S0-BZF-077

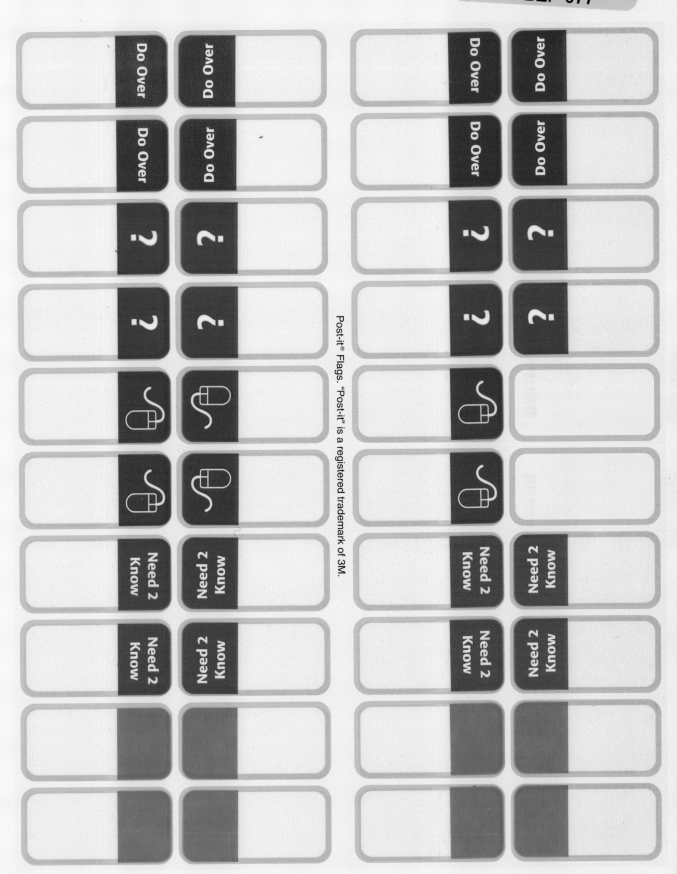

Post-it® Flags. "Post-it" is a registered trademark of 3M.

Tab it. Do it. Ace it.

Do Over

Do you need to review something? Try again? Work it out on your own after class? Tab it.

?

Got a question for office hours? Do you need to review an example on your own to get a full understanding? Do you need to look something up before moving on? Tab it.

Do you need to see the video? Check out an online source? Complete your online homework? Tab it.

Need 2 Know

Is this going to be on the test? Need to mark a key formula? Do you need to memorize these steps? Tab it.

Do you have your own study system? Do you need to make a note? Do you want to express yourself? Tab it.

Tab it. Do it. Ace it.

ISBN 13: 978-0-495-55855-2
ISBN 10: 0-4955-5855-9

Applied Mathematics for the Managerial, Life & Social Sciences

University of Rhode Island

Soo T. Tan

CENGAGE
Learning·

Australia · Brazil · Japan · Korea · Mexico · Singapore · Spain · United Kingdom · United States

CENGAGE
Learning

Applied Mathematics for the Managerial, Life & Social Sciences: University of Rhode Island

Senior Manager, Student Engagement:

Linda deStefano

Janey Moeller

Manager, Student Engagement:

Julie Dierig

Marketing Manager:

Rachael Kloos

Manager, Production Editorial:

Kim Fry

Manager, Intellectual Property Project Manager:

Brian Methe

Senior Manager, Production and Manufacturing:

Donna M. Brown

Manager, Production:

Terri Daley

Printed in the United States of America

Source:

Applied Mathematics for the Managerial, Life, and Social Sciences, 6th Edition
Soo T. Tan
© 2013, 2010 Cengage Learning. All rights reserved.

For product information and technology assistance, contact us at
Cengage Learning Customer & Sales Support, 1-800-354-9706
For permission to use material from this text or product,
submit all requests online at **cengage.com/permissions**
Further permissions questions can be emailed to
permissionrequest@cengage.com

This book contains select works from existing Cengage Learning resources and was produced by Cengage Learning Custom Solutions for collegiate use. As such, those adopting and/or contributing to this work are responsible for editorial content accuracy, continuity and completeness.

Compilation © 2014 Cengage Learning

ISBN-13: 978-1-305-01426-8

ISBN-10: 1-305-01426-X

WCN: 01-100-101

Cengage Learning

5191 Natorp Boulevard
Mason, Ohio 45040
USA

Cengage Learning is a leading provider of customized learning solutions with office locations around the globe, including Singapore, the United Kingdom, Australia, Mexico, Brazil, and Japan. Locate your local office at:
international.cengage.com/region.

Cengage Learning products are represented in Canada by Nelson Education, Ltd. For your lifelong learning solutions, visit **www.cengage.com/custom.**
Visit our corporate website at **www.cengage.com.**

TO PAT, BILL, AND MICHAEL

CONTENTS

CHAPTER 4

Mathematics of Finance   189

CHAPTER 5

Systems of Linear Equations and Matrices   247

CHAPTER 6

Linear Programming   329

PREFACE

Math is an integral part of our increasingly complex life. *Applied Mathematics for the Managerial, Life, and Social Sciences* attempts to illustrate this point with its applied approach to mathematics. In preparing the Sixth Edition, I have kept in mind three longstanding goals: (1) to write an applied text that motivates students while providing the background in the quantitative techniques necessary to better understand and appreciate the courses normally taken in undergraduate training, (2) to lay the foundation for more advanced courses, such as statistics and operations research, and (3) to make the book a useful teaching tool for instructors.

One of the most important lessons I learned from my experience teaching mathematics to non-mathematics majors is that many of the students come into these courses with some degree of apprehension. This awareness led to the intuitive approach I have adopted in all of my texts. As you will see, I try to introduce each abstract mathematical concept through an example drawn from a common, real-life experience. Once the idea has been conveyed, I then proceed to make it precise, thereby assuring that no mathematical rigor is lost in this intuitive treatment of the subject.

Another lesson I learned from my students is that they have a much greater appreciation of the material if the applications are drawn from their fields of interest and from situations that occur in the real world. This is one reason you will see so many exercises in my texts that are modeled on data gathered from newspapers, magazines, journals, and other media. Whether it be the market for cholesterol-reducing drugs, financing a home, bidding for cable rights, broadband Internet households, or Starbucks' annual sales, I weave topics of current interest into my examples and exercises to keep the book relevant to all of my readers.

This text offers more than enough material for a two-semester or three-semester course. The following chart on chapter dependency is provided to help the instructor design a course that is most suitable for the intended audience.

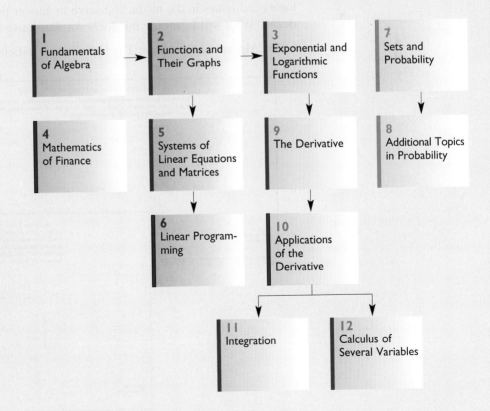

THE APPROACH

Level of Presentation

My approach is intuitive, and I state the results informally. However, I have taken special care to ensure that this approach does not compromise the mathematical content and accuracy.

Problem-Solving Approach

A problem-solving approach is stressed throughout the book. Numerous examples and applications illustrate each new concept and result. Special emphasis is placed on helping students formulate, solve, and interpret the results of the problems involving applications. Because students often have difficulty setting up and solving word problems, extra care has been taken to help students master these skills:

- Very early on in the text, students are given practice in solving word problems (see Example 7, Section 1.8).
- Guidelines are given to help students formulate and solve word problems (see Section 2.7).
- One entire section is devoted to modeling and setting up linear programming problems (see Section 6.2).
- In Chapter 10, optimization problems are covered in two sections. First, students solve problems in which the function to be optimized is given (Section 10.4); second, students solve problems in which the function to be optimized must first be formulated (Section 10.5).

Intuitive Introduction to Concepts

Mathematical concepts are introduced with concrete, real-life examples, wherever appropriate. These examples and other applications have been chosen from current topics and issues in the media and serve to answer the question often posed by students: "What will I ever use this for?" An illustrative list follows.

- **The Algebra of Functions:** The U.S. Budget Deficit

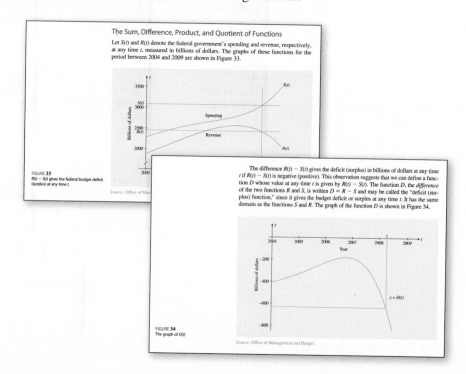

The Sum, Difference, Product, and Quotient of Functions

Let $S(t)$ and $R(t)$ denote the federal government's spending and revenue, respectively, at any time t, measured in billions of dollars. The graphs of these functions for the period between 2004 and 2009 are shown in Figure 33.

FIGURE 33
$R(t) - S(t)$ gives the federal budget deficit (surplus) at any time t.

Source: Office of Man...

The difference $R(t) - S(t)$ gives the deficit (surplus) in billions of dollars at any time t if $R(t) - S(t)$ is negative (positive). This observation suggests that we can define a function D whose value at any time t is given by $R(t) - S(t)$. The function D, the *difference* of the two functions R and S, is written $D = R - S$ and may be called the "deficit (surplus) function," since it gives the budget deficit or surplus at any time t. It has the same domain as the functions S and R. The graph of the function D is shown in Figure 34.

FIGURE 34
The graph of $D(t)$

Source: Office of Management and Budget.

- **Mathematical Modeling:** Social Security Trust Fund Assets
- **Limits:** The Motion of a Maglev
- **The Chain Rule:** The Population of Americans Aged 55 Years and Older
- **Increasing and Decreasing Functions:** The Fuel Economy of a Car
- **Concavity:** U.S. and World Population Growth
- **Inflection Points:** The Point of Diminishing Returns
- **Curve Sketching:** The Dow-Jones Industrial Average on "Black Monday"
- **Exponential functions:** Income Distribution of American Families
- **Area Between Two Curves:** Petroleum Saved with Conservation Measures

Connections

One example (the maglev) is used as a common thread throughout the development of calculus—from limits through integration. The goal here is to show students the connections between the concepts presented: limits, continuity, rates of change, the derivative, the definite integral, and so on.

Motivation

Illustrating the practical value of mathematics in applied areas is an objective of my approach. Many of the applications are based on mathematical models (functions) that I have constructed using data drawn from various sources, including current newspapers, magazines, and the Internet. Sources are given in the text for these applied problems.

Modeling

I believe that one of the important skills that a student should acquire is the ability to translate a real problem into a mathematical model that can provide insight into the problem. In Section 2.7, the modeling process is discussed, and students are asked to use models (functions) constructed from real-life data to answer questions. Students get hands-on experience constructing these models in the Using Technology sections.

NEW TO THIS EDITION

Motivating Real-World Applications

Among the many new and updated applied examples and exercises are global warming, the solvency of the U.S. Social Security trust fund, the Case-Shiller Home Price Index, treasury bills, Roth IRAs, smartphone sales, bounced-check charges, home affordability, and sleeping with cell phones.

24. **CASE-SHILLER HOME PRICE INDEX** The following graph shows the change in the S&P/Case-Shiller Home Price Index based on a 20-city average from June 2001 ($t = \frac{1}{2}$) through June 2008 ($t = 7\frac{1}{2}$), adjusted for inflation.

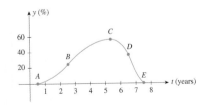

a. What do the points $A(\frac{1}{2}, 0)$ and $E(7\frac{1}{2}, 0)$ tell you about the change in the Case-Shiller Home Price Index over the period under consideration?
b. What does the point $C(5\frac{1}{3}, 56)$ tell you about the Case-Shiller Home Price Index?
c. Give an interpretation of the inflection points $B(2\frac{1}{2}, 24)$ and $D(6\frac{1}{2}, 36)$.

Source: New York Times.

Modeling with Data

As in the previous edition, Modeling with Data exercises are found in many of the Using Technology sections throughout the text. Here students can actually see how some of the functions found in the exercises are constructed. Many of these applications have been updated and some new exercises have been added. (See Social Security Trust Fund Assets, Example 2, page 139, and the corresponding exercise in which the model is derived in Exercise 14, page 152.)

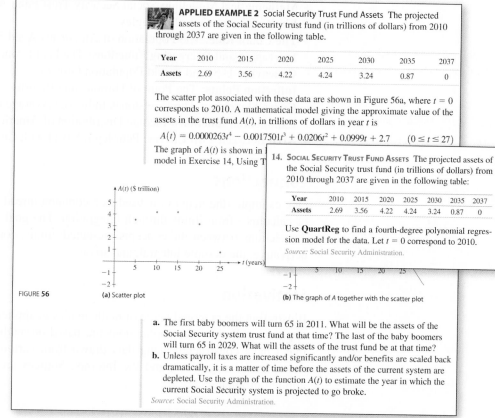

APPLIED EXAMPLE 2 Social Security Trust Fund Assets The projected assets of the Social Security trust fund (in trillions of dollars) from 2010 through 2037 are given in the following table.

Year	2010	2015	2020	2025	2030	2035	2037
Assets	2.69	3.56	4.22	4.24	3.24	0.87	0

The scatter plot associated with these data are shown in Figure 56a, where $t = 0$ corresponds to 2010. A mathematical model giving the approximate value of the assets in the trust fund $A(t)$, in trillions of dollars in year t is

$$A(t) = 0.0000263t^4 - 0.0017501t^3 + 0.0206t^2 + 0.0999t + 2.7 \qquad (0 \le t \le 27)$$

The graph of $A(t)$ is shown in [...] model in Exercise 14, Using T[...]

FIGURE 56   (a) Scatter plot

14. SOCIAL SECURITY TRUST FUND ASSETS The projected assets of the Social Security trust fund (in trillions of dollars) from 2010 through 2037 are given in the following table:

Year	2010	2015	2020	2025	2030	2035	2037
Assets	2.69	3.56	4.22	4.24	3.24	0.87	0

Use **QuartReg** to find a fourth-degree polynomial regression model for the data. Let $t = 0$ correspond to 2010.
Source: Social Security Administration.

(b) The graph of *A* together with the scatter plot

a. The first baby boomers will turn 65 in 2011. What will be the assets of the Social Security system trust fund at that time? The last of the baby boomers will turn 65 in 2029. What will the assets of the trust fund be at that time?
b. Unless payroll taxes are increased significantly and/or benefits are scaled back dramatically, it is a matter of time before the assets of the current system are depleted. Use the graph of the function $A(t)$ to estimate the year in which the current Social Security system is projected to go broke.
Source: Social Security Administration.

Making Connections with Technology

All of the art in the Using Technology sections has been redone. The graphing calculator screens now show a numbered scale for both axes, making it easier for students to use and understand these graphs. In addition, Microsoft Excel 2010 examples and exercises now appear in the technology sections. Step-by-step instructions (including keystrokes), all new dialog boxes, graphs, and spreadsheets for Excel 2010 are given in the text. Instructions for solving these examples and exercises with Microsoft Excel 2007 are provided on CourseMate.

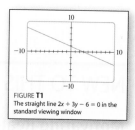

FIGURE T1
The straight line $2x + 3y - 6 = 0$ in the standard viewing window

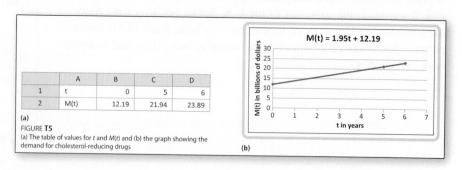

	A	B	C	D
1	t	0	5	6
2	M(t)	12.19	21.94	23.89

(a)
FIGURE T5
(a) The table of values for *t* and *M(t)* and (b) the graph showing the demand for cholesterol-reducing drugs

(b)

Variety of Problem Types

Additional rote questions, true or false questions, and concept questions have been added throughout the text to enhance the exercise sets. (See, for example, Concept Question 2.4, Exercise 1, page 109.)

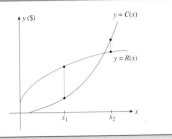

2.4   Concept Questions

1. The figure opposite shows the graphs of a total cost function and a total revenue function. Let P, defined by $P(x) = R(x) - C(x)$, denote the total profit function.
   a. Find an expression for $P(x_1)$. Explain its significance.
   b. Find an expression for $P(x_2)$. Explain its significance.

Carefully Crafted Solutions

The solutions manual has been completely revamped. All new art has been created for the manual, and the solutions have been copyedited and streamlined for ease of use. As in previous editions, the solutions to all of the exercises have been written by the author.

3.3   Exponential Functions as Mathematical Models

Problem-Solving Tips

Four mathematical models were introduced in this section:

1. **Exponential growth:** $Q(t) = Q_0 e^{kt}$ describes a quantity $Q(t)$ that is initially present in the amount $Q(0) = Q_0$ and whose rate of growth at any time t is directly proportional to the amount of the quantity present at time t.

2. **Exponential decay:** $Q(t) = Q_0 e^{-kt}$ describes a quantity $Q(t)$ that is initially present in the amount $Q(0) = Q_0$ and decreases at a rate that is directly proportional to its size.

3. **Learning curves:** $Q(t) = C - Ae^{-kt}$ describes a quantity $Q(t)$, where $Q(0) = C - A$, and $Q(t)$ increases and approaches the number C as t increases without bound.

4. **Logistic growth functions:** $Q(t) = \dfrac{A}{1 + Be^{-kt}}$ describes a quantity $Q(t)$, where $Q(0) = \dfrac{A}{1+B}$. Note that $Q(t)$ increases rapidly for small values of t but the rate of growth of $Q(t)$ decreases quickly as t increases. $Q(t)$ approaches the number A as t increases without bound.

Try to familiarize yourself with the examples and graphs for each of these models before you work through the applied problems in this section.

Concept Questions

1. $Q(t) = Q_0 e^{kt}$ where $k > 0$ represents exponential growth and $k < 0$ represents exponential decay. The larger the magnitude of k, the more quickly the former grows and the more quickly the latter decays.

3. $Q(t) = \dfrac{A}{1 + Be^{-kt}}$, where A, B, and k are positive constants. Q increases rapidly for small values of t but the rate of increase slows down as Q (always increasing) approaches the number A.

Enhanced Graphics

The three-dimensional artwork in Chapter 12 has been redone with an emphasis on making it easier for students to see the concepts being described in 3-space. For example, Figure 7 in Section 12.1 now shows the trace of the graph of $z = f(x, y)$ and the plane $z = k$ and its projection onto the xy-plane (Figure 7a) and the corresponding level curve (Figure 7b).

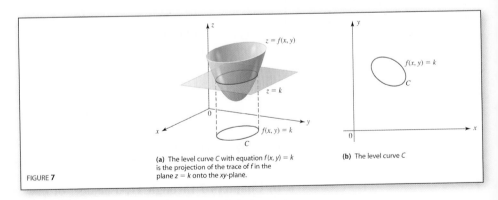

FIGURE 7

(a) The level curve C with equation $f(x, y) = k$ is the projection of the trace of f in the plane $z = k$ onto the xy-plane.

(b) The level curve C

Specific Content Changes

- **Chapter 1:** Sixteen new applications have been added to the exercise sets and one applied example has been added to Section 1.4.
- **Chapter 2:** A new model has been constructed for Applied Example 4 in the Using Technology in Section 2.3. The budget-deficit graphs that are used as motivation to introduce Section 2.4, "The Algebra of Functions," have been redone to reflect the current deficit figures. A new graphical concept exercise has been added to Section 2.4. In Section 2.5, a new applied example, Bounced-Check Charges, has been added. In Section 2.7, "Functions and Mathematical Models," new models for three applied examples—Global Warming, Social Security Trust Fund Assets, and Driving Costs—have been added.
- **Chapter 4:** The formulas given in this chapter form the basis for understanding many of the complex financial instruments in the market today. New examples were added in this chapter to help students better understand how these formulas can be applied to solve problems involving investments, mortgages, and retirement planning. In Section 4.1, a new example illustrating an application of simple interest to treasury bills has been added. In Section 4.2, an example has been added to show the after-tax effect of saving in tax-sheltered accounts. Another example that illustrates the difference between Roth IRAs and traditional IRAs has also been added. In Section 4.3, a new example has been added comparing mortgage payments for a 7/1 ARM and an interest-only loan. Here, students can see how the interest-only loans were used to make homes appear more affordable. Another application of sinking funds was also added; in Example 8, Section 4.3, the monthly contributions needed to meet a retirement goal of monthly withdrawals of a fixed amount are computed.
- **Chapter 5:** In Section 5.2, the pivoting process is now explained in greater detail. In Section 5.3, an introductory example using a 2×2 system of equations with infinitely many solutions has been added. In Section 5.4, an example illustrating the use of the associative and commutative laws for matrices has been added.
- **Chapter 6:** In Section 6.5, a new example has been added in which a primal problem and its associated dual problem are first solved graphically. Then these problems are solved algebraically using the simplex method. This provides students with a graphical justification of the method of duality.

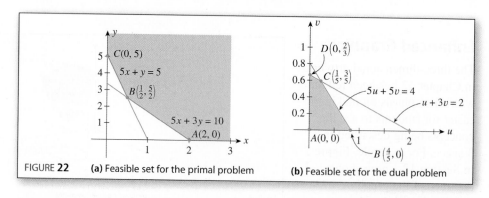

FIGURE 22   **(a)** Feasible set for the primal problem   **(b)** Feasible set for the dual problem

- **In Chapters 7 through 11,** a number of new and unique applications have been added to the exercise sets. Among these are Pharmaceutical Theft, Federal Lobbying, Total Knee Replacement Procedures, Mexico's Hedging Tactics, and the British Deficit.
- **Chapter 10:** The budget curves used to motivate relative extrema were updated to reflect the current deficit. Seven new graphical exercises were added to Exercise Set 10.2, including the applications Rumors of a Run on a Bank and the Case-Shiller

Home Price Index. The Average Age of Cars example that is used to motivate the concept of absolute extrema in Section 10.4 was updated.

- **Chapter 12:** The three-dimensional artwork in this chapter has been redone. The new graphics make it easier for students to see the concepts being described in 3-space.
- Many new rote exercises have been added throughout the text.

TRUSTED FEATURES

In addition to the new features, we have retained many of the following hallmarks that have made this series so usable and well-received in past editions:

- Review material to reinforce prerequisite algebra skills
- Section exercises to help students understand and apply concepts
- Optional technology sections to explore mathematical ideas and solve problems
- End-of-chapter review sections to assess understanding and problem-solving skills
- Features to motivate further exploration

Algebra Review Where Students Need It Most

Well-placed algebra review notes, keyed to the review chapter, appear where students need them most throughout the text. These are indicated by the (x^2) icon. See this feature in action on pages 556 and 591.

EXAMPLE 6 Evaluate

$$\lim_{h\to 0} \frac{\sqrt{1+h}-1}{h}$$

Solution   Letting h approach zero, we obtain the indeterminate form $0/0$. Next, we rationalize the numerator of the quotient by multiplying both the numerator and the denominator by the expression $(\sqrt{1+h}+1)$, obtaining

$$\frac{\sqrt{1+h}-1}{h} = \frac{(\sqrt{1+h}-1)(\sqrt{1+h}+1)}{h(\sqrt{1+h}+1)} \quad (x^2) \text{ See page 42.}$$

$$= \frac{1+h-1}{h(\sqrt{1+h}+1)} \quad (\sqrt{a}-\sqrt{b})(\sqrt{a}+\sqrt{b})=a-b$$

$$= \frac{h}{h(\sqrt{1+h}+1)}$$

$$= \frac{1}{\sqrt{1+h}+1}$$

Therefore,

$$\lim_{h\to 0}\frac{\sqrt{1+h}-1}{h}=\lim_{h\to 0}\frac{1}{\sqrt{1+h}+1}=\frac{1}{\sqrt{1}+1}=\frac{1}{2}$$

Self-Check Exercises

Offering students immediate feedback on key concepts, these exercises begin each end-of-section exercise set. Fully worked-out solutions can be found at the end of each exercise section.

7.1   Self-Check Exercises

1. Let $U = \{1, 2, 3, 4, 5, 6, 7\}$, $A = \{1, 2, 3\}$, $B = \{3, 4, 5, 6\}$, and $C = \{2, 3, 4\}$. Find the following sets:
   a. A^c b. $A \cup B$ c. $B \cap C$
   d. $(A \cup B) \cap C$ e. $(A \cap B) \cup C$ f. $A^c \cap (B \cup C)^c$

2. Let U denote the set of all members of the House of Representatives. Let
   $$D = \{x \in U \mid x \text{ is a Democrat}\}$$
   $$R = \{x \in U \mid x \text{ is a Republican}\}$$

$$F = \{x \in U \mid x \text{ is a female}\}$$
$$L = \{x \in U \mid x \text{ is a lawyer by training}\}$$

Describe each of the following sets in words.
a. $D \cap F$ b. $F^c \cap R$ c. $D \cap F \cap L^c$

Solutions to Self-Check Exercises 7.1 can be found on page 415.

Concept Questions

Designed to test students' understanding of the basic concepts discussed in the section, these questions encourage students to explain learned concepts in their own words.

7.1 Concept Questions

1. **a.** What is a set? Give an example.
   b. When are two sets equal? Give an example of two equal sets.
   c. What is the empty set?

2. What can you say about two sets A and B such that
   a. $A \cup B \subseteq A$ **b.** $A \cup B = \varnothing$
   c. $A \cap B = B$ **d.** $A \cap B = \varnothing$

3. **a.** If $A \subset B$, what can you say about the relationship between A^c and B^c?
   b. If $A^c = \varnothing$, what can you say about A?

Exercises

Each exercise section contains an ample set of problems of a routine computational nature followed by an extensive set of application-oriented problems.

7.1 Exercises

In Exercises 1–4, write the set in set-builder notation.

1. The set of gold medalists in the 2010 Winter Olympic Games

2. The set of football teams in the NFL

3. $\{3, 4, 5, 6, 7\}$

4. $\{1, 3, 5, 7, 9, 11, \ldots, 39\}$

In Exercises 5–8, list the elements of the set in roster notation.

5. $\{x \mid x \text{ is a digit in the number } 352{,}646\}$

6. $\{x \mid x \text{ is a letter in the word } HIPPOPOTAMUS\}$

In Exercises 15 and 16, let $A = \{1, 2, 3, 4, 5\}$. Determine whether the statements are true or false.

15. **a.** $2 \in A$ **b.** $A \subseteq \{2, 4, 6\}$

16. **a.** $0 \in A$ **b.** $\{1, 3, 5\} \in A$

17. Let $A = \{1, 2, 3\}$. Which of the following sets are equal to A?
    a. $\{2, 1, 3\}$ **b.** $\{3, 2, 1\}$
    c. $\{0, 1, 2, 3\}$

18. Let $A = \{a, e, l, t, r\}$. Which of the following sets are equal to A?
    a. $\{x \mid x \text{ is a letter of the word } later\}$

Using Technology

These optional features appear after the section exercises. They can be used in the classroom if desired or as material for self-study by the student. Here, the graphing calculator and Microsoft Excel 2010 are used as tools to solve problems. (Instructions for Microsoft Excel 2007 are online on CourseMate.) These sections are written in the traditional example-exercise format, with answers given at the back of the book. Illustrations showing graphing calculator screens and spreadsheets are extensively used. In keeping with the theme of motivation through real-life examples, many sourced applications are again included. Students can construct their own models using real-life data in many of the Using Technology sections. These include models for growth of the Indian gaming industry, health-care spending, worldwide PC shipments, and the federal debt, among others.

A *How-To Technology Index* is included at the back of the book for easy reference.

USING TECHNOLOGY

Although the proof is outside the scope of this book, it can be proved that an exponential function of the form $f(x) = b^x$, where $b > 1$, will ultimately grow faster than the power function $g(x) = x^n$ for *any* positive real number n. To give a visual demonstration of this result for the special case of the exponential function $f(x) = e^x$, we can use a graphing utility to plot the graphs of both f and g (for selected values of n) on the same set of axes in an appropriate viewing window and observe that the graph of f ultimately lies above that of g.

EXAMPLE 1 Use a graphing utility to plot the graphs of (a) $f(x) = e^x$ and $g(x) = x^3$ on the same set of axes in the viewing window $[0, 6] \times [0, 250]$ and (b) $f(x) = e^x$ and $g(x) = x^5$ in the viewing window $[0, 20] \times [0, 1{,}000{,}000]$.

Solution

a. The graphs of $f(x) = e^x$ and $g(x) = x^3$ in the viewing window $[0, 6] \times [0, 250]$ are shown in Figure T1a.

b. The graphs of $f(x) = e^x$ and $g(x) = x^5$ in the viewing window $[0, 20] \times [0, 1{,}000{,}000]$ are shown in Figure T1b.

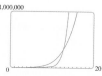

(a) The graphs of $f(x) = e^x$ and $g(x) = x^3$ in the viewing window $[0, 6] \times [0, 250]$

(b) The graphs of $f(x) = e^x$ and $g(x) = x^5$ in the viewing window $[0, 20] \times [0, 1{,}000{,}000]$

FIGURE **T1**

In the exercises that follow, you are asked to use a graphing utility to reveal the properties of exponential functions.

Exploring with Technology

Designed to explore mathematical concepts and to shed further light on examples in the text, these optional questions appear throughout the main body of the text and serve to enhance the student's understanding of the concepts and theory presented. Often the solution of an example in the text is augmented with a graphical or numerical solution. Complete solutions to these exercises are given in Solution Builder.

> **Exploring with TECHNOLOGY**
>
> To obtain a visual confirmation of the fact that the expression $(1 + 1/m)^m$ approaches the number $e = 2.71828\ldots$ as m gets larger and larger, plot the graph of $f(x) = (1 + 1/x)^x$ in a suitable viewing window, and observe that $f(x)$ approaches $2.71828\ldots$ as x gets larger and larger. Use ZOOM and TRACE to find the value of $f(x)$ for large values of x.

Summary of Principal Formulas and Terms

Each review section begins with the Summary highlighting important formulas and terms with page numbers given for quick review.

> **CHAPTER 3** Summary of Principal Formulas and Terms
>
> FORMULAS
>
> | | |
> |---|---|
> | 1. Exponential function with base b | $y = b^x$, where $b > 0$ and $b \neq 1$ |
> | 2. The number e | $e = 2.7182818\ldots$ |
> | 3. Exponential function with base e | $y = e^x$ |
> | 4. Logarithmic function with base b | $y = \log_b x \quad (x > 0)$ |
> | 5. Logarithmic function with base e | $y = \ln x \quad (x > 0)$ |
> | 6. Inverse properties of $\ln x$ and e^x | $\ln e^x = x$ and $e^{\ln x} = x$ |
>
> TERMS
>
> common logarithm (166) growth constant (175) half-life of a radioactive substance
> natural logarithm (166) exponential decay (177) (177)
> exponential growth (175) decay constant (177) logistic growth function (180)

Concept Review Questions

These questions give students a chance to check their knowledge of the basic definitions and concepts given in each chapter.

> **CHAPTER 3** Concept Review Questions
>
> Fill in the blanks.
>
> 1. The function $f(x) = x^b$ (b, a real number) is called a/an ____ function, whereas the function $g(x) = b^x$, where $b >$ ____ and $b \neq$ ____, is called a/an ____ function.
>
> 2. a. The domain of the function $y = 3^x$ is ____, and its range is ____.
>   b. The graph of the function $y = 0.3^x$ passes through the point ____ and falls from ____ to ____.
>
> 3. a. If $b > 0$ and $b \neq 1$, then the logarithmic function $y = \log_b x$ has domain ____ and range ____; its graph passes through the point ____.
>   b. The graph of $y = \log_b x$ ____ from left to right if $b < 1$ and ____ from left to right if $b > 1$.
>
> 4. a. If $x > 0$, then $e^{\ln x} =$ ____.
>   b. If x is any real number, then $\ln e^x =$ ____.

Review Exercises

Offering a solid review of the chapter material, the Review Exercises contain routine computational exercises followed by applied problems.

> **CHAPTER 3** Review Exercises
>
> In Exercises 1–4, sketch the graph of the function.
>
> 1. $f(x) = 5^x$
> 2. $y = \left(\dfrac{1}{5}\right)^x$
> 3. $f(x) = \log_4 x$
> 4. $y = \log_{1/4} x$
>
> In Exercises 5–8, express each equation in logarithmic form.
>
> 5. $3^4 = 81$
> 6. $9^{1/2} = 3$
> 7. $\left(\dfrac{2}{3}\right)^{-3} = \dfrac{27}{8}$
> 8. $16^{-3/4} = 0.125$
>
> 34. $\dfrac{20}{1 + 2e^{0.2x}} = 4$
> 35. $\dfrac{30}{1 + 2e^{-0.1x}} = 5$
>
> 36. Sketch the graph of the function $y = \log_2(x + 3)$.
>
> 37. Sketch the graph of the function $y = \log_3(x + 1)$.
>
> 38. GROWTH OF BACTERIA A culture of bacteria that initially contained 2000 bacteria has a count of 18,000 bacteria after 2 hr.
>   a. Determine the function $Q(t)$ that expresses the exponential growth of the number of cells of this bacterium as a function of time t (in minutes).
>   b. Find the number of bacteria present after 4 hr.

Before Moving On . . .

Found at the end of each chapter review, these exercises give students a chance to see if they have mastered the basic computational skills developed in each chapter. If students need step-by-step help, they can use the CourseMate tutorials that are keyed to text and work out similar problems at their own pace.

CHAPTER 3 Before Moving On . . .

1. Simplify the expression $(2x^{-2})^2(9x^{-4})^{1/2}$.

2. Solve $e^{2x} - e^x - 6 = 0$ for x.
   Hint: Let $u = e^x$.

3. Solve $\log_2(x^2 - 8x + 1) = 0$.

4. Solve the equation $\dfrac{100}{1 + 2e^{0.3t}} = 40$ for t.

5. The temperature of a cup of coffee at time t (in minutes) is

   $$T(t) = 70 + ce^{-kt}$$

   Initially, the temperature of the coffee was 200°F. Three minutes later, it was 180°. When will the temperature of the coffee be 150°F?

Explore & Discuss

These optional questions can be discussed in class or assigned as homework. These questions generally require more thought and effort than the usual exercises. They may also be used to add a writing component to the class or as team projects. Complete solutions to these exercises are given in Solution Builder and in the instructor's *Complete Solutions Manual*.

Explore & Discuss

The average price of gasoline at the pump over a 3-month period, during which there was a temporary shortage of oil, is described by the function f defined on the interval [0, 3]. During the first month, the price was increasing at an increasing rate. Starting with the second month, the good news was that the rate of increase was slowing down, although the price of gas was still increasing. This pattern continued until the end of the second month. The price of gas peaked at $t = 2$ and began to fall at an increasing rate until $t = 3$.

1. Describe the signs of $f'(t)$ and $f''(t)$ over each of the intervals (0, 1), (1, 2), and (2, 3).

2. Make a sketch showing a plausible graph of f over [0, 3].

Portfolios

The real-life experiences of a variety of professionals who use mathematics in the workplace are related in these interviews. Among those interviewed are the Senior Vice-President of Supply at Earthbound Farm and a software engineer at Iron Mountain who uses statistics and calculus to develop document management tools.

PORTFOLIO   **Todd Kodet**

TITLE   Senior Vice-President of Supply
INSTITUTION   Earthbound Farm

Earthbound Farm is America's largest grower of organic produce, offering more than 100 varieties of organic salads, vegetables, fruits, and herbs on 34,000 crop acres. As Senior Vice-President of Supply, I am responsible for getting our products into and out of Earthbound Farm. A major part of my work is scheduling plantings for upcoming seasons, matching projected supply to projected demand for any given day and season. I use applied mathematics in every step of my planning to create models for predicting supply and demand.

After the sales department provides me with information about projected demand, I take their estimates, along with historical data for expected yields, to determine how much of each organic product we need to plant. There are several factors that I have to think about when I make these determinations. For example, I not only have to consider gross yield per acre of farmland, but also have to calculate average trimming waste per acre, to arrive at net pounds needed per customer.

Some of the other variables I consider are the amount of organic land available, the location of the farms, seasonal information (because days to maturity for each of our crops varies greatly depending on the weather), and historical information relating to weeds, pests, and diseases.

I emphasize the importance of understanding the mathematics that drives our business plans when I work with my team to analyze the reports they have generated. They need to recognize when the information they have gathered does not make sense so that they can spot errors that could skew our projections. With a sound understanding of mathematics, we are able to create more accurate predictions to help us meet our company's goals.

Alli Pura, Earthbound Farm; (inset) © istockphoto.com/Dan Moore

Example Videos

Available through CourseMate and Enhanced WebAssign, these videos offer hours of instruction from award-winning teacher Deborah Upton of Molloy College. Watch as she walks students through key examples from the text, step by step, giving them a foundation in the skills that they need to have. Each example available online is identified by the video icon located in the margin.

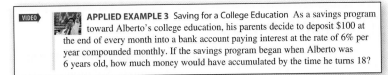

VIDEO ► **APPLIED EXAMPLE 3** Saving for a College Education As a savings program toward Alberto's college education, his parents decide to deposit $100 at the end of every month into a bank account paying interest at the rate of 6% per year compounded monthly. If the savings program began when Alberto was 6 years old, how much money would have accumulated by the time he turns 18?

Action-Oriented Study Tabs

Convenient color-coded study tabs, similar to Post-it® flags, make it easy for students to tab pages that they want to return to later, whether it be for additional review, exam preparation, online exploration, or identifying a topic to be discussed with the instructor.

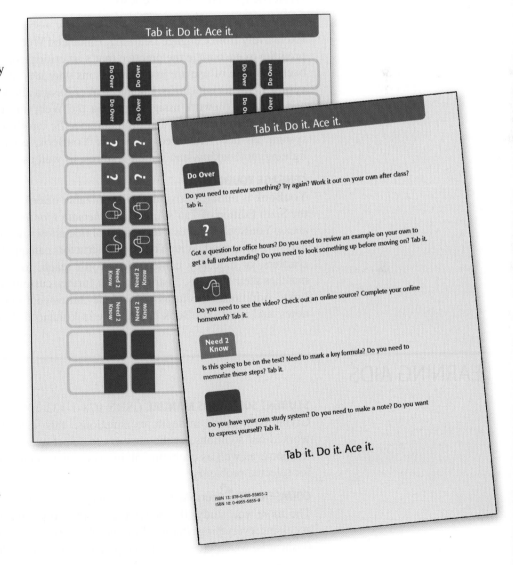

TEACHING AIDS

COMPLETE SOLUTIONS MANUAL (ISBN 978-1-133-36434-4) by Soo T. Tan
The *Complete Solutions Manual* contains fully worked-out solutions to all section exercises in the text, including Technology Exercises in Using Technology, Explore & Discuss activities, and Exploring with Technology activities.

SOLUTION BUILDER (www.cengage.com/solutionbuilder)
This online instructor database offers complete worked-out solutions to all exercises in the text, including Exploring with Technology and Explore & Discuss questions. Solution Builder allows you to create customized, secure solutions printouts (in PDF format) matched exactly to the problems you assign in class.

POWERLECTURE (ISBN 978-1-133-49041-8)
This comprehensive CD-ROM includes Solution Builder, PowerPoint slides, and ExamView® Computerized Testing featuring algorithmically generated questions to create, deliver, and customize tests.

ENHANCED WEBASSIGN WebAssign (www.webassign.net)
Exclusively from Cengage Learning, Enhanced WebAssign offers an extensive online program to encourage the practice that's so critical for concept mastery. Instant feedback and ease of use are just two reasons why it is the most widely used homework system in higher education. Enhanced WebAssign allows you to assign, collect, grade, and record homework assignments via the Web, and includes links to text-specific content, video examples, and problem-specific tutorials. Now this proven homework system has been enhanced to include YouBook, a customizable eBook with highlighting, note-taking, and search features, as well as links to multimedia resources.

CENGAGE YOUBOOK
YouBook is an interactive and customizable eBook. Containing all the content from the Sixth Edition of Tan's *Applied Mathematics for the Managerial, Life, and Social Sciences*, YouBook features a text edit tool that allows instructors to modify the textbook narrative as needed. With YouBook instructors can quickly reorder entire sections and chapters, or hide any content they don't teach, to create an eBook that perfectly matches their syllabus. Instructors can further customize the text by publishing Web links. Additional media assets include animated figures, video clips, highlighting, notes, and more. YouBook is available in Enhanced WebAssign.

LEARNING AIDS

STUDENT SOLUTIONS MANUAL (ISBN 978-1-133-10932-7) by Soo T. Tan
Giving you more in-depth explanations, this insightful resource includes fully worked-out solutions for the answers to select exercises included at the back of the textbook, as well as problem-solving strategies, additional algebra steps, and review for selected problems.

COURSEMATE CourseMate (www.cengagebrain.com)
The more you study, the better the results. Make the most of your study time by accessing everything you need in one place. Read your textbook, take notes, review flashcards, watch videos, and take practice quizzes—online with CourseMate.

CENGAGEBRAIN.COM
Visit **www.cengagebrain.com** to access additional course materials and companion resources. At the CengageBrain.com home page, search for the ISBN of your title (from the back cover of your book) using the search box at the top of the page. This will take you to the product page where free companion resources can be found.

ACKNOWLEDGMENTS

I wish to express my personal appreciation to each of the following reviewers of the Sixth Edition, whose many suggestions have helped make a much improved book.

Richard Baslaw
York College of CUNY

Denis Bell
University of North Florida

Debra Carney
University of Denver

James Eby
Blinn College—Bryan Campus

Edna Greenwood
Tarrant County College—Northwest Campus

Velma Hill
York College of CUNY

Jiashi Hou
Norfolk State University

Xingde Jia
Texas State University

Kristi Karber
University of Central Oklahoma

Rebecca Leefers
Michigan State University

James Liu
James Madison University

Theresa Manns
Salisbury University

Michael Paulding
Kapiolani Community College

Armando Perez
Laredo Community College

Donald Stengel
California State University—Fresno

Beimnet Teclezghi
New Jersey City University

I also thank the following reviewers whose comments and suggestions for the previous edition have greatly shaped the current form of this edition.

Paul Abraham
Kent State University—Stark

James Adair
Missouri Valley College

Jill Britton
Camosun College

Debra D. Bryant
Tennessee Technological University

Michelle Dedeo
University of North Florida

Scott L. Dennison
University of Wisconsin—Oshkosh

Christine Devena
Miles Community College

Andrew Diener
Christian Brothers University

Mike Everett
Santa Ana College

Kevin Ferland
Bloomsburg University

Tao Guo
Rock Valley College

Mark Jacobson
Montana State University—Billings

Sarah Kilby
North Country Community College

Murray Lieb
New Jersey Institute of Technology

Lia Liu
University of Illinois at Chicago

Rebecca Lynn
Colorado State University

Mary T. McMahon
North Central College

Daniela Mihai
University of Pittsburgh

Kathy Nickell
College of DuPage

Carol Overdeep
Saint Martin's University

Mohammed Rajah
Miracosta College

Dennis H. Risher
Loras College

Brian Rodas
Santa Monica College

Dr. Arthur Rosenthal
Salem State College

Abdelrida Saleh
Miami Dade College

Stephanie Anne Salomone
University of Portland

Mohammed Siddique
Virginia Union University

Ray Toland
Clarkson University

Jennifer Strehler
Oakton Community College

Justin Wyss-Gallifent
University of Maryland at College Park

I also wish to thank Tao Guo for the superb job he did as the accuracy checker for this text and on the *Complete Solutions Manual* that accompanies the text. I also thank the editorial, production, and marketing staffs of Brooks/Cole—Richard Stratton, Laura Wheel, Haeree Chang, Andrew Coppola, Cheryll Linthicum, Vernon Boes, and Barb Bartoszek—for all of their help and support during the development and production of this edition. I also thank Martha Emry and Barbara Willette who both did an excellent job of ensuring the accuracy and readability of this edition. Simply stated, the team I have been working with is outstanding, and I truly appreciate all their hard work and efforts.

S. T. Tan

ABOUT THE AUTHOR

SOO T. TAN received his S.B. degree from Massachusetts Institute of Technology, his M.S. degree from the University of Wisconsin–Madison, and his Ph.D. from the University of California at Los Angeles. He has published numerous papers in optimal control theory, numerical analysis, and mathematics of finance. He is also the author of a series of calculus textbooks.

1 FUNDAMENTALS OF ALGEBRA

How much money is needed to purchase at least 100,000 shares of the Starr Communications Company? Corbyco, a giant conglomerate, wishes to purchase a minimum of 100,000 shares of the company. In Applied Example 8, page 59, you will see how Corbyco's management determines how much money they will need for the acquisition.

THIS CHAPTER CONTAINS a brief review of the algebra you will use in this course. In the process of solving many practical problems, you will need to solve algebraic equations. You will also need to simplify algebraic expressions. This chapter also contains a short review of inequalities and absolute value; their uses range from describing the domains of functions to formulating applied problems.

1.1   Real Numbers

The Set of Real Numbers

We use *real numbers* every day to describe various quantities such as temperature, salary, annual percentage rate, shoe size, and grade point average. Some of the symbols we use to represent real numbers are

$$3, \quad -17, \quad \sqrt{2}, \quad 0.666\ldots, \quad 113, \quad 3.9, \quad 0.12875$$

To construct the set of real numbers, we start with the set of **natural numbers** (also called counting numbers)

$$N = \{1, 2, 3, \ldots\}$$

and adjoin other numbers to it. The set

$$W = \{0, 1, 2, 3, \ldots\}$$

of **whole numbers** is obtained by adjoining the single number 0 to N. By adjoining the negatives of the natural numbers to the set W, we obtain the set of **integers**

$$I = \{\ldots, -3, -2, -1, 0, 1, 2, 3, \ldots\}$$

Next, we consider the set Q of **rational numbers,** numbers of the form $\frac{a}{b}$, where a and b are integers, with $b \neq 0$. Using set notation, we write

$$Q = \{\tfrac{a}{b} \mid a \text{ and } b \text{ are integers, } b \neq 0\}$$

Observe that I is contained in Q, since each integer may be written in the form $\frac{a}{b}$, with $b = 1$. For example, the integer 6 may be written in the form $\frac{6}{1}$. Symbolically, we express the fact that I is contained in Q by writing

$$I \subset Q$$

However, Q is not contained in I, since fractions such as $\frac{1}{2}$ and $\frac{23}{25}$ are not integers. To show the relationships of the sets N, W, I, and Q, we write

$$N \subset W \subset I \subset Q$$

This says that N is a proper subset of W, W is a proper subset of I, and so on.*

Finally, we obtain the set of real numbers by adjoining the set of rational numbers to the set of **irrational numbers** (Ir)—numbers that cannot be expressed in the form $\frac{a}{b}$, where a, b are integers ($b \neq 0$). Examples of irrational numbers are $\sqrt{2}$, $\sqrt{3}$, π, and so on. Thus, the set

$$R = Q \cup Ir$$

comprising all rational numbers and irrational numbers is called the set of **real numbers.** (See Figure 1.)

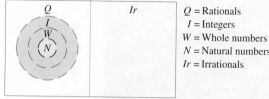

Q = Rationals
I = Integers
W = Whole numbers
N = Natural numbers
Ir = Irrationals

FIGURE 1
The set of all real numbers consists of the set of rational numbers plus the set of irrational numbers.

---

*A set A is a proper subset of a set B if every element of a set A is also an element of a set B and there exists at least one element in B that is not in A.

Representing Real Numbers as Decimals

Every real number can be written as a decimal. A rational number can be represented as either a repeating decimal or a terminating decimal. For example, $\frac{2}{3}$ is represented by the repeating decimal

$$0.66666666\ldots \qquad \text{Repeating decimal; note that the integer 6 repeats.}$$

which may also be written $0.\overline{6}$, where the bar above the 6 indicates that the 6 repeats indefinitely. The number $\frac{1}{2}$ is represented by the terminating decimal

$$0.5 \qquad \text{Terminating decimal}$$

When an irrational number is represented as a decimal, it neither terminates nor repeats. For example,

$$\sqrt{2} = 1.41421\ldots \quad \text{and} \quad \pi = 3.14159\ldots$$

Table 1 summarizes this classification of real numbers.

TABLE 1

The Set of Real Numbers

Set	Description	Examples	Decimal Representation
Natural numbers	Counting numbers	$1, 2, 3, \ldots$	Terminating decimals
Whole numbers	Counting numbers and 0	$0, 1, 2, 3, \ldots$	Terminating decimals
Integers	Natural numbers, their negatives, and 0	$\ldots, -3, -2, -1, 0, 1, 2, 3, \ldots$	Terminating decimals
Rational numbers	Numbers that can be written in the form $\frac{a}{b}$, where a and b are integers and $b \neq 0$	$-3, -\frac{3}{4}, -0.2\overline{22}, 0, \frac{5}{6}, 2, 4.31\overline{11}$	Terminating or repeating decimals
Irrational numbers	Numbers that cannot be written in the form $\frac{a}{b}$, where a and b are integers and $b \neq 0$	$\sqrt{2}, \sqrt{3}, \pi$ $1.414213\ldots,$ $1.732050\ldots$	Nonterminating, non-repeating decimals
Real numbers	Rational and irrational numbers	All of the above	All types of decimals

Representing Real Numbers on a Number Line

Real numbers may be represented geometrically by points on a line. This *real number*, or *coordinate*, *line* is constructed as follows: Arbitrarily select a point on a straight line to represent the number 0. This point is called the *origin*. If the line is horizontal, then choose a point at a convenient distance to the right of the origin to represent the number 1. This determines the scale for the number line.

The point representing each positive real number x lies x units to the right of 0, and the point representing each negative real number x lies $-x$ units to the left of 0. Thus, real numbers may be represented by points on a line in such a way that corresponding to each real number there is exactly one point on a line, and vice versa. In this way, a *one-to-one correspondence* is set up between the set of real numbers and the set of points on the number line, with all the positive numbers lying to the right of the origin and all the negative numbers lying to the left of the origin (Figure 2).

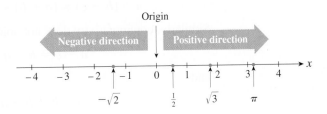

FIGURE 2
The real number line

Operations with Real Numbers

Two real numbers may be combined to obtain a real number. The operation of addition, written $+$, enables us to combine any two numbers a and b to obtain their *sum,* denoted by $a + b$. Another operation, multiplication, written $\cdot$, enables us to combine any two real numbers a and b to form their product, the number $a \cdot b$ (more simply written ab). These two operations are subject to the rules of operation given in Table 2.

TABLE 2

Rules of Operation for Real Numbers

Rule		**Illustration**
Under addition		
1. $a + b = b + a$	Commutative law of addition	$2 + 3 = 3 + 2$
2. $a + (b + c) = (a + b) + c$	Associative law of addition	$4 + (2 + 3) = (4 + 2) + 3$
3. $a + 0 = a$	Identity law of addition	$6 + 0 = 6$
4. $a + (-a) = 0$	Inverse law of addition	$5 + (-5) = 0$
Under multiplication		
1. $ab = ba$	Commutative law of multiplication	$3 \cdot 2 = 2 \cdot 3$
2. $a(bc) = (ab)c$	Associative law of multiplication	$4(3 \cdot 2) = (4 \cdot 3)2$
3. $a \cdot 1 = 1 \cdot a$	Identity law of multiplication	$4 \cdot 1 = 1 \cdot 4$
4. $a\left(\dfrac{1}{a}\right) = 1 \quad (a \neq 0)$	Inverse law of multiplication	$3\left(\dfrac{1}{3}\right) = 1$
Under addition and multiplication		
1. $a(b + c) = ab + ac$	Distributive law for multiplication with respect to addition	$3(4 + 5) = 3 \cdot 4 + 3 \cdot 5$

The operation of subtraction is defined in terms of addition. Thus,

$$a + (-b)$$

where $-b$ is the additive inverse of b, may be written in the more familiar form $a - b$, and we say that b is subtracted from a. Similarly, the operation of division is defined in terms of multiplication. Recall that the multiplicative inverse of a nonzero real number b is $\frac{1}{b}$, also written b^{-1}. Then,

$$a\left(\frac{1}{b}\right)$$

is written $\frac{a}{b}$, and we say that a is divided by b. Thus, $4\left(\frac{1}{3}\right) = \frac{4}{3}$. Remember, zero does not have a multiplicative inverse since division by zero is not defined.

Do the operations of associativity and commutativity hold for subtraction and division? Looking first at associativity, we see that the answer is no, since

$$a - (b - c) \neq (a - b) - c \qquad 7 - (4 - 2) \neq (7 - 4) - 2, \text{ or } 5 \neq 1$$

and

$$a \div (b \div c) \neq (a \div b) \div c \qquad 8 \div (4 \div 2) \neq (8 \div 4) \div 2, \text{ or } 4 \neq 1$$

Similarly, commutativity does not hold because

$$a - b \neq b - a \qquad 7 - 4 \neq 4 - 7, \text{ or } 3 \neq -3$$

and

$$a \div b \neq b \div a \qquad 8 \div 4 \neq 4 \div 8, \text{ or } 2 \neq \frac{1}{2}$$

VIDEO **EXAMPLE 1** State the real number property that justifies each statement.

Statement	Property
a. $4 + (x - 2) = 4 + (-2 + x)$	Commutative law of addition
b. $(a + 2b) + c = a + (2b + c)$	Associative law of addition
c. $x(y - z + 2) = (y - z + 2)x$	Commutative law of multiplication
d. $4(xy^2) = (4x)y^2$	Associative law of multiplication
e. $x(y - 2) = xy - 2x$	Distributive law for multiplication under addition

Using the properties of real numbers listed earlier, we can derive all other algebraic properties of real numbers. Some of the more important properties are given in Tables 3–5.

TABLE 3

Properties of Negatives

Property	Illustration
1. $-(-a) = a$	$-(-6) = 6$
2. $(-a)b = -(ab) = a(-b)$	$(-3)4 = -(3 \cdot 4) = 3(-4)$
3. $(-a)(-b) = ab$	$(-3)(-4) = 3 \cdot 4$
4. $(-1)a = -a$	$(-1)5 = -5$

TABLE 4

Properties Involving Zero

Property
1. $a \cdot 0 = 0$
2. If $ab = 0$, then $a = 0$, $b = 0$, or both.

TABLE 5

Properties of Quotients

Property		Illustration
1. $\dfrac{a}{b} = \dfrac{c}{d}$ if $ad = bc$	$(b, d \neq 0)$	$\dfrac{3}{4} = \dfrac{9}{12}$ because $3 \cdot 12 = 9 \cdot 4$
2. $\dfrac{ca}{cb} = \dfrac{a}{b}$	$(b, c \neq 0)$	$\dfrac{4 \cdot 3}{4 \cdot 8} = \dfrac{3}{8}$
3. $\dfrac{a}{-b} = \dfrac{-a}{b} = -\dfrac{a}{b}$	$(b \neq 0)$	$\dfrac{4}{-3} = \dfrac{-4}{3} = -\dfrac{4}{3}$
4. $\dfrac{a}{b} \cdot \dfrac{c}{d} = \dfrac{ac}{bd}$	$(b, d \neq 0)$	$\dfrac{3}{4} \cdot \dfrac{5}{2} = \dfrac{15}{8}$
5. $\dfrac{a}{b} \div \dfrac{c}{d} = \dfrac{a}{b} \cdot \dfrac{d}{c} = \dfrac{ad}{bc}$	$(b, c, d \neq 0)$	$\dfrac{3}{4} \div \dfrac{5}{2} = \dfrac{3}{4} \cdot \dfrac{2}{5} = \dfrac{3}{10}$
6. $\dfrac{a}{b} + \dfrac{c}{d} = \dfrac{ad + bc}{bd}$	$(b, d \neq 0)$	$\dfrac{3}{4} + \dfrac{5}{2} = \dfrac{3 \cdot 2 + 4 \cdot 5}{8} = \dfrac{13}{4}$
7. $\dfrac{a}{b} - \dfrac{c}{d} = \dfrac{ad - bc}{bd}$	$(b, d \neq 0)$	$\dfrac{3}{4} - \dfrac{5}{2} = \dfrac{3 \cdot 2 - 4 \cdot 5}{8} = -\dfrac{7}{4}$

Note In the rest of this book, we will assume that all variables are restricted so that division by zero is excluded.

VIDEO **EXAMPLE 2** State the real number property that justifies each statement.

Statement	Property
a. $-(-4) = 4$	Property 1 of negatives
b. If $(4x - 1)(x + 3) = 0$, then $x = \dfrac{1}{4}$ or $x = -3$.	Property 2 of zero properties
c. $\dfrac{(x - 1)(x + 1)}{(x - 1)(x - 3)} = \dfrac{x + 1}{x - 3}$	Property 2 of quotients
d. $\dfrac{x - 1}{y} \div \dfrac{y + 1}{x} = \dfrac{x - 1}{y} \cdot \dfrac{x}{y + 1} = \dfrac{x(x - 1)}{y(y + 1)}$	Property 5 of quotients
e. $\dfrac{x}{y} + \dfrac{x}{y + 1} = \dfrac{x(y + 1) + xy}{y(y + 1)}$	Property 6 of quotients
$= \dfrac{xy + x + xy}{y(y + 1)} = \dfrac{2xy + x}{y(y + 1)}$	Distributive law ▪

1.1   Self-Check Exercises

State the property (or properties) that justify each statement.

1. $(3v + 2) - w = 3v + (2 - w)$

2. $(3s)(4t) = 3[s(4t)]$

3. $-(-s + t) = s - t$

4. $\dfrac{2}{-(u - v)} = -\dfrac{2}{u - v}$

Solutions to Self-Check Exercises 1.1 can be found on page 7.

1.1   Concept Questions

1. What is a natural number? A whole number? An integer? A rational number? An irrational number? A real number? Give examples.

2. a. The associative law of addition states that
$a + (b + c) = $ _____.
   b. The distributive law states that $ab + ac = $ _____.

3. What can you say about a and b if $ab \neq 0$? How about a, b, and c if $abc \neq 0$?

1.1   Exercises

In Exercises 1–10, classify the number as to type. (For example, $\frac{1}{2}$ is rational and real, whereas $\sqrt{5}$ is irrational and real.)

1. -3 **2.** -420 **3.** $\dfrac{3}{8}$ **4.** $-\dfrac{4}{125}$

5. $\sqrt{11}$ **6.** $-\sqrt{5}$ **7.** $\dfrac{\pi}{2}$ **8.** $\dfrac{2}{\pi}$

9. $2.\overline{421}$ **10.** $2.71828\ldots$

In Exercises 11–16, indicate whether the statement is true or false.

11. Every integer is a whole number.

12. Every integer is a rational number.

13. Every natural number is an integer.

14. Every rational number is a real number.

15. Every natural number is an irrational number.

16. Every irrational number is a real number.

In Exercises 17–36, state the real number property that justifies the statement.

17. $(2x + y) + z = z + (2x + y)$

18. $3x + (2y + z) = (3x + 2y) + z$

19. $u(3v + w) = (3v + w)u$

20. $a^2(b^2c) = (a^2b^2)c$

21. $u(2v + w) = 2uv + uw$

22. $(2u + v)w = 2uw + vw$

23. $(2x + 3y) + (x + 4y) = 2x + [3y + (x + 4y)]$

24. $(a + 2b)(a - 3b) = a(a - 3b) + 2b(a - 3b)$

25. $a - [-(c + d)] = a + (c + d)$

26. $-(2x + y)[-(3x + 2y)] = (2x + y)(3x + 2y)$

27. $0(2a + 3b) = 0$

28. If $(x - y)(x + y) = 0$, then $x = y$ or $x = -y$.

29. If $(x - 2)(2x + 5) = 0$, then $x = 2$ or $x = -\frac{5}{2}$.

30. If $x(2x - 9) = 0$, then $x = 0$ or $x = \frac{9}{2}$.

31. $\dfrac{(x + 1)(x - 3)}{(2x + 1)(x - 3)} = \dfrac{x + 1}{2x + 1}$

32. $\dfrac{(2x + 1)(x + 3)}{(2x - 1)(x + 3)} = \dfrac{2x + 1}{2x - 1}$

33. $\dfrac{a + b}{b} \div \dfrac{a - b}{ab} = \dfrac{a(a + b)}{a - b}$

34. $\dfrac{x + 2y}{3x + y} \div \dfrac{x}{6x + 2y} = \dfrac{x + 2y}{3x + y} \cdot \dfrac{2(3x + y)}{x} = \dfrac{2(x + 2y)}{x}$

35. $\dfrac{a}{b + c} + \dfrac{c}{b} = \dfrac{ab + bc + c^2}{b(b + c)}$

36. $\dfrac{x + y}{x + 1} - \dfrac{y}{x} = \dfrac{x^2 - y}{x(x + 1)}$

In Exercises 37–42, indicate whether the statement is true or false.

37. If $ab = 1$, then $a = 1$ or $b = 1$.

38. If $ab = 0$ and $a \neq 0$, then $b = 0$.

39. $a - b = b - a$

40. $a \div b = b \div a$

41. $(a - b) - c = a - (b - c)$

42. $a \div (b \div c) = (a \div b) \div c$

1.1 Solutions to Self-Check Exercises

1. Associative law of addition: $a + (b + c) = (a + b) + c$

2. Associative law of multiplication: $a(bc) = (ab)c$

3. Distributive law for multiplication: $a(b + c) = ab + ac$
   Properties 1 and 4 of negatives: $-(-a) = a; (-1)a = -a$

4. Property 3 of quotients: $\dfrac{a}{-b} = \dfrac{-a}{b} = -\dfrac{a}{b}$

1.2 Polynomials

Exponents

Expressions such as 2^5, $(-3)^2$, and $\left(\frac{1}{4}\right)^4$ are exponential expressions. More generally, if n is a natural number and a is a real number, then a^n represents the product of the real number a and itself n times.

Exponential Notation

If a is a real number and n is a natural number, then

$$a^n = \underbrace{a \cdot a \cdot a \cdots \cdot a}_{n \text{ factors}} \qquad \underbrace{3^4 = 3 \cdot 3 \cdot 3 \cdot 3}_{4 \text{ factors}}$$

The natural number n is called the **exponent**, and the real number a is called the **base.**

EXAMPLE 1

a. $4^4 = (4)(4)(4)(4) = 256$

b. $(-5)^3 = (-5)(-5)(-5) = -125$

c. $\left(\dfrac{1}{2}\right)^3 = \left(\dfrac{1}{2}\right)\left(\dfrac{1}{2}\right)\left(\dfrac{1}{2}\right) = \dfrac{1}{8}$

d. $\left(-\dfrac{1}{3}\right)^2 = \left(-\dfrac{1}{3}\right)\left(-\dfrac{1}{3}\right) = \dfrac{1}{9}$

When we evaluate expressions such as $3^2 \cdot 3^3$, we use the following property of exponents to write the product in exponential form.

Property 1

If m and n are natural numbers and a is any real number, then

$$a^m \cdot a^n = a^{m+n} \qquad 3^2 \cdot 3^3 = 3^{2+3} = 3^5$$

To verify that Property 1 follows from the definition of an exponential expression, we note that the total number of factors in the exponential expression

$$a^m \cdot a^n = \underbrace{a \cdot a \cdot a \cdots \cdots a}_{m \text{ factors}} \cdot \underbrace{a \cdot a \cdot a \cdots \cdots a}_{n \text{ factors}}$$

is $m + n$.

EXAMPLE 2

a. $3^2 \cdot 3^3 = 3^{2+3} = 3^5 = 243$

b. $(-2)^2 \cdot (-2)^5 = (-2)^{2+5} = (-2)^7 = -128$

c. $(3x) \cdot (3x)^3 = (3x)^{1+3} = (3x)^4 = 81x^4$

⚠ Be careful to apply the exponent to the indicated base only. Note that

$$4 \cdot x^2 = 4x^2 \neq (4x)^2 = 4^2 \cdot x^2 = 16x^2$$

The exponent applies to $4x$.
The exponent applies only to the base x.

and

$$-3^2 = -9 \neq (-3)^2 = 9$$

The exponent applies to -3.
The exponent applies only to the base 3.

Polynomials

Recall that a *variable* is a letter that is used to represent any element of a given set. However, unless specified otherwise, variables in this text will represent real numbers. Sometimes physical considerations impose restrictions on the values a variable may assume. For example, if the variable x denotes the number of LCD television sets sold daily in an appliance store, then x must be a nonnegative integer. At other times, restrictions must be imposed on x in order for an expression to make sense. For example, in the expression $\frac{1}{x+2}$, x cannot take on the value -2, since division by 0 is not permitted. We call the set of all real numbers that a variable is allowed to assume the *domain of the variable*.

In contrast to a variable, a *constant* is a fixed number or letter whose value remains fixed throughout a particular discussion. For example, in the expression $\frac{1}{2}gt^2$, which gives the distance in feet covered by a free-falling body near the surface of the earth, t seconds from rest, the letter g represents the constant of acceleration due to gravity (approximately 32 feet/second/second), whereas the letter t is a variable with domain consisting of nonnegative real numbers.

By combining constants and variables through addition, subtraction, multiplication, division, exponentiation, and root extraction, we obtain *algebraic expressions.* Examples of algebraic expressions are

$$3x - 4y \qquad 2x^2 - y + \frac{1}{xy} \qquad \frac{ax - b}{1 - x^2} \qquad \frac{3xy^{-2} + \pi}{x^2 + y^2 + z^2}$$

where a and b are constants and x, y, and z are variables. Intimidating as some of these expressions might be, remember that they are just real numbers. For example, if $x = 1$ and $y = 4$, then the second expression represents the number

$$2(1)^2 - 4 + \frac{1}{(1)(4)}$$

or $-\frac{7}{4}$, obtained by replacing x and y in the expression by the appropriate values.

Polynomials are an important class of algebraic expressions. The simplest polynomials are those involving *one* variable.

Polynomial in One Variable

A **polynomial** in x is an expression of the form

$$a_n x^n + a_{n-1} x^{n-1} + \cdots + a_1 x + a_0$$

where n is a nonnegative integer and $a_0, a_1, \ldots, a_n$ are real numbers, with $a_n \neq 0$.

The expressions $a_k x^k$ in the sum are called the *terms* of the polynomial. The numbers $a_0, a_1, \ldots, a_n$ are called the *coefficients* of 1, x, x^2, $\ldots$, x^n, respectively. The coefficient a_n of x^n (the highest power in x) is called the *leading coefficient* of the polynomial. The nonnegative integer n gives the *degree* of the polynomial. For example, consider the polynomial

$$-2x^5 + 8x^3 - 6x^2 + 3x + 1$$

1. The terms of the polynomial are $-2x^5$, $8x^3$, $-6x^2$, $3x$, and 1.
2. The coefficients of 1, x, x^2, x^3, x^4, and x^5 are 1, 3, -6, 8, 0, and -2, respectively.
3. The leading coefficient of the polynomial is -2.
4. The degree of the polynomial is 5.

A polynomial that has just one term (such as $2x^3$) is called a *monomial*; a polynomial that has exactly two terms (such as $x^3 + x$) is called a *binomial*; and a polynomial that has only three terms (such as $-2x^3 + x - 8$) is called a *trinomial.* Also, a polynomial consisting of one (constant) term a_0 (such as the monomial -8) is called a *constant polynomial.* Observe that the degree of a constant polynomial a_0, with $a_0 \neq 0$, is 0 because we can write $a_0 = a_0 x^0$ and see that $n = 0$ in this situation. If all the coefficients of a polynomial are 0, it is called the *zero polynomial* and is denoted by 0. The zero polynomial is not assigned a degree.

Most of the terminology used for a polynomial in one variable carries over to the discussion of polynomials in several variables. But the *degree of a term* in a polynomial in several variables is obtained by adding the powers of all variables in the term, and the *degree of the polynomial* is given by the highest degree of all its terms. For example, the polynomial

$$2x^2y^5 - 3xy^3 + 8xy^2 - 3y + 4$$

is a polynomial in the two variables x and y. It has five terms with degrees 7, 4, 3, 1, and 0, respectively. Accordingly, the degree of the polynomial is 7.

Adding and Subtracting Polynomials

Constant terms and terms that have the same variable and exponent are called *like* or *similar* terms. Like terms may be combined by adding or subtracting their numerical coefficients. For example,

$$3x + 7x = (3 + 7)x = 10x \qquad \text{Add like terms.}$$

and

$$\frac{1}{2}m^2 - 3m^2 = \left(\frac{1}{2} - 3\right)m^2 = -\frac{5}{2}m^2 \qquad \text{Subtract like terms.}$$

The distributive property of the real number system,

$$ab + ac = a(b + c)$$

is used to justify this procedure.

To add or subtract two or more polynomials, first remove the parentheses and then combine like terms. The resulting expression is then written in order of decreasing degree from left to right.

EXAMPLE 3

a. $(3x^3 + 2x^2 - 4x + 5) + (-2x^3 - 2x^2 - 2)$

$\qquad = 3x^3 + 2x^2 - 4x + 5 - 2x^3 - 2x^2 - 2 \qquad$ Remove parentheses.

$\qquad = 3x^3 - 2x^3 + 2x^2 - 2x^2 - 4x + 5 - 2 \qquad$ Group like terms together.

$\qquad = x^3 - 4x + 3 \qquad$ Combine like terms.

b. $(2x^4 + 3x^3 + 4x + 6) - (3x^4 + 9x^3 + 3x^2)$

$\qquad = 2x^4 + 3x^3 + 4x + 6 - 3x^4 - 9x^3 - 3x^2 \qquad$ Remove parentheses. Note that the minus sign preceding the second polynomial changes the sign of each term of that polynomial.

$\qquad = 2x^4 - 3x^4 + 3x^3 - 9x^3 - 3x^2 + 4x + 6 \qquad$ Group like terms.

$\qquad = -x^4 - 6x^3 - 3x^2 + 4x + 6 \qquad$ Combine like terms.   ■

Multiplying Polynomials

To find the product of two polynomials, we again use the distributive property for real numbers. For example, to compute the product $3x(4x - 2)$, we use the distributive law to obtain

$$3x(4x - 2) = (3x)(4x) + (3x)(-2) \qquad a(b + c) = ab + ac$$
$$= 12x^2 - 6x$$

Observe that each term of one polynomial is multiplied by each term of the other. The resulting expression is then simplified by combining like terms. In general, an algebraic expression is *simplified* if none of its terms are similar.

EXAMPLE 4 Find the product of $(3x + 5)(2x - 3)$.

Solution

$$
\begin{aligned}
(3x + 5)(2x - 3) &= 3x(2x - 3) + 5(2x - 3) &&\text{Distributive property} \\
&= (3x)(2x) + (3x)(-3) &&\text{Distributive property} \\
&\quad + (5)(2x) + (5)(-3) \\
&= 6x^2 - 9x + 10x - 15 &&\text{Multiply terms.} \\
&= 6x^2 + x - 15 &&\text{Combine like terms.} \quad\blacksquare
\end{aligned}
$$

EXAMPLE 5 Find the product of $(2t^2 - t + 3)(2t^2 - 1)$.

Solution

$$
\begin{aligned}
(2t^2 - t &+ 3)(2t^2 - 1) \\
&= 2t^2(2t^2 - 1) - t(2t^2 - 1) + 3(2t^2 - 1) &&\text{Distributive property} \\
&= (2t^2)(2t^2) + (2t^2)(-1) + (-t)(2t^2) &&\text{Distributive property} \\
&\quad + (-t)(-1) + (3)(2t^2) + (3)(-1) \\
&= 4t^4 - 2t^2 - 2t^3 + t + 6t^2 - 3 &&\text{Multiply terms.} \\
&= 4t^4 - 2t^3 + 4t^2 + t - 3 &&\text{Combine terms.}
\end{aligned}
$$

Alternative Solution We can also find the product by arranging the polynomials vertically and multiplying:

$$
\begin{array}{r}
2t^2 - t + 3 \\
2t^2 - 1 \\
\hline
4t^4 - 2t^3 + 6t^2 \\
- 2t^2 + t - 3 \\
\hline
4t^4 - 2t^3 + 4t^2 + t - 3
\end{array}
$$

$\blacksquare$

The polynomials in Examples 4 and 5 are polynomials in one variable. The operations of addition, subtraction, and multiplication are performed on polynomials of more than one variable in the same way as they are for polynomials in one variable.

▶ VIDEO **EXAMPLE 6** Multiply $(3x - y)(4x^2 - 2y)$.

Solution

$$
\begin{aligned}
(3x - y)(4x^2 - 2y) &= 3x(4x^2 - 2y) - y(4x^2 - 2y) &&\text{Distributive property} \\
&= 12x^3 - 6xy - 4x^2y + 2y^2 &&\text{Distributive property} \\
&= 12x^3 - 4x^2y - 6xy + 2y^2 &&\text{\small Arrange terms in order of} \\
&&&\text{\small descending powers of } x. \quad\blacksquare
\end{aligned}
$$

Several commonly used products of polynomials are summarized in Table 6. Since products of this type occur so frequently, you will find it helpful to memorize these formulas.

TABLE 6

Special Products

Formula	Illustration
1. $(a + b)^2 = a^2 + 2ab + b^2$	$(2x + 3y)^2 = (2x)^2 + 2(2x)(3y) + (3y)^2$ $= 4x^2 + 12xy + 9y^2$
2. $(a - b)^2 = a^2 - 2ab + b^2$	$(4x - 2y)^2 = (4x)^2 - 2(4x)(2y) + (2y)^2$ $= 16x^2 - 16xy + 4y^2$
3. $(a + b)(a - b) = a^2 - b^2$	$(2x + y)(2x - y) = (2x)^2 - (y)^2$ $= 4x^2 - y^2$

EXAMPLE 7 Use the special product formulas to compute:

a. $(2x + y)^2$ **b.** $(3a - 4b)^2$ **c.** $\left(\frac{1}{2}x - 1\right)\left(\frac{1}{2}x + 1\right)$

Solution

a. $(2x + y)^2 = (2x)^2 + 2(2x)(y) + y^2$ Formula 1
$= 4x^2 + 4xy + y^2$

b. $(3a - 4b)^2 = (3a)^2 - 2(3a)(4b) + (4b)^2$ Formula 2
$= 9a^2 - 24ab + 16b^2$

c. $\left(\frac{1}{2}x - 1\right)\left(\frac{1}{2}x + 1\right) = \left(\frac{1}{2}x\right)^2 - 1 = \frac{1}{4}x^2 - 1$ Formula 3

Order of Operations

The common steps in Examples 1–7 have been to remove parentheses and combine like terms. If more than one grouping symbol is present, the innermost symbols are removed first. As you work through Examples 8 and 9, note the order in which the grouping symbols are removed: parentheses ( ) first, brackets [ ] second, and finally braces { }. Also, note that the operations of multiplication and division take precedence over addition and subtraction.

VIDEO **EXAMPLE 8** Perform the indicated operations:

$$2t^3 - \{t^2 - [t - (2t - 1)] + 4\}$$

Solution

$2t^3 - \{t^2 - [t - (2t - 1)] + 4\}$
$= 2t^3 - \{t^2 - [t - 2t + 1] + 4\}$ Remove parentheses.
$= 2t^3 - \{t^2 - [-t + 1] + 4\}$ Combine like terms within the brackets.
$= 2t^3 - \{t^2 + t - 1 + 4\}$ Remove brackets.
$= 2t^3 - \{t^2 + t + 3\}$ Combine like terms within the braces.
$= 2t^3 - t^2 - t - 3$ Remove braces.

EXAMPLE 9 Simplify $2\{3 - 2[x - 2x(3 - x)]\}$.

Solution

$2\{3 - 2[x - 2x(3 - x)]\} = 2\{3 - 2[x - 6x + 2x^2]\}$ Remove parentheses.
$= 2\{3 - 2[-5x + 2x^2]\}$ Combine like terms.
$= 2\{3 + 10x - 4x^2\}$ Remove brackets.
$= 6 + 20x - 8x^2$ Remove braces.
$= -8x^2 + 20x + 6$ Write answer in order of descending powers of x.

1.2 Self-Check Exercises

1. Find the product of $(2x + 3y)(3x - 2y)$.

2. Simplify $3x - 2\{2x - [x - 2(x - 2)] + 1\}$.

Solutions to Self-Check Exercises 1.2 can be found on page 14.

1.2 Concept Questions

1. Describe a polynomial of degree n in x. Give an example of a polynomial of degree 4 in x.

2. Without looking at the formulas in the text, complete the following:
   a. $(1 + b)^2 =$ _____
   b. $(a - b)^2 =$ _____
   c. $(a + b)(a - b) =$ _____

1.2 Exercises

In Exercises 1–12, evaluate the expression.

1. 3^4

2. $(-2)^5$

3. $\left(\dfrac{2}{3}\right)^3$

4. $\left(-\dfrac{3}{4}\right)^2$

5. -3^4

6. $-\left(-\dfrac{4}{5}\right)^3$

7. $-3\left(\dfrac{3}{5}\right)^3$

8. $\left(-\dfrac{2}{3}\right)^2\left(-\dfrac{3}{4}\right)^3$

9. $2^3 \cdot 2^5$

10. $(-3)^2 \cdot (-3)^3$

11. $(3y)^2(3y)^3$

12. $(-2x)^3(-2x)^2$

In Exercises 13–56, perform the indicated operations and simplify.

13. $(2x + 3) + (4x - 6)$

14. $(-3x + 2) - (4x - 3)$

15. $(7x^2 - 2x + 5) + (2x^2 + 5x - 4)$

16. $(3x^2 + 5xy + 2y) + (4 - 3xy - 2x^2)$

17. $(5y^2 - 2y + 1) - (y^2 - 4y - 8)$

18. $(2x^2 - 3x + 4) - (-x^2 + 2x - 6)$

19. $(2.4x^3 - 3x^2 + 1.7x - 6.2) - (1.2x^3 + 1.2x^2 - 0.8x + 2)$

20. $(1.4x^3 - 1.2x^2 + 3.2) - (-0.8x^3 - 2.1x - 1.8)$

21. $(3x^2)(2x^3)$

22. $(-2rs^2)(4r^2s^2)(2s)$

23. $-2x(x^2 - 2) + 4x^3$

24. $xy(2y - 3x)$

25. $2m(3m - 4) + m(m - 1)$

26. $-3x(2x^2 + 3x - 5) + 2x(x^2 - 3)$

27. $3(2a - b) - 4(b - 2a)$

28. $2(3m - 1) - 3(-4m + 2n)$

29. $(2x + 3)(3x - 2)$

30. $(3r - 1)(2r + 5)$

31. $(2x - 3y)(3x + 2y)$

32. $(5m - 2n)(5m + 3n)$

33. $(3r + 2s)(4r - 3s)$

34. $(2m + 3n)(3m - 2n)$

35. $(0.2x + 1.2y)(0.3x - 2.1y)$

36. $(3.2m - 1.7n)(4.2m + 1.3n)$

37. $(2x - y)(3x^2 + 2y)$

38. $(3m - 2n^2)(2m^2 + 3n)$

39. $(2x + 3y)^2$

40. $(3m - 2n)^2$

41. $(2u - v)(2u + v)$

42. $(3r + 4s)(3r - 4s)$

43. $(2x - 1)^2 + 3x - 2(x^2 + 1) + 3$

44. $(3m + 2)^2 - 2m(1 - m) - 4$

45. $(2x + 3y)^2 - (2y + 1)(3x - 2) + 2(x - y)$

46. $(x - 2y)(y + 3x) - 2xy + 3(x + y - 1)$

47. $(t^2 - 2t + 4)(2t^2 + 1)$

48. $(3m^2 - 1)(2m^2 + 3m - 4)$

49. $2x - \{3x - [x - (2x - 1)]\}$

50. $3m - 2\{m - 3[2m - (m - 5)] + 4\}$

51. $x - \{2x - [-x - (1 + x)]\}$

52. $3x^2 - \{x^2 + 1 - x[x - (2x - 1)]\} + 2$

53. $(2x - 3)^2 - 3(x + 4)(x - 4) + 2(x - 4) + 1$

54. $(x - 2y)^2 + 2(x + y)(x - 3y) + x(2x + 3y + 2)$

55. $2x\{3x[2x - (3 - x)] + (x + 1)(2x - 3)\}$

56. $-3[(x + 2y)^2 - (3x - 2y)^2 + (2x - y)(2x + y)]$

57. **PROFIT OF A COMPANY** The total revenue realized in the sale of x units of the LectroCopy photocopying machine is
$$-0.04x^2 + 2000x$$
dollars/week, and the total cost incurred in manufacturing x units of the machines is
$$0.000002x^3 - 0.02x^2 + 1000x + 120{,}000$$
dollars/week ($0 \le x \le 50{,}000$). Find an expression giving the total weekly profit of the company.
Hint: The profit is revenue minus cost.

58. **PROFIT OF A COMPANY** A manufacturer of tennis rackets finds that the total cost of manufacturing x rackets/day is given by
$$0.0001x^2 + 4x + 400$$
dollars. Each racket can be sold at a price of p dollars, where
$$p = -0.0004x + 10$$
Find an expression giving the daily profit for the manufacturer, assuming that all the rackets manufactured can be sold.
Hint: The total revenue is given by the total number of rackets sold multiplied by the price of each racket. The profit is given by revenue minus cost.

59. REVENUE OF A COMPANY Jake owns two gas stations in town. The total revenue of the first gas station for the next 12 months is projected to be

$$0.2t^2 + 150t \qquad (0 \le t \le 12)$$

thousand dollars t months from now. The total revenue of the second gas station for the next 12 months is projected to be

$$0.5t^2 + 200t \qquad (0 \le t \le 12)$$

thousand dollars t months from now. Find an expression that gives the total revenue realized by Jake's gas stations in month t $(0 \le t \le 12)$.

60. REVENUE OF A COMPANY Refer to Exercise 59. Find an expression that gives the amount by which the revenue of the second gas station will exceed that of the first gas stations in month t $(0 < t \le 12)$.

61. PRISON OVERCROWDING The 1980s saw a trend toward old-fashioned punitive deterrence as opposed to the more liberal penal policies and community-based corrections popular in the 1960s and early 1970s. As a result, prisons became more crowded, and the gap between the number of people in prison and prison capacity widened. Based on figures from the U.S. Department of Justice, the number of prisoners (in thousands) in federal and state prisons is approximately

$$3.5t^2 + 26.7t + 436.2 \qquad (0 \le t \le 10)$$

and the number of inmates (in thousands) for which prisons were designed is given by

$$24.3t + 365 \qquad (0 \le t \le 10)$$

where t is measured in years and $t = 0$ corresponds to 1984. Find an expression giving the gap between the number of prisoners and the number for which the prisons were designed at any time t.

Source: U.S. Department of Justice.

62. HEALTH-CARE SPENDING Health-care spending per person (in dollars) by the private sector includes payments by individuals, corporations, and their insurance companies and is approximated by

$$2.5t^2 + 18.5t + 509 \qquad (0 \le t \le 6)$$

where t is measured in years and $t = 0$ corresponds to the beginning of 1994. The corresponding government spending (in dollars), including expenditures for Medicaid and other federal, state, and local government public health care, is

$$-1.1t^2 + 29.1t + 429 \qquad (0 \le t \le 6)$$

where t has the same meaning as before. Find an expression for the difference between private and government expenditures per person at any time t. What was the difference between private and government expenditures per person at the beginning of 1998? At the beginning of 2000?

Source: Health Care Financing Administration.

In Exercises 63–66, determine whether the statement is true or false. If it is true, explain why it is true. If it is false, give an example to show why it is false.

63. If m and n are natural numbers and a and b are real numbers, then $a^m \cdot b^n = (ab)^{m+n}$.

64. $a^{16} - b^{16} = (a^8 + b^8)(a^4 + b^4)(a^2 + b^2)(a + b)(a - b)$

65. The degree of the product of a polynomial of degree m and a polynomial of degree n is mn.

66. Suppose p and q are polynomials of degree n. Then $p + q$ is a polynomial of degree n.

67. Suppose p is a polynomial of degree m and q is a polynomial of degree n, where $m > n$. What is the degree of $p - q$?

1.2 Solutions to Self-Check Exercises

1. $(2x + 3y)(3x - 2y) = 2x(3x - 2y) + 3y(3x - 2y)$

$\qquad\qquad\qquad\qquad = 6x^2 - 4xy + 9xy - 6y^2$

$\qquad\qquad\qquad\qquad = 6x^2 + 5xy - 6y^2$

2. $3x - 2\{2x - [x - 2(x - 2)] + 1\}$

$\qquad = 3x - 2\{2x - [x - 2x + 4] + 1\}$

$\qquad = 3x - 2\{2x - [-x + 4] + 1\}$

$\qquad = 3x - 2\{2x + x - 4 + 1\}$

$\qquad = 3x - 2\{3x - 3\}$

$\qquad = 3x - 6x + 6$

$\qquad = -3x + 6$

1.3   Factoring Polynomials

Factoring

Factoring a polynomial is the process of expressing it as a product of two or more polynomials. For example, by applying the distributive property, we may write

$$3x^2 - x = x(3x - 1)$$

and we say that x and $3x - 1$ are factors of $3x^2 - x$.

How do we know whether a polynomial is completely factored? Recall that an integer greater than 1 is *prime* if its only positive integer factors are itself and 1. For example, the number 3 is prime because its only factors are 3 and 1. In the same way, a polynomial is said to be prime over the set of integral coefficients if it cannot be expressed as a product of two or more polynomials of positive degree with integral coefficients. For example, $x^2 + 2x + 2$ is a prime polynomial relative to the set of integers, whereas $x^2 - 9$ is not a prime polynomial, since $x^2 - 9 = (x + 3)(x - 3)$. Finally, a polynomial is said to be *completely factored* over the set of integers if it is expressed as a product of prime polynomials with integral coefficients.

Note   Unless otherwise mentioned, we will consider only factorization over the set of integers in this text. Hence, when the term *factor* is used, it will be understood that the factorization is to be completed over the set of integers.   ∎

Common Factors

The first step in factoring a polynomial is to check whether it contains any common factors. If it does, the common factor of highest degree is then factored out. For example, the greatest common factor of $2a^2x + 4ax + 6a$ is $2a$ because

$$2a^2x + 4ax + 6a = 2a \cdot ax + 2a \cdot 2x + 2a \cdot 3$$
$$= 2a(ax + 2x + 3)$$

EXAMPLE 1   Factor out the greatest common factor.

a. $-3t^2 + 3t$ **b.** $6a^4b^4c - 9a^2b^2$

Solution

a. Since $3t$ is a common factor of each term, we have

$$-3t^2 + 3t = 3t(-t + 1) = -3t(t - 1)$$

b. Since $3a^2b^2$ is the common factor of highest degree, we have

$$6a^4b^4c - 9a^2b^2 = 3a^2b^2(2a^2b^2c - 3)$$ ∎

Some Important Formulas

Having checked for common factors, the next step in factoring a polynomial is to express the polynomial as the product of a constant and/or one or more prime polynomials. The formulas given in Table 7 for factoring polynomials should be memorized.

TABLE 7

Factoring Formulas

Formula	Illustration
Difference of two squares $a^2 - b^2 = (a + b)(a - b)$	$x^2 - 36 = (x + 6)(x - 6)$ $8x^2 - 2y^2 = 2(4x^2 - y^2) = 2[(2x)^2 - y^2]$ $\qquad\qquad = 2(2x + y)(2x - y)$ $9 - a^6 = 3^2 - (a^3)^2 = (3 + a^3)(3 - a^3)$
Perfect square trinomial $a^2 + 2ab + b^2 = (a + b)^2$ $a^2 - 2ab + b^2 = (a - b)^2$	$x^2 + 8x + 16 = (x + 4)^2$ $4x^2 - 4xy + y^2 = (2x)^2 - 2(2x)(y) + y^2$ $\qquad\qquad\qquad = (2x - y)^2$
Sum of two cubes $a^3 + b^3 = (a + b)(a^2 - ab + b^2)$	$z^3 + 27 = z^3 + (3)^3$ $\qquad\quad = (z + 3)(z^2 - 3z + 9)$
Difference of two cubes $a^3 - b^3 = (a - b)(a^2 + ab + b^2)$	$8x^3 - y^6 = (2x)^3 - (y^2)^3$ $\qquad\qquad = (2x - y^2)(4x^2 + 2xy^2 + y^4)$

Note   Observe that a formula is given for factoring the sum of two cubes, but none is given for factoring the sum of two squares, since $x^2 + a^2$ is prime over the set of integers.   ■

VIDEO   **EXAMPLE 2** Factor:

a. $x^2 - 9$ **b.** $16x^2 - 81y^4$ **c.** $(a - b)^2 - (a^2 + b)^2$

Solution   Observe that each of the polynomials in parts (a)–(c) is the difference of two squares. Using the formula given in Table 7, we have

a. $x^2 - 9 = x^2 - 3^2 = (x + 3)(x - 3)$
b. $16x^2 - 81y^4 = (4x)^2 - (9y^2)^2 = (4x + 9y^2)(4x - 9y^2)$
c. $(a - b)^2 - (a^2 + b)^2 = [(a - b) + (a^2 + b)][(a - b) - (a^2 + b)]$
$\qquad\qquad\qquad\quad = [a - b + a^2 + b][a - b - a^2 - b]$ Remove parentheses.
$\qquad\qquad\qquad\quad = (a + a^2)(-a^2 + a - 2b)$ Combine like terms.
$\qquad\qquad\qquad\quad = a(1 + a)(-a^2 + a - 2b)$ ■

EXAMPLE 3 Factor:

a. $x^2 + 4xy + 4y^2$ **b.** $4a^2 - 12ab + 9b^2$

Solution   Recognizing each of these polynomials as a perfect square trinomial, we use a formula given in Table 7 to factor each polynomial. Thus,

a. $x^2 + 4xy + 4y^2 = x^2 + 2x(2y) + (2y)^2 = (x + 2y)(x + 2y) = (x + 2y)^2$
b. $4a^2 - 12ab + 9b^2 = (2a)^2 - 2(2a)(3b) + (3b)^2$
$\qquad\qquad\qquad\quad = (2a - 3b)(2a - 3b) = (2a - 3b)^2$ ■

EXAMPLE 4 Factor:

a. $x^3 + 8y^3$ **b.** $27a^3 - 64b^3$

Solution

a. This polynomial is the sum of two cubes. Using the formula given in Table 7, we have

$$x^3 + 8y^3 = x^3 + (2y)^3 = (x + 2y)[x^2 - x(2y) + (2y)^2]$$
$$= (x + 2y)(x^2 - 2xy + 4y^2)$$

b. Using the formula for the difference of two cubes given in Table 7, we have

$$27a^3 - 64b^3 = [(3a)^3 - (4b)^3]$$
$$= (3a - 4b)[(3a)^2 + (3a)(4b) + (4b)^2]$$
$$= (3a - 4b)(9a^2 + 12ab + 16b^2)$$

∎

Trial-and-Error Factorization

The factors of the second-degree polynomial $px^2 + qx + r$, where p, q, and r are integers, have the form

$$(ax + b)(cx + d)$$

where $ac = p$, $ad + bc = q$, and $bd = r$. Since only a limited number of choices are possible, we use a trial-and-error method to factor polynomials having this form.

For example, to factor $x^2 - 2x - 3$, we first observe that the only possible first-degree terms in each factor in the product are

$$(x \qquad)(x \qquad) \qquad \text{Since the coefficient of } x^2 \text{ is } 1$$

Next, we observe that the product of the constant terms is (-3). This gives us the following possible factors:

$$(x - 1)(x + 3)$$
$$(x + 1)(x - 3)$$

Looking once again at the polynomial $x^2 - 2x - 3$, we see that the coefficient of x is -2. Checking to see which set of factors yields -2 for the coefficient of x, we find that

Coefficients of inner terms
Coefficients of outer terms

$$(-1)(1) + (1)(3) = 2$$

Factors
Outer terms
$$(x - 1)(x + 3)$$
Inner terms

Coefficients of inner terms
Coefficients of outer terms

$$(1)(1) + (1)(-3) = -2$$

Outer terms
$$(x + 1)(x - 3)$$
Inner terms

and we conclude that the correct factorization is

$$x^2 - 2x - 3 = (x + 1)(x - 3)$$

With practice, you will soon find that you can perform many of these steps mentally, and you will no longer need to write out each step.

EXAMPLE 5 Factor:

a. $3x^2 + 4x - 4$ **b.** $3x^2 - 6x - 24$

Solution

a. Using trial and error, we find that the correct factorization is

$$3x^2 + 4x - 4 = (3x - 2)(x + 2)$$

b. Since each term has the common factor 3, we have

$$3x^2 - 6x - 24 = 3(x^2 - 2x - 8)$$

Using the trial-and-error method of factorization, we find that

$$x^2 - 2x - 8 = (x - 4)(x + 2)$$

Thus, we have

$$3x^2 - 6x - 24 = 3(x - 4)(x + 2)$$

∎

Factoring by Regrouping

Sometimes a polynomial may be factored by regrouping and rearranging terms so that a common term can be factored out. This technique is illustrated in Example 6.

EXAMPLE 6 Factor:

a. $x^3 + x + x^2 + 1$ **b.** $2ax + 2ay + bx + by$

Solution

a. We begin by rearranging the terms in order of descending powers of x. Thus,

$$x^3 + x + x^2 + 1 = x^3 + x^2 + x + 1$$
$$= x^2(x + 1) + x + 1 \quad \text{Factor the first two terms.}$$
$$= (x + 1)(x^2 + 1) \quad \text{Factor the common term } x + 1.$$

b. First, factor the common term $2a$ from the first two terms and the common term b from the last two terms. Thus,

$$2ax + 2ay + bx + by = 2a(x + y) + b(x + y)$$

Since $(x + y)$ is common to both terms of the polynomial on the right, we can factor it out. Hence,

$$2a(x + y) + b(x + y) = (x + y)(2a + b)$$

More Examples on Factoring

EXAMPLE 7 Factor:

a. $4x^6 - 4x^2$ **b.** $18x^4 - 3x^3 - 6x^2$

Solution

a. $4x^6 - 4x^2 = 4x^2(x^4 - 1)$ Common factor
$= 4x^2(x^2 - 1)(x^2 + 1)$ Difference of two squares
$= 4x^2(x - 1)(x + 1)(x^2 + 1)$ Difference of two squares

b. $18x^4 - 3x^3 - 6x^2 = 3x^2(6x^2 - x - 2)$ Common factor
$= 3x^2(3x - 2)(2x + 1)$ Trial-and-error factorization

EXAMPLE 8 Factor:

a. $3x^2y + 9x^2 - 12y - 36$ **b.** $(x - at)^3 - (x + at)^3$

Solution

a. $3x^2y + 9x^2 - 12y - 36 = 3(x^2y + 3x^2 - 4y - 12)$ Common factor
$= 3[x^2(y + 3) - 4(y + 3)]$ Regrouping
$= 3(y + 3)(x^2 - 4)$ Common factor
$= 3(y + 3)(x - 2)(x + 2)$ Difference of two squares

b. $(x - at)^3 - (x + at)^3$
$= [(x - at) - (x + at)]$
$\quad \cdot [(x - at)^2 + (x - at)(x + at) + (x + at)^2]$ Difference of two cubes
$= -2at(x^2 - 2atx + a^2t^2 + x^2 - a^2t^2 + x^2 + 2atx + a^2t^2)$
$= -2at(3x^2 + a^2t^2)$

Be sure you become familiar with the factorization methods discussed in this chapter because we will be using them throughout the text. As with many other algebraic techniques, you will find yourself becoming more proficient at factoring as you work through the exercises.

1.3 ## Self-Check Exercises

1. Factor: **a.** $4x^3 - 2x^2$ **b.** $3(a^2 + 2b^2) + 4(a^2 + 2b^2)^2$

2. Factor: **a.** $6x^2 - x - 12$ **b.** $4x^2 + 10x - 6$

Solutions to Self-Check Exercises 1.3 can be found on page 20.

Concept Questions

1. What is meant by the expression *factor a polynomial*? Illustrate the process with an example.

2. Without looking at the formulas in the text, complete the following formulas:
   a. $a^3 + b^3 = $ _____   **b.** $a^3 - b^3 = $ _____

Exercises

In Exercises 1–10, factor out the greatest common factor.

1. $6m^2 - 4m$

2. $4t^4 - 12t^3$

3. $9ab^2 - 6a^2b$

4. $12x^3y^5 + 16x^2y^3$

5. $10m^2n - 15mn^2 + 20mn$

6. $6x^4y - 4x^2y^2 + 2x^2y^3$

7. $3x(2x + 1) - 5(2x + 1)$

8. $2u(3v^2 + w) + 5v(3v^2 + w)$

9. $(3a + b)(2c - d) + 2a(2c - d)^2$

10. $4uv^2(2u - v) + 6u^2v(v - 2u)$

In Exercises 11–54, factor the polynomial. If the polynomial is prime, state it.

11. $2m^2 - 11m - 6$

12. $6x^2 - x - 1$

13. $x^2 - xy - 6y^2$

14. $2u^2 + 5uv - 12v^2$

15. $x^2 - 3x - 1$

16. $m^2 + 2m + 3$

17. $4a^2 - b^2$

18. $12x^2 - 3y^2$

19. $u^2v^2 - w^2$

20. $4a^2b^2 - 25c^2$

21. $z^2 + 4$

22. $u^2 + 25v^2$

23. $x^2 + 6xy + y^2$

24. $4u^2 - 12uv + 9v^2$

25. $x^2 + 3x - 4$

26. $3m^3 + 3m^2 - 18m$

27. $12x^2y - 10xy - 12y$

28. $12x^2y - 2xy - 24y$

29. $35r^2 + r - 12$

30. $6uv^2 + 9uv - 6v$

31. $9x^3y - 4xy^3$

32. $4u^4v - 9u^2v^3$

33. $x^4 - 16y^2$

34. $16u^4v - 9v^3$

35. $(a - 2b)^2 - (a + 2b)^2$

36. $2x(x + y)^2 - 8x(x + y^2)^2$

37. $8m^3 + 1$

38. $27m^3 - 8$

39. $8r^3 - 27s^3$

40. $x^3 + 64y^3$

41. $u^2v^6 - 8u^2$

42. $r^6s^6 + 8s^3$

43. $2x^3 + 6x + x^2 + 3$

44. $2u^4 - 4u^2 + 2u^2 - 4$

45. $3ax + 6ay + bx + 2by$

46. $6ux - 4uy + 3vx - 2vy$

47. $u^4 - v^4$

48. $u^4 - u^2v^2 - 6v^4$

49. $4x^3 - 9xy^2 + 4x^2y - 9y^3$

50. $4u^4 + 11u^2v^2 - 3v^4$

51. $x^4 + 3x^3 - 2x - 6$

52. $a^2 - b^2 + a + b$

53. $au^2 + (a + c)u + c$

54. $ax^2 - (1 + ab)xy + by^2$

55. SIMPLE INTEREST The accumulated amount after t years when a deposit of P dollars is made in a bank and earning interest at the rate of r/year is $A = P + Prt$. Factor the expression on the right-hand side of this equation.

56. WORKER EFFICIENCY An efficiency study conducted by Elektra Electronics showed that the number of Space Commander walkie-talkies assembled by the average worker t hr after starting work at 8 A.M. is

$$-t^3 + 6t^2 + 15t \qquad (0 \le t \le 4)$$

Factor the expression.

57. REVENUE Williams Commuter Air Service realizes a monthly revenue of

$$8000x - 100x^2 \qquad (0 \le x \le 80)$$

dollars when the price charged per passenger is x dollars. Factor the expression.

58. CHEMICAL REACTION In an autocatalytic chemical reaction, the product that is formed acts as a catalyst for the reaction. If Q is the amount of the original substrate present initially and x is the amount of catalyst formed, then the rate of change of the chemical reaction with respect to the amount of catalyst present in the reaction is

$$R = kQx - kx^2 \qquad (0 \le x \le Q)$$

where k is a constant. Factor the expression on the right-hand side of the equation.

59. SPREAD OF AN EPIDEMIC The incidence (number of new cases/day) of a contagious disease spreading in a population of M people, where k is a positive constant and x denotes the number of people already infected, is given by $kMx - kx^2$. Factor this expression.

60. REVENUE The total revenue realized by the Apollo Company from the sale of x smartphones is given by $R = -0.1x^2 + 500x$ dollars. Factor the expression on the right-hand side of this equation.

61. CHARLES' LAW Charles' Law for gases states that if the pressure remains constant, then the volume V that a gas occupies is related to its temperature T in degrees Celsius by the equation

$$V = V_0 + \frac{V_0}{273}T$$

Factor the expression on the right-hand side of the equation.

62. REACTION TO A DRUG The strength of a human body's reaction to a dosage D of a certain drug, where k is a positive constant, is given by

$$\frac{kD^2}{2} - \frac{D^3}{3}$$

Factor this expression.

1.3 Solutions to Self-Check Exercises

1. a. The common factor is $2x^2$. Therefore,

$$4x^3 - 2x^2 = 2x^2(2x - 1)$$

b. The common factor is $a^2 + 2b^2$. Therefore,

$$3(a^2 + 2b^2) + 4(a^2 + 2b^2)^2 = (a^2 + 2b^2)[3 + 4(a^2 + 2b^2)]$$
$$= (a^2 + 2b^2)(3 + 4a^2 + 8b^2)$$

2. a. Using the trial-and-error method of factorization, we find that

$$6x^2 - x - 12 = (3x + 4)(2x - 3)$$

b. We first factor out the common factor 2. Thus,

$$4x^2 + 10x - 6 = 2(2x^2 + 5x - 3)$$

Using the trial-and-error method of factorization, we find that

$$2x^2 + 5x - 3 = (2x - 1)(x + 3)$$

and, consequently,

$$4x^2 + 10x - 6 = 2(2x - 1)(x + 3)$$

1.4 Rational Expressions

Quotients of polynomials are called **rational expressions**. Examples of rational expressions are

$$\frac{6x - 1}{2x + 3} \quad \text{and} \quad \frac{3x^2y^3 - 2xy}{4x - y}$$

Because division by zero is not allowed, the denominator of a rational expression must not be equal to zero. Thus, in the first example, $x \neq -\frac{3}{2}$, and in the second example, $y \neq 4x$.

Since rational expressions are quotients in which the variables represent real numbers, the properties of real numbers apply to rational expressions as well. For this reason, operations with rational fractions are performed in the same way as operations with arithmetic fractions.

Simplifying Rational Expressions

A rational expression is *simplified,* or reduced to lowest terms, if its numerator and denominator have no common factors other than 1 and -1. If a rational expression does contain common factors, we use the properties of the real number system to write

$$\frac{ac}{bc} = \frac{a}{b} \cdot \frac{c}{c} = \frac{a}{b} \cdot 1 = \frac{a}{b} \qquad (a, b, c \text{ are real numbers, and } bc \neq 0.)$$

This process is often called "canceling common factors." To indicate this process, we often write

$$\frac{a\cancel{c}}{b\cancel{c}} = \frac{a}{b}$$

where a slash is shown through the common factors. As another example, the rational expression

$$\frac{(x+2)(x-3)}{(x-2)(x-3)} \qquad (x \neq 2, 3)$$

is simplified by canceling the common factors $(x-3)$ and writing

$$\frac{(x+2)(x\!\!\!\!\diagup3)}{(x-2)(x\!\!\!\!\diagup3)} = \frac{x+2}{x-2}$$

⚠ $\dfrac{3+4x}{3} = 1 + 4x$ is an example of incorrect cancellation. Instead, we write

$$\frac{3+4x}{3} = \frac{3}{3} + \frac{4x}{3} = 1 + \frac{4x}{3}$$

EXAMPLE 1 Simplify:

a. $\dfrac{x^2 + 2x - 3}{x^2 + 4x + 3}$ b. $\dfrac{3 - 4x - 4x^2}{2x - 1}$ c. $\dfrac{(k+4)^2(k-1)}{k^2 - 16}$

Solution

a. $\dfrac{x^2 + 2x - 3}{x^2 + 4x + 3} = \dfrac{(x+3)(x-1)}{(x+3)(x+1)} = \dfrac{x-1}{x+1}$ Factor numerator and denominator, and cancel common factors.

b. $\dfrac{3 - 4x - 4x^2}{2x - 1} = \dfrac{(1 - 2x)(3 + 2x)}{2x - 1}$

$\qquad = -\dfrac{(2x - 1)(2x + 3)}{2x - 1}$ Rewrite the term $1 - 2x$ in the equivalent form $-(2x - 1)$.

$\qquad = -(2x + 3)$ Cancel common factors.

c. $\dfrac{(k+4)^2(k-1)}{k^2 - 16} = \dfrac{(k+4)^2(k-1)}{(k+4)(k-4)} = \dfrac{(k+4)(k-1)}{k-4}$ ■

Multiplication and Division

The operations of multiplication and division are performed with rational expressions in the same way they are with arithmetic fractions (Table 8).

TABLE 8		
Multiplication and Division of Rational Expressions		
Operation		**Illustration**
If $P, Q, R,$ and S are polynomials, then		
Multiplication		
$\dfrac{P}{Q} \cdot \dfrac{R}{S} = \dfrac{PR}{QS} \qquad (Q, S \neq 0)$		$\dfrac{2x}{y} \cdot \dfrac{(x+1)}{(y-1)} = \dfrac{2x(x+1)}{y(y-1)}$
Division		
$\dfrac{P}{Q} \div \dfrac{R}{S} = \dfrac{P}{Q} \cdot \dfrac{S}{R} = \dfrac{PS}{QR} \qquad (Q, R, S \neq 0)$		$\dfrac{x^2 + 3}{y} \div \dfrac{y^2 + 1}{x} = \dfrac{x^2 + 3}{y} \cdot \dfrac{x}{y^2 + 1} = \dfrac{x(x^2 + 3)}{y(y^2 + 1)}$

When the operations of multiplication and division are performed on rational expressions, the resulting expression should be simplified.

VIDEO ▶   **EXAMPLE 2** Perform the indicated operations and simplify.

a. $\dfrac{2x-8}{x+2} \cdot \dfrac{x^2+4x+4}{x^2-16}$ **b.** $\dfrac{x^2-6x+9}{3x+12} \div \dfrac{x^2-9}{6x^2+18x}$

Solution

a. $\dfrac{2x-8}{x+2} \cdot \dfrac{x^2+4x+4}{x^2-16}$

$\qquad = \dfrac{2(x-4)}{x+2} \cdot \dfrac{(x+2)^2}{(x+4)(x-4)}$ Factor numerators and denominators.

$\qquad = \dfrac{2(x-4)(x+2)(x+2)}{(x+2)(x+4)(x-4)}$

$\qquad = \dfrac{2(x+2)}{x+4}$ Cancel the common factors $(x+2)(x-4)$.

b. $\dfrac{x^2-6x+9}{3x+12} \div \dfrac{x^2-9}{6x^2+18x} = \dfrac{x^2-6x+9}{3x+12} \cdot \dfrac{6x^2+18x}{x^2-9}$

$\qquad = \dfrac{(x-3)^2}{3(x+4)} \cdot \dfrac{6x(x+3)}{(x+3)(x-3)}$

$\qquad = \dfrac{(x-3)(x-3)(6x)(x+3)}{3(x+4)(x+3)(x-3)}$

$\qquad = \dfrac{2x(x-3)}{x+4}$ ■

Addition and Subtraction

For rational expressions the operations of addition and subtraction are performed by finding a common denominator for the fractions and then adding or subtracting the fractions. Table 9 shows the rules for fractions with common denominators.

TABLE 9		
Adding and Subtracting Fractions with Common Denominators		
Operation		**Illustration**
If P, Q, and R are polynomials, then		
Addition		
$\dfrac{P}{R}+\dfrac{Q}{R}=\dfrac{P+Q}{R}$ $(R \neq 0)$		$\dfrac{2x}{x+2}+\dfrac{6x}{x+2}=\dfrac{2x+6x}{x+2}=\dfrac{8x}{x+2}$
Subtraction		
$\dfrac{P}{R}-\dfrac{Q}{R}=\dfrac{P-Q}{R}$ $(R \neq 0)$		$\dfrac{3y}{y-x}-\dfrac{y}{y-x}=\dfrac{3y-y}{y-x}=\dfrac{2y}{y-x}$

To add or subtract fractions that have different denominators, first find the least common denominator (LCD). To find the LCD of two or more rational expressions, follow these steps:

1. *Find the prime factors* of each denominator.
2. *Form the product of the different prime factors* that occur in the denominators. Raise each prime factor in this product to the highest power of that factor appearing in the denominators.

After finding the LCD, carry out the indicated operations following the procedure for adding and subtracting fractions with common denominators.

EXAMPLE 3 Simplify:

a. $\dfrac{3x + 4}{4x} + \dfrac{4y - 2}{3y}$ **b.** $\dfrac{2x}{x^2 - 1} + \dfrac{3x + 1}{2x^2 - x - 1}$ **c.** $\dfrac{1}{x + h} - \dfrac{1}{x}$

Solution

a. $\dfrac{3x + 4}{4x} + \dfrac{4y - 2}{3y} = \dfrac{3x + 4}{4x} \cdot \dfrac{3y}{3y} + \dfrac{4y - 2}{3y} \cdot \dfrac{4x}{4x}$ $\text{LCD} = (4x)(3y) = 12xy$

$$= \dfrac{9xy + 12y}{12xy} + \dfrac{16xy - 8x}{12xy}$$

$$= \dfrac{25xy - 8x + 12y}{12xy}$$

b. $\dfrac{2x}{x^2 - 1} + \dfrac{3x + 1}{2x^2 - x - 1} = \dfrac{2x}{(x + 1)(x - 1)} + \dfrac{3x + 1}{(2x + 1)(x - 1)}$

$$= \dfrac{2x(2x + 1) + (3x + 1)(x + 1)}{(x + 1)(x - 1)(2x + 1)} \quad \begin{array}{l} \text{LCD} = (2x + 1) \\ \cdot (x + 1)(x - 1) \end{array}$$

$$= \dfrac{4x^2 + 2x + 3x^2 + 3x + x + 1}{(x + 1)(x - 1)(2x + 1)}$$

$$= \dfrac{7x^2 + 6x + 1}{(x + 1)(x - 1)(2x + 1)}$$

c. $\dfrac{1}{x + h} - \dfrac{1}{x} = \dfrac{1}{x + h} \cdot \dfrac{x}{x} - \dfrac{1}{x} \cdot \dfrac{x + h}{x + h}$ $\text{LCD} = x(x + h)$

$$= \dfrac{x}{x(x + h)} - \dfrac{x + h}{x(x + h)}$$

$$= \dfrac{x - x - h}{x(x + h)}$$

$$= -\dfrac{h}{x(x + h)}$$

Complex Fractions

A fractional expression that contains fractions in its numerator and/or denominator is called a **complex fraction.** The techniques used to simplify rational expressions may be used to simplify these fractions.

VIDEO ▶ **EXAMPLE 4** Simplify:

a. $\dfrac{1 + \dfrac{1}{x + 1}}{x - \dfrac{4}{x}}$ **b.** $\dfrac{\dfrac{1}{x} + \dfrac{1}{y}}{\dfrac{1}{x^2} - \dfrac{1}{y^2}}$

Solution

a. We first express the numerator and denominator of the given expression as a single quotient. Thus,

$$\frac{1 + \dfrac{1}{x+1}}{x - \dfrac{4}{x}} = \frac{1 \cdot \dfrac{x+1}{x+1} + \dfrac{1}{x+1}}{x \cdot \dfrac{x}{x} - \dfrac{4}{x}}$$

The LCD for the fraction in the numerator is $x + 1$, and the LCD for the fraction in the denominator is x.

$$= \frac{\dfrac{x+1+1}{x+1}}{\dfrac{x^2-4}{x}}$$

$$= \frac{\dfrac{x+2}{x+1}}{\dfrac{x^2-4}{x}}$$

We then multiply the numerator by the reciprocal of the denominator, obtaining

$$\frac{x+2}{x+1} \cdot \frac{x}{x^2-4} = \frac{x+2}{x+1} \cdot \frac{x}{(x-2)(x+2)}$$

Factor the denominator of the second fraction.

$$= \frac{x}{(x+1)(x-2)}$$

Cancel the common factors.

b. As before, we first write the numerator and denominator of the given expression as a single quotient and then simplify the resulting fraction.

$$\frac{\dfrac{1}{x} + \dfrac{1}{y}}{\dfrac{1}{x^2} - \dfrac{1}{y^2}} = \frac{\dfrac{y+x}{xy}}{\dfrac{y^2-x^2}{x^2y^2}}$$

The LCD for the fractions in the numerator is xy, and the LCD for the fractions in the denominator is x^2y^2.

$$= \frac{y+x}{xy} \cdot \frac{x^2y^2}{y^2-x^2}$$

$$\frac{a}{b} \div \frac{c}{d} = \frac{a}{b} \cdot \frac{d}{c}$$

$$= \frac{y+x}{xy} \cdot \frac{x^2y^2}{(y+x)(y-x)}$$

$$= \frac{xy}{y-x}$$

Cancel common factors.   ∎

 APPLIED EXAMPLE 5 Optics The equation

$$\frac{1}{f} = \frac{1}{p} + \frac{1}{q}$$

sometimes called a **lens-maker's equation,** gives the relationship between the focal length f of a thin lens, the distance p of the object from the lens, and the distance q of its image from the lens.

a. Write the right-hand side of the equation as a single fraction.
b. Use the result of part (a) to find the focal length of the lens.

Solution

a. $\dfrac{1}{p} + \dfrac{1}{q} = \dfrac{q + p}{pq} = \dfrac{p + q}{pq}$

b. Using the result of part (a), we have

$$\frac{1}{f} = \frac{1}{p} + \frac{1}{q} = \frac{p + q}{pq}$$

from which we see that

$$f = \frac{pq}{p + q}$$

1.4 Self-Check Exercises

1. Simplify $\dfrac{3a^2b^3}{2ab^2 + 4ab} \cdot \dfrac{b^2 + 4b + 4}{6a^2b^5}$.

2. Simplify $\dfrac{\dfrac{x}{y} - \dfrac{y}{x}}{\dfrac{x^2 + 2xy + y^2}{x^2 - y^2}}$.

Solutions to Self-Check Exercises 1.4 can be found on page 27.

1.4 Concept Questions

1. **a.** What is a rational expression? Give an example.
   b. Explain why a polynomial is a rational expression but not vice versa.

2. **a.** If P, Q, R, and S are polynomials and $Q \neq 0$ and $S \neq 0$, what is $\left(\frac{P}{Q}\right)\left(\frac{R}{S}\right)$? What is $\left(\frac{P}{Q}\right) \div \left(\frac{R}{S}\right)$ if $Q \neq 0$, $R \neq 0$, and $S \neq 0$?
   b. If P, Q, and R are polynomials with $R \neq 0$, what are $\left(\frac{P}{R}\right) + \left(\frac{Q}{R}\right)$ and $\left(\frac{P}{R}\right) - \left(\frac{Q}{R}\right)$?

1.4 Exercises

In Exercises 1–12, simplify the expression.

1. $\dfrac{28x^2}{7x^3}$

2. $\dfrac{3y^4}{18y^2}$

3. $\dfrac{4x + 12}{5x + 15}$

4. $\dfrac{12m - 6}{18m - 9}$

5. $\dfrac{6x^2 - 3x}{6x^2}$

6. $\dfrac{8y^2}{4y^3 - 4y^2 + 8y}$

7. $\dfrac{x^2 + x - 2}{x^2 + 3x + 2}$

8. $\dfrac{2y^2 - y - 3}{2y^2 + y - 1}$

9. $\dfrac{x^2 - 9}{2x^2 - 5x - 3}$

10. $\dfrac{6y^2 + 11y + 3}{4y^2 - 9}$

11. $\dfrac{x^3 + y^3}{x^2 - xy + y^2}$

12. $\dfrac{8r^3 - s^3}{2r^2 + rs - s^2}$

In Exercises 13–46, perform the indicated operations and simplify.

13. $\dfrac{6x^3}{32} \cdot \dfrac{8}{3x^2}$

14. $\dfrac{25y^4}{12y} \cdot \dfrac{3y^2}{5y^3}$

15. $\dfrac{3x^3}{8x^2} \div \dfrac{15x^4}{16x^5}$

16. $\dfrac{6x^5}{21x^2} \div \dfrac{4x}{7x^3}$

17. $\dfrac{3x}{x + 2y} \cdot \dfrac{5x + 10y}{6}$

18. $\dfrac{4y + 12}{y + 2} \cdot \dfrac{3y + 6}{2y - 1}$

19. $\dfrac{2m + 6}{3} \div \dfrac{3m + 9}{6}$

20. $\dfrac{3y - 6}{4y + 6} \div \dfrac{6y + 24}{8y + 12}$

21. $\dfrac{6r^2 - r - 2}{2r + 4} \cdot \dfrac{6r + 12}{4r + 2}$

22. $\dfrac{x^2 - x - 6}{2x^2 + 7x + 6} \cdot \dfrac{2x^2 - x - 6}{x^2 + x - 6}$

23. $\dfrac{k^2 - 2k - 3}{k^2 - k - 6} \div \dfrac{k^2 - 6k + 8}{k^2 - 2k - 8}$

24. $\dfrac{6y^2 - 5y - 6}{6y^2 + 13y + 6} \div \dfrac{6y^2 - 13y + 6}{9y^2 - 12y + 4}$

25. $\dfrac{2}{2x + 3} + \dfrac{3}{2x - 1}$
26. $\dfrac{2x - 1}{x + 2} - \dfrac{x + 3}{x - 1}$

27. $\dfrac{3}{x^2 - x - 6} + \dfrac{2}{x^2 + x - 2}$

28. $\dfrac{4}{x^2 - 9} - \dfrac{5}{x^2 - 6x + 9}$

29. $\dfrac{2m}{2m^2 - 2m - 1} + \dfrac{3}{2m^2 - 3m + 3}$

30. $\dfrac{t}{t^2 + t - 2} - \dfrac{2t - 1}{2t^2 + 3t - 2}$

31. $\dfrac{x}{1 - x} + \dfrac{2x + 3}{x^2 - 1}$

32. $2 + \dfrac{1}{a + 2} - \dfrac{2a}{a - 2}$
33. $x - \dfrac{x^2}{x + 2} + \dfrac{2}{x - 2}$

34. $\dfrac{y}{y^2 - 1} + \dfrac{y - 1}{y + 1} - \dfrac{2y}{1 - y}$

35. $\dfrac{x}{x^2 + 5x + 6} + \dfrac{2}{x^2 - 4} - \dfrac{3}{x^2 + 3x + 2}$

36. $\dfrac{2x + 1}{2x^2 - x - 1} - \dfrac{x + 1}{2x^2 + 3x + 1} + \dfrac{4}{x^2 + 2x - 3}$

37. $\dfrac{x}{ax - ay} + \dfrac{y}{by - bx}$
38. $\dfrac{ax + by}{ax - bx} + \dfrac{ay - bx}{by - ay}$

39. $\dfrac{1 + \dfrac{1}{x}}{1 - \dfrac{1}{x}}$
40. $\dfrac{2 + \dfrac{2}{x}}{x - \dfrac{2}{x}}$

41. $\dfrac{\dfrac{1}{x} + \dfrac{1}{y}}{1 - \dfrac{1}{xy}}$
42. $\dfrac{1 + \dfrac{x}{y}}{1 - \dfrac{x^2}{y^2}}$

43. $\dfrac{\dfrac{1}{x^2} - \dfrac{1}{y^2}}{x + y}$
44. $\dfrac{\dfrac{1}{x^3} - \dfrac{1}{y^3}}{\dfrac{1}{x} - \dfrac{1}{y}}$

45. $\dfrac{\dfrac{1}{2(x + h)} - \dfrac{1}{2x}}{h}$
46. $\dfrac{\dfrac{1}{(x + h)^2} - \dfrac{1}{x^2}}{h}$

47. **AVERAGE COST** The cost incurred by Herald Records in pressing x DVDs is

$$2.2 + \frac{2500}{x}$$

dollars per disc.
   a. Write the cost per disc as a single fraction.
   b. What is the total cost incurred by Herald Records in pressing x discs?

48. **INVENTORY CONTROL** The equation

$$A = \frac{km}{q} + cm + \frac{hq}{2}$$

gives the annual cost of ordering and storing (as yet unsold) merchandise. Here, q is the size of each order, k is the cost of placing each order, c is the unit cost of the product, m is the number of units of the product sold per year, and h is the annual cost for storing each unit. Write the right-hand side of this equation as a single fraction.

49. **AMORTIZATION FORMULA** The periodic payment R on a loan of P dollars to be amortized (paid off gradually by periodic payments of principal and interest) over n periods with interest charged at the rate of i per period is found by solving the equation

$$P = \frac{R}{i} - \frac{R}{i(1 + i)^n}$$

for R. Write the right-hand side of this equation as a single fraction.

50. **CYLINDER PRESSURE** The pressure P, volume V, and temperature T of a gas in a cylinder are related by the van der Waals equation

$$P = \frac{kT}{V - b} + \frac{ab}{V^2(V - b)} - \frac{a}{V(V - b)}$$

where a, b, and k are constants. Write the pressure P as a single fraction.

51. **AIR POLLUTION** The amount of nitrogen dioxide, a brown gas that impairs breathing, present in the atmosphere on a certain May day in the city of Long Beach is approximated by

$$A = \frac{136}{1 + 0.25(t - 4.5)^2} + 28 \qquad (0 \le t \le 11)$$

where A is measured in pollutant standard index (PSI) and t is measured in hours, with $t = 0$ corresponding to 7 A.M. Express A as a single fraction.
Source: Los Angeles Times.

1.4   Solutions to Self-Check Exercises

1. Factoring the numerator and denominator of each expression, we have

$$\frac{3a^2b^3}{2ab^2 + 4ab} \cdot \frac{b^2 + 4b + 4}{6a^2b^5} = \frac{3a^2b^3}{2ab(b + 2)} \cdot \frac{(b + 2)^2}{(3a^2b^3)(2b^2)}$$

$$= \frac{b + 2}{2ab(2b^2)} \qquad \text{Cancel common factors.}$$

$$= \frac{b + 2}{4ab^3}$$

2. Writing the numerator of the given expression as a single quotient, we have

$$\frac{\dfrac{x}{y} - \dfrac{y}{x}}{\dfrac{x^2 + 2xy + y^2}{x^2 - y^2}} = \frac{\dfrac{x^2 - y^2}{xy}}{\dfrac{x^2 + 2xy + y^2}{x^2 - y^2}} \qquad \begin{array}{l}\text{The LCD for the fractions in the numerator is } xy.\end{array}$$

$$= \frac{\dfrac{(x + y)(x - y)}{xy}}{\dfrac{(x + y)(x + y)}{(x + y)(x - y)}} \qquad \text{Factor.}$$

$$= \frac{(x + y)(x - y)}{xy} \cdot \frac{(x + y)(x - y)}{(x + y)(x + y)} \qquad \dfrac{a}{b} \div \dfrac{c}{d} = \dfrac{a}{b} \cdot \dfrac{d}{c}$$

$$= \frac{(x - y)^2}{xy} \qquad \begin{array}{l}\text{Cancel common factors.}\end{array}$$

1.5   Integral Exponents

Exponents

We begin by recalling the definition of the exponential expression a^n, where a is a real number and n is a positive integer.

Exponential Expressions

If a is any real number and n is a natural number, then the expression a^n (read "a to the power n") is defined as the number

$$a^n = \underbrace{a \cdot a \cdot a \cdots \cdot a}_{n \text{ factors}}$$

Recall that the number a is the *base* and the superscript n is the *exponent*, or *power*, to which the base is raised.

EXAMPLE 1 Write each of the following numbers without using exponents.

a. 2^5 **b.** $\left(\dfrac{2}{3}\right)^3$

Solution

a. $2^5 = 2 \cdot 2 \cdot 2 \cdot 2 \cdot 2 = 32$

b. $\left(\dfrac{2}{3}\right)^3 = \left(\dfrac{2}{3}\right)\left(\dfrac{2}{3}\right)\left(\dfrac{2}{3}\right) = \dfrac{8}{27}$

Next, we extend our definition of a^n to include $n = 0$; that is, we define the expression a^0. Observe that if a is any real number and m and n are positive integers, then we have the rule

$$a^m a^n = \underbrace{(a \cdot a \cdot \cdots \cdot a)}_{m \text{ factors}}\underbrace{(a \cdot a \cdot \cdots \cdot a)}_{n \text{ factors}} = \underbrace{a \cdot a \cdot \cdots \cdot a}_{(m + n) \text{ factors}} = a^{m+n}$$

Now, if we require that this rule hold for the zero exponent as well, then we must have, upon setting $m = 0$,

$$a^0 a^n = a^{0+n} = a^n \quad \text{or} \quad a^0 a^n = a^n$$

Therefore, if $a \neq 0$, we can divide both sides of this last equation by a^n to obtain $a^0 = 1$. This motivates the following definition.

> **Zero Exponent**
>
> For any nonzero real number a,
>
> $$a^0 = 1$$
>
> The expression 0^0 is not defined.

EXAMPLE 2

a. $2^0 = 1$ **b.** $(-2)^0 = 1$ **c.** $(\pi)^0 = 1$ **d.** $\left(\dfrac{1}{3}\right)^0 = 1$ ∎

Next, we extend our definition to include expressions of the form a^n, where the exponent is a negative integer. Once again, we use the rule

$$a^m a^n = a^{m+n}$$

where n is a positive integer. Now, if we require that this rule hold for negative integral exponents as well, upon setting $m = -n$, we have

$$a^{-n} a^n = a^{-n+n} = a^0 = 1 \quad \text{or} \quad a^{-n} a^n = 1$$

Therefore, if $a \neq 0$, we can divide both sides of this last equation by a^n to obtain $a^{-n} = 1/a^n$. This motivates the following definition.

> **Exponential Expressions with Negative Exponents**
>
> If a is any nonzero real number and n is a positive integer, then
>
> $$a^{-n} = \frac{1}{a^n}$$

VIDEO **EXAMPLE 3** Write each of the following numbers without using exponents.

a. 4^{-2} **b.** 3^{-1} **c.** -2^{-3} **d.** $\left(\dfrac{2}{3}\right)^{-1}$ **e.** $\left(\dfrac{3}{2}\right)^{-3}$

Solution

a. $4^{-2} = \dfrac{1}{4^2} = \dfrac{1}{16}$ **b.** $3^{-1} = \dfrac{1}{3^1} = \dfrac{1}{3}$ **c.** $-2^{-3} = -\dfrac{1}{2^3} = -\dfrac{1}{8}$

d. $\left(\dfrac{2}{3}\right)^{-1} = \dfrac{1}{\left(\frac{2}{3}\right)^1} = \dfrac{1}{\frac{2}{3}} = \dfrac{3}{2}$ **e.** $\left(\dfrac{3}{2}\right)^{-3} = \dfrac{1}{\left(\frac{3}{2}\right)^3} = \dfrac{2^3}{3^3} = \dfrac{8}{27}$ ∎

Note   In Example 3d and 3e, the intermediate steps may be omitted by observing that

$$\left(\frac{a}{b}\right)^{-n} = \left(\frac{b}{a}\right)^{n}$$ For example, $\left(\frac{2}{3}\right)^{-1} = \left(\frac{3}{2}\right)^{1}$

since

$$\left(\frac{a}{b}\right)^{-n} = \frac{1}{\left(\frac{a}{b}\right)^{n}} = \frac{1}{\frac{a^{n}}{b^{n}}} = 1 \cdot \frac{b^{n}}{a^{n}} = \left(\frac{b}{a}\right)^{n}$$

Five basic properties of exponents are given in Table 10.

TABLE 10

Properties of Exponents

Property	**Illustration**
1. $a^{m} \cdot a^{n} = a^{m+n}$	$x^{2} \cdot x^{3} = x^{2+3} = x^{5}$
2. $\dfrac{a^{m}}{a^{n}} = a^{m-n}$	$\dfrac{x^{7}}{x^{4}} = x^{7-4} = x^{3}$
3. $(a^{m})^{n} = a^{mn}$	$(x^{4})^{3} = x^{4 \cdot 3} = x^{12}$
4. $(ab)^{n} = a^{n} \cdot b^{n}$	$(2x)^{4} = 2^{4} \cdot x^{4} = 16x^{4}$
5. $\left(\dfrac{a}{b}\right)^{n} = \dfrac{a^{n}}{b^{n}}$ $(b \neq 0)$	$\left(\dfrac{x}{2}\right)^{3} = \dfrac{x^{3}}{2^{3}} = \dfrac{x^{3}}{8}$

It can be shown that these laws are valid for any real numbers a and b and any integers m and n.

Simplifying Exponential Expressions

The next two examples illustrate the use of the laws of exponents.

EXAMPLE 4 Simplify the expression and write your answer using positive exponents only.

a. $(2x^{3})(3x^{5})$ **b.** $\dfrac{2x^{5}}{3x^{4}}$ **c.** $(x^{-2})^{-3}$ **d.** $(2u^{-1}v^{3})^{3}$ **e.** $\left(\dfrac{2m^{3}n^{4}}{m^{5}n^{3}}\right)^{-1}$

Solution

a. $(2x^{3})(3x^{5}) = 6x^{3+5} = 6x^{8}$ Property 1

b. $\dfrac{2x^{5}}{3x^{4}} = \dfrac{2}{3}x^{5-4} = \dfrac{2}{3}x$ Property 2

c. $(x^{-2})^{-3} = x^{(-2)(-3)} = x^{6}$ Property 3

d. $(2u^{-1}v^{3})^{3} = 2^{3}u^{(-1)(3)}v^{3(3)} = 8u^{-3}v^{9} = \dfrac{8v^{9}}{u^{3}}$ Property 4

e. $\left(\dfrac{2m^{3}n^{4}}{m^{5}n^{3}}\right)^{-1} = (2m^{3-5}n^{4-3})^{-1}$ Property 2

$= (2m^{-2}n)^{-1}$ Property 1

$= \dfrac{1}{2m^{-2}n} = \dfrac{m^{2}}{2n}$

VIDEO ▶ **EXAMPLE 5** Simplify the expression and write your answer using positive exponents only.

    a. $(2^2)^3 - (3^2)^2$ **b.** $(x^{-1} + y^{-1})^{-1}$ **c.** $\dfrac{2^{-4} \cdot (2^{-1})^2}{(2^0 + 1)^{-1}}$

Solution

    a. $(2^2)^3 - (3^2)^2 = 2^6 - 3^4 = 64 - 81 = -17$

    b. $(x^{-1} + y^{-1})^{-1} = \left(\dfrac{1}{x} + \dfrac{1}{y}\right)^{-1} = \left(\dfrac{y + x}{xy}\right)^{-1} = \dfrac{xy}{y + x}$

    c. $\dfrac{2^{-4} \cdot (2^{-1})^2}{(2^0 + 1)^{-1}} = \dfrac{2^{-4} \cdot 2^{-2}}{(2)^{-1}} = 2^{-4-2+1} = 2^{-5} = \dfrac{1}{2^5} = \dfrac{1}{32}$

                                                                 ■

1.5   Self-Check Exercises

1. Simplify the expression and write your answer using positive exponents only.

    a. $(3a^4)(4a^3)$ **b.** $\left(\dfrac{u^{-3}}{u^{-5}}\right)^{-2}$

2. Simplify the expression and write your answer using positive exponents only.

    a. $(x^2 y^3)^3 (x^5 y)^{-2}$ **b.** $\left(\dfrac{a^2 b^{-1} c^3}{a^3 b^{-2}}\right)^2$

Solutions to Self-Check Exercises 1.5 can be found on page 31.

1.5   Concept Questions

1. Explain the meaning of the expression a^n. What restrictions, if any, are placed on a and n? What is a^0 if a is a nonzero real number? What is a^{-n} if n is a positive integer and $a \neq 0$?

2. Write all the properties of exponents, and illustrate with examples.

1.5   Exercises

In Exercises 1–20, rewrite the number without using exponents.

1. $(-2)^3$

2. $\left(-\dfrac{2}{3}\right)^4$

3. 7^{-2}

4. $\left(\dfrac{3}{4}\right)^{-2}$

5. $-\left(-\dfrac{1}{4}\right)^{-2}$

6. -4^2

7. $2^{-2} + 3^{-1}$

8. $-3^{-2} - \left(-\dfrac{2}{3}\right)^2$

9. $(0.03)^2$

10. $(-0.3)^{-2}$

11. 1996^0

12. $(18 + 25)^0$

13. $(ab^2)^0$, where $a, b \neq 0$ **14.** $(3x^2 y^3)^0$, where $x, y \neq 0$

15. $\dfrac{2^3 \cdot 2^5}{2^4 \cdot 2^9}$

16. $\dfrac{6 \cdot 10^4}{3 \cdot 10^2}$

17. $\dfrac{2^{-3} \cdot 2^{-4}}{2^{-5} \cdot 2^{-2}}$

18. $\dfrac{4 \cdot 2^{-3}}{2 \cdot 4^{-2}}$

19. $\left(\dfrac{3^4 \cdot 3^{-3}}{3^{-2}}\right)^{-1}$

20. $\left(\dfrac{5^{-2} \cdot 5^{-2}}{5^{-5}}\right)^{-2}$

In Exercises 21–54, simplify the expression and write your answer using positive exponents only.

21. $(2x^3)\left(\dfrac{1}{8}x^2\right)$

22. $(-2x^2)(3x^{-4})$

23. $\dfrac{3x^3}{2x^4}$

24. $\dfrac{(3x^2)(4x^3)}{2x^4}$

25. $(a^{-2})^3$

26. $(-a^2)^{-3}$

27. $(2x^{-2}y^2)^3$

28. $(3u^{-1}v^{-2})^{-3}$

29. $(4x^2y^{-3})(2x^{-3}y^2)$

30. $\left(\dfrac{1}{2}u^{-2}v^3\right)(4v^3)$

31. $(-x^2y)^3\left(\dfrac{2y^2}{x^4}\right)$

32. $\left(-\dfrac{1}{2}x^2y\right)^{-2}$

33. $\left(\dfrac{2u^2v^3}{3uv}\right)^{-1}$

34. $\left(\dfrac{a^{-2}}{2b^2}\right)^{-3}$

35. $(3x^{-2})^3(2x^2)^5$

36. $(2^{-1}r^3)^{-2}(3s^{-1})^2$

37. $\dfrac{3^0 \cdot 4x^{-2}}{16 \cdot (x^2)^3}$

38. $\dfrac{5x^2(3x^{-2})}{(4x^{-1})(x^3)^{-2}}$

39. $\dfrac{2^2u^{-2}(v^{-1})^3}{3^2(u^{-3}v)^2}$

40. $\dfrac{(3a^{-1}b^2)^{-2}}{(2a^2b^{-1})^{-3}}$

41. $(-2x)^{-2}(3y)^{-3}(4z)^{-2}$

42. $(3x^{-1})^2(4y^{-1})^3(2z)^{-2}$

43. $(a^2b^{-3})^2(a^{-2}b^2)^{-3}$

44. $(5u^2v^{-3})^{-1} \cdot 3(2u^2v^2)^{-2}$

45. $\left[\left(\dfrac{a^{-2}b^{-2}}{3a^{-1}b^2}\right)^2\right]^{-1}$

46. $\left[\left(\dfrac{x^2y^{-3}z^{-4}}{x^{-2}y^{-1}z^2}\right)^{-2}\right]^3$

47. $\left(\dfrac{3^2u^{-2}v^2}{2^2u^3v^{-3}}\right)^{-2}\left(\dfrac{3^2v^5}{4^2u}\right)^2$

48. $\left[\left(-\dfrac{2^2x^{-2}y^0}{3^2x^3y^{-2}}\right)^{-2}\right]^{-2}$

49. $\dfrac{x^{-1}-1}{x^{-1}+1}$

50. $\dfrac{x^{-1}-y^{-1}}{x^{-1}+y^{-1}}$

51. $\dfrac{u^{-1}-v^{-1}}{v-u}$

52. $\dfrac{(uv)^{-1}}{u^{-1}+v^{-1}}$

53. $\left(\dfrac{a^{-1}-b^{-1}}{a^{-1}+b^{-1}}\right)^{-1}$

54. $[(a^{-1}+b^{-1})(a^{-1}-b^{-1})]^{-2}$

In Exercises 55–57, determine whether the statement is true or false. If it is true, explain why it is true. If it is false, give an example to show why it is false.

55. If a and b are real numbers and m and n are natural numbers, then $a^mb^n = (ab)^{mn}$.

56. If a and b are real numbers $(b \neq 0)$ and m and n are natural numbers, then

$$\frac{a^m}{b^n} = \left(\frac{a}{b}\right)^{m-n}$$

57. If a and b are real numbers and n is a natural number, then $(a+b)^n = a^n + b^n$.

1.5   Solutions to Self-Check Exercises

1. a. $(3a^4)(4a^3) = 3 \cdot a^4 \cdot 4 \cdot a^3 = 12a^{4+3} = 12a^7$

   b. $\left(\dfrac{u^{-3}}{u^{-5}}\right)^{-2} = \dfrac{u^{(-3)(-2)}}{u^{(-5)(-2)}} = \dfrac{u^6}{u^{10}} = u^{6-10} = u^{-4} = \dfrac{1}{u^4}$

2. a. $(x^2y^3)^3(x^5y)^{-2} = x^{2\cdot3}y^{3\cdot3}x^{5(-2)}y^{-2} = x^6y^9x^{-10}y^{-2}$

         $= x^{6-10}y^{9-2} = x^{-4}y^7$

         $= \dfrac{y^7}{x^4}$

   b. $\left(\dfrac{a^2b^{-1}c^3}{a^3b^{-2}}\right)^2 = \dfrac{a^{2\cdot2}b^{(-1)(2)}c^{3\cdot2}}{a^{3\cdot2}b^{(-2)(2)}} = \dfrac{a^4b^{-2}c^6}{a^6b^{-4}}$

         $= a^{4-6}b^{-2+4}c^6 = a^{-2}b^2c^6$

         $= \dfrac{b^2c^6}{a^2}$

1.6   Solving Equations

Equations

An **equation** is a statement that two mathematical expressions are equal.

EXAMPLE 1 The following are examples of equations.

   a. $2x + 3 = 7$ b. $3(2x + 3) = 4(x - 1) + 4$

   c. $\dfrac{y}{y-2} = \dfrac{3y+1}{3y-4}$ d. $\sqrt{z - 1} = 2$

In Example 1, the letters x, y, and z are called variables. A **variable** is a letter that stands for a number belonging to a set of (real) numbers.

A **solution of an equation** involving one variable is a number that renders the equation a true statement when the number is substituted for the variable. For example, replacing the variable x in the equation $2x + 3 = 7$ by the number 2 gives

$$2(2) + 3 = 7$$

$$4 + 3 = 7$$

which is true. This shows that the number 2 is a solution of $2x + 3 = 7$. The set of all solutions of an equation is called the **solution set.** To *solve* an equation is synonymous with finding its solution set.

The standard procedure for solving an equation is to transform the given equation, using an appropriate operation, into an *equivalent* equation—that is, one having exactly the same solution(s) as the original equation. The transformations are repeated if necessary until the solution(s) are easily read off. The following properties of real numbers can be used to produce equivalent equations.

Equality Properties of Real Numbers

Let a, b, and c be real numbers.

1. If $a = b$, then $a + c = b + c$ and $a - c = b - c$.   Addition and subtraction properties

2. If $a = b$ and $c \neq 0$, then $ca = cb$ and $\dfrac{a}{c} = \dfrac{b}{c}$.   Multiplication and division properties

---

Thus, adding or subtracting the same number to both sides of an equation leads to an equivalent equation. Also, multiplying or dividing both sides of an equation by a *nonzero* number leads to an equivalent equation. Let's apply the procedure to the solution of some linear equations.

Linear Equations

A **linear equation** in the variable x is an equation that can be written in the form $ax + b = 0$, where a and b are constants with $a \neq 0$. A linear equation in x is also called a **first-degree equation in x** or an **equation of degree 1 in x.**

EXAMPLE 2 Solve the linear equation $8x - 3 = 2x + 9$.

Solution   We use the equality properties of real numbers to obtain the following equivalent equations, in which the aim is to isolate x.

$$8x - 3 = 2x + 9$$
$$8x - 3 - 2x = 2x + 9 - 2x \quad \text{Subtract } 2x \text{ from both sides.}$$
$$6x - 3 = 9$$
$$6x - 3 + 3 = 9 + 3 \quad \text{Add 3 to both sides.}$$
$$6x = 12$$
$$\frac{1}{6}(6x) = \frac{1}{6}(12) \quad \text{Multiply both sides by } \frac{1}{6}.$$
$$x = 2$$

and so the required solution is 2.

EXAMPLE 3 Solve the linear equation $3p + 2(p - 1) = -2p - 4$.

Solution

$$3p + 2(p - 1) = -2p - 4$$

$$3p + 2p - 2 = -2p - 4 \qquad \text{Use the distributive property.}$$

$$5p - 2 = -2p - 4 \qquad \text{Simplify.}$$

$$5p - 2 + 2p = -2p - 4 + 2p \qquad \text{Add } 2p \text{ to both sides.}$$

$$7p - 2 = -4$$

$$7p - 2 + 2 = -4 + 2 \qquad \text{Add 2 to both sides.}$$

$$7p = -2$$

$$\frac{1}{7}(7p) = \frac{1}{7}(-2) \qquad \text{Multiply both sides by } \frac{1}{7}.$$

$$p = -\frac{2}{7}$$

EXAMPLE 4 Solve the linear equation $\dfrac{2k + 1}{3} - \dfrac{k - 1}{4} = 1$.

Solution First multiply both sides of the given equation by 12, the LCD. Thus,

$$12\left(\frac{2k + 1}{3} - \frac{k - 1}{4}\right) = 12(1)$$

$$12 \cdot \frac{2k + 1}{3} - 12 \cdot \frac{k - 1}{4} = 12 \qquad \text{Use the distributive property.}$$

$$4(2k + 1) - 3(k - 1) = 12 \qquad \text{Simplify.}$$

$$8k + 4 - 3k + 3 = 12 \qquad \text{Use the distributive property.}$$

$$5k + 7 = 12 \qquad \text{Simplify.}$$

$$5k = 5 \qquad \text{Subtract 7 from both sides.}$$

$$k = 1 \qquad \text{Multiply both sides by } \frac{1}{5}.$$

Some Special Nonlinear Equations

The solution(s) of some nonlinear equations are found by solving a related linear equation as the following examples show.

VIDEO   **EXAMPLE 5** Solve $\dfrac{2}{3(x + 1)} - \dfrac{x}{2(x + 1)} = \dfrac{1}{3}$.

Solution We multiply both sides of the given equation by $6(x + 1)$, the LCD. Thus,

$$6(x + 1) \cdot \frac{2}{3(x + 1)} - 6(x + 1) \cdot \frac{x}{2(x + 1)} = 6(x + 1) \cdot \frac{1}{3}$$

which, upon simplification, yields

$$4 - 3x = 2(x + 1)$$
$$4 - 3x = 2x + 2$$
$$4 - 3x - 2x = 2x + 2 - 2x \qquad \text{Subtract } 2x \text{ from both sides.}$$
$$4 - 5x = 2$$
$$4 - 5x - 4 = 2 - 4 \qquad \text{Subtract 4 from both sides.}$$
$$-5x = -2$$
$$x = \frac{2}{5} \qquad \text{Multiply both sides by } -\frac{1}{5}.$$

We can verify that $x = \frac{2}{5}$ is a solution of the original equation by substituting $\frac{2}{5}$ into the left-hand side of the equation. Thus,

$$\frac{2}{3\left(\frac{2}{5} + 1\right)} - \frac{\frac{2}{5}}{2\left(\frac{2}{5} + 1\right)} = \frac{2}{3\left(\frac{7}{5}\right)} - \frac{\frac{2}{5}}{2\left(\frac{7}{5}\right)}$$
$$= \frac{10}{21} - \frac{1}{7} = \frac{7}{21} = \frac{1}{3}$$

which is equal to the right-hand side.   ■

⚠ When we solve an equation in x, we sometimes multiply both sides of the equation by an expression in x. The resulting equation may contain solution(s) that are not solution(s) of the original equation. Such a solution is called an **extraneous solution.** For example, the solution of the equation $3x = 0$ is of course 0. But multiplying both sides of this equation by the expression $(x - 2)$ leads to the equation $3x(x - 2) = 0$ whose solutions are 0 and 2. The solution 2 is not a solution of the original equation. Thus, whenever you need to multiply both sides of an equation by an expression that involves a variable, it is a good idea to check whether each solution of the modified equation is indeed a solution of the given equation.

EXAMPLE 6 Solve $\dfrac{x + 1}{x} - \dfrac{x - 1}{x + 1} = \dfrac{1}{x^2 + x}$.

Solution   Multiplying both sides of the equation by the LCD, $x(x + 1)$, we obtain

$$(x + 1)^2 - x(x - 1) = 1 \qquad \text{Note: } x^2 + x = x(x + 1)$$
$$x^2 + 2x + 1 - x^2 + x = 1$$
$$3x + 1 = 1$$
$$3x = 0$$
$$x = 0$$

Since the original equation is not defined for $x = 0$ (division by 0 is not permitted), we see that 0 is an extraneous solution of the given equation and conclude, accordingly, that the given equation has no solution.   ■

Solving for a Specified Variable

Equations involving more than one variable occur frequently in practical applications. In these situations, we may be interested in solving for one of the variables in terms of the others. To obtain such a solution, we think of all the variables other than the one we are solving for as constants. This technique is illustrated in the next example.

EXAMPLE 7 The equation $A = P + Prt$ gives the relationship between the value A of an investment of P dollars after t years when the investment earns simple interest at the rate of r percent per year. Solve the equation for (a) P, (b) t, and (c) r.

Solution

a.
$$A = P + Prt$$
$$A = P(1 + rt) \qquad \text{Factor.}$$
$$\frac{A}{1 + rt} = P \qquad \text{Multiply both sides by } \frac{1}{1 + rt}.$$

b.
$$A = P + Prt$$
$$A - P = Prt \qquad \text{Subtract } P \text{ from both sides.}$$
$$\frac{A - P}{Pr} = t \qquad \text{Multiply both sides by } \frac{1}{Pr}.$$

c.
$$A = P + Prt$$
$$A - P = Prt$$
$$\frac{A - P}{Pt} = r$$

1.6 Self-Check Exercises

1. Solve $2\left(\dfrac{x - 1}{4}\right) - \dfrac{2x}{3} = \dfrac{4 - 3x}{12}$.

2. Solve $\dfrac{k}{2k + 1} = \dfrac{3}{8}$.

Solutions to Self-Check Exercises 1.6 can be found on page 37.

1.6 Concept Questions

1. What is an equation? What is a solution of an equation? What is the solution set of an equation? Give examples.

2. Write the equality properties of real numbers. Illustrate with examples.

3. What is a linear equation in x? Give an example of one and solve it.

1.6 Exercises

In Exercises 1–32, solve the given equation.

1. $3x = 12$

2. $2x = 0$

3. $0.3y = 2$

4. $2x + 5 = 11$

5. $3x + 4 = 2$

6. $2 - 3y = 8$

7. $-2y + 3 = -7$

8. $\dfrac{1}{3}k + 1 = \dfrac{1}{4}k - 2$

9. $\dfrac{1}{5}p - 3 = -\dfrac{1}{3}p + 5$

10. $3.1m + 2 = 3 - 0.2m$

11. $0.4 - 0.3p = 0.1(p + 4)$

12. $\dfrac{1}{3}k + 4 = -2\left(k + \dfrac{1}{3}\right)$

13. $\dfrac{3}{5}(k + 1) = \dfrac{1}{4}(2k + 4)$

14. $3\left(\dfrac{3m}{4} - 1\right) + \dfrac{m}{5} = \dfrac{42 - m}{4}$

15. $\dfrac{2x - 1}{3} + \dfrac{3x + 4}{4} = \dfrac{7(x + 3)}{10}$

16. $\dfrac{w - 1}{3} + \dfrac{w + 1}{4} = -\dfrac{w + 1}{6}$

17. $\dfrac{1}{2}[2x - 3(x - 4)] = \dfrac{2}{3}(x - 5)$

18. $\dfrac{1}{3}[2 - 3(x + 2)] = \dfrac{1}{4}\left[(-3x + 1) + \dfrac{1}{2}x\right]$

19. $(2x + 1)^2 - (3x - 2)^2 = 5x(2 - x)$

20. $x[(2x - 3)^2 + 5x^2] = 3x^2(3x - 4) + 18$

21. $\dfrac{8}{x} = 24$

22. $\dfrac{1}{x} + \dfrac{2}{x} = 6$

23. $\dfrac{2}{y - 1} = 4$

24. $\dfrac{1}{x + 3} = 0$

25. $\dfrac{2x - 3}{x + 1} = \dfrac{2}{5}$

26. $\dfrac{r}{3r - 1} = 4$

27. $\dfrac{2}{q - 1} = \dfrac{3}{q - 2}$

28. $\dfrac{y}{3} - \dfrac{2}{y + 1} = \dfrac{1}{3}(y - 3)$

29. $\dfrac{3k - 2}{4} - \dfrac{3k}{4} = \dfrac{k + 3}{k}$

30. $\dfrac{2x - 1}{3x + 2} = \dfrac{2x + 1}{3x + 1}$

31. $\dfrac{m - 2}{m} + \dfrac{2}{m} = \dfrac{m + 3}{m - 3}$

32. $\dfrac{4}{x(x - 2)} = \dfrac{2}{x - 2}$

In Exercises 33–46, solve the equation for the indicated variable.

33. $I = Prt; \ r$

34. $ax + by + c = 0; \ y$

35. $p = -3q + 1; \ q$

36. $w = \dfrac{kuv}{s^2}; \ u$

37. $S = R\left[\dfrac{(1 + i)^n - 1}{i}\right]; \ R$

38. $S = R(1 + i)\left[\dfrac{(1 + i)^n - 1}{i}\right]; \ R$

39. $V = \dfrac{ax}{x + b}; \ x$

40. $V = C\left(1 - \dfrac{n}{N}\right); \ n$

41. $r = \dfrac{2mI}{B(n + 1)}; \ m$

42. $p = \dfrac{x + 10}{x + 4}; \ x$

43. $r = \dfrac{2mI}{B(n + 1)}; \ n$

44. $y = 10\left(1 - \dfrac{1}{1 + 2x}\right); \ x$

45. $\dfrac{1}{f} = \dfrac{1}{p} + \dfrac{1}{q}; \ p$

46. $\dfrac{1}{R} = \dfrac{1}{R_1} + \dfrac{1}{R_2} + \dfrac{1}{R_3}; \ R_3$

47. SIMPLE INTEREST The simple interest I (in dollars) earned when P dollars is invested for a term of t years is given by $I = Prt$, where r is the (simple) interest rate/year. Solve for t in terms of I, P, and r. If Susan invests $1000 in a bank paying interest at the rate of 6%/year, how long must she leave it in the bank before it earns an interest of $90?

48. TEMPERATURE CONVERSION The relationship between the temperature in degrees Fahrenheit (°F) and the temperature in degrees Celsius (°C) is $F = \frac{9}{5}C + 32$. Solve for C in terms of F. Then use the result to find the temperature in degrees Celsius corresponding to a temperature of 70°F.

49. WEISS'S LAW According to Weiss's law of excitation of tissue, the strength S of an electric current is related to the time t the current takes to excite tissue by the formula

$$S = \dfrac{a}{t} + b \qquad (t > 0)$$

where a and b are positive constants. Solve this equation for t.

50. SPEED OF A CHEMICAL REACTION Certain proteins, known as enzymes, serve as catalysts for chemical reactions in living things. In 1913, Leonor Michaelis and L. M. Menten discovered the following formula:

$$V = \dfrac{ax}{x + b}$$

where a and b are positive constants, giving the initial speed V, in miles per liter per second, at which a reaction begins in terms of the amount of substrate x (the substance being acted upon), measured in moles per liters. Solve this equation for x.

51. LINEAR DEPRECIATION Suppose an asset has an original value of $\$C$ and is depreciated linearly over N years with a scrap value of $\$S$. Then the book value V (in dollars) of the asset at the end of t years is given by

$$V = C - \left(\dfrac{C - S}{N}\right)t$$

a. Solve for C in terms of V, S, N, and t.
b. A speed boat is being depreciated linearly over 5 years. If the scrap value of the boat is $40,000 and the book value of the boat at the end of 3 years is $70,000, what was its original value?

52. MOTION OF A CAR The distance s (in feet) covered by a car traveling along a straight road is related to its initial speed u (in ft/sec), its final speed v (in ft/sec), and its (constant) acceleration a (in ft/sec^2) by the equation $v^2 = u^2 + 2as$.
a. Solve the equation for a in terms of the other variables.
b. A car starting from rest and accelerating at a constant rate reaches a speed of 88 ft/sec after traveling $\frac{1}{4}$ mile (1320 ft). What is its acceleration?

53. COWLING'S RULE Cowling's Rule is a method for calculating pediatric drug dosages. If a denotes the adult dosage (in milligrams) and if t is the child's age (in years), then the child's dosage (in milligrams) is given by

$$c = \left(\dfrac{t + 1}{24}\right)a$$

a. Solve the equation for t in terms of a and c.
b. If the adult dose of a drug is 500 mg and a child received a dose of 125 mg, how old was the child?

54. AMOUNT OF RAINFALL The total amount of rain (in inches) after t hr during a rainfall is given by

$$T = \dfrac{0.8t}{t + 4.1}$$

a. Solve for t in terms of T.
b. How long does it take for the total amount of rain to reach 0.4 in.?

1.6 Solutions to Self-Check Exercises

1. $2\left(\dfrac{x-1}{4}\right) - \dfrac{2x}{3} = \dfrac{4-3x}{12}$

$\quad 6(x-1) - 8x = 4 - 3x$ Multiply both sides by 12, the LCD.

$\quad\quad 6x - 6 - 8x = 4 - 3x$

$\quad\quad\quad -6 - 2x = 4 - 3x$

$\quad\quad\quad\quad -6 + x = 4$ Add $3x$ to both sides.

$\quad\quad\quad\quad\quad\quad x = 10$

2. $\dfrac{k}{2k+1} = \dfrac{3}{8}$

$\quad 8k = 3(2k+1)$ Multiply both sides by $8(2k+1)$, the LCD.

$\quad 8k = 6k + 3$

$\quad 2k = 3$

$\quad k = \dfrac{3}{2}$

If we substitute this value of k into the original equation, we find

$$\frac{\frac{3}{2}}{2(\frac{3}{2}) + 1} = \frac{\frac{3}{2}}{3 + 1} = \frac{3}{8}$$

which is equal to the right-hand side. So $k = \frac{3}{2}$ is the solution of the given equation.

1.7 Rational Exponents and Radicals

nth Roots of Real Numbers

Thus far, we have described the expression a^n, where a is a real number and n is an integer. We now direct our attention to a closely related topic: roots of real numbers. As we will soon see, expressions of the form a^n for fractional (rational) powers of n may be defined in terms of the roots of a.

> **nth Root of a Real Number**
>
> If n is a natural number and a and b are real numbers such that
>
> $$a^n = b$$
>
> then we say that a is the **nth root** of b.

For $n = 2$ and $n = 3$, the roots are commonly referred to as the **square roots** and **cube roots**, respectively. Some examples of roots follow:

- -2 and 2 are square roots of 4 because $(-2)^2 = 4$ and $2^2 = 4$.
- -3 and 3 are fourth roots of 81 because $(-3)^4 = 81$ and $3^4 = 81$.
- -4 is a cube root of -64 because $(-4)^3 = -64$.
- $\frac{1}{2}$ is a fifth root of $\frac{1}{32}$ because $\left(\frac{1}{2}\right)^5 = \frac{1}{32}$.

How many real roots does a real number b have?

1. *When n is even, the real nth roots of a positive real number b must come in pairs—one positive and the other negative.* For example, the real fourth roots of 81 include -3 and 3.
2. *When n is even and b is a negative real number, there are no real nth roots of b.* For example, if $b = -9$ and the real number a is a square root of b, then by definition, $a^2 = -9$. But this is a contradiction, since the square of a real number cannot be negative, and we conclude that b has no real roots in this case.
3. *When n is odd, then there is only one real nth root of b.* For example, the cube root of -64 is -4.

As you can see from the first statement, given a number b, there might be more than one real root. So to avoid ambiguity, we define the *principal* nth root of a positive real number, when n is even, to be the positive root. Thus, the principal square root of a real number, when n is even, is the positive root. For example, the principal square root of 4 is 2, and the principal fourth root of 81 is 3. Of course, the principal nth root of any real number b, when n is odd, is given by the (unique) nth root of b. For example, the principal cube root of -64 is -4, and the principal fifth root of $\frac{1}{32}$ is $\frac{1}{2}$.

A summary of the number of roots of a real number b is given in Table 11.

TABLE 11

Number of Roots of a Real Number b

Index	b	Number of Roots
n even	$b > 0$	Two real roots (one principal root)
	$b < 0$	No real roots
	$b = 0$	One real root
n odd	$b > 0$	One real root
	$b < 0$	One real root
	$b = 0$	One real root

We use the notation $\sqrt[n]{b}$, called a **radical**, to denote the principal nth root of b. The symbol $\sqrt{}$ is called a **radical sign**, and the number b within the radical sign is called the **radicand**. The positive integer n is called the **index** of the radical. For square roots ($n = 2$), we write $\sqrt{b}$ instead of $\sqrt[2]{b}$.

⚠️ A common mistake is to write $\sqrt[4]{16} = \pm 2$. This is wrong because $\sqrt[4]{16}$ denotes the principal fourth root of 16, which is the positive root 2. Of course, the negative of the fourth root of 16 is $-\sqrt[4]{16} = -(2) = -2$.

EXAMPLE 1 Determine the number of roots of each real number.

a. $\sqrt{25}$ **b.** $\sqrt[5]{0}$ **c.** $\sqrt[3]{-27}$ **d.** $\sqrt{-27}$

Solution

a. Here, $b > 0$, n is even, and there is one principal root. Thus, $\sqrt{25} = 5$.
b. Here, $b = 0$, n is odd, and there is one root. Thus, $\sqrt[5]{0} = 0$.
c. Here, $b < 0$, n is odd, and there is one root. Thus, $\sqrt[3]{-27} = -3$.
d. Here, $b < 0$, n is even, and no real root exists. Thus, $\sqrt{-27}$ is not defined. ∎

⚠️ Note that $(-81)^{1/4}$ does not exist because n is even and $b < 0$, but $-81^{1/4} = -(81)^{1/4} = -3$. The first expression is "the fourth root of -81," whereas the second expression is "the negative of the fourth root of 81."

EXAMPLE 2 Evaluate the following radicals.

a. $\sqrt[6]{64}$ **b.** $\sqrt[5]{-32}$ **c.** $\sqrt[3]{\dfrac{8}{27}}$ **d.** $-\sqrt{\dfrac{4}{25}}$

Solution

a. $\sqrt[6]{64} = 2$ because $2^6 = 64$.
b. $\sqrt[5]{-32} = -2$ because $(-2)^5 = -32$.
c. $\sqrt[3]{\dfrac{8}{27}} = \dfrac{2}{3}$ because $\left(\dfrac{2}{3}\right)^3 = \dfrac{8}{27}$.
d. $-\sqrt{\dfrac{4}{25}} = -\dfrac{2}{5}$ because $\sqrt{\dfrac{4}{25}} = \dfrac{2}{5}$, so $-\sqrt{\dfrac{4}{25}} = -\dfrac{2}{5}$. ∎

Rational Exponents and Radicals

In Section 1.5, we defined expressions such as 2^{-3}, $\left(\frac{1}{2}\right)^2$, and $1/\pi^3$ involving integral exponents. But how do we evaluate expressions such as $8^{1/3}$, where the exponent is a rational number? From the definition of the nth root of a real number, we know that $\sqrt[3]{8} = 2$. Using rational exponents, we write this same result in the form $8^{1/3} = 2$. More generally, we have the following definitions.

Rational Exponents	Illustration
1. If n is a natural number and b is a real number, then $$b^{1/n} = \sqrt[n]{b}$$ (If $b < 0$ and n is even, $b^{1/n}$ is not defined.)	$9^{1/2} = \sqrt{9} = 3$ $(-8)^{1/3} = \sqrt[3]{-8} = -2$
2. If m/n is a rational number reduced to lowest terms (m, n natural numbers), then $$b^{m/n} = (b^{1/n})^m$$ or, equivalently, $$b^{m/n} = \sqrt[n]{b^m}$$ whenever it exists.	$(27)^{2/3} = (27^{1/3})^2 = 3^2 = 9$ $(27)^{2/3} = [(27^2)]^{1/3} = (729)^{1/3} = 9$ $(-27)^{2/3} = (-27^{1/3})^2$ $\qquad = (-3)^2 = 9$

EXAMPLE 3

a. $(64)^{1/3} = \sqrt[3]{64} = 4$
b. $(81)^{3/4} = (81^{1/4})^3 = 3^3 = 27$
c. $(-8)^{5/3} = (-8^{1/3})^5 = (-2)^5 = -32$
d. $\left(\dfrac{1}{27}\right)^{2/3} = \left[\left(\dfrac{1}{27}\right)^{1/3}\right]^2 = \left(\dfrac{1}{3}\right)^2 = \dfrac{1}{9}$

Expressions involving *negative* rational exponents are taken care of by the following definition.

Negative Exponents
$$a^{-m/n} = \frac{1}{a^{m/n}} \qquad (a \neq 0)$$

EXAMPLE 4

a. $4^{-5/2} = \dfrac{1}{4^{5/2}} = \dfrac{1}{(4^{1/2})^5} = \dfrac{1}{2^5} = \dfrac{1}{32}$ **b.** $(-8)^{-1/3} = \dfrac{1}{(-8)^{1/3}} = \dfrac{1}{-2} = -\dfrac{1}{2}$

All the properties of integral exponents listed in Table 10 (on page 29) hold for rational exponents. Examples 5 and 6 illustrate the use of these properties.

EXAMPLE 5

a. $\dfrac{16^{5/4}}{16^{1/2}} = 16^{5/4-1/2} = 16^{5/4-2/4} = 16^{3/4} = (16^{1/4})^3 = 2^3 = 8$

b. $(6^{2/3})^3 = 6^{(2/3)\cdot 3} = 6^{6/3} = 6^2 = 36$

c. $\left(\dfrac{16}{81}\right)^{3/4} = \left[\left(\dfrac{16}{81}\right)^{1/4}\right]^3 = \left(\dfrac{16^{1/4}}{81^{1/4}}\right)^3 = \left(\dfrac{2}{3}\right)^3 = \dfrac{8}{27}$

VIDEO▶ **EXAMPLE 6** Evaluate:

a. $2x^{1/2}(x^{2/3} - x^{1/4})$ **b.** $(x^{1/3} - y^{2/3})^2$

Solution

a. $2x^{1/2}(x^{2/3} - x^{1/4}) = 2x^{1/2}(x^{2/3}) - 2x^{1/2}(x^{1/4})$
$$= 2x^{1/2 + 2/3} - 2x^{1/2 + 1/4} = 2x^{3/6 + 4/6} - 2x^{2/4 + 1/4}$$
$$= 2x^{7/6} - 2x^{3/4}$$

b. $(x^{1/3} - y^{2/3})^2 = (x^{1/3})^2 - 2x^{1/3}y^{2/3} + (y^{2/3})^2$
$$= x^{2/3} - 2x^{1/3}y^{2/3} + y^{4/3}$$

Simplifying Radicals

The properties of radicals given in Table 12 follow directly from the properties of exponents discussed earlier (Table 10, page 29).

TABLE 12	
Properties of Radicals	
Property	**Illustration**
If m and n are natural numbers and a and b are real numbers for which the indicated roots exist, then	
1. $(\sqrt[n]{a})^n = a$	$(\sqrt[3]{2})^3 = (2^{1/3})^3 = 2^1 = 2$
2. $\sqrt[n]{ab} = \sqrt[n]{a} \cdot \sqrt[n]{b}$	$\sqrt[3]{216} = \sqrt[3]{27 \cdot 8} = \sqrt[3]{27} \cdot \sqrt[3]{8} = 3 \cdot 2 = 6$
3. $\sqrt[n]{\dfrac{a}{b}} = \dfrac{\sqrt[n]{a}}{\sqrt[n]{b}} \quad (b \neq 0)$	$\sqrt[3]{\dfrac{8}{64}} = \dfrac{\sqrt[3]{8}}{\sqrt[3]{64}} = \dfrac{2}{4} = \dfrac{1}{2}$
4. $\sqrt[m]{\sqrt[n]{a}} = \sqrt[mn]{a}$	$\sqrt[3]{\sqrt[2]{64}} = \sqrt[2 \cdot 3]{64} = \sqrt[6]{64} = 2$
5. If n is even: $\sqrt[n]{a^n} = |a|$.	$\sqrt{(-3)^2} = |-3| = 3$
If n is odd: $\sqrt[n]{a^n} = a$.	$\sqrt[3]{-8} = -2$

⚠ A common error is to write $\sqrt[n]{a^n} = a$. This is not true if a is negative (see Illustration 5 in Table 12). Thus, unless the variable a is known to be nonnegative, the correct answer is given by Property 5.

When we work with algebraic expressions involving radicals, we usually express the radical in simplified form.

> **Simplifying Radicals**
>
> An expression involving radicals is simplified if the following conditions are satisfied:
>
> **1.** The powers of all factors under the radical sign are less than the index of the radical.
> **2.** The index of the radical has been reduced as far as possible.
> **3.** No radical appears in a denominator.
> **4.** No fraction appears within a radical.

EXAMPLE 7 Determine whether each radical is in simplified form. If not, state which condition is violated.

a. $\dfrac{1}{\sqrt[4]{x^5}}$ **b.** $\sqrt[6]{y^2}$ **c.** $\sqrt{\dfrac{5}{4}}$

Solution   None of the three radicals are in simplified form. The radical in part (a) violates conditions 1 and 3; that is, the power of x is 5, which is greater than 4, the index of the radical, and a radical appears in the denominator. The radical in part (b) violates condition 2, since the index of the radical can be reduced; that is,

$$\sqrt[6]{y^2} = \sqrt[3 \cdot 2]{y^2} = \sqrt[3]{y}$$

The radical in part (c) violates condition 4, since there is a fraction within the radical. Rewriting the radical, we see that

$$\sqrt{\frac{5}{4}} = \frac{\sqrt{5}}{\sqrt{4}} = \frac{\sqrt{5}}{2} = \frac{1}{2}\sqrt{5}$$ ◼

EXAMPLE 8 Simplify:

a. $\sqrt[3]{375}$ **b.** $\sqrt[3]{8x^3y^6z^9}$ **c.** $\sqrt[6]{81x^4y^2}$

Solution

a. $\sqrt[3]{375} = \sqrt[3]{3 \cdot 125} = \sqrt[3]{3} \cdot \sqrt[3]{5^3} = 5\sqrt[3]{3}$ $\sqrt[3]{5^3} = 5$

b. $\sqrt[3]{8x^3y^6z^9} = \sqrt[3]{2^3 \cdot (xy^2z^3)^3}$
$$= \sqrt[3]{2^3} \cdot \sqrt[3]{(xy^2z^3)^3}$$
$$= 2xy^2z^3 \qquad \sqrt[3]{2^3} = 2, \quad \sqrt[3]{(xy^2z^3)^3} = xy^2z^3$$

c. $\sqrt[6]{81x^4y^2} = \sqrt[6]{9^2(x^2y)^2}$
$$= \sqrt[3 \cdot 2]{9^2} \cdot \sqrt[3 \cdot 2]{(x^2y)^2}$$
$$= \sqrt[3]{9} \cdot \sqrt[3]{x^2y} \qquad \sqrt[3 \cdot 2]{9^2} = \sqrt[3]{9}, \quad \sqrt[3 \cdot 2]{(x^2y)^2} = \sqrt[3]{x^2y}$$
$$= \sqrt[3]{9x^2y}$$ ◼

  As was mentioned earlier, a simplified rational expression should not have radicals in its denominator. For example, $3/\sqrt{5}$ is *not* in simplified form, whereas $3\sqrt{5}/5$ is, since the denominator of the latter is free of radicals. How do we get rid of the radical $\sqrt{5}$ in the fraction $3/\sqrt{5}$? Obviously, multiplying by $\sqrt{5}$ does the job, since $(\sqrt{5})(\sqrt{5}) = \sqrt{25} = 5$! But we cannot multiply the denominator of a fraction by any number other than 1 without changing the fraction. So the solution is to multiply *both* the numerator and the denominator of $3/\sqrt{5}$ by $\sqrt{5}$. Equivalently, we multiply $3/\sqrt{5}$ by $\sqrt{5}/\sqrt{5}$ (which is equal to 1). Thus,

$$\frac{3}{\sqrt{5}} = \frac{3}{\sqrt{5}} \cdot \frac{\sqrt{5}}{\sqrt{5}} = \frac{3\sqrt{5}}{\sqrt{25}} = \frac{3\sqrt{5}}{5}$$

This process of eliminating a radical from the denominator of an algebraic expression is referred to as *rationalizing the denominator* and is illustrated in Examples 9 and 10.

EXAMPLE 9 Rationalize the denominator.

a. $\dfrac{1}{\sqrt{2}}$ **b.** $\dfrac{3x}{2\sqrt{x}}$ **c.** $\dfrac{x}{\sqrt[3]{y}}$

Solution

a. $\dfrac{1}{\sqrt{2}} \cdot \dfrac{\sqrt{2}}{\sqrt{2}} = \dfrac{\sqrt{2}}{2} = \dfrac{1}{2}\sqrt{2}$

b. $\dfrac{3x}{2\sqrt{x}} \cdot \dfrac{\sqrt{x}}{\sqrt{x}} = \dfrac{3x\sqrt{x}}{2x} = \dfrac{3}{2}\sqrt{x}$

c. $\dfrac{x}{\sqrt[3]{y}} \cdot \dfrac{\sqrt[3]{y^2}}{\sqrt[3]{y^2}} = \dfrac{x\sqrt[3]{y^2}}{\sqrt[3]{y^3}} = \dfrac{x\sqrt[3]{y^2}}{y}$ ◾

Note that in Example 9c, we multiplied the denominator by $\sqrt[3]{y^2}$ so that we would obtain $\sqrt[3]{y^3}$, which is equal to y. In general, to rationalize a denominator involving an nth root, we multiply the numerator and the denominator by a factor that will yield a product in the denominator involving an nth power. For example,

$$\frac{1}{\sqrt[5]{x^3}} = \frac{1}{\sqrt[5]{x^3}} \cdot \frac{\sqrt[5]{x^2}}{\sqrt[5]{x^2}} = \frac{\sqrt[5]{x^2}}{\sqrt[5]{x^5}} = \frac{\sqrt[5]{x^2}}{x} \qquad \text{Since } \sqrt[5]{x^3} \cdot \sqrt[5]{x^2} = x^{3/5} \cdot x^{2/5} = x^{5/5} = x$$

EXAMPLE 10 Rationalize the denominator.

a. $\sqrt[3]{\dfrac{8}{3}}$ **b.** $\sqrt[3]{\dfrac{x}{y^2}}$

Solution

a. $\sqrt[3]{\dfrac{8}{3}} = \dfrac{\sqrt[3]{8}}{\sqrt[3]{3}} = \dfrac{2}{\sqrt[3]{3}} \cdot \dfrac{\sqrt[3]{3^2}}{\sqrt[3]{3^2}} = \dfrac{2\sqrt[3]{3^2}}{3} = \dfrac{2}{3}\sqrt[3]{9}$

b. $\sqrt[3]{\dfrac{x}{y^2}} = \dfrac{\sqrt[3]{x}}{\sqrt[3]{y^2}} = \dfrac{\sqrt[3]{x}}{\sqrt[3]{y^2}} \cdot \dfrac{\sqrt[3]{y}}{\sqrt[3]{y}} = \dfrac{\sqrt[3]{xy}}{\sqrt[3]{y^3}} = \dfrac{\sqrt[3]{xy}}{y}$ ◾

How do we rationalize the denominator of a fraction like $\dfrac{1}{1 - \sqrt{3}}$? Rather than multiplying by $\dfrac{\sqrt{3}}{\sqrt{3}}$ (which does not eliminate the radical in the denominator), we multiply by $\dfrac{1 + \sqrt{3}}{1 + \sqrt{3}}$, obtaining

$$\frac{1}{1 - \sqrt{3}} \cdot \frac{1 + \sqrt{3}}{1 + \sqrt{3}} = \frac{1 + \sqrt{3}}{1 - (\sqrt{3})^2} \qquad (a - b)(a + b) = a^2 - b^2$$

$$= \frac{1 + \sqrt{3}}{1 - 3} = \frac{1 + \sqrt{3}}{-2}$$

$$= -\frac{1 + \sqrt{3}}{2}$$

In general, to rationalize a denominator of the form $a + \sqrt{b}$, we multiply by $\dfrac{a - \sqrt{b}}{a - \sqrt{b}}$.

Similarly, to rationalize a denominator of the form $a - \sqrt{b}$, we multiply by $\dfrac{a + \sqrt{b}}{a + \sqrt{b}}$.

We refer to the quantities $a + \sqrt{b}$ and $a - \sqrt{b}$ as **conjugates** of each other.

VIDEO   **EXAMPLE 11** Rationalize the denominator in the expression $\dfrac{3}{2 - \sqrt{5}}$.

Solution   The conjugate of $2 - \sqrt{5}$ is $2 + \sqrt{5}$. Therefore, we multiply the given fraction by $\dfrac{2 + \sqrt{5}}{2 + \sqrt{5}}$, obtaining

$$\frac{3}{2 - \sqrt{5}} \cdot \frac{2 + \sqrt{5}}{2 + \sqrt{5}} = \frac{3(2 + \sqrt{5})}{(2 - \sqrt{5})(2 + \sqrt{5})} = \frac{3(2 + \sqrt{5})}{2^2 - (\sqrt{5})^2}$$

$$= \frac{3(2 + \sqrt{5})}{-1}$$

$$= -3(2 + \sqrt{5})$$

Sometimes you may find it easier to convert an expression containing a radical to one involving rational exponents before evaluating the expression.

EXAMPLE 12 Evaluate:

a. $\sqrt[3]{3^2} \cdot \sqrt[4]{9^3}$ **b.** $\sqrt[3]{x^2 y} \cdot \sqrt{xy}$

Solution

a. $\sqrt[3]{3^2} \cdot \sqrt[4]{9^3} = 3^{2/3} \cdot 9^{3/4} = 3^{2/3} \cdot 3^{6/4} = 3^{2/3 + 3/2} = 3^{4/6 + 9/6} = 3^{13/6}$

b. $\sqrt[3]{x^2 y} \cdot \sqrt{xy} = x^{2/3} y^{1/3} \cdot x^{1/2} y^{1/2} = x^{2/3 + 1/2} y^{1/3 + 1/2} = x^{4/6 + 3/6} y^{2/6 + 3/6} = x^{7/6} y^{5/6}$

Note how much easier it is to work with rational exponents in this case.

Solving Equations Involving Radicals

The following examples show how to solve equations involving radicals.

EXAMPLE 13 Solve $\sqrt{2x + 5} = 3$.

Solution   This equation is not a linear equation. To solve it, we square both sides of the equation, obtaining

$$(\sqrt{2x + 5})^2 = 3^2$$
$$2x + 5 = 9$$
$$2x = 4$$
$$x = 2$$

Substituting $x = 2$ into the left-hand side of the original equation yields

$$\sqrt{2(2) + 5} = \sqrt{9} = 3$$

which is the same as the number on the right-hand side of the equation. Therefore, the required solution is 3.

VIDEO   **EXAMPLE 14** Solve $\sqrt{k^2 - 4} = k - 4$.

Solution   Squaring both sides of the equation leads to

$$k^2 - 4 = (k - 4)^2$$
$$k^2 - 4 = k^2 - 8k + 16$$
$$-4 = -8k + 16$$
$$-20 = -8k$$
$$\frac{5}{2} = k$$

Substituting this value of k into the left-hand side of the original equation gives

$$\sqrt{\left(\frac{5}{2}\right)^2 - 4} = \sqrt{\frac{25}{4} - 4} = \sqrt{\frac{9}{4}} = \frac{3}{2}$$

But this is not equal to $\frac{5}{2} - 4 = -\frac{3}{2}$, which is the result obtained if the value of k is substituted into the right-hand side of the original equation. We conclude that the given equation has no solution.

1.7 Self-Check Exercises

1. Simplify:

   a. $\dfrac{8^{2/3} \cdot 8^{4/3}}{8^{3/2}}$ **b.** $\left(\dfrac{3y^{-4}y^4}{z^{-2}}\right)^3$ **c.** $\sqrt{5} \cdot \sqrt{45}$

2. Rationalize the denominator:

   a. $\dfrac{1}{\sqrt[3]{xy}}$ **b.** $\dfrac{4}{3 + \sqrt{8}}$

   Solutions to Self-Check Exercises 1.7 can be found on page 45.

1.7 Concept Questions

1. What is the nth root of a real number? Give an example.

2. What is the principal nth root of a positive real number? Give an example.

3. What is meant by the statement "rationalize the denominator of an algebraic expression"? Illustrate the process with an example.

1.7 Exercises

In Exercises 1–20, rewrite the number without radicals or exponents.

1. $\sqrt{81}$

2. $\sqrt[3]{-27}$

3. $\sqrt[4]{256}$

4. $\sqrt[5]{-32}$

5. $16^{1/2}$

6. $625^{1/4}$

7. $8^{2/3}$

8. $32^{2/5}$

9. $-25^{1/2}$

10. $-16^{3/2}$

11. $(-8)^{2/3}$

12. $(-32)^{3/5}$

13. $\left(\dfrac{4}{9}\right)^{1/2}$

14. $\left(\dfrac{9}{25}\right)^{3/2}$

15. $\left(\dfrac{27}{8}\right)^{2/3}$

16. $\left(-\dfrac{8}{125}\right)^{1/3}$

17. $8^{-2/3}$

18. $81^{-1/4}$

19. $-\left(\dfrac{27}{8}\right)^{-1/3}$

20. $-\left(-\dfrac{8}{27}\right)^{-2/3}$

In Exercises 21–40, carry out the indicated operation and write your answer using positive exponents only.

21. $3^{1/3} \cdot 3^{5/3}$

22. $2^{6/5} \cdot 2^{-1/5}$

23. $\dfrac{3^{1/2}}{3^{5/2}}$

24. $\dfrac{3^{-5/4}}{3^{-1/4}}$

25. $\dfrac{2^{-1/2} \cdot 3^{2/3}}{2^{3/2} \cdot 3^{-1/3}}$

26. $\dfrac{4^{1/3} \cdot 4^{-2/5}}{4^{2/3}}$

27. $(2^{3/2})^4$

28. $[(-3)^{1/3}]^2$

29. $x^{2/5} \cdot x^{-1/5}$

30. $y^{-3/8} \cdot y^{1/4}$

31. $\dfrac{x^{3/4}}{x^{-1/4}}$

32. $\dfrac{x^{7/3}}{x^{-2}}$

33. $\left(\dfrac{x^3}{-27x^{-6}}\right)^{-2/3}$

34. $\left(\dfrac{27x^{-3}y^2}{8x^{-2}y^{-5}}\right)^{1/3}$

35. $\left(\dfrac{x^{-3}}{y^{-2}}\right)^{1/2}\left(\dfrac{y}{x}\right)^{3/2}$

36. $\left(\dfrac{r^n}{r^{5-2n}}\right)^4$

37. $x^{2/5}(x^2 - 2x^3)$

38. $s^{1/3}(2s - s^{1/4})$

39. $2p^{3/2}(2p^{1/2} - p^{-1/2})$

40. $3y^{1/3}(y^{2/3} - 1)^2$

In Exercises 41–52, write the expression in simplest radical form. (Assume that all variables are nonnegative.)

41. $\sqrt{32}$

42. $\sqrt{45}$

43. $\sqrt[3]{-54}$

44. $-\sqrt[4]{48}$

45. $\sqrt{16x^2y^3}$

46. $\sqrt{40a^3b^4}$

47. $\sqrt[3]{m^6n^3p^{12}}$

48. $\sqrt[3]{-27p^2q^3r^4}$

49. $\sqrt[3]{\sqrt{9}}$

50. $\sqrt[5]{\sqrt[3]{9}}$

51. $\sqrt[3]{\sqrt{x}}$

52. $\sqrt[3]{-\sqrt[4]{x^3}}$

In Exercises 53–68, rationalize the denominator of the expression.

53. $\dfrac{2}{\sqrt{3}}$ **54.** $\dfrac{3}{\sqrt{5}}$ **55.** $\dfrac{3}{2\sqrt{x}}$

56. $\dfrac{3}{\sqrt{xy}}$ **57.** $\dfrac{2y}{\sqrt{3y}}$ **58.** $\dfrac{5x^2}{\sqrt{3x}}$

59. $\dfrac{1}{\sqrt[3]{x}}$ **60.** $\sqrt{\dfrac{2x}{y}}$

61. $\dfrac{2}{1+\sqrt{3}}$ **62.** $\dfrac{3}{1-\sqrt{2}}$

63. $\dfrac{1+\sqrt{2}}{1-\sqrt{2}}$ **64.** $\dfrac{9+\sqrt{2}}{3-\sqrt{2}}$

65. $\dfrac{q}{\sqrt{q}-1}$ **66.** $\dfrac{xy}{\sqrt{x}+\sqrt{y}}$

67. $\dfrac{y}{\sqrt[3]{x^2z}}$ **68.** $\dfrac{2x}{\sqrt[3]{xy^2}}$

In Exercises 69–76, simplify the expression. (Assume that all variables are positive.)

69. $\sqrt{\dfrac{16}{3}}$ **70.** $-\sqrt{\dfrac{8}{3}}$ **71.** $\sqrt[3]{\dfrac{2}{3}}$ **72.** $\sqrt[3]{\dfrac{81}{4}}$

73. $\sqrt{\dfrac{3}{2x^2}}$ **74.** $\sqrt{\dfrac{x^3y^5}{4}}$ **75.** $\sqrt[3]{\dfrac{2y^2}{3}}$ **76.** $\sqrt[3]{\dfrac{3a^3}{b^2}}$

In Exercises 77–84, simplify the expression.

77. $\dfrac{1}{\sqrt{a}}+\sqrt{a}$ **78.** $\dfrac{x}{\sqrt{x-y}}-\sqrt{x-y}$

79. $\dfrac{\sqrt{x}}{\sqrt{x}+\sqrt{y}}+\dfrac{\sqrt{y}}{\sqrt{x}-\sqrt{y}}$

80. $\dfrac{a}{\sqrt{a^2-b^2}}-\dfrac{\sqrt{a^2-b^2}}{a}$

81. $(x+1)^{1/2}+\dfrac{1}{2}x(x+1)^{-1/2}$

82. $\dfrac{1}{2}x^{-1/2}(x+y)^{1/3}+\dfrac{1}{3}x^{1/2}(x+y)^{-2/3}$

83. $\dfrac{\frac{1}{2}(1+x^{1/3})x^{-1/2}-\frac{1}{3}x^{1/2}\cdot x^{-2/3}}{(1+x^{1/3})^2}$

84. $\dfrac{\frac{1}{2}x^{-1/2}(x+y)^{1/2}-\frac{1}{2}x^{1/2}(x+y)^{-1/2}}{x+y}$

In Exercises 85–90, solve the given equation.

85. $\sqrt{3x+1}=2$

86. $\sqrt{2x-3}-3=0$

87. $\sqrt{k^2-4}=4-k$

88. $\sqrt{4k^2-3}=2k+1$

89. $\sqrt{k+1}+\sqrt{k}=3\sqrt{k}$

90. $\sqrt{x+1}-\sqrt{x}=\sqrt{4x-3}$

91. DEMAND FOR TIRES The management of Titan Tire Company has determined that x thousand Super Titan tires will be sold each week if the price per tire is p dollars, where p and x are related by the equation

$$x=\sqrt{144-p}\qquad(0<p\le144)$$

Solve the equation for p in terms of x.

92. DEMAND FOR WATCHES The equation

$$x=10\sqrt{\dfrac{50-p}{p}}\qquad(0<p\le50)$$

gives the relationship between the number x, in thousands, of the Sicard sports watch demanded per week and the unit price p in dollars. Solve the equation for p in terms of x.

In Exercises 93–96, determine whether the statement is true or false. If it is true, explain why it is true. If it is false, explain why or give an example to show why it is false.

93. If a is a real number, then $\sqrt{a^2}=|a|$.

94. If a is a real number, then $-(a^2)^{1/4}$ is not defined.

95. If n is a natural number and a is a positive real number, then $(a^{1/n})^n=a$.

96. If a and b are positive real numbers, then $\sqrt{a^2+b^2}=a+b$.

1.7 Solutions to Self-Check Exercises

1. a. $\dfrac{8^{2/3}\cdot8^{4/3}}{8^{3/2}}=8^{2/3+4/3-3/2}=8^{6/3-3/2}=8^{1/2}=\sqrt{8}=2\sqrt{2}$

b. $\left(\dfrac{3y^{-4}y^4}{z^{-2}}\right)^3=\left(\dfrac{3y^{-4+4}}{z^{-2}}\right)^3=\left(\dfrac{3y^0}{z^{-2}}\right)^3$
$=(3z^2)^3=3^3z^{2\cdot3}=27z^6$

c. $\sqrt{5}\cdot\sqrt{45}=\sqrt{5\cdot45}=\sqrt{225}=15$

2. a. $\dfrac{1}{\sqrt[3]{xy}}\cdot\dfrac{\sqrt[3]{x^2y^2}}{\sqrt[3]{x^2y^2}}=\dfrac{\sqrt[3]{x^2y^2}}{\sqrt[3]{x^3y^3}}=\dfrac{\sqrt[3]{x^2y^2}}{xy}$

b. $\dfrac{4}{3+\sqrt{8}}\cdot\dfrac{3-\sqrt{8}}{3-\sqrt{8}}=\dfrac{4(3-\sqrt{8})}{3^2-(\sqrt{8})^2}$
$=\dfrac{4(3-\sqrt{8})}{9-8}=4(3-\sqrt{8})$

1.8   Quadratic Equations

The equation

$$2x^2 + 3x + 1 = 0$$

is an example of a quadratic equation. In general, a **quadratic equation** in the variable x is any equation that can be written in the form

$$ax^2 + bx + c = 0$$

where a, b, and c are constants and $a \neq 0$. We refer to this form as the *standard form*. Equations such as

$$3x^2 + 4x = 1 \quad \text{and} \quad 2t^2 = t + 1$$

are quadratic equations in nonstandard form, but they can be easily transformed into standard form. For example, adding -1 to both sides of the first equation leads to $3x^2 + 4x - 1 = 0$, which is in standard form. Similarly, by subtracting $(t + 1)$ from both sides of the second equation, we obtain $2t^2 - t - 1 = 0$, which is also in standard form.

Solving by Factoring

We solve a quadratic equation in x by finding its roots. The *roots* of a quadratic equation in x are precisely the values of x that satisfy the equation. The method of solving quadratic equations by *factoring* relies on the following zero-product property of real numbers, which we restate here.

> **Zero-Product Property of Real Numbers**
>
> If a and b are real numbers and $ab = 0$, then $a = 0$, or $b = 0$, or both $a, b = 0$.

Simply stated, this property says that the product of two real numbers is equal to zero if and only if one (or both) of the factors is equal to zero.

EXAMPLE 1 Solve $x^2 - 3x + 2 = 0$ by factoring.

Solution Factoring the given equation, we find that

$$x^2 - 3x + 2 = (x - 2)(x - 1) = 0$$

By the zero-product property of real numbers, we have

$$x - 2 = 0 \quad \text{or} \quad x - 1 = 0$$

from which we see that $x = 2$ or $x = 1$ are the roots of the equation.   ∎

If a quadratic equation is not in standard form, we first rewrite it in standard form and then factor the equation to find its roots.

EXAMPLE 2 Solve by factoring.

a. $2x^2 - 7x = -6$ **b.** $4x^2 = 3x$ **c.** $2x^2 = 6x - 4$

Solution

a. Rewriting the equation in standard form, we have

$$2x^2 - 7x + 6 = 0$$

Factoring this equation, we obtain

$$(2x - 3)(x - 2) = 0$$

Therefore,

$$2x - 3 = 0 \quad \text{or} \quad x - 2 = 0$$
$$x = \frac{3}{2} \qquad\qquad x = 2$$

b. We first write the equation in standard form:

$$4x^2 - 3x = 0$$

Factoring this equation, we have

$$x(4x - 3) = 0$$

Therefore,

$$x = 0 \quad \text{or} \quad 4x - 3 = 0$$
$$x = 0 \qquad\qquad x = \frac{3}{4}$$

(Observe that if we had divided $4x^2 - 3x = 0$ by x before factoring the original equation, we would have lost the solution $x = 0$.)

c. Rewriting the equation in standard form, we have

$$2x^2 - 6x + 4 = 0$$

Factoring, we have

$$2(x^2 - 3x + 2) = 0 \qquad \text{2 is a common factor.}$$
$$2(x - 2)(x - 1) = 0$$

and

$$x = 2 \quad \text{or} \quad x = 1$$

Solving by Completing the Square

The method of solution by factoring works well for equations that are easily factored. But what about equations such as $x^2 - 2x - 2 = 0$ that are not easily factored? Equations of this type may be solved by using the method of *completing the square*.

The Method of Completing the Square	Illustration
1. Write the equation $ax^2 + bx + c = 0$ in the form $$x^2 + \frac{b}{a}x = -\frac{c}{a}$$ Coefficient of x ⌐ └ Constant term where the coefficient of x^2 is 1 and the constant term is on the right side of the equation.	$x^2 - 2x - 2 = 0$ $x^2 - 2x = 2$ Coefficient ⌐ └ Constant of x term
2. Square half of the coefficient of x.	$\left(-\frac{2}{2}\right)^2 = 1$
3. Add the number obtained in step 2 to both sides of the equation, factor, and solve for x.	$x^2 - 2x + 1 = 2 + 1$ $(x - 1)^2 = 3$ $x - 1 = \pm\sqrt{3}$ $x = 1 \pm \sqrt{3}$

EXAMPLE 3 Solve by completing the square.

a. $4x^2 - 3x - 2 = 0$ **b.** $6x^2 - 27 = 0$

Solution

a. Step 1   First, write

$$x^2 - \frac{3}{4}x - \frac{1}{2} = 0 \qquad \text{Divide the original equation}$$
by 4, the coefficient of x^2.

$$x^2 - \frac{3}{4}x = \frac{1}{2} \qquad \text{Add } \frac{1}{2} \text{ to both sides of the equation so that}$$
the constant term is on the right side.

Step 2   Square half of the coefficient of x, obtaining

$$\left(\frac{-\frac{3}{4}}{2}\right)^2 = \left(-\frac{3}{8}\right)^2 = \frac{9}{64}$$

Step 3   Add $\frac{9}{64}$ to both sides of the equation:

$$x^2 - \frac{3}{4}x + \frac{9}{64} = \frac{1}{2} + \frac{9}{64}$$

$$= \frac{41}{64}$$

Factoring the left side of the equation, we have

$$\left(x - \frac{3}{8}\right)^2 = \frac{41}{64}$$

$$x - \frac{3}{8} = \pm\frac{\sqrt{41}}{8}$$

$$x = \frac{3}{8} \pm \frac{\sqrt{41}}{8} = \frac{1}{8}(3 \pm \sqrt{41})$$

b. This equation is much easier to solve because the coefficient of x is 0. As before, we write the equation in the form

$$6x^2 = 27 \quad \text{or} \quad x^2 = \frac{9}{2}$$

Taking the square root of both sides of the equation, we have

$$x = \pm\sqrt{\frac{9}{2}} = \pm\frac{3}{\sqrt{2}} = \pm\frac{3\sqrt{2}}{2}$$ ■

Using the Quadratic Formula

By using the method of completing the square to solve the general quadratic equation,

$$ax^2 + bx + c = 0 \qquad (a \neq 0)$$

we obtain the quadratic formula (but we will not derive the formula here). This formula can be used to solve any quadratic equation.

The Quadratic Formula

The solutions of $ax^2 + bx + c = 0 \ (a \neq 0)$ are given by

$$x = \frac{-b \pm \sqrt{b^2 - 4ac}}{2a}$$

VIDEO **EXAMPLE 4** Use the quadratic formula to solve the following:

a. $2x^2 + 5x - 12 = 0$ **b.** $x^2 + 165x + 6624 = 0$ **c.** $x^2 = -3x + 8$

Solution

a. The equation is in standard form, with $a = 2$, $b = 5$, and $c = -12$. Using the quadratic formula, we find

$$x = \frac{-b \pm \sqrt{b^2 - 4ac}}{2a} = \frac{-5 \pm \sqrt{5^2 - 4(2)(-12)}}{2(2)}$$

$$= \frac{-5 \pm \sqrt{121}}{4} = \frac{-5 \pm 11}{4}$$

$$= -4 \quad \text{or} \quad \frac{3}{2}$$

Observe that this equation can also be solved by factoring. Thus,

$$2x^2 + 5x - 12 = (2x - 3)(x + 4) = 0$$

from which we see that the desired roots are $x = \frac{3}{2}$ or $x = -4$, as obtained earlier.

b. The equation is in standard form, with $a = 1$, $b = 165$, and $c = 6624$. Using the quadratic formula, we find

$$x = \frac{-b \pm \sqrt{b^2 - 4ac}}{2a} = \frac{-165 \pm \sqrt{165^2 - 4(1)(6624)}}{2(1)}$$

$$= \frac{-165 \pm \sqrt{729}}{2}$$

$$= \frac{-165 \pm 27}{2} = -96 \quad \text{or} \quad -69$$

In this case, using the quadratic formula is preferable to factoring the quadratic equation.

c. We first rewrite the given equation in the standard form $x^2 + 3x - 8 = 0$, from which we see that $a = 1$, $b = 3$, and $c = -8$. Using the quadratic formula, we find

$$x = \frac{-b \pm \sqrt{b^2 - 4ac}}{2a} = \frac{-3 \pm \sqrt{3^2 - 4(1)(-8)}}{2(1)} = \frac{-3 \pm \sqrt{41}}{2}$$

That is, the solutions are

$$\frac{-3 + \sqrt{41}}{2} \approx 1.7 \quad \text{or} \quad \frac{-3 - \sqrt{41}}{2} \approx -4.7$$

In this case, the quadratic formula proves quite handy! ∎

EXAMPLE 5 Use the quadratic formula to solve $9x^2 - 12x + 4 = 0$.

Solution   Using the quadratic formula with $a = 9$, $b = -12$, and $c = 4$, we find that

$$x = \frac{-b \pm \sqrt{b^2 - 4ac}}{2a} = \frac{-(-12) \pm \sqrt{(-12)^2 - 4(9)(4)}}{2(9)}$$

$$= \frac{12 \pm \sqrt{144 - 144}}{18} = \frac{12}{18} = \frac{2}{3}$$

Here, the only solution is $x = \frac{2}{3}$. We refer to $\frac{2}{3}$ as a *double root*. (This equation could also be solved by factoring. Try it!) ∎

EXAMPLE 6 Solve the equation $\sqrt{2x-1} - \sqrt{x+3} + 1 = 0$.

Solution   We proceed as follows:

$$\sqrt{2x-1} = \sqrt{x+3} - 1 \qquad \text{Add } \sqrt{x+3} - 1 \text{ to both sides.}$$
$$2x - 1 = (\sqrt{x+3} - 1)^2 \qquad \text{Square both sides.}$$
$$2x - 1 = x + 3 - 2\sqrt{x+3} + 1$$
$$x - 5 = -2\sqrt{x+3} \qquad \text{Simplify.}$$
$$(x-5)^2 = (-2\sqrt{x+3})^2 \qquad \text{Square both sides.}$$
$$x^2 - 10x + 25 = 4x + 12$$
$$x^2 - 14x + 13 = 0$$
$$(x-1)(x-13) = 0$$
$$x = 1 \quad \text{or} \quad 13$$

Next, we need to verify that these solutions of the quadratic equation are indeed the solutions of the original equation. Recall that squaring an equation could introduce extraneous solutions. Now, substituting $x = 1$ into the original equation gives

$$\sqrt{2-1} - \sqrt{1+3} + 1 = 1 - 2 + 1 = 0$$

so $x = 1$ is a solution. On the other hand, if $x = 13$, we have

$$\sqrt{26-1} - \sqrt{16} + 1 = 5 - 4 + 1 = 2 \neq 0$$

so $x = 13$ is an extraneous solution. Therefore, the required solution is $x = 1$.   ∎

The following example gives an application involving quadratic equations.

VIDEO ▶

APPLIED EXAMPLE 7 Book Design   A production editor at a textbook publishing house decided that the pages of a book should have 1-inch margins at the top and bottom and $\frac{1}{2}$-inch margins on the sides. She further stipulated that the length of a page should be $1\frac{1}{2}$ times its width and have a printed area of exactly 51 square inches. Find the dimensions of a page of the book.

Solution   Let x denote the width of a page of the book (Figure 3). Then the length of the page is $\frac{3}{2}x$. The dimensions of the printed area of the page are $\left(\frac{3}{2}x - 2\right)$ inches by $(x - 1)$ inches, so its area is $\left(\frac{3}{2}x - 2\right)(x - 1)$ square inches. Since the printed area is to be exactly 51 square inches, we must have

$$\left(\frac{3}{2}x - 2\right)(x - 1) = 51$$

Expanding the left-hand side of the equation gives

$$\frac{3}{2}x^2 - \frac{3}{2}x - 2x + 2 = 51$$
$$\frac{3}{2}x^2 - \frac{7}{2}x - 49 = 0$$
$$3x^2 - 7x - 98 = 0 \qquad \text{Multiply both sides by 2.}$$

Factoring, we have

$$(3x + 14)(x - 7) = 0$$

so $x = -\frac{14}{3}$ or $x = 7$. Since x must be positive, we reject the negative root and conclude that the required solution is $x = 7$. Therefore, the dimensions of the page are 7 inches by $\frac{3}{2}(7)$, or $10\frac{1}{2}$, inches.

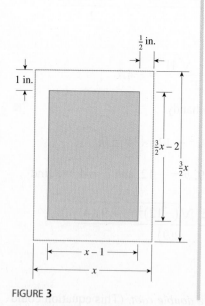

$\frac{1}{2}$ in.

1 in.

$\frac{3}{2}x - 2$

$\frac{3}{2}x$

$x - 1$

x

FIGURE 3

TABLE 13

Solutions of a Quadratic Equation

Discriminant $b^2 - 4ac$	Number of Solutions
Positive	Two real and distinct solutions
Equal to 0	One real solution
Negative	No real solution

The quantity $b^2 - 4ac$, which appears under the radical sign in the quadratic formula, is called the **discriminant.** We use the discriminant to determine the number of solutions of a quadratic equation. If the discriminant is positive, the equation has two real and distinct roots (Example 4a–c). If the discriminant is equal to 0, the equation has one double root (Example 5). Finally, if the discriminant is negative, the equation has no real roots. These results are summarized in Table 13.

EXAMPLE 8 Use the discriminant to determine the number of real solutions of each equation.

a. $x^2 - 7x + 4 = 0$ **b.** $2x^2 - 3x + 4 = 0$

Solution

a. Here, $a = 1$, $b = -7$, and $c = 4$. Therefore,

$$b^2 - 4ac = (-7)^2 - 4(1)(4) = 49 - 16$$
$$= 33$$

and we conclude that the equation has two real and distinct solutions.

b. Here, $a = 2$, $b = -3$, and $c = 4$. Therefore,

$$b^2 - 4ac = (-3)^2 - 4(2)(4) = 9 - 32$$
$$= -23$$

Since the discriminant is negative, we conclude that the equation has no real solution. ■

1.8 Self-Check Exercises

1. Solve by factoring:
   a. $x^2 - 5x + 6 = 0$
   b. $4t^2 - 4t = 3$

2. Solve the equation $2x^2 - 4x - 8 = 0$:
   a. By completing the square
   b. By using the quadratic formula

   Solutions to Self-Check Exercises 1.8 can be found on page 54.

1.8 Concept Questions

1. What is a *quadratic equation* in x? Give an example.

2. Explain the method of completing the square. Illustrate with an example.

3. State the quadratic formula. Illustrate its use with an example.

1.8 Exercises

In Exercises 1–16, solve the equation by factoring, if required.

1. $(x + 2)(x - 3) = 0$

2. $(y - 3)(y - 4) = 0$

3. $x^2 - 4 = 0$

4. $2m^2 - 32 = 0$

5. $x^2 + x - 12 = 0$

6. $3x^2 - x - 4 = 0$

7. $4t^2 + 2t - 2 = 0$

8. $-6x^2 + x + 12 = 0$

9. $\frac{1}{4}x^2 - x + 1 = 0$

10. $\frac{1}{2}a^2 + a - 12 = 0$

11. $2m^2 - 7m = -6$

12. $6x^2 = -5x + 6$

13. $4x^2 - 9 = 0$

14. $8m^2 + 64m = 0$

15. $z(2z + 1) = 6$

16. $13m = -5 - 6m^2$

In Exercises 17–26, solve the equation by completing the square.

17. $x^2 + 2x - 8 = 0$

18. $x^2 - x - 6 = 0$

19. $6x^2 - 12x = 3$

20. $2x^2 - 6x = 20$

21. $m^2 - 3 = -m$

22. $p^2 - 4 = -2p$

23. $3x - 4 = -2x^2$

24. $10x - 5 = 4x^2$

25. $4x^2 - 13 = 0$

26. $7p^2 - 20 = 0$

In Exercises 27–36, solve the equation by using the quadratic formula.

27. $2x^2 - x - 6 = 0$

28. $6x^2 - 7x - 3 = 0$

29. $m^2 = 4m - 1$

30. $2x^2 = 8x - 3$

31. $8x + 3 = 8x^2$

32. $6p - 6 = p^2$

33. $4x = -2x^2 + 3$

34. $15 - 2y^2 = 7y$

35. $2.1x^2 - 4.7x - 6.2 = 0$

36. $0.2m^2 + 1.6m + 1.2 = 0$

In Exercises 37–44, solve the equation.

37. $x^4 - 5x^2 + 6 = 0$
Hint: Let $m = x^2$. Then solve the quadratic equation in m.

38. $m^4 - 13m^2 + 36 = 0$
Hint: Let $x = m^2$. Then solve the quadratic equation in x.

39. $y^4 - 7y^2 + 10 = 0$
Hint: Let $x = y^2$.

40. $4x^4 - 21x^2 + 5 = 0$
Hint: Let $y = x^2$.

41. $6(x + 2)^2 + 7(x + 2) - 3 = 0$
Hint: Let $y = x + 2$.

42. $8(2m + 3)^2 + 14(2m + 3) - 15 = 0$
Hint: Let $x = 2m + 3$.

43. $6w - 13\sqrt{w} + 6 = 0$
Hint: Let $x = \sqrt{w}$.

44. $\left(\dfrac{t}{t-1}\right)^2 - \dfrac{2t}{t-1} - 3 = 0$

Hint: Let $x = \dfrac{t}{t-1}$.

In Exercises 45–64, solve the equation.

Hint: Be sure to check for extraneous solutions.

45. $\dfrac{2}{x+3} - \dfrac{4}{x} = 4$

46. $\dfrac{3y-1}{4} + \dfrac{4}{y+1} = \dfrac{5}{2}$

47. $x + 2 - \dfrac{3}{2x-1} = 0$

48. $\dfrac{x^2}{x-1} = \dfrac{3-2x}{x-1}$

49. $2 - \dfrac{7}{2y} - \dfrac{15}{y^2} = 0$

50. $6 + \dfrac{1}{k} - \dfrac{2}{k^2} = 0$

51. $\dfrac{3}{x^2-1} + \dfrac{2x}{x+1} = \dfrac{7}{3}$

52. $\dfrac{m}{m-2} - \dfrac{27}{7} = \dfrac{2}{m^2-m-2}$

53. $\dfrac{3x}{x-2} + \dfrac{4}{x+2} = \dfrac{24}{x^2-4}$

54. $\dfrac{3x}{x+1} + \dfrac{2}{x} + 5 = \dfrac{3}{x^2+x}$

55. $\dfrac{2t+1}{t-2} - \dfrac{t}{t+1} = -1$

56. $\dfrac{x}{x+1} - \dfrac{3}{x-2} + \dfrac{2}{x^2-x-2} = 0$

57. $\sqrt{u^2+u-5} = 1$ **58.** $\sqrt{6x^2-5x} - 2 = 0$

59. $\sqrt{2r+3} = r$ **60.** $\sqrt{3-4x} + 2x = 0$

61. $\sqrt{s-2} - \sqrt{s+3} + 1 = 0$

62. $\sqrt{x+1} - \sqrt{2x-5} + 1 = 0$

63. $\dfrac{1}{(x-3)^2} - \dfrac{10}{x-3} + 21 = 0$

64. $\dfrac{2}{(2x-1)^2} - \dfrac{5}{2x-1} + 3 = 0$

In Exercises 65–72, use the discriminant to determine the number of real solutions of the equation.

65. $x^2 - 6x + 5 = 0$ **66.** $2m^2 + 5m + 3 = 0$

67. $3y^2 - 4y + 5 = 0$ **68.** $2p^2 + 5p + 6 = 0$

69. $4x^2 + 12x + 9 = 0$ **70.** $25x^2 - 80x + 64 = 0$

71. $\dfrac{6}{k^2} + \dfrac{1}{k} - 2 = 0$

72. $(2p + 1)^2 - 3(2p + 1) + 4 = 0$

73. MOTION OF A BALL A person standing on the balcony of a building throws a ball directly upward. The height of the ball (in feet) as measured from the ground after t sec is given by $h = -16t^2 + 64t + 768$. When does the ball reach the ground?

74. MOTION OF A MODEL ROCKET A model rocket is launched vertically upward so that its height (measured in feet) t sec after launch is given by

$$h(t) = -16t^2 + 384t + 4$$

a. Find the time(s) when the rocket is at a height of 1284 ft.
b. How long is the rocket in flight?

75. MOTION OF A CYCLIST A cyclist riding along a straight path has a speed of u ft/sec as she passes a tree. Accelerating at a ft/sec^2, she reaches a speed of v ft/sec t sec later, where $v = ut + at^2$. If the cyclist was traveling at 10 ft/sec and she began accelerating at a rate of 4 ft/sec^2 as she passed the tree, how long did it take her to reach a speed of 22 ft/sec?

76. **PROFIT OF A VINEYARD** Phillip, the proprietor of a vineyard, estimates that the profit (in dollars) from producing and selling $(x + 10,000)$ bottles of wine is $P = -0.0002x^2 + 3x + 50,000$. Find the level(s) of production that will yield a profit of $60,800.

77. **DEMAND FOR SMOKE ALARMS** The quantity demanded x (measured in units of a thousand) of the Sentinel smoke alarm/week is related to its unit price p (in dollars) by the equation

$$p = \frac{30}{0.02x^2 + 1} \qquad (0 \le x \le 10)$$

If the unit price is set at $10, what is the quantity demanded?

78. **DEMAND FOR COMMODITIES** The quantity demanded x (measured in units of a thousand) of a certain commodity when the unit price is set at $\$p$ is given by the equation

$$p = \sqrt{-x^2 + 100}$$

If the unit price is set at $6, what is the quantity demanded?

79. **SUPPLY OF SATELLITE RADIOS** The quantity x of satellite radios that a manufacturer will make available in the marketplace is related to the unit price p (in dollars) by the equation

$$p = \frac{1}{10}\sqrt{x} + 10$$

How many satellite radios will the manufacturer make available in the marketplace if the unit price is $30?

80. **SUPPLY OF DESK LAMPS** The supplier of the Luminar desk lamp will make x thousand units of the lamp available in the marketplace if its unit price is p dollars, where p and x are related by the equation

$$p = 0.1x^2 + 0.5x + 15$$

If the unit price of the lamp is set at $20, how many units will the supplier make available in the marketplace?

81. **OXYGEN CONTENT OF A POND** When organic waste is dumped into a pond, the oxidation process that takes place reduces the pond's oxygen content. However, given time, nature will restore the oxygen content to its natural level. Suppose the oxygen content t days after organic waste has been dumped into the pond is given by

$$P = 100\left(\frac{t^2 + 10t + 100}{t^2 + 20t + 100}\right)$$

percent of its normal level. Find t corresponding to an oxygen content of 80%, and interpret your results.

82. **THE GOLDEN RATIO** Consider a rectangle of width x and height y (see the accompanying figure). The ratio $r = \frac{x}{y}$ satisfying the equation

$$\frac{x}{y} = \frac{x + y}{x}$$

is called the *golden ratio*. Show that

$$r = \left(\frac{1}{2}\right)(1 + \sqrt{5}) \approx 1.6$$

Note: A structure or a picture with a ratio of width to height equal to the golden ratio is especially pleasing to the eye. In fact, this golden ratio was used by the ancient Greeks in designing their beautiful temples and public buildings such as the Parthenon (see photo below).

© Jim Winkley/Ecoscene/Corbis

83. **CONSTRUCTING A BOX** By cutting away identical squares from each corner of a rectangular piece of cardboard and folding up the resulting flaps, an open box may be made (see the accompanying figure). If the cardboard is 16 in. long and 10 in. wide, find the dimensions of the resulting box if it is to have a total surface area of 144 in.^{2}.

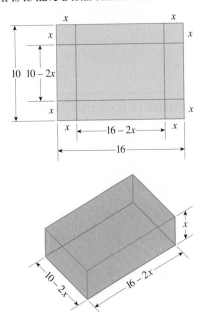

84. ENCLOSING AN AREA Carmen wishes to put up a fence around a proposed rectangular garden in her backyard. The length of the garden is to be twice its width, and the area of the garden is to be 200 ft². How many feet of fencing does she need?

85. ENCLOSING AN AREA George has 120 ft of fencing. He wishes to cut it into two pieces, with the purpose of enclosing two square regions. If the sum of the areas of the regions enclosed is 562.5 ft², how long should each piece of fencing be?

86. WIDTH OF A SIDEWALK A rectangular garden of length 40 ft and width 20 ft is surrounded by a path of uniform width. If the area of the walkway is 325 ft², what is its width?

87. ENCLOSING AN AREA The owner of the Rancho los Feliz has 3000 yd of fencing to enclose a rectangular piece of grazing land along the straight portion of a river. If an area of 1,125,000 yd² is to be enclosed, what will be the dimensions of the fenced area?

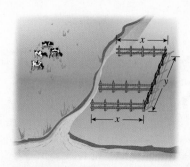

88. RADIUS OF A CYLINDRICAL CAN The surface area of a right circular cylinder is given by $S = 2\pi r^2 + 2\pi rh$, where r is the radius of the cylinder and h is its height. What is the radius of a cylinder of surface area 100 in.² and height 3 in.?

89. DESIGNING A METAL CONTAINER A metal container consists of a right circular cylinder with hemispherical ends. The surface area of the container is $S = 2\pi rl + 4\pi r^2$, where l is the length of the cylinder and r is the radius of the hemisphere. If the length of the cylinder is 4 ft and the surface area of the container is 28π ft², what is the radius of each hemisphere?

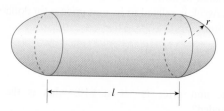

90. OIL SPILLS In calm waters, the oil spilling from the ruptured hull of a grounded oil tanker spreads in all directions. The area polluted at a certain instant of time was circular with a radius of 100 ft. A little later, the area, still circular, had increased by 4400π ft². By how much had the radius increased?

In Exercises 91–94, determine whether the statement is true or false. If it is true, explain why it is true. If it is false, explain why or give an example to show why it is false.

91. If a and b are real numbers and $ab \neq 0$, then $a \neq 0$ or $b \neq 0$.

92. If a, b, and c are real numbers and $abc \neq 0$, then $\frac{a+b}{c}$ is a real number but $\frac{a}{b+c}$ might not be a real number.

93. If $b^2 - 4ac > 0$ and $a \neq 0$, then the roots of $ax^2 - bx + c = 0$ are the negatives of the roots of $ax^2 + bx + c = 0$.

94. If $b^2 - 4ac \neq 0$ and $a \neq 0$, then $ax^2 + bx + c = 0$ has two distinct real roots, or it has no real roots at all.

1.8   Solutions to Self-Check Exercises

1. a. Factoring the given equation, we have

$$x^2 - 5x + 6 = (x - 3)(x - 2) = 0$$

and $x = 3$ or $x = 2$.

b. Rewriting the given equation, we have

$$4t^2 - 4t - 3 = 0 \qquad \text{Add } -3 \text{ to both sides of the equation.}$$

Factoring this equation gives

$$(2t - 3)(2t + 1) = 0$$

and $t = \frac{3}{2}$ or $t = -\frac{1}{2}$.

2. a. Step 1   First write

$$x^2 - 2x - 4 = 0 \qquad \text{Divide the original equation by 2, the coefficient of } x^2.$$

$$x^2 - 2x = 4 \qquad \text{Add 4 to both sides so that the constant term is on the right side.}$$

Step 2   Square half of the coefficient of x, obtaining

$$\left(\frac{-2}{2}\right)^2 = 1$$

Step 3 Add 1 to both sides of the equation:

$$x^2 - 2x + 1 = 5$$

Factoring, we have

$$(x - 1)^2 = 5$$
$$x - 1 = \pm\sqrt{5}$$
$$x = 1 \pm \sqrt{5}$$

b. Using the quadratic formula, with $a = 2$, $b = -4$, and $c = -8$, we obtain

$$x = \frac{-b \pm \sqrt{b^2 - 4ac}}{2a} = \frac{-(-4) \pm \sqrt{(-4)^2 - 4(2)(-8)}}{2(2)}$$

$$= \frac{4 \pm \sqrt{80}}{4} = \frac{4 \pm 4\sqrt{5}}{4}$$

$$= 1 \pm \sqrt{5}$$

1.9 Inequalities and Absolute Value

Intervals

We described the system of real numbers and its properties in Section 1.1. Often, we will restrict our attention to certain subsets of the set of real numbers. For example, if x denotes the number of cars rolling off an assembly line each day in an automobile assembly plant, then x must be nonnegative; that is, $x \geq 0$. Taking this example one step further, suppose management decides that the daily production must not exceed 200 cars. Then x must satisfy the inequality $0 \leq x \leq 200$.

More generally, we will be interested in certain subsets of real numbers called finite intervals and infinite intervals. **Finite intervals** are open, closed, or half-open. The set of all real numbers that lie *strictly* between two fixed numbers a and b is called an **open interval** (a, b). It consists of all real numbers x that satisfy the inequalities $a < x < b$; it is called "open" because neither of its endpoints is included in the interval. A **closed interval** contains both of its endpoints. Thus, the set of all real numbers x that satisfy the inequalities $a \leq x \leq b$ is the closed interval $[a, b]$. Notice that brackets are used to indicate that the endpoints are included in this interval. **Half-open intervals** (also called *half-closed intervals*) contain only *one* of their endpoints. The interval $[a, b)$ is the set of all real numbers x that satisfy $a \leq x < b$, whereas the interval $(a, b]$ is described by the inequalities $a < x \leq b$. Examples of these finite intervals are illustrated in Table 14.

TABLE 14

Finite Intervals

Interval	Graph	Example	
Open: (a, b)		$(-2, 1)$	
Closed: $[a, b]$		$[-1, 2]$	
Half-open: $(a, b]$		$\left(\frac{1}{2}, 3\right]$	
Half-open: $[a, b)$		$\left[-\frac{1}{2}, 3\right)$	

Infinite intervals include the half-lines (a, ∞), $[a, \infty)$, $(-\infty, a)$, and $(-\infty, a]$, defined by the set of all real numbers that satisfy $x > a$, $x \geq a$, $x < a$, and $x \leq a$, respectively. The symbol ∞, called *infinity*, is not a real number. It is used here only for notational purposes in conjunction with the definition of infinite intervals. The notation $(-\infty, \infty)$ is used for the set of real numbers x, since, by definition, the inequalities $-\infty < x < \infty$ hold for any real number x. These infinite intervals are illustrated in Table 15.

TABLE 15

Infinite Intervals

Interval	Graph	Example	
(a, ∞)		$(2, \infty)$	
$[a, \infty)$		$[-1, \infty)$	
$(-\infty, a)$		$(-\infty, 1)$	
$(-\infty, a]$		$(-\infty, -\frac{1}{2}]$	

Inequalities

In practical applications, intervals are often found by solving one or more inequalities involving a variable. To solve these inequalities, we use the properties listed in Table 16.

TABLE 16

Properties of Inequalities

Property	Illustration
Let a, b, and c be any real numbers.	
1. If $a < b$ and $b < c$, then $a < c$.	$2 < 3$ and $3 < 8$, so $2 < 8$.
2. If $a < b$, then $a + c < b + c$.	$-5 < -3$, so $-5 + 2 < -3 + 2$; that is, $-3 < -1$.
3. If $a < b$ and $c > 0$, then $ac < bc$.	$-5 < -3$ and $2 > 0$, so $(-5)(2) < (-3)(2)$; that is, $-10 < -6$.
4. If $a < b$ and $c < 0$, then $ac > bc$.	$-2 < 4$ and $-3 < 0$, so $(-2)(-3) > (4)(-3)$; that is, $6 > -12$.

Similar properties hold if each inequality sign, $<$, between a and b is replaced by $\geq$, $>$, or $\leq$.

A real number is a *solution of an inequality* involving a variable if a true statement is obtained when the variable is replaced by that number. The set of all real numbers satisfying the inequality is called the *solution set*.

EXAMPLE 1 Solve $3x - 2 < 7$.

Solution   Add 2 to each side of the inequality, obtaining

$$3x - 2 + 2 < 7 + 2$$
$$3x < 9$$

Next, multiply each side of the inequality by $\frac{1}{3}$, obtaining

$$\frac{1}{3}(3x) < \frac{1}{3}(9)$$

$$x < 3$$

The solution is the set of all values of x in the interval $(-\infty, 3)$.

EXAMPLE 2 Solve $-1 \le 2x - 5 < 7$ and graph the solution set.

Solution   Add 5 to each member of the double inequality, obtaining

$$4 \le 2x < 12$$

Next multiply each member of the resulting double inequality by $\frac{1}{2}$, yielding

$$2 \le x < 6$$

Thus, the solution is the set of all values of x lying in the interval $[2, 6)$. The graph of the solution set is shown in Figure 4.

FIGURE **4**
The graph of the solution set for
$-1 \le 2x - 5 < 7$

Solving Inequalities by Factoring

The method of factoring can be used to solve inequalities that involve polynomials of degree 2 or higher. This method relies on the principle that a polynomial changes sign only at a point where its value is 0. To find the values of x where the polynomial is equal to 0, we set the polynomial equal to 0 and then solve for x. The values obtained can then be used to help us solve the given inequality. In Examples 3 and 4, detailed steps are provided for this technique.

EXAMPLE 3 Solve $x^2 - 5x + 6 > 0$.

Solution

Step 1   *Set the polynomial in the inequality equal to 0:*

$$x^2 - 5x + 6 = 0$$

Step 2   *Factor the polynomial:*

$$(x - 3)(x - 2) = 0$$

Step 3   *Construct a sign diagram for the factors of the polynomial.* We use a $+$ to indicate that a factor is positive for a given value of x, a $-$ to indicate that it is negative, and a 0 to indicate that it is equal to 0. Now, $x - 3 < 0$ if $x < 3$, $x - 3 > 0$ if $x > 3$, and $x - 3 = 0$ if $x = 3$. Similarly, $x - 2 < 0$ if $x < 2$, $x - 2 > 0$ if $x > 2$, and $x - 2 = 0$ if $x = 2$. Using this information, we construct the sign diagram shown in Figure 5.

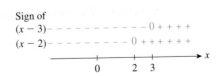

FIGURE **5**

Step 4   *Determine the intervals that satisfy the given inequality.* Since $x^2 - 5x + 6 > 0$, we require that the product of the two factors be positive—that is, that both factors have the same sign. From the sign diagram, we see that the two factors have the same sign when $x < 2$ or $x > 3$. Thus, the solution set is $(-\infty, 2) \cup (3, \infty)$.

EXAMPLE 4 Solve $x^2 + 2x - 8 < 0$.

Solution

Step 1 $x^2 + 2x - 8 = 0$

Step 2 $(x + 4)(x - 2) = 0$, so $x = -4$ or $x = 2$.

Step 3 $x + 4 > 0$ when $x > -4$, $x + 4 < 0$ when $x < -4$, and $x + 4 = 0$ when $x = -4$. Similarly, $x - 2 > 0$ when $x > 2$, $x - 2 < 0$ when $x < 2$, and $x - 2 = 0$ when $x = 2$. Using these results, we construct the sign diagram for the factors of $x^2 + 2x - 8$ (Figure 6).

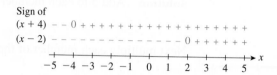

FIGURE **6**

Step 4   Since $x^2 + 2x - 8 < 0$, the product of the two factors must be negative; that is, the signs of the two factors must differ. From the sign diagram, we see that the two factors $x + 4$ and $x - 2$ have opposite signs when x lies strictly between -4 and 2. Therefore, the required solution is the interval $(-4, 2)$. ∎

Solving Inequalities Involving a Quotient

The next two examples show how an inequality involving the quotient of two algebraic expressions is solved.

EXAMPLE 5 Solve $\dfrac{x + 1}{x - 1} \geq 0$.

Solution   The quotient $(x + 1)/(x - 1)$ is positive (greater than 0) when the numerator and denominator have the *same* sign. The signs of $x + 1$ and $x - 1$ are shown in Figure 7.

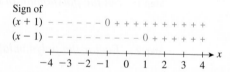

FIGURE **7**

From the sign diagram, we see that $x + 1$ and $x - 1$ have the same sign when $x < -1$ or $x > 1$. The quotient $(x + 1)/(x - 1)$ is equal to 0 when $x = -1$. It is undefined at $x = 1$, since the denominator is 0 at that point. Therefore, the required solution is the set of all x in the intervals $(-\infty, -1]$ and $(1, \infty)$. ∎

EXAMPLE 6 Solve $\dfrac{2x - 1}{x - 2} \geq 1$.

Solution   We rewrite the given inequality so that the right side is equal to 0:

$$\frac{2x - 1}{x - 2} - 1 \geq 0$$

$$\frac{2x - 1 - (x - 2)}{x - 2} \geq 0$$

$$\frac{2x - 1 - x + 2}{x - 2} \geq 0$$

$$\frac{x + 1}{x - 2} \geq 0$$

Next we construct the sign diagram for the factors in the numerator and the denominator (Figure 8).

Sign of
$(x + 1)$ $- - - - 0 + + + + + + + + + +$
$(x - 2)$ $- - - - - - - - - - 0 + + + +$

$$\xrightarrow{\hspace{3cm}} x$$
$$-1 \quad 0 \qquad 2$$

FIGURE **8**

Since the quotient of these two factors must be positive or equal to 0, we require that the sign of each factor be the same or that the quotient of the two factors be equal to 0. From the sign diagram, we see that the solution set is given by $(-\infty, -1]$ and $(2, \infty)$. Note that $x = 2$ is not included in the second interval, since division by 0 is not allowed. ∎

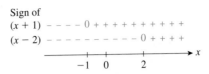

APPLIED EXAMPLE 7 Gross Domestic Product The gross domestic product (GDP) of a certain country is projected to be $t^2 + 2t + 50$ billion dollars t years from now. Find the time t when the GDP of the country will first equal or exceed \$58 billion.

Solution The GDP of the country will equal or exceed \$58 billion when

$$t^2 + 2t + 50 \geq 58$$

To solve this inequality for t, we first write it in the form

$$t^2 + 2t - 8 \geq 0$$
$$(t + 4)(t - 2) \geq 0$$

The sign diagram for the factors of $t^2 + 2t - 8$ is shown in Figure 9.

Sign of
$(t + 4)$ $- - - 0 + + + + + + + + + + +$
$(t - 2)$ $- - - - - - - - - - - 0 + + + +$

$$\xrightarrow{\hspace{3cm}} t$$
$$-4 \qquad 0 \quad 2$$

FIGURE **9**

From the sign diagram, we see that the solution set is $(-\infty, -4] \cup [2, \infty)$. Since t must be nonnegative for the problem to be meaningful, we see that the GDP of the country is greater than or equal to \$58 billion when $t \geq 2$; that is, the GDP will first equal or exceed \$58 billion when $t = 2$, or 2 years from now. ∎

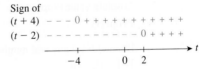

APPLIED EXAMPLE 8 Stock Purchase The management of Corbyco, a giant conglomerate, has estimated that x thousand dollars is needed to purchase

$$100,000(-1 + \sqrt{1 + 0.001x})$$

shares of common stock of the Starr Communications Company. Determine how much money Corbyco needs in order to purchase at least 100,000 shares of Starr's stock.

Solution The amount of cash Corbyco needs to purchase at least 100,000 shares is found by solving the inequality

$$100,000(-1 + \sqrt{1 + 0.001x}) \geq 100,000$$

Proceeding, we find

$$-1 + \sqrt{1 + 0.001x} \geq 1$$
$$\sqrt{1 + 0.001x} \geq 2$$
$$1 + 0.001x \geq 4 \quad \text{Square both sides.}$$
$$0.001x \geq 3$$
$$x \geq 3000$$

so Corbyco needs at least $3,000,000. (Remember, x is measured in thousands of dollars.) ∎

Absolute Value

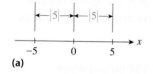

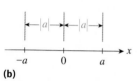

FIGURE **10**

> ### Absolute Value
> The **absolute value** of a number a is denoted by $|a|$ and is defined by
> $$|a| = \begin{cases} a & \text{if } a \geq 0 \\ -a & \text{if } a < 0 \end{cases}$$

Since $-a$ is a positive number when a is negative, it follows that the absolute value of a number is always nonnegative. For example, $|5| = 5$ and $|-5| = -(-5) = 5$. Geometrically, $|a|$ is the distance between the origin and the point on the number line that represents the number a (Figure 10a and b).

The absolute value properties are given in Table 17. Property 4 is called the **triangle inequality.**

TABLE 17

Absolute Value Properties

Property	Illustration
If a and b are any real numbers, then	
1. $\|-a\| = \|a\|$	$\|-3\| = -(-3) = 3 = \|3\|$
2. $\|ab\| = \|a\|\|b\|$	$\|(2)(-3)\| = \|-6\| = 6$
	$= \|2\|\|-3\|$
3. $\left\|\dfrac{a}{b}\right\| = \dfrac{\|a\|}{\|b\|}$	$\left\|\dfrac{-3}{-4}\right\| = \left\|\dfrac{3}{4}\right\| = \dfrac{3}{4} = \dfrac{\|-3\|}{\|-4\|}$
4. $\|a + b\| \leq \|a\| + \|b\|$	$\|8 + (-5)\| = \|3\| = 3$
	$\leq \|8\| + \|-5\| = 13$

EXAMPLE 9 Evaluate:

a. $|\pi - 5| + 3$ **b.** $|\sqrt{3} - 2| + |2 - \sqrt{3}|$

Solution

a. Since $\pi - 5 < 0$, we see that $|\pi - 5| = -(\pi - 5)$. Therefore,

$$|\pi - 5| + 3 = -(\pi - 5) + 3 = -\pi + 5 + 3 = 8 - \pi$$

b. Since $\sqrt{3} - 2 < 0$, we see that $|\sqrt{3} - 2| = -(\sqrt{3} - 2)$. Next observe that $2 - \sqrt{3} > 0$, so $|2 - \sqrt{3}| = 2 - \sqrt{3}$. Therefore,

$$|\sqrt{3} - 2| + |2 - \sqrt{3}| = -(\sqrt{3} - 2) + (2 - \sqrt{3}) = -\sqrt{3} + 2 + 2 - \sqrt{3}$$
$$= 4 - 2\sqrt{3} = 2(2 - \sqrt{3})$$

EXAMPLE 10 Solve the inequalities $|x| \leq 5$ and $|x| \geq 5$.

Solution   We first consider the inequality $|x| \leq 5$. If $x > 0$, then $|x| = x$, so $|x| \leq 5$ implies $x \leq 5$ in this case. However, if $x < 0$, then $|x| = -x$, so $|x| \leq 5$ implies that $-x \leq 5$, or $x \geq -5$. Thus, $|x| \leq 5$ means $-5 \leq x \leq 5$ (Figure 11a). Alternatively, observe that $|x|$ is the distance from the point x to 0, so the inequality $|x| \leq 5$ implies immediately that $-5 \leq x \leq 5$.

Next, the inequality $|x| \geq 5$ states that the distance from x to 0 is greater than or equal to 5. This observation yields the result $x \geq 5$ or $x \leq -5$ (Figure 11b).

FIGURE **11**

(a) $|x| \leq 5$ (b) $|x| \geq 5$

The results of Example 10 may be generalized. Thus, if $k > 0$, then $|x| \leq k$ is equivalent to $-k \leq x \leq k$, and $|x| \geq k$ is equivalent to $x \geq k$ or $x \leq -k$.

EXAMPLE 11 Solve the inequality $|2x - 3| \leq 1$.

Solution   The inequality $|2x - 3| \leq 1$ is equivalent to the inequalities $-1 \leq 2x - 3 \leq 1$ (see Example 10). Then

$$2 \leq 2x \leq 4 \qquad \text{Add 3 to each member of the inequality.}$$

and

$$1 \leq x \leq 2 \qquad \text{Multiply each member of the inequality by } \frac{1}{2}.$$

FIGURE **12**
$|2x - 3| \leq 1$

Therefore, the solution is given by the set of all x in the interval [1, 2] (Figure 12).

EXAMPLE 12 Solve the inequality $|5x + 7| \geq 18$.

Solution   Referring to Example 10 once again, we see that $|5x + 7| \geq 18$ is equivalent to

$$5x + 7 \leq -18 \quad \text{or} \quad 5x + 7 \geq 18$$

That is,

$$5x \leq -25 \quad \text{or} \quad 5x \geq 11$$
$$x \leq -5 \qquad\qquad x \geq \frac{11}{5}$$

FIGURE **13**
$|5x + 7| \geq 18$

Therefore, the solution is given by the set of all x in the interval $(-\infty, -5]$ or the interval $\left[\frac{11}{5}, \infty\right)$ (Figure 13).

1.9 Self-Check Exercises

1. Solve $-1 < 2x - 1 \leq 5$, and graph the solution set.

2. Solve $6x^2 - 5x - 4 \leq 0$.

Solutions to Self-Check Exercises 1.9 can be found on page 64.

1.9   Concept Questions

1. State the properties of inequalities. Illustrate with examples.

2. What is the absolute value of a number a? Can $|a|$ be negative? Explain.

3. State the absolute value properties. Illustrate with examples.

1.9   Exercises

In Exercises 1–4, determine whether the statement is true or false.

1. $-3 < -20$

2. $-5 \leq -5$

3. $\dfrac{2}{3} > \dfrac{5}{6}$

4. $-\dfrac{5}{6} < -\dfrac{11}{12}$

In Exercises 5–10, show the interval on a number line.

5. $(3, 6)$

6. $(-2, 5]$

7. $[-1, 4)$

8. $\left[-\dfrac{6}{5}, -\dfrac{1}{2} \right]$

9. $(0, \infty)$

10. $(-\infty, 5]$

In Exercises 11–28, find the values of x that satisfy the inequalities.

11. $2x + 2 < 8$

12. $-6 > 4 + 5x$

13. $-4x \geq 20$

14. $-12 \leq -3x$

15. $-6 < x - 2 < 4$

16. $0 \leq x + 1 \leq 4$

17. $x + 1 > 4$ or $x + 2 < -1$

18. $x + 1 > 2$ or $x - 1 < -2$

19. $x + 3 > 1$ and $x - 2 < 1$

20. $x - 4 \leq 1$ and $x + 3 > 2$

21. $(x + 3)(x - 5) \leq 0$

22. $(2x - 4)(x + 2) \geq 0$

23. $(2x - 3)(x - 1) \leq 0$

24. $(3x - 4)(2x + 2) \leq 0$

25. $\dfrac{x + 3}{x - 2} \geq 0$

26. $\dfrac{2x - 3}{x + 1} \geq 4$

27. $\dfrac{x - 2}{x - 1} \leq 2$

28. $\dfrac{2x - 1}{x + 2} \leq 4$

In Exercises 29–38, evaluate the expression.

29. $|-6 + 2|$

30. $4 + |-4|$

31. $\dfrac{|-12 + 4|}{|16 - 12|}$

32. $\left| \dfrac{0.2 - 1.4}{1.6 - 2.4} \right|$

33. $\sqrt{3}\,|-2| + 3\,|-\sqrt{3}|$

34. $|-1| + \sqrt{2}\,|-2|$

35. $|\pi - 1| + 2$

36. $|\pi - 6| - 3$

37. $|\sqrt{2} - 1| + |3 - \sqrt{2}|$

38. $|2\sqrt{3} - 3| - |\sqrt{3} - 4|$

In Exercises 39–44, suppose a and b are real numbers other than 0 and $a > b$. State whether the inequality is true or false.

39. $b - a > 0$

40. $\dfrac{a}{b} > 1$

41. $a^2 > b^2$

42. $\dfrac{1}{a} > \dfrac{1}{b}$

43. $a^3 > b^3$

44. $-a < -b$

45. Write the inequality $|x - a| < b$ without using absolute values.

46. Write the inequality $|x - a| \geq b$ without using absolute values.

In Exercises 47–52, determine whether the statement is true for all real numbers a and b.

47. $|-a| = a$

48. $|b^2| = b^2$

49. $|a - 4| = |4 - a|$

50. $|a + 1| = |a| + 1$

51. $|a + b| = |a| + |b|$

52. $|a - b| = |a| - |b|$

53. Find the minimum cost C (in dollars) given that
$$5(C - 25) \geq 1.75 + 2.5C$$

54. Find the maximum profit P (in dollars) given that
$$6(P - 2500) \leq 4(P + 2400)$$

55. **DRIVING RANGE OF A CAR**   An advertisement for a certain car states that the EPA fuel economy is 20 mpg city and 27 mpg highway and that the car's fuel-tank capacity is 18.1 gal. Assuming ideal driving conditions, determine the driving range for the car from the foregoing data.

56. **CELSIUS AND FAHRENHEIT TEMPERATURES**   The relationship between Celsius (°C) and Fahrenheit (°F) temperatures is given by the formula
$$C = \dfrac{5}{9}(F - 32)$$

a. If the temperature range for Montreal during the month of January is $-15° < C° < -5°$, find the range in degrees Fahrenheit in Montreal for the same period.

b. If the temperature range for New York City during the month of June is $63° < F° < 80°$, find the range in degrees Celsius in New York City for the same period.

57. Meeting Sales Targets A salesman's monthly commission is 15% on all sales over $12,000. If his goal is to make a commission of at least $6000/month, what minimum monthly sales figures must he attain?

58. Markup on a Car The markup on a used car was at least 30% of its current wholesale price. If the car was sold for $11,200, what was the maximum wholesale price?

59. Meeting Profit Goals A manufacturer of a certain commodity has estimated that her profit (in thousands of dollars) is given by the expression

$$-6x^2 + 30x - 10$$

where x (in thousands) is the number of units produced. What production range will enable the manufacturer to realize a profit of at least $14,000 on the commodity?

60. Concentration of a Drug in the Bloodstream The concentration (in milligrams/cubic centimeter) of a certain drug in a patient's bloodstream t hr after injection is given by

$$\frac{0.2t}{t^2 + 1}$$

Find the interval of time when the concentration of the drug is greater than or equal to 0.08 mg/cc.

61. Cost of Removing Toxic Pollutants A city's main well was recently found to be contaminated with trichloroethylene (a cancer-causing chemical) as a result of an abandoned chemical dump that leached chemicals into the water. A proposal submitted to the city council indicated that the cost, in millions of dollars, of removing $x\%$ of the toxic pollutants is

$$\frac{0.5x}{100 - x}$$

If the city could raise between $25 and $30 million inclusive for the purpose of removing the toxic pollutants, what is the range of pollutants that could be expected to be removed?

62. Average Speed of a Vehicle The average speed of a vehicle in miles per hour on a stretch of route 134 between 6 A.M. and 10 A.M. on a typical weekday is approximated by the expression

$$20t - 40\sqrt{t} + 50 \qquad (0 \le t \le 4)$$

where t is measured in hours, with $t = 0$ corresponding to 6 A.M. Over what interval of time is the average speed of a vehicle less than or equal to 35 mph?

63. Effect of Bactericide The number of bacteria in a certain culture t min after an experimental bactericide is introduced is given by

$$\frac{10,000}{t^2 + 1} + 2000$$

Find the time when the number of bacteria will have dropped below 4000.

64. Air Pollution Nitrogen dioxide is a brown gas that impairs breathing. The amount of nitrogen dioxide present in the atmosphere on a certain May day in the city of Long Beach measured in PSI (pollutant standard index) at time t, where t is measured in hours, and $t = 0$ corresponds to 7 A.M., is approximated by

$$\frac{136}{1 + 0.25(t - 4.5)^2} + 28 \qquad (0 \le t \le 11)$$

Find the time of the day when the amount of nitrogen dioxide is greater than or equal to 128 PSI.

Source: Los Angeles Times.

65. A ball is thrown straight up so that its height after t sec is

$$128t - 16t^2 + 4$$

ft. Determine the length of time the ball stays at or above a height of 196 ft.

66. Distribution of Income The distribution of income in a certain city can be described by the mathematical model $y = (5.6 \cdot 10^{11})(x)^{-1.5}$, where y is the number of families with an income of x or more dollars.

a. How many families in this city have an income of $30,000 or more?

b. How many families have an income of $60,000 or more?

c. How many families have an income of $150,000 or more?

67. Quality Control PAR Manufacturing Company manufactures steel rods. Suppose the rods ordered by a customer are manufactured to a specification of 0.5 in. and are acceptable only if they are within the *tolerance limits* of 0.49 in. and 0.51 in. Letting x denote the diameter of a rod, write an inequality using absolute value to express a criterion involving x that must be satisfied in order for a rod to be acceptable.

68. Quality Control The diameter x (in inches) of a batch of ball bearings manufactured by PAR Manufacturing satisfies the inequality

$$|x - 0.1| \le 0.01$$

What is the smallest diameter a ball bearing in the batch can have? The largest diameter?

1.9   Solutions to Self-Check Exercises

1. $-1 < 2x - 1 \le 5$

   $-1 + 1 < 2x - 1 + 1 \le 5 + 1$ Add 1 to each member of
                                        the inequality.

   $0 < 2x \le 6$ Combine like terms.

   $0 < x \le 3$ Multiply each member of
                                        the inequality by $\frac{1}{2}$.

We conclude that the solution set is $(0, 3]$. The graph of the solution set is shown in the following figure:

2. Step 1 $6x^2 - 5x - 4 \le 0$

   Step 2 $(3x - 4)(2x + 1) \le 0$

   Step 3 $3x - 4 > 0$ when $x > \frac{4}{3}$, $3x - 4 = 0$ when $x = \frac{4}{3}$, and $3x - 4 < 0$ when $x < \frac{4}{3}$. Similarly, $2x + 1 > 0$ when $x > -\frac{1}{2}$, $2x + 1 = 0$ when $x = -\frac{1}{2}$, and

$2x + 1 < 0$ when $x < -\frac{1}{2}$. Using these results, we construct the following sign diagram for the factors of $6x^2 - 5x - 4$:

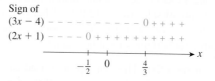

Step 4   Since $6x^2 - 5x - 4 \le 0$, the signs of the two factors must differ or be equal to 0. From the sign diagram, we see that x must lie between $-\frac{1}{2}$ and $\frac{4}{3}$, inclusive. Therefore, the required solution is $[-\frac{1}{2}, \frac{4}{3}]$.

CHAPTER 1   Summary of Principal Formulas and Terms

FORMULAS

1. Product formula	$(a + b)^2 = a^2 + 2ab + b^2$ $(a - b)^2 = a^2 - 2ab + b^2$ $(a + b)(a - b) = a^2 - b^2$
2. Quadratic formula	$x = \dfrac{-b \pm \sqrt{b^2 - 4ac}}{2a}$
3. Difference of two squares	$a^2 - b^2 = (a + b)(a - b)$
4. Perfect square trinomial	$a^2 + 2ab + b^2 = (a + b)^2$ $a^2 - 2ab + b^2 = (a - b)^2$
5. Sum of two cubes	$a^3 + b^3 = (a + b)(a^2 - ab + b^2)$
6. Difference of two cubes	$a^3 - b^3 = (a - b)(a^2 + ab + b^2)$

TERMS

natural number (2)	equation (31)	index (38)
whole number (2)	variable (31)	conjugate (42)
integer (2)	solution of an equation (32)	quadratic equation (46)
rational number (2)	solution set (32)	discriminant (51)
irrational number (2)	linear equation (32)	finite interval (55)
real number (2)	extraneous solution (34)	open interval (55)
exponent (7)	nth root (37)	closed interval (55)
base (7)	square root (37)	half-open interval (55)
polynomial (9)	cube root (37)	infinite interval (55)
rational expression (20)	radical (38)	absolute value (60)
complex fraction (23)	radical sign (38)	triangle inequality (60)
zero exponent (28)	radicand (38)	

CHAPTER 1 Concept Review Questions

Fill in the blanks.

1. a. A number of the form $\frac{a}{b}$, where a and b are integers, with $b \neq 0$ is called a/an _____ number. A rational number can be represented by either a/an _____ or _____ decimal.

  b. A real number that is not rational is called _____. When such a number is represented by a decimal, it neither _____ nor _____.

2. a. Under addition, we have $a + b =$ _____, $a + (b + c) =$ _____, $a + 0 =$ _____, and $a + (-a) =$ _____.

  b. Under multiplication, we have $ab =$ _____, $a(bc) =$ _____, $a \cdot 1 =$ _____, and $a\left(\frac{1}{a}\right) =$ _____ $(a \neq 0)$.

  c. Under addition and multiplication, we have $a(b + c) =$ _____.

3. a. If a and b are numbers, then $-(-a) =$ _____, $(-a)b =$ _____, $(-a)(-b) =$ _____, $(-1)a = -a$, and $a \cdot 0 = 0$.

  b. If $ab = 0$, then $a =$ _____, or $b =$ _____, or both.

4. a. An expression of the form $a_n x^n + a_{n-1}x^{n-1} + \cdots + a_0$ is called a/an _____ in _____; the nonnegative integer n is called its _____; the expression $a_k x^k$ is called the _____ of the _____; and a_k is called the _____ of x_k.

  b. To add or subtract two polynomials, we add or subtract _____ terms.

5. To factor a polynomial, we express it as a/an _____ of two or more _____ polynomials. For example, $x^3 + x^2 - 2x =$ _____.

6. a. A rational expression is a quotient of _____.

  b. A rational expression is simplified or reduced to lowest terms if the _____ and the _____ have no common _____ other than _____ and _____.

  c. To add or subtract rational expressions, first find the least common _____ of the expressions, if necessary. Then follow the procedure for adding and subtracting _____ with common denominators.

7. A rational expression that contains fractions in its numerator or denominator is called a/an _____ fraction. An example of a compound fraction is _____.

8. a. If a is any real number and n is a natural number, then $a^n =$ _____. The number a is the _____, and the superscript n is called the _____, or _____.

  b. For any nonzero real number a, $a^0 =$ _____. The expression 0^0 is _____ _____.

  c. If a is any nonzero number and n is a positive integer, then $a^{-n} =$ _____.

9. a. A statement that two mathematical statements are equal is called a/an _____.

  b. A variable is a letter that stands for a/an _____ belonging to a set of real numbers.

  c. A linear equation in the variable x is an equation that can be written in the form _____; a linear equation in x has degree _____ in x.

10. a. If n is a natural number and a and b are real numbers, we say that a is the nth root of b if _____.

  b. If n is even, the real nth roots of a positive number b must come in _____.

  c. If n is even and b is negative, then there are _____ real roots.

  d. If n is odd, then there is only one _____ _____ of b.

11. a. If n is a natural number and b is a real number, then $\sqrt[n]{b}$ is called a/an _____; also, $\sqrt[n]{b} =$ _____.

  b. To rationalize a denominator of an algebraic expression means to eliminate a/an _____ from the denominator.

12. a. A quadratic equation is an equation in x that can be written in the form _____.

  b. A quadratic equation can be solved by _____, by _____ _____, or by using the quadratic formula. The quadratic formula is _____.

CHAPTER 1 Review Exercises

In Exercises 1–6, classify the number as to type.

1. $\dfrac{7}{8}$ **2.** $\sqrt{13}$ **3.** -2π

4. 0   **5.** $2.\overline{71}$ **6.** $3.14159\ldots$

In Exercises 7–14, evaluate the expression.

7. $\left(\dfrac{9}{4}\right)^{3/2}$ **8.** $\dfrac{5^6}{5^4}$

9. $(3 \cdot 4)^{-2}$ **10.** $(-8)^{5/3}$

11. $\left(\dfrac{16}{9}\right)^{3/2}$ **12.** $\dfrac{(3 \cdot 2^{-3})(4 \cdot 3^5)}{2 \cdot 9^3}$

13. $\sqrt[3]{\dfrac{27}{125}}$ **14.** $\dfrac{3\sqrt[3]{54}}{\sqrt[3]{18}}$

In Exercises 15–22, simplify the expression. (Assume that all variables are positive.)

15. $\dfrac{4(x^2 + y)^3}{x^2 + y}$ **16.** $\dfrac{a^6 b^{-5}}{(a^3 b^{-2})^{-3}}$

17. $\dfrac{\sqrt[4]{16x^5yz}}{\sqrt[4]{81xyz^5}}$

18. $(2x^3)(-3x^{-2})\left(\dfrac{1}{6}x^{-1/2}\right)$

19. $\left(\dfrac{3xy^2}{4x^3y}\right)^{-2}\left(\dfrac{3xy^3}{2x^2}\right)^3$

20. $(-3a^2b^3)^2(2a^{-1}b^{-2})^{-1}$

21. $\sqrt[3]{81x^5y^{10}}\ \sqrt[3]{9xy^2}$

22. $\left(\dfrac{-x^{1/2}y^{2/3}}{x^{1/3}y^{3/4}}\right)^6$

In Exercises 23–30, perform the indicated operations and simplify the expression.

23. $(3x^4 + 10x^3 + 6x^2 + 10x + 3) + (2x^4 + 10x^3 + 6x^2 + 4x)$

24. $(3x - 4)(3x^2 - 2x + 3)$

25. $(2x + 3y)^2 - (3x + 1)(2x - 3)$

26. $2(3a + b) - 3[(2a + 3b) - (a + 2b)]$

27. $\dfrac{(t + 6)(60) - (60t + 180)}{(t + 6)^2}$

28. $\dfrac{6x}{2(3x^2 + 2)} + \dfrac{1}{4(x + 2)}$

29. $\dfrac{2}{3}\left(\dfrac{4x}{2x^2 - 1}\right) + 3\left(\dfrac{3}{3x - 1}\right)$

30. $\dfrac{-2x}{\sqrt{x + 1}} + 4\sqrt{x + 1}$

In Exercises 31–40, factor the expression.

31. $-2\pi^2r^3 + 100\pi r^2$

32. $2v^3w + 2vw^3 + 2u^2vw$

33. $16 - x^2$

34. $12t^3 - 6t^2 - 18t$

35. $-2x^2 - 4x + 6$

36. $12x^2 - 92x + 120$

37. $9a^2 - 25b^2$

38. $8u^6v^3 + 27u^3$

39. $6a^4b^4c - 3a^3b^2c - 9a^2b^2$

40. $6x^2 - xy - y^2$

In Exercises 41–46, perform the indicated operations and simplify the expression.

41. $\dfrac{2x^2 + 3x - 2}{2x^2 + 5x - 3}$

42. $\dfrac{[(t^2 + 4)(2t - 4)] - (t^2 - 4t + 4)(2t)}{(t^2 + 4)^2}$

43. $\dfrac{2x - 6}{x + 3} \cdot \dfrac{x^2 + 6x + 9}{x^2 - 9}$

44. $\dfrac{3x}{x^2 + 2} + \dfrac{3x^2}{x^3 + 1}$

45. $\dfrac{1 + \dfrac{1}{x + 2}}{x - \dfrac{9}{x}}$

46. $\dfrac{x(3x^2 + 1)}{x - 1} \cdot \dfrac{3x^3 - 5x^2 + x}{x(x - 1)(3x^2 + 1)^{1/2}}$

In Exercises 47–54, solve the equation.

47. $8x^2 + 2x - 3 = 0$

48. $-6x^2 - 10x + 4 = 0$

49. $2x^2 - 3x - 4 = 0$

50. $x^2 + 5x + 3 = 0$

51. $2y^2 - 3y + 1 = 0$

52. $0.3m^2 - 2.1m - 3.2 = 0$

53. $-x^3 - 2x^2 + 3x = 0$

54. $2x^4 + x^2 = 1$

In Exercises 55–62, solve the equation.

55. $\dfrac{1}{4}x + 2 = \dfrac{3}{4}x - 5$

56. $\dfrac{3p + 1}{2} - \dfrac{2p - 1}{3} = \dfrac{5p}{12}$

57. $(x + 2)^2 - 3x(1 - x) = (x - 2)^2$

58. $\dfrac{3(2q + 1)}{4q - 3} = \dfrac{3q + 1}{2q + 1}$

59. $\sqrt{k - 1} = \sqrt{2k - 3}$

60. $\sqrt{x} - \sqrt{x - 1} = \sqrt{4x - 3}$

61. Solve $C = \dfrac{20x}{100 - x}$ for x.

62. Solve $r = \dfrac{2mI}{B(n + 1)}$ for I.

In Exercises 63–66, find the values of x that satisfy the inequalities.

63. $-x + 3 \le 2x + 9$

64. $-2 \le 3x + 1 \le 7$

65. $x - 3 > 2$ or $x + 3 < -1$

66. $2x^2 > 50$

In Exercises 67–70, evaluate the expression.

67. $|-5 + 7| + |-2|$

68. $\left|\dfrac{5 - 12}{-4 - 3}\right|$

69. $|2\pi - 6| - \pi$

70. $|\sqrt{3} - 4| + |4 - 2\sqrt{3}|$

In Exercises 71–76, find the value(s) of x that satisfy the expression.

71. $2x^2 + 3x - 2 \le 0$

72. $x^2 + x - 12 \le 0$

73. $\dfrac{1}{x + 2} > 2$

74. $|2x - 3| < 5$

75. $|3x - 4| \le 2$

76. $\left|\dfrac{x + 1}{x - 1}\right| = 5$

77. Rationalize the numerator:

$$\dfrac{\sqrt{x} - 1}{x - 1}$$

78. Rationalize the numerator:

$$\sqrt[3]{\dfrac{x^2}{yz^3}}$$

79. Rationalize the denominator:

$$\frac{\sqrt{x} - 1}{2\sqrt{x}}$$

80. Rationalize the denominator:

$$\frac{3}{1 + 2\sqrt{x}}$$

In Exercises 81 and 82, use the quadratic formula to solve the quadratic equation.

81. $x^2 - 2x - 5 = 0$ **82.** $2x^2 + 8x + 7 = 0$

83. Find the minimum cost C (in dollars) given that

$$2(1.5C + 80) \le 2(2.5C - 20)$$

84. Find the maximum revenue R (in dollars) given that

$$12(2R - 320) \le 4(3R + 240)$$

The problem-solving skills that you learn in each chapter are building blocks for the rest of the course. Therefore, it is a good idea to make sure that you have mastered these skills before moving on to the next chapter. The Before Moving On exercises that follow are designed for that purpose. After taking this test, you can see where your weaknesses, if any, are. Then you can log in at www.cengagebrain.com where you will find a link to CourseMate. Here, you can access additional Concept Quiz and Concept Review modules.

If you feel that you need additional help with these exercises, at this Website you can also use Tutorial Example Videos.

CHAPTER 1   Before Moving On . . .

1. Perform the indicated operations and simplify:

$$2(3x - 2)^2 - 3x(x + 1) + 4$$

2. Factor:
 a. $x^4 - x^3 - 6x^2$
 b. $(a - b)^2 - (a^2 + b)^2$

3. Perform the indicated operation and simplify:

$$\frac{2x}{3x^2 - 5x - 2} + \frac{x - 1}{x^2 - x - 2}$$

4. Simplify $\left(\dfrac{8x^2y^{-3}}{9x^{-3}y^2}\right)^{-1}\left(\dfrac{2x^2}{3y^3}\right)^2$.

5. Solve $2s = \dfrac{r}{s + r}$ for r.

6. Rationalize the denominator in the expression

$$\frac{2 - \sqrt{3}}{2 + \sqrt{3}}$$

7. **a.** Solve $2x^2 + 5x - 12 = 0$ by factoring.
 b. Solve $m^2 - 3m - 2 = 0$.

8. Solve $\sqrt{x + 4} - \sqrt{x - 5} - 1 = 0$.

9. Find the values of x that satisfy $(3x + 2)(2x - 3) \le 0$.

10. Find the values of x that satisfy $|2x + 3| \le 1$.

2

FUNCTIONS AND THEIR GRAPHS

Unless changes are made, when is the current Social Security system expected to go broke? In Example 2, page 139, we use a mathematical model constructed from data from the Social Security Administration to predict the year in which the assets of the current system will be depleted.

THIS CHAPTER INTRODUCES the Cartesian coordinate system, a system that allows us to represent points in the plane in terms of ordered pairs of real numbers. This in turn enables us to study geometry, using algebraic methods. Specifically, we will see how straight lines in the plane can be represented by algebraic equations. Next, we study functions, which are special relationships between two quantities. These relationships, or mathematical models, can be found in fields of study as diverse as business, economics, the social sciences, physics, and medicine. We study in detail two special classes of functions: linear functions and quadratic functions. More general functions require the use of the tools of calculus and will be studied in later chapters. Finally, we look at the process of solving real-world problems using mathematics, a process called mathematical modeling.

2.1 The Cartesian Coordinate System and Straight Lines

In Section 1.1, we saw how a one-to-one correspondence between the set of real numbers and the points on a straight line leads to a coordinate system on a line (a one-dimensional space).

A similar representation for points in a plane (a two-dimensional space) is realized through the **Cartesian coordinate system,** which we construct as follows: Take two perpendicular lines, one of which is normally chosen to be horizontal. These lines intersect at a point O, called the **origin** (Figure 1). The horizontal line is called the **x-axis,** and the vertical line is called the **y-axis.** A number scale is set up along the x-axis, with the positive numbers lying to the right of the origin and the negative numbers lying to the left of it. Similarly, a number scale is set up along the y-axis, with the positive numbers lying above the origin and the negative numbers lying below it.

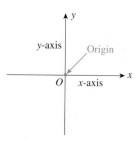

FIGURE **1**
The Cartesian coordinate system

Note   The number scales on the two axes need not be the same. Indeed, in many applications, different quantities are represented by x and y. For example, x may represent the number of smartphones sold, and y may represent the total revenue resulting from the sales. In such cases, it is often desirable to choose different number scales to represent the different quantities. Note, however, that the zeros of both number scales coincide at the origin of the two-dimensional coordinate system. ■

We can represent a point in the plane that is uniquely in this coordinate system by an **ordered pair** of numbers—that is, a pair (x, y) in which x is the first number and y is the second. To see this, let P be any point in the plane (Figure 2). Draw perpendicular lines from P to the x-axis and y-axis, respectively. Then the number x is precisely the number that corresponds to the point on the x-axis at which the perpendicular line through P hits the x-axis. Similarly, y is the number that corresponds to the point on the y-axis at which the perpendicular line through P crosses the y-axis.

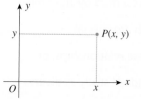

FIGURE **2**
An ordered pair in the coordinate plane

Conversely, given an ordered pair (x, y) with x as the first number and y as the second, a point P in the plane is uniquely determined as follows: Locate the point on the x-axis represented by the number x, and draw a line through that point perpendicular to the x-axis. Next, locate the point on the y-axis represented by the number y, and draw a line through that point perpendicular to the y-axis. The point of intersection of these two lines is the point P (Figure 2).

In the ordered pair (x, y), x is called the **abscissa,** or **x-coordinate;** y is called the **ordinate,** or **y-coordinate;** and x and y together are referred to as the **coordinates** of the point P. The point P with x-coordinate equal to a and y-coordinate equal to b is often written $P(a, b)$.

The points $A(2, 3)$, $B(-2, 3)$, $C(-2, -3)$, $D(2, -3)$, $E(3, 2)$, $F(4, 0)$, and $G(0, -5)$ are plotted in Figure 3.

Note   In general, $(x, y) \neq (y, x)$. This is illustrated by the points A and E in Figure 3.

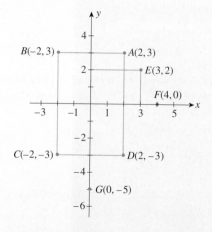

FIGURE **3**
Several points in the coordinate plane

The axes divide the plane into four quadrants. Quadrant I consists of the points P with coordinates x and y, denoted by $P(x, y)$, satisfying $x > 0$ and $y > 0$; Quadrant II consists of the points $P(x, y)$ where $x < 0$ and $y > 0$; Quadrant III consists of the points $P(x, y)$ where $x < 0$ and $y < 0$; and Quadrant IV consists of the points $P(x, y)$ where $x > 0$ and $y < 0$ (Figure 4).

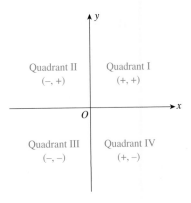

FIGURE 4
The four quadrants in the coordinate plane

Straight Lines

In computing income tax, business firms are allowed by law to depreciate certain assets such as buildings, machines, furniture, and automobiles over a period of time. *Linear depreciation*, or the *straight-line method*, is often used for this purpose. The graph of the straight line shown in Figure 5 describes the book value V of a network server that has an initial value of $10,000 and that is being depreciated linearly over 5 years with a scrap value of $3000. Note that only the solid portion of the straight line is of interest here.

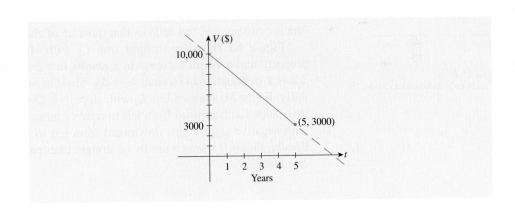

FIGURE 5
Linear depreciation of an asset

The book value of the server at the end of year t, where t lies between 0 and 5, can be read directly from the graph. But there is one shortcoming in this approach: The result depends on how accurately you draw and read the graph. A better and more accurate method is based on finding an *algebraic* representation of the depreciation line. (We continue our discussion of the linear depreciation problem in Section 2.5.)

To see how a straight line in the xy-plane may be described algebraically, we need first to recall certain properties of straight lines.

Slope of a Line

Let L denote the unique straight line that passes through the two distinct points (x_1, y_1) and (x_2, y_2). If $x_1 \neq x_2$, then we define the slope of L as follows.

Slope of a Nonvertical Line

If (x_1, y_1) and (x_2, y_2) are any two distinct points on a nonvertical line L, then the slope m of L is given by

$$m = \frac{\Delta y}{\Delta x} = \frac{y_2 - y_1}{x_2 - x_1} \tag{1}$$

(Figure 6).

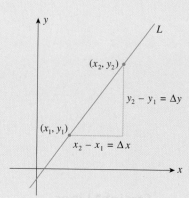

FIGURE **6**

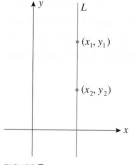

FIGURE **7**
The slope is undefined if $x_1 = x_2$.

If $x_1 = x_2$, then L is a vertical line (Figure 7). Its slope is undefined since the denominator in Equation (1) will be zero and division by zero is proscribed.

Observe that the slope of a straight line is a constant whenever it is defined. The number $\Delta y = y_2 - y_1$ (Δy is read "delta y") is a measure of the vertical change in y, and $\Delta x = x_2 - x_1$ is a measure of the horizontal change in x as shown in Figure 6. From this figure we can see that the slope m of a straight line L is a measure of the *rate of change of y with respect to x*. Furthermore, the slope of a nonvertical straight line is constant, and this tells us that this rate of change is constant.

Figure 8a shows a straight line L_1 with slope 2. Observe that L_1 has the property that a 1-unit increase in x results in a 2-unit increase in y. To see this, let $\Delta x = 1$ in Equation (1) so that $m = \Delta y$. Since $m = 2$, we conclude that $\Delta y = 2$. Similarly, Figure 8b shows a line L_2 with slope -1. Observe that a straight line with positive slope slants upward from left to right (y increases as x increases), whereas a line with negative slope slants downward from left to right (y decreases as x increases). Finally, Figure 9 shows a family of straight lines passing through the origin with indicated slopes.

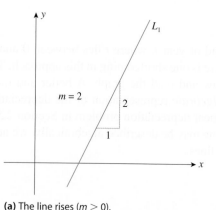

(a) The line rises ($m > 0$).
FIGURE **8**

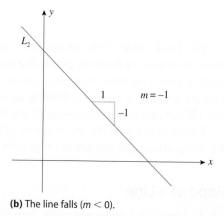

(b) The line falls ($m < 0$).

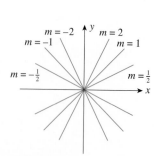

FIGURE **9**
A family of straight lines

Explore & Discuss

Show that the slope of a nonvertical line is independent of the two distinct points used to compute it.

Hint: Suppose we pick two other distinct points, $P_3(x_3, y_3)$ and $P_4(x_4, y_4)$ lying on L. Draw a picture, and use similar triangles to demonstrate that using P_3 and P_4 gives the same value as that obtained by using P_1 and P_2.

VIDEO▶ **EXAMPLE 1** Sketch the straight line that passes through the point $(-2, 5)$ and has slope $-\frac{4}{3}$.

Solution   First, plot the point $(-2, 5)$ (Figure 10). Next, recall that a slope of $-\frac{4}{3}$ indicates that an increase of 1 unit in the x-direction produces a *decrease* of $\frac{4}{3}$ units in the y-direction, or equivalently, a 3-unit increase in the x-direction produces a $3(\frac{4}{3})$, or 4-unit, decrease in the y-direction. Using this information, we plot the point $(1, 1)$ and draw the line through the two points.

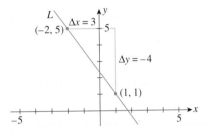

FIGURE **10**
L has slope $-\frac{4}{3}$ and passes through $(-2, 5)$.

EXAMPLE 2 Find the slope m of the line that passes through the points $(-1, 1)$ and $(5, 3)$.

Solution   Choose (x_1, y_1) to be the point $(-1, 1)$ and (x_2, y_2) to be the point $(5, 3)$. Then, with $x_1 = -1$, $y_1 = 1$, $x_2 = 5$, and $y_2 = 3$, we find, using Equation (1),

$$m = \frac{y_2 - y_1}{x_2 - x_1} = \frac{3 - 1}{5 - (-1)} = \frac{2}{6} = \frac{1}{3}$$

(Figure 11). You may verify that the result obtained would be the same had we chosen the point $(-1, 1)$ to be (x_2, y_2) and the point $(5, 3)$ to be (x_1, y_1).

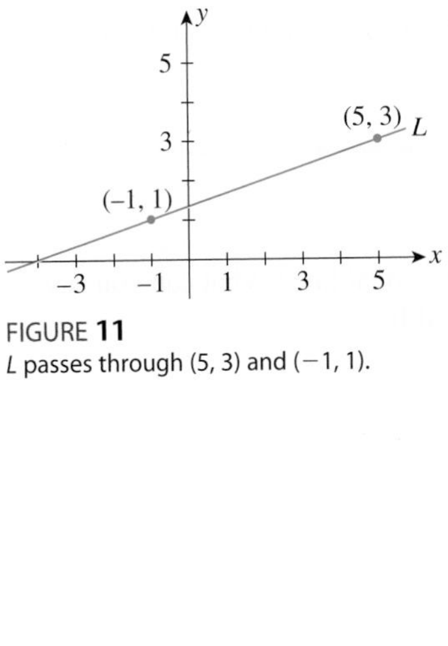

FIGURE **11**
L passes through $(5, 3)$ and $(-1, 1)$.

EXAMPLE 3 Find the slope of the line that passes through the points $(-2, 5)$ and $(3, 5)$.

Solution   The slope of the required line is given by

$$m = \frac{5 - 5}{3 - (-2)} = \frac{0}{5} = 0$$

(Figure 12).

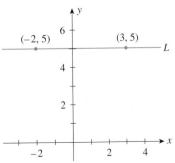

FIGURE **12**
The slope of the horizontal line L is zero.

Note   The slope of a horizontal line is zero.

We can use the slope of a straight line to determine whether a line is parallel to another line.

> **Parallel Lines**
> Two distinct lines are **parallel** if and only if their slopes are equal or their slopes are undefined.

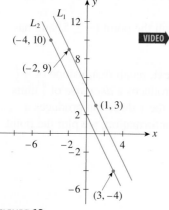

EXAMPLE 4 Let L_1 be a line that passes through the points $(-2, 9)$ and $(1, 3)$, and let L_2 be the line that passes through the points $(-4, 10)$ and $(3, -4)$. Determine whether L_1 and L_2 are parallel.

Solution The slope m_1 of L_1 is given by

$$m_1 = \frac{3 - 9}{1 - (-2)} = -2$$

The slope m_2 of L_2 is given by

$$m_2 = \frac{-4 - 10}{3 - (-4)} = -2$$

Since $m_1 = m_2$, the lines L_1 and L_2 are parallel (Figure 13).

FIGURE 13
L_1 and L_2 have the same slope and hence are parallel.

2.1 Self-Check Exercise

Determine the number a such that the line passing through the points $(a, 2)$ and $(3, 6)$ is parallel to a line with slope 4.

The solution to Self-Check Exercise 2.1 can be found on page 76.

2.1 Concept Questions

1. What can you say about the signs of a and b if the point $P(a, b)$ lies in (a) the second quadrant? (b) The third quadrant? (c) The fourth quadrant?

2. What is the slope of a nonvertical line? What can you say about the slope of a vertical line?

2.1 Exercises

In Exercises 1–6, refer to the accompanying figure and determine the coordinates of the point and the quadrant in which it is located.

1. A **2.** B **3.** C

4. D **5.** E **6.** F

In Exercises 7–12, refer to the accompanying figure.

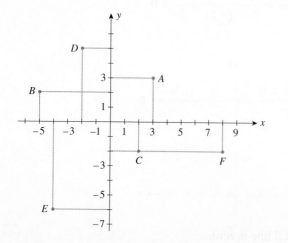

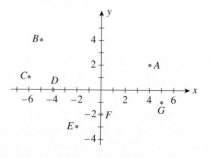

7. Which point has coordinates $(4, 2)$?

8. What are the coordinates of point B?

9. Which points have negative y-coordinates?

10. Which point has a negative x-coordinate and a negative y-coordinate?

11. Which point has an x-coordinate that is equal to zero?

12. Which point has a y-coordinate that is equal to zero?

In Exercises 13–20, sketch a set of coordinate axes and then plot the point.

13. $(-2, 5)$ 14. $(1, 3)$

15. $(3, -1)$ 16. $(3, -4)$

17. $\left(8, -\frac{7}{2}\right)$ 18. $\left(-\frac{5}{2}, \frac{3}{2}\right)$

19. $(4.5, -4.5)$ 20. $(1.2, -3.4)$

In Exercises 21–24, find the slope of the line shown in each figure.

21.

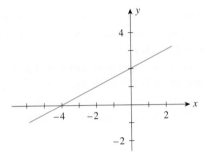

22.

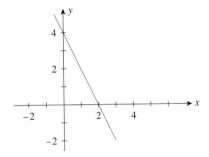

23.

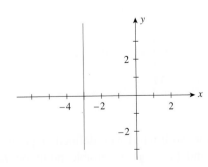

24.

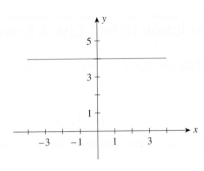

In Exercises 25–30, find the slope of the line that passes through the given pair of points.

25. $(4, 3)$ and $(5, 8)$ 26. $(4, 5)$ and $(3, 8)$

27. $(-2, 3)$ and $(4, 8)$ 28. $(-2, -2)$ and $(4, -4)$

29. (a, b) and (c, d)

30. $(-a + 1, b - 1)$ and $(a + 1, -b)$

31. Given the equation $y = 4x - 3$, answer the following questions.
    a. If x increases by 1 unit, what is the corresponding change in y?
    b. If x decreases by 2 units, what is the corresponding change in y?

32. Given the equation $2x + 3y = 4$, answer the following questions.
    a. Is the slope of the line described by this equation positive or negative?
    b. As x increases in value, does y increase or decrease?
    c. If x decreases by 2 units, what is the corresponding change in y?

In Exercises 33 and 34, determine whether the lines through the pairs of points are parallel.

33. $A(1, -2)$, $B(-3, -10)$ and $C(1, 5)$, $D(-1, 1)$

34. $A(2, 3)$, $B(2, -2)$ and $C(-2, 4)$, $D(-2, 5)$

35. If the line passing through the points $(1, a)$ and $(4, -2)$ is parallel to the line passing through the points $(2, 8)$ and $(-7, a + 4)$, what is the value of a?

36. If the line passing through the points $(a, 1)$ and $(5, 8)$ is parallel to the line passing through the points $(4, 9)$ and $(a + 2, 1)$, what is the value of a?

37. Is there a difference between the statements "The slope of a straight line is zero" and "The slope of a straight line does not exist (is not defined)"? Explain your answer.

2.1   Solution to Self-Check Exercise

The slope of the line that passes through the points $(a, 2)$ and $(3, 6)$ is

$$m = \frac{6 - 2}{3 - a} = \frac{4}{3 - a}$$

Since this line is parallel to a line with slope 4, m must be equal to 4; that is,

$$\frac{4}{3 - a} = 4$$

or, upon multiplying both sides of the equation by $3 - a$,

$$4 = 4(3 - a)$$
$$4 = 12 - 4a$$
$$4a = 8$$
$$a = 2$$

2.2   Equations of Lines

Point-Slope Form

We now show that every straight line lying in the xy-plane may be represented by an equation involving the variables x and y. One immediate benefit of this is that problems involving straight lines may be solved algebraically.

Let L be a straight line parallel to the y-axis (perpendicular to the x-axis) (Figure 14). Then L crosses the x-axis at some point $(a, 0)$ with the x-coordinate given by $x = a$, where a is some real number. Any other point on L has the form (a, y), where y is an appropriate number. Therefore, the vertical line L is described by the sole condition

$$x = a$$

and this is accordingly an equation of L. For example, the equation $x = -2$ represents a vertical line 2 units to the left of the y-axis, and the equation $x = 3$ represents a vertical line 3 units to the right of the y-axis (Figure 15).

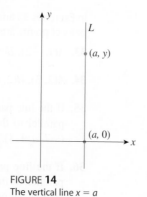

FIGURE 14
The vertical line $x = a$

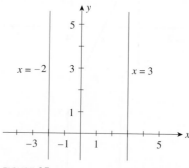

FIGURE 15
The vertical lines $x = -2$ and $x = 3$

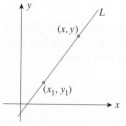

FIGURE 16
L passes through (x_1, y_1) and has slope m.

Next, suppose L is a nonvertical line, so it has a well-defined slope m. Suppose (x_1, y_1) is a fixed point lying on L and (x, y) is a variable point on L distinct from (x_1, y_1) (Figure 16). Using Equation (1) with the point $(x_2, y_2) = (x, y)$, we find that the slope of L is given by

$$m = \frac{y - y_1}{x - x_1}$$

Upon multiplying both sides of the equation by $x - x_1$, we obtain Equation (2).

> **Point-Slope Form of an Equation of a Line**
>
> An equation of the line that has slope m and passes through the point (x_1, y_1) is given by
>
> $$y - y_1 = m(x - x_1) \qquad (2)$$

Equation (2) is called the *point-slope form* of an equation of a line because it uses a given point (x_1, y_1) on a line and the slope m of the line.

VIDEO **EXAMPLE 1** Find an equation of the line that passes through the point $(1, 3)$ and has slope 2.

Solution   Using the point-slope form of the equation of a line with the point $(1, 3)$ and $m = 2$, we obtain

$$y - 3 = 2(x - 1) \qquad {\scriptstyle y - y_1 = m(x - x_1)}$$

which, when simplified, becomes

$$2x - y + 1 = 0$$

(Figure 17).

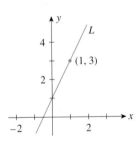

FIGURE 17
L passes through $(1, 3)$ and has slope 2.

EXAMPLE 2 Find an equation of the line that passes through the points $(-3, 2)$ and $(4, -1)$.

Solution   The slope of the line is given by

$$m = \frac{-1 - 2}{4 - (-3)} = -\frac{3}{7}$$

Using the point-slope form of the equation of a line with the point $(4, -1)$ and the slope $m = -\frac{3}{7}$, we have

$$y + 1 = -\frac{3}{7}(x - 4) \qquad {\scriptstyle y - y_1 = m(x - x_1)}$$

$$7y + 7 = -3x + 12$$

$$3x + 7y - 5 = 0$$

(Figure 18).

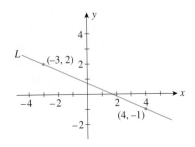

FIGURE 18
L passes through $(-3, 2)$ and $(4, -1)$.

We can use the slope of a straight line to determine whether a line is perpendicular to another line.

> ### Perpendicular Lines
>
> If L_1 and L_2 are two distinct nonvertical lines that have slopes m_1 and m_2, respectively, then L_1 is **perpendicular** to L_2 (written $L_1 \perp L_2$) if and only if
>
> $$m_1 = -\frac{1}{m_2}$$

If the line L_1 is vertical (so that its slope is undefined), then L_1 is perpendicular to another line, L_2, if and only if L_2 is horizontal (so that its slope is zero). For a proof of these results, see Exercise 74, page 85.

EXAMPLE 3 Find an equation of the line that passes through the point (3, 1) and is perpendicular to the line of Example 1.

Solution Since the slope of the line in Example 1 is 2, it follows that the slope of the required line is given by $m = -\frac{1}{2}$, the negative reciprocal of 2. Using the point-slope form of the equation of a line, we obtain

$$y - 1 = -\frac{1}{2}(x - 3) \qquad y - y_1 = m(x - x_1)$$

$$2y - 2 = -x + 3$$

$$x + 2y - 5 = 0$$

(Figure 19).

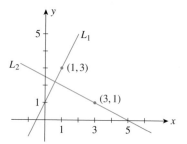

FIGURE 19
L_2 is perpendicular to L_1 and passes through (3, 1).

> ### Exploring with
> ### TECHNOLOGY
>
> 1. Use a graphing utility to plot the straight lines L_1 and L_2 with equations $2x + y - 5 = 0$ and $41x + 20y - 11 = 0$ on the same set of axes, using the standard viewing window.
>   **a.** Can you tell whether the lines L_1 and L_2 are parallel to each other?
>   **b.** Verify your observations by computing the slopes of L_1 and L_2 algebraically.
> 2. Use a graphing utility to plot the straight lines L_1 and L_2 with equations $x + 2y - 5 = 0$ and $5x - y + 5 = 0$ on the same set of axes, using the standard viewing window.
>   **a.** Can you tell whether the lines L_1 and L_2 are perpendicular to each other?
>   **b.** Verify your observation by computing the slopes of L_1 and L_2 algebraically.

Slope-Intercept Form

A straight line L that is neither horizontal nor vertical cuts the x-axis and the y-axis at, say, points $(a, 0)$ and $(0, b)$, respectively (Figure 20). The numbers a and b are called the x-**intercept** and y-**intercept,** respectively, of L.

Now, let L be a line with slope m and y-intercept b. Using Equation (2), the point-slope form of the equation of a line, with the point given by $(0, b)$ and slope m, we have

$$y - b = m(x - 0)$$

$$y = mx + b$$

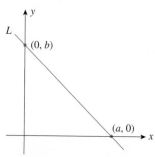

FIGURE 20
The line L has x-intercept a and y-intercept b.

Slope-Intercept Form of an Equation of a Line

An equation of the line that has slope m and intersects the y-axis at the point $(0, b)$ is given by

$$y = mx + b \qquad\qquad (3)$$

EXAMPLE 4 Find an equation of the line that has slope 3 and y-intercept -4.

Solution Using Equation (3) with $m = 3$ and $b = -4$, we obtain the required equation:

$$y = 3x - 4$$

EXAMPLE 5 Determine the slope and y-intercept of the line whose equation is $3x - 4y = 8$.

Solution Rewrite the given equation in the slope-intercept form. Thus,

$$3x - 4y = 8$$
$$-4y = -3x + 8$$
$$y = \frac{3}{4}x - 2$$

Comparing this result with Equation (3), we find $m = \frac{3}{4}$ and $b = -2$, and we conclude that the slope and y-intercept of the given line are $\frac{3}{4}$ and -2, respectively.

Exploring with TECHNOLOGY

1. Use a graphing utility to plot the straight lines with equations $y = -2x + 3$, $y = -x + 3$, $y = x + 3$, and $y = 2.5x + 3$ on the same set of axes, using the standard viewing window. What effect does changing the coefficient m of x in the equation $y = mx + b$ have on its graph?
2. Use a graphing utility to plot the straight lines with equations $y = 2x - 2$, $y = 2x - 1$, $y = 2x$, $y = 2x + 1$, and $y = 2x + 4$ on the same set of axes, using the standard viewing window. What effect does changing the constant b in the equation $y = mx + b$ have on its graph?
3. Describe in words the effect of changing both m and b in the equation $y = mx + b$.

VIDEO **APPLIED EXAMPLE 6** Sales of a Sporting Goods Store The sales manager of a local sporting goods store plotted sales versus time for the last 5 years and found the points to lie approximately along a straight line. (See Figure 21 on the next page.) By using the points corresponding to the first and fifth years, find an equation of the *trend line*. What sales figure can be predicted for the sixth year?

Solution Using Equation (1) with the points $(1, 20)$ and $(5, 60)$, we find that the slope of the required line is given by

$$m = \frac{60 - 20}{5 - 1} = 10$$

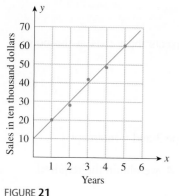

FIGURE 21
Sales of a sporting goods store `VIDEO`

Next, using the point-slope form of the equation of a line with the point $(1, 20)$ and $m = 10$, we obtain

$$y - 20 = 10(x - 1) \qquad y - y_1 = m(x - x_1)$$
$$y = 10x + 10$$

as the required equation.

The sales figure for the sixth year is obtained by letting $x = 6$ in the last equation, giving

$$y = 10(6) + 10 = 70$$

or $700,000. (Note that the sales are measured in ten thousand dollars.) ∎

APPLIED EXAMPLE 7 Appreciation in Value of an Art Object Suppose an art object purchased for $50,000 is expected to appreciate in value at a constant rate of $5000 per year for the next 5 years. Use Equation (3) to write an equation predicting the value of the art object in the next several years. What will be its value 3 years from the purchase date?

Solution Let x denote the time (in years) that has elapsed since the purchase date and let y denote the object's value (in dollars). Then, $y = 50,000$ when $x = 0$. Furthermore, the slope of the required equation is given by $m = 5000$, since each unit increase in x (1 year) implies an increase of 5000 units (dollars) in y. Using Equation (3) with $m = 5000$ and $b = 50,000$, we obtain

$$y = 5000x + 50,000 \qquad y = mx + b$$

Three years from the purchase date, the value of the object will be given by

$$y = 5000(3) + 50,000$$

or $65,000. ∎

Explore & Discuss

Refer to Example 7. Can the equation predicting the value of the art object be used to predict long-term growth?

General Form of an Equation of a Line

We have considered several forms of the equation of a straight line in the plane. These different forms of the equation are equivalent to each other. In fact, each is a special case of the following equation.

General Form of a Linear Equation

The equation

$$Ax + By + C = 0 \tag{4}$$

where A, B, and C are constants and A and B are not both zero, is called the general form of a linear equation in the variables x and y.

We now state (without proof) an important result concerning the algebraic representation of straight lines in the plane.

THEOREM 1

An equation of a straight line is a linear equation; conversely, every linear equation represents a straight line.

This result justifies the use of the adjective *linear* in describing Equation (4).

EXAMPLE 8 Sketch the straight line represented by the equation

$$3x - 4y - 12 = 0$$

Solution   Since every straight line is uniquely determined by two distinct points, we need to find only two points through which the line passes in order to sketch it. For convenience, let's compute the points at which the line crosses the x- and y-axes. Setting $y = 0$, we find $x = 4$, the x-intercept, so the line crosses the x-axis at the point $(4, 0)$. Setting $x = 0$ gives $y = -3$, the y-intercept, so the line crosses the y-axis at the point $(0, -3)$. A sketch of the line appears in Figure 22.

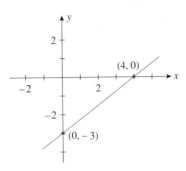

FIGURE **22**
To sketch $3x - 4y - 12 = 0$, first find the x-intercept, 4, and the y-intercept, -3.

Here is a summary of the common forms of the equations of straight lines discussed in this section.

Equations of Straight Lines

Vertical line:	$x = a$
Horizontal line:	$y = b$
Point-slope form:	$y - y_1 = m(x - x_1)$
Slope-intercept form:	$y = mx + b$
General form:	$Ax + By + C = 0$

2.2   Self-Check Exercises

1. Find an equation of the line that passes through the point $(3, -1)$ and is perpendicular to a line with slope $-\frac{1}{2}$.

2. Does the point $(3, -3)$ lie on the line with equation $2x - 3y - 12 = 0$? Sketch the graph of the line.

3. **SATELLITE TV SUBSCRIBERS** The following table gives the number of satellite TV subscribers in the United States (in millions) from 2004 through 2008 ($t = 0$ corresponds to the beginning of 2004):

Year, t	0	1	2	3	4
Number, y	22.5	24.8	27.1	29.1	30.7

a. Plot the number of satellite TV subscribers in the United States (y) versus the year (t).
b. Draw the line L through the points $(0, 22.5)$ and $(4, 30.7)$.
c. Find an equation of the line L.
d. Assuming that this trend continued, estimate the number of satellite TV subscribers in the United States in 2010.

Sources: National Cable & Telecommunications Association and the Federal Communications Commission.

Solutions to Self-Check Exercises 2.2 can be found on page 85.

2.2   Concept Questions

1. Give (a) the point-slope form, (b) the slope-intercept form, and (c) the general form of an equation of a line.

2. Let L_1 have slope m_1 and let L_2 have slope m_2. State the conditions on m_1 and m_2 if (a) L_1 is parallel to L_2 and (b) L_1 is perpendicular to L_2.

2.2 Exercises

In Exercises 1–6, match the statement with one of the graphs (a)–(f).

1. The slope of the line is zero.

2. The slope of the line is undefined.

3. The slope of the line is positive, and its *y*-intercept is positive.

4. The slope of the line is positive, and its *y*-intercept is negative.

5. The slope of the line is negative, and its *x*-intercept is negative.

6. The slope of the line is negative, and its *x*-intercept is positive.

(a)

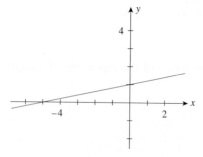

(b)

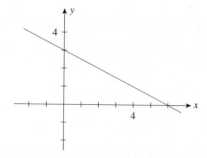

(c)

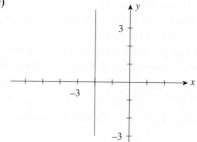

(d)

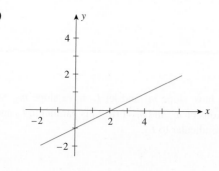

(e)

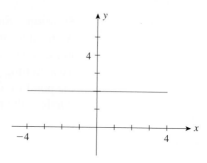

(f)

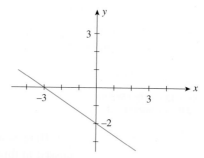

In Exercises 7 and 8, determine whether the lines through the pairs of points are perpendicular.

7. $A(-2, 5)$, $B(4, 2)$ and $C(-1, -2)$, $D(3, 6)$

8. $A(2, 0)$, $B(1, -2)$ and $C(4, 2)$, $D(-8, 4)$

9. Find an equation of the horizontal line that passes through $(-4, -5)$.

10. Find an equation of the vertical line that passes through $(0, 5)$.

In Exercises 11–14, find an equation of the line that passes through the point and has the indicated slope *m*.

11. $(3, -4)$; $m = 2$ **12.** $(2, 4)$; $m = -1$

13. $(-3, 2)$; $m = 0$ **14.** $(1, 2)$; $m = -\dfrac{1}{2}$

In Exercises 15–18, find an equation of the line that passes through the given points.

15. $(2, 4)$ and $(3, 7)$ **16.** $(2, 1)$ and $(2, 5)$

17. $(1, 2)$ and $(-3, -2)$ **18.** $(-1, -2)$ and $(3, -4)$

In Exercises 19–22, find an equation of the line that has slope *m* and *y*-intercept *b*.

19. $m = 3$; $b = 5$ **20.** $m = -2$; $b = -1$

21. $m = 0$; $b = 5$ **22.** $m = -\dfrac{1}{2}$; $b = \dfrac{3}{4}$

In Exercises 23–28, write the equation in the slope-intercept form and then find the slope and y-intercept of the corresponding line.

23. $x - 2y = 0$

24. $y - 2 = 0$

25. $2x - 3y - 9 = 0$

26. $3x - 4y + 8 = 0$

27. $2x + 4y = 14$

28. $5x + 8y - 24 = 0$

29. Find an equation of the line that passes through the point $(-2, 2)$ and is parallel to the line $2x - 4y - 8 = 0$.

30. Find an equation of the line that passes through the point $(-1, 3)$ and is parallel to the line passing through the points $(-2, -3)$ and $(2, 5)$.

31. Find an equation of the line that passes through the point $(2, 4)$ and is perpendicular to the line $3x + 4y - 22 = 0$.

32. Find an equation of the line that passes through the point $(1, -2)$ and is perpendicular to the line passing through the points $(-2, -1)$ and $(4, 3)$.

In Exercises 33–38, find an equation of the line that satisfies the given condition.

33. The line parallel to the x-axis and 6 units below it

34. The line passing through the origin and parallel to the line passing through the points $(2, 4)$ and $(4, 7)$

35. The line passing through the point (a, b) with slope equal to zero

36. The line passing through $(-3, 4)$ and parallel to the x-axis

37. The line passing through $(-5, -4)$ and parallel to the line passing through $(-3, 2)$ and $(6, 8)$

38. The line passing through (a, b) with undefined slope

39. Given that the point $P(-3, 5)$ lies on the line $kx + 3y + 9 = 0$, find k.

40. Given that the point $P(2, -3)$ lies on the line $-2x + ky + 10 = 0$, find k.

In Exercises 41–46, sketch the straight line defined by the linear equation by finding the x- and y-intercepts.

Hint: See Example 8.

41. $3x - 2y + 6 = 0$

42. $2x - 5y + 10 = 0$

43. $x + 2y - 4 = 0$

44. $2x + 3y - 15 = 0$

45. $y + 5 = 0$

46. $-2x - 8y + 24 = 0$

47. Show that an equation of a line through the points $(a, 0)$ and $(0, b)$ with $a \neq 0$ and $b \neq 0$ can be written in the form

$$\frac{x}{a} + \frac{y}{b} = 1$$

(Recall that the numbers a and b are the x- and y-intercepts, respectively, of the line. This form of an equation of a line is called the **intercept form.**)

In Exercises 48–51, use the results of Exercise 47 to find an equation of a line with the x- and y-intercepts.

48. x-intercept 3; y-intercept 4

49. x-intercept -2; y-intercept -4

50. x-intercept $-\dfrac{1}{2}$; y-intercept $\dfrac{3}{4}$

51. x-intercept 4; y-intercept $-\dfrac{1}{2}$

In Exercises 52 and 53, determine whether the points lie on a straight line.

52. $A(-1, 7)$, $B(2, -2)$, and $C(5, -9)$

53. $A(-2, 1)$, $B(1, 7)$, and $C(4, 13)$

54. **TEMPERATURE CONVERSION** The relationship between the temperature in degrees Fahrenheit (°F) and the temperature in degrees Celsius (°C) is

$$F = \frac{9}{5}C + 32$$

   a. Sketch the line with the given equation.
   b. What is the slope of the line? What does it represent?
   c. What is the F-intercept of the line? What does it represent?

55. **NUCLEAR PLANT UTILIZATION** The United States is not building many nuclear plants, but the ones it has are running at nearly full capacity. The output (as a percentage of total capacity) of nuclear plants is described by the equation

$$y = 1.9467t + 70.082$$

where t is measured in years, with $t = 0$ corresponding to the beginning of 1990.

   a. Sketch the line with the given equation.
   b. What are the slope and the y-intercept of the line found in part (a)?
   c. Give an interpretation of the slope and the y-intercept of the line found in part (a).
   d. If the utilization of nuclear power continued to grow at the same rate and the total capacity of nuclear plants in the United States remained constant, by what year were the plants generating at maximum capacity?

Source: Nuclear Energy Institute.

56. **SOCIAL SECURITY CONTRIBUTIONS** For wages less than the maximum taxable wage base, Social Security contributions (including those for Medicare) by employees are 7.65% of the employee's wages.

   a. Find an equation that expresses the relationship between the wages earned (x) and the Social Security taxes paid (y) by an employee who earns less than the maximum taxable wage base.
   b. For each additional dollar that an employee earns, by how much is his or her Social Security contribution increased? (Assume that the employee's wages are less than the maximum taxable wage base.)
   c. What Social Security contributions will an employee who earns $65,000 (which is less than the maximum taxable wage base) be required to make?

Source: Social Security Administration.

57. **COLLEGE ADMISSIONS** Using data compiled by the Admissions Office at Faber University, college admissions officers estimate that 55% of the students who are offered admission to the freshman class at the university will actually enroll.

   a. Find an equation that expresses the relationship between the number of students who actually enroll (y) and the number of students who are offered admission to the university (x).

   b. If the desired freshman class size for the upcoming academic year is 1100 students, how many students should be admitted?

58. **WEIGHT OF WHALES** The equation $W = 3.51L - 192$, expressing the relationship between the length L (in feet) and the expected weight W (in British tons) of adult blue whales, was adopted in the late 1960s by the International Whaling Commission.

   a. What is the expected weight of an 80-ft blue whale?

   b. Sketch the straight line that represents the equation.

59. **THE NARROWING GENDER GAP** Since the founding of the Equal Employment Opportunity Commission and the passage of equal-pay laws, the gulf between men's and women's earnings has continued to close gradually. At the beginning of 1990 ($t = 0$), women's wages were 68% of men's wages, and by the beginning of 2000 ($t = 10$), women's wages were 80% of men's wages. If this gap between women's and men's wages continued to narrow *linearly*, then women's wages were what percentage of men's wages at the beginning of 2004?

   Source: Journal of Economic Perspectives.

60. **SALES OF NAVIGATION SYSTEMS** The projected number of navigation systems (in millions) installed in vehicles in North America, Europe, and Japan from 2002 through 2006 are shown in the following table ($x = 0$ corresponds to 2002):

Year, x	0	1	2	3	4
Systems Installed, y	3.9	4.7	5.8	6.8	7.8

   a. Plot the annual sales (y) versus the year (x).

   b. Draw a straight line L through the points corresponding to 2002 and 2006.

   c. Derive an equation of the line L.

   d. Use the equation found in part (c) to estimate the number of navigation systems installed in 2005. Compare this figure with the sales for that year.

   Source: ABI Research.

61. **SALES OF GPS EQUIPMENT** The annual sales (in billions of dollars) of global positioning systems (GPS) equipment from 2000 through 2006 are shown in the following table ($x = 0$ corresponds to 2000):

Year, x	0	1	2	3	4	5	6
Annual Sales, y	7.9	9.6	11.5	13.3	15.2	17	18.8

   a. Plot the annual sales (y) versus the year (x).

   b. Draw a straight line L through the points corresponding to 2000 and 2006.

   c. Derive an equation of the line L.

   d. Use the equation found in part (c) to estimate the annual sales of GPS equipment in 2005. Compare this figure with the projected sales for that year.

   Source: ABI Research.

62. **IDEAL HEIGHTS AND WEIGHTS FOR WOMEN** The Venus Health Club for Women provides its members with the following table, which gives the average desirable weight (in pounds) for women of a given height (in inches):

Height, x	60	63	66	69	72
Weight, y	108	118	129	140	152

   a. Plot the weight (y) versus the height (x).

   b. Draw a straight line L through the points corresponding to heights of 5 ft and 6 ft.

   c. Derive an equation of the line L.

   d. Using the equation of part (c), estimate the average desirable weight for a woman who is 5 ft, 5 in. tall.

63. **COST OF A COMMODITY** A manufacturer obtained the following data relating the cost y (in dollars) to the number of units (x) of a commodity produced:

Units Produced, x	0	20	40	60	80	100
Cost in Dollars, y	200	208	222	230	242	250

   a. Plot the cost (y) versus the quantity produced (x).

   b. Draw a straight line through the points (0, 200) and (100, 250).

   c. Derive an equation of the straight line of part (b).

   d. Taking this equation to be an approximation of the relationship between the cost and the level of production, estimate the cost of producing 54 units of the commodity.

64. **DIGITAL TV SERVICES** The percentage of homes with digital TV services stood at 5% at the beginning of 1999 ($t = 0$) and was projected to grow linearly so that, at the beginning of 2003 ($t = 4$), the percentage of such homes was 25%.

   a. Derive an equation of the line passing through the points $A(0, 5)$ and $B(4, 25)$.

   b. Plot the line with the equation found in part (a).

   c. Using the equation found in part (a), find the percentage of homes with digital TV services at the beginning of 2001.

   Source: Paul Kagan Associates.

65. **SALES GROWTH** Metro Department Store's annual sales (in millions of dollars) during the past 5 years were

Annual Sales, y	5.8	6.2	7.2	8.4	9.0
Year, x	1	2	3	4	5

a. Plot the annual sales (y) versus the year (x).

b. Draw a straight line L through the points corresponding to the first and fifth years.

c. Derive an equation of the line L.

d. Using the equation found in part (c), estimate Metro's annual sales 4 years from now ($x = 9$).

In Exercises 66–72, determine whether the statement is true or false. If it is true, explain why it is true. If it is false, give an example to show why it is false.

66. Suppose the slope of a line L is $-\frac{1}{2}$ and P is a given point on L. If Q is the point on L lying 4 units to the left of P, then Q is situated 2 units above P.

67. The point $(-1, 1)$ lies on the line with equation $3x + 7y = 5$.

68. The point $(1, k)$ lies on the line with equation $3x + 4y = 12$ if and only if $k = \frac{9}{4}$.

69. The line with equation $Ax + By + C = 0$ ($B \neq 0$) and the line with equation $ax + by + c = 0$ ($b \neq 0$) are parallel if $Ab - aB = 0$.

70. If the slope of the line L_1 is positive, then the slope of a line L_2 perpendicular to L_1 may be positive or negative.

71. The lines with equations $ax + by + c_1 = 0$ and $bx - ay + c_2 = 0$, where $a \neq 0$ and $b \neq 0$, are perpendicular to each other.

72. If L is the line with equation $Ax + By + C = 0$, where $A \neq 0$, then L crosses the x-axis at the point $(-C/A, 0)$.

73. Show that two distinct lines with equations $a_1x + b_1y + c_1 = 0$ and $a_2x + b_2y + c_2 = 0$, respectively, are parallel if and only if $a_1b_2 - b_1a_2 = 0$.
Hint: Write each equation in the slope-intercept form and compare.

74. Prove that if a line L_1 with slope m_1 is perpendicular to a line L_2 with slope m_2, then $m_1m_2 = -1$.
Hint: Refer to the accompanying figure. Show that $m_1 = b$ and $m_2 = c$. Next, apply the Pythagorean Theorem and the distance formula to the triangles OAC, OCB, and OBA to show that $1 = -bc$.

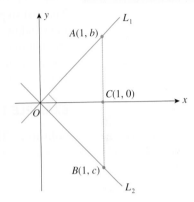

2.2 Solutions to Self-Check Exercises

1. Since the required line L is perpendicular to a line with slope $-\frac{1}{2}$, the slope of L is

$$m = -\frac{1}{-\frac{1}{2}} = 2$$

Next, using the point-slope form of the equation of a line, we have

$$y - (-1) = 2(x - 3)$$
$$y + 1 = 2x - 6$$
$$y = 2x - 7$$

2. Substituting $x = 3$ and $y = -3$ into the left-hand side of the given equation, we find

$$2(3) - 3(-3) - 12 = 3$$

which is not equal to zero (the right-hand side). Therefore, $(3, -3)$ does not lie on the line with equation $2x - 3y - 12 = 0$. (See the accompanying figure.)

Setting $x = 0$, we find $y = -4$, the y-intercept. Next, setting $y = 0$ gives $x = 6$, the x-intercept. We now draw the line passing through the points $(0, -4)$ and $(6, 0)$, as shown.

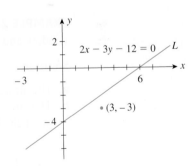

3. a and **b.** See the following figure.

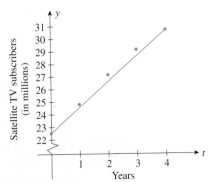

c. The slope of L is

$$m = \frac{30.7 - 22.5}{4 - 0} = 2.05$$

Using the point-slope form of the equation of a line with the point $(0, 22.5)$, we find

$$y - 22.5 = 2.05(t - 0)$$
$$y = 2.05t + 22.5$$

d. Here the year 2010 corresponds to $t = 6$, so the estimated number of satellite TV subscribers in the United States in 2010 is

$$y = 2.05(6) + 22.5 = 34.8$$

or 34.8 million.

USING TECHNOLOGY

Graphing a Straight Line

Graphing Utility

The first step in plotting a straight line with a graphing utility is to select a suitable viewing window. We usually do this by experimenting. For example, you might first plot the straight line using the **standard viewing window** $[-10, 10] \times [-10, 10]$. If necessary, you then might adjust the viewing window by enlarging it or reducing it to obtain a sufficiently complete view of the line or at least the portion of the line that is of interest.

EXAMPLE 1 Plot the straight line $2x + 3y - 6 = 0$ in the standard viewing window.

Solution   The straight line in the standard viewing window is shown in Figure T1.

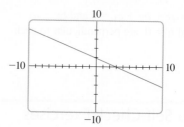

FIGURE **T1**
The straight line $2x + 3y - 6 = 0$ in the standard viewing window

EXAMPLE 2 Plot the straight line $2x + 3y - 30 = 0$ in (a) the standard viewing window and (b) the viewing window $[-5, 20] \times [-5, 20]$.

Solution

a. The straight line in the standard viewing window is shown in Figure T2a.
b. The straight line in the viewing window $[-5, 20] \times [-5, 20]$ is shown in Figure T2b. This figure certainly gives a more complete view of the straight line.

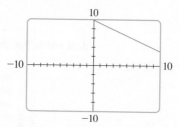

(a) The graph of $2x + 3y - 30 = 0$ in the standard viewing window

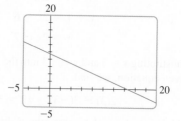

(b) The graph of $2x + 3y - 30 = 0$ in the viewing window $[-5, 20] \times [-5, 20]$

FIGURE **T2**

Excel

In the examples and exercises that follow, we assume that you are familiar with the basic features of Microsoft Excel. Please consult your Excel manual or use Excel's Help features to answer questions regarding the standard commands and operating instructions for Excel. Here we use Microsoft Excel 2010.*

---

*Instructions for solving these examples and exercises using Microsoft Excel 2007 are given on CourseMate.

EXAMPLE 3 Plot the graph of the straight line $2x + 3y - 6 = 0$ over the interval $[-10, 10]$.

Solution

1. *Write the equation in the slope-intercept form:*

$$y = -\frac{2}{3}x + 2$$

2. *Create a table of values.* First, enter the input values: Enter the values of the endpoints of the interval over which you are graphing the straight line. (Recall that we need only two distinct data points to draw the graph of a straight line. In general, we select the endpoints of the interval over which the straight line is to be drawn as our data points.) In this case, we enter -10 in cell B1 and 10 in cell C1.

    Second, enter the formula for computing the *y*-values: Here, we enter

$$= -(2/3)*B1+2$$

    in cell B2 and then press $\boxed{\textbf{Enter}}$.

    Third, evaluate the function at the other input value: To extend the formula to cell C2, move the pointer to the small black box at the lower right corner of cell B2 (the cell containing the formula). Observe that the pointer now appears as a black + (plus sign). Drag this pointer through cell C2 and then release it. The *y*-value, -4.66667, corresponding to the *x*-value in cell C1(10) will appear in cell C2 (Figure T3).

	A	B	C
1	x	-10	10
2	y	8.666667	-4.66667

FIGURE T3
Table of values for *x* and *y*

3. *Graph the straight line determined by these points.* First, highlight the numerical values in the table. Here we highlight cells B1:B2 and C1:C2.

    Step 1   Click on the $\boxed{\textbf{Insert}}$ ribbon tab and then select $\boxed{\textbf{Scatter}}$ from the Charts group. Select the chart subtype in the first row and second column. A chart will then appear on your worksheet.

    Step 2   From the Chart Tools group that now appears at the end of the ribbon, click the $\boxed{\textbf{Layout}}$ tab and then select $\boxed{\textbf{Chart Title}}$ from the Labels group followed by $\boxed{\textbf{Above Chart}}$. Type `y =-(2/3)x + 2` and press $\boxed{\textbf{Enter}}$. Click $\boxed{\textbf{Axis Titles}}$ from the Labels group and select $\boxed{\textbf{Primary Horizontal Axis Title}}$ followed by $\boxed{\textbf{Title Below Axis}}$. Type `x` and then press $\boxed{\textbf{Enter}}$. Next, click $\boxed{\textbf{Axis Titles}}$ again and select $\boxed{\textbf{Primary Vertical Axis Title}}$ followed by $\boxed{\textbf{Vertical Title}}$. Type `y` and press $\boxed{\textbf{Enter}}$.

    Step 3   Click $\boxed{\textbf{Series1}}$ which appears on the right side of the graph and press $\boxed{\textbf{Delete}}$.

Note: Boldfaced words/characters enclosed in a box (for example, $\boxed{\textbf{Enter}}$) indicate that an action (click, select, or press) is required. Words/characters printed blue (for example, Chart Type) indicate words/characters that appear on the screen. Words/characters printed in a monospace font (for example, `=(-2/3)*A2+2`) indicate words/characters that need to be typed and entered.

(continued)

The graph shown in Figure T4 will appear.

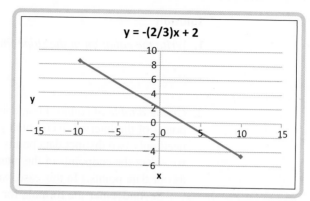

FIGURE **T4**
The graph of $y = -\frac{2}{3}x + 2$ over the interval $[-10, 10]$

If the interval over which the straight line is to be plotted is not specified, then you might have to experiment to find an appropriate interval for the x-values in your graph. For example, you might first plot the straight line over the interval $[-10, 10]$. If necessary you then might adjust the interval by enlarging it or reducing it to obtain a sufficiently complete view of the line or at least the portion of the line that is of interest.

EXAMPLE 4 Plot the straight line $2x + 3y - 30 = 0$ over the intervals
(a) $[-10, 10]$ and (b) $[-5, 20]$.

Solution   **a** and **b.** We first cast the equation in the slope-intercept form, obtaining $y = -\frac{2}{3}x + 10$. Following the procedure given in Example 3, we obtain the graphs shown in Figure T5.

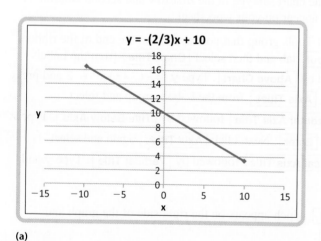

(a)

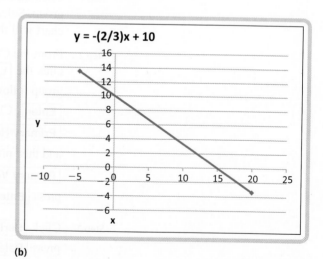

(b)

FIGURE **T5**
The graph of $y = -\frac{2}{3}x + 10$ over the intervals (a) $[-10, 10]$ and (b) $[-5, 20]$

Observe that the graph in Figure T5b includes the x- and y-intercepts. This figure certainly gives a more complete view of the straight line.

TECHNOLOGY EXERCISES

Graphing Utility

In Exercises 1–4, plot the straight line with the equation in the standard viewing window.

1. $3.2x + 2.1y - 6.72 = 0$ 2. $2.3x - 4.1y - 9.43 = 0$

3. $1.6x + 5.1y = 8.16$ 4. $-3.2x + 2.1y = 6.72$

In Exercises 5–8, plot the straight line with the equation in (a) the standard viewing window and (b) the indicated viewing window.

5. $12.1x + 4.1y - 49.61 = 0$; $[-10, 10] \times [-10, 20]$

6. $4.1x - 15.2y - 62.32 = 0$; $[-10, 20] \times [-10, 10]$

7. $20x + 16y = 300$; $[-10, 20] \times [-10, 30]$

8. $32.2x + 21y = 676.2$; $[-10, 30] \times [-10, 40]$

In Exercises 9–12, plot the straight line with the equation in an appropriate viewing window. (*Note:* The answer is *not* unique.)

9. $20x + 30y = 600$

10. $30x - 20y = 600$

11. $22.4x + 16.1y - 352 = 0$

12. $18.2x - 15.1y = 274.8$

Excel

In Exercises 1–4, plot the straight line with the equation over the interval $[-10, 10]$.

1. $3.2x + 2.1y - 6.72 = 0$ 2. $2.3x - 4.1y - 9.43 = 0$

3. $1.6x + 5.1y = 8.16$ 4. $-3.2x + 2.1y = 6.72$

In Exercises 5–8, plot the straight line with the equation over the given interval.

5. $12.1x + 4.1y - 49.61 = 0$; $[-10, 10]$

6. $4.1x - 15.2y - 62.32 = 0$; $[-10, 20]$

7. $20x + 16y = 300$; $[-10, 20]$

8. $32.2x + 21y = 676.2$; $[-10, 30]$

In Exercises 9–12, plot the straight line with the equation. (*Note:* The answer is *not* unique.)

9. $20x + 30y = 600$

10. $30x - 20y = 600$

11. $22.4x + 16.1y - 352 = 0$

12. $18.2x - 15.1y = 274.8$

2.3 Functions and Their Graphs

Functions

A manufacturer would like to know how his company's profit is related to its production level; a biologist would like to know how the size of the population of a certain culture of bacteria will change over time; a psychologist would like to know the relationship between the learning time of an individual and the length of a vocabulary list; and a chemist would like to know how the initial speed of a chemical reaction is related to the amount of substrate used. In each instance, we are concerned with the same question: How does one quantity depend upon another? The relationship between two quantities is conveniently described in mathematics by using the concept of a function.

> **Function**
> A **function** is a rule that assigns to each element in a set A one and only one element in a set B.

The set A is called the **domain** of the function. It is customary to denote a function by a letter of the alphabet, such as the letter f. If x is an element in the domain of a function f, then the element in B that f associates with x is written $f(x)$ (read "f of x") and

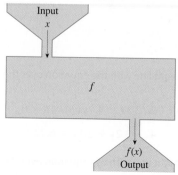

FIGURE 23
A function machine

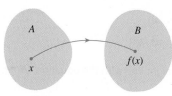

FIGURE 24
The function *f* viewed as a mapping

is called the value of *f* at *x*. The set comprising all the values assumed by $y = f(x)$ as *x* takes on all possible values in its domain is called the **range** of the function *f*.

We can think of a function *f* as a machine. The domain is the set of inputs (raw material) for the machine, the rule describes how the input is to be processed, and the values of the function are the outputs of the machine (Figure 23).

We can also think of a function *f* as a mapping in which an element *x* in the domain of *f* is mapped onto a unique element $f(x)$ in *B* (Figure 24).

Notes

1. The output $f(x)$ associated with an input *x* is unique. To appreciate the importance of this uniqueness property, consider a rule that associates with each item *x* in a department store its selling price *y*. Then, each *x* must correspond to *one and only one y*. Notice, however, that different *x*'s may be associated with the same *y*. In the context of the present example, this says that different items may have the same price.

2. Although the sets *A* and *B* that appear in the definition of a function may be quite arbitrary, in this book they will denote sets of real numbers. ∎

An example of a function may be taken from the familiar relationship between the area of a circle and its radius. Letting *x* and *y* denote the radius and area of a circle, respectively, we have, from elementary geometry,

$$y = \pi x^2 \tag{5}$$

Equation (5) defines *y* as a function of *x* since for each admissible value of *x* (that is, for each nonnegative number representing the radius of a certain circle), there corresponds precisely one number $y = \pi x^2$ that gives the area of the circle. The rule defining this "area function" may be written as

$$f(x) = \pi x^2 \tag{6}$$

To compute the area of a circle of radius 5 inches, we simply replace *x* in Equation (6) with the number 5. Thus, the area of the circle is

$$f(5) = \pi \, 5^2 = 25\pi$$

or 25π square inches.

In general, to evaluate a function at a specific value of *x*, we replace *x* with that value, as illustrated in Examples 1 and 2.

[VIDEO] **EXAMPLE 1** Let the function *f* be defined by the rule $f(x) = 2x^2 - x + 1$. Find:

a. $f(1)$ **b.** $f(-2)$ **c.** $f(a)$ **d.** $f(a + h)$

Solution

a. $f(1) = 2(1)^2 - (1) + 1 = 2 - 1 + 1 = 2$
b. $f(-2) = 2(-2)^2 - (-2) + 1 = 8 + 2 + 1 = 11$
c. $f(a) = 2(a)^2 - (a) + 1 = 2a^2 - a + 1$
d. $f(a + h) = 2(a + h)^2 - (a + h) + 1 = 2a^2 + 4ah + 2h^2 - a - h + 1$ ∎

APPLIED EXAMPLE 2 Profit Functions ThermoMaster manufactures an indoor–outdoor thermometer at its Mexican subsidiary. Management estimates that the profit (in dollars) realizable by ThermoMaster in the manufacture and sale of *x* thermometers per week is

$$P(x) = -0.001x^2 + 8x - 5000$$

Find ThermoMaster's weekly profit if its level of production is (a) 1000 thermometers per week and (b) 2000 thermometers per week.

Solution

a. The weekly profit when the level of production is 1000 units per week is found by evaluating the profit function P at $x = 1000$. Thus,

$$P(1000) = -0.001(1000)^2 + 8(1000) - 5000 = 2000$$

or $2000.

b. When the level of production is 2000 units per week, the weekly profit is given by

$$P(2000) = -0.001(2000)^2 + 8(2000) - 5000 = 7000$$

or $7000.   ∎

Determining the Domain of a Function

Suppose we are given the function $y = f(x)$.* Then, the variable x is called the **independent variable.** The variable y, whose value depends on x, is called the **dependent variable.**

To determine the domain of a function, we need to find what restrictions, if any, are to be placed on the independent variable x. In many practical applications, the domain of a function is dictated by the nature of the problem, as illustrated in Example 3.

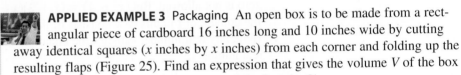

APPLIED EXAMPLE 3 Packaging An open box is to be made from a rectangular piece of cardboard 16 inches long and 10 inches wide by cutting away identical squares (x inches by x inches) from each corner and folding up the resulting flaps (Figure 25). Find an expression that gives the volume V of the box as a function of x. What is the domain of the function?

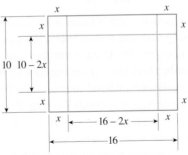

(a) The box is constructed by cutting x" by x" squares from each corner.

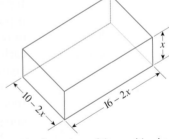

(b) The dimensions of the resulting box are $(10 - 2x)$" by $(16 - 2x)$" by x".

FIGURE **25**

Solution The dimensions of the box are $(10 - 2x)$ inches by $(16 - 2x)$ inches by x inches, so its volume (in cubic inches) is given by

$$V = f(x) = (16 - 2x)(10 - 2x)x \qquad \text{Length} \cdot \text{width} \cdot \text{height}$$
$$= (160 - 52x + 4x^2)x$$
$$= 4x^3 - 52x^2 + 160x$$

Since the length of each side of the box must be greater than or equal to zero, we see that

$$16 - 2x \geq 0 \qquad 10 - 2x \geq 0 \qquad x \geq 0$$

---

*It is customary to refer to a function f as $f(x)$ or by the equation $y = f(x)$ defining it.

simultaneously; that is,

$$x \leq 8 \qquad x \leq 5 \qquad x \geq 0$$

All three inequalities are satisfied simultaneously provided that $0 \leq x \leq 5$. Thus, the domain of the function f is the interval $[0, 5]$. ■

In general, if a function is defined by a rule relating x to $f(x)$ without specific mention of its domain, it is understood that the domain will consist of all values of x for which $f(x)$ is a real number. In this connection, you should keep in mind that (1) division by zero is not permitted and (2) the even root of a negative number is not a real number.

EXAMPLE 4 Find the domain of each function.

a. $f(x) = \sqrt{x - 1}$ **b.** $f(x) = \dfrac{1}{x^2 - 4}$ **c.** $f(x) = x^2 + 3$

Solution

a. Since the square root of a negative number is not a real number, it is necessary that $x - 1 \geq 0$. The inequality is satisfied by the set of real numbers $x \geq 1$. Thus, the domain of f is the interval $[1, \infty)$.
b. The only restriction on x is that $x^2 - 4$ be different from zero, since division by zero is not allowed. But $(x^2 - 4) = (x + 2)(x - 2) = 0$ if $x = -2$ or $x = 2$. Thus, the domain of f in this case consists of the intervals $(-\infty, -2)$, $(-2, 2)$, and $(2, \infty)$.
c. Here, any real number satisfies the equation, so the domain of f is the set of all real numbers. ■

Graphs of Functions

If f is a function with domain A, then corresponding to each real number x in A, there is precisely one real number $f(x)$. We can also express this fact by using ordered pairs of real numbers. Write each number x in A as the first member of an ordered pair and each number $f(x)$ corresponding to x as the second member of the ordered pair. This gives exactly one ordered pair $(x, f(x))$ for each x in A.

Observe that the condition that there be one and only one number $f(x)$ corresponding to each number x in A translates into the requirement that *no two ordered pairs have the same first number.*

Since ordered pairs of real numbers correspond to points in the plane, we have found a way to exhibit a function graphically.

> **Graph of a Function of One Variable**
> The **graph of a function** f is the set of all points (x, y) in the xy-plane such that x is in the domain of f and $y = f(x)$.

Figure 26 shows the graph of a function f. Observe that the y-coordinate of the point (x, y) on the graph of f gives the height of that point (the distance above the x-axis), if $f(x)$ is positive. If $f(x)$ is negative, then $-f(x)$ gives the depth of the point (x, y) (the distance below the x-axis). Also, observe that the domain of f is a set of real numbers lying on the x-axis, whereas the range of f lies on the y-axis.

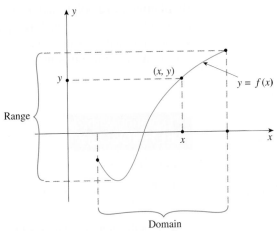

FIGURE **26**
The graph of f

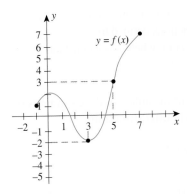

FIGURE **27**
The graph of f

EXAMPLE 5 The graph of a function f is shown in Figure 27.

a. What is the value of $f(3)$? The value of $f(5)$?
b. What is the height or depth of the point $(3, f(3))$ from the x-axis? The point $(5, f(5))$ from the x-axis?
c. What is the domain of f? The range of f?

Solution

a. From the graph of f, we see that $y = -2$ when $x = 3$, and we conclude that $f(3) = -2$. Similarly, we see that $f(5) = 3$.
b. Since the point $(3, -2)$ lies below the x-axis, we see that the depth of the point $(3, f(3))$ is $-f(3) = -(-2) = 2$ units below the x-axis. The point $(5, f(5))$ lies above the x-axis and is located at a height of $f(5)$, or 3 units above the x-axis.
c. Observe that x may take on all values between $x = -1$ and $x = 7$, inclusive, and so the domain of f is $[-1, 7]$. Next, observe that as x takes on all values in the domain of f, $f(x)$ takes on all values between -2 and 7, inclusive. (You can easily see this by running your index finger along the x-axis from $x = -1$ to $x = 7$ and observing the corresponding values assumed by the y-coordinate of each point of the graph of f.) Therefore, the range of f is $[-2, 7]$. ∎

   Much information about the graph of a function can be gained by plotting a few points on its graph. Later on, we will develop more systematic and sophisticated techniques for graphing functions.

EXAMPLE 6 Sketch the graph of the function defined by the equation $y = x^2 + 1$. What is the range of f?

Solution   The domain of the function is the set of all real numbers. By assigning several values to the variable x and computing the corresponding values for y, we obtain the following solutions to the equation $y = x^2 + 1$:

x	-3	-2	-1	0	1	2	3
y	10	5	2	1	2	5	10

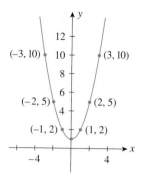

FIGURE **28**
The graph of $y = x^2 + 1$ is a parabola.

By plotting these points and then connecting them with a smooth curve, we obtain the graph of $y = f(x)$, which is a parabola (Figure 28). To determine the range of f, we observe that $x^2 \geq 0$ if x is any real number, and so $x^2 + 1 \geq 1$ for all real numbers x. We conclude that the range of f is $[1, \infty)$. The graph of f confirms this result visually. ■

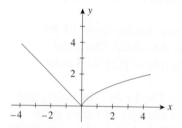

Exploring with
TECHNOLOGY

Let $f(x) = x^2$.

1. Plot the graphs of $F(x) = x^2 + c$ on the same set of axes for $c = -2, -1, -\frac{1}{2}, 0, \frac{1}{2}, 1, 2$.

2. Plot the graphs of $G(x) = (x + c)^2$ on the same set of axes for $c = -2, -1, -\frac{1}{2}, 0, \frac{1}{2}, 1, 2$.

3. Plot the graphs of $H(x) = cx^2$ on the same set of axes for $c = -2, -1, -\frac{1}{2}, -\frac{1}{4}, 0, \frac{1}{4}, \frac{1}{2}, 1, 2$.

4. Study the family of graphs in parts 1–3, and describe the relationship between the graph of a function f and the graphs of the functions defined by
(a) $y = f(x) + c$, (b) $y = f(x + c)$, and (c) $y = cf(x)$, where c is a constant.

Sometimes a function is defined by giving different formulas for different parts of its domain. Such a function is said to be a **piecewise-defined function.**

EXAMPLE 7 Sketch the graph of the function f defined by

$$f(x) = \begin{cases} -x & \text{if } x < 0 \\ \sqrt{x} & \text{if } x \geq 0 \end{cases}$$

FIGURE **29**
The graph of $y = f(x)$ is obtained by graphing $y = -x$ over $(-\infty, 0)$ and $y = \sqrt{x}$ over $[0, \infty)$.

Solution The function f is defined in a piecewise fashion on the set of all real numbers. In the subdomain $(-\infty, 0)$, the rule for f is given by $f(x) = -x$. The equation $y = -x$ is a linear equation in the slope-intercept form (with slope -1 and intercept 0). Therefore, the graph of f corresponding to the subdomain $(-\infty, 0)$ is the half-line shown in Figure 29. Next, in the subdomain $[0, \infty)$, the rule for f is given by $f(x) = \sqrt{x}$. The values of $f(x)$ corresponding to $x = 0, 1, 2, 3,$ and 4 are shown in the following table:

x	0	1	2	3	4
$f(x)$	0	1	$\sqrt{2}$	$\sqrt{3}$	2

Using these values, we sketch the graph of the function f as shown in Figure 29. ■

VIDEO▶

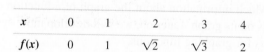

APPLIED EXAMPLE 8 Bank Deposits Madison Finance Company plans to open two branch offices 2 years from now in two separate locations: an industrial complex and a newly developed commercial center in the city. As a result of these expansion plans, Madison's total deposits during the next 5 years are expected to grow in accordance with the rule

$$f(x) = \begin{cases} \sqrt{2x} + 20 & \text{if } 0 \leq x \leq 2 \\ \dfrac{1}{2}x^2 + 20 & \text{if } 2 < x \leq 5 \end{cases}$$

where $y = f(x)$ gives the total amount of money (in millions of dollars) on deposit with Madison in year x ($x = 0$ corresponds to the present). Sketch the graph of the function f.

Solution The function f is defined in a piecewise fashion on the interval $[0, 5]$. In the subdomain $[0, 2]$, the rule for f is given by $f(x) = \sqrt{2x} + 20$. The values of $f(x)$ corresponding to $x = 0, 1,$ and 2 may be tabulated as follows:

x	0	1	2
$f(x)$	20	21.4	22

Next, in the subdomain $(2, 5]$, the rule for f is given by $f(x) = \frac{1}{2}x^2 + 20$. The values of $f(x)$ corresponding to $x = 3, 4,$ and 5 are shown in the following table:

x	3	4	5
$f(x)$	24.5	28	32.5

Using the values of $f(x)$ in this table, we sketch the graph of the function f as shown in Figure 30. ∎

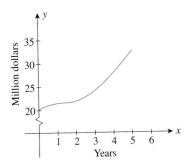

FIGURE **30**
We obtain the graph of the function $y = f(x)$ by graphing $y = \sqrt{2x} + 20$ over $[0, 2]$ and $y = \frac{1}{2}x^2 + 20$ over $(2, 5]$.

The Vertical Line Test

Although it is true that every function f of a variable x has a graph in the xy-plane, it is not true that every curve in the xy-plane is the graph of a function. For example, consider the curve depicted in Figure 31. This is the graph of the equation $y^2 = x$. In general, the **graph of an equation** is the set of all ordered pairs (x, y) that satisfy the given equation. Observe that the points $(9, -3)$ and $(9, 3)$ both lie on the curve. This implies that the number $x = 9$ is associated with *two* numbers: $y = -3$ and $y = 3$. But this clearly violates the uniqueness property of a function. Thus, we conclude that the curve under consideration cannot be the graph of a function.

This example suggests the following **Vertical Line Test** for determining whether a curve is the graph of a function.

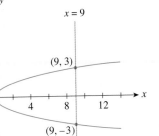

FIGURE **31**
Since a vertical line passes through the curve at more than one point, we deduce that the curve is *not* the graph of a function.

> **Vertical Line Test**
> A curve in the xy-plane is the graph of a function $y = f(x)$ if and only if each vertical line intersects it in at most one point.

EXAMPLE 9 Determine which of the curves shown in Figure 32 (see next page) are the graphs of functions of x.

Solution The curves depicted in Figure 32a, c, and d are graphs of functions because each curve satisfies the requirement that each vertical line intersects the curve in at most one point. Note that the vertical line shown in Figure 32c does *not* intersect the graph because the point on the x-axis through which this line passes does not lie in the domain of the function. The curve depicted in Figure 32b is *not* the graph of a function of x because the vertical line shown there intersects the graph at three points.

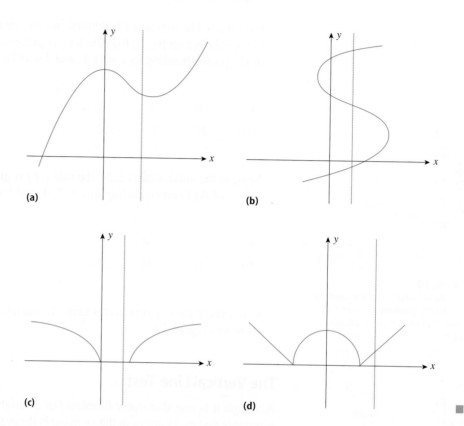

FIGURE 32
The Vertical Line Test can be used to determine which of these curves are graphs of functions.

2.3 Self-Check Exercises

1. Let f be the function defined by

$$f(x) = \frac{\sqrt{x + 1}}{x}$$

a. Find the domain of f.   **b.** Compute $f(3)$.
c. Compute $f(a + h)$.

2. Let

$$f(x) = \begin{cases} -x + 1 & \text{if } -1 \le x < 1 \\ \sqrt{x - 1} & \text{if } 1 \le x \le 5 \end{cases}$$

a. Find $f(0)$ and $f(2)$.
b. Sketch the graph of f.

3. Let $f(x) = \sqrt{2x + 1} + 2$. Determine whether the point $(4, 6)$ lies on the graph of f.

Solutions to Self-Check Exercises 2.3 can be found on page 101.

2.3 Concept Questions

1. a. What is a function?
 b. What is the domain of a function? The range of a function?
 c. What is an independent variable? A dependent variable?

2. a. What is the graph of a function? Use a drawing to illustrate the graph, the domain, and the range of a function.
 b. If you are given a curve in the xy-plane, how can you tell whether the graph is that of a function f defined by $y = f(x)$?

3. Are the following graphs of functions? Explain.

a.

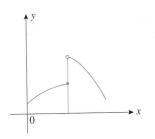

b.

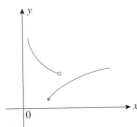

c.

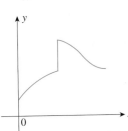

d.

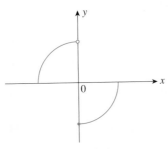

4. What are the domain and range of the function f with the following graph?

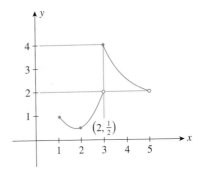

$\left(2, \frac{1}{2}\right)$

2.3 Exercises

1. Let f be the function defined by $f(x) = 5x + 6$. Find $f(3)$, $f(-3), f(a), f(-a)$, and $f(a + 3)$.

2. Let f be the function defined by $f(x) = 4x - 3$. Find $f(4)$, $f(\frac{1}{4})$, $f(0), f(a)$, and $f(a + 1)$.

3. Let g be the function defined by $g(x) = 3x^2 - 6x + 3$. Find $g(0), g(-1), g(a), g(-a)$, and $g(x + 1)$.

4. Let h be the function defined by $h(x) = x^3 - x^2 + x + 1$. Find $h(-5), h(0), h(a)$, and $h(-a)$.

5. Let f be the function defined by $f(x) = 2x + 5$. Find $f(a + h), f(-a), f(a^2), f(a - 2h)$, and $f(2a - h)$.

6. Let g be the function defined by $g(x) = -x^2 + 2x$. Find $g(a + h), g(-a), g(\sqrt{a}), a + g(a)$, and $\dfrac{1}{g(a)}$.

7. Let s be the function defined by $s(t) = \dfrac{2t}{t^2 - 1}$. Find $s(4), s(0), s(a), s(2 + a)$, and $s(t + 1)$.

8. Let g be the function defined by $g(u) = (3u - 2)^{3/2}$. Find $g(1), g(6), g(\frac{11}{3})$, and $g(u + 1)$.

9. Let f be the function defined by $f(t) = \dfrac{2t^2}{\sqrt{t - 1}}$. Find $f(2)$, $f(a), f(x + 1)$, and $f(x - 1)$.

10. Let f be the function defined by $f(x) = 2 + 2\sqrt{5 - x}$. Find $f(-4), f(1), f(\frac{11}{4})$, and $f(x + 5)$.

11. Let f be the function defined by

$$f(x) = \begin{cases} x^2 + 1 & \text{if } x \le 0 \\ \sqrt{x} & \text{if } x > 0 \end{cases}$$

Find $f(-3), f(0)$, and $f(1)$.

12. Let g be the function defined by

$$g(x) = \begin{cases} -\dfrac{1}{2}x + 1 & \text{if } x < 2 \\ \sqrt{x - 2} & \text{if } x \ge 2 \end{cases}$$

Find $g(-2), g(0), g(2)$, and $g(4)$.

13. Let f be the function defined by

$$f(x) = \begin{cases} -\dfrac{1}{2}x^2 + 3 & \text{if } x < 1 \\ 2x^2 + 1 & \text{if } x \ge 1 \end{cases}$$

Find $f(-1), f(0), f(1)$, and $f(2)$.

14. Let f be the function defined by

$$f(x) = \begin{cases} 2 + \sqrt{1 - x} & \text{if } x \le 1 \\ \dfrac{1}{1 - x} & \text{if } x > 1 \end{cases}$$

Find $f(0), f(1)$, and $f(2)$.

(continued)

15. Refer to the graph of the function f in the following figure.

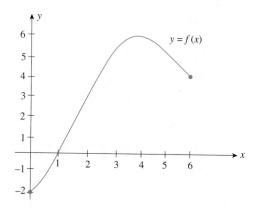

a. Find the value of $f(0)$.
b. Find the value of x for which (i) $f(x) = 3$ and (ii) $f(x) = 0$.
c. Find the domain of f.
d. Find the range of f.

16. Refer to the graph of the function f in the following figure.

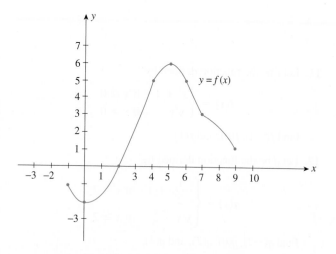

a. Find the value of $f(7)$.
b. Find the values of x corresponding to the point(s) on the graph of f located at a height of 5 units from the x-axis.
c. Find the point on the x-axis at which the graph of f crosses it. What is the value of $f(x)$ at this point?
d. Find the domain and range of f.

In Exercises 17–20, determine whether the point lies on the graph of the function.

17. $(2, \sqrt{3})$; $g(x) = \sqrt{x^2 - 1}$

18. $(3, 3)$; $f(x) = \dfrac{x + 1}{\sqrt{x^2 + 7}} + 2$

19. $(-2, -3)$; $f(t) = \dfrac{|t - 1|}{t + 1}$

20. $\left(-3, -\dfrac{1}{13}\right)$; $h(t) = \dfrac{|t + 1|}{t^3 + 1}$

In Exercises 21 and 22, find the value of c such that the point $P(a, b)$ lies on the graph of the function f.

21. $f(x) = 2x^2 - 4x + c$; $P(1, 5)$

22. $f(x) = x\sqrt{9 - x^2} + c$; $P(2, 4)$

In Exercises 23–36, find the domain of the function.

23. $f(x) = x^2 + 3$ **24.** $f(x) = 7 - x^2$

25. $f(x) = \dfrac{3x + 1}{x^2}$ **26.** $g(x) = \dfrac{2x + 1}{x - 1}$

27. $f(x) = \sqrt{x^2 + 1}$ **28.** $f(x) = \sqrt{x - 5}$

29. $f(x) = \sqrt{5 - x}$ **30.** $g(x) = \sqrt{2x^2 + 3}$

31. $f(x) = \dfrac{x}{x^2 - 4}$ **32.** $f(x) = \dfrac{1}{x^2 + x - 2}$

33. $f(x) = (x + 3)^{3/2}$ **34.** $g(x) = 2(x - 1)^{5/2}$

35. $f(x) = \dfrac{\sqrt{1 - x}}{x^2 - 4}$ **36.** $f(x) = \dfrac{\sqrt{x - 1}}{(x + 2)(x - 3)}$

37. Let f be the function defined by the rule $f(x) = x^2 - x - 6$.
a. Find the domain of f.
b. Compute $f(x)$ for $x = -3, -2, -1, 0, \frac{1}{2}, 1, 2, 3$.
c. Use the results obtained in parts (a) and (b) to sketch the graph of f.

38. Let f be the function defined by the rule $f(x) = 2x^2 + x - 3$.
a. Find the domain of f.
b. Compute $f(x)$ for $x = -3, -2, -1, -\frac{1}{2}, 0, 1, 2, 3$.
c. Use the results obtained in parts (a) and (b) to sketch the graph of f.

In Exercises 39–50, sketch the graph of the function f. Find the domain and range of the function.

39. $f(x) = 2x^2 + 1$ **40.** $f(x) = 9 - x^2$

41. $f(x) = 2 + \sqrt{x}$ **42.** $g(x) = 4 - \sqrt{x}$

43. $f(x) = \sqrt{1 - x}$ **44.** $f(x) = \sqrt{x - 1}$

45. $f(x) = |x| - 1$ **46.** $f(x) = |x| + 1$

47. $f(x) = \begin{cases} x & \text{if } x < 0 \\ 2x + 1 & \text{if } x \geq 0 \end{cases}$

48. $f(x) = \begin{cases} 4 - x & \text{if } x < 2 \\ 2x - 2 & \text{if } x \geq 2 \end{cases}$

49. $f(x) = \begin{cases} -x + 1 & \text{if } x \leq 1 \\ x^2 - 1 & \text{if } x > 1 \end{cases}$

50. $f(x) = \begin{cases} -x - 1 & \text{if } x < -1 \\ 0 & \text{if } -1 \leq x \leq 1 \\ x + 1 & \text{if } x > 1 \end{cases}$

In Exercises 51–58, use the Vertical Line Test to determine whether the graph represents y as a function of x.

51.

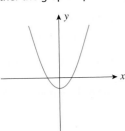

52.

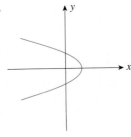

53.

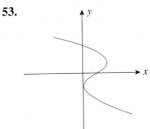

54.

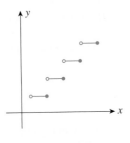

55.

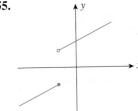

56.

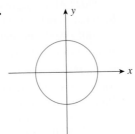

57.

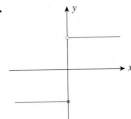

58.

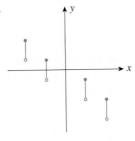

59. The circumference of a circle is given by

$$C(r) = 2\pi r$$

where r is the radius of the circle. What is the circumference of a circle with a 5-in. radius?

60. The volume of a sphere of radius r is given by

$$V(r) = \frac{4}{3}\pi r^3$$

Compute $V(2.1)$ and $V(2)$. What does the quantity $V(2.1) - V(2)$ measure?

61. GROWTH OF A CANCEROUS TUMOR The volume of a spherical cancerous tumor is given by the function

$$V(r) = \frac{4}{3}\pi r^3$$

where r is the radius of the tumor in centimeters. By what factor is the volume of the tumor increased if its radius is doubled?

62. LIFE EXPECTANCY AFTER AGE 65 The average life expectancy after age 65 is soaring, putting pressure on the Social Security Administration's resources. According to the Social Security Trustees, the average life expectancy after age 65 is given by

$$L(t) = 0.056t + 18.1 \qquad (0 \le t \le 7)$$

where t is measured in years, with $t = 0$ corresponding to 2003.
  a. How fast is the average life expectancy after age 65 changing at any time during the period under consideration?
  b. What was the average life expectancy after age 65 in 2010?

Source: Social Security Trustees.

63. CLOSING THE GENDER GAP IN EDUCATION The following graph shows the ratio of the number of bachelor's degrees earned by women to that of men from 1960 through 1990.

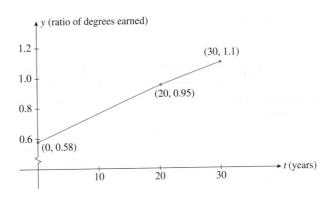

  a. Write the rule for the function f giving the ratio of the number of bachelor's degrees earned by women to that of men in year t, with $t = 0$ corresponding to 1960.
  Hint: The function f is defined piecewise and is linear over each of two subintervals.
  b. How fast was the ratio changing in the period from 1960 to 1980? From 1980 to 1990?
  c. In what year (approximately) was the number of bachelor's degrees earned by women equal for the first time to the number earned by men?

Source: Department of Education.

64. THE GENDER GAP The following graph shows the ratio of women's earnings to men's from 1960 through 2000.

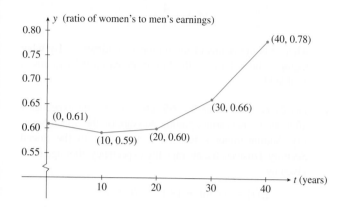

a. Write the rule for the function f giving the ratio of women's earnings to men's in year t, with $t = 0$ corresponding to 1960.
   Hint: The function f is defined piecewise and is linear over each of four subintervals.
b. In what decade(s) was the gender gap expanding? Shrinking?
c. Refer to part (b). How fast was the gender gap expanding or shrinking in each of these decades?
Source: U.S. Bureau of Labor Statistics.

65. WORKER EFFICIENCY An efficiency study conducted for Elektra Electronics showed that the number of Space Commander walkie-talkies assembled by the average worker t hr after starting work at 8 A.M. is given by

$$N(t) = -t^3 + 6t^2 + 15t \qquad (0 \leq t \leq 4)$$

How many walkie-talkies can an average worker be expected to assemble between 8 and 9 A.M.? Between 9 and 10 A.M.?

66. POLITICS Political scientists have discovered the following empirical rule, known as the "cube rule," which gives the relationship between the proportion of seats in the House of Representatives won by Democratic candidates $s(x)$ and the proportion of popular votes x received by the Democratic presidential candidate:

$$s(x) = \frac{x^3}{x^3 + (1 - x)^3} \qquad (0 \leq x \leq 1)$$

Compute $s(0.6)$, and interpret your result.

67. U.S. HEALTH-CARE INFORMATION TECHNOLOGY SPENDING As health-care costs increase, payers are turning to technology and outsourced services to keep a lid on expenses. The amount of health-care information technology (IT) spending by payer is projected to be

$$S(t) = -0.03t^3 + 0.2t^2 + 0.23t + 5.6 \qquad (0 \leq t \leq 4)$$

where $S(t)$ is measured in billions of dollars and t is measured in years, with $t = 0$ corresponding to 2004. What was the amount spent by payers on health-care IT in 2004? Assuming that the projection held true, what amount was spent by payers in 2008?
Source: U.S. Department of Commerce.

68. HOTEL RATES The average daily rate of U.S. hotels from 2006 through 2009 is approximated by the function

$$f(t) = \begin{cases} 0.88t^2 + 3.21t + 96.75 & \text{if } 0 \leq t < 2 \\ -5.58t + 117.85 & \text{if } 2 \leq t \leq 3 \end{cases}$$

where $f(t)$ is measured in dollars and $t = 0$ corresponds to 2006.
a. What was the average daily rate of U.S. hotels in 2006? In 2007? In 2008?
b. Sketch the graph of f.
Source: Smith Travel Research.

69. INVESTMENTS IN HEDGE FUNDS Investments in hedge funds have increased along with their popularity. The assets of hedge funds (in trillions of dollars) from 2002 through 2007 are modeled by the function

$$f(t) = \begin{cases} 0.6 & \text{if } 0 \leq t < 1 \\ 0.6t^{0.43} & \text{if } 1 \leq t \leq 5 \end{cases}$$

where t is measured in years, with $t = 0$ corresponding to the beginning of 2002.
a. What were the assets in hedge funds at the beginning of 2002? At the beginning of 2003?
b. What were the assets in hedge funds at the beginning of 2005? At the beginning of 2007?
Source: Hennessee Group.

70. RISING MEDIAN AGE Increased longevity and the aging of the baby boom generation—those born between 1946 and 1965—are the primary reasons for a rising median age. The median age (in years) of the U.S. population from 1900 through 2000 is approximated by the function

$$f(t) = \begin{cases} 1.3t + 22.9 & \text{if } 0 \leq t \leq 3 \\ -0.7t^2 + 7.2t + 11.5 & \text{if } 3 < t \leq 7 \\ 2.6t + 9.4 & \text{if } 7 < t \leq 10 \end{cases}$$

where t is measured in decades, with $t = 0$ corresponding to the beginning of 1900.
a. What was the median age of the U.S. population at the beginning of 1900? At the beginning of 1950? At the beginning of 1990?
b. Sketch the graph of f.
Source: U.S. Census Bureau.

71. HARBOR CLEANUP The amount of solids discharged from the MWRA (Massachusetts Water Resources Authority) sewage treatment plant on Deer Island (in Boston Harbor) is given by the function

$$f(t) = \begin{cases} 130 & \text{if } 0 \leq t \leq 1 \\ -30t + 160 & \text{if } 1 < t \leq 2 \\ 100 & \text{if } 2 < t \leq 4 \\ -5t^2 + 25t + 80 & \text{if } 4 < t \leq 6 \\ 1.25t^2 - 26.25t + 162.5 & \text{if } 6 < t \leq 10 \end{cases}$$

where $f(t)$ is measured in tons per day and t is measured in years, with $t = 0$ corresponding to 1989.
a. What amount of solids were discharged per day in 1989? In 1992? In 1996?
b. Sketch the graph of f.
Source: Metropolitan District Commission.

In Exercises 72–76, determine whether the statement is true or false. If it is true, explain why it is true. If it is false, give an example to show why it is false.

72. If $a = b$, then $f(a) = f(b)$.

73. If $f(a) = f(b)$, then $a = b$.

74. If f is a function, then $f(a + b) = f(a) + f(b)$.

75. A vertical line must intersect the graph of $y = f(x)$ at exactly one point.

76. The domain of $f(x) = \sqrt{x + 2} + \sqrt{2 - x}$ is $[-2, 2]$.

2.3 Solutions to Self-Check Exercises

1. a. The expression under the radical sign must be nonnegative, so $x + 1 \geq 0$ or $x \geq -1$. Also, $x \neq 0$ because division by zero is not permitted. Therefore, the domain of f is $[-1, 0) \cup (0, \infty)$.

b. $f(3) = \dfrac{\sqrt{3 + 1}}{3} = \dfrac{\sqrt{4}}{3} = \dfrac{2}{3}$

c. $f(a + h) = \dfrac{\sqrt{(a + h) + 1}}{a + h} = \dfrac{\sqrt{a + h + 1}}{a + h}$

2. a. The function f is defined in a piecewise fashion. For $x = 0$, the rule is $f(x) = -x + 1$, and so $f(0) = 1$. For $x = 2$, the rule is $f(x) = \sqrt{x - 1}$, and so $f(2) = \sqrt{2 - 1} = 1$.

b. In the subdomain $[-1, 1)$, the graph of f is the line segment $y = -x + 1$, which is a linear equation with slope -1 and y-intercept 1. In the subdomain $[1, 5]$, the graph of f is given by the rule $f(x) = \sqrt{x - 1}$. From the table below,

x	1	2	3	4	5
$f(x)$	0	1	$\sqrt{2}$	$\sqrt{3}$	2

we obtain the following graph of f.

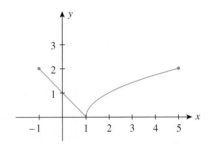

3. A point (x, y) lies on the graph of the function f if the coordinates satisfy the equation $y = f(x)$. Now,

$$f(4) = \sqrt{2(4) + 1} + 2 = \sqrt{9} + 2 = 5 \neq 6$$

and we conclude that the given point does *not* lie on the graph of f.

USING TECHNOLOGY

Graphing a Function

Most of the graphs of functions in this book can be plotted with the help of a graphing utility. Furthermore, a graphing utility can be used to analyze the nature of a function. However, the amount and accuracy of the information obtained by using a graphing utility depend on the experience and sophistication of the user. As you progress through this book, you will see that the more knowledge of mathematics you gain, the more effective the graphing utility will prove to be as a tool in problem solving.

(continued)

EXAMPLE 1 Plot the graph of $f(x) = 2x^2 - 4x - 5$ in the standard viewing window.

Solution The graph of f, shown in Figure T1a, is a parabola. From our previous work (Example 6, Section 2.3), we know that the figure does give a good view of the graph.

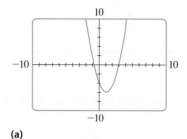 

FIGURE T1
(a) The graph of $f(x) = 2x^2 - 4x - 5$ on $[-10, 10] \times [-10, 10]$; (b) the TI-83/84 window screen for (a); (c) the TI-83/84 equation screen

EXAMPLE 2 Let $f(x) = x^3(x - 3)^4$.

a. Plot the graph of f in the standard viewing window.
b. Plot the graph of f in the window $[-1, 5] \times [-40, 40]$.

Solution

a. The graph of f in the standard viewing window is shown in Figure T2a. Since the graph does not appear to be complete, we need to adjust the viewing window.

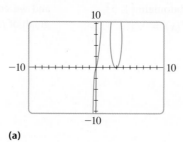

FIGURE T2
(a) An incomplete sketch of $f(x) = x^3(x - 3)^4$ on $[-10, 10] \times [-10, 10]$; (b) the TI-83/84 equation screen

b. The graph of f in the window $[-1, 5] \times [-40, 40]$, shown in Figure T3a, is an improvement over the previous graph. (Later we will be able to show that the figure does in fact give a rather complete view of the graph of f.)

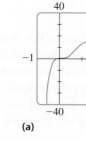

FIGURE T3
(a) A complete sketch of $f(x) = x^3(x - 3)^4$ is shown using the window $[-1, 5] \times [-40, 40]$; (b) the TI-83/84 window screen

Evaluating a Function

A graphing utility can be used to find the value of a function with minimal effort, as the next example shows.

EXAMPLE 3 Let $f(x) = x^3 - 4x^2 + 4x + 2$.

a. Plot the graph of f in the standard viewing window.
b. Find $f(3)$ and verify your result by direct computation.
c. Find $f(4.215)$.

Solution

a. The graph of f is shown in Figure T4a.

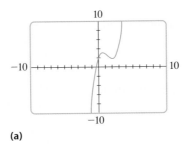

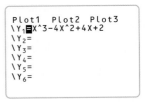

FIGURE **T4**
(a) The graph of $f(x) = x^3 - 4x^2 + 4x + 2$ in the standard viewing window; (b) the TI-83/84 equation screen

(a)   (b)

b. Using the evaluation function of the graphing utility and the value 3 for x, we find $y = 5$. This result is verified by computing

$$f(3) = 3^3 - 4(3^2) + 4(3) + 2 = 27 - 36 + 12 + 2 = 5$$

c. Using the evaluation function of the graphing utility and the value 4.215 for x, we find $y = 22.679738$. Thus, $f(4.215) = 22.679738$. The efficacy of the graphing utility is clearly demonstrated here! ∎

 APPLIED EXAMPLE 4 Number of Alzheimer's Patients The number of Alzheimer's patients in the United States is approximated by

$$f(t) = 0.142t^3 - 0.557t^2 + 1.340t + 3.8 \qquad (0 \le t \le 4)$$

where $f(t)$ is measured in millions and t is measured in decades, with $t = 0$ corresponding to the beginning of 1990.

a. Use a graphing utility to plot the graph of f in the viewing window $[0, 4] \times [0, 12]$.
b. What is the projected number of Alzheimer's patients in the United States at the beginning of 2020 ($t = 3$)?

Source: Alzheimer's Association.

Solution

a. The graph of f in the viewing window $[0, 4] \times [0, 12]$ is shown in Figure T5a.

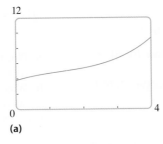

FIGURE **T5**
(a) The graph of f in the viewing window $[0, 4] \times [0, 12]$; (b) the TI-83/84 equation screen

(a)   (b)

(continued)

b. Using the evaluation function of the graphing utility and the value 3 for x, we see that the anticipated number of Alzheimer's patients at the beginning of 2020 is given by $f(3) \approx 6.64$, or approximately 6.6 million.   ∎

TECHNOLOGY EXERCISES

In Exercises 1–4, plot the graph of the function f in (a) the standard viewing window and (b) the indicated window.

1. $f(x) = x^4 - 2x^2 + 8; [-2, 2] \times [6, 10]$

2. $f(x) = x^3 - 20x^2 + 8x - 10; [-20, 20] \times [-1200, 100]$

3. $f(x) = x\sqrt{4 - x^2}; [-3, 3] \times [-2, 2]$

4. $f(x) = \dfrac{4}{x^2 - 8}; [-5, 5] \times [-5, 5]$

In Exercises 5–8, plot the graph of the function f in an appropriate viewing window. (*Note:* The answer is *not* unique.)

5. $f(x) = 2x^4 - 3x^3 + 5x^2 - 20x + 40$

6. $f(x) = -2x^4 + 5x^2 - 4$

7. $f(x) = \dfrac{x^3}{x^3 + 1}$ **8.** $f(x) = \dfrac{2x^4 - 3x}{x^2 - 1}$

In Exercises 9–12, use the evaluation function of your graphing utility to find the value of f at the indicated value of x. Express your answer accurate to four decimal places.

9. $f(x) = 3x^3 - 2x^2 + x - 4; x = 2.145$

10. $f(x) = 5x^4 - 2x^2 + 8x - 3; x = 1.28$

11. $f(x) = \dfrac{2x^3 - 3x + 1}{3x - 2}; x = 2.41$

12. $f(x) = \sqrt{2x^2 + 1} + \sqrt{3x^2 - 1}; x = 0.62$

13. LOBBYISTS' SPENDING Lobbyists try to persuade legislators to propose, pass, or defeat legislation or to change existing laws. The amount (in billions of dollars) spent by lobbyists from 2003 through 2009, where $t = 0$ corresponds to 2003, is given by

$$f(t) = -0.0056t^3 + 0.112t^2 + 0.51t + 8 \quad (0 \le t \le 6)$$

a. Plot the graph of f in the viewing window $[0, 6] \times [0, 15]$.

b. What amount was spent by lobbyists in the year 2005? In 2009?

Source: OpenSecrets.org.

14. SURVEILLANCE CAMERAS Research reports indicate that surveillance cameras at major intersections dramatically reduce the number of drivers who barrel through red lights. The cameras automatically photograph vehicles that drive into intersections after the light turns red. Vehicle owners are then mailed citations instructing them to pay a fine or sign an affidavit that they weren't driving at the time. The function

$$N(t) = 6.08t^3 - 26.79t^2 + 53.06t + 69.5 \quad (0 \le t \le 4)$$

gives the number, $N(t)$, of U.S. communities using surveillance cameras at intersections in year t, with $t = 0$ corresponding to 2003.

a. Plot the graph of N in the viewing window $[0, 4] \times [0, 250]$.

b. How many communities used surveillance cameras at intersections in 2004? In 2006?

Source: Insurance Institute for Highway Safety.

15. KEEPING WITH THE TRAFFIC FLOW By driving at a speed to match the prevailing traffic speed, you decrease the chances of an accident. According to data obtained in a university study, the number of accidents per 100 million vehicle miles, y, is related to the deviation from the mean speed, x, in miles per hour by

$$y = 1.05x^3 - 21.95x^2 + 155.9x - 327.3 \quad (6 \le x \le 11)$$

a. Plot the graph of y in the viewing window $[6, 11] \times [20, 150]$.

b. What is the number of accidents per 100 million vehicle miles if the deviation from the mean speed is 6 mph, 8 mph, and 11 mph?

Source: University of Virginia School of Engineering and Applied Science.

16. SAFE DRIVERS The fatality rate in the United States (per 100 million miles traveled) by age of driver (in years) is given by the function

$$f(x) = 0.00000304x^4 - 0.0005764x^3 + 0.04105x^2 - 1.30366x + 16.579 \quad (18 \le x \le 82)$$

a. Plot the graph of f in the viewing window $[18, 82] \times [0, 8]$.

b. What is the fatality rate for 18-year-old drivers? For 50-year-old drivers? For 80-year-old drivers?

Source: National Highway Traffic Safety Administration.

The Algebra of Functions

The Sum, Difference, Product, and Quotient of Functions

Let $S(t)$ and $R(t)$ denote the federal government's spending and revenue, respectively, at any time t, measured in billions of dollars. The graphs of these functions for the period between 2004 and 2009 are shown in Figure 33.

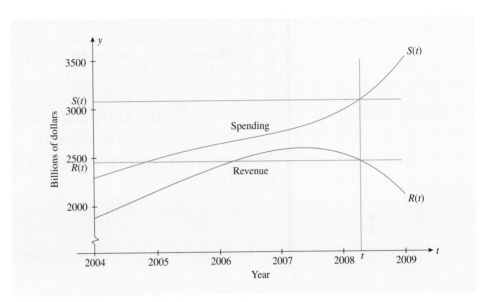

FIGURE 33
$R(t) - S(t)$ gives the federal budget deficit (surplus) at any time t.

Source: Office of Management and Budget.

The difference $R(t) - S(t)$ gives the deficit (surplus) in billions of dollars at any time t if $R(t) - S(t)$ is negative (positive). This observation suggests that we can define a function D whose value at any time t is given by $R(t) - S(t)$. The function D, the *difference* of the two functions R and S, is written $D = R - S$ and may be called the "deficit (surplus) function," since it gives the budget deficit or surplus at any time t. It has the same domain as the functions S and R. The graph of the function D is shown in Figure 34.

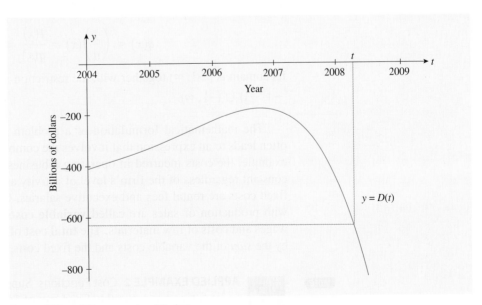

FIGURE 34
The graph of $D(t)$

Source: Office of Management and Budget.

Most functions are built up from other, generally simpler, functions. For example, we may view the function $f(x) = 2x + 4$ as the sum of the two functions $g(x) = 2x$ and $h(x) = 4$. The function $g(x) = 2x$ may in turn be viewed as the product of the functions $p(x) = 2$ and $q(x) = x$.

In general, given the functions f and g, we define the sum $f + g$, the difference $f - g$, the product fg, and the quotient f/g of f and g as follows.

The Sum, Difference, Product, and Quotient of Functions

Let f and g be functions with domains A and B, respectively. Then the **sum** $f + g$, **difference** $f - g$, and **product** fg of f and g are functions with domain $A \cap B$* and rule given by

$$(f + g)(x) = f(x) + g(x) \quad \text{Sum}$$
$$(f - g)(x) = f(x) - g(x) \quad \text{Difference}$$
$$(fg)(x) = f(x)g(x) \quad \text{Product}$$

The **quotient** f/g of f and g has domain $A \cap B$ excluding all numbers x such that $g(x) = 0$ and rule given by

$$\left(\frac{f}{g}\right)(x) = \frac{f(x)}{g(x)} \quad \text{Quotient}$$

*$A \cap B$ is read "A intersected with B" and denotes the set of all points common to both A and B.

EXAMPLE 1 Let $f(x) = \sqrt{x + 1}$ and $g(x) = 2x + 1$. Find the sum s, the difference d, the product p, and the quotient q of the functions f and g.

Solution   Since the domain of f is $A = [-1, \infty)$ and the domain of g is $B = (-\infty, \infty)$, we see that the domain of s, d, and p is $A \cap B = [-1, \infty)$. The rules follow.

$$s(x) = (f + g)(x) = f(x) + g(x) = \sqrt{x + 1} + 2x + 1$$
$$d(x) = (f - g)(x) = f(x) - g(x) = \sqrt{x + 1} - (2x + 1) = \sqrt{x + 1} - 2x - 1$$
$$p(x) = (fg)(x) = f(x)g(x) = \sqrt{x + 1}(2x + 1) = (2x + 1)\sqrt{x + 1}$$

The quotient function q has rule

$$q(x) = \left(\frac{f}{g}\right)(x) = \frac{f(x)}{g(x)} = \frac{\sqrt{x + 1}}{2x + 1}$$

Its domain is $[-1, \infty)$ together with the restriction $x \neq -\frac{1}{2}$. We denote this by $\left[-1, -\frac{1}{2}\right) \cup \left(-\frac{1}{2}, \infty\right)$. ∎

The mathematical formulation of a problem arising from a practical situation often leads to an expression that involves the combination of functions. Consider, for example, the costs incurred in operating a business. Costs that remain more or less constant regardless of the firm's level of activity are called **fixed costs.** Examples of fixed costs are rental fees and executive salaries. On the other hand, costs that vary with production or sales are called **variable costs.** Examples of variable costs are wages and costs of raw materials. The **total cost** of operating a business is thus given by the *sum* of the variable costs and the fixed costs, as illustrated in the next example.

VIDEO ▶   **APPLIED EXAMPLE 2** Cost Functions Suppose Puritron, a manufacturer of water filters, has a monthly fixed cost of \$10,000 and a variable cost of

$$-0.0001x^2 + 10x \quad (0 \leq x \leq 40,000)$$

dollars, where x denotes the number of filters manufactured per month. Find a function C that gives the total monthly cost incurred by Puritron in the manufacture of x filters.

Solution   Puritron's monthly fixed cost is always $10,000, regardless of the level of production, and it is described by the constant function $F(x) = 10,000$. Next, the variable cost is described by the function $V(x) = -0.0001x^2 + 10x$. Since the total cost incurred by Puritron at any level of production is the sum of the variable cost and the fixed cost, we see that the required total cost function is given by

$$C(x) = V(x) + F(x)$$
$$= -0.0001x^2 + 10x + 10,000 \qquad (0 \le x \le 40,000) \qquad \blacksquare$$

Next, the **total profit** realized by a firm in operating a business is the *difference* between the total revenue realized and the total cost incurred; that is,

$$P(x) = R(x) - C(x)$$

APPLIED EXAMPLE 3 Profit Functions Refer to Example 2. Suppose the total revenue realized by Puritron from the sale of x water filters is given by the total revenue function

$$R(x) = -0.0005x^2 + 20x \qquad (0 \le x \le 40,000)$$

a. Find the total profit function—that is, the function that describes the total profit Puritron realizes in manufacturing and selling x water filters per month.
b. What is the profit when the level of production is 10,000 filters per month?

Solution

a. The total profit realized by Puritron in manufacturing and selling x water filters per month is the difference between the total revenue realized and the total cost incurred. Thus, the required total profit function is given by

$$P(x) = R(x) - C(x)$$
$$= (-0.0005x^2 + 20x) - (-0.0001x^2 + 10x + 10,000)$$
$$= -0.0004x^2 + 10x - 10,000$$

b. The profit realized by Puritron when the level of production is 10,000 filters per month is

$$P(10,000) = -0.0004(10,000)^2 + 10(10,000) - 10,000 = 50,000$$

or $50,000 per month. $\qquad \blacksquare$

Composition of Functions

Another way to build up a function from other functions is through a process known as the *composition of functions*. Consider, for example, the function h, whose rule is given by $h(x) = \sqrt{x^2 - 1}$. Let f and g be functions defined by the rules $f(x) = x^2 - 1$ and $g(x) = \sqrt{x}$. Evaluating the function g at the point $f(x)$ [remember that for each real number x in the domain of f, $f(x)$ is simply a real number], we find that

$$g(f(x)) = \sqrt{f(x)} = \sqrt{x^2 - 1}$$

which is just the rule defining the function h!

In general, the composition of a function g with a function f is defined as follows.

> ### The Composition of Two Functions
>
> Let f and g be functions. Then the composition of g and f is the function $g \circ f$ defined by
>
> $$(g \circ f)(x) = g(f(x))$$
>
> The domain of $g \circ f$ is the set of all x in the domain of f such that $f(x)$ lies in the domain of g.

The function $g \circ f$ (read "g circle f") is also called a **composite function.** The interpretation of the function $h = g \circ f$ as a machine is illustrated in Figure 35, and its interpretation as a mapping is shown in Figure 36.

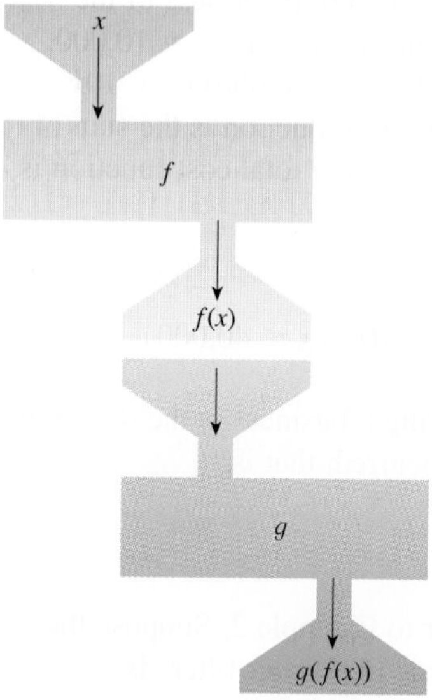

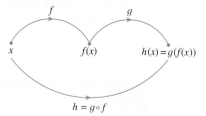

FIGURE 36
The function $h = g \circ f$ viewed as a mapping

FIGURE 35
The composite function $h = g \circ f$ viewed as a machine

EXAMPLE 4 Let $f(x) = x^2 - 1$ and $g(x) = \sqrt{x} + 1$. Find:

a. The rule for the composite function $g \circ f$.
b. The rule for the composite function $f \circ g$.

Solution

a. To find the rule for the composite function $g \circ f$, evaluate the function g at $f(x)$. We obtain

$$(g \circ f)(x) = g(f(x)) = \sqrt{f(x)} + 1 = \sqrt{x^2 - 1} + 1$$

b. To find the rule for the composite function $f \circ g$, evaluate the function f at $g(x)$. Thus,

$$(f \circ g)(x) = f(g(x)) = (g(x))^2 - 1 = (\sqrt{x} + 1)^2 - 1$$
$$= x + 2\sqrt{x} + 1 - 1 = x + 2\sqrt{x}$$

Example 4 shows us that in general $g \circ f$ is different from $f \circ g$, so care must be taken in finding the rule for a composite function.

Explore & Discuss

Let $f(x) = \sqrt{x} + 1$ for $x \geq 0$, and let $g(x) = (x - 1)^2$ for $x \geq 1$.

1. Show that $(g \circ f)(x) = x$ and $(f \circ g)(x) = x$. (*Note:* The function g is said to be the *inverse* of f and vice versa.)

2. Plot the graphs of f and g together with the straight line $y = x$. Describe the relationship between the graphs of f and g.

VIDEO ▶ **APPLIED EXAMPLE 5** Automobile Pollution An environmental impact study conducted for the city of Oxnard indicates that under existing environmental protection laws, the level of carbon monoxide (CO) present in the air due to pollution from automobile exhaust will be $0.01x^{2/3}$ parts per million when the number of motor vehicles is x thousand. A separate study conducted by a state government agency estimates that t years from now, the number of motor vehicles in Oxnard will be $0.2t^2 + 4t + 64$ thousand.

a. Find an expression for the concentration of CO in the air due to automobile exhaust t years from now.

b. What will be the level of concentration 5 years from now?

Solution

a. The level of CO present in the air due to pollution from automobile exhaust is described by the function $g(x) = 0.01x^{2/3}$, where x is the number (in thousands) of motor vehicles. But the number of motor vehicles x (in thousands) t years from now may be estimated by the rule $f(t) = 0.2t^2 + 4t + 64$. Therefore, the concentration of CO due to automobile exhaust t years from now is given by

$$C(t) = (g \circ f)(t) = g(f(t)) = 0.01(0.2t^2 + 4t + 64)^{2/3}$$

parts per million.

b. The level of concentration 5 years from now will be

$$C(5) = 0.01[0.2(5)^2 + 4(5) + 64]^{2/3}$$
$$= (0.01)89^{2/3} \approx 0.20$$

or approximately 0.20 parts per million. ∎

2.4 Self-Check Exercises

1. Let f and g be functions defined by the rules

$$f(x) = \sqrt{x} + 1 \quad \text{and} \quad g(x) = \frac{x}{1 + x}$$

respectively. Find the rules for
a. The sum s, the difference d, the product p, and the quotient q of f and g.
b. The composite functions $f \circ g$ and $g \circ f$.

2. HEALTH-CARE SPENDING Health-care spending per person by the private sector includes payments by individuals, corporations, and their insurance companies and is approximated by the function

$$f(t) = 2.48t^2 + 18.47t + 509 \quad (0 \le t \le 6)$$

where $f(t)$ is measured in dollars and t is measured in years, with $t = 0$ corresponding to the beginning of 1994. The corresponding government spending—including expenditures for Medicaid, Medicare, and other federal, state, and local government public health care—is

$$g(t) = -1.12t^2 + 29.09t + 429 \quad (0 \le t \le 6)$$

where t has the same meaning as before.
a. Find a function that gives the difference between private and government health-care spending per person at any time t.
b. What was the difference between private and government expenditures per person at the beginning of 1995? At the beginning of 2000?

Source: Health Care Financing Administration.

Solutions to Self-Check Exercises 2.4 can be found on page 112.

2.4 Concept Questions

1. The figure opposite shows the graphs of a total cost function and a total revenue function. Let P, defined by $P(x) = R(x) - C(x)$, denote the total profit function.
a. Find an expression for $P(x_1)$. Explain its significance.
b. Find an expression for $P(x_2)$. Explain its significance.

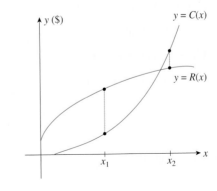

2. a. Explain what is meant by the sum, difference, product, and quotient of the functions f and g with domains A and B, respectively.

  b. If $f(2) = 3$ and $g(2) = -2$, what is $(f + g)(2)$? $(f - g)(2)$? $(fg)(2)$? $(f/g)(2)$?

3. Let f and g be functions, and suppose that (x, y) is a point on the graph of h. What is the value of y for $h = f + g$? $h = f - g$? $h = fg$? $h = f/g$?

4. a. What is the composition of the functions f and g? The functions g and f?

  b. If $f(2) = 3$ and $g(3) = 8$, what is $(g \circ f)(2)$? Can you conclude from the given information what $(f \circ g)(3)$ is? Explain.

5. Let f be a function with domain A, and let g be a function whose domain contains the range of f. If a is any number in A, must $(g \circ f)(a)$ be defined? Explain with an example.

2.4 Exercises

In Exercises 1–8, let $f(x) = x^3 + 5$, $g(x) = x^2 - 2$, and $h(x) = 2x + 4$. Find the rule for each function.

1. $f + g$ **2.** $f - g$ **3.** fg **4.** gf

5. $\dfrac{f}{g}$ **6.** $\dfrac{f - g}{h}$ **7.** $\dfrac{fg}{h}$ **8.** fgh

In Exercises 9–18, let $f(x) = x - 1$, $g(x) = \sqrt{x + 1}$, and $h(x) = 2x^3 - 1$. Find the rule for each function.

9. $f + g$ **10.** $g - f$ **11.** fg **12.** gf

13. $\dfrac{g}{h}$ **14.** $\dfrac{h}{g}$ **15.** $\dfrac{fg}{h}$ **16.** $\dfrac{fh}{g}$

17. $\dfrac{f - h}{g}$ **18.** $\dfrac{gh}{g - f}$

In Exercises 19–24, find the functions $f + g$, $f - g$, fg, and f/g.

19. $f(x) = x^2 + 5$; $g(x) = \sqrt{x} - 2$

20. $f(x) = \sqrt{x - 1}$; $g(x) = x^3 + 1$

21. $f(x) = \sqrt{x + 3}$; $g(x) = \dfrac{1}{x - 1}$

22. $f(x) = \dfrac{1}{x^2 + 1}$; $g(x) = \dfrac{1}{x^2 - 1}$

23. $f(x) = \dfrac{x + 1}{x - 1}$; $g(x) = \dfrac{x + 2}{x - 2}$

24. $f(x) = x^2 + 1$; $g(x) = \sqrt{x + 1}$

In Exercises 25–30, find the rules for the composite functions $f \circ g$ and $g \circ f$.

25. $f(x) = x^2 + x + 1$; $g(x) = x^2 + 1$

26. $f(x) = 3x^2 + 2x + 1$; $g(x) = x + 3$

27. $f(x) = \sqrt{x} + 1$; $g(x) = x^2 - 1$

28. $f(x) = 2\sqrt{x} + 3$; $g(x) = x^2 + 1$

29. $f(x) = \dfrac{x}{x^2 + 1}$; $g(x) = \dfrac{1}{x}$

30. $f(x) = \sqrt{x + 1}$; $g(x) = \dfrac{1}{x - 1}$

In Exercises 31–34, evaluate $h(2)$, where $h = g \circ f$.

31. $f(x) = x^2 + x + 1$; $g(x) = x^2$

32. $f(x) = \sqrt[3]{x^2 - 1}$; $g(x) = 3x^3 + 1$

33. $f(x) = \dfrac{1}{2x + 1}$; $g(x) = \sqrt{x}$

34. $f(x) = \dfrac{1}{x - 1}$; $g(x) = x^2 + 1$

In Exercises 35–42, find functions f and g such that $h = g \circ f$. (*Note:* The answer is *not* unique.)

35. $h(x) = (2x^3 + x^2 + 1)^5$ **36.** $h(x) = (3x^2 - 4)^{-3}$

37. $h(x) = \sqrt{x^2 - 1}$ **38.** $h(x) = (2x - 3)^{3/2}$

39. $h(x) = \dfrac{1}{x^2 - 1}$ **40.** $h(x) = \dfrac{1}{\sqrt{x^2 - 4}}$

41. $h(x) = \dfrac{1}{(3x^2 + 2)^{3/2}}$ **42.** $h(x) = \dfrac{1}{\sqrt{2x + 1}} + \sqrt{2x + 1}$

In Exercises 43–46, find $f(a + h) - f(a)$ for each function. Simplify your answer.

43. $f(x) = 3x + 4$ **44.** $f(x) = -\dfrac{1}{2}x + 3$

45. $f(x) = 4 - x^2$ **46.** $f(x) = x^2 - 2x + 1$

In Exercises 47–52, find and simplify

$$\frac{f(a + h) - f(a)}{h} \qquad (h \neq 0)$$

for each function.

47. $f(x) = 2x^2 + 1$ **48.** $f(x) = 2x^2 - x + 1$

49. $f(x) = x^3 - x$ **50.** $f(x) = 2x^3 - x^2 + 1$

51. $f(x) = \dfrac{1}{x}$ **52.** $f(x) = \sqrt{x}$

53. RESTAURANT REVENUE Nicole owns and operates two restaurants. The revenue of the first restaurant at time t is $f(t)$ dollars, and the revenue of the second restaurant at time t is $g(t)$ dollars. What does the function $F(t) = f(t) + g(t)$ represent?

54. BIRTHRATE OF ENDANGERED SPECIES The birthrate of an endangered species of whales in year t is $f(t)$ whales/year. This species of whales is dying at the rate of $g(t)$ whales/year in year t. What does the function $F(t) = f(t) - g(t)$ represent?

55. VALUE OF AN INVESTMENT The number of IBM shares that Nancy owns is given by $f(t)$. The price per share of the stock of IBM at time t is $g(t)$ dollars. What does the function $f(t)g(t)$ represent?

56. PRODUCTION COSTS The total cost incurred by time t in the production of a certain commodity is $f(t)$ dollars. The number of products produced by time t is $g(t)$ units. What does the function $f(t)/g(t)$ represent?

57. CARBON MONOXIDE POLLUTION The number of cars running in the business district of a town at time t is given by $f(t)$. Carbon monoxide pollution coming from these cars is given by $g(x)$ parts per million, where x is the number of cars being operated in the district. What does the function $g \circ f$ represent?

58. EFFECT OF ADVERTISING ON REVENUE The revenue of Leisure Travel is given by $f(x)$ dollars, where x is the dollar amount spent by the company on advertising. The amount spent by Leisure at time t on advertising is given by $g(t)$ dollars. What does the function $f \circ g$ represent?

59. MANUFACTURING COSTS TMI, a manufacturer of blank DVDs, has a monthly fixed cost of $12,100 and a variable cost of $0.60/disc. Find a function C that gives the total cost incurred by TMI in the manufacture of x discs/month.

60. SPAM MESSAGES The total number of email messages per day (in billions) between 2003 and 2007 is approximated by

$$f(t) = 1.54t^2 + 7.1t + 31.4 \qquad (0 \le t \le 4)$$

where t is measured in years, with $t = 0$ corresponding to 2003. Over the same period, the total number of spam messages per day (in billions) is approximated by

$$g(t) = 1.21t^2 + 6t + 14.5 \qquad (0 \le t \le 4)$$

a. Find the rule for the function $D = f - g$. Compute $D(4)$ and explain what it measures.
b. Find the rule for the function $P = f/g$. Compute $P(4)$ and explain what it means.
Source: Technology Review.

61. PUBLIC TRANSPORTATION BUDGET DEFICIT According to the Massachusetts Bay Transportation Authority (MBTA), the projected cumulative MBTA budget deficit (in billions of dollars) with a $160 million rescue package is given by

$$D_1(t) = 0.0275t^2 + 0.081t + 0.07 \qquad (0 \le t \le 3)$$

and the budget deficit without the rescue package is given by

$$D_2(t) = 0.035t^2 + 0.21t + 0.24 \qquad (0 \le t \le 3)$$

Find the function $D = D_2 - D_1$, and interpret your result.
Source: MBTA Review.

62. MOTORCYCLE DEATHS Suppose the fatality rate (deaths per 100 million miles traveled) of motorcyclists is given by $g(x)$, where x is the percentage of motorcyclists who wear helmets. Next, suppose the percent of motorcyclists who wear helmets at time t (t measured in years) is $f(t)$, with $t = 0$ corresponding to 2000.
a. If $f(0) = 0.64$ and $g(0.64) = 26$, find $(g \circ f)(0)$ and interpret your result.

b. If $f(6) = 0.51$ and $g(0.51) = 42$, find $(g \circ f)(6)$ and interpret your result.
c. Comment on the results of parts (a) and (b).
Source: National Highway Traffic Safety Administration.

63. FIGHTING CRIME Suppose the reported serious crimes (crimes that include homicide, rape, robbery, aggravated assault, burglary, and car theft) that end in arrests or in the identification of suspects is $g(x)$ percent, where x denotes the total number of detectives. Next, suppose the total number of detectives in year t is $f(t)$, with $t = 0$ corresponding to 2001.
a. If $f(1) = 406$ and $g(406) = 23$, find $(g \circ f)(1)$ and interpret your result.
b. If $f(6) = 326$ and $g(326) = 18$, find $(g \circ f)(6)$ and interpret your result.
c. Comment on the results of parts (a) and (b).
Source: Boston Police Department.

64. COST OF PRODUCING SMARTPHONES Apollo manufactures smartphones at a variable cost of

$$V(x) = 0.000003x^3 - 0.03x^2 + 200x$$

dollars, where x denotes the number of units manufactured per month. The monthly fixed cost attributable to the division that produces them is $100,000. Find a function C that gives the total cost incurred by the manufacture of x smartphones. What is the total cost incurred in producing 2000 units/month?

65. PROFIT FROM SALE OF SMARTPHONES Refer to Exercise 64. Suppose the total revenue realized by Apollo from the sale of x smartphones is given by the total revenue function

$$R(x) = -0.1x^2 + 500x \qquad (0 \le x \le 5000)$$

where $R(x)$ is measured in dollars.
a. Find the total profit function.
b. What is the profit when 1500 units are produced and sold each month?

66. PROFIT FROM SALE OF PAGERS A division of Chapman Corporation manufactures a pager. The weekly fixed cost for the division is $20,000, and the variable cost for producing x pagers/week is

$$V(x) = 0.000001x^3 - 0.01x^2 + 50x$$

dollars. The company realizes a revenue of

$$R(x) = -0.02x^2 + 150x \qquad (0 \le x \le 7500)$$

dollars from the sale of x pagers/week.
a. Find the total cost function.
b. Find the total profit function.
c. What is the profit for the company if 2000 units are produced and sold each week?

67. OVERCROWDING OF PRISONS The 1980s saw a trend toward old-fashioned punitive deterrence as opposed to the more liberal penal policies and community-based corrections popular in the 1960s and early 1970s. As a result, prisons became more crowded, and the gap between the number of people in prison and the prison capacity widened. The number of prisoners (in thousands) in federal and state prisons is approximated by the function

$$N(t) = 3.5t^2 + 26.7t + 436.2 \qquad (0 \le t \le 10)$$

where t is measured in years, with $t = 0$ corresponding to 1983. The number of inmates for which prisons were designed is given by

$$C(t) = 24.3t + 365 \qquad (0 \le t \le 10)$$

where $C(t)$ is measured in thousands and t has the same meaning as before.

a. Find an expression that shows the gap between the number of prisoners and the number of inmates for which the prisons were designed at any time t.

b. Find the gap at the beginning of 1983 and at the beginning of 1986.

Source: U.S. Department of Justice.

68. EFFECT OF MORTGAGE RATES ON HOUSING STARTS A study prepared for the National Association of Realtors estimated that the number of housing starts per year over the next 5 years will be

$$N(r) = \frac{7}{1 + 0.02r^2}$$

million units, where r (percent) is the mortgage rate. Suppose the mortgage rate t months from now will be

$$r(t) = \frac{5t + 75}{t + 10} \qquad (0 \le t \le 24)$$

percent/year.

a. Find an expression for the number of housing starts per year as a function of t, t months from now.

b. Using the result from part (a), determine the number of housing starts at present, 12 months from now, and 18 months from now.

69. HOTEL OCCUPANCY RATE The occupancy rate of the all-suite Wonderland Hotel, located near an amusement park, is given by the function

$$r(t) = \frac{10}{81}t^3 - \frac{10}{3}t^2 + \frac{200}{9}t + 55 \qquad (0 \le t \le 11)$$

percent, where t is measured in months and $t = 0$ corresponds to the beginning of January. Management has estimated that the monthly revenue (in thousands of dollars) is approximated by the function

$$R(r) = -\frac{3}{5000}r^3 + \frac{9}{50}r^2 \qquad (0 \le r \le 100)$$

where r (percent) is the occupancy rate.

a. What is the hotel's occupancy rate at the beginning of January? At the beginning of June?

b. What is the hotel's monthly revenue at the beginning of January? At the beginning of June?

Hint: Compute $R(r(0))$ and $R(r(5))$.

70. HOUSING STARTS AND CONSTRUCTION JOBS The president of a major housing construction firm reports that the number of construction jobs (in millions) created is given by

$$N(x) = 1.42x$$

where x denotes the number of housing starts. Suppose the number of housing starts in the next t months is expected to be

$$x(t) = \frac{7(t + 10)^2}{(t + 10)^2 + 2(t + 15)^2}$$

million units. Find an expression for the number of jobs created per year in the next t months. How many jobs/year will have been created 6 months and 12 months from now?

71. a. Let f, g, and h be functions. How would you define the "sum" of f, g, and h?

b. Give a real-life example involving the sum of three functions. (*Note:* The answer is not unique.)

72. a. Let f, g, and h be functions. How would you define the "composition" of h, g, and f, in that order?

b. Give a real-life example involving the composition of these functions. (*Note:* The answer is not unique.)

In Exercises 73–76, determine whether the statement is true or false. If it is true, explain why it is true. If it is false, give an example to show why it is false.

73. If f and g are functions with domain D, then $f + g = g + f$.

74. If $g \circ f$ is defined at $x = a$, then $f \circ g$ must also be defined at $x = a$.

75. If f and g are functions, then $f \circ g = g \circ f$.

76. If f is a function, then $(f \circ f)(x) = [f(x)]^2$.

2.4 Solutions to Self-Check Exercises

1. a. $s(x) = f(x) + g(x) = \sqrt{x} + 1 + \dfrac{x}{1 + x}$

$d(x) = f(x) - g(x) = \sqrt{x} + 1 - \dfrac{x}{1 + x}$

$p(x) = f(x)g(x) = (\sqrt{x} + 1) \cdot \dfrac{x}{1 + x} = \dfrac{x(\sqrt{x} + 1)}{1 + x}$

$q(x) = \dfrac{f(x)}{g(x)} = \dfrac{\sqrt{x} + 1}{\dfrac{x}{1 + x}} = \dfrac{(\sqrt{x} + 1)(1 + x)}{x}$

b. $(f \circ g)(x) = f(g(x)) = \sqrt{\dfrac{x}{1 + x} + 1}$

$(g \circ f)(x) = g(f(x)) = \dfrac{\sqrt{x} + 1}{1 + (\sqrt{x} + 1)} = \dfrac{\sqrt{x} + 1}{\sqrt{x} + 2}$

2. a. The difference between private and government health-care spending per person at any time t is given by the function d with the rule

$$d(t) = f(t) - g(t) = (2.48t^2 + 18.47t + 509)$$
$$- (-1.12t^2 + 29.09t + 429)$$
$$= 3.6t^2 - 10.62t + 80$$

b. The difference between private and government expenditures per person at the beginning of 1995 is given by

$$d(1) = 3.6(1)^2 - 10.62(1) + 80$$

or \$72.98/person.

The difference between private and government expenditures per person at the beginning of 2000 is given by

$$d(6) = 3.6(6)^2 - 10.62(6) + 80$$

or \$145.88/person.

2.5 Linear Functions

We now focus our attention on an important class of functions known as linear functions. Recall that a linear equation in x and y has the form $Ax + By + C = 0$, where A, B, and C are constants and A and B are not both zero. If $B \neq 0$, the equation can always be solved for y in terms of x; in fact, as we saw in Section 2.2, the equation may be cast in the slope-intercept form:

$$y = mx + b \qquad (m, b \text{ constants}) \tag{7}$$

Equation (7) defines y as a function of x. The domain and range of this function are the set of all real numbers. Furthermore, the graph of this function, as we saw in Section 2.2, is a straight line in the plane. For this reason, the function $f(x) = mx + b$ is called a linear function.

> **Linear Function**
> The function f defined by
> $$f(x) = mx + b$$
> where m and b are constants, is called a **linear function.**

Linear functions play an important role in the quantitative analysis of business and economic problems. First, many problems that arise in these and other fields are linear in nature or are linear in the intervals of interest and thus can be formulated in terms of linear functions. Second, because linear functions are relatively easy to work with, assumptions involving linearity are often made in the formulation of problems. In many cases, these assumptions are justified, and acceptable mathematical models are obtained that approximate real-life situations.

We now look at several applications that can be modeled by using linear functions.

APPLIED EXAMPLE 1 Bounced-Check Charges Overdraft fees have become an important piece of a bank's total fee income. The following table gives the bank revenue from overdraft fees (in billions of dollars) from 2004 through 2009.

Year, t	0	1	2	3	4	5
Revenue, y	27.5	29	31	34	36	38

where t is measured in years, with $t = 0$ corresponding to 2004. A mathematical model giving the approximate projected bank revenue from overdraft fees over the period under consideration is given by the linear function

$$f(t) = 2.19t + 27.12 \qquad (0 \leq t \leq 5)$$

a. Plot the six data points and sketch the graph of the function f on the same set of axes.

b. Assuming that the projection held and the trend continued, what was the projected bank revenue from overdraft fees in 2010 ($t = 6$)?

c. What was the rate of increase of the bank revenue from overdraft fees over the period from 2004 through 2009?

Source: New York Times.

Solution

a. The graph of f is shown in Figure 37.

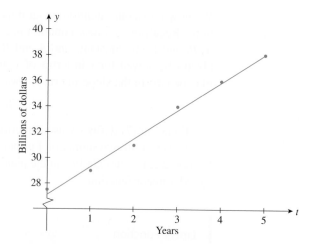

FIGURE **37**
Bank revenue from overdraft fees from 2004 to 2009

b. The projected bank revenue from overdraft fees in 2010 was

$$f(6) = 2.19(6) + 27.12 = 40.26$$

or $40.26 billion.

c. The rate of increase of the bank revenue from overdraft fees over the period from 2004 through 2009 is $2.19 billion per year. ∎

Simple Depreciation

We first discussed linear depreciation in Section 2.1 as a real-world application of straight lines. The following example illustrates how to derive an equation describing the book value of an asset that is being depreciated linearly.

APPLIED EXAMPLE 2 Linear Depreciation A network server has an original value of $10,000 and is to be depreciated linearly over 5 years with a $3000 scrap value. Find an expression giving the book value at the end of year t. What will be the book value of the server at the end of the second year? What is the rate of depreciation of the server?

Solution   Let $V(t)$ denote the network server's book value at the end of the tth year. Since the depreciation is linear, V is a linear function of t. Equivalently, the graph of the function is a straight line. Now, to find an equation of the straight line, observe that $V = 10,000$ when $t = 0$; this tells us that the line passes through the point $(0, 10,000)$. Similarly, the condition that $V = 3000$ when $t = 5$ says that the line also passes through the point $(5, 3000)$. The slope of the line is given by

$$m = \frac{10,000 - 3000}{0 - 5} = -\frac{7000}{5} = -1400$$

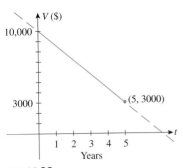

FIGURE 38
Linear depreciation of an asset

Using the point-slope form of an equation of a line with the point $(0, 10{,}000)$ and the slope $m = -1400$, we have

$$V - 10{,}000 = -1400(t - 0)$$
$$V = -1400t + 10{,}000$$

the required expression. The book value at the end of the second year is given by

$$V(2) = -1400(2) + 10{,}000 = 7200$$

or $7200. The rate of depreciation of the server is given by the negative of the slope of the depreciation line. Since the slope of the line is $m = -1400$, the rate of depreciation is $1400 per year. The graph of $V = -1400t + 10{,}000$ is sketched in Figure 38. ∎

Linear Cost, Revenue, and Profit Functions

Whether a business is a sole proprietorship or a large corporation, the owner or chief executive must constantly keep track of operating costs, revenue resulting from the sale of products or services, and, perhaps most important, the profits realized. Three functions provide management with a measure of these quantities: the total cost function, the revenue function, and the profit function.

> **Cost, Revenue, and Profit Functions**
>
> Let x denote the number of units of a product manufactured or sold. Then, the **total cost function** is
>
> $$C(x) = \text{Total cost of manufacturing } x \text{ units of the product}$$
>
> The **revenue function** is
>
> $$R(x) = \text{Total revenue realized from the sale of } x \text{ units of the product}$$
>
> The **profit function** is
>
> $$P(x) = \text{Total profit realized from manufacturing and selling } x \text{ units of the product}$$

Generally speaking, the total cost, revenue, and profit functions associated with a company will probably be nonlinear (these functions are best studied using the tools of calculus). But *linear* cost, revenue, and profit functions do arise in practice, and we will consider such functions in this section. Before deriving explicit forms of these functions, we need to recall some common terminology.

The costs that are incurred in operating a business are usually classified into two categories. Costs that remain more or less constant regardless of the firm's activity level are called **fixed costs.** Examples of fixed costs are rental fees and executive salaries. Costs that vary with production or sales are called **variable costs.** Examples of variable costs are wages and costs for raw materials.

Suppose a firm has a fixed cost of F dollars, a production cost of c dollars per unit, and a selling price of s dollars per unit. Then the *cost function* $C(x)$, the *revenue function* $R(x)$, and the *profit function* $P(x)$ for the firm are given by

$$C(x) = cx + F$$
$$R(x) = sx$$
$$P(x) = R(x) - C(x) \qquad \text{Revenue } - \text{ cost}$$
$$= (s - c)x - F$$

where x denotes the number of units of the commodity produced and sold. The functions C, R, and P are linear functions of x.

VIDEO ▶

APPLIED EXAMPLE 3 Profit Functions Puritron, a manufacturer of water filters, has a monthly fixed cost of $20,000, a production cost of $20 per unit, and a selling price of $30 per unit. Find the cost function, the revenue function, and the profit function for Puritron.

Solution   Let x denote the number of units produced and sold. Then

$$C(x) = 20x + 20{,}000$$
$$R(x) = 30x$$
$$\begin{aligned} P(x) &= R(x) - C(x) \\ &= 30x - (20x + 20{,}000) \\ &= 10x - 20{,}000 \end{aligned}$$

Intersection of Straight Lines

The solution of certain practical problems involves finding the point of intersection of two straight lines. To see how such a problem may be solved algebraically, suppose we are given two straight lines L_1 and L_2 with equations

$$y = m_1 x + b_1 \quad \text{and} \quad y = m_2 x + b_2$$

(where m_1, b_1, m_2, and b_2 are constants) that intersect at the point $P(x_0, y_0)$ (Figure 39).
The point $P(x_0, y_0)$ lies on the line L_1, so it satisfies the equation $y = m_1 x + b_1$. It also lies on the line L_2, so it satisfies the equation $y = m_2 x + b_2$. Therefore, to find the point of intersection $P(x_0, y_0)$ of the lines L_1 and L_2, we solve the system composed of the two equations

$$y = m_1 x + b_1 \quad \text{and} \quad y = m_2 x + b_2$$

for x and y.

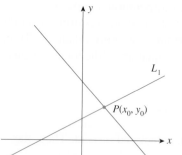

FIGURE 39
L_1 and L_2 intersect at the point $P(x_0, y_0)$.

EXAMPLE 4 Find the point of intersection of the straight lines that have equations $y = x + 1$ and $y = -2x + 4$.

Solution   We solve the given simultaneous equations. Substituting the value y as given in the first equation into the second, we obtain

$$x + 1 = -2x + 4$$
$$3x = 3$$
$$x = 1$$

Substituting this value of x into either one of the given equations yields $y = 2$. Therefore, the required point of intersection is $(1, 2)$ (Figure 40).

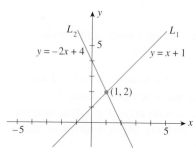

FIGURE 40
The point of intersection of L_1 and L_2 is $(1, 2)$.

Exploring with
TECHNOLOGY

1. Use a graphing utility to plot the straight lines L_1 and L_2 with equations $y = 3x - 2$ and $y = -2x + 3$, respectively, on the same set of axes in the standard viewing window. Then use TRACE and ZOOM to find the point of intersection of L_1 and L_2. Repeat using the "intersection" function of your graphing utility.

2. Find the point of intersection of L_1 and L_2 algebraically.

3. Comment on the effectiveness of each method.

We now turn to some applications involving the intersections of pairs of straight lines.

Break-Even Analysis

Consider a firm with (linear) cost function $C(x)$, revenue function $R(x)$, and profit function $P(x)$ given by

$$C(x) = cx + F$$
$$R(x) = sx$$
$$P(x) = R(x) - C(x) = (s - c)x - F$$

where c denotes the unit cost of production, s the selling price per unit, F the fixed cost incurred by the firm, and x the level of production and sales. The level of production at which the firm neither makes a profit nor sustains a loss is called the **break-even level of operation** and may be determined by solving the equations $y = C(x)$ and $y = R(x)$ simultaneously. At the level of production x_0, the profit is zero, so

$$P(x_0) = R(x_0) - C(x_0) = 0$$
$$R(x_0) = C(x_0)$$

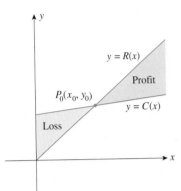

FIGURE 41
P_0 is the break-even point.

The point $P_0(x_0, y_0)$, the solution of the simultaneous equations $y = R(x)$ and $y = C(x)$, is referred to as the **break-even point;** the number x_0 and the number y_0 are called the **break-even quantity** and the **break-even revenue,** respectively.

Geometrically, the break-even point $P_0(x_0, y_0)$ is just the point of intersection of the straight lines representing the cost and revenue functions, respectively. This follows because $P_0(x_0, y_0)$, being the solution of the simultaneous equations $y = R(x)$ and $y = C(x)$, must lie on both these lines simultaneously (Figure 41).

Note that if $x < x_0$, then $R(x) < C(x)$, so $P(x) = R(x) - C(x) < 0$; thus, the firm sustains a loss at this level of production. On the other hand, if $x > x_0$, then $P(x) > 0$, and the firm operates at a profitable level.

VIDEO

APPLIED EXAMPLE 5 Break-Even Level Prescott manufactures its products at a cost of $4 per unit and sells them for $10 per unit. If the firm's fixed cost is $12,000 per month, determine the firm's break-even point.

Solution The cost function C and the revenue function R are given by $C(x) = 4x + 12,000$ and $R(x) = 10x$, respectively (Figure 42).
Setting $R(x) = C(x)$, we obtain

$$10x = 4x + 12,000$$
$$6x = 12,000$$
$$x = 2000$$

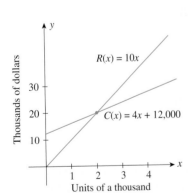

FIGURE 42
The point at which $R(x) = C(x)$ is the break-even point.

Substituting this value of x into $R(x) = 10x$ gives

$$R(2000) = (10)(2000) = 20,000$$

So for a break-even operation, the firm should manufacture 2000 units of its product, resulting in a break-even revenue of $20,000 per month.

APPLIED EXAMPLE 6 Break-Even Analysis Using the data given in Example 5, answer the following questions:

a. What is the loss sustained by the firm if only 1500 units are produced and sold each month?
b. What is the profit if 3000 units are produced and sold each month?
c. How many units should the firm produce to realize a minimum monthly profit of $9000?

Solution   The profit function P is given by the rule

$$P(x) = R(x) - C(x)$$
$$= 10x - (4x + 12,000)$$
$$= 6x - 12,000$$

a. If 1500 units are produced and sold each month, we have

$$P(1500) = 6(1500) - 12,000 = -3000$$

so the firm will sustain a loss of $3000 per month.

b. If 3000 units are produced and sold each month, we have

$$P(3000) = 6(3000) - 12,000 = 6000$$

or a monthly profit of $6000.

c. Substituting 9000 for $P(x)$ in the equation $P(x) = 6x - 12,000$, we obtain

$$9000 = 6x - 12,000$$
$$6x = 21,000$$
$$x = 3500$$

Thus, the firm should produce at least 3500 units to realize a $9000 minimum monthly profit.

APPLIED EXAMPLE 7 Decision Analysis The management of Robertson Controls must decide between two manufacturing processes for its model C electronic thermostat. The monthly cost of the first process is given by $C_1(x) = 20x + 10,000$ dollars, where x is the number of thermostats produced; the monthly cost of the second process is given by $C_2(x) = 10x + 30,000$ dollars. If the projected monthly sales are 800 thermostats at a unit price of $40, which process should management choose in order to maximize the company's profit?

Solution   The break-even level of operation using the first process is obtained by solving the equation

$$40x = 20x + 10,000$$
$$20x = 10,000$$
$$x = 500$$

giving an output of 500 units. Next, we solve the equation

$$40x = 10x + 30,000$$
$$30x = 30,000$$
$$x = 1000$$

giving an output of 1000 units for a break-even operation using the second process. Since the projected sales are 800 units, we conclude that management should choose the first process, which will give the firm a profit.

APPLIED EXAMPLE 8 Decision Analysis Referring to Example 7, decide which process Robertson's management should choose if the projected monthly sales are (a) 1500 units and (b) 3000 units.

Solution   In both cases, the production is past the break-even level. Since the revenue is the same regardless of which process is employed, the decision will be based on how much each process costs.

a. If $x = 1500$, then

$$C_1(x) = (20)(1500) + 10,000 = 40,000$$
$$C_2(x) = (10)(1500) + 30,000 = 45,000$$

Hence, management should choose the first process.
b. If $x = 3000$, then

$$C_1(x) = (20)(3000) + 10,000 = 70,000$$
$$C_2(x) = (10)(3000) + 30,000 = 60,000$$

In this case, management should choose the second process.

Exploring with TECHNOLOGY

1. Use a graphing utility to plot the straight lines L_1 and L_2 with equations $y = 2x - 1$ and $y = 2.1x + 3$, respectively, on the same set of axes, using the standard viewing window. Do the lines appear to intersect?

2. Plot the straight lines L_1 and L_2, using the viewing window $[-100, 100] \times [-100, 100]$. Do the lines appear to intersect? Can you find the point of intersection using TRACE and ZOOM? Using the "intersection" function of your graphing utility?

3. Find the point of intersection of L_1 and L_2 algebraically.

4. Comment on the effectiveness of the solution methods in parts 2 and 3.

2.5 Self-Check Exercises

1. A manufacturer has a monthly fixed cost of $60,000 and a production cost of $10 for each unit produced. The product sells for $15/unit.
   a. What is the cost function?
   b. What is the revenue function?
   c. What is the profit function?
   d. Compute the profit (loss) corresponding to production levels of 10,000 and 14,000 units/month.

2. Find the point of intersection of the straight lines with equations $2x + 3y = 6$ and $x - 3y = 4$.

Solutions to Self-Check Exercises 2.5 can be found on page 122.

2.5 Concept Questions

1. **a.** What is a *linear function*? Give an example.
   b. What is the domain of a linear function? The range?
   c. What is the graph of a linear function?

2. What is the general form of a linear cost function? A linear revenue function? A linear profit function?

3. Explain the meaning of each term:
   a. Break-even point
   b. Break-even quantity
   c. Break-even revenue

2.5 Exercises

In Exercises 1–10, determine whether the equation defines y as a linear function of x. If so, write it in the form $y = mx + b$.

1. $2x + 3y = 6$

2. $-2x + 4y = 7$

3. $x = 2y - 4$

4. $2x = 3y + 8$

5. $2x - 4y + 9 = 0$

6. $3x - 6y + 7 = 0$

7. $2x^2 - 8y + 4 = 0$ **8.** $3\sqrt{x} + 4y = 0$

9. $2x - 3y^2 + 8 = 0$ **10.** $2x + \sqrt{y} - 4 = 0$

11. A manufacturer has a monthly fixed cost of $40,000 and a production cost of $8 for each unit produced. The product sells for $12/unit.
  a. What is the cost function?
  b. What is the revenue function?
  c. What is the profit function?
  d. Compute the profit (loss) corresponding to production levels of 8000 and 12,000 units.

12. A manufacturer has a monthly fixed cost of $100,000 and a production cost of $14 for each unit produced. The product sells for $20/unit.
  a. What is the cost function?
  b. What is the revenue function?
  c. What is the profit function?
  d. Compute the profit (loss) corresponding to production levels of 12,000 and 20,000 units.

13. Find the constants m and b in the linear function $f(x) = mx + b$ such that $f(0) = 4$ and $f(3) = -2$.

14. Find the constants m and b in the linear function $f(x) = mx + b$ such that $f(2) = 4$ and the straight line represented by f has slope -1.

In Exercises 15–20, find the point of intersection of each pair of straight lines.

15. $y = 3x + 4$
$\quad\ y = -2x + 19$

16. $\quad\ y = -4x - 7$
$\quad\ -y = 5x + 10$

17. $2x - 3y = 6$
$\quad 3x + 6y = 16$

18. $\quad 2x + 4y = 11$
$\quad -5x + 3y = 5$

19. $y = \dfrac{1}{4}x - 5$
$\quad 2x - \dfrac{3}{2}y = 1$

20. $y = \dfrac{2}{3}x - 4$
$\quad x + 3y + 3 = 0$

In Exercises 21–24, find the break-even point for the firm whose cost function C and revenue function R are given.

21. $C(x) = 5x + 10,000; R(x) = 15x$

22. $C(x) = 15x + 12,000; R(x) = 21x$

23. $C(x) = 0.2x + 120; R(x) = 0.4x$

24. $C(x) = 150x + 20,000; R(x) = 270x$

25. LINEAR DEPRECIATION An office building worth $1 million when completed in 2005 is being depreciated linearly over 50 years. What was the book value of the building in 2010? What will it be in 2015? (Assume that the scrap value is $0.)

26. LINEAR DEPRECIATION An automobile purchased for use by the manager of a firm at a price of $24,000 is to be depreciated using the straight-line method over 5 years. What will be the book value of the automobile at the end of 3 years? (Assume that the scrap value is $0.)

27. SOCIAL SECURITY BENEFITS Social Security recipients receive an automatic cost-of-living adjustment (COLA) once each year. Their monthly benefit is increased by the same percentage that consumer prices have increased during the preceding year. Suppose consumer prices have increased by 3.3% during the preceding year.
  a. Express the adjusted monthly benefit of a Social Security recipient as a function of his or her current monthly benefit.
  b. If Carlos Garcia's monthly Social Security benefit is now $1520, what will be his adjusted monthly benefit?

28. PROFIT FUNCTIONS AutoTime, a manufacturer of 24-hr variable timers, has a monthly fixed cost of $48,000 and a production cost of $8 for each timer manufactured. The timers sell for $14 each.
  a. What is the cost function?
  b. What is the revenue function?
  c. What is the profit function?
  d. Compute the profit (loss) corresponding to production levels of 4000, 6000, and 10,000 timers, respectively.

29. PROFIT FUNCTIONS The management of TMI finds that the monthly fixed costs attributable to the production of their 100-watt light bulbs is $12,100.00. If the cost of producing each twin-pack of light bulbs is $0.60 and each twin-pack sells for $1.15, find the company's cost function, revenue function, and profit function.

30. LINEAR DEPRECIATION In 2007, National Textile installed a new machine in one of its factories at a cost of $250,000. The machine is depreciated linearly over 10 years with a scrap value of $10,000.
  a. Find an expression for the machine's book value in the tth year of use ($0 \le t \le 10$).
  b. Sketch the graph of the function of part (a).
  c. Find the machine's book value in 2011.
  d. Find the rate at which the machine is being depreciated.

31. LINEAR DEPRECIATION A workcenter system purchased at a cost of $60,000 in 2010 has a scrap value of $12,000 at the end of 4 years. If the straight-line method of depreciation is used,
  a. Find the rate of depreciation.
  b. Find the linear equation expressing the system's book value at the end of t years.
  c. Sketch the graph of the function of part (b).
  d. Find the system's book value at the end of the third year.

32. LINEAR DEPRECIATION Suppose an asset has an original value of $\$C$ and is depreciated linearly over N years with a scrap value of $\$S$. Show that the asset's book value at the end of the tth year is described by the function

$$V(t) = C - \left(\frac{C - S}{N}\right)t$$

Hint: Find an equation of the straight line passing through the points $(0, C)$ and (N, S). (Why?)

33. Rework Exercise 25 using the formula derived in Exercise 32.

34. Rework Exercise 26 using the formula derived in Exercise 32.

35. DRUG DOSAGES A method sometimes used by pediatricians to calculate the dosage of medicine for children is based on the child's surface area. If a denotes the adult dosage (in milligrams) and if S is the child's surface area (in square meters), then the child's dosage is given by

$$D(S) = \frac{Sa}{1.7}$$

a. Show that D is a linear function of S.

Hint: Think of D as having the form $D(S) = mS + b$. What are the slope m and the y-intercept b?

b. If the adult dose of a drug is 500 mg, how much should a child whose surface area is 0.4 m² receive?

36. DRUG DOSAGES Cowling's Rule is a method for calculating pediatric drug dosages. If a denotes the adult dosage (in milligrams) and if t is the child's age (in years), then the child's dosage is given by

$$D(t) = \left(\frac{t + 1}{24} \right)a$$

a. Show that D is a linear function of t.

Hint: Think of $D(t)$ as having the form $D(t) = mt + b$. What is the slope m and the y-intercept b?

b. If the adult dose of a drug is 500 mg, how much should a 4-year-old child receive?

37. BROADBAND INTERNET HOUSEHOLDS The number of U.S. broadband Internet households stood at 20 million at the beginning of 2002 and was projected to grow at the rate of 6.5 million households per year for the next 8 years.

a. Find a linear function $f(t)$ giving the projected number of U.S. broadband Internet households (in millions) in year t, where $t = 0$ corresponds to the beginning of 2002.

b. What was the projected number of U.S. broadband Internet households at the beginning of 2010?

Source: Jupiter Research.

38. DIAL-UP INTERNET HOUSEHOLDS The number of U.S. dial-up Internet households stood at 42.5 million at the beginning of 2004 and was projected to decline at the rate of 3.9 million households per year for the next 6 years.

a. Find a linear function f giving the projected U.S. dial-up Internet households (in millions) in year t, where $t = 0$ corresponds to the beginning of 2004.

b. What was the projected number of U.S. dial-up Internet households at the beginning of 2010?

Source: Strategy Analytics, Inc.

39. CELSIUS AND FAHRENHEIT TEMPERATURES The relationship between temperature measured on the Celsius scale and on the Fahrenheit scale is linear. The freezing point is 0°C and 32°F, and the boiling point is 100°C and 212°F.

a. Find an equation giving the relationship between the temperature F measured on the Fahrenheit scale and the temperature C measured on the Celsius scale.

b. Find F as a function of C and use this formula to determine the temperature in Fahrenheit corresponding to a temperature of 20°C.

c. Find C as a function of F and use this formula to determine the temperature in Celsius corresponding to a temperature of 70°F.

40. CRICKET CHIRPING AND TEMPERATURE Entomologists have discovered that a linear relationship exists between the rate of chirping of crickets of a certain species and the air temperature. When the temperature is 70°F, the crickets chirp at the rate of 120 chirps/min, and when the temperature is 80°F, they chirp at the rate of 160 chirps/min.

a. Find an equation giving the relationship between the air temperature T and the number of chirps per minute N of the crickets.

b. Find N as a function of T, and use this function to determine the rate at which the crickets chirp when the temperature is 102°F.

41. BREAK-EVEN ANALYSIS AutoTime, a manufacturer of 24-hr variable timers, has a monthly fixed cost of $48,000 and a production cost of $8 for each timer manufactured. The units sell for $14 each.

a. Sketch the graphs of the cost function and the revenue function and thereby find the break-even point graphically.

b. Find the break-even point algebraically.

c. Sketch the graph of the profit function.

d. At what point does the graph of the profit function cross the x-axis? Interpret your result.

42. BREAK-EVEN ANALYSIS A division of Carter Enterprises produces "Personal Income Tax" diaries. Each diary sells for $8. The monthly fixed costs incurred by the division are $25,000, and the variable cost of producing each diary is $3.

a. Find the break-even point for the division.

b. What should be the level of sales in order for the division to realize a 15% profit over the cost of making the diaries?

43. BREAK-EVEN ANALYSIS A division of the Gibson Corporation manufactures bicycle pumps. Each pump sells for $9, and the variable cost of producing each unit is 40% of the selling price. The monthly fixed costs incurred by the division are $50,000. What is the break-even point for the division?

44. LEASING Ace Truck Leasing Company leases a certain size truck for $30/day and $0.50/mi, whereas Acme Truck Leasing Company leases the same size truck for $25/day and $0.60/mi.

a. Find the functions describing the daily cost of leasing from each company.

b. Sketch the graphs of the two functions on the same set of axes.

c. If a customer plans to drive at most 70 mi, from which company should he rent a truck for a single day?

45. DECISION ANALYSIS A product may be made by using Machine I or Machine II. The manufacturer estimates that the monthly fixed costs of using Machine I are $18,000, whereas the monthly fixed costs of using Machine II are $15,000. The variable costs of manufacturing 1 unit of the product using Machine I and Machine II are $15 and $20, respectively. The product sells for $50 each.

   a. Find the cost functions associated with using each machine.

   b. Sketch the graphs of the cost functions of part (a) and the revenue functions on the same set of axes.

   c. Which machine should management choose in order to maximize their profit if the projected sales are 450 units? 550 units? 650 units?

   d. What is the profit for each case in part (c)?

46. ANNUAL SALES The annual sales of Crimson Drug Store are expected to be given by $S = 2.3 + 0.4t$ million dollars t years from now, whereas the annual sales of Cambridge Drug Store are expected to be given by $S = 1.2 + 0.6t$ million dollars t years from now. When will Cambridge's annual sales first surpass Crimson's annual sales?

47. LCDs VERSUS CRTs The global shipments of traditional cathode-ray tube monitors (CRTs) is approximated by the equation

$$y = -12t + 88 \qquad (0 \le t \le 3)$$

where y is measured in millions and t in years, with $t = 0$ corresponding to the beginning of 2001. The equation

$$y = 18t + 13.4 \qquad (0 \le t \le 3)$$

gives the approximate number (in millions) of liquid crystal displays (LCDs) over the same period. When did the global shipments of LCDs first overtake the global shipments of CRTs?

Source: International Data Corporation.

48. DIGITAL VERSUS FILM CAMERAS The sales of digital cameras (in millions of units) in year t is given by the function

$$f(t) = 3.05t + 6.85 \qquad (0 \le t \le 3)$$

where $t = 0$ corresponds to 2001. Over that same period, the sales of film cameras (in millions of units) is given by

$$g(t) = -1.85t + 16.58 \qquad (0 \le t \le 3)$$

   a. Show that more film cameras than digital cameras were sold in 2001.

   b. When did the sales of digital cameras first exceed those of film cameras?

Source: Popular Science.

49. U.S. FINANCIAL TRANSACTIONS The percentage of U.S. transactions by check between the beginning of 2001 ($t = 0$) and the beginning of 2010 ($t = 9$) is approximated by

$$f(t) = -\frac{11}{9}t + 43 \qquad (0 \le t \le 9)$$

whereas the percentage of transactions done electronically during the same period is approximated by

$$g(t) = \frac{11}{3}t + 23 \qquad (0 \le t \le 9)$$

   a. Sketch the graphs of f and g on the same set of axes.

   b. Find the time when transactions done electronically first exceeded those done by check.

Source: Foreign Policy.

50. BROADBAND VERSUS DIAL-UP The number of U.S. broadband Internet households (in millions) between the beginning of 2004 ($t = 0$) and the beginning of 2008 ($t = 4$) is approximated by

$$f(t) = 6.5t + 33 \qquad (0 \le t \le 4)$$

Over the same period, the number of U.S. dial-up Internet households (in millions) is approximated by

$$g(t) = -3.9t + 42.5 \qquad (0 \le t \le 4)$$

   a. Sketch the graphs of f and g on the same set of axes.

   b. Solve the equation $f(t) = g(t)$, and interpret your result.

Source: Strategic Analytics, Inc.

In Exercises 51 and 52, determine whether the statement is true or false. If it is true, explain why it is true. If it is false, give an example to show why it is false.

51. If the book value V at the end of the year t of an asset being depreciated linearly is given by $V = -at + b$, where a and b are positive constants, then the rate of depreciation of the asset is a units per year.

52. Suppose $C(x) = cx + F$ and $R(x) = sx$ are the cost and revenue functions of a certain firm. Then the firm is operating at a break-even level of production if its level of production is $F/(s - c)$.

2.5   Solutions to Self-Check Exercises

1. Let x denote the number of units produced and sold. Then

   a. $C(x) = 10x + 60,000$

   b. $R(x) = 15x$

   c. $P(x) = R(x) - C(x) = 15x - (10x + 60,000)$
$$= 5x - 60,000$$

   d. $P(10,000) = 5(10,000) - 60,000$
$$= -10,000$$
   or a loss of $10,000 per month.

$$P(14,000) = 5(14,000) - 60,000$$
$$= 10,000$$
   or a profit of $10,000 per month.

2. The point of intersection of the two straight lines is found by solving the system of linear equations

$$2x + 3y = 6$$
$$x - 3y = 4$$

Solving the first equation for y in terms of x, we obtain

$$y = -\frac{2}{3}x + 2$$

Substituting this expression for y into the second equation, we obtain

$$x - 3\left(-\frac{2}{3}x + 2\right) = 4$$
$$x + 2x - 6 = 4$$
$$3x = 10$$

or $x = \frac{10}{3}$. Substituting this value of x into the expression for y obtained earlier, we find

$$y = -\frac{2}{3}\left(\frac{10}{3}\right) + 2 = -\frac{2}{9}$$

Therefore, the point of intersection is $\left(\frac{10}{3}, -\frac{2}{9}\right)$.

USING TECHNOLOGY

Linear Functions

Graphing Utility

A graphing utility can be used to find the value of a function f at a given point with minimal effort. However, to find the value of y for a given value of x in a linear equation such as $Ax + By + C = 0$, the equation must first be cast in the slope-intercept form $y = mx + b$, thus revealing the desired rule $f(x) = mx + b$ for y as a function of x.

EXAMPLE 1 Consider the equation $2x + 5y = 7$.

a. Plot the straight line with the given equation in the standard viewing window.
b. Find the value of y when $x = 2$ and verify your result by direct computation.
c. Find the value of y when $x = 1.732$.

Solution

a. The straight line with equation $2x + 5y = 7$ or, equivalently, $y = -\frac{2}{5}x + \frac{7}{5}$ in the standard viewing window is shown in Figure T1.

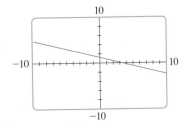

FIGURE **T1**
The straight line $2x + 5y = 7$ in the standard viewing window

b. Using the evaluation function of the graphing utility and the value of 2 for x, we find $y = 0.6$. This result is verified by computing

$$y = -\frac{2}{5}(2) + \frac{7}{5} = -\frac{4}{5} + \frac{7}{5} = \frac{3}{5} = 0.6$$

when $x = 2$.
c. Once again using the evaluation function of the graphing utility, this time with the value 1.732 for x, we find $y = 0.7072$.

⚠ When evaluating $f(x)$ at $x = a$, remember that the number a must lie between xMin and xMax.

(continued)

APPLIED EXAMPLE 2 Market for Cholesterol-Reducing Drugs In a study conducted in early 2000, experts projected a rise in the market for cholesterol-reducing drugs. The U.S. market (in billions of dollars) for such drugs from 1999 through 2004 is approximated by

$$M(t) = 1.95t + 12.19$$

where t is measured in years, with $t = 0$ corresponding to 1999.

a. Plot the graph of the function M in the viewing window $[0, 5] \times [0, 25]$.
b. What was the estimated market for cholesterol-reducing drugs in 2005 ($t = 6$)?
c. What was the rate of increase of the market for cholesterol-reducing drugs over the period in question?

Source: S. G. Cowen.

Solution

a. The graph of M is shown in Figure T2.
b. The market in 2005 for cholesterol-reducing drugs was approximately

$$M(6) = 1.95(6) + 12.19 = 23.89$$

or $23.89 billion.

c. The function M is linear; hence, we see that the rate of increase of the market for cholesterol-reducing drugs is given by the slope of the straight line represented by M, which is approximately $1.95 billion per year. ∎

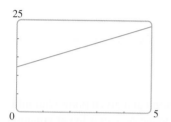

25
0   5
FIGURE **T2**
The graph of M in the viewing window
$[0, 5] \times [0, 25]$

Excel

Excel can be used to find the value of a function at a given value with minimal effort. However, to find the value of y for a given value of x in a linear equation such as $Ax + By + C = 0$, the equation must first be cast in the slope-intercept form $y = mx + b$, thus revealing the desired rule $f(x) = mx + b$ for y as a function of x.

EXAMPLE 3 Consider the equation $2x + 5y = 7$.

a. Find the value of y for $x = 0$, 5, and 10.
b. Plot the straight line with the given equation over the interval $[0, 10]$.

Solution

a. Since this is a linear equation, we first cast the equation in slope-intercept form:

$$y = -\frac{2}{5}x + \frac{7}{5}$$

Next, we create a table of values (Figure T3), following the same procedure outlined in Example 3, pages 87–88. In this case we use the formula `=(-2/5)*B1+7/5` for the y-values.

	A	B	C	D
1	x	0	5	10
2	y	1.4	-0.6	-2.6

FIGURE **T3**
Table of values for x and y

Note: Words/characters printed in a monospace font (for example, `=(-2/3)*A2+2`) indicate words/characters that need to be typed and entered.

b. Following the procedure outlined in Example 3, we obtain the graph shown in Figure T4.

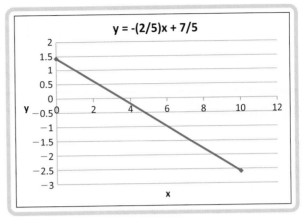

FIGURE **T4**
The graph of $y = -\frac{2}{5}x + \frac{7}{5}$ over the interval [0, 10]

APPLIED EXAMPLE 4 Market for Cholesterol-Reducing Drugs In a study conducted in early 2000, experts projected a rise in the market for cholesterol-reducing drugs. The U.S. market (in billions of dollars) for such drugs from 1999 through 2004 is approximated by

$$M(t) = 1.95t + 12.19$$

where t is measured in years, with $t = 0$ corresponding to 1999.

a. Plot the graph of the function M over the interval [0, 6].
b. What was the estimated market for cholesterol-reducing drugs in 2005 ($t = 6$)?
c. What was the rate of increase of the market for cholesterol-reducing drugs over the period in question?
Source: S. G. Cowen.

Solution

a. Following the instructions given in Example 3, pages 87–88, we obtain the spreadsheet and graph shown in Figure T5. [*Note:* We have made the appropriate entries for the title and x- and y-axis labels. In particular, for Primary Vertical Axis Title, select [Rotated Title] and type M(t) in billions of dollars].

	A	B	C	D
1	t	0	5	6
2	M(t)	12.19	21.94	23.89

(a)

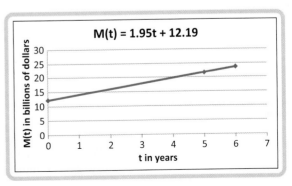

(b)

FIGURE **T5**
(a) The table of values for t and $M(t)$ and (b) the graph showing the demand for cholesterol-reducing drugs

(*continued*)

b. From the table of values, we see that

$$M(6) = 1.95(6) + 12.19 = 23.89$$

or $23.89 billion.

c. The function M is linear; hence, we see that the rate of increase of the market for cholesterol-reducing drugs is given by the slope of the straight line represented by M, which is approximately $1.95 billion per year.   ■

TECHNOLOGY EXERCISES

Find the value of *y* corresponding to the given value of *x*.

1. $3.1x + 2.4y - 12 = 0$; $x = 2.1$

2. $1.2x - 3.2y + 8.2 = 0$; $x = 1.2$

3. $2.8x + 4.2y = 16.3$; $x = 1.5$

4. $-1.8x + 3.2y - 6.3 = 0$; $x = -2.1$

5. $22.1x + 18.2y - 400 = 0$; $x = 12.1$

6. $17.1x - 24.31y - 512 = 0$; $x = -8.2$

7. $2.8x = 1.41y - 2.64$; $x = 0.3$

8. $0.8x = 3.2y - 4.3$; $x = -0.4$

2.6 Quadratic Functions

Quadratic Functions

A **quadratic function** is one of the form

$$f(x) = ax^2 + bx + c$$

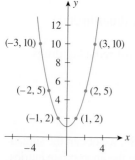

FIGURE **43**
The graph of $f(x) = x^2 + 1$ is a parabola.

where a, b, and c are constants and $a \neq 0$. For example, the function $f(x) = 2x^2 - 3x + 4$ is quadratic (here, $a = 2$, $b = -3$, and $c = 4$). Also, the function $f(x) = x^2 + 1$ of Example 6, Section 2.3, is quadratic (here, $a = 1$, $b = 0$, and $c = 1$). Its graph illustrates the general shape of the graph of a quadratic function, which is called a *parabola* (Figure 43).

In general, the graph of a quadratic function is a parabola that opens upward or downward (Figure 44). Furthermore, the parabola is symmetric with respect to a vertical line called the *axis of symmetry* (shown dashed in Figure 44). This line also passes through the lowest point or the highest point of the parabola. The point of intersection of the parabola with its axis of symmetry is called the *vertex* of the parabola.

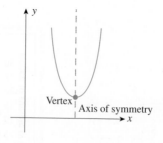

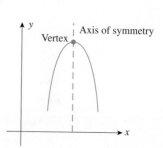

FIGURE **44**
Graphs of quadratic functions are parabolas.

We can use these properties to help us sketch the graph of a quadratic function. For example, suppose we want to sketch the graph of

$$f(x) = 2x^2 - 4x + 1$$

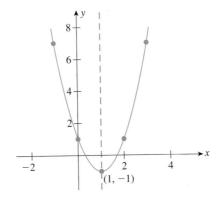

If we complete the square in x, we obtain

$$f(x) = 2(x^2 - 2x) + 1 \qquad \text{Factor out the coefficient of } x^2 \\ \text{from the first two terms.}$$

$$\text{Adding and subtracting 2} \\ \downarrow \qquad\qquad \downarrow$$

$$= 2\left[x^2 - 2x + (-1)^2\right] + 1 - 2 \qquad \begin{array}{l}\text{Because of the 2 outside the brackets, we} \\ \text{have added } 2(1) \text{ and must therefore} \\ \text{subtract 2.}\end{array}$$

$$= 2(x - 1)^2 - 1 \qquad \text{Factor the terms within the brackets.}$$

Observe that the first term, $2(x - 1)^2$, is nonnegative. In fact, it is equal to zero when $x = 1$ and is greater than zero if $x \neq 1$. Consequently, we see that $f(x) \geq -1$ for all values of x. This tells us that the vertex (in this case, the lowest point) of the parabola is the point $(1, -1)$. The axis of symmetry of the parabola is the vertical line $x = 1$. Finally, plotting the vertex and a few additional points on either side of the axes of symmetry of the parabola, we obtain the graph shown in Figure 45.

x	y
-1	7
0	1
2	1
3	7

FIGURE 45
The graph of $f(x) = 2x^2 - 4x + 1$

The x-intercepts of f, the x-coordinates of the points at which the parabola intersects the x-axis, can be found by solving the equation $f(x) = 0$. Here, we use the quadratic formula with $a = 2$, $b = -4$, and $c = 1$ to find that

$$x = \frac{-(-4) \pm \sqrt{(-4)^2 - 4(2)(1)}}{2(2)} = \frac{4 \pm \sqrt{8}}{4} = \frac{4 \pm 2\sqrt{2}}{4} = 1 \pm \frac{\sqrt{2}}{2}$$

Therefore, the x-intercepts are $1 + \sqrt{2}/2 \approx 1.71$ and $1 - \sqrt{2}/2 \approx 0.29$. The y-intercept of f (obtained by setting $x = 0$) is $f(0) = 1$.

The technique that we used to analyze $f(x) = 2x^2 - 4x + 1$ can be used to study the general quadratic function

$$f(x) = ax^2 + bx + c \qquad (a \neq 0)$$

If we complete the square in x, we find

$$f(x) = a\left(x + \frac{b}{2a}\right)^2 + \frac{4ac - b^2}{4a} \qquad \text{See Exercise 52.}$$

From this equation, we obtain the following properties of the quadratic function f.

Properties of the Quadratic Function
$$f(x) = ax^2 + bx + c \qquad (a \neq 0)$$

1. The domain of f is the set of all real numbers, and the graph of f is a parabola.

2. If $a > 0$, the parabola opens upward, and if $a < 0$, it opens downward.

(continued)

3. The vertex of the parabola is $\left(-\dfrac{b}{2a}, f\left(-\dfrac{b}{2a}\right)\right)$.

4. An equation of the axis of symmetry of the parabola is $x = -\dfrac{b}{2a}$.

5. The x-intercepts (if any) are found by solving $f(x) = 0$. The y-intercept is $f(0) = c$.

VIDEO   **EXAMPLE 1** Given the quadratic function

$$f(x) = -2x^2 + 5x - 2$$

a. Find the vertex of the parabola.
b. Find the x-intercepts (if any) of the parabola.
c. Sketch the parabola.

Solution

a. Comparing $f(x) = -2x^2 + 5x - 2$ with the general form of the quadratic equation, we find that $a = -2$, $b = 5$, and $c = -2$. Therefore, the x-coordinate of the vertex of the parabola is

$$-\frac{b}{2a} = -\frac{5}{2(-2)} = \frac{5}{4}$$

Next, to find the y-coordinate of the vertex, we evaluate f at $x = \frac{5}{4}$, obtaining

$$f\left(\frac{5}{4}\right) = -2\left(\frac{5}{4}\right)^2 + 5\left(\frac{5}{4}\right) - 2$$

$$= -\frac{25}{8} + \frac{25}{4} - 2 \qquad -2\left(\frac{25}{16}\right) = -\frac{25}{8}$$

$$= \frac{9}{8} \qquad -\frac{25}{8} + \frac{50}{8} - \frac{16}{8} = \frac{9}{8}$$

b. To find the x-intercepts of the parabola, we solve the equation

$$-2x^2 + 5x - 2 = 0$$

using the quadratic formula with $a = -2$, $b = 5$, and $c = -2$. We find

$$x = \frac{-5 \pm \sqrt{25 - 4(-2)(-2)}}{2(-2)} \qquad x = \frac{-b \pm \sqrt{b^2 - 4ac}}{2a}$$

$$= \frac{-5 \pm \sqrt{9}}{-4}$$

$$= \frac{-5 \pm 3}{-4}$$

$$= \frac{1}{2} \quad \text{or} \quad 2$$

Thus, the x-intercepts of the parabola are $\frac{1}{2}$ and 2.

c. Since $a = -2 < 0$, the parabola opens downward. The vertex of the parabola $\left(\frac{5}{4}, \frac{9}{8}\right)$ is therefore the highest point on the curve. The parabola crosses the x-axis at the points $\left(\frac{1}{2}, 0\right)$ and $(2, 0)$. Setting $x = 0$ gives -2 as the y-intercept of the curve. Finally, using this information, we sketch the parabola shown in Figure 46.

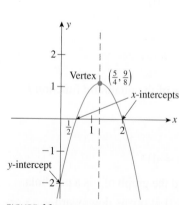

FIGURE 46
The graph of $f(x) = -2x^2 + 5x - 2$

APPLIED EXAMPLE 2 Effect of Advertising on Profit The quarterly profit (in thousands of dollars) of Cunningham Realty is given by

$$P(x) = -\frac{1}{3}x^2 + 7x + 30 \qquad (0 \le x \le 50)$$

where x (in thousands of dollars) is the amount of money Cunningham spends on advertising per quarter. Find the amount of money Cunningham should spend on advertising to realize a maximum quarterly profit. What is the maximum quarterly profit realizable by Cunningham?

Solution The profit function P is a quadratic function, so its graph is a parabola. Furthermore, the coefficient of x^2 is $a = -\frac{1}{3} < 0$, so the parabola opens downward. The x-coordinate of the vertex of the parabola is

$$-\frac{b}{2a} = -\frac{7}{2(-\frac{1}{3})} = \frac{21}{2} = 10.5 \qquad a = -\frac{1}{3} \text{ and } b = 7$$

The corresponding y-coordinate is

$$f\left(\frac{21}{2}\right) = -\frac{1}{3}\left(\frac{21}{2}\right)^2 + 7\left(\frac{21}{2}\right) + 30 = \frac{267}{4} = 66.75$$

Therefore, the vertex of the parabola is $(\frac{21}{2}, \frac{267}{4})$. Since the parabola opens downward, the vertex of the parabola is the highest point on the parabola. Accordingly, the y-coordinate of the vertex gives the maximum value of P. This implies that the maximum quarterly profit of \$66,750 [remember that $P(x)$ is measured in thousands of dollars] is realized if Cunningham spends \$10,500 per quarter on advertising. The graph of P is shown in Figure 47.

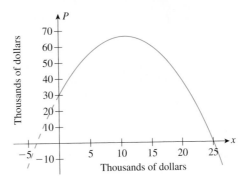

FIGURE **47**
The graph of the profit function
$P(x) = -\frac{1}{3}x^2 + 7x + 30$

Demand and Supply Curves

In a free-market economy, consumer demand for a particular commodity depends on the commodity's unit price. A *demand equation* expresses the relationship between the unit price and the quantity demanded. The graph of a demand equation is called a *demand curve*. In general, the quantity demanded of a commodity decreases as the commodity's unit price increases, and vice versa. Accordingly, a **demand function**, defined by $p = f(x)$, where p measures the unit price and x measures the number of units of the commodity in question, is generally characterized as a decreasing function of x; that is, $p = f(x)$ decreases as x increases. Since both x and p assume only nonnegative values, the demand curve is that part of the graph of $f(x)$ that lies in the first quadrant (Figure 48).

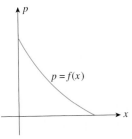

FIGURE **48**
A demand curve

Todd Kodet

TITLE   Senior Vice-President of Supply
INSTITUTION   Earthbound Farm

Earthbound Farm is America's largest grower of organic produce, offering more than 100 varieties of organic salads, vegetables, fruits, and herbs on 34,000 crop acres. As Senior Vice-President of Supply, I am responsible for getting our products into and out of Earthbound Farm. A major part of my work is scheduling plantings for upcoming seasons, matching projected supply to projected demand for any given day and season. I use applied mathematics in every step of my planning to create models for predicting supply and demand.

After the sales department provides me with information about projected demand, I take their estimates, along with historical data for expected yields, to determine how much of each organic product we need to plant. There are several factors that I have to think about when I make these determinations. For example, I not only have to consider gross yield per acre of farmland, but also have to calculate aver-

age trimming waste per acre, to arrive at net pounds needed per customer.

Some of the other variables I consider are the amount of organic land available, the location of the farms, seasonal information (because days to maturity for each of our crops varies greatly depending on the weather), and historical information relating to weeds, pests, and diseases.

I emphasize the importance of understanding the mathematics that drives our business plans when I work with my team to analyze the reports they have generated. They need to recognize when the information they have gathered does not make sense so that they can spot errors that could skew our projections. With a sound understanding of mathematics, we are able to create more accurate predictions to help us meet our company's goals.

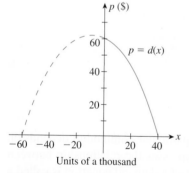

 APPLIED EXAMPLE 3 Demand for Bluetooth Headsets The demand function for a certain brand of bluetooth wireless headsets is given by

$$p = d(x) = -0.025x^2 - 0.5x + 60$$

where p is the wholesale unit price in dollars and x is the quantity demanded each month, measured in units of a thousand. Sketch the corresponding demand curve. Above what price will there be no demand? What is the maximum quantity demanded per month?

Solution   The given function is quadratic, and its graph may be sketched using the methods just developed (Figure 49). The p-intercept, 60, gives the wholesale unit price above which there will be no demand. To obtain the maximum quantity demanded, set $p = 0$, which gives

$$-0.025x^2 - 0.5x + 60 = 0$$
$$x^2 + 20x - 2400 = 0 \qquad \text{Upon multiplying both sides of the equation by } -40$$
$$(x + 60)(x - 40) = 0$$

That is, $x = -60$ or $x = 40$. Since x must be nonnegative, we reject the root $x = -60$. Thus, the maximum number of headsets demanded per month is 40,000. ∎

In a competitive market, a relationship also exists between the unit price of a commodity and the commodity's availability in the market. In general, an increase in the commodity's unit price induces the producer to increase the supply of the commodity. Conversely, a decrease in the unit price generally leads to a drop in the supply. The equation that expresses the relation between the unit price and the quantity supplied is called a *supply equation*, and its graph is called a *supply curve*. A **supply function**, defined by $p = f(x)$, is generally characterized as an increasing function of

FIGURE 49
The demand curve $p = d(x)$

Units of a thousand

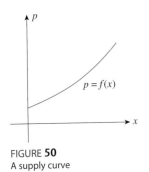

FIGURE 50
A supply curve

x; that is, $p = f(x)$ increases as x increases. Since both x and p assume only nonnegative values, the supply curve is the part of the graph of $f(x)$ that lies in the first quadrant (Figure 50).

APPLIED EXAMPLE 4 Supply of Bluetooth Headsets The supply function for a certain brand of bluetooth wireless headsets is given by

$$p = s(x) = 0.02x^2 + 0.6x + 20$$

where p is the unit wholesale price in dollars and x stands for the quantity in units of a thousand) that will be made available in the market by the supplier. Sketch the corresponding supply curve. What is the lowest price at which the supplier will make the headsets available in the market?

Solution A sketch of the supply curve appears in Figure 51. The p-intercept, 20, gives the lowest price at which the supplier will make the headsets available in the market.

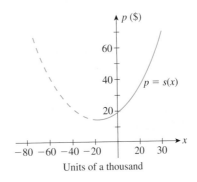

FIGURE 51
The supply curve $p = s(x)$

Market Equilibrium

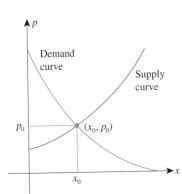

FIGURE 52
Market equilibrium corresponds to (x_0, p_0), the point at which the supply and demand curves intersect.

Under pure competition, the price of a commodity will eventually settle at a level dictated by the following condition: The supply of the commodity will be equal to the demand for it. If the price is too high, the consumer will not buy, and if the price is too low, the supplier will not produce. **Market equilibrium** prevails when the quantity produced is equal to the quantity demanded. The quantity produced at market equilibrium is called the *equilibrium quantity,* and the corresponding price is called the *equilibrium price.*

Market equilibrium corresponds to the point at which the demand curve and the supply curve intersect. In Figure 52, x_0 represents the equilibrium quantity, and p_0 represents the equilibrium price. The point (x_0, p_0) lies on the supply curve and therefore satisfies the supply equation. At the same time, it also lies on the demand curve and therefore satisfies the demand equation. Thus, to find the point (x_0, p_0), and hence the equilibrium quantity and price, we solve the demand and supply equations simultaneously for x and p. For meaningful solutions, x and p must both be positive.

APPLIED EXAMPLE 5 Market Equilibrium Refer to Examples 3 and 4. The demand function for a certain brand of bluetooth wireless headsets is given by

$$p = d(x) = -0.025x^2 - 0.5x + 60$$

and the corresponding supply function is given by

$$p = s(x) = 0.02x^2 + 0.6x + 20$$

where p is expressed in dollars and x is measured in units of a thousand. Find the equilibrium quantity and price.

Solution   We solve the following system of equations:

$$p = -0.025x^2 - 0.5x + 60$$
$$p = 0.02x^2 + 0.6x + 20$$

Substituting the first equation into the second yields

$$-0.025x^2 - 0.5x + 60 = 0.02x^2 + 0.6x + 20$$

which is equivalent to

$$0.045x^2 + 1.1x - 40 = 0$$
$$45x^2 + 1100x - 40{,}000 = 0 \qquad \text{Multiply by 1000.}$$
$$9x^2 + 220x - 8000 = 0 \qquad \text{Divide by 5.}$$
$$(9x + 400)(x - 20) = 0$$

Thus, $x = -\frac{400}{9}$ or $x = 20$. Since x must be nonnegative, the root $x = -\frac{400}{9}$ is rejected. Therefore, the equilibrium quantity is 20,000 headsets. The equilibrium price is given by

$$p = 0.02(20)^2 + 0.6(20) + 20 = 40$$

or $40 per headset (Figure 53).

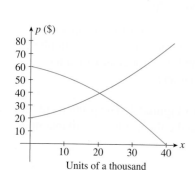

FIGURE **53**
The supply curve and the demand curve intersect at the point (20, 40).

Given the quadratic function

$$f(x) = 2x^2 - 3x - 3$$

1. Find the vertex of the parabola.

2. Find the x-intercepts (if any) of the parabola.
3. Sketch the parabola.

Solutions to Self-Check Exercises 2.6 can be found on page 135.

---

1. Consider the quadratic function
   $f(x) = ax^2 + bx + c \ (a \neq 0)$.
   a. What is the domain of f?
   b. What can you say about the parabola if $a > 0$?
   c. What is the vertex of the parabola in terms of a and b?
   d. What is the axis of symmetry of the parabola?

2. a. What is a demand function? A supply function?
   b. What is market equilibrium?
   c. What are the equilibrium quantity and equilibrium price? How do you determine these quantities?

In Exercises 1–18, find the vertex, the x-intercepts (if any), and sketch the parabola.

1. $f(x) = x^2 + x - 6$

2. $f(x) = 3x^2 - 5x - 2$

3. $f(x) = x^2 - 4x + 4$

4. $f(x) = x^2 + 6x + 9$

5. $f(x) = -x^2 + 5x - 6$

6. $f(x) = -4x^2 + 4x + 3$

7. $f(x) = 3x^2 - 5x + 1$

8. $f(x) = -2x^2 + 6x - 3$

9. $f(x) = 2x^2 - 3x + 3$

10. $f(x) = 3x^2 - 4x + 2$

11. $f(x) = x^2 - 4$

12. $f(x) = 2x^2 + 3$

13. $f(x) = 16 - x^2$

14. $f(x) = 5 - x^2$

15. $f(x) = \dfrac{3}{8}x^2 - 2x + 2$

16. $f(x) = \dfrac{3}{4}x^2 - \dfrac{1}{2}x + 1$

17. $f(x) = 1.2x^2 + 3.2x - 1.2$

18. $f(x) = 2.3x^2 - 4.1x + 3$

In Exercises 19–24, find the points of intersection of the graphs of the functions.

19. $f(x) = -x^2 + 4$; $g(x) = x - 2$

20. $f(x) = x^2 - 5x + 6$; $g(x) = \dfrac{1}{2}x + \dfrac{3}{2}$

21. $f(x) = -x^2 + 2x + 6$; $g(x) = x^2 - 6$

22. $f(x) = x^2 - 2x - 2$; $g(x) = -x^2 - x + 1$

23. $f(x) = 2x^2 - 5x - 8$; $g(x) = -3x^2 + x + 5$

24. $f(x) = 0.2x^2 - 1.2x - 4$; $g(x) = -0.3x^2 + 0.7x + 8.2$

For the demand equations in Exercises 25 and 26, where x represents the quantity demanded in units of a thousand and p is the unit price in dollars, (a) sketch the demand curve and (b) determine the quantity demanded when the unit price is set at $\$p$.

25. $p = -x^2 + 36$; $p = 11$ **26.** $p = -x^2 + 16$; $p = 7$

For the supply equations in Exercises 27 and 28, where x is the quantity supplied in units of a thousand and p is the unit price in dollars, (a) sketch the supply curve and (b) determine the price at which the supplier will make 2000 units of the commodity available in the market.

27. $p = 2x^2 + 18$ **28.** $p = x^2 + 16x + 40$

In Exercises 29–32, for each pair of supply and demand equations, where x represents the quantity demanded in units of a thousand and p the unit price in dollars, find the equilibrium quantity and the equilibrium price.

29. $p = -2x^2 + 80$ and $p = 15x + 30$

30. $p = -x^2 - 2x + 100$ and $p = 8x + 25$

31. $11p + 3x - 66 = 0$ and $2p^2 + p - x = 10$

32. $p = 60 - 2x^2$ and $p = x^2 + 9x + 30$

33. CANCER SURVIVORS The number of living Americans who have had a cancer diagnosis has increased drastically since 1971. In part, this is due to more testing for cancer and better treatment for some cancers. In part, it is because the population is older, and cancer is largely a disease of the elderly. The number of cancer survivors (in millions) between 1975 ($t = 0$) and 2000 ($t = 25$) is approximately

$$N(t) = 0.0031t^2 + 0.16t + 3.6 \qquad (0 \le t \le 25)$$

a. How many living Americans had a cancer diagnosis in 1975? In 2000?

b. Assuming the trend continued, how many cancer survivors were there in 2005?

Source: National Cancer Institute.

34. PREVALENCE OF ALZHEIMER'S PATIENTS Based on a study conducted in 1997, the percent of the U.S. population by age afflicted with Alzheimer's disease is given by the function

$$P(x) = 0.0726x^2 + 0.7902x + 4.9623 \qquad (0 \le x \le 25)$$

where x is measured in years, with $x = 0$ corresponding to age 65. What percent of the U.S. population at age 65 is expected to have Alzheimer's disease? At age 90?

Source: Alzheimer's Association.

35. MOTION OF A STONE A stone is thrown straight up from the roof of an 80-ft building. The distance of the stone from the ground at any time t (in seconds) is given by

$$h(t) = -16t^2 + 64t + 80$$

a. Sketch the graph of h.

b. At what time does the stone reach the highest point? What is the stone's maximum height from the ground?

36. MAXIMIZING PROFIT Lynbrook West, an apartment complex, has 100 two-bedroom units. The monthly profit realized from renting out x apartments is given by

$$P(x) = -10x^2 + 1760x - 50{,}000$$

dollars. How many units should be rented out to maximize the monthly rental profit? What is the maximum monthly profit realizable?

37. MAXIMIZING PROFIT The estimated monthly profit realizable by the Cannon Precision Instruments Corporation for manufacturing and selling x units of its model M1 cameras is

$$P(x) = -0.04x^2 + 240x - 10{,}000$$

dollars. Determine how many cameras Cannon should produce per month to maximize its profits.

38. EFFECT OF ADVERTISING ON PROFIT The relationship between Northwood Realty's quarterly profit, $P(x)$, and the amount of money x spent on advertising per quarter is described by the function

$$P(x) = -\frac{1}{8}x^2 + 7x + 30 \qquad (0 \le x \le 50)$$

where both $P(x)$ and x are measured in thousands of dollars.

a. Sketch the graph of P.

b. Find the amount of money the company should spend on advertising per quarter to maximize its quarterly profits.

39. MAXIMIZING REVENUE The monthly revenue R (in hundreds of dollars) realized in the sale of Royal electric shavers is related to the unit price p (in dollars) by the equation

$$R(p) = -\frac{1}{2}p^2 + 30p$$

a. Sketch the graph of R.

b. At what unit price is the monthly revenue maximized?

40. SUPPLY FUNCTIONS The supply function for the Luminar daylight LED desk lamp is given by

$$p = 0.1x^2 + 0.5x + 15$$

where x is the quantity supplied (in thousands) and p is the unit price in dollars. Sketch the graph of the supply function. What unit price will induce the supplier to make 5000 lamps available in the marketplace?

41. Market Equilibrium The weekly demand and supply functions for Sportsman 5×7 tents are given by

$$p = -0.1x^2 - x + 40$$
$$p = 0.1x^2 + 2x + 20$$

respectively, where p is measured in dollars and x is measured in units of a hundred. Find the equilibrium quantity and price.

42. Market Equilibrium The management of the Titan Tire Company has determined that the weekly demand and supply functions for their Super Titan tires are given by

$$p = 144 - x^2$$
$$p = 48 + \frac{1}{2}x^2$$

respectively, where p is measured in dollars and x is measured in units of a thousand. Find the equilibrium quantity and price.

43. Poiseuille's Law According to a law discovered by the nineteenth century physician Poiseuille, the velocity (in centimeters/second) of blood r cm from the central axis of an artery is given by

$$v(r) = k(R^2 - r^2)$$

where k is a constant and R is the radius of the artery. Suppose that for a certain artery, $k = 1000$ and $R = 0.2$ so that $v(r) = 1000(0.04 - r^2)$.
a. Sketch the graph of v.
b. For what value of r is $v(r)$ largest? Smallest? Interpret your results.

44. Motion of a Ball A ball is thrown straight upward from the ground and attains a height of $s(t) = -16t^2 + 128t + 4$ ft above the ground after t sec. When does the ball reach the maximum height? What is the maximum height?

45. Designing a Norman Window A Norman window has the shape of a rectangle surmounted by a semicircle (see the accompanying figure). If a Norman window is to have a perimeter of 28 ft, what should be its dimensions in order to allow the maximum amount of light through the window?

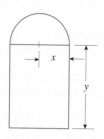

46. Distance of Water Flow A cylindrical tank of height h ft is filled to the top with water. If a hole is punched into the lateral side of the tank, the stream of water flowing out of the tank will reach the ground at a distance of x ft from the base of the tank where $x = 2\sqrt{y(h - y)}$ (see the accompanying figure). Find the location of the hole so that x is a maximum. What is this maximum value of x?
Hint: It suffices to maximize the expression for x^2. (Why?)

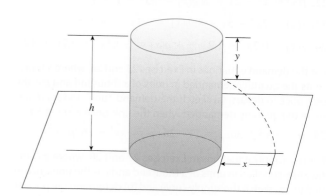

In Exercises 47–51, determine whether the statement is true or false. If it is true, explain why it is true. If it is false, explain why or give an example to show why it is false.

47. If $f(x) = ax^2 + bx + c \ (a \neq 0)$, then

$$f\left(\frac{-b + \sqrt{b^2 - 4ac}}{2a}\right) = 0$$

48. The quadratic function $f(x) = ax^2 + bx + c \ (a \neq 0)$ has no x-intercepts if $b^2 - 4ac > 0$.

49. If a and c have opposite signs, then the parabola with equation $y = ax^2 + bx + c$ intersects the x-axis at two distinct points.

50. If $b^2 = 4ac$, then the graph of the quadratic function $f(x) = ax^2 + bx + c \ (a \neq 0)$ touches the x-axis at exactly one point.

51. If a profit function is given by $P(x) = ax^2 + bx + c$, where x is the number of units produced and sold, then the level of production that yields the maximum profit is $-\dfrac{b}{2a}$ units.

52. Let $f(x) = ax^2 + bx + c \ (a \neq 0)$. By completing the square in x, show that

$$f(x) = a\left(x + \frac{b}{2a}\right)^2 + \frac{4ac - b^2}{4a}$$

2.6 Solutions to Self-Check Exercises

1. Here, $a = 2$, $b = -3$, and $c = -3$. The x-coordinate of the vertex is

$$-\frac{b}{2a} = -\frac{(-3)}{2(2)} = \frac{3}{4}$$

The corresponding y-coordinate is

$$f\left(\frac{3}{4}\right) = 2\left(\frac{3}{4}\right)^2 - 3\left(\frac{3}{4}\right) - 3 = \frac{9}{8} - \frac{9}{4} - 3 = -\frac{33}{8}$$

Therefore, the vertex of the parabola is $\left(\frac{3}{4}, -\frac{33}{8}\right)$.

2. Solving the equation $2x^2 - 3x - 3 = 0$, we find

$$x = \frac{-(-3) \pm \sqrt{(-3)^2 - 4(2)(-3)}}{2(2)} = \frac{3 \pm \sqrt{33}}{4}$$

So the x-intercepts are $\dfrac{3}{4} - \dfrac{\sqrt{33}}{4} \approx -0.7$ and

$$\frac{3}{4} + \frac{\sqrt{33}}{4} \approx 2.2.$$

3. Since $a = 2 > 0$, the parabola opens upward. The y-intercept is -3. The graph of the parabola is shown in the accompanying figure.

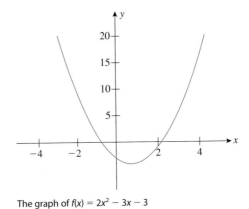

The graph of $f(x) = 2x^2 - 3x - 3$

USING TECHNOLOGY

Finding the Points of Intersection of Two Graphs

A graphing utility can be used to find the point(s) of intersection of the graphs of two functions.

EXAMPLE 1 Find the points of intersection of the graphs of

$$f(x) = 0.3x^2 - 1.4x - 3 \quad \text{and} \quad g(x) = -0.4x^2 + 0.8x + 6.4$$

Solution The graphs of both f and g in the standard viewing window are shown in Figure T1a. Using the function for finding the points of intersection of two graphs on a graphing utility, we find the point(s) of intersection, accurate to four decimal places, to be $(-2.4158, 2.1329)$ (Figure T1b) and $(5.5587, -1.5125)$ (Figure T1c). To access this function on the TI-83/84, select **5: intersect** on the Calc menu.

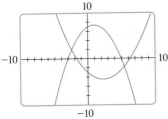

(a)

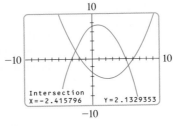

(b)

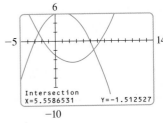

(c)

FIGURE T1
(a) The graphs of f and g in the standard viewing window; (b) and (c) the TI-83/84 intersection screens

(continued)

EXAMPLE 2 Consider the demand and supply functions

$$p = d(x) = -0.01x^2 - 0.2x + 8 \quad \text{and} \quad p = s(x) = 0.01x^2 + 0.1x + 3$$

a. Plot the graphs of d and s in the viewing window $[0, 15] \times [0, 10]$.
b. Verify that the equilibrium point is $(10, 5)$.

Solution

a. The graphs of d and s are shown in Figure T2a.

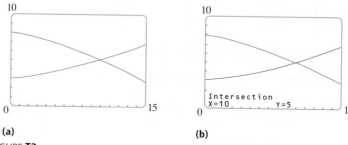

(a)   **(b)**
FIGURE **T2**
(a) The graphs of d and s in the window $[0, 15] \times [0, 10]$; (b) the TI-83/84 intersection screen

b. Using the function for finding the point of intersection of two graphs, we see that $x = 10$ and $y = 5$ (Figure T2b), so the equilibrium point is $(10, 5)$. ∎

TECHNOLOGY EXERCISES

In Exercises 1–6, find the points of intersection of the graphs of the functions. Express your answer accurate to four decimal places.

1. $f(x) = 1.2x + 3.8$; $g(x) = -0.4x^2 + 1.2x + 7.5$

2. $f(x) = 0.2x^2 - 1.3x - 3$; $g(x) = -1.3x + 2.8$

3. $f(x) = 0.3x^2 - 1.7x - 3.2$; $g(x) = -0.4x^2 + 0.9x + 6.7$

4. $f(x) = -0.3x^2 + 0.6x + 3.2$; $g(x) = 0.2x^2 - 1.2x - 4.8$

5. $f(x) = -1.8x^2 + 2.1x - 2$; $g(x) = 2.1x - 4.2$

6. $f(x) = 1.2x^2 - 1.2x + 2$; $g(x) = -0.2x^2 + 0.8x + 2.1$

7. MARKET EQUILIBRIUM The monthly demand and supply functions for a certain brand of wall clocks are given by

$$p = -0.2x^2 - 1.2x + 50$$
$$p = 0.1x^2 + 3.2x + 25$$

respectively, where p is measured in dollars and x is measured in units of a hundred.
a. Plot the graphs of both functions in an appropriate viewing window.
b. Find the equilibrium quantity and price.

8. MARKET EQUILIBRIUM The quantity demanded x (in units of a hundred) of Mikado miniature cameras per week is related to the unit price p (in dollars) by

$$p = -0.2x^2 + 80$$

The quantity x (in units of a hundred) that the supplier is willing to make available in the market is related to the unit price p (in dollars) by

$$p = 0.1x^2 + x + 40$$

a. Plot the graphs of both functions in an appropriate viewing window.
b. Find the equilibrium quantity and price.

2.7 Functions and Mathematical Models

Mathematical Models

One of the fundamental goals in this book is to show how mathematics and, in particular, calculus can be used to solve real-world problems such as those arising from the world of business and the social, life, and physical sciences. You have already seen some of these problems earlier. Here are a few more examples of real-world phenomena that we will analyze in this and ensuing chapters.

- Global warming (page 138)
- The solvency of the U.S. Social Security trust fund (page 139)
- The total cost of the health-care bill (page 146)
- Pharmaceutical theft (page 653)
- The Case-Shiller Home Price Index (page 707)
- Mexico's hedging tactic (page 823)

Regardless of the field from which the real-world problem is drawn, the problem is analyzed by using a process called **mathematical modeling.** The four steps in this process are illustrated in Figure 54.

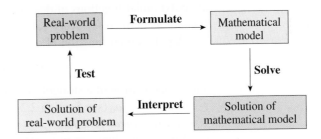

FIGURE **54**

1. **Formulate** Given a real-world problem, our first task is to formulate the problem, using the language of mathematics. The many techniques used in constructing mathematical models range from theoretical consideration of the problem on the one extreme to an interpretation of data associated with the problem on the other. For example, the mathematical model giving the accumulated amount at any time when a certain sum of money is deposited in the bank can be derived theoretically (see Chapter 4). On the other hand, many of the mathematical models in this book are constructed by studying the data associated with the problem (see Using Technology, pages 149–152). In calculus, we are primarily concerned with how one (dependent) variable depends on one or more (independent) variables. Consequently, most of our mathematical models will involve functions of one or more variables or equations defining these functions (implicitly).
2. **Solve** Once a mathematical model has been constructed, we can use the appropriate mathematical techniques, which we will develop throughout the book, to solve the problem.
3. **Interpret** Bearing in mind that the solution obtained in Step 2 is just the solution of the mathematical model, we need to interpret these results in the context of the original real-world problem.
4. **Test** Some mathematical models of real-world applications describe the situations with complete accuracy. For example, the model describing a deposit in a bank account gives the exact accumulated amount in the account at any time. But other mathematical models give, at best, an approximate description of the real-world problem. In this case, we need to test the accuracy of the model by observing how well it describes the original real-world problem and how well it predicts past

and/or future behavior. If the results are unsatisfactory, then we may have to reconsider the assumptions made in the construction of the model or, in the worst case, return to Step 1.

Many real-world phenomena, including those mentioned at the beginning of this section, are modeled by an appropriate function.

In what follows, we will recall some familiar functions and give examples of real-world phenomena that are modeled by using these functions.

Polynomial Functions

A **polynomial function** of degree n is a function of the form

$$f(x) = a_n x^n + a_{n-1} x^{n-1} + \cdots + a_2 x^2 + a_1 x + a_0 \qquad (a_n \neq 0)$$

where n is a nonnegative integer and the numbers $a_0, a_1, \ldots, a_n$ are constants, called the **coefficients** of the polynomial function. For example, the functions

$$f(x) = 2x^5 - 3x^4 + \frac{1}{2}x^3 + \sqrt{2}x^2 - 6$$

$$g(x) = 0.001x^3 - 0.2x^2 + 10x + 200$$

are polynomial functions of degrees 5 and 3, respectively. Observe that a polynomial function is defined for every value of x and so its domain is $(-\infty, \infty)$.

A polynomial function of degree 1 ($n = 1$) has the form

$$y = f(x) = a_1 x + a_0 \qquad (a_1 \neq 0)$$

and is an equation of a straight line in the slope-intercept form with slope $m = a_1$ and y-intercept $b = a_0$ (see Section 2.2). For this reason, a polynomial function of degree 1 is called a *linear function*.

Linear functions are used extensively in mathematical modeling for two important reasons. First, some models are *linear* by nature. For example, the formula for converting temperature from Celsius (°C) to Fahrenheit (°F) is $F = \frac{9}{5}C + 32$, and F is a linear function of C. Second, some natural phenomena exhibit linear characteristics over a small range of values and can therefore be modeled by a linear function restricted to a small interval.

A polynomial function of degree 2 has the form

$$y = f(x) = a_2 x^2 + a_1 x + a_0 \qquad (a_2 \neq 0)$$

or, more simply, $y = ax^2 + bx + c$, and is called a *quadratic function*.

Quadratic functions serve as mathematical models for many phenomena, as Example 1 shows.

APPLIED EXAMPLE 1   Global Warming The increase in carbon dioxide (CO_2) in the atmosphere is a major cause of global warming. The Keeling curve, named after Charles David Keeling, a professor at Scripps Institution of Oceanography, gives the average amount of CO_2, measured in parts per million volume (ppmv), in the atmosphere from 1958 through 2010. Even though data were available for every year in this time interval, we'll construct the curve based only on the following randomly selected data points.

Year	1958	1970	1974	1978	1985	1991	1998	2003	2007	2010
Amount	315	325	330	335	345	355	365	375	380	390

The **scatter plot** associated with these data is shown in Figure 55a. A mathematical model giving the approximate amount of CO_2 in the atmosphere during this period is given by

$$A(t) = 0.012313t^2 + 0.7545t + 313.9 \qquad (1 \le t \le 53)$$

where t is measured in years, with $t = 1$ corresponding to 1958. The graph of A is shown in Figure 55b.

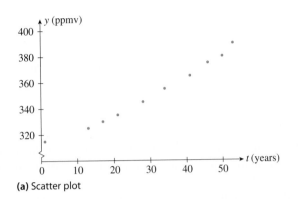

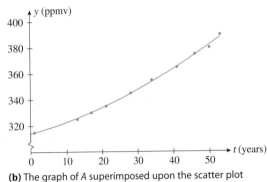

FIGURE 55 (a) Scatter plot (b) The graph of A superimposed upon the scatter plot

a. Use the model to estimate the average amount of atmospheric CO_2 in 1980 ($t = 23$).
b. Assume that the trend continued, and use the model to predict the average amount of atmospheric CO_2 in 2013.
Source: Scripps Institution of Oceanography.

Solution

a. The average amount of atmospheric carbon dioxide in 1980 is given by

$$A(23) = 0.012313(23)^2 + 0.7545(23) + 313.9 \approx 337.77$$

or approximately 338 ppmv.
b. Assuming that the trend continued, the average amount of atmospheric CO_2 in 2013 will be

$$A(56) = 0.012313(56)^2 + 0.7545(56) + 313.9 \approx 394.77$$

or approximately 395 ppmv. ∎

The next example uses a polynomial of degree 4 to help us construct a model that describes the projected assets of the Social Security trust fund.

APPLIED EXAMPLE 2 Social Security Trust Fund Assets The projected assets of the Social Security trust fund (in trillions of dollars) from 2010 through 2037 are given in the following table.

Year	2010	2015	2020	2025	2030	2035	2037
Assets	2.69	3.56	4.22	4.24	3.24	0.87	0

The scatter plot associated with these data are shown in Figure 56a, where $t = 0$ corresponds to 2010. A mathematical model giving the approximate value of the assets in the trust fund $A(t)$, in trillions of dollars in year t is

$$A(t) = 0.0000263t^4 - 0.0017501t^3 + 0.0206t^2 + 0.0999t + 2.7 \qquad (0 \le t \le 27)$$

The graph of $A(t)$ is shown in Figure 56b. (You will be asked to construct this model in Exercise 14, Using Technology Exercises 2.7.)

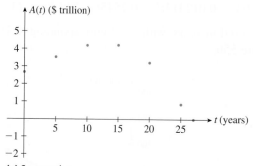

FIGURE 56

(a) Scatter plot

(b) The graph of A together with the scatter plot

a. The first baby boomers will turn 65 in 2011. What will be the assets of the Social Security system trust fund at that time? The last of the baby boomers will turn 65 in 2029. What will the assets of the trust fund be at that time?

b. Unless payroll taxes are increased significantly and/or benefits are scaled back dramatically, it is a matter of time before the assets of the current system are depleted. Use the graph of the function $A(t)$ to estimate the year in which the current Social Security system is projected to go broke.

Source: Social Security Administration.

Solution

a. The assets of the Social Security trust fund in 2011 ($t = 1$) will be

$$A(1) = 0.0000263(1)^4 - 0.0017501(1)^3 + 0.0206(1)^2 + 0.0999(1) + 2.7 \approx 2.82$$

or approximately $2.82 trillion. The assets of the trust fund in 2029 ($t = 19$) will be

$$A(19) = 0.0000263(19)^4 - 0.0017501(19)^3 + 0.0206(19)^2$$
$$+ 0.0999(19) + 2.7 \approx 3.46$$

or approximately $3.46 trillion.

b. From Figure 56b, we see that the graph of A crosses the t-axis at approximately $t = 27$. So unless the current system is changed, it is projected to go broke in 2037. (At this time, the first of the baby boomers will be 91, and the last of the baby boomers will be 73.)

Rational and Power Functions

Another important class of functions is rational functions. A **rational function** is simply the quotient of two polynomials. Examples of rational functions are

$$F(x) = \frac{3x^3 + x^2 - x + 1}{x - 2}$$

$$G(x) = \frac{x^2 + 1}{x^2 - 1}$$

In general, a rational function has the form

$$R(x) = \frac{f(x)}{g(x)}$$

where $f(x)$ and $g(x)$ are polynomial functions. Since division by zero is not allowed, we conclude that the domain of a rational function is the set of all real numbers except the zeros of g—that is, the roots of the equation $g(x) = 0$. Thus, the domain of the function F is the set of all numbers except $x = 2$, whereas the domain of the function G is the set of all numbers except those that satisfy $x^2 - 1 = 0$, or $x = \pm 1$.

Functions of the form

$$f(x) = x^r$$

where r is any real number, are called **power functions**. We encountered examples of power functions earlier in our work. For example, the functions

$$f(x) = \sqrt{x} = x^{1/2} \quad \text{and} \quad g(x) = \frac{1}{x^2} = x^{-2}$$

are power functions.

Many of the functions that we encounter later will involve combinations of the functions introduced here. For example, the following functions may be viewed as combinations of such functions:

$$f(x) = \sqrt{\frac{1 - x^2}{1 + x^2}}$$

$$g(x) = \sqrt{x^2 - 3x + 4}$$

$$h(x) = (1 + 2x)^{1/2} + \frac{1}{(x^2 + 2)^{3/2}}$$

As with polynomials of degree 3 or greater, analyzing the properties of these functions is facilitated by using the tools of calculus, to be developed later.

In the next example, we use a power function to construct a model that describes the driving costs of a car.

APPLIED EXAMPLE 3 Driving Costs A study of driving costs based on a 2008 medium-sized sedan found the following average costs (car payments, gas, insurance, upkeep, and depreciation), measured in cents per mile.

Miles/year	5000	10,000	15,000	20,000
Cost/mile, y (¢)	147.52	71.9	55.2	46.9

A mathematical model (using least-squares techniques) giving the average cost in cents per mile is

$$C(x) = \frac{1735.2}{x^{1.72}} + 38.6$$

where x (in thousands) denotes the number of miles the car is driven in each year. The scatter plot associated with these data and the graph of C are shown in Figure 57. Using this model, estimate the average cost of driving a 2008 medium-sized sedan 8000 miles per year and 18,000 miles per year.

Source: American Automobile Association.

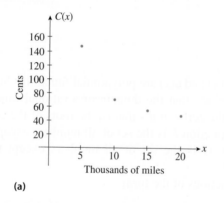

(a)

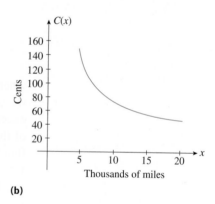

(b)

FIGURE **57**
(a) The scatter plot and (b) the graph of the model for driving costs

Solution   The average cost for driving a car 8000 miles per year is

$$C(8) = \frac{1735.2}{8^{1.72}} + 38.6 \approx 87.1$$

or approximately 87.1¢/mile. The average cost for driving it 18,000 miles per year is

$$C(18) = \frac{1735.2}{18^{1.72}} + 38.6 \approx 50.6$$

or approximately 50.6¢/mile.

Constructing Mathematical Models

We close this section by showing how some mathematical models can be constructed by using elementary geometric and algebraic arguments.

The following guidelines can be used to construct mathematical models.

Guidelines for Constructing Mathematical Models

1. Assign a letter to each variable mentioned in the problem. If appropriate, draw and label a figure.
2. Find an expression for the quantity sought.
3. Use the conditions given in the problem to write the quantity sought as a function *f* of one variable. Note any restrictions to be placed on the domain of *f* from physical considerations of the problem.

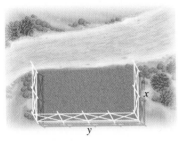

FIGURE 58
The rectangular grazing land has width *x* and length *y*.

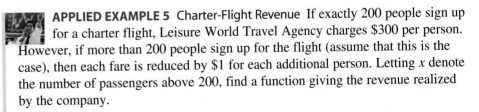

APPLIED EXAMPLE 4 Enclosing an Area The owner of Rancho Los Feliz has 3000 yards of fencing with which to enclose a rectangular piece of grazing land along the straight portion of a river. Fencing is not required along the river. Letting *x* denote the width of the rectangle, find a function *f* in the variable *x* giving the area of the grazing land if she uses all of the fencing (Figure 58).

Solution

1. This information was given.
2. The area of the rectangular grazing land is $A = xy$. Next, observe that the amount of fencing is $2x + y$ and this must be equal to 3000, since all the fencing is used; that is,

$$2x + y = 3000$$

3. From the equation, we see that $y = 3000 - 2x$. Substituting this value of y into the expression for A gives

$$A = xy = x(3000 - 2x) = 3000x - 2x^2$$

Finally, observe that both x and y must be nonnegative, since they represent the width and length of a rectangle, respectively. Thus, $x \geq 0$ and $y \geq 0$. But the latter is equivalent to $3000 - 2x \geq 0$, or $x \leq 1500$. So the required function is $f(x) = 3000x - 2x^2$ with domain $0 \leq x \leq 1500$. ■

Note Observe that if we view the function $f(x) = 3000x - 2x^2$ strictly as a mathematical entity, then its domain is the set of all real numbers. But physical considerations dictate that its domain should be restricted to the interval [0, 1500]. ■

APPLIED EXAMPLE 5 Charter-Flight Revenue If exactly 200 people sign up for a charter flight, Leisure World Travel Agency charges $300 per person. However, if more than 200 people sign up for the flight (assume that this is the case), then each fare is reduced by $1 for each additional person. Letting *x* denote the number of passengers above 200, find a function giving the revenue realized by the company.

Solution

1. This information was given.
2. If there are x passengers above 200, then the number of passengers signing up for the flight is $200 + x$. Furthermore, the fare will be $(300 - x)$ dollars per passenger.
3. The revenue will be

$$R = (200 + x)(300 - x) \qquad \text{Number of passengers · the fare per passenger}$$
$$= -x^2 + 100x + 60,000$$

Clearly, x must be nonnegative, and $300 - x \geq 0$, or $x \leq 300$. So the required function is $f(x) = -x^2 + 100x + 60,000$ with domain [0, 300]. ■

2.7 Self-Check Exercise

The Cunningham Day Care Center wants to enclose a playground of rectangular shape having an area of 500 ft^2 with a wooden fence. Find a function f giving the amount of fencing required in terms of the width x of the rectangular playground.

The solution to Self-Check Exercise 2.7 can be found on page 149.

2.7 Concept Questions

1. Describe mathematical modeling in your own words.

2. Define (a) a polynomial function and (b) a rational function. Give an example of each.

2.7 Exercises

In Exercises 1–6, determine whether the given function is a polynomial function, a rational function, or some other function. State the degree of each polynomial function.

1. $f(x) = 3x^6 - 2x^2 + 1$

2. $f(x) = \dfrac{x^2 - 9}{x - 3}$

3. $G(x) = 2(x^2 - 3)^3$

4. $H(x) = 2x^{-3} + 5x^{-2} + 6$

5. $f(t) = 2t^2 + 3\sqrt{t}$

6. $f(r) = \dfrac{6r}{(r^3 - 8)}$

7. INSTANT MESSAGING ACCOUNTS The number of enterprise instant messaging (IM) accounts is projected to grow according to the function

$$N(t) = 2.96t^2 + 11.37t + 59.7 \qquad (0 \le t \le 5)$$

where $N(t)$ is measured in millions and t in years, with $t = 0$ corresponding to 2006.
a. How many enterprise IM accounts were there in 2006?
b. What was the expected number of enterprise IM accounts in 2010?

Source: The Radical Group.

8. SOLAR POWER More and more businesses and homeowners are installing solar panels on their roofs to draw energy from the sun's rays. According to the U.S. Department of Energy, the solar cell kilowatt-hour use in the United States (in millions) is projected to be

$$S(t) = 0.73t^2 + 15.8t + 2.7 \qquad (0 \le t \le 8)$$

in year t, with $t = 0$ corresponding to 2000. What was the projected solar cell kilowatt-hours used in the United States for 2006? For 2008?

Source: U.S. Department of Energy.

9. AVERAGE SINGLE-FAMILY PROPERTY TAX Based on data from 298 of 351 cities and towns in Massachusetts, the average single-family tax bill from 1997 through 2010 is approximated by the function

$$T(t) = 7.26t^2 + 91.7t + 2360 \qquad (0 \le t \le 13)$$

where $T(t)$ is measured in dollars and t in years, with $t = 0$ corresponding to 1997.
a. What was the average property tax on a single-family home in Massachusetts in 1997?
b. If the trend continued, what was the average property tax in 2010?

Source: Massachusetts Department of Revenue.

10. REVENUE OF POLO RALPH LAUREN Citing strong sales and benefits from a new arm that will design lifestyle brands for department and specialty stores, the company projects revenue (in billions of dollars) to be

$$R(t) = -0.06t^2 + 0.69t + 3.25 \qquad (0 \le t \le 3)$$

in year t, where $t = 0$ corresponds to 2005.
a. What was the revenue of the company in 2005?
b. Find $R(1)$, $R(2)$, and $R(3)$ and interpret your results.
c. Sketch the graph of R.

Source: Company reports.

11. BABY BOOMERS AND SOCIAL SECURITY BENEFITS Aging baby boomers will put a strain on Social Security benefits unless Congress takes action. The Social Security benefits to be paid out from 2010 through 2040 are projected to be

$$S(t) = 0.1375t^2 + 0.5185t + 0.72 \qquad (0 \le t \le 3)$$

where $S(t)$ is measured in trillions of dollars and t is measured in decades with $t = 0$ corresponding to 2010.
a. What was the amount of the Social Security benefits paid out in 2010?
b. What is the amount of the Social Security benefits projected to be paid out in 2040?

Source: Social Security and Medicare Trustees' 2010 Report.

12. BABY BOOMERS AND MEDICARE BENEFITS Aging baby boomers will put a strain on Medicare benefits unless Congress takes action. The Medicare benefits to be paid out from 2010 through 2040 are projected to be

$$B(t) = 0.09t^2 + 0.102t + 0.25 \qquad (0 \le t \le 3)$$

where $B(t)$ is measured in trillions of dollars and t is measured in decades with $t = 0$ corresponding to 2010.
   a. What was the amount of Medicare benefits paid out in 2010?
   b. What is the amount of Medicare benefits projected to be paid out in 2040?
   Source: Social Security and Medicare Trustees' 2010 Report.

13. **REACTION OF A FROG TO A DRUG** Experiments conducted by A. J. Clark suggest that the response $R(x)$ of a frog's heart muscle to the injection of x units of acetylcholine (as a percent of the maximum possible effect of the drug) may be approximated by the rational function

$$R(x) = \frac{100x}{b + x} \qquad (x \geq 0)$$

where b is a positive constant that depends on the particular frog.
   a. If a concentration of 40 units of acetylcholine produces a response of 50% for a certain frog, find the "response function" for this frog.
   b. Using the model found in part (a), find the response of the frog's heart muscle when 60 units of acetylcholine are administered.

14. **AGING DRIVERS** The number of fatalities due to car crashes, based on the number of miles driven, begins to climb after the driver is past age 65. Aside from declining ability as one ages, the older driver is more fragile. The number of fatalities per 100 million vehicle miles driven is approximately

$$N(x) = 0.0336x^3 - 0.118x^2 + 0.215x + 0.7 \qquad (0 \leq x \leq 7)$$

where x denotes the age group of drivers, with $x = 0$ corresponding to those aged 50–54, $x = 1$ corresponding to those aged 55–59, $x = 2$ corresponding to those aged 60–64, . . . , and $x = 7$ corresponding to those aged 85–89. What is the fatality rate per 100 million vehicle miles driven for an average driver in the 50–54 age group? In the 85–89 age group?
   Source: U.S. Department of Transportation.

15. **TOTAL GLOBAL MOBILE DATA TRAFFIC** In a 2009 report, equipment maker Cisco forecast the total global mobile data traffic to be

$$f(t) = 0.021t^3 + 0.015t^2 + 0.12t + 0.06 \qquad (0 \leq t \leq 5)$$

million terabytes per month in year t, where $t = 0$ corresponds to 2009.
   a. What was the total global mobile data traffic in 2009?
   b. According to Cisco, what will the total global mobile data traffic per month be in 2012?
   Source: Cisco.

16. **GIFT CARDS** Gift cards have increased in popularity in recent years. Consumers appreciate gift cards because they get to select the present they like. The U.S. sales of gift cards (in billions of dollars) is approximated by

$$S(t) = -0.6204t^3 + 4.671t^2 + 3.354t + 47.4 \qquad (0 \leq t \leq 5)$$

in year t, where $t = 0$ corresponds to 2003.
   a. What were the sales of gift cards for 2003?
   b. What were the sales of gift cards for 2008?
   Source: The Tower Group.

17. **BLACKBERRY SUBSCRIBERS** According to a study conducted in 2004, the number of subscribers of BlackBerry, the handheld email devices manufactured by Research in Motion Ltd., is approximated by

$$N(t) = -0.0675t^4 + 0.5083t^3 - 0.893t^2 + 0.66t + 0.32$$
$$(0 \leq t \leq 4)$$

where $N(t)$ is measured in millions and t in years, with $t = 0$ corresponding to the beginning of 2002.
   a. How many BlackBerry subscribers were there at the beginning of 2002?
   b. How many BlackBerry subscribers were there at the beginning of 2006?
   Source: ThinkEquity Partners.

18. **INFANT MORTALITY RATES IN MASSACHUSETTS** The deaths of children less than 1 year old per 1000 live births is modeled by the function

$$R(t) = 162.8t^{-3.025} \qquad (1 \leq t \leq 3)$$

where t is measured in 50-year intervals, with $t = 1$ corresponding to 1900.
   a. Find $R(1)$, $R(2)$, and $R(3)$ and use your result to sketch the graph of the function R over the domain [1, 3].
   b. What was the infant mortality rate in 1900? in 1950? in 2000?
   Source: Massachusetts Department of Public Health.

19. **ONLINE VIDEO VIEWERS** As broadband Internet grows more popular, video services such as YouTube will continue to expand. The number of online video viewers (in millions) in year t is projected to grow according to the rule

$$N(t) = 52t^{0.531} \qquad (1 \leq t \leq 10)$$

where $t = 1$ corresponds to 2003.
   a. Sketch the graph of N.
   b. How many online video viewers will there be in 2012?
   Source: eMarketer.com.

20. **CHIP SALES** The worldwide sales of flash memory chip (in billions of dollars) is approximated by

$$S(t) = 4.3(t + 2)^{0.94} \qquad (0 \leq t \leq 6)$$

where t is measured in years, with $t = 0$ corresponding to 2002. Flash chips are used in cell phones, digital cameras, and other products.
   a. What were the worldwide flash memory chip sales in 2002?
   b. What were the estimated sales for 2010?
   Source: Web-Feet Research, Inc.

(continued)

21. Outsourcing of Jobs According to a study conducted in 2003, the total number of U.S. jobs (in millions) that are projected to leave the country by year t, where $t = 0$ corresponds to 2000, is

$$N(t) = 0.0018425(t + 5)^{2.5} \qquad (0 \le t \le 15)$$

What was the projected number of outsourced jobs for 2005 ($t = 5$)? For 2010 ($t = 10$)?

Source: Forrester Research.

22. Immigration to the United States Immigration to the United States from Europe, as a percentage of total immigration, is approximately

$$P(t) = 0.767t^3 - 0.636t^2 - 19.17t + 52.7 \qquad (0 \le t \le 4)$$

where t is measured in decades, with $t = 0$ corresponding to the decade of the 1950s.
a. Complete the following table:

t	0	1	2	3	4
$P(t)$					

b. Use the result of part (a) to sketch the graph of P.
c. Use the result of part (b) to estimate the decade when immigration, as a percentage of total immigration, was the greatest and the decade when it was the smallest.

Source: Jeffrey Williamson, Harvard University.

23. Selling Price of DVD Recorders The rise of digital music and the improvement to the DVD format are part of the reasons why the average selling price of stand-alone DVD recorders will drop in the coming years. The function

$$A(t) = \frac{699}{(t + 1)^{0.94}} \qquad (0 \le t \le 5)$$

gives the average selling price (in dollars) of stand-alone DVD recorders in year t, where $t = 0$ corresponds to the beginning of 2002. What was the average selling price of stand-alone DVD recorders at the beginning of 2002? at the beginning of 2007?

Source: Consumer Electronics Association.

24. Walking Versus Running The oxygen consumption (in milliliter per pound per minute) for a person walking at x mph is approximated by the function

$$f(x) = \frac{5}{3}x^2 + \frac{5}{3}x + 10 \qquad (0 \le x \le 9)$$

whereas the oxygen consumption for a runner at x mph is approximated by the function

$$g(x) = 11x + 10 \qquad (4 \le x \le 9)$$

a. Sketch the graphs of f and g.
b. At what speed is the oxygen consumption the same for a walker as it is for a runner? What is the level of oxygen consumption at that speed?
c. What happens to the oxygen consumption of the walker and the runner at speeds beyond that found in part (b)?

Source: William McArdley, Frank Katch, and Victor Katch, *Exercise Physiology.*

25. Price of Automobile Parts For years, automobile manufacturers had a monopoly on the replacement-parts market, particularly for sheet metal parts such as fenders, doors, and hoods, the parts most often damaged in a crash. Beginning in the late 1970s, however, competition appeared on the scene. In a report conducted by an insurance company to study the effects of the competition, the price of an OEM (original equipment manufacturer) fender for a particular 1983 model car was found to be

$$f(t) = \frac{110}{\frac{1}{2}t + 1} \qquad (0 \le t \le 2)$$

where $f(t)$ is measured in dollars and t is in years. Over the same period of time, the price of a non-OEM fender for the car was found to be

$$g(t) = 26\left(\frac{1}{4}t^2 - 1\right)^2 + 52 \qquad (0 \le t \le 2)$$

where $g(t)$ is also measured in dollars. Find a function $h(t)$ that gives the difference in price between an OEM fender and a non-OEM fender. Compute $h(0)$, $h(1)$, and $h(2)$. What does the result of your computation seem to say about the price gap between OEM and non-OEM fenders over the 2 years?

26. Obese Children in the United States The percentage of obese children aged 12–19 in the United States is approximately

$$P(t) = \begin{cases} 0.04t + 4.6 & \text{if } 0 \le t < 10 \\ -0.01005t^2 + 0.945t - 3.4 & \text{if } 10 \le t \le 30 \end{cases}$$

where t is measured in years, with $t = 0$ corresponding to the beginning of 1970. What was the percentage of obese children aged 12–19 at the beginning of 1970? At the beginning of 1985? At the beginning of 2000?

Source: Centers for Disease Control and Prevention.

27. Cost of the Health-Care Bill The Congressional Budget Office estimates that the health-care bill passed by the Senate in November 2009, combined with a package of revisions known as the reconciliation bill, will result in a cost by year t (in billions of dollars) of

$$f(t) = \begin{cases} 5 & \text{if } 0 \le t < 2 \\ -0.5278t^3 + 3.012t^2 + 49.23t - 103.29 & \text{if } 2 \le t \le 8 \end{cases}$$

where t is measured in years, with $t = 0$ corresponding to 2010. What will be the cost of the health-care bill by 2011? By 2015?

Source: U.S. Congressional Budget Office.

28. Price of Ivory According to the World Wildlife Fund, a group in the forefront of the fight against illegal ivory trade, the price of ivory (in dollars per kilo) compiled from a variety of legal and black market sources is approximated by the function

$$f(t) = \begin{cases} 8.37t + 7.44 & \text{if } 0 \le t \le 8 \\ 2.84t + 51.68 & \text{if } 8 < t \le 30 \end{cases}$$

where t is measured in years, with $t = 0$ corresponding to the beginning of 1970.
a. Sketch the graph of the function f.
b. What was the price of ivory at the beginning of 1970? At the beginning of 1990?

Source: World Wildlife Fund.

29. **WORKING-AGE POPULATION** The ratio of working-age population to the elderly in the United States in year t (including projections after 2000) is given by

$$f(t) = \begin{cases} 4.1 & \text{if } 0 \le t < 5 \\ -0.03t + 4.25 & \text{if } 5 \le t < 15 \\ -0.075t + 4.925 & \text{if } 15 \le t \le 35 \end{cases}$$

with $t = 0$ corresponding to the beginning of 1995.
a. Sketch the graph of f.
b. What was the ratio at the beginning of 2005? What will the ratio be at the beginning of 2020?
c. Over what years is the ratio constant?
d. Over what years is the decline of the ratio greatest?

Source: U.S. Census Bureau.

30. **DEMAND FOR SMOKE ALARMS** The demand function for the Sentinel smoke alarm is given by

$$p = \frac{30}{0.02x^2 + 1} \qquad (0 \le x \le 10)$$

where x (measured in units of a thousand) is the quantity demanded per week and p is the unit price in dollars.
a. Sketch the graph of the demand function.
b. What is the unit price that corresponds to a quantity demanded of 10,000 units?

31. **DEMAND FOR COMMODITIES** Assume that the demand function for a certain commodity has the form

$$p = \sqrt{-ax^2 + b} \qquad (a \ge 0, b \ge 0)$$

where x is the quantity demanded, measured in units of a thousand and p is the unit price in dollars. Suppose the quantity demanded is 6000 ($x = 6$) when the unit price is $8.00 and 8000 ($x = 8$) when the unit price is $6.00. Determine the demand equation. What is the quantity demanded when the unit price is set at $7.50?

32. **SUPPLY OF SATELLITE RADIOS** Suppliers of satellite radios will market 10,000 units when the unit price is $20 and 62,500 units when the unit price is $35. Determine the supply function if it is known to have the form

$$p = a\sqrt{x} + b \qquad (a > 0, b > 0)$$

where x is the quantity supplied and p is the unit price in dollars. Sketch the graph of the supply function. What unit price will induce the supplier to make 40,000 satellite radios available in the marketplace?

33. Suppose the demand and supply equations for a certain commodity are given by $p = ax + b$ and $p = cx + d$, respectively, where $a < 0$, $c > 0$, and $b > d > 0$ (see the following figure).

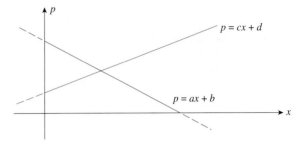

a. Find the equilibrium quantity and equilibrium price in terms of a, b, c, and d.
b. Use part (a) to determine what happens to the market equilibrium if c is increased while a, b, and d remain fixed. Interpret your answer in economic terms.
c. Use part (a) to determine what happens to the market equilibrium if b is decreased while a, c, and d remain fixed. Interpret your answer in economic terms.

34. **ENCLOSING AN AREA** Patricia wishes to have a rectangular garden in her backyard. She has 80 ft of fencing with which to enclose her garden. Letting x denote the width of the garden, find a function f in the variable x giving the area of the garden. What is its domain?

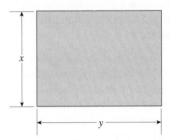

35. **ENCLOSING AN AREA** Juanita wishes to have a rectangular garden in her backyard. But Juanita wants her garden to have an area of 250 ft². Letting x denote the width of the garden, find a function f in the variable x giving the length of the fencing required to construct the garden. What is the domain of the function?

Hint: Refer to the figure for Exercise 34. The amount of fencing required is equal to the perimeter of the rectangle, which is twice the width plus twice the length of the rectangle.

36. **PACKAGING** By cutting away identical squares from each corner of a rectangular piece of cardboard and folding up the resulting flaps, an open box can be made. If the cardboard is 15 in. long and 8 in. wide and the square cutaways have dimensions of x in. by x in., find a function giving the volume of the resulting box.

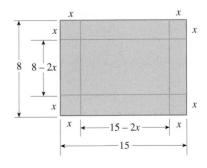

37. Construction Costs A rectangular box is to have a square base and a volume of 20 ft³. The material for the base costs 30¢/ft², the material for the sides costs 10¢/ft², and the material for the top costs 20¢/ft². Letting x denote the length of one side of the base, find a function in the variable x giving the cost of constructing the box.

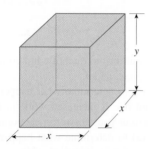

38. Oil Spills The oil spilling from the ruptured hull of a grounded tanker spreads in all directions in calm waters. Suppose the area polluted is a circle of radius r ft and the radius is increasing at the rate of 2 ft/sec.
a. Find a function f giving the area polluted in terms of r.
b. Find a function g giving the radius of the polluted area in terms of t.
c. Find a function h giving the area polluted in terms of t.
d. What is the size of the polluted area 30 sec after the hull was ruptured?

39. Yield of an Apple Orchard An apple orchard has an average yield of 36 bushels of apples/tree if tree density is 22 trees/acre. For each unit increase in tree density, the yield decreases by 2 bushels/tree. Letting x denote the number of trees beyond 22/acre, find a function in x that gives the yield of apples.

40. Book Design A book designer has decided that the pages of a book should have 1-in. margins at the top and bottom and $\frac{1}{2}$-in. margins on the sides. She further stipulated that each page should have a total area of 50 in.². Find a function in the variable x, giving the area of the printed part of the page. What is the domain of the function?

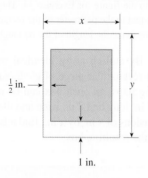

41. Profit of a Vineyard Phillip, the proprietor of a vineyard, estimates that if 10,000 bottles of wine were produced this season, then the profit would be $5/bottle. But if more than 10,000 bottles were produced, then the profit per bottle for the entire lot would drop by $0.0002 for each additional bottle sold. Assume that at least 10,000 bottles of wine are produced and sold, and let x denote the number of bottles produced and sold in excess of 10,000.
a. Find a function P giving the profit in terms of x.
b. What is the profit Phillip can expect from the sale of 16,000 bottles of wine from his vineyard?

42. Charter Revenue The owner of a luxury motor yacht that sails among the 4000 Greek islands charges $600/person/day if exactly 20 people sign up for the cruise. However, if more than 20 people sign up for the cruise (up to the maximum capacity of 90), the fare for all the passengers is reduced by $4 per person for each additional passenger. Assume that at least 20 people sign up for the cruise, and let x denote the number of passengers above 20.
a. Find a function R giving the revenue per day realized from the charter.
b. What is the revenue per day if 60 people sign up for the cruise?
c. What is the revenue per day if 80 people sign up for the cruise?

In Exercises 43–46, determine whether the statement is true or false. If it is true, explain why it is true. If it is false, give an example to show why it is false.

43. A polynomial function is a sum of constant multiples of power functions.

44. A polynomial function is a rational function, but the converse is false.

45. If $r > 0$, then the power function $f(x) = x^r$ is defined for all values of x.

46. The function $f(x) = 2^x$ is a power function.

2.7   Solution to Self-Check Exercise

Let the length of the rectangular playground be y ft (see the figure).

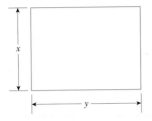

Then, the amount of fencing required is $L = 2x + 2y$. But the requirement that the area of the rectangular playground be 500 ft^2 implies that $xy = 500$, or upon solving for y, $y = 500/x$. Therefore, the amount of fencing required is

$$L = f(x) = 2x + 2\left(\frac{500}{x}\right) = 2x + \frac{1000}{x}$$

with domain $(0, \infty)$.

USING TECHNOLOGY

Constructing Mathematical Models from Raw Data

A graphing utility can sometimes be used to construct mathematical models from sets of data. For example, if the points corresponding to the given data are scattered about a straight line, then use **LinReg(ax+b)** (linear regression) from the statistical calculations menu of the graphing utility to obtain a function (model) that approximates the data at hand. If the points seem to be scattered along a parabola (the graph of a quadratic function), then use **QuadReg** (second-degree polynomial regression), and so on. (These are functions on the TI-83/84 calculator.)

 APPLIED EXAMPLE 1 Indian Gaming Industry The following table gives the estimated gross revenues (in billions of dollars) from the Indian gaming industries from 2000 ($t = 0$) to 2008 ($t = 8$).

Year	0	1	2	3	4	5	6	7	8
Revenue	11.0	12.8	14.7	16.8	19.5	22.7	25.1	26.4	26.8

a. Use a graphing utility to find a polynomial function f of degree 4 that models the data.
b. Plot the graph of the function f, using the viewing window $[0, 8] \times [0, 30]$.
c. Use the function evaluation capability of the graphing utility to compute $f(0)$, $f(1), \ldots, f(8)$, and compare these values with the original data.
d. If the trend continued, what was the gross revenue for 2009 ($t = 9$)?
Source: National Indian Gaming Association.

Solution

a. First, enter the data using the statistical menu. Then choose **QuartReg** (fourth-degree polynomial regression) from the statistical calculations menu of a graphing utility. We find

$$f(t) = -0.00737t^4 + 0.0655t^3 - 0.008t^2 + 1.61t + 11$$

b. The graph of f is shown in Figure T1.

(continued)

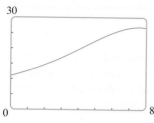

FIGURE **T1**
The graph of *f* in the viewing window
[0, 8] × [0, 30]

c. The required values, which compare favorably with the given data, follow:

t	0	1	2	3	4	5	6	7	8
f(t)	11.0	12.7	14.6	16.9	19.6	22.4	25.0	26.6	26.7

d. The gross revenue for 2009 (*t* = 9) is given by

$$f(9) = -0.00737(9)^4 + 0.0655(9)^3 - 0.008(9)^2 + 1.61(9) + 11 \approx 24.24$$

or approximately $24.2 billion. ∎

TECHNOLOGY EXERCISES

In Exercises 1–14, use the statistical calculations menu to construct a mathematical model associated with the given data.

1. CONSUMPTION OF BOTTLED WATER The annual per-capita consumption of bottled water (in gallons) and the scatter plot for these data follow:

Year	2001	2002	2003	2004	2005	2006
Consumption	18.8	20.9	22.4	24	26.1	28.3

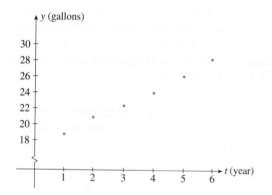

a. Use **LinReg(ax+b)** to find a first-degree (linear) polynomial regression model for the data. Let *t* = 1 correspond to 2001.
b. Plot the graph of the function *f* found in part (a), using the viewing window [1, 6] × [0, 30].
c. Compute the values for *t* = 1, 2, 3, 4, 5, and 6. How do your figures compare with the given data?
d. If the trend continued, what was the annual per-capita consumption of bottled water in 2008?

Source: Beverage Marketing Corporation.

2. WEB CONFERENCING Web conferencing is a big business, and it's growing rapidly. The amount (in billions of dollars) spent on Web conferencing from the beginning of 2003 through 2010, and the scatter plot for these data follow:

Year	2003	2004	2005	2006	2007	2008	2009	2010
Amount	0.50	0.63	0.78	0.92	1.16	1.38	1.60	1.90

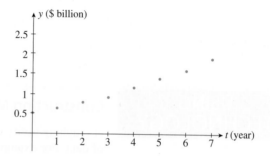

a. Let *t* = 0 correspond to the beginning of 2003 and use **QuadReg** to find a second-degree polynomial regression model based on the given data.
b. Plot the graph of the function *f* found in part (a) using the window [0, 7] × [0, 2].
c. Compute *f*(0), *f*(3), *f*(6), and *f*(7). Compare these values with the given data.

Source: Gartner Dataquest.

3. STUDENT POPULATION The projected total number of students in elementary schools, secondary schools, and colleges (in millions) from 1995 through 2015 is given in the following table:

Year	1995	2000	2005	2010	2015
Number	64.8	68.7	72.6	74.8	78

a. Use **QuadReg** to find a second-degree polynomial regression model for the data. Let *t* be measured in 5-year intervals, with *t* = 0 corresponding to the beginning of 1995.
b. Plot the graph of the function *f* found in part (a), using the viewing window [0, 4] × [0, 85].
c. Using the model found in part (a), what will be the projected total number of students (all categories) enrolled in 2015?

Source: U.S. National Center for Education Statistics.

4. DIGITAL TV SHIPMENTS The estimated number of digital TV shipments between the year 2000 and 2006 (in millions of units) and the scatter plot for these data follow:

Year	2000	2001	2002	2003	2004	2005	2006
Units Shipped	0.63	1.43	2.57	4.1	6	8.1	10

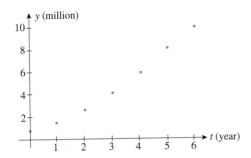

a. Use **CubicReg** to find a third-degree polynomial regression model for the data. Let $t = 0$ correspond to the beginning of 2000.

b. Plot the graph of the function f found in part (a), using the viewing window $[0, 6] \times [0, 11]$.

c. Compute the values of $f(t)$ for $t = 0, 1, 2, 3, 4, 5,$ and 6.

Source: Consumer Electronics Manufacturers Association.

5. **HEALTH-CARE SPENDING** Health-care spending by business (in billions of dollars) from the year 2000 through 2006 is summarized below:

Year	2000	2001	2002	2003	2004	2005	2006
Number	185	235	278	333	389	450	531

a. Plot the scatter diagram for the above data. Let $t = 0$ correspond to the beginning of 2000.

b. Use **QuadReg** to find a second-degree polynomial regression model for the data.

c. If the trend continued, what was the spending in 2010?

Source: Centers for Medicine and Medicaid Services.

6. **TIVO OWNERS** The projected number of households (in millions) with digital video recorders that allow viewers to record shows onto a server and skip commercials are given in the following table:

Year	2006	2007	2008	2009	2010
Households	31.2	49.0	71.6	97.0	130.2

a. Let $t = 0$ correspond to the beginning of 2006, and use **QuadReg** to find a second-degree polynomial regression model based on the given data.

b. Obtain the scatter plot and the graph of the function f found in part (a), using the viewing window $[0, 4] \times [0, 140]$.

Source: Strategy Analytics.

7. **TELECOMMUNICATIONS INDUSTRY REVENUE** Telecommunications industry revenue is expected to grow in the coming years, fueled by the demand for broadband and high-speed data services. The worldwide revenue for the industry (in trillions of dollars) and the scatter plot for these data follow:

Year	2000	2002	2004	2006	2008	2010
Revenue	1.7	2.0	2.5	3.0	3.6	4.2

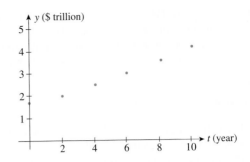

a. Let $t = 0$ correspond to the beginning of 2000 and use **CubicReg** to find a third-degree polynomial regression model based on the given data.

b. Plot the graph of the function f found in part (a), using the viewing window $[0, 10] \times [0, 5]$.

c. Find the worldwide revenue for the industry in 2001 and 2005 and find the projected revenue for 2010.

Source: Telecommunication Industry Association.

8. **POPULATION GROWTH IN CLARK COUNTY** Clark County in Nevada—dominated by greater Las Vegas—is one of the fastest-growing metropolitan areas in the United States. The population of the county from 1970 through 2000 is given in the following table:

Year	1970	1980	1990	2000
Population	273,288	463,087	741,459	1,375,765

a. Use **CubicReg** to find a third-degree polynomial regression model for the data. Let t be measured in decades, with $t = 0$ corresponding to the beginning of 1970.

b. Plot the graph of the function f found in part (a), using the viewing window $[0, 3] \times [0, 1,500,000]$.

c. Compare the values of f at $t = 0, 1, 2,$ and 3, with the given data.

Source: U.S. Census Bureau.

9. **LOBBYISTS' SPENDING** Lobbyists try to persuade legislators to propose, pass, or defeat legislation or to change existing laws. The amount (in billions of dollars) spent by lobbyists from 2003 through 2009 is shown in the following table:

Year	2003	2004	2005	2006	2007	2008	2009
Amount	8.0	8.5	9.7	10.2	11.3	12.9	13.8

a. Use **CubicReg** to find a third-degree polynomial regression model for the data, letting $t = 0$ correspond to 2003.

b. Plot the scatter diagram and the graph of the function f found in part (a), using the viewing window $[0, 6] \times [0, 15]$.

c. Compare the values of f at $t = 0, 3,$ and 6 with the given data.

Source: Center for Public Integrity.

(continued)

10. **MOBILE ENTERPRISE IM ACCOUNTS** The projected number of mobile enterprise instant messaging (IM) accounts (in millions) from 2006 through 2010 is given in the following table ($t = 0$ corresponds to the beginning of 2006):

Year	0	1	2	3	4
Accounts	2.3	3.6	5.8	8.7	14.9

a. Use **CubicReg** to find a third-degree polynomial regression model based on the given data.
b. Plot the graph of the function f found in part (a), using the viewing window $[0, 5] \times [0, 16]$.
c. Compute $f(0), f(1), f(2), f(3)$, and $f(4)$.
Source: The Radical Group.

11. **MEASLES DEATHS** Measles is still a leading cause of vaccine-preventable death among children, but because of improvements in immunizations, measles deaths have dropped globally. The following table gives the number of measles deaths (in thousands) in sub-Saharan Africa from 1999 through 2005:

Year	1999	2001	2003	2005
Amount	506	338	250	126

a. Use **CubicReg** to find a third-degree polynomial regression model for the data, letting $t = 0$ correspond to the beginning of 1999.
b. Plot the scatter diagram and the graph of the function f found in part (a).
c. Compute the values of f for $t = 0, 2$, and 6.
Source: Centers for Disease Control and Prevention and World Health Organization.

12. **OFFICE VACANCY RATE** The total vacancy rate (as a percent) of offices in Manhattan from 2000 through 2006 is shown in the following table:

Year	2000	2001	2002	2003	2004	2005	2006
Vacancy Rate	3.8	8.9	12	12.5	11	8.4	6.7

a. Use **CubicReg** to find a third-degree polynomial regression model for the data, letting $t = 0$ correspond to the beginning of 2000.
b. Plot the scatter diagram and the graph of the function f found in part (a).
c. Compute the values for $t = 1, 2, 3, 4, 5$, and 6.
Source: Cushman and Wakefield.

13. **NICOTINE CONTENT OF CIGARETTES** Even as measures to discourage smoking have been growing more stringent in recent years, the nicotine content of cigarettes has been rising, making it more difficult for smokers to quit. The following table gives the average amount of nicotine in cigarette smoke from 1999 through 2004:

Year	1999	2000	2001	2002	2003	2004
Yield per Cigarette (mg)	1.71	1.81	1.85	1.84	1.83	1.89

a. Use **QuartReg** to find a fourth-degree polynomial regression model for the data. Let $t = 0$ correspond to the beginning of 1999.
b. Plot the graph of the function f found in part (a), using the viewing window $[0, 5] \times [0, 2]$.
c. Compute the values of $f(t)$ for $t = 0, 1, 2, 3, 4$, and 5.
d. If the trend continued, what was the average amount of nicotine in cigarettes in 2005?
Source: Massachusetts Tobacco Control Program.

14. **SOCIAL SECURITY TRUST FUND ASSETS** The projected assets of the Social Security trust fund (in trillions of dollars) from 2010 through 2037 are given in the following table:

Year	2010	2015	2020	2025	2030	2035	2037
Assets	2.69	3.56	4.22	4.24	3.24	0.87	0

Use **QuartReg** to find a fourth-degree polynomial regression model for the data. Let $t = 0$ correspond to 2010.
Source: Social Security Administration.

CHAPTER 2   Summary of Principal Formulas and Terms

FORMULAS

1. Slope of a line	$m = \dfrac{y_2 - y_1}{x_2 - x_1}$
2. Equation of a vertical line	$x = a$
3. Equation of a horizontal line	$y = b$
4. Point-slope form of the equation of a line	$y - y_1 = m(x - x_1)$
5. Slope-intercept form of the equation of a line	$y = mx + b$
6. General equation of a line	$Ax + By + C = 0$

TERMS

Cartesian coordinate system (70)

origin (70)

ordered pair (70)

coordinates (70)

parallel lines (74)

perpendicular lines (78)

x-intercept (78)

y-intercept (78)

function (89)

domain (89)

range (90)

independent variable (91)

dependent variable (91)

graph of a function (92)

piecewise-defined function (94)

graph of an equation (95)

vertical line test (95)

composite function (108)

linear function (113)

total cost function (115)

revenue function (115)

profit function (115)

break-even point (117)

quadratic function (126)

demand function (129)

supply function (130)

market equilibrium (131)

polynomial function (138)

rational function (140)

power function (141)

CHAPTER 2   Concept Review Questions

Fill in the blanks.

1. A point in the plane can be represented uniquely by a/an
_____ pair of numbers. The first number of the pair is called
the _____, and the second number of the pair is called the
_____.

2. **a.** The point $P(a, 0)$ lies on the _____-axis, and the point
$P(0, b)$ lies on the _____-axis.
   b. If the point $P(a, b)$ lies in the fourth quadrant, then the
point $P(-a, b)$ lies in the _____ quadrant.

3. **a.** If $P_1(x_1, y_1)$ and $P_2(x_2, y_2)$ are any two distinct points on
a nonvertical line L, then the slope of L is $m =$ _____.
   b. The slope of a vertical line is _____.
   c. The slope of a horizontal line is _____.
   d. The slope of a line that slants upward from left to right
is _____.

4. If L_1 and L_2 are nonvertical lines with slopes m_1 and m_2,
respectively, then L_1 is parallel to L_2 if and only if _____
and L_1 is perpendicular to L_2 if and only if _____.

5. **a.** An equation of the line passing through the point
$P(x_1, y_1)$ and having slope m is _____. This form of the
equation of a line is called the _____ _____.
   b. An equation of the line that has slope m and y-intercept
b is _____. It is called the _____ form of an equation
of a line.

6. **a.** The general form of an equation of a line is _____.
   b. If a line has equation $ax + by + c = 0$ ($b \neq 0$), then its
slope is _____.

7. If f is a function from the set A to the set B, then A is called
the _____ of f, and the set of all values of $f(x)$ as x takes

on all possible values in A is called the _____ of f. The
range of f is contained in the set _____.

8. The graph of a function is the set of all points (x, y) in the
xy-plane such that x is in the _____ of f and $y =$ _____.
The Vertical Line Test states that a curve in the xy-plane is
the graph of a function $y = f(x)$ if and only if each _____
line intersects it in at most one _____.

9. If f and g are functions with domains A and B, respectively,
then (a) $(f \pm g)(x) =$ _____, (b) $(fg)(x) =$ _____, and
(c) $\left(\dfrac{f}{g}\right)(x) =$ _____. The domain of $f + g$ is _____. The
domain of $\dfrac{f}{g}$ is _____ with the additional condition that
$g(x)$ is never _____.

10. The composition of g and f is the function with rule
$(g \circ f)(x) =$ _____. Its domain is the set of all x in the
domain of _____ such that _____ lies in the domain of
_____.

11. A quadratic function has the form $f(x) =$ _____. Its graph
is a/an _____ that opens _____ if $a > 0$ and _____ if
$a < 0$. Its highest point or lowest point is called its _____.
The x-coordinate of its vertex is _____, and an equation
of its axis of symmetry is _____.

12. **a.** A polynomial function of degree n is a function of the
form _____.
   b. A polynomial function of degree 1 is called a/an _____
function; one of degree 2 is called a/an _____ function.
   c. A rational function is a/an _____ of two _____.
   d. A power function has the form $f(x) =$ _____.

In Exercises 1–6, find an equation of the line L that passes through the point $(-2, 4)$ and satisfies the given condition.

1. L is a vertical line.

2. L is a horizontal line.

3. L passes through the point $(3, \frac{7}{2})$.

4. The x-intercept of L is 3.

5. L is parallel to the line $5x - 2y = 6$.

6. L is perpendicular to the line $4x + 3y = 6$.

7. Find an equation of the line with slope $-\frac{1}{2}$ and y-intercept -3.

8. Find the slope and y-intercept of the line with equation $3x - 5y = 6$.

9. Find an equation of the line passing through the point $(2, 3)$ and parallel to the line with equation $3x + 4y - 8 = 0$.

10. Find an equation of the line passing through the point $(-1, 3)$ and parallel to the line joining the points $(-3, 4)$ and $(2, 1)$.

11. Find an equation of the line passing through the point $(-2, -4)$ that is perpendicular to the line with equation $2x - 3y - 24 = 0$.

In Exercises 12 and 13, sketch the graph of the equation.

12. $3x - 4y = 24$　　　　**13.** $-2x + 5y = 15$

In Exercises 14 and 15, find the domain of the function.

14. $f(x) = \sqrt{9 - x}$

15. $f(x) = \dfrac{x + 3}{2x^2 - x - 3}$

16. Let $f(x) = 3x^2 + 5x - 2$. Find:
  a. $f(-2)$
  b. $f(a + 2)$
  c. $f(2a)$
  d. $f(a + h)$

17. **SALES OF PRERECORDED MUSIC** The following graphs show the sales y of prerecorded music (in billions of dollars) by format as a function of time t (in years), with $t = 0$ corresponding to 1985.

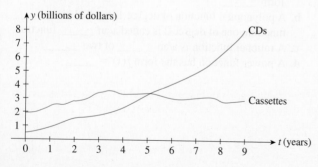

a. In what years were the sales of prerecorded cassettes greater than those of prerecorded CDs?
b. In what years were the sales of prerecorded CDs greater than those of prerecorded cassettes?
c. In what year were the sales of prerecorded cassettes the same as those of prerecorded CDs? Estimate the level of sales in each format at that time.

Source: Recording Industry Association of America.

18. Let $y^2 = 2x + 1$.
  a. Sketch the graph of this equation.
  b. Is y a function of x? Why?
  c. Is x a function of y? Why?

19. Sketch the graph of the function defined by

$$f(x) = \begin{cases} x + 1 & \text{if } x < 1 \\ -x^2 + 4x - 1 & \text{if } x \geq 1 \end{cases}$$

20. Let $f(x) = \frac{1}{x}$ and $g(x) = 2x + 3$. Find:
  a. $f(x)g(x)$　　　　**b.** $f(x)/g(x)$
  c. $f(g(x))$　　　　**d.** $g(f(x))$

In Exercises 21 and 22, find the vertex and the x-intercepts of the parabola and sketch the parabola.

21. $y = 6x^2 - 11x - 10$　　**22.** $y = -4x^2 + 4x + 3$

In Exercises 23 and 24, find the point of intersection of the lines with the given equations.

23. $3x + 4y = -6$ and $2x + 5y = -11$

24. $y = \dfrac{3}{4}x + 6$ and $3x - 2y + 3 = 0$

25. Find the point of intersection of the two straight lines having the equations $7x + 9y = -11$ and $3x = 6y - 8$.

26. The cost and revenue functions for a certain firm are given by $C(x) = 12x + 20{,}000$ and $R(x) = 20x$, respectively. Find the company's break-even point.

27. **DEMAND FOR CLOCK RADIOS** In the accompanying figure, L_1 is the demand curve for the model A clock radios manufactured by Ace Radio, and L_2 is the demand curve for their model B clock radios. Which line has the greater slope? Interpret your results.

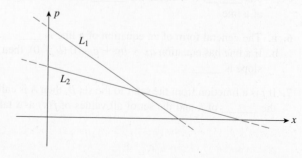

28. **SUPPLY OF CLOCK RADIOS** In the accompanying figure, L_1 is the supply curve for the model A clock radios manufactured by Ace Radio, and L_2 is the supply curve for their model B clock radios. Which line has the greater slope? Interpret your results.

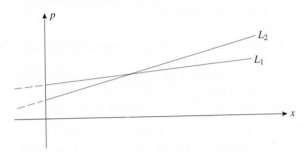

29. **SALES OF MP3 PLAYERS** Sales of a certain brand of MP3 player are approximated by the relationship

$$S(x) = 6000x + 30,000 \qquad (0 \le x \le 5)$$

where $S(x)$ denotes the number of MP3 players sold in year x ($x = 0$ corresponds to 2007). Find the number of MP3 players expected to be sold in 2012.

30. **COMPANY SALES** A company's total sales (in millions of dollars) are approximately linear as a function of time (in years). Sales in 2006 ($x = 0$) were \$2.4 million, whereas sales in 2011 amounted to \$7.4 million.
    a. Find an equation giving the company's sales as a function of time.
    b. What were the sales in 2009?

31. **PROFIT FUNCTIONS** A company has a fixed cost of \$30,000 and a production cost of \$6 for each CD it manufactures. Each CD sells for \$10.
    a. What is the cost function?
    b. What is the revenue function?
    c. What is the profit function?
    d. Compute the profit (loss) corresponding to production levels of 6000, 8000, and 12,000 units, respectively.

32. **LINEAR DEPRECIATION** An office building worth \$6 million when it was completed in 2008 is being depreciated linearly over 30 years.
    a. What is the rate of depreciation?
    b. What will be the book value of the building in 2018?

33. **DEMAND EQUATIONS** There is no demand for a certain commodity when the unit price is \$200 or more, but for each \$10 decrease in price below \$200, the quantity demanded increases by 200 units. Find the demand equation and sketch its graph.

34. **SUPPLY EQUATIONS** Bicycle suppliers will make 200 bicycles available in the market per month when the unit price is \$50 and 2000 bicycles available per month when the unit price is \$100. Find the supply equation if it is known to be linear.

35. **CLARK'S RULE** Clark's Rule is a method for calculating pediatric drug dosages based on a child's weight. If a denotes the adult dosage (in milligrams) and w is the weight of the child (in pounds), then the child's dosage is given by

$$D(w) = \frac{aw}{150}$$

If the adult dose of a substance is 500 mg, how much should a child who weighs 35 lb receive?

36. **COLLEGE ADMISSIONS** The accompanying data were compiled by the Admissions Office of Carter College during the past 5 years. The data relate the number of college brochures and follow-up letters (x) sent to a preselected list of high school juniors who took the PSAT and the number of completed applications (y) received from these students (both measured in thousands).

Brochures Sent, x	1.8	2	3.2	4	4.8
Applications Completed, y	0.4	0.5	0.7	1	1.3

   a. Derive an equation of the straight line L that passes through the points $(2, 0.5)$ and $(4, 1)$.
   b. Use this equation to predict the number of completed applications that might be expected if 6400 brochures and follow-up letters are sent out during the next year.

37. **REVENUE FUNCTIONS** The monthly revenue R (in hundreds of dollars) realized in the sale of Royal electric shavers is related to the unit price p (in dollars) by the equation

$$R(p) = -\frac{1}{2}p^2 + 30p$$

Find the revenue when an electric shaver is priced at \$30.

38. **HEALTH CLUB MEMBERSHIP** The membership of the newly opened Venus Health Club is approximated by the function

$$N(x) = 200(4 + x)^{1/2} \qquad (1 \le x \le 24)$$

where $N(x)$ denotes the number of members x months after the club's grand opening. Find $N(0)$ and $N(12)$, and interpret your results.

39. **POPULATION GROWTH** A study prepared for a Sunbelt town's Chamber of Commerce projected that the population of the town in the next 3 years will grow according to the rule

$$P(x) = 50,000 + 30x^{3/2} + 20x$$

where $P(x)$ denotes the population x months from now. By how much will the population increase during the next 9 months? During the next 16 months?

40. **THURSTONE LEARNING CURVE** Psychologist L. L. Thurstone discovered the following model for the relationship between the learning time T and the length of a list n:

$$T = f(n) = An\sqrt{n - b}$$

where A and b are constants that depend on the person and the task. Suppose that, for a certain person and a certain task, $A = 4$ and $b = 4$. Compute $f(4), f(5), \ldots, f(12)$, and use this information to sketch the graph of the function f. Interpret your results.

41. TESTOSTERONE USE Fueled by the promotion of testosterone as an antiaging elixir, use of the hormone by middle-age and older men grew dramatically. The total number of prescriptions for testosterone from 1999 through 2002 is given by

$$N(t) = -35.8t^3 + 202t^2 + 87.8t + 648 \qquad (0 \le t \le 3)$$

where $N(t)$ is measured in thousands and t is measured in years, with $t = 0$ corresponding to the beginning of 1999. Find the total number of prescriptions for testosterone in 1999, 2000, 2001, and 2002.

Source: IMS Health.

42. U.S. NUTRITIONAL SUPPLEMENTS MARKET The size of the U.S. nutritional supplements market from 1999 through 2003 is approximated by the function

$$A(t) = 16.4(t + 1)^{0.1} \qquad (0 \le t \le 4)$$

where $A(t)$ is measured in billions of dollars and t is measured in years, with $t = 0$ corresponding to the beginning of 1999.

a. Compute $A(0)$, $A(1)$, $A(2)$, $A(3)$, and $A(4)$. Interpret your results.

b. Use the results of part (a) to sketch the graph of A.

Source: Nutrition Business Journal.

43. GLOBAL SUPPLY OF PLUTONIUM The global stockpile of plutonium for military applications between 1990 ($t = 0$) and 2003 ($t = 13$) stood at a constant 267 tons. On the other hand, the global stockpile of plutonium for civilian use was

$$2t^2 + 46t + 733$$

tons in year t over the same period.

a. Find the function f giving the global stockpile of plutonium for military use from 1990 through 2003 and the function g giving the global stockpile of plutonium for civilian use over the same period.

b. Find the function h giving the total global stockpile of plutonium between 1990 and 2003.

c. What was the total global stockpile of plutonium in 2003?

Source: Institute for Science and International Security.

44. MARKET EQUILIBRIUM The monthly demand and supply functions for the Luminar desk lamp are given by

$$p = d(x) = -1.1x^2 + 1.5x + 40$$
$$p = s(x) = 0.1x^2 + 0.5x + 15$$

respectively, where p is measured in dollars and x in units of a thousand. Find the equilibrium quantity and price.

45. INFLATING A BALLOON A spherical balloon is being inflated at a rate of $\frac{9}{2}\pi$ ft³/min.

a. Find a function f giving the radius r of the balloon in terms of its volume V.

Hint: $V = \frac{4}{3}\pi r^3$

b. Find a function g giving the volume of the balloon in terms of time t.

c. Find a function h giving the radius of the balloon in terms of time.

d. What is the radius of the balloon after 8 min?

46. HOTEL OCCUPANCY RATE A forecast released by PricewaterhouseCoopers in June of 2004 predicted the occupancy rate of U.S. hotels between 2001 ($t = 0$) and 2005 ($t = 4$) to be

$$P(t) = \begin{cases} -0.9t + 59.8 & \text{if } 0 \le t < 1 \\ 0.3t + 58.6 & \text{if } 1 \le t < 2 \\ 56.79t^{0.06} & \text{if } 2 \le t \le 4 \end{cases}$$

percent.

a. Compute $P(0)$, $P(1)$, $P(2)$, $P(3)$, and $P(4)$.

b. Sketch the graph of P.

c. What was the predicted occupancy rate of hotels for 2004?

Source: PricewaterhouseCoopers LLP Hospitality & Leisure Research.

47. PACKAGING By cutting away identical squares from each corner of a 20-in. × 20-in. piece of cardboard and folding up the resulting flaps, an open box can be made. Denoting the length of a side of a cutaway by x, find a function of x giving the volume of the resulting box.

48. CONSTRUCTION COSTS The length of a rectangular box is to be twice its width, and its volume is to be 30 ft³. The material for the base costs 30¢/ft², the material for the sides costs 15¢/ft², and the material for the top costs 20¢/ft². Letting x denote the width of the box, find a function in the variable x giving the cost of constructing the box.

CHAPTER 2   Before Moving On . . .

1. Find an equation of the line that passes through $(-1, -2)$ and $(4, 5)$.

2. Find an equation of the line that has slope $-\frac{1}{3}$ and y-intercept $\frac{4}{3}$.

3. Let

$$f(x) = \begin{cases} -2x + 1 & \text{if } -1 \le x < 0 \\ x^2 + 2 & \text{if } 0 \le x \le 2 \end{cases}$$

Find (a) $f(-1)$, (b) $f(0)$, and (c) $f(\frac{3}{2})$.

4. Let $f(x) = \dfrac{1}{x + 1}$ and $g(x) = x^2 + 1$. Find the rules for

(a) $f + g$, (b) fg, (c) $f \circ g$, and (d) $g \circ f$.

5. Postal regulations specify that a parcel sent by parcel post may have a combined length and girth of no more than 108 in. Suppose a rectangular package that has a square cross section of x in. × x in. is to have a combined length and girth of exactly 108 in. Find a function in terms of x giving the volume of the package.

Hint: The length plus the girth is $4x + h$ (see the accompanying figure).

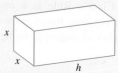

3

EXPONENTIAL AND LOGARITHMIC FUNCTIONS

THE EXPONENTIAL FUNCTION is without doubt the most important function in mathematics and its applications. After a brief introduction to the exponential function and its *inverse,* the logarithmic function, we explore some of the many applications involving exponential functions, such as the growth rate of a bacteria population in the laboratory, the way in which radioactive matter decays, the rate at which a factory worker learns a certain process, and the rate at which a communicable disease is spread over time. Exponential functions also play an important role in computing interest earned in a bank account, a topic to be discussed in Chapter 4.

How many cameras can a new employee at Eastman Optical assemble after completing the basic training program, and how many cameras can he assemble after being on the job for 6 months? In Example 5, page 179, you will see how to answer these questions.

3.1   Exponential Functions

Exponential Functions and Their Graphs

Suppose you deposit a sum of $1000 in an account earning interest at the rate of 10% per year *compounded continuously* (the way most financial institutions compute interest). Then, the accumulated amount at the end of t years ($0 \leq t \leq 20$) is described by the function f, whose graph appears in Figure 1.* This function is called an *exponential function*. Observe that the graph of f rises rather slowly at first but very rapidly as time goes by. For purposes of comparison, we have also shown the graph of the function $y = g(t) = 1000(1 + 0.10t)$, giving the accumulated amount for the same principal ($1000) but earning *simple* interest at the rate of 10% per year. The moral of the story: It is never too early to save.

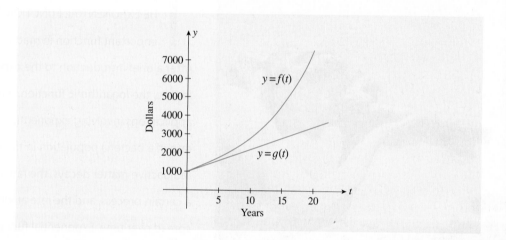

FIGURE 1
Under continuous compounding, a sum of money grows exponentially.

Exponential functions play an important role in many real-world applications, as you will see throughout this chapter.

Recall that whenever b is a positive number and n is any real number, the expression b^n is a real number. This enables us to define an exponential function as follows:

> **Exponential Function**
>
> The function defined by
>
> $$f(x) = b^x \qquad (b > 0, b \neq 1)$$
>
> is called an **exponential function with base b and exponent x.** The domain of f is the set of all real numbers.

For example, the exponential function with base 2 is the function

$$f(x) = 2^x$$

with domain $(-\infty, \infty)$. The values of $f(x)$ for selected values of x follow:

$$f(3) = 2^3 = 8 \qquad f\left(\frac{3}{2}\right) = 2^{3/2} = 2 \cdot 2^{1/2} = 2\sqrt{2} \qquad f(0) = 2^0 = 1$$

$$f(-1) = 2^{-1} = \frac{1}{2} \qquad f\left(-\frac{2}{3}\right) = 2^{-2/3} = \frac{1}{2^{2/3}} = \frac{1}{\sqrt[3]{4}}$$

---

*We will derive the rule for f in Section 4.1.

Computations involving exponentials are facilitated by the laws of exponents. These laws were stated in Section 1.5, and you might want to review the material there. For convenience, however, we will restate these laws.

Laws of Exponents

Let a and b be positive numbers and let x and y be real numbers. Then,

1. $b^x \cdot b^y = b^{x+y}$ **4.** $(ab)^x = a^x b^x$

2. $\dfrac{b^x}{b^y} = b^{x-y}$ **5.** $\left(\dfrac{a}{b}\right)^x = \dfrac{a^x}{b^x}$

3. $(b^x)^y = b^{xy}$

The use of the laws of exponents is illustrated in the next two examples.

EXAMPLE 1

a. $16^{7/4} \cdot 16^{-1/2} = 16^{7/4 - 1/2} = 16^{5/4} = 2^5 = 32$ Law 1

b. $\dfrac{8^{5/3}}{8^{-1/3}} = 8^{5/3 - (-1/3)} = 8^2 = 64$ Law 2

c. $(64^{4/3})^{-1/2} = 64^{(4/3)(-1/2)} = 64^{-2/3}$

$$= \frac{1}{64^{2/3}} = \frac{1}{(64^{1/3})^2} = \frac{1}{4^2} = \frac{1}{16}$$ Law 3

d. $(16 \cdot 81)^{-1/4} = 16^{-1/4} \cdot 81^{-1/4} = \dfrac{1}{16^{1/4}} \cdot \dfrac{1}{81^{1/4}} = \dfrac{1}{2} \cdot \dfrac{1}{3} = \dfrac{1}{6}$ Law 4

e. $\left(\dfrac{3^{1/2}}{2^{1/3}}\right)^4 = \dfrac{3^{4/2}}{2^{4/3}} = \dfrac{9}{2^{4/3}}$ Law 5

EXAMPLE 2 Let $f(x) = 2^{2x-1}$. Find the value of x for which $f(x) = 16$.

Solution We want to solve the equation

$$2^{2x-1} = 16 = 2^4$$

But this equation holds if and only if

$$2x - 1 = 4 \quad b^m = b^n \Rightarrow m = n$$

giving $x = \frac{5}{2}$.

Exponential functions play an important role in mathematical analysis. Because of their special characteristics, they are some of the most useful functions and are found in virtually every field in which mathematics is applied. To mention a few examples: Under ideal conditions, the number of bacteria present at any time t in a culture may be described by an exponential function of t; radioactive substances decay over time in accordance with an "exponential" law of decay; money left on fixed deposit and earning compound interest grows exponentially; and some of the most important distribution functions encountered in statistics are exponential.

Let's begin our investigation into the properties of exponential functions by studying their graphs.

EXAMPLE 3 Sketch the graph of the exponential function $y = 2^x$.

Solution   First, as was discussed earlier, the domain of the exponential function $y = f(x) = 2^x$ is the set of real numbers. Next, putting $x = 0$ gives $y = 2^0 = 1$, the y-intercept of f. There is no x-intercept, since there is no value of x for which $y = 0$. To find the range of f, consider the following table of values:

x	-5	-4	-3	-2	-1	0	1	2	3	4	5
y	$\frac{1}{32}$	$\frac{1}{16}$	$\frac{1}{8}$	$\frac{1}{4}$	$\frac{1}{2}$	1	2	4	8	16	32

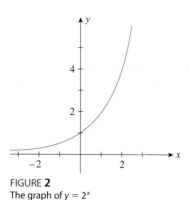

FIGURE **2**
The graph of $y = 2^x$

We see from these computations that 2^x decreases and approaches zero as x decreases without bound and that 2^x increases without bound as x increases without bound. Thus, the range of f is the interval $(0, \infty)$—that is, the set of positive real numbers. Finally, we sketch the graph of $y = f(x) = 2^x$ in Figure 2. ∎

VIDEO **EXAMPLE 4** Sketch the graph of the exponential function $y = (1/2)^x$.

Solution   The domain of the exponential function $y = (1/2)^x$ is the set of all real numbers. The y-intercept is $(1/2)^0 = 1$; there is no x-intercept, since there is no value of x for which $y = 0$. From the following table of values

x	-5	-4	-3	-2	-1	0	1	2	3	4	5
y	32	16	8	4	2	1	$\frac{1}{2}$	$\frac{1}{4}$	$\frac{1}{8}$	$\frac{1}{16}$	$\frac{1}{32}$

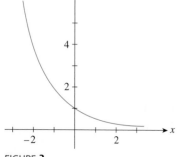

FIGURE **3**
The graph of $y = \left(\dfrac{1}{2}\right)^x$

we deduce that $(1/2)^x = 1/2^x$ increases without bound as x decreases without bound and that $(1/2)^x$ decreases and approaches zero as x increases without bound. Thus, the range of f is the interval $(0, \infty)$. The graph of $y = f(x) = (1/2)^x$ is sketched in Figure 3. ∎

The functions $y = 2^x$ and $y = (1/2)^x$, whose graphs you studied in Examples 3 and 4, are special cases of the exponential function $y = f(x) = b^x$, obtained by setting $b = 2$ and $b = 1/2$, respectively. In general, the exponential function $y = b^x$ with $b > 1$ has a graph similar to that of $y = 2^x$, whereas the graph of $y = b^x$ for $0 < b < 1$ is similar to that of $y = (1/2)^x$ (Exercises 23 and 24 on page 162). When $b = 1$, the function $y = b^x$ reduces to the constant function $y = 1$. For comparison, the graphs of all three functions are sketched in Figure 4.

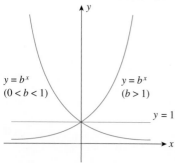

FIGURE **4**
$y = b^x$ is an increasing function of x if $b > 1$, a constant function if $b = 1$, and a decreasing function if $0 < b < 1$.

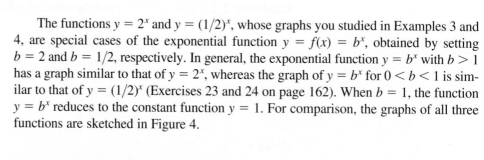

Properties of the Exponential Function

The exponential function $y = b^x$ $(b > 0, b \neq 1)$ has the following properties:

1. Its domain is $(-\infty, \infty)$.
2. Its range is $(0, \infty)$.
3. Its graph passes through the point $(0, 1)$.
4. Its graph is an unbroken curve devoid of holes or jumps.
5. Its graph rises from left to right if $b > 1$ and falls from left to right if $b < 1$.

TABLE 1	
m	$\left(1 + \dfrac{1}{m}\right)^m$
10	2.59374
100	2.70481
1000	2.71692
10,000	2.71815
100,000	2.71827
1,000,000	2.71828

The Base e

It can be shown, although we will not do so here, that as m gets larger and larger, the value of the expression

$$\left(1 + \frac{1}{m}\right)^m$$

approaches the irrational number 2.7182818, which we denote by e. You may convince yourself of the plausibility of this definition of the number e by examining Table 1, which may be constructed with the help of a calculator. (Also, see the Exploring with Technology exercise that follows.)

Exploring with TECHNOLOGY

To obtain a visual confirmation of the fact that the expression $(1 + 1/m)^m$ approaches the number $e = 2.71828.\ldots$ as m gets larger and larger, plot the graph of $f(x) = (1 + 1/x)^x$ in a suitable viewing window, and observe that $f(x)$ approaches 2.71828. . . as x gets larger and larger. Use ZOOM and TRACE to find the value of $f(x)$ for large values of x.

VIDEO **EXAMPLE 5** Sketch the graph of the function $y = e^x$.

Solution Since $e > 1$, it follows from our previous discussion that the graph of $y = e^x$ is similar to the graph of $y = 2^x$ (see Figure 2). With the aid of a calculator, we obtain the following table:

x	-3	-2	-1	0	1	2	3
y	0.05	0.14	0.37	1	2.72	7.39	20.09

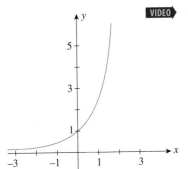

FIGURE **5**
The graph of $y = e^x$

The graph of $y = e^x$ is sketched in Figure 5.

Next, we consider another exponential function to the base e that is closely related to the previous function and is particularly useful in constructing models that describe "exponential decay."

EXAMPLE 6 Sketch the graph of the function $y = e^{-x}$.

Solution Since $e > 1$, it follows that $0 < 1/e < 1$, so $f(x) = e^{-x} = 1/e^x = (1/e)^x$ is an exponential function with base less than 1. Therefore, it has a graph similar to that of the exponential function $y = (1/2)^x$. As before, we construct the following table of values of $y = e^{-x}$ for selected values of x:

x	-3	-2	-1	0	1	2	3
y	20.09	7.39	2.72	1	0.37	0.14	0.05

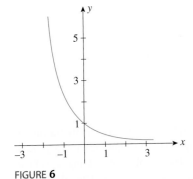

FIGURE **6**
The graph of $y = e^{-x}$

Using this table, we sketch the graph of $y = e^{-x}$ in Figure 6.

3.1   Self-Check Exercises

1. Solve the equation $2^{2x+1} \cdot 2^{-3} = 2^{x-1}$.

2. Sketch the graph of $y = e^{0.4x}$.

Solutions to Self-Check Exercises 3.1 can be found on page 164.

3.1   Concept Questions

1. Define the exponential function f with base b and exponent x. What restrictions, if any, are placed on b?

2. For the exponential function $y = b^x$ ($b > 0$, $b \neq 1$), state (a) its domain and range, (b) its y-intercept, and (c) where its graph rises and where it falls for the case $b > 1$ and the case $b < 1$.

3.1   Exercises

In Exercises 1–6, evaluate the expression.

1. **a.** $4^{-3} \cdot 4^5$ **b.** $3^{-3} \cdot 3^6$

2. **a.** $(2^{-1})^3$ **b.** $(3^{-2})^3$

3. **a.** $9(9)^{-1/2}$ **b.** $5(5)^{-1/2}$

4. **a.** $\left[\left(-\dfrac{1}{2}\right)^3\right]^{-2}$ **b.** $\left[\left(-\dfrac{1}{3}\right)^2\right]^{-3}$

5. **a.** $\dfrac{(-3)^4(-3)^5}{(-3)^8}$ **b.** $\dfrac{(2^{-4})(2^6)}{2^{-1}}$

6. **a.** $3^{1/4} \cdot 9^{-5/8}$ **b.** $2^{3/4} \cdot 4^{-3/2}$

In Exercises 7–12, simplify the expression.

7. **a.** $(64x^9)^{1/3}$ **b.** $(25x^3y^4)^{1/2}$

8. **a.** $(2x^3)(-4x^{-2})$ **b.** $(4x^{-2})(-3x^5)$

9. **a.** $\dfrac{6a^{-4}}{3a^{-3}}$ **b.** $\dfrac{4b^{-4}}{12b^{-6}}$

10. **a.** $y^{-3/2}y^{5/3}$ **b.** $x^{-3/5}x^{8/3}$

11. **a.** $(2x^3y^2)^3$ **b.** $(4x^2y^2z^3)^2$

12. **a.** $\dfrac{5^0}{(2^{-3}x^{-3}y^2)^2}$ **b.** $\dfrac{(x+y)(x-y)}{(x-y)^0}$

In Exercises 13–22, solve the equation for x.

13. $6^{2x} = 6^6$ 14. $5^{-x} = 5^3$

15. $3^{3x-4} = 3^5$ 16. $10^{2x-1} = 10^{x+3}$

17. $(2.1)^{x+2} = (2.1)^5$ 18. $(1.3)^{x-2} = (1.3)^{2x+1}$

19. $8^x = \left(\dfrac{1}{32}\right)^{x-2}$ 20. $3^{x-x^2} = \dfrac{1}{9^x}$

21. $3^{2x} - 12 \cdot 3^x + 27 = 0$

22. $2^{2x} - 4 \cdot 2^x + 4 = 0$

In Exercises 23–32, sketch the graphs of the given functions on the same axes.

23. $y = 2^x$, $y = 3^x$, and $y = 4^x$

24. $y = \left(\dfrac{1}{2}\right)^x$, $y = \left(\dfrac{1}{3}\right)^x$, and $y = \left(\dfrac{1}{4}\right)^x$

25. $y = 2^{-x}$, $y = 3^{-x}$, and $y = 4^{-x}$

26. $y = 4^{0.5x}$ and $y = 4^{-0.5x}$

27. $y = 4^{0.5x}$, $y = 4^x$, and $y = 4^{2x}$

28. $y = e^x$, $y = 2e^x$, and $y = 3e^x$

29. $y = e^{0.5x}$, $y = e^x$, and $y = e^{1.5x}$

30. $y = e^{-0.5x}$, $y = e^{-x}$, and $y = e^{-1.5x}$

31. $y = 0.5e^{-x}$, $y = e^{-x}$, and $y = 2e^{-x}$

32. $y = 1 - e^{-x}$ and $y = 1 - e^{-0.5x}$

33. A function f has the form $f(x) = Ae^{kx}$. Find f if it is known that $f(0) = 100$ and $f(1) = 120$.
    Hint: $e^{kx} = (e^k)^x$

34. If $f(x) = Axe^{-kx}$, find $f(3)$ if $f(1) = 5$ and $f(2) = 7$.
    Hint: $e^{kx} = (e^k)^x$

35. If
$$f(t) = \frac{1000}{1 + Be^{-kt}}$$
    find $f(5)$ given that $f(0) = 20$ and $f(2) = 30$.
    Hint: $e^{kx} = (e^k)^x$

36. **TRACKING WITH GPS** Employers are increasingly turning to GPS (global positioning system) technology to keep track of their fleet vehicles. The number of automatic vehicle trackers installed on fleet vehicles in the United States is approximated by
$$N(t) = 0.6e^{0.17t} \qquad (0 \leq t \leq 5)$$

where $N(t)$ is measured in millions and t is measured in years, with $t = 0$ corresponding to 2000.

a. How many automatic vehicle trackers were installed in the year 2000? How many were installed in 2005?

b. Sketch the graph of N.

Source: C. J. Driscoll Associates.

37. DISABILITY RATES Because of medical technology advances, the disability rates for people over 65 years old have been dropping rather dramatically. The function

$$R(t) = 26.3e^{-0.016t} \qquad (0 \le t \le 18)$$

gives the disability rate $R(t)$, in percent, for people over age 65 from 1982 ($t = 0$) through 2000, where t is measured in years.

a. What was the disability rate in 1982? In 1986? In 1994? In 2000?

b. Sketch the graph of R.

Source: Frost and Sullivan.

38. MARRIED HOUSEHOLDS The percentage of families that were married households in the United States between 1970 and 2000 is approximately

$$P(t) = 86.9e^{-0.05t} \qquad (0 \le t \le 3)$$

where t is measured in decades, with $t = 0$ corresponding to 1970.

a. What percentage of families in the United States were married households in 1970? In 1980? In 1990? In 2000?

b. Sketch the graph of P.

Source: U.S. Census Bureau.

39. GROWTH IN NUMBER OF WEBSITES According to a study conducted in 2000, the projected number of Web addresses (in billions) is approximated by the function

$$N(t) = 0.45e^{0.5696t} \qquad (0 \le t \le 5)$$

where t is measured in years, with $t = 0$ corresponding to 1997.

a. Complete the following table by finding the number of Web addresses in each year:

Year	0	1	2	3	4	5
Number of Web Addresses (billions)						

b. Sketch the graph of N.

40. INTERNET USERS IN CHINA The number of Internet users in China is approximated by

$$N(t) = 94.5e^{0.2t} \qquad (1 \le t \le 6)$$

where $N(t)$ is measured in millions and t is measured in years, with $t = 1$ corresponding to 2005.

a. How many Internet users were there in 2005? In 2006?

b. How many Internet users were there expected to be in 2010?

c. Sketch the graph of N.

Source: C. E. Unterberg.

41. ALTERNATIVE MINIMUM TAX The alternative minimum tax was created in 1969 to prevent the very wealthy from using creative deductions and shelters to avoid having to pay anything to the Internal Revenue Service. But it has increasingly hit the middle class. The number of taxpayers subjected to an alternative minimum tax is projected to be

$$N(t) = \frac{35.5}{1 + 6.89e^{-0.8674t}} \qquad (0 \le t \le 6)$$

where $N(t)$ is measured in millions and t is measured in years, with $t = 0$ corresponding to 2004. What was the projected number of taxpayers subjected to an alternative minimum tax in 2010?

Source: Brookings Institution.

42. ABSORPTION OF DRUGS The concentration of a drug in an organ at any time t (in seconds) is given by

$$C(t) = \begin{cases} 0.3t - 18(1 - e^{-t/60}) & \text{if } 0 \le t \le 20 \\ 18e^{-t/60} - 12e^{-(t-20)/60} & \text{if } t > 20 \end{cases}$$

where $C(t)$ is measured in grams per cubic centimeter (g/cm^3).

a. What is the initial concentration of the drug in the organ?

b. What is the concentration of the drug in the organ after 10 sec?

c. What is the concentration of the drug in the organ after 30 sec?

43. ABSORPTION OF DRUGS The concentration of a drug in an organ at any time t (in seconds) is given by

$$x(t) = 0.08 + 0.12(1 - e^{-0.02t})$$

where $x(t)$ is measured in grams per cubic centimeter (g/cm^3).

a. What is the initial concentration of the drug in the organ?

b. What is the concentration of the drug in the organ after 20 sec?

44. ABSORPTION OF DRUGS Jane took 100 mg of a drug in the morning and another 100 mg of the same drug at the same time the following morning. The amount of the drug in her body t days after the first dose was taken is given by

$$A(t) = \begin{cases} 100e^{-1.4t} & \text{if } 0 \le t < 1 \\ 100(1 + e^{1.4})e^{-1.4t} & \text{if } t \ge 1 \end{cases}$$

What was the amount of drug in Jane's body immediately after taking the second dose? After 2 days?

In Exercises 45–48, determine whether the statement is true or false. If it is true, explain why it is true. If it is false, give an example to show why it is false.

45. $(x^2 + 1)^3 = x^6 + 1$

46. $e^{xy} = e^x e^y$

47. If $x < y$, then $e^x < e^y$.

48. If $0 < b < 1$ and $x < y$, then $b^x > b^y$.

3.1 Solutions to Self-Check Exercises

1. $2^{2x+1} \cdot 2^{-3} = 2^{x-1}$

$2^{(2x+1)+(-3)} = 2^{x-1}$ Laws of exponents

$2x + 1 - 3 = x - 1$ $b^m = b^n \Rightarrow m = n$

$x = 1$

This is true if and only if $x - 1 = 0$ or $x = 1$.

2. We first construct the following table of values:

Next, we plot these points and join them by a smooth curve to obtain the graph of f shown in the accompanying figure.

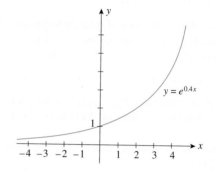

x	-3	-2	-1	0	1	2	3	4
$y = e^{0.4x}$	0.3	0.4	0.7	1	1.5	2.2	3.3	5

USING TECHNOLOGY

Although the proof is outside the scope of this book, it can be proved that an exponential function of the form $f(x) = b^x$, where $b > 1$, will ultimately grow faster than the power function $g(x) = x^n$ for *any* positive real number n. To give a visual demonstration of this result for the special case of the exponential function $f(x) = e^x$, we can use a graphing utility to plot the graphs of both f and g (for selected values of n) on the same set of axes in an appropriate viewing window and observe that the graph of f ultimately lies above that of g.

EXAMPLE 1 Use a graphing utility to plot the graphs of (a) $f(x) = e^x$ and $g(x) = x^3$ on the same set of axes in the viewing window $[0, 6] \times [0, 250]$ and (b) $f(x) = e^x$ and $g(x) = x^5$ in the viewing window $[0, 20] \times [0, 1,000,000]$.

Solution

a. The graphs of $f(x) = e^x$ and $g(x) = x^3$ in the viewing window $[0, 6] \times [0, 250]$ are shown in Figure T1a.

b. The graphs of $f(x) = e^x$ and $g(x) = x^5$ in the viewing window $[0, 20] \times [0, 1,000,000]$ are shown in Figure T1b.

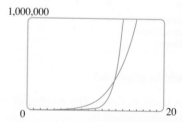

FIGURE **T1**

(a) The graphs of $f(x) = e^x$ and $g(x) = x^3$ in the viewing window $[0, 6] \times [0, 250]$

(b) The graphs of $f(x) = e^x$ and $g(x) = x^5$ in the viewing window $[0, 20] \times [0, 1,000,000]$

In the exercises that follow, you are asked to use a graphing utility to reveal the properties of exponential functions.

TECHNOLOGY EXERCISES

In Exercises 1 and 2, plot the graphs of the functions f and g on the same set of axes in the specified viewing window.

1. $f(x) = e^x$ and $g(x) = x^2$; $[0, 4] \times [0, 30]$

2. $f(x) = e^x$ and $g(x) = x^4$; $[0, 15] \times [0, 20{,}000]$

In Exercises 3 and 4, plot the graphs of the functions f and g on the same set of axes in an appropriate viewing window to demonstrate that f ultimately grows faster than g. (*Note:* Your answer will *not* be unique.)

3. $f(x) = 2^x$ and $g(x) = x^{2.5}$

4. $f(x) = 3^x$ and $g(x) = x^3$

5. Plot the graphs of $f(x) = 2^x$, $g(x) = 3^x$, and $h(x) = 4^x$ on the same set of axes in the viewing window $[0, 5] \times [0, 100]$. Comment on the relationship between the base b and the growth of the function $f(x) = b^x$.

6. Plot the graphs of $f(x) = (1/2)^x$, $g(x) = (1/3)^x$, and $h(x) = (1/4)^x$ on the same set of axes in the viewing window $[0, 4] \times [0, 1]$. Comment on the relationship between the base b and the growth of the function $f(x) = b^x$.

7. Plot the graphs of $f(x) = e^x$, $g(x) = 2e^x$, and $h(x) = 3e^x$ on the same set of axes in the viewing window $[-3, 3] \times [0, 10]$. Comment on the role played by the constant k in the graph of $f(x) = ke^x$.

8. Plot the graphs of $f(x) = -e^x$, $g(x) = -2e^x$, and $h(x) = -3e^x$ on the same set of axes in the viewing window $[-3, 3] \times [-10, 0]$. Comment on the role played by the constant k in the graph of $f(x) = ke^x$.

9. Plot the graphs of $f(x) = e^{0.5x}$, $g(x) = e^x$, and $h(x) = e^{1.5x}$ on the same set of axes in the viewing window $[-2, 2] \times [0, 4]$. Comment on the role played by the constant k in the graph of $f(x) = e^{kx}$.

10. Plot the graphs of $f(x) = e^{-0.5x}$, $g(x) = e^{-x}$, and $h(x) = e^{-1.5x}$ on the same set of axes in the viewing window $[-2, 2] \times [0, 4]$. Comment on the role played by the constant k in the graph of $f(x) = e^{kx}$.

11. **ABSORPTION OF DRUGS** The concentration of a drug in an organ at any time t (in seconds) is given by

$$x(t) = 0.08 + 0.12(1 - e^{-0.02t})$$

where $x(t)$ is measured in grams per cubic centimeter (g/cm^3).

a. Plot the graph of the function x in the viewing window $[0, 200] \times [0, 0.2]$.

b. What is the initial concentration of the drug in the organ?

c. What is the concentration of the drug in the organ after 20 sec?

12. **ABSORPTION OF DRUGS** Jane took 100 mg of a drug in the morning and another 100 mg of the same drug at the same time the following morning. The amount of the drug in her body t days after the first dosage was taken is given by

$$A(t) = \begin{cases} 100e^{-1.4t} & \text{if } 0 \le t < 1 \\ 100(1 + e^{1.4})e^{-1.4t} & \text{if } t \ge 1 \end{cases}$$

a. Plot the graph of the function A in the viewing window $[0, 5] \times [0, 140]$.

b. Verify the results of Exercise 44, page 163.

13. **ABSORPTION OF DRUGS** The concentration of a drug in an organ at any time t (in seconds) is given by

$$C(t) = \begin{cases} 0.3t - 18(1 - e^{-t/60}) & \text{if } 0 \le t \le 20 \\ 18e^{-t/60} - 12e^{-(t-20)/60} & \text{if } t > 20 \end{cases}$$

where $C(t)$ is measured in grams per cubic centimeter (g/cm^3).

a. Plot the graph of the function C in the viewing window $[0, 120] \times [0, 1]$.

b. How long after the drug is first introduced will it take for the concentration of the drug to reach a peak?

c. How long after the concentration of the drug has peaked will it take for the concentration of the drug to fall back to 0.5 g/cm^3?
Hint: Plot the graphs of $y_1 = C(x)$ and $y_2 = 0.5$, and use the ISECT function of your graphing utility.

14. **MODELING WITH DATA** The estimated number of Internet users in China (in millions) from 2005 through 2010 are shown in the following table:

Year	2005	2006	2007	2008	2009	2010
Number	116.1	141.9	169.0	209.0	258.1	314.8

a. Use **ExpReg** to find an exponential regression model for the data. Let $t = 1$ correspond to 2005.
Hint: $a^x = e^{x \ln a}$

b. Plot the scatter diagram and the graph of the function f found in part (a).

3.2 Logarithmic Functions

Logarithms

You are already familiar with exponential equations of the form

$$b^y = x \qquad (b > 0, b \neq 1)$$

where the variable x is expressed in terms of a real number b and a variable y. But what about solving this same equation for y? You may recall from your study of algebra that the number y is called the **logarithm of x to the base b** and is denoted by $\log_b x.$ It is the power to which the base b must be raised to obtain the number x.

> **Logarithm of x to the Base b**
>
> $$y = \log_b x \quad \text{if and only if} \quad x = b^y \qquad (b > 0, b \neq 1, \text{ and } x > 0)$$

⚠️ Observe that the logarithm $\log_b x$ is defined only for positive values of x.

EXAMPLE 1

a. $\log_{10} 100 = 2$ since $100 = 10^2$
b. $\log_5 125 = 3$ since $125 = 5^3$

c. $\log_3 \dfrac{1}{27} = -3$ since $\dfrac{1}{27} = \dfrac{1}{3^3} = 3^{-3}$

d. $\log_{20} 20 = 1$ since $20 = 20^1$ ∎

EXAMPLE 2 Solve each of the following equations for x.

a. $\log_3 x = 4$ **b.** $\log_{16} 4 = x$ **c.** $\log_x 8 = 3$

Solution

a. By definition, $\log_3 x = 4$ implies $x = 3^4 = 81$.
b. $\log_{16} 4 = x$ is equivalent to $4 = 16^x = (4^2)^x = 4^{2x}$, or $4^1 = 4^{2x}$, from which we deduce that

$$2x = 1 \qquad b^m = b^n \Rightarrow m = n$$
$$x = \frac{1}{2}$$

c. Referring once again to the definition, we see that the equation $\log_x 8 = 3$ is equivalent to

$$8 = 2^3 = x^3$$
$$x = 2 \qquad a^m = b^m \Rightarrow a = b$$ ∎

The two most widely used systems of logarithms are the system of **common logarithms,** which uses the number 10 as its base, and the system of **natural logarithms,** which uses the irrational number $e = 2.71828\ldots$ as its base. Also, it is standard practice to write **log** for $\log_{10}$ and **ln** for $\log_e$.

> **Logarithmic Notation**
>
> $$\log x = \log_{10} x \qquad \text{Common logarithm}$$
> $$\ln x = \log_e x \qquad \text{Natural logarithm}$$

The system of natural logarithms is widely used in theoretical work. Using natural logarithms rather than logarithms to other bases often leads to simpler expressions.

Laws of Logarithms

Computations involving logarithms are facilitated by the following **laws of logarithms.**

> **Laws of Logarithms**
> If m and n are positive numbers and $b > 0$, $b \neq 1$, then
>
> **1.** $\log_b mn = \log_b m + \log_b n$
>
> **2.** $\log_b \dfrac{m}{n} = \log_b m - \log_b n$
>
> **3.** $\log_b m^n = n \log_b m$
>
> **4.** $\log_b 1 = 0$
>
> **5.** $\log_b b = 1$

⚠ Do not confuse the expression $\log m/n$ (Law 2) with the expression $\log m/\log n$. For example,

$$\log \frac{100}{10} = \log 100 - \log 10 = 2 - 1 = 1 \neq \frac{\log 100}{\log 10} = \frac{2}{1} = 2$$

You will be asked to prove these laws in Exercises 74–76 on page 174. Their derivations are based on the definition of a logarithm and the corresponding laws of exponents. The following examples illustrate the properties of logarithms.

EXAMPLE 3

a. $\log(2 \cdot 3) = \log 2 + \log 3$ **b.** $\ln \dfrac{5}{3} = \ln 5 - \ln 3$

c. $\log \sqrt{7} = \log 7^{1/2} = \dfrac{1}{2} \log 7$ **d.** $\log_5 1 = 0$

e. $\log_{45} 45 = 1$

EXAMPLE 4 Given that $\log 2 \approx 0.3010$, $\log 3 \approx 0.4771$, and $\log 5 \approx 0.6990$, use the laws of logarithms to find

a. $\log 15$ **b.** $\log 7.5$ **c.** $\log 81$ **d.** $\log 50$

Solution

a. Note that $15 = 3 \cdot 5$, so by Law 1 for logarithms,

$$
\begin{aligned}
\log 15 &= \log 3 \cdot 5 \\
&= \log 3 + \log 5 \\
&\approx 0.4771 + 0.6990 \\
&= 1.1761
\end{aligned}
$$

b. Observing that $7.5 = 15/2 = (3 \cdot 5)/2$, we apply Laws 1 and 2, obtaining

$$\log 7.5 = \log \frac{(3)(5)}{2}$$
$$= \log 3 + \log 5 - \log 2$$
$$\approx 0.4771 + 0.6990 - 0.3010$$
$$= 0.8751$$

c. Since $81 = 3^4$, we apply Law 3 to obtain

$$\log 81 = \log 3^4$$
$$= 4 \log 3$$
$$\approx 4(0.4771)$$
$$= 1.9084$$

d. We write $50 = 5 \cdot 10$ and find

$$\log 50 = \log(5)(10)$$
$$= \log 5 + \log 10$$
$$\approx 0.6990 + 1 \qquad \text{Use Law 5}$$
$$= 1.6990$$

VIDEO   **EXAMPLE 5** Expand and simplify the following expressions:

a. $\log_3 x^2 y^3$ **b.** $\log_2 \dfrac{x^2 + 1}{2^x}$ **c.** $\ln \dfrac{x^2 \sqrt{x^2 - 1}}{e^x}$

Solution

a. $\log_3 x^2 y^3 = \log_3 x^2 + \log_3 y^3 \qquad$ Law 1
$$= 2 \log_3 x + 3 \log_3 y \qquad \text{Law 3}$$

b. $\log_2 \dfrac{x^2 + 1}{2^x} = \log_2(x^2 + 1) - \log_2 2^x \qquad$ Law 2

$$= \log_2(x^2 + 1) - x \log_2 2 \qquad \text{Law 3}$$
$$= \log_2(x^2 + 1) - x \qquad \text{Law 5}$$

c. $\ln \dfrac{x^2 \sqrt{x^2 - 1}}{e^x} = \ln \dfrac{x^2(x^2 - 1)^{1/2}}{e^x} \qquad$ Rewrite

$$= \ln x^2 + \ln(x^2 - 1)^{1/2} - \ln e^x \qquad \text{Laws 1 and 2}$$
$$= 2 \ln x + \frac{1}{2} \ln(x^2 - 1) - x \ln e \qquad \text{Law 3}$$
$$= 2 \ln x + \frac{1}{2} \ln(x^2 - 1) - x \qquad \text{Law 5}$$

Examples 6 and 7 illustrate how the properties of logarithms are used to solve equations.

EXAMPLE 6 Solve $\log_3(x + 1) - \log_3(x - 1) = 1$ for x.

Solution Using the properties of logarithms, we obtain

$$\log_3(x + 1) - \log_3(x - 1) = 1$$

$$\log_3 \frac{x + 1}{x - 1} = 1 \qquad \text{Law 2}$$

$$\frac{x + 1}{x - 1} = 3^1 = 3 \qquad \text{Definition of logarithms}$$

So

$$x + 1 = 3(x - 1)$$
$$x + 1 = 3x - 3$$
$$4 = 2x$$
$$x = 2$$

EXAMPLE 7 Solve $\log x + \log(2x - 1) = \log 6$.

Solution We have

$$\log x + \log(2x - 1) = \log 6$$

$$\log x + \log(2x - 1) - \log 6 = 0$$

$$\log\left[\frac{x(2x - 1)}{6}\right] = 0 \qquad \text{Laws 1 and 2}$$

$$\frac{x(2x - 1)}{6} = 10^0 = 1 \qquad \text{Definition of logarithms}$$

So

$$x(2x - 1) = 6$$
$$2x^2 - x - 6 = 0$$
$$(2x + 3)(x - 2) = 0$$
$$x = -\frac{3}{2} \quad \text{or} \quad 2$$

Because $(2x - 1)$ is defined for $2x - 1 > 0$, or $x > \frac{1}{2}$, the domain of $\log(2x - 1)$ is the interval $(\frac{1}{2}, \infty)$. So we reject the root $-\frac{3}{2}$ of the quadratic equation and conclude that the solution of the given equation is $x = 2$.

Note   Using the fact that $\log a = \log b$ if and only if $a = b$, we can also solve the equation of Example 7 as follows:

$$\log x + \log(2x - 1) = \log 6$$
$$\log x(2x - 1) = \log 6$$
$$x(2x - 1) = 6$$

The rest of the solution is the same as that in Example 7.

Logarithmic Functions and Their Graphs

The definition of a logarithm implies that if b and n are positive numbers and b is different from 1, then the expression $\log_b n$ is a real number. This enables us to define a logarithmic function as follows.

> **Logarithmic Function**
> The function defined by
> $$f(x) = \log_b x \qquad (b > 0, b \neq 1)$$
> is called the **logarithmic function with base** b. The domain of f is the set of all positive numbers.

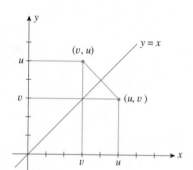

FIGURE 7
The points (u, v) and (v, u) are mirror reflections of each other.

One way to obtain the graph of the logarithmic function $y = \log_b x$ is to construct a table of values of the logarithm (base b). However, another method—and a more instructive one—is based on exploiting the intimate relationship between logarithmic and exponential functions.

If a point (u, v) lies on the graph of $y = \log_b x$, then

$$v = \log_b u$$

But we can also write this equation in exponential form as

$$u = b^v$$

So the point (v, u) also lies on the graph of the function $y = b^x$. Let's look at the relationship between the points (u, v) and (v, u) and the line $y = x$ (Figure 7). If we think of the line $y = x$ as a mirror, then the point (v, u) is the mirror reflection of the point (u, v). Similarly, the point (u, v) is the mirror reflection of the point (v, u). We can take advantage of this relationship to help us draw the graph of logarithmic functions. For example, if we wish to draw the graph of $y = \log_b x$, where $b > 1$, then we need only draw the mirror reflection of the graph of $y = b^x$ with respect to the line $y = x$ (Figure 8).

You may discover the following properties of the logarithmic function by taking the reflection of the graph of an appropriate exponential function (Exercises 47 and 48 on page 173).

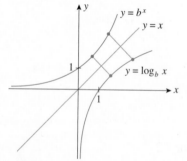

FIGURE 8
The graphs of $y = b^x$ and $y = \log_b x$ are mirror reflections of each other.

> **Properties of the Logarithmic Function**
> The logarithmic function $y = \log_b x$ $(b > 0, b \neq 1)$ has the following properties:
> **1.** Its domain is $(0, \infty)$.
> **2.** Its range is $(-\infty, \infty)$.
> **3.** Its graph passes through the point $(1, 0)$.
> **4.** Its graph is an unbroken curve devoid of holes or jumps.
> **5.** Its graph rises from left to right if $b > 1$ and falls from left to right if $b < 1$.

EXAMPLE 8 Sketch the graph of the function $y = \ln x$.

Solution   We first sketch the graph of $y = e^x$. Then, the required graph is obtained by tracing the mirror reflection of the graph of $y = e^x$ with respect to the line $y = x$ (Figure 9). ∎

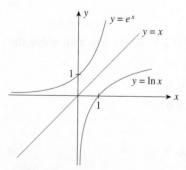

FIGURE 9
The graph of $y = \ln x$ is the mirror reflection of the graph of $y = e^x$.

Properties Relating the Exponential and Logarithmic Functions

We made use of the relationship that exists between the exponential function $f(x) = e^x$ and the logarithmic function $g(x) = \ln x$ when we sketched the graph of g in Example 8. This relationship is further described by the following properties, which are an immediate consequence of the definition of the logarithm of a number.

Properties Relating e^x and $\ln x$

$$e^{\ln x} = x \qquad \text{(for } x > 0\text{)} \tag{1}$$

$$\ln e^x = x \qquad \text{(for any real number } x\text{)} \tag{2}$$

(Try to verify these properties.)

From Properties 1 and 2, we conclude that the composite function satisfies

$$(f \circ g)(x) = f[g(x)]$$
$$= e^{\ln x} = x \qquad \text{(for all } x > 0\text{)}$$

$$(g \circ f)(x) = g[f(x)]$$
$$= \ln e^x = x \qquad \text{(for all } x > 0\text{)}$$

Any two functions f and g that satisfy this relationship are said to be **inverses** of each other. Note that the function f undoes what the function g does, and vice versa, so the composition of the two functions in any order results in the identity function $H(x) = (f \circ g)(x) = (g \circ f)(x)$.

The relationships expressed in Equations (1) and (2) are useful in solving equations that involve exponentials and logarithms.

Exploring with TECHNOLOGY

You can demonstrate the validity of Properties 1 and 2, which state that the exponential function $f(x) = e^x$ and the logarithmic function $g(x) = \ln x$ are inverses of each other, as follows:

1. Sketch the graph of $(f \circ g)(x) = e^{\ln x}$, using the viewing window $[0, 10] \times [0, 10]$. Interpret the result.
2. Sketch the graph of $(g \circ f)(x) = \ln e^x$, using the standard viewing window. Interpret the result.

VIDEO **EXAMPLE 9** Solve the equation $2e^{x+2} = 5$.

Explore & Discuss

Consider the equation $y = y_0 b^{kx}$, where y_0 and k are positive constants and $b > 0$, $b \neq 1$. Suppose we want to express y in the form $y = y_0 e^{px}$. Use the laws of logarithms to show that $p = k \ln b$ and hence that $y = y_0 e^{(k \ln b)x}$ is an alternative form of $y = y_0 b^{kx}$ using the base e.

Solution We first divide both sides of the equation by 2 to obtain

$$e^{x+2} = \frac{5}{2} = 2.5$$

Next, taking the natural logarithm of each side of the equation and using Equation (2), we have

$$\ln e^{x+2} = \ln 2.5$$
$$x + 2 = \ln 2.5$$
$$x = -2 + \ln 2.5$$
$$\approx -1.08$$

EXAMPLE 10 Solve the equation $5 \ln x + 3 = 0$.

Solution   Adding -3 to both sides of the equation leads to

$$5 \ln x = -3$$

$$\ln x = -\frac{3}{5} = -0.6$$

and so

$$e^{\ln x} = e^{-0.6}$$

Using Equation (1), we conclude that

$$x = e^{-0.6}$$

$$\approx 0.55$$

3.2 Self-Check Exercises

1. Sketch the graph of $y = 3^x$ and $y = \log_3 x$ on the same set of axes.

2. Solve the equation $3e^{x+1} - 2 = 4$.

Solutions to Self-Check Exercises 3.2 can be found on page 175.

3.2 Concept Questions

1. Define the logarithmic function f with base b. What restrictions, if any, are placed on b?

2. For the logarithmic function $y = \log_b x$ ($b > 0$, $b \neq 1$), state (a) its domain and range, (b) its x-intercept, (c) where its graph rises and where it falls for the case $b > 1$ and the case $b < 1$.

3. a. If $x > 0$, what is $e^{\ln x}$?
   b. If x is any real number, what is $\ln e^x$?

3.2 Exercises

In Exercises 1–10, express each equation in logarithmic form.

1. $2^6 = 64$

2. $3^5 = 243$

3. $4^{-2} = \dfrac{1}{16}$

4. $5^{-3} = \dfrac{1}{125}$

5. $\left(\dfrac{1}{3}\right)^1 = \dfrac{1}{3}$

6. $\left(\dfrac{1}{2}\right)^{-4} = 16$

7. $32^{4/5} = 16$

8. $81^{3/4} = 27$

9. $10^{-3} = 0.001$

10. $16^{-1/4} = 0.5$

In Exercises 11–16, given that $\log 3 \approx 0.4771$ and $\log 4 \approx 0.6021$, find each quantity using the laws of logarithms. (Do not use a calculator.)

11. $\log 12$

12. $\log \dfrac{3}{4}$

13. $\log 16$

14. $\log \sqrt{3}$

15. $\log 48$

16. $\log \dfrac{1}{300}$

In Exercises 17–20, write the expression as the logarithm of a single quantity.

17. $2 \ln a + 3 \ln b$

18. $\dfrac{1}{2} \ln x + 2 \ln y - 3 \ln z$

19. $\ln 3 + \dfrac{1}{2} \ln x + \ln y - \dfrac{1}{3} \ln z$

20. $\ln 2 + \dfrac{1}{2} \ln(x + 1) - 2 \ln(1 + \sqrt{x})$

In Exercises 21–28, use the laws of logarithms to expand and simplify the expression.

21. $\log x(x + 1)^4$

22. $\log x(x^2 + 1)^{-1/2}$

23. $\log \dfrac{\sqrt{x + 1}}{x^2 + 1}$

24. $\ln \dfrac{e^x}{1 + e^x}$

25. $\ln xe^{-x^2}$

26. $\ln x(x + 1)(x + 2)$

27. $\ln \dfrac{x^{1/2}}{x^2 \sqrt{1 + x^2}}$

28. $\ln \dfrac{x^2}{\sqrt{x}(1 + x)^2}$

In Exercises 29–42, use the laws of logarithms to solve the equation.

29. $\log_2 x = 3$

30. $\log_3 x = 2$

31. $\log_2 8 = x$

32. $\log_3 27 = 2x$

33. $\log_x 10^3 = 3$

34. $\log_x \dfrac{1}{16} = -2$

35. $\log_2(2x + 5) = 4$

36. $\log_4(5x - 4) = 2$

37. $\log_2 x - \log_2(x - 2) = 3$

38. $\log x - \log(x + 6) = -1$

39. $\log_5(2x + 1) - \log_5(x - 2) = 1$

40. $\log(x + 7) - \log(x - 2) = 1$

41. $\log x + \log(2x - 5) = \log 3$

42. $\log_3(x + 1) + \log_3(2x - 3) = 1$

In Exercises 43–46, sketch the graph of the equation.

43. $y = \log_3 x$

44. $y = \log_{1/3} x$

45. $y = \ln 2x$

46. $y = \ln \dfrac{1}{2} x$

In Exercises 47 and 48, sketch the graphs of the equations on the same coordinate axes.

47. $y = 2^x$ and $y = \log_2 x$

48. $y = e^{3x}$ and $y = \dfrac{1}{3} \ln x$

In Exercises 49–58, use logarithms to solve the equation for t.

49. $e^{0.4t} = 8$

50. $\dfrac{1}{3} e^{-3t} = 0.9$

51. $5e^{-2t} = 6$

52. $4e^{t-1} = 4$

53. $2e^{-0.2t} - 4 = 6$

54. $12 - e^{0.4t} = 3$

55. $\dfrac{50}{1 + 4e^{0.2t}} = 20$

56. $\dfrac{200}{1 + 3e^{-0.3t}} = 100$

57. $A = Be^{-t/2}$

58. $\dfrac{A}{1 + Be^{t/2}} = C$

59. A function f has the form $f(x) = a + b \ln x$. Find f if it is known that $f(1) = 2$ and $f(2) = 4$.

60. AVERAGE LIFE SPAN One reason for the increase in human life span over the years has been the advances in medical technology. The average life span for American women from 1907 through 2007 is given by

$$W(t) = 49.9 + 17.1 \ln t \qquad (1 \le t \le 6)$$

where $W(t)$ is measured in years and t is measured in 20-year intervals, with $t = 1$ corresponding to 1907.
   a. What was the average life expectancy for women in 1907?
   b. If the trend continues, what will be the average life expectancy for women in 2027 ($t = 7$)?

Source: American Association of Retired Persons (AARP).

61. BLOOD PRESSURE A normal child's systolic blood pressure may be approximated by the function

$$p(x) = m(\ln x) + b$$

where $p(x)$ is measured in millimeters of mercury, x is measured in pounds, and m and b are constants. Given that $m = 19.4$ and $b = 18$, determine the systolic blood pressure of a child who weighs 92 lb.

62. MAGNITUDE OF EARTHQUAKES On the Richter scale, the magnitude R of an earthquake is given by the formula

$$R = \log \dfrac{I}{I_0}$$

where I is the intensity of the earthquake being measured and I_0 is the standard reference intensity.
   a. Express the intensity I of an earthquake of magnitude $R = 5$ in terms of the standard intensity I_0.
   b. Express the intensity I of an earthquake of magnitude $R = 8$ in terms of the standard intensity I_0. How many times greater is the intensity of an earthquake of magnitude 8 than one of magnitude 5?
   c. In modern times, the greatest loss of life attributable to an earthquake occurred in eastern China in 1976. Known as the Tangshan earthquake, it registered 8.2 on the Richter scale. How does the intensity of this earthquake compare with the intensity of an earthquake of magnitude $R = 5$?

63. SOUND INTENSITY The relative loudness of a sound D of intensity I is measured in decibels (db), where

$$D = 10 \log \frac{I}{I_0}$$

and I_0 is the standard threshold of audibility.

a. Express the intensity I of a 30-db sound (the sound level of normal conversation) in terms of I_0.

b. Determine how many times greater the intensity of an 80-db sound (rock music) is than that of a 30-db sound.

c. Prolonged noise above 150 db causes permanent deafness. How does the intensity of a 150-db sound compare with the intensity of an 80-db sound?

64. BAROMETRIC PRESSURE Halley's Law states that the barometric pressure (in inches of mercury) at an altitude of x mi above sea level is approximated by the equation

$$p(x) = 29.92 e^{-0.2x} \qquad (x \geq 0)$$

If the barometric pressure as measured by a hot-air balloonist is 20 in. of mercury, what is the balloonist's altitude?

65. HEIGHT OF TREES The height (in feet) of a certain kind of tree is approximated by

$$h(t) = \frac{160}{1 + 240 e^{-0.2t}}$$

where t is the age of the tree in years. Estimate the age of an 80-ft tree.

66. NEWTON'S LAW OF COOLING The temperature of a cup of coffee t min after it is poured is given by

$$T = 70 + 100 e^{-0.0446t}$$

where T is measured in degrees Fahrenheit.

a. What was the temperature of the coffee when it was poured?

b. When will the coffee be cool enough to drink (say, 120°F)?

67. LENGTHS OF FISH The length (in centimeters) of a typical Pacific halibut t years old is approximately

$$f(t) = 200(1 - 0.956 e^{-0.18t})$$

Suppose a Pacific halibut caught by Mike measures 140 cm. What is its approximate age?

68. ABSORPTION OF DRUGS The concentration of a drug in an organ t seconds after it has been administered is given by

$$x(t) = 0.08(1 - e^{-0.02t})$$

where $x(t)$ is measured in grams per cubic centimeter (g/cm³).

a. How long would it take for the concentration of the drug in the organ to reach 0.02 g/cm³?

b. How long would it take for the concentration of the drug in the organ to reach 0.04 g/cm³?

69. ABSORPTION OF DRUGS The concentration of a drug in an organ t seconds after it has been administered is given by

$$x(t) = 0.08 + 0.12 e^{-0.02t}$$

where $x(t)$ is measured in grams per cubic centimeter (g/cm³).

a. How long would it take for the concentration of the drug in the organ to reach 0.18 g/cm³?

b. How long would it take for the concentration of the drug in the organ to reach 0.16 g/cm³?

70. FORENSIC SCIENCE Forensic scientists use the following law to determine the time of death of accident or murder victims. If T denotes the temperature of a body t hr after death, then

$$T = T_0 + (T_1 - T_0)(0.97)^t$$

where T_0 is the air temperature and T_1 is the body temperature at the time of death. John Doe was found murdered at midnight in his house; the room temperature was 70°F, and his body temperature was 80°F when he was found. When was he killed? Assume that the normal body temperature is 98.6°F.

In Exercises 71 and 72, determine whether the statement is true or false. If it is true, explain why it is true. If it is false, give an example to show why it is false.

71. $(\ln x)^3 = 3 \ln x$ for all x in $(0, \infty)$.

72. $\ln a - \ln b = \ln(a - b)$ for all positive real numbers a and b.

73. a. Given that $2^x = e^{kx} \ (x \neq 0)$, find k.

b. Show that, in general, if b is a positive real number, then any equation of the form $y = b^x$ may be written in the form $y = e^{kx}$, for some real number k.

74. Use the definition of a logarithm to prove

a. $\log_b mn = \log_b m + \log_b n$

b. $\log_b \dfrac{m}{n} = \log_b m - \log_b n$

  Hint: Let $\log_b m = p$ and $\log_b n = q$. Then, $b^p = m$ and $b^q = n$.

75. Use the definition of a logarithm to prove

$$\log_b m^n = n \log_b m$$

76. Use the definition of a logarithm to prove

a. $\log_b 1 = 0$

b. $\log_b b = 1$

Solutions to Self-Check Exercises

1. First, sketch the graph of $y = 3^x$ with the help of the following table of values:

x	-3	-2	-1	0	1	2	3
$y = 3^x$	$\frac{1}{27}$	$\frac{1}{9}$	$\frac{1}{3}$	1	3	9	27

Next, take the mirror reflection of this graph with respect to the line $y = x$ to obtain the graph of $y = \log_3 x$.

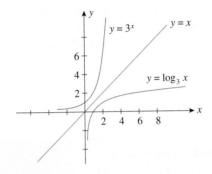

2.
$$3e^{x+1} - 2 = 4$$
$$3e^{x+1} = 6$$
$$e^{x+1} = 2$$
$$\ln e^{x+1} = \ln 2 \qquad \text{Take the logarithm of both sides.}$$
$$(x + 1)\ln e = \ln 2 \qquad \text{Law 3}$$
$$x + 1 = \ln 2 \qquad \text{Law 5}$$
$$x = \ln 2 - 1$$
$$\approx -0.3069$$

3.3 Exponential Functions as Mathematical Models

Exponential Growth

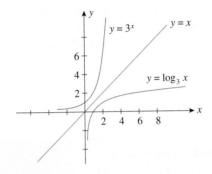

Many problems arising from practical situations can be described mathematically in terms of exponential functions or functions closely related to the exponential function. In this section, we look at some applications involving exponential functions from the fields of the life and social sciences.

In Section 3.1, we saw that the exponential function $f(x) = b^x$ is an increasing function when $b > 1$. In particular, the function $f(x) = e^x$ has this property. Suppose that $Q(t)$ represents a quantity at time t; then one may deduce that the function $Q(t) = Q_0 e^{kt}$, where Q_0 and k are positive constants, has the following properties:

1. $Q(0) = Q_0$
2. $Q(t)$ increases "rapidly" without bound as t increases without bound (Figure 10).

FIGURE 10
Exponential growth

Property 1 follows from the computation
$$Q(0) = Q_0 e^0 = Q_0$$

The exponential function
$$Q(t) = Q_0 e^{kt} \qquad (0 \le t < \infty) \tag{3}$$

provides us with a mathematical model of a quantity $Q(t)$ that is initially present in the amount of $Q(0) = Q_0$ and whose rate of growth at any time t is directly proportional to the amount of the quantity present at time t (see Example 5, Section 9.7). Such a quantity is said to exhibit unrestricted **exponential growth,** and the constant k of proportionality is called the **growth constant.** Interest earned on a fixed deposit when compounded continuously exhibits exponential growth (see Chapter 4). Other examples of unrestricted exponential growth follow.

Carol A. Reeb, Ph.D.

TITLE   Research Associate
INSTITUTION   Hopkins Marine Station, Stanford University

Historically, the world's oceans were thought to provide an unlimited source of inexpensive seafood. However, in a world in which the human population now exceeds six billion people, overfishing has pushed one third of all marine fishery stocks toward a state of collapse.

As a fishery geneticist at Hopkins Marine Station, I study commercially harvested marine populations and use exponential models in my work. The equation for determining the size of a population that grows or declines exponentially is $x_t = x_0 e^{rt}$, where x_0 is the initial population, t is time, and r is the growth or decay constant (positive for growth, negative for decay).

This equation can be used to estimate the population in the past as well as in the future. We know that the demand for seafood increased as the human population grew, eventually causing fish populations to decline. Because genetic diversity is linked to population size, the exponential function is useful to model change in fishery populations and their gene pools over time.

Interestingly, exponential functions can also be used to model the increase in the market value of seafood in the United States over the past 60 years. In general, the price of seafood has increased exponentially, although the price did stabilize briefly in 1995.

Although exponential curves are important to my work, they are not always the best fit. Exponential curves are best applied across short time frames when environments or markets are unlimited. Over longer periods, the logistic growth function is more suitable. In my research, selecting the most accurate model requires examining many possibilities.

© Joe Wible/Hopkins Marine Station; (inset) © Rich Carey/Shutterstock.com

APPLIED EXAMPLE 1   Growth of Bacteria   Under ideal laboratory conditions, the number of bacteria in a culture grows in accordance with the law $Q(t) = Q_0 e^{kt}$, where Q_0 denotes the number of bacteria initially present in the culture, k is a constant determined by the strain of bacteria under consideration and other factors, and t is the elapsed time measured in hours. Suppose 10,000 bacteria are present initially in the culture and 60,000 are present 2 hours later. How many bacteria will there be in the culture at the end of 4 hours?

Solution   We are given that $Q(0) = Q_0 = 10{,}000$, so $Q(t) = 10{,}000 e^{kt}$. Next, the fact that 60,000 bacteria are present 2 hours later translates into $Q(2) = 60{,}000$. Thus,

$$60{,}000 = 10{,}000 e^{2k}$$
$$e^{2k} = 6$$

Taking the natural logarithm on both sides of the equation, we obtain

$$\ln e^{2k} = \ln 6$$
$$2k = \ln 6 \qquad \text{Since } \ln e = 1$$
$$k = \frac{\ln 6}{2}$$
$$k \approx 0.8959$$

Thus, the number of bacteria present at any time t is given by

$$Q(t) \approx 10{,}000 e^{0.8959t}$$

In particular, the number of bacteria present in the culture at the end of 4 hours is given by

$$Q(4) \approx 10{,}000e^{0.8959(4)}$$
$$\approx 360{,}000$$

∎

Exponential Decay

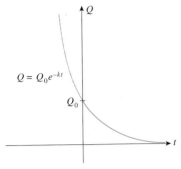

$$Q = Q_0 e^{-kt}$$

Q_0

FIGURE 11
Exponential decay

In contrast to exponential growth, a quantity exhibits **exponential decay** if it decreases at a rate that is directly proportional to its size. Such a quantity may be described by the exponential function

$$Q(t) = Q_0 e^{-kt} \qquad (0 \le t < \infty) \tag{4}$$

where the positive constant Q_0 measures the amount present initially ($t = 0$) and k is some suitable positive number, called the **decay constant.** The choice of this number is determined by the nature of the substance under consideration and other factors. The graph of this function is sketched in Figure 11.

APPLIED EXAMPLE 2 Radioactive Decay Radioactive substances decay exponentially. For example, the amount of radium present at any time t obeys the law $Q(t) = Q_0 e^{-kt}$, where Q_0 is the initial amount present and k is a specific positive constant. The **half-life of a radioactive substance** is the time required for a given amount to be reduced by one-half. Now, it is known that the half-life of radium is approximately 1600 years. Suppose initially there are 200 milligrams of pure radium. Find the amount left after t years. What is the amount left after 800 years?

Solution The initial amount of radium present is 200 milligrams, so $Q(0) = Q_0 = 200$. Thus, $Q(t) = 200e^{-kt}$. Next, the datum concerning the half-life of radium implies that $Q(1600) = 100$, and this gives

$$100 = 200e^{-1600k}$$

$$e^{-1600k} = \frac{1}{2}$$

Taking the natural logarithm on both sides of this equation yields

$$-1600k \ln e = \ln \frac{1}{2}$$

$$-1600k = \ln \frac{1}{2} \qquad \text{\scriptsize ln } e = 1$$

$$k = -\frac{1}{1600} \ln \left(\frac{1}{2} \right) \approx 0.0004332$$

Therefore, the amount of radium left after t years is

$$Q(t) = 200e^{-0.0004332t}$$

In particular, the amount of radium left after 800 years is

$$Q(800) = 200e^{-0.0004332(800)} \approx 141.42$$

or approximately 141 milligrams.

∎

APPLIED EXAMPLE 3 Radioactive Decay Carbon 14, a radioactive isotope of carbon, has a half-life of 5730 years. What is its decay constant?

Solution   We have $Q(t) = Q_0 e^{-kt}$. Since the half-life of the element is 5730 years, half of the substance is left at the end of that period; that is,

$$Q(5730) = Q_0 e^{-5730k} = \frac{1}{2} Q_0$$

$$e^{-5730k} = \frac{1}{2}$$

Taking the natural logarithm on both sides of this equation, we have

$$\ln e^{-5730k} = \ln \frac{1}{2}$$

$$-5730k \approx -0.693147$$

$$k \approx 0.000121$$

Carbon-14 dating is a well-known method used by anthropologists to establish the age of animal and plant fossils. This method assumes that the proportion of carbon 14 (C-14) present in the atmosphere has remained constant over the past 50,000 years. Professor Willard Libby, recipient of the Nobel Prize in chemistry in 1960, proposed this theory.

The amount of C-14 in the tissues of a living plant or animal is constant. However, when an organism dies, it stops absorbing new quantities of C-14, and the amount of C-14 in the remains diminishes because of the natural decay of the radioactive substance. Therefore, the approximate age of a plant or animal fossil can be determined by measuring the amount of C-14 present in the remains.

APPLIED EXAMPLE 4 Carbon-14 Dating A skull from an archeological site has one tenth the amount of C-14 that it originally contained. Determine the approximate age of the skull.

Solution   Here,
$$Q(t) = Q_0 e^{-kt}$$
$$= Q_0 e^{-0.000121t}$$

where Q_0 is the amount of C-14 present originally and k, the decay constant, is equal to 0.000121 (see Example 3). Since $Q(t) = (1/10)Q_0$, we have

$$\frac{1}{10} Q_0 = Q_0 e^{-0.000121t}$$

$$\ln \frac{1}{10} = -0.000121t \qquad \text{Take the natural logarithm on both sides.}$$

$$t = \frac{\ln \frac{1}{10}}{-0.000121}$$

$$\approx 19,030$$

or approximately 19,030 years.

Learning Curves

The next example shows how the exponential function may be applied to describe certain types of learning processes. Consider the function

$$Q(t) = C - Ae^{-kt}$$

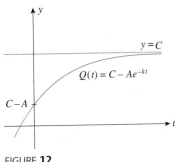

FIGURE **12**
A learning curve

where C, A, and k are positive constants. The graph of the function Q is shown in Figure 12, where that part of the graph corresponding to the negative values of t is drawn with a gray line since, in practice, one normally restricts the domain of the function to the interval $[0, \infty)$. Observe that starting at $t = 0$, $Q(t)$ increases rather rapidly but then the rate of increase slows down considerably after a while. The value of $Q(t)$) never exceeds C.

This behavior of the graph of the function Q closely resembles the learning pattern experienced by workers engaged in highly repetitive work. For example, the productivity of an assembly-line worker increases very rapidly in the early stages of the training period. This productivity increase is a direct result of the worker's training and accumulated experience. But the rate of increase of productivity slows as time goes by, and the worker's productivity level approaches some fixed level due to the limitations of the worker and the machine. Because of this characteristic, the graph of the function $Q(t) = C - Ae^{-kt}$ is often called a **learning curve.**

APPLIED EXAMPLE 5 Assembly Time The Camera Division of Eastman Optical produces a compact digital camera. Eastman's training department determines that after completing the basic training program, a new, previously inexperienced employee will be able to assemble

$$Q(t) = 50 - 30e^{-0.5t}$$

Model F cameras per day t months after the employee starts work on the assembly line.

a. How many Model F cameras can a new employee assemble per day after basic training?
b. How many Model F cameras can an employee with 1 month of experience assemble per day? An employee with 2 months of experience? An employee with 6 months of experience?
c. How many Model F cameras can the average experienced employee assemble per day?

Solution

a. The number of Model F cameras a new employee can assemble is given by

$$Q(0) = 50 - 30 = 20$$

b. The number of Model F cameras that an employee with 1 month of experience, 2 months of experience, and 6 months of experience can assemble per day is given by

$$Q(1) = 50 - 30e^{-0.5} \approx 31.80$$
$$Q(2) = 50 - 30e^{-1} \approx 38.96$$
$$Q(6) = 50 - 30e^{-3} \approx 48.51$$

or approximately 32, 39, and 49, respectively.
c. As t gets larger and larger, $Q(t)$ approaches 50. Hence, the average experienced employee can ultimately be expected to assemble 50 Model F cameras per day.

Other applications of the learning curve are found in models that describe the dissemination of information about a product or the velocity of an object dropped into a viscous medium.

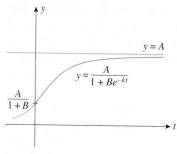

$$y = \frac{A}{1 + Be^{-kt}}$$

FIGURE **13**
A logistic curve

Logistic Growth Functions

Our last example of an application of exponential functions to the description of natural phenomena involves the **logistic** (also called the **S-shaped**, or **sigmoidal**) **curve,** which is the graph of the function

$$Q(t) = \frac{A}{1 + Be^{-kt}}$$

where A, B, and k are positive constants. The function Q is called a **logistic growth function.** The graph of the function Q is sketched in Figure 13.

Observe that $Q(t)$ increases slowly at first but more rapidly as t increases. In fact, for small positive values of t, the logistic curve resembles an exponential growth curve. However, the *rate of growth* of $Q(t)$ decreases quite rapidly as t increases and $Q(t)$ approaches the number A as t gets larger and larger, but $Q(t)$ never exceeds A.

Thus, the logistic curve exhibits both the property of rapid growth of the exponential growth curve as well as the "saturation" property of the learning curve. Because of these characteristics, the logistic curve serves as a suitable mathematical model for describing many natural phenomena. For example, if a small number of rabbits were introduced to a tiny island in the South Pacific, the rabbit population might be expected to grow very rapidly at first, but the growth rate would decrease quickly as overcrowding, scarcity of food, and other environmental factors affected it. The population would eventually stabilize at a level compatible with the life-support capacity of the environment. This level, given by A, is called the *carrying capacity* of the environment. Models describing the spread of rumors and epidemics are other examples of the application of the logistic curve.

VIDEO

APPLIED EXAMPLE 6 Spread of Flu The number of soldiers at Fort MacArthur who contracted influenza after t days during a flu epidemic is approximated by the exponential model

$$Q(t) = \frac{5000}{1 + 1249e^{-kt}}$$

If 40 soldiers contracted the flu by day 7, find how many soldiers contracted the flu by day 15.

Solution   The given information implies that

$$Q(7) = \frac{5000}{1 + 1249e^{-7k}} = 40$$

Thus,

$$40(1 + 1249e^{-7k}) = 5000$$

$$1 + 1249e^{-7k} = \frac{5000}{40} = 125$$

$$e^{-7k} = \frac{124}{1249}$$

$$-7k = \ln \frac{124}{1249}$$

$$k = -\frac{\ln \frac{124}{1249}}{7} \approx 0.33$$

Therefore, the number of soldiers who contracted the flu after t days is given by

$$Q(t) = \frac{5000}{1 + 1249e^{-0.33t}}$$

In particular, the number of soldiers who contracted the flu by day 15 is given by

$$Q(15) = \frac{5000}{1 + 1249e^{-15(0.33)}}$$
$$\approx 508$$

or approximately 508 soldiers. ■

Exploring with
TECHNOLOGY

Refer to Example 6.

1. Use a graphing utility to plot the graph of the function Q, using the viewing window $[0, 40] \times [0, 5000]$.

2. How long will it take for the first 1000 soldiers to contract the flu?
   Hint: Plot the graphs of $y_1 = Q(t)$ and $y_2 = 1000$, and find the point of intersection of the two graphs.

3.3 Self-Check Exercise

IMMIGRATION AND POPULATION GROWTH Suppose the population (in millions) of a country grows in accordance with the rule

$$P = \left(P_0 + \frac{I}{k}\right)e^{kt} - \frac{I}{k}$$

where P denotes the population (in millions) at any time t (in years), k is a constant reflecting the natural growth rate of the population, I is a constant giving the (constant) rate of immi-

gration into the country, and P_0 is the total population of the country at time $t = 0$. The population of the United States in 1980 ($t = 0$) was 226.5 million. If the natural growth rate is 0.8% annually ($k = 0.008$) and net immigration is allowed at the rate of half a million people per year ($I = 0.5$), what was the expected population of the United States in 2010?

The solution to Self-Check Exercise 3.3 can be found on page 184.

3.3 Concept Questions

1. Give the model for unrestricted exponential growth and the model for exponential decay. What effect does the magnitude of the growth (decay) constant have on the growth (decay) of a quantity?

2. What is the half-life of a radioactive substance?

3. What is the logistic growth function? What are its characteristics?

3.3 Exercises

1. EXPONENTIAL GROWTH Given that a quantity $Q(t)$ is described by the exponential growth function

$$Q(t) = 300e^{0.02t}$$

where t is measured in minutes, answer the following questions:

a. What is the growth constant?
b. What quantity is present initially?
c. Complete the following table of values:

t	0	10	20	100	1000
Q					

2. **EXPONENTIAL DECAY** Given that a quantity $Q(t)$ exhibiting exponential decay is described by the function

$$Q(t) = 2000e^{-0.06t}$$

where t is measured in years, answer the following questions:

   a. What is the decay constant?
   b. What quantity is present initially?
   c. Complete the following table of values:

t	0	5	10	20	100
Q					

3. **GROWTH OF BACTERIA** The growth rate of the bacterium *Escherichia coli*, a common bacterium found in the human intestine, is proportional to its size. Under ideal laboratory conditions, when this bacterium is grown in a nutrient broth medium, the number of cells in a culture doubles approximately every 20 min.

   a. If the initial cell population is 100, determine the function $Q(t)$ that expresses the exponential growth of the number of cells of this bacterium as a function of time t (in minutes).
   b. How long will it take for a colony of 100 cells to increase to a population of 1 million?
   c. If the initial cell population were 1000, how would this alter our model?

4. **WORLD POPULATION** The world population at the beginning of 1990 was 5.3 billion. Assume that the population continues to grow at the rate of approximately 2%/year and find the function $Q(t)$ that expresses the world population (in billions) as a function of time t (in years), with $t = 0$ corresponding to the beginning of 1990. Using this function, complete the following table of values and sketch the graph of the function Q.

Year	1990	1995	2000	2005
World Population				

Year	2010	2015	2020	2025
World Population				

5. **WORLD POPULATION** Refer to Exercise 4.

   a. If the world population continues to grow at the rate of approximately 2%/year, find the length of time t_0 required for the world population to triple in size.
   b. Using the time t_0 given in Exercise 4, what would be the world population if the growth rate were reduced to 1.8%/year?

6. **RESALE VALUE** Garland Mills purchased a certain piece of machinery 3 years ago for $500,000. Its present resale value is $320,000. Assuming that the machine's resale value decreases exponentially, what will it be 4 years from now?

7. **ATMOSPHERIC PRESSURE** If the temperature is constant, then the atmospheric pressure P (in pounds per square inch) varies with the altitude above sea level h in accordance with the law

$$P = p_0 e^{-kh}$$

where p_0 is the atmospheric pressure at sea level and k is a constant. If the atmospheric pressure is 15 lb/in.² at sea level and 12.5 lb/in.² at 4000 ft, find the atmospheric pressure at an altitude of 12,000 ft.

8. **RADIOACTIVE DECAY** The radioactive element polonium decays according to the law

$$Q(t) = Q_0 \cdot 2^{-(t/140)}$$

where Q_0 is the initial amount in milligrams and the time t is measured in days. If the amount of polonium left after 280 days is 20 mg, what was the initial amount present?

9. **RADIOACTIVE DECAY** Phosphorus 32 (P-32) has a half-life of 14.2 days. If 100 g of this substance are present initially, find the amount present after t days. What amount will be left after 7.1 days?

10. **NUCLEAR FALLOUT** Strontium 90 (Sr-90), a radioactive isotope of strontium, is present in the fallout resulting from nuclear explosions. It is especially hazardous to animal life, including humans, because, upon ingestion of contaminated food, it is absorbed into the bone structure. Its half-life is 27 years. If the amount of Sr-90 in a certain area is found to be four times the "safe" level, find how much time must elapse before the safe level is reached.

11. **CARBON-14 DATING** Wood deposits recovered from an archeological site contain 20% of the C-14 they originally contained. How long ago did the tree from which the wood was obtained die?

12. **CARBON-14 DATING** The skeletal remains of the so-called Pittsburgh Man, unearthed in Pennsylvania, had lost 82% of the C-14 they originally contained. Determine the approximate age of the bones.

13. **LEARNING CURVES** The American Court Reporting Institute finds that the average student taking Advanced Machine Shorthand, an intensive 20-week course, progresses according to the function

$$Q(t) = 120(1 - e^{-0.05t}) + 60 \qquad (0 \le t \le 20)$$

where $Q(t)$ measures the number of words (per minute) of dictation that the student can take in machine shorthand after t weeks in the course. Sketch the graph of the function Q and answer the following questions:

   a. What is the beginning shorthand speed for the average student in this course?
   b. What shorthand speed does the average student attain halfway through the course?
   c. How many words per minute can the average student take after completing this course?

14. **PEOPLE LIVING WITH HIV** On the basis of data compiled by WHO, the number of people living with HIV (human immunodeficiency virus) worldwide from 1985 through 2006 is approximated by

$$N(t) = \frac{39.88}{1 + 18.94e^{-0.2957t}} \qquad (0 \le t \le 21)$$

where $N(t)$ is measured in millions and t in years, with $t = 0$ corresponding to the beginning of 1985.
   a. How many people were living with HIV worldwide at the beginning of 1985? At the beginning of 2005?
   b. Assuming that the trend continued, how many people were living with HIV worldwide at the beginning of 2008?

   Source: World Health Organization.

15. **FEDERAL DEBT** According to data obtained from the CBO, the total federal debt (in trillions of dollars) from 2001 through 2006 is given by

$$f(t) = 5.37e^{0.078t} \qquad (1 \le t \le 6)$$

where t is measured in years, with $t = 1$ corresponding to 2001. What was the total federal debt in 2001? In 2006?

   Source: Congressional Budget Office.

16. **EFFECT OF ADVERTISING ON SALES** Metro Department Store found that t weeks after the end of a sales promotion the volume of sales was given by

$$S(t) = B + Ae^{-kt} \qquad (0 \le t \le 4)$$

where $B = 50,000$ and is equal to the average weekly volume of sales before the promotion. The sales volumes at the end of the first and third weeks were \$83,515 and \$65,055, respectively. Assume that the sales volume is decreasing exponentially.
   a. Find the decay constant k.
   b. Find the sales volume at the end of the fourth week.

17. **DEMAND FOR TABLET COMPUTERS** Universal Instruments found that the monthly demand for its new line of Galaxy tablet computers t months after placing the line on the market was given by

$$D(t) = 2000 - 1500e^{-0.05t} \qquad (t > 0)$$

Graph this function and answer the following questions:
   a. What is the demand after 1 month? After 1 year? After 2 years? After 5 years?
   b. At what level is the demand expected to stabilize?

18. **RELIABILITY OF COMPUTER CHIPS** The percentage of a certain brand of computer chips that will fail after t years of use is estimated to be

$$P(t) = 100(1 - e^{-0.1t})$$

What percentage of this brand of computer chips are expected to be usable after 3 years?

19. **LENGTHS OF FISH** The length (in centimeters) of a typical Pacific halibut t years old is approximately

$$f(t) = 200(1 - 0.956e^{-0.18t})$$

What is the length of a typical 6-year-old Pacific halibut?

20. **SPREAD OF AN EPIDEMIC** During a flu epidemic, the number of children in the Woodbridge Community School System who contracted influenza after t days was given by

$$Q(t) = \frac{1000}{1 + 199e^{-0.8t}}$$

   a. How many children were stricken by the flu after the first day?
   b. How many children had the flu after 10 days?

21. **LAY TEACHERS AT ROMAN CATHOLIC SCHOOLS** The change from religious to lay teachers at Roman Catholic schools has been attributed partly to the decline in the number of women and men entering religious orders. The percentage of teachers who are lay teachers is given by

$$f(t) = \frac{98}{1 + 2.77e^{-t}} \qquad (0 \le t \le 4)$$

where t is measured in decades, with $t = 0$ corresponding to the beginning of 1960. What percentage of teachers were lay teachers at the beginning of 1990?

   Sources: National Catholic Education Association and the U.S. Department of Education.

22. **GROWTH OF A FRUIT FLY POPULATION** On the basis of data collected during an experiment, a biologist found that the growth of a fruit fly population (*Drosophila*) with a limited food supply could be approximated by

$$N(t) = \frac{400}{1 + 39e^{-0.16t}}$$

where t denotes the number of days since the beginning of the experiment.
   a. What was the initial fruit fly population in the experiment?
   b. What was the population of the fruit fly colony on the 20th day?

23. **DEMOGRAPHICS** The number of citizens aged 45–64 years is approximated by

$$P(t) = \frac{197.9}{1 + 3.274e^{-0.0361t}} \qquad (0 \le t \le 25)$$

where $P(t)$ is measured in millions and t is measured in years, with $t = 0$ corresponding to 1990. People belonging to this age group are the targets of insurance companies that want to sell them annuities. What is the expected population of citizens aged 45–64 years in 2010? In 2015?

   Source: K. G. Securities.

24. **POPULATION GROWTH IN THE TWENTY-FIRST CENTURY** The U.S. population is approximated by the function

$$P(t) = \frac{616.5}{1 + 4.02e^{-0.5t}}$$

where $P(t)$ is measured in millions of people and t is measured in 30-year intervals, with $t = 0$ corresponding to 1930. What is the expected population of the United States in 2020 ($t = 3$)?

25. DISSEMINATION OF INFORMATION Three hundred students attended the dedication ceremony of a new building on a college campus. The president of the traditionally female college announced a new expansion program, which included plans to make the college coeducational. The number of students who learned of the new program t hr later is given by the function

$$f(t) = \frac{3000}{1 + Be^{-kt}}$$

If 600 students on campus had heard about the new program 2 hr after the ceremony, how many students had heard about the policy after 4 hr?

26. RADIOACTIVE DECAY A radioactive substance decays according to the formula

$$Q(t) = Q_0 e^{-kt}$$

where $Q(t)$ denotes the amount of the substance present at time t (measured in years), Q_0 denotes the amount of the substance present initially, and k (a positive constant) is the decay constant.

a. Show that half-life of the substance is $\bar{t} = (\ln 2)/k$.
b. Suppose a radioactive substance decays according to the formula

$$Q(t) = 20e^{-0.0001238t}$$

How long will it take for the substance to decay to half the original amount?

27. LOGISTIC GROWTH FUNCTION Consider the logistic growth function

$$Q(t) = \frac{A}{1 + Be^{-kt}}$$

Suppose the population is Q_1 when $t = t_1$ and Q_2 when $t = t_2$. Show that the value of k is

$$k = \frac{1}{t_2 - t_1} \ln \left[\frac{Q_2(A - Q_1)}{Q_1(A - Q_2)} \right]$$

28. LOGISTIC GROWTH FUNCTION The carrying capacity of a colony of fruit flies (*Drosophila*) is 600. The population of fruit flies after 14 days is 76, and the population after 21 days is 167. What is the value of the growth constant k?
Hint: Use the result of Exercise 27.

3.3 Solution to Self-Check Exercise

We are given that $P_0 = 226.5$, $k = 0.008$, and $I = 0.5$. So

$$P = \left(226.5 + \frac{0.5}{0.008} \right) e^{0.008t} - \frac{0.5}{0.008}$$

$$= 289 e^{0.008t} - 62.5$$

Therefore, the expected population in 2010 is given by

$$P(30) = 289 e^{0.24} - 62.5$$

$$\approx 304.9$$

or approximately 304.9 million.

USING TECHNOLOGY

Analyzing Mathematical Models

We can use a graphing utility to analyze the mathematical models encountered in this section.

APPLIED EXAMPLE 1 Internet-Gaming Sales The estimated growth in global Internet-gaming revenue (in billions of dollars), as predicted by industry analysts, is given in the following table:

Year	2001	2002	2003	2004	2005	2006	2007	2008	2009	2010
Revenue	3.1	3.9	5.6	8.0	11.8	15.2	18.2	20.4	22.7	24.5

a. Use **Logistic** to find a regression model for the data. Let $t = 0$ correspond to 2001.
b. Plot the scatter diagram and the graph of the function f found in part (a) using the viewing window $[0, 9] \times [0, 35]$.

Source: Christiansen Capital/Advisors.

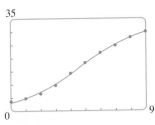

FIGURE **T1**
The graph of *f* in the viewing window
[0, 9] × [0, 35]

Solution

a. Using **Logistic** we find

$$f(t) = \frac{27.11}{1 + 9.64e^{-0.49t}} \qquad (0 \le t \le 9)$$

b. The scatter plot for the data, and the graph of *f* in the viewing window
[0, 9] × [0, 35], are shown in Figure T1.   ■

TECHNOLOGY EXERCISES

1. **ONLINE BANKING** In a study prepared in 2000, the percentage of households using online banking was projected to be

$$f(t) = 1.5e^{0.78t} \qquad (0 \le t \le 4)$$

where *t* is measured in years, with *t* = 0 corresponding to the beginning of 2000. Plot the graph of *f*, using the viewing window [0, 4] × [0, 40].

Source: Online Banking Report.

2. **NEWTON'S LAW OF COOLING** The temperature of a cup of coffee *t* min after it is poured is given by

$$T = 70 + 100e^{-0.0446t}$$

where *T* is measured in degrees Fahrenheit.
a. Plot the graph of *T*, using the viewing window [0, 30] × [0, 200].
b. When will the coffee be cool enough to drink (say, 120°)?
Hint: Use the ISECT function.

3. **AIR TRAVEL** Air travel has been rising dramatically in the past 30 years. In a study conducted in 2000, the FAA projected further exponential growth for air travel through 2010. The function

$$f(t) = 666e^{0.0413t} \qquad (0 \le t \le 10)$$

gives the number of passengers (in millions) in year *t*, with *t* = 0 corresponding to 2000.
a. Plot the graph of *f*, using the viewing window [0, 10] × [0, 1000].
b. How many air passengers were there in 2000? What was the projected number of air passengers for 2010?
Source: Federal Aviation Administration.

4. **COMPUTER GAME SALES** The total number of Starr Communication's newest game, Laser Beams, sold *t* months after its release is given by

$$N(t) = -20(t + 20)e^{-0.05t} + 400$$

thousand units. Plot the graph of *N*, using the viewing window [0, 500] × [0, 500].

5. **POPULATION GROWTH IN THE TWENTY-FIRST CENTURY** The U.S. population is approximated by the function

$$P(t) = \frac{616.5}{1 + 4.02e^{-0.5t}}$$

where *P(t)* is measured in millions of people and *t* is measured in 30-year intervals, with *t* = 0 corresponding to 1930.
a. Plot the graph of *P*, using the viewing window [0, 4] × [0, 650].
b. What is the expected population of the United States in 2020 (*t* = 3)?
Source: U.S. Census Bureau.

6. **TIME RATE OF GROWTH OF A TUMOR** The rate at which a tumor grows, with respect to time, is given by

$$R = Ax \ln \frac{B}{x} \qquad (\text{for } 0 < x < B)$$

where *A* and *B* are positive constants and *x* is the radius of the tumor. Plot the graph of *R* for the case *A* = *B* = 10.

7. **ABSORPTION OF DRUGS** The concentration of a drug in an organ at any time *t* (in seconds) is given by

$$C(t) = \begin{cases} 0.3t - 18(1 - e^{-t/60}) & \text{if } 0 \le t \le 20 \\ 18e^{-t/60} - 12e^{-(t-20)/60} & \text{if } t > 20 \end{cases}$$

where *C(t)* is measured in grams per cubic centimeter (g/cm³).
a. Plot the graph of *C*, using the viewing window [0, 120] × [0, 1].
b. What is the initial concentration of the drug in the organ?
c. What is the concentration of the drug in the organ after 10 sec?
d. What is the concentration of the drug in the organ after 30 sec?

(continued)

8. **MODELING WITH DATA** The snowfall accumulation at Logan Airport (in inches), t hr after the beginning of a 33-hr snowstorm in Boston on a certain day, follows:

Hour	0	3	6	9	12	15	18	21	24	27	30	33
Inches	0.1	0.4	3.6	6.5	9.1	14.4	19.5	22	23.6	24.8	26.6	27

Here, $t = 0$ corresponds to noon of February 6.
a. Use **Logistic** to find a regression model for the data.
b. Plot the scatter diagram and the graph of the function f found in part (a), using the viewing window $[0, 33] \times [0, 30]$.

Source: Boston Globe.

9. **WORLDWIDE PC SHIPMENTS** Estimated worldwide PC shipments (in millions of units) from 2005 through 2009 are given in the following table:

Year	2005	2006	2007	2008	2009
PCs	207.1	226.2	252.9	283.3	302.4

a. Use **Logistic** to find a regression model for the data. Let $t = 0$ correspond to 2005.
b. Plot the graph of the function f found in part (a), using the viewing window $[0, 4] \times [200, 300]$.

Source: International Data Corporation.

10. **FEDERAL DEBT** According to data obtained from the CBO, the total federal debt (in trillions of dollars) from 2001 through 2006 is given in the following table:

Year	2001	2002	2003	2004	2005	2006
Debt	5.81	6.23	6.78	7.40	7.93	8.51

a. Use **ExpReg** to find a regression model for the data. Let $t = 1$ correspond to 2001.
b. Plot the graph of the function f found in part (a), using the viewing window $[1, 6] \times [4, 10]$.

Source: Congressional Budget Office.

CHAPTER 3   Summary of Principal Formulas and Terms

FORMULAS

1. Exponential function with base b	$y = b^x$, where $b > 0$ and $b \neq 1$
2. The number e	$e = 2.7182818\ldots$
3. Exponential function with base e	$y = e^x$
4. Logarithmic function with base b	$y = \log_b x \quad (x > 0)$
5. Logarithmic function with base e	$y = \ln x \quad (x > 0)$
6. Inverse properties of $\ln x$ and e^x	$\ln e^x = x \quad$ and $\quad e^{\ln x} = x$

TERMS

common logarithm (166)
natural logarithm (166)
exponential growth (175)

growth constant (175)
exponential decay (177)
decay constant (177)

half-life of a radioactive substance (177)
logistic growth function (180)

CHAPTER 3   Concept Review Questions

Fill in the blanks.

1. The function $f(x) = x^b$ (b, a real number) is called a/an _____ function, whereas the function $g(x) = b^x$, where $b > $ _____ and $b \neq$ _____, is called a/an _____ function.

2. a. The domain of the function $y = 3^x$ is _____, and its range is _____.
   b. The graph of the function $y = 0.3^x$ passes through the point _____ and falls from _____ to _____.

3. a. If $b > 0$ and $b \neq 1$, then the logarithmic function $y = \log_b x$ has domain _____ and range _____; its graph passes through the point _____.
   b. The graph of $y = \log_b x$ _____ from left to right if $b < 1$ and _____ from left to right if $b > 1$.

4. a. If $x > 0$, then $e^{\ln x} = $ _____.
   b. If x is any real number, then $\ln e^x = $ _____.

5. a. In the unrestricted exponential growth model $Q = Q_0 e^{kt}$, Q_0 represents the quantity present _____, and k is called the _____ constant.

  b. In the exponential decay model $Q = Q_0 e^{-kt}$, k is called the _____ constant.

  c. The half-life of a radioactive substance is the _____ required for a substance to decay to _____ _____ of its original amount.

6. a. The model $Q(t) = C - Ae^{-kt}$ is called a/an _____ _____. The value of $Q(t)$ never exceeds _____.

  b. The model $Q(t) = \dfrac{A}{1 + Be^{-kt}}$, $y = A$, is called a/an _____ _____ _____ of the graph of Q. If the quantity $Q(t)$ is initially smaller than A, then $Q(t)$ will eventually approach _____ as t increases; the number A, represents the life-support capacity of the environment and is called the _____ _____ of the environment.

CHAPTER 3   Review Exercises

In Exercises 1–4, sketch the graph of the function.

1. $f(x) = 5^x$

2. $y = \left(\dfrac{1}{5}\right)^x$

3. $f(x) = \log_4 x$

4. $y = \log_{1/4} x$

In Exercises 5–8, express each equation in logarithmic form.

5. $3^4 = 81$

6. $9^{1/2} = 3$

7. $\left(\dfrac{2}{3}\right)^{-3} = \dfrac{27}{8}$

8. $16^{-3/4} = 0.125$

In Exercises 9–12, given that $\ln 2 \approx 0.6931$, $\ln 3 \approx 1.0986$, and $\ln 5 \approx 1.6094$, find the value of the expression using the laws of logarithms.

9. $\ln 30$

10. $\ln 9$

11. $\ln 3.6$

12. $\ln 75$

In Exercises 13–15, given that $\ln 2 = x$, $\ln 3 = y$, and $\ln 5 = z$, express each of the given logarithmic values in terms of x, y, and z.

13. $\ln 30$

14. $\ln 3.6$

15. $\ln 75$

In Exercises 16–21, solve for x without using a calculator.

16. $2^{2x-3} = 8$

17. $e^{x^2+x} = e^2$

18. $3^{x-1} = 9^{x+2}$

19. $2^{x^2+x} = 4^{x^2-3}$

20. $\log_4(2x + 1) = 2$

21. $\ln(x - 1) + \ln 4 = \ln(2x + 4) - \ln 2$

In Exercises 22–35, solve for x, giving your answer accurate to four decimal places.

22. $4^x = 5$

23. $3^{-2x} = 8$

24. $3 \cdot 2^{-x} = 17$

25. $2e^{-x} = 7$

26. $0.2e^x = 3.4$

27. $e^{2x-1} = 14$

28. $5^{3x+1} = 16$

29. $2^{3x+1} = 3^{2x-3}$

30. $2^{x^2} = 12$

31. $3e^{\sqrt{x}} = 15$

32. $4e^{-0.1x} - 2 = 8$

33. $8 - e^{0.2x} = 2$

34. $\dfrac{20}{1 + 2e^{0.2x}} = 4$

35. $\dfrac{30}{1 + 2e^{-0.1x}} = 5$

36. Sketch the graph of the function $y = \log_2(x + 3)$.

37. Sketch the graph of the function $y = \log_3(x + 1)$.

38. GROWTH OF BACTERIA A culture of bacteria that initially contained 2000 bacteria has a count of 18,000 bacteria after 2 hr.

  a. Determine the function $Q(t)$ that expresses the exponential growth of the number of cells of this bacterium as a function of time t (in minutes).

  b. Find the number of bacteria present after 4 hr.

39. RADIOACTIVE DECAY The radioactive element radium has a half-life of 1600 years. What is its decay constant?

40. DEMAND FOR DVD PLAYERS VCA Television found that the monthly demand for its new line of DVD players t months after placing the players on the market is given by:

$$D(t) = 4000 - 3000e^{-0.06t} \qquad (t \geq 0)$$

Graph this function and answer the following questions:

  a. What was the demand after 1 month? After 1 year? After 2 years?

  b. At what level is the demand expected to stabilize?

41. FLU EPIDEMIC During a flu epidemic, the number of students at a certain university who contracted influenza after t days could be approximated by the exponential model

$$Q(t) = \dfrac{3000}{1 + 499e^{-kt}}$$

If 90 students contracted the flu by day 10, how many students contracted the flu by day 20?

42. U.S. INFANT MORTALITY RATE The U.S. infant mortality rate (per 1000 live births) is approximated by the function

$$N(t) = 12.5e^{-0.0294t} \qquad (0 \leq t \leq 21)$$

where t is measured in years, with $t = 0$ corresponding to 1980. What was the mortality rate in 1980? In 1990? In 2000?

Source: U.S. Department of Health and Human Services.

43. **ABSORPTION OF DRUGS** The concentration of a drug in an organ at any time t (in seconds) is given by

$$x(t) = 0.08(1 - e^{-0.02t})$$

where $x(t)$ is measured in grams/cubic centimeter (g/cm³).

a. What is the initial concentration of the drug in the organ?

b. What is the concentration of the drug in the organ after 30 sec?

CHAPTER 3 Before Moving On . . .

1. Simplify the expression $(2x^{-2})^2(9x^{-4})^{1/2}$.

2. Solve $e^{2x} - e^x - 6 = 0$ for x.
   Hint: Let $u = e^x$.

3. Solve $\log_2(x^2 - 8x + 1) = 0$.

4. Solve the equation $\dfrac{100}{1 + 2e^{0.3t}} = 40$ for t.

5. The temperature of a cup of coffee at time t (in minutes) is

$$T(t) = 70 + ce^{-kt}$$

Initially, the temperature of the coffee was 200°F. Three minutes later, it was 180°. When will the temperature of the coffee be 150°F?

4

MATHEMATICS OF FINANCE

INTEREST THAT IS periodically added to the principal and thereafter itself earns interest is called *compound interest*. We begin this chapter by deriving the *compound interest formula*, which gives the amount of money accumulated when an initial amount of money is invested in an account for a fixed term and earns compound interest.

An *annuity* is a sequence of payments made at regular intervals. We derive formulas giving the *future value of an annuity* (what you end up with) and the *present value of an annuity* (the lump sum that, when invested now, will yield the same future value as that of the annuity). Then, using these formulas, we answer questions involving the amortization of certain types of installment loans and questions involving *sinking funds* (funds that are set up to be used for a specific purpose at a future date).

How much can the Jacksons afford to borrow from the bank for the purchase of a home? They have determined that after making a down payment they can afford a monthly payment of $2000. In Example 4, page 223, we learn how to determine the maximum amount they can afford to borrow if they secure a 30-year fixed mortgage at the current rate.

4.1 Compound Interest

Simple Interest

A natural application of linear functions to the business world is found in the computation of **simple interest**—interest that is computed on the original principal only. Thus, if I denotes the interest on a principal P (in dollars) at an interest rate of r per year for t years, we have

$$I = Prt$$

The **accumulated amount** A, the sum of the principal and interest after t years, is given by

$$A = P + I = P + Prt$$
$$= P(1 + rt)$$

and is a linear function of t (see Exercise 46). In business applications, we are normally interested only in the case in which t is positive, so only the part of the line that lies in Quadrant I is of interest to us (Figure 1).

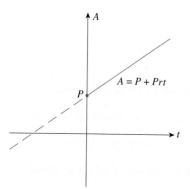

FIGURE **1**
The accumulated amount is a linear function of t.

$A = P + Prt$

Simple Interest Formulas

Interest:	$I = Prt$	**(1a)**
Accumulated amount:	$A = P(1 + rt)$	**(1b)**

EXAMPLE 1 A bank pays simple interest at the rate of 8% per year for certain deposits. If a customer deposits $1000 and makes no withdrawals for 3 years, what is the total amount on deposit at the end of 3 years? What is the interest earned in that period of time?

Solution Using Formula (1b) with $P = 1000$, $r = 0.08$, and $t = 3$, we see that the total amount on deposit at the end of 3 years is given by

$$A = P(1 + rt)$$
$$= 1000[1 + (0.08)(3)] = 1240$$

or $1240.

The interest earned over the 3-year period is given by

$$I = Prt \quad \text{Use Formula (1a).}$$
$$= 1000(0.08)(3) = 240$$

or $240. ∎

Exploring with TECHNOLOGY

Refer to Example 1. Use a graphing utility to plot the graph of the function $A = 1000(1 + 0.08t)$, using the viewing window $[0, 10] \times [0, 2000]$.

1. What is the A-intercept of the straight line, and what does it represent?
2. What is the slope of the straight line, and what does it represent? (See Exercise 46.)

APPLIED EXAMPLE 2 Trust Funds An amount of $2000 is invested in a 10-year trust fund that pays 6% annual simple interest. What is the total amount of the trust fund at the end of 10 years?

Solution   The total amount of the trust fund at the end of 10 years is given by

$$A = P(1 + rt)$$
$$= 2000[1 + (0.06)(10)] = 3200$$

or $3200.

A Treasury Bill (T-Bill) is a short-term debt obligation (less than or equal to 1 year) backed by the U.S. government. Rather than paying fixed interest payments, T-Bills are sold at a discount from face value. The appreciation of a T-Bill (face value minus purchase price) provides the investment return to the holder.

APPLIED EXAMPLE 3 T-Bills Suppose that Jane buys a 26-week T-Bill with a maturity value of $10,000. If she pays $9850 for the T-Bill, what will be the rate of return on her investment?

Solution   We use Formula (1b) with $A = 10{,}000$, $P = 9850$, and $t = \frac{26}{52} = \frac{1}{2}$. We obtain

$$10{,}000 = 9850\left(1 + \frac{1}{2}r\right)$$

$$= 9850 + 4925r$$

Solving for r, we find

$$4925r = 150$$

$$r = \frac{150}{4925} \approx 0.0305$$

So Jane's investment will earn simple interest at the rate of approximately 3.05% per year.

Compound Interest

In contrast to simple interest, **compound interest** is earned interest that is periodically added to the principal and thereafter itself earns interest at the same rate. To find a formula for the accumulated amount, let's consider a numerical example. Suppose $1000 (the principal) is deposited in a bank for a term of 3 years, earning interest at the rate of 8% per year (called the **nominal,** or **stated, rate**) compounded annually. Then, using Formula (1b) with $P = 1000$, $r = 0.08$, and $t = 1$, we see that the accumulated amount A_1 at the end of the first year is

$$A_1 = P(1 + rt)$$
$$= 1000[1 + (0.08)(1)] = 1000(1.08) = 1080$$

or $1080.

To find the accumulated amount A_2 at the end of the second year, we use Formula (1b) once again, this time with $P = A_1$. (Remember, the principal *and* interest now earn interest over the second year.) We obtain

$$A_2 = P(1 + rt) = A_1(1 + rt)$$
$$= 1000[1 + 0.08(1)][1 + 0.08(1)]$$
$$= 1000[1 + 0.08]^2 = 1000(1.08)^2 = 1166.40$$

or $1166.40.

Finally, the accumulated amount A_3 at the end of the third year is found using (1b) with $P = A_2$, giving

$$
\begin{aligned}
A_3 = P(1 + rt) &= A_2(1 + rt) \\
&= 1000[1 + 0.08(1)]^2[1 + 0.08(1)] \\
&= 1000[1 + 0.08]^3 = 1000(1.08)^3 \approx 1259.71
\end{aligned}
$$

or approximately \$1259.71.

If you reexamine our calculations, you will see that the accumulated amounts at the end of each year have the following form.

First year: $A_1 = 1000(1 + 0.08)$, or $A_1 = P(1 + r)$
Second year: $A_2 = 1000(1 + 0.08)^2$, or $A_2 = P(1 + r)^2$
Third year: $A_3 = 1000(1 + 0.08)^3$, or $A_3 = P(1 + r)^3$

These observations suggest the following general result: If P dollars is invested over a term of t years, earning interest at the rate of r per year compounded annually, then the accumulated amount is

$$A = P(1 + r)^t \tag{2}$$

Equation (2) was derived under the assumption that interest was compounded *annually*. In practice, however, interest is usually compounded more than once a year. The interval of time between successive interest calculations is called the **conversion period.**

If interest at a nominal rate of r per year is compounded m times a year on a principal of P dollars, then the simple interest rate per conversion period is

$$i = \frac{r}{m} \qquad \begin{array}{l} \text{Annual interest rate} \\ \hline \text{Periods per year} \end{array}$$

For example, if the nominal interest rate is 8% per year ($r = 0.08$) and interest is compounded quarterly ($m = 4$), then

$$i = \frac{r}{m} = \frac{0.08}{4} = 0.02$$

or 2% per period.

To find a general formula for the accumulated amount when a principal of P dollars is deposited in a bank for a term of t years and earns interest at the (nominal) rate of r per year compounded m times per year, we proceed as before, using Formula (1b) repeatedly with the interest rate $i = \frac{r}{m}$. We see that the accumulated amount at the end of each period is as follows:

First period: $A_1 = P(1 + i)$
Second period: $A_2 = A_1(1 + i)$ $= [P(1 + i)](1 + i) = P(1 + i)^2$
Third period: $A_3 = A_2(1 + i)$ $= [P(1 + i)^2](1 + i) = P(1 + i)^3$

$$\vdots \qquad\qquad\qquad \vdots$$

nth period: $A_n = A_{n-1}(1 + i) = [P(1 + i)^{n-1}](1 + i) = P(1 + i)^n$

There are $n = mt$ periods in t years (number of conversion periods per year times the term in years). Hence, the accumulated amount at the end of t years is given by

$$A = P(1 + i)^n$$

Compound Interest Formula (Accumulated Amount)

$$A = P(1 + i)^n \qquad (3)$$

where $i = \dfrac{r}{m}$, $n = mt$, and

A = Accumulated amount at the end of n conversion periods
P = Principal
r = Nominal interest rate per year
m = Number of conversion periods per year
t = Term (number of years)

Exploring with
TECHNOLOGY

Let $A_1(t)$ denote the accumulated amount of $100 earning simple interest at the rate of 6% per year over t years, and let $A_2(t)$ denote the accumulated amount of $100 earning interest at the rate of 6% per year compounded monthly over t years.

1. Find expressions for $A_1(t)$ and $A_2(t)$.
2. Use a graphing utility to plot the graphs of A_1 and A_2 on the same set of axes, using the viewing window $[0, 20] \times [0, 400]$.
3. Comment on the growth of $A_1(t)$ and $A_2(t)$ by referring to the graphs of A_1 and A_2.

EXAMPLE 4 Find the accumulated amount after 3 years if $1000 is invested at 8% per year compounded (a) annually, (b) semiannually, (c) quarterly, (d) monthly, and (e) daily (assume a 365-day year).

Solution

a. Here, $P = 1000$, $r = 0.08$, and $m = 1$. Thus, $i = r = 0.08$ and $n = 3$, so Formula (3) gives

$$A = 1000(1 + 0.08)^3$$
$$\approx 1259.71$$

or $1259.71.

b. Here, $P = 1000$, $r = 0.08$, and $m = 2$. Thus, $i = \frac{0.08}{2}$ and $n = (3)(2) = 6$, so Formula (3) gives

$$A = 1000\left(1 + \frac{0.08}{2}\right)^6$$
$$\approx 1265.32$$

or $1265.32.

c. In this case, $P = 1000$, $r = 0.08$, and $m = 4$. Thus, $i = \frac{0.08}{4}$ and $n = (3)(4) = 12$, so Formula (3) gives

$$A = 1000\left(1 + \frac{0.08}{4}\right)^{12}$$
$$\approx 1268.24$$

or $1268.24.

d. Here, $P = 1000$, $r = 0.08$, and $m = 12$. Thus, $i = \frac{0.08}{12}$ and $n = (3)(12) = 36$, so Formula (3) gives

$$A = 1000\left(1 + \frac{0.08}{12}\right)^{36}$$

$$\approx 1270.24$$

or $1270.24.

e. Here, $P = 1000$, $r = 0.08$, $m = 365$, and $t = 3$. Thus, $i = \frac{0.08}{365}$ and $n = (3)(365) = 1095$, so Formula (3) gives

$$A = 1000\left(1 + \frac{0.08}{365}\right)^{1095}$$

$$\approx 1271.22$$

or $1271.22. These results are summarized in Table 1.

TABLE 1

Nominal Rate, r	Conversion Period	Interest Rate/ Conversion Period	Initial Investment	Accumulated Amount
8%	Annually ($m = 1$)	8%	$1000	$1259.71
8	Semiannually ($m = 2$)	4	1000	1265.32
8	Quarterly ($m = 4$)	2	1000	1268.24
8	Monthly ($m = 12$)	2/3	1000	1270.24
8	Daily ($m = 365$)	8/365%	1000	1271.22

Exploring with TECHNOLOGY

Investments that are allowed to grow over time can increase in value surprisingly fast. Consider the potential growth of $10,000 if earnings are reinvested. More specifically, suppose $A_1(t)$, $A_2(t)$, $A_3(t)$, $A_4(t)$, and $A_5(t)$ denote the accumulated values of an investment of $10,000 over a term of t years and earning interest at the rate of 4%, 6%, 8%, 10%, and 12% per year compounded annually.

1. Find expressions for $A_1(t)$, $A_2(t)$, . . . , $A_5(t)$.
2. Use a graphing utility to plot the graphs of $A_1, A_2, \ldots, A_5$ on the same set of axes, using the viewing window $[0, 20] \times [0, 100{,}000]$.
3. Use TRACE to find $A_1(20)$, $A_2(20)$, . . . , $A_5(20)$ and then interpret your results.

Continuous Compounding of Interest

One question that arises naturally in the study of compound interest is: What happens to the accumulated amount over a fixed period of time if the interest is computed more and more frequently?

Intuition suggests that the more often interest is compounded, the larger the accumulated amount will be. This is confirmed by the results of Example 4, where we found that the accumulated amounts did in fact increase when we increased the number of conversion periods per year.

This leads us to another question: Does the accumulated amount keep growing without bound, or does it approach a fixed number when the interest is computed more and more frequently over a fixed period of time?

To answer this question, let's look again at the compound interest formula:

$$A = P(1 + i)^n = P\left(1 + \frac{r}{m}\right)^{mt} \tag{4}$$

Recall that m is the number of conversion periods per year. So to find an answer to our question, we should let m get larger and larger in Formula (4). If we let $u = \frac{m}{r}$ so that $m = ru$, then (4) becomes

$$A = P\left(1 + \frac{1}{u}\right)^{urt} \qquad \frac{r}{m} = \frac{1}{u}$$

$$= P\left[\left(1 + \frac{1}{u}\right)^u\right]^{rt} \qquad \text{Since } a^{xy} = (a^x)^y$$

Now observe that u gets larger and larger as m gets larger and larger. But from our work in Section 3.1, we know that $(1 + 1/u)^u$ approaches e as u gets larger and larger. Using this result, we can see that, as m gets larger and larger, A approaches $P(e)^{rt} = Pe^{rt}$. In this situation, we say that interest is *compounded continuously*. Let's summarize this important result.

Continuous Compound Interest Formula

$$A = Pe^{rt} \tag{5}$$

where

P = Principal

r = Nominal interest rate compounded continuously

t = Time in years

A = Accumulated amount at the end of t years

EXAMPLE 5 Find the accumulated amount after 3 years if $1000 is invested at 8% per year compounded (a) daily (assume a 365-day year) and (b) continuously.

Solution

a. Use Formula (3) with $P = 1000$, $r = 0.08$, $m = 365$, and $t = 3$. Thus, $i = \frac{0.08}{365}$ and $n = (365)(3) = 1095$, so

$$A = 1000\left(1 + \frac{0.08}{365}\right)^{1095} \approx 1271.22$$

or $1271.22.

b. Here, we use Formula (5) with $P = 1000$, $r = 0.08$, and $t = 3$, obtaining

$$A = 1000e^{(0.08)(3)}$$
$$\approx 1271.25$$

or $1271.25.

Observe that the accumulated amounts corresponding to interest compounded daily and interest compounded continuously differ by very little. The continuous compound interest formula is a very important tool in theoretical work in financial analysis.

Exploring with
TECHNOLOGY

In the opening paragraph of Section 3.1, we pointed out that the accumulated amount of an account earning interest *compounded continuously* will eventually outgrow by far the accumulated amount of an account earning interest at the same nominal rate but earning simple interest. Illustrate this fact using the following example.

Suppose you deposit $1000 in Account I, earning interest at the rate of 10% per year compounded continuously so that the accumulated amount at the end of t years is $A_1(t) = 1000e^{0.1t}$. Suppose you also deposit $1000 in Account II, earning simple interest at the rate of 10% per year so that the accumulated amount at the end of t years is $A_2(t) = 1000(1 + 0.1t)$. Use a graphing utility to sketch the graphs of the functions A_1 and A_2 in the viewing window $[0, 20] \times [0, 10,000]$ to see the accumulated amounts $A_1(t)$ and $A_2(t)$ over a 20-year period.

Effective Rate of Interest

Examples 4 and 5 showed that the interest actually earned on an investment depends on the frequency with which the interest is compounded. Thus, the stated, or nominal, rate of 8% per year does not reflect the actual rate at which interest is earned. This suggests that we need to find a common basis for comparing interest rates. One such way of comparing interest rates is provided by the use of the *effective rate*. The **effective rate** is the *simple* interest rate that would produce the same accumulated amount in 1 year as the nominal rate compounded m times a year. The effective rate is also called the **annual percentage yield,** often abbreviated APY.

To derive a relationship between the nominal interest rate, r per year compounded m times, and its corresponding effective rate, R per year, let's assume an initial investment of P dollars. Then the accumulated amount after 1 year at a simple interest rate of R per year is

$$A = P(1 + R)$$

Also, the accumulated amount after 1 year at an interest rate of r per year compounded m times a year is

$$A = P(1 + i)^n = P\left(1 + \frac{r}{m}\right)^m \qquad \text{Since } i = \frac{r}{m} \text{ and } t = 1$$

Equating the two expressions gives

$$P(1 + R) = P\left(1 + \frac{r}{m}\right)^m$$

$$1 + R = \left(1 + \frac{r}{m}\right)^m \qquad \text{Divide both sides by } P.$$

If we solve the preceding equation for R, we obtain the following formula for computing the effective rate of interest.

> **Effective Rate of Interest Formula**
>
> $$r_{\text{eff}} = \left(1 + \frac{r}{m}\right)^m - 1 \qquad \qquad \textbf{(6)}$$
>
> where
>
> r_{eff} = Effective rate of interest
> r = Nominal interest rate per year
> m = Number of conversion periods per year

VIDEO ▶ **EXAMPLE 6** Find the effective rate of interest corresponding to a nominal rate of 8% per year compounded (a) annually, (b) semiannually, (c) quarterly, (d) monthly, and (e) daily.

Solution

a. The effective rate of interest corresponding to a nominal rate of 8% per year compounded annually is, of course, given by 8% per year. This result is also confirmed by using Formula (6) with $r = 0.08$ and $m = 1$. Thus,

$$r_{\text{eff}} = (1 + 0.08) - 1 = 0.08$$

b. Let $r = 0.08$ and $m = 2$. Then Formula (6) yields

$$r_{\text{eff}} = \left(1 + \frac{0.08}{2}\right)^2 - 1$$
$$= (1.04)^2 - 1$$
$$= 0.0816$$

so the effective rate is 8.16% per year.

c. Let $r = 0.08$ and $m = 4$. Then Formula (6) yields

$$r_{\text{eff}} = \left(1 + \frac{0.08}{4}\right)^4 - 1$$
$$= (1.02)^4 - 1$$
$$\approx 0.08243$$

so the corresponding effective rate in this case is 8.243% per year.

d. Let $r = 0.08$ and $m = 12$. Then Formula (6) yields

$$r_{\text{eff}} = \left(1 + \frac{0.08}{12}\right)^{12} - 1$$
$$\approx 0.08300$$

so the corresponding effective rate in this case is 8.3% per year.

e. Let $r = 0.08$ and $m = 365$. Then Formula (6) yields

$$r_{\text{eff}} = \left(1 + \frac{0.08}{365}\right)^{365} - 1$$
$$\approx 0.08328$$

so the corresponding effective rate in this case is 8.328% per year. ▪

Explore & Discuss

Recall the effective rate of interest formula:

$$r_{\text{eff}} = \left(1 + \frac{r}{m}\right)^m - 1$$

1. Show that

$$r = m\left[(1 + r_{\text{eff}})^{1/m} - 1\right]$$

2. A certificate of deposit (CD) is known to have an effective rate of 5.3%. If interest is compounded monthly, find the nominal rate of interest by using the result of part 1.

If the effective rate of interest r_{eff} is known, then the accumulated amount after t years on an investment of P dollars may be more readily computed by using the formula

$$A = P(1 + r_{\text{eff}})^t$$

The 1968 Truth in Lending Act passed by Congress requires that the effective rate of interest be disclosed in all contracts involving interest charges. The passage of this act has benefited consumers because they now have a common basis for comparing the various nominal rates quoted by different financial institutions. Furthermore, knowing the effective rate enables consumers to compute the actual charges involved in a transaction. Thus, if the effective rates of interest found in Example 6 were known, then the accumulated values of Example 4 could have been readily found (see Table 2).

TABLE 2

Nominal Rate, r	Frequency of Interest Payment	Effective Rate	Initial Investment	Accumulated Amount after 3 Years	
8%	Annually	8%	$1000	$1000(1 + 0.08)^3$	$\approx \$1259.71$
8	Semiannually	8.16	1000	$1000(1 + 0.0816)^3$	≈ 1265.32
8	Quarterly	8.243	1000	$1000(1 + 0.08243)^3$	≈ 1268.23
8	Monthly	8.300	1000	$1000(1 + 0.08300)^3$	≈ 1270.24
8	Daily	8.328	1000	$1000(1 + 0.08328)^3$	≈ 1271.22

Present Value

Let's return to the compound interest Formula (3), which expresses the accumulated amount at the end of n periods when interest at the rate of r is compounded m times a year. The principal P in (3) is often referred to as the **present value**, and the accumulated value A is called the **future value**, since it is realized at a future date. In certain instances, an investor might wish to determine how much money he should invest now, at a fixed rate of interest, so that he will realize a certain sum at some future date. This problem may be solved by expressing P in terms of A. Thus, from Formula (3), we find

$$P = A(1 + i)^{-n}$$

Here, as before, $i = \frac{r}{m}$, where m is the number of conversion periods per year.

> **Present Value Formula for Compound Interest**
> $$P = A(1 + i)^{-n} \tag{7}$$

EXAMPLE 7 How much money should be deposited in a bank paying interest at the rate of 6% per year compounded monthly so that at the end of 3 years, the accumulated amount will be $20,000?

Solution Here, $r = 0.06$ and $m = 12$, so $i = \frac{0.06}{12}$ and $n = (3)(12) = 36$. Thus, the problem is to determine P given that $A = 20,000$. Using Formula (7), we obtain

$$P = 20{,}000\left(1 + \frac{0.06}{12}\right)^{-36}$$

$$\approx 16{,}713$$

or $16,713.

EXAMPLE 8 Find the present value of $49,158.60 due in 5 years at an interest rate of 10% per year compounded quarterly.

Solution Using Formula (7) with $r = 0.1$, $m = 4$, and $t = 5$ so that $i = \frac{0.1}{4}$, $n = mt = (4)(5) = 20$, and $A = 49,158.6$, we obtain

$$P = (49,158.6)\left(1 + \frac{0.1}{4}\right)^{-20} \approx 30,000.07$$

or approximately $30,000.

If we solve Formula (5) for P, we have

$$P = Ae^{-rt}$$

which gives the present value in terms of the future (accumulated) value for the case of continuous compounding.

> **Present Value Formula for Continuous Compound Interest**
> $$P = Ae^{-rt} \tag{8}$$

Using Logarithms to Solve Problems in Finance

The next two examples show how logarithms can be used to solve problems involving compound interest.

EXAMPLE 9 How long will it take $10,000 to grow to $15,000 if the investment earns an interest rate of 12% per year compounded quarterly?

Solution Using Formula (3) with $A = 15,000$, $P = 10,000$, $r = 0.12$, and $m = 4$, we obtain

$$15,000 = 10,000\left(1 + \frac{0.12}{4}\right)^{4t}$$

$$(1.03)^{4t} = \frac{15,000}{10,000} = 1.5$$

Taking the logarithm on each side of the equation gives

$$\ln(1.03)^{4t} = \ln 1.5$$

$$4t \ln 1.03 = \ln 1.5 \qquad \log_b m^n = n \log_b m$$

$$4t = \frac{\ln 1.5}{\ln 1.03}$$

$$t = \frac{\ln 1.5}{4 \ln 1.03} \approx 3.43$$

So it will take approximately 3.4 years for the investment to grow from $10,000 to $15,000.

EXAMPLE 10 Find the interest rate needed for an investment of $10,000 to grow to an amount of $18,000 in 5 years if the interest is compounded monthly.

Solution Use Formula (3) with $A = 18,000$, $P = 10,000$, $m = 12$, and $t = 5$. Thus, $i = \frac{r}{12}$ and $n = (12)(5) = 60$, so

$$18,000 = 10,000\left(1 + \frac{r}{12}\right)^{12(5)}$$

Dividing both sides of the equation by 10,000 gives

$$\frac{18,000}{10,000} = \left(1 + \frac{r}{12}\right)^{60}$$

or, upon simplification,

$$\left(1 + \frac{r}{12}\right)^{60} = 1.8$$

Now we take the logarithm on each side of the equation, obtaining

$$\ln\left(1 + \frac{r}{12}\right)^{60} = \ln 1.8$$

$$60 \ln\left(1 + \frac{r}{12}\right) = \ln 1.8$$

$$\ln\left(1 + \frac{r}{12}\right) = \frac{\ln 1.8}{60} \approx 0.009796$$

$$\left(1 + \frac{r}{12}\right) \approx e^{0.009796} \qquad \ln e^x = x$$

$$\approx 1.009844$$

and

$$\frac{r}{12} = 1.009844 - 1$$

$$r = 0.118128$$

or approximately 11.81% per year.   ∎

APPLIED EXAMPLE 11 Real Estate Investment Blakely Investment Company owns an office building located in the commercial district of a city. As a result of the continued success of an urban renewal program, local business is enjoying a miniboom. The market value of Blakely's property is

$$V(t) = 300,000e^{\sqrt{t}/2}$$

where $V(t)$ is measured in dollars and t is the time in years from the present. If the expected rate of appreciation is 9% per year compounded continuously for the next 10 years, find an expression for the present value $P(t)$ of the market price of the property that will be valid for the next 10 years. Compute $P(7)$, $P(8)$, and $P(9)$, and then interpret your results.

Solution   Using Formula (8) with $A = V(t)$ and $r = 0.09$, we find that the present value of the market price of the property t years from now is

$$P(t) = V(t)e^{-0.09t}$$

$$= 300,000e^{-0.09t + \sqrt{t}/2} \qquad (0 \le t \le 10)$$

Letting $t = 7$, 8, and 9, we find

$$P(7) = 300,000e^{-0.09(7) + \sqrt{7}/2} \approx 599,837, \text{ or } \$599,837$$

$$P(8) = 300,000e^{-0.09(8) + \sqrt{8}/2} \approx 600,640, \text{ or } \$600,640$$

$$P(9) = 300,000e^{-0.09(9) + \sqrt{9}/2} \approx 598,115, \text{ or } \$598,115$$

respectively. From the results of these computations, we see that the present value of the property's market price seems to decrease after a certain period of growth. This suggests that there is an optimal time for the owners to sell. Later we will show that the highest present value of the property's market value is approximately $600,779 and that it occurs at time $t \approx 7.72$ years.   ■

The returns on certain investments such as zero coupon certificates of deposit (CDs) and zero coupon bonds are compared by quoting the time it takes for each investment to triple, or even quadruple. These calculations make use of the compound interest Formula (3).

VIDEO ▶

APPLIED EXAMPLE 12 Investment Options Jane has narrowed her investment options down to two:

1. Purchase a CD that matures in 12 years and pays interest upon maturity at the rate of 10% per year compounded daily (assume 365 days in a year).
2. Purchase a zero coupon CD that will triple her investment in the same period.

Which option will optimize Jane's investment?

Solution Let's compute the accumulated amount under option 1. Here,

$$r = 0.10 \qquad m = 365 \qquad t = 12$$

so $n = 12(365) = 4380$ and $i = \frac{0.10}{365}$. The accumulated amount at the end of 12 years (after 4380 conversion periods) is

$$A = P\left(1 + \frac{0.10}{365}\right)^{4380} \approx 3.32P$$

or $3.32P$. If Jane chooses option 2, the accumulated amount of her investment after 12 years will be $3P$. Therefore, she should choose option 1.   ■

APPLIED EXAMPLE 13 IRAs Moesha has an Individual Retirement Account (IRA) with a brokerage firm. Her money is invested in a money market mutual fund that pays interest on her investment. Over a 2-year period in which no deposits or withdrawals were made, her account grew from $4500 to $5268.24. Find the effective rate at which Moesha's account was earning interest over that period.

Solution Let r_{eff} denote the required effective rate of interest. We have

$$5268.24 = 4500(1 + r_{\text{eff}})^2$$
$$(1 + r_{\text{eff}})^2 = 1.17072$$
$$1 + r_{\text{eff}} \approx 1.081998 \qquad \text{Take the square root on both sides.}$$

or $r_{\text{eff}} \approx 0.081998$. Therefore, the effective rate was 8.20% per year.   ■

4.1 Self-Check Exercises

1. Find the present value of $20,000 due in 3 years at an interest rate of 5.4%/year compounded monthly.

2. Paul is a retiree living on Social Security and the income from his investment. Currently, his $100,000 investment in a 1-year CD is yielding 4.6% interest compounded daily. If he reinvests the principal ($100,000) on the due date of the

CD in another 1-year CD paying 3.2% interest compounded daily, find the net decrease in his yearly income from his investment. (Use a 365-day year.)

Solutions to Self-Check Exercises 4.1 can be found on page 205.

4.1   Concept Questions

1. Explain the difference between simple interest and compound interest.

2. What is the difference between the accumulated amount (future value) and the present value of an investment?

3. What is the effective rate of interest?

4.1   Exercises

1. Find the simple interest on a $500 investment made for 2 years at an interest rate of 8%/year. What is the accumulated amount?

2. Find the simple interest on a $1000 investment made for 3 years at an interest rate of 5%/year. What is the accumulated amount?

3. Find the accumulated amount at the end of 9 months on an $800 deposit in a bank paying simple interest at a rate of 6%/year.

4. Find the accumulated amount at the end of 8 months on a $1200 bank deposit paying simple interest at a rate of 7%/year.

5. If the accumulated amount is $1160 at the end of 2 years and the simple rate of interest is 8%/year, what is the principal?

6. A bank deposit paying simple interest at the rate of 5%/year grew to a sum of $3100 in 10 months. Find the principal.

7. How many days will it take for a sum of $1000 to earn $20 interest if it is deposited in a bank paying ordinary simple interest at the rate of 3%/year? (Use a 365-day year.)

8. How many days will it take for a sum of $1500 to earn $25 interest if it is deposited in a bank paying 5%/year? (Use a 365-day year.)

9. A bank deposit paying simple interest grew from an initial sum of $1000 to a sum of $1075 in 9 months. Find the interest rate.

10. Determine the simple interest rate at which $1200 will grow to $1250 in 8 months.

In Exercises 11–20, find the accumulated amount A if the principal P is invested at the interest rate of r/year for t years.

11. $P = \$1000$, $r = 7\%$, $t = 8$, compounded annually

12. $P = \$1000$, $r = 8\frac{1}{2}\%$, $t = 6$, compounded annually

13. $P = \$2500$, $r = 7\%$, $t = 10$, compounded semiannually

14. $P = \$2500$, $r = 9\%$, $t = 10\frac{1}{2}$, compounded semiannually

15. $P = \$12,000$, $r = 8\%$, $t = 10\frac{1}{2}$, compounded quarterly

16. $P = \$42,000$, $r = 7\frac{3}{4}\%$, $t = 8$, compounded quarterly

17. $P = \$150,000$, $r = 14\%$, $t = 4$, compounded monthly

18. $P = \$180,000$, $r = 9\%$, $t = 6\frac{1}{4}$, compounded monthly

19. $P = \$150,000$, $r = 12\%$, $t = 3$, compounded daily

20. $P = \$200,000$, $r = 8\%$, $t = 4$, compounded daily

In Exercises 21–24, find the effective rate corresponding to the given nominal rate.

21. 10%/year compounded semiannually

22. 9%/year compounded quarterly

23. 8%/year compounded monthly

24. 8%/year compounded daily

In Exercises 25–28, find the present value of $40,000 due in 4 years at the given rate of interest.

25. 6%/year compounded semiannually

26. 8%/year compounded quarterly

27. 7%/year compounded monthly

28. 9%/year compounded daily

29. Find the accumulated amount after 4 years if $5000 is invested at 8%/year compounded continuously.

30. Find the accumulated amount after 6 years if $6500 is invested at 7%/year compounded continuously.

In Exercises 31–38, use logarithms to solve each problem.

31. How long will it take $5000 to grow to $6500 if the investment earns interest at the rate of 12%/year compounded monthly?

32. How long will it take $12,000 to grow to $15,000 if the investment earns interest at the rate of 8%/year compounded monthly?

33. How long will it take an investment of $2000 to double if the investment earns interest at the rate of 9%/year compounded monthly?

34. How long will it take an investment of $5000 to triple if the investment earns interest at the rate of 8%/year compounded daily?

35. Find the interest rate needed for an investment of $5000 to grow to an amount of $6000 in 3 years if interest is compounded continuously.

36. Find the interest rate needed for an investment of $4000 to double in 5 years if interest is compounded continuously.

37. How long will it take an investment of $6000 to grow to $7000 if the investment earns interest at the rate of $7\frac{1}{2}$%/year compounded continuously?

38. How long will it take an investment of $8000 to double if the investment earns interest at the rate of 8%/year compounded continuously?

39. CONSUMER DECISIONS Mitchell has been given the option of either paying his $300 bill now or settling it for $306 after 1 month (30 days). If he chooses to pay after 1 month, find the simple interest rate at which he would be charged.

40. COURT JUDGMENT Jennifer was awarded damages of $150,000 in a successful lawsuit she brought against her employer 5 years ago. Interest (simple) on the judgment accrues at the rate of 12%/year from the date of filing. If the case were settled today, how much would Jennifer receive in the final judgment?

41. BRIDGE LOANS To help finance the purchase of a new house, the Abdullahs have decided to apply for a short-term loan (a bridge loan) in the amount of $120,000 for a term of 3 months. If the bank charges simple interest at the rate of 10%/year, how much will the Abdullahs owe the bank at the end of the term?

42. CORPORATE BONDS David owns $20,000 worth of 10-year bonds of Ace Corporation. These bonds pay interest every 6 months at the rate of 7%/year (simple interest). How much income will David receive from this investment every 6 months? How much interest will David receive over the life of the bonds?

43. MUNICIPAL BONDS Maya paid $10,000 for a 7-year bond issued by a city. She received interest amounting to $3500 over the life of the bonds. What rate of (simple) interest did the bond pay?

44. TREASURY BILLS Isabella purchased $20,000 worth of 13-week T-Bills for $19,875. What will be the rate of return on her investment?

45. TREASURY BILLS Maxwell purchased $15,000 worth of 52-week T-Bills for $14,650. What will be the rate of return on his investment?

46. Write Equation (1b) in the slope-intercept form, and interpret the meaning of the slope and the A-intercept in terms of r and P.
Hint: Refer to Figure 1.

47. HOSPITAL COSTS If the cost of a semiprivate room in a hospital was $580/day 5 years ago and it has risen at the rate of 8%/year since that time, what rate would you expect to pay for a semiprivate room today?

48. FAMILY FOOD EXPENDITURE Today a typical family of four spends $880/month for food. If inflation occurs at the rate of 3%/year over the next 6 years, how much should the typical family of four expect to spend for food 6 years from now?

49. HOUSING APPRECIATION The Kwans are planning to buy a house 4 years from now. Housing experts in their area have estimated that the cost of a home will increase at a rate of 5%/year during that period. If this economic prediction holds true, how much can the Kwans expect to pay for a house that currently costs $210,000?

50. ELECTRICITY CONSUMPTION A utility company in a western city of the United States expects the consumption of electricity to increase by 8%/year during the next decade, owing mainly to the expected increase in population. If consumption does increase at this rate, find the amount by which the utility company will have to increase its generating capacity in order to meet the needs of the area at the end of the decade.

51. PENSION FUNDS The managers of a pension fund have invested $1.5 million in U.S. government certificates of deposit that pay interest at the rate of 5.5%/year compounded semiannually over a period of 10 years. At the end of this period, how much will the investment be worth?

52. RETIREMENT FUNDS Five and a half years ago, Chris invested $10,000 in a retirement fund that grew at the rate of 10.82%/year compounded quarterly. What is his account worth today?

53. MUTUAL FUNDS Jodie invested $15,000 in a mutual fund 4 years ago. If the fund grew at the rate of 9.8%/year compounded monthly, what would Jodie's account be worth today?

54. TRUST FUNDS A young man is the beneficiary of a trust fund established for him 21 years ago at his birth. If the original amount placed in trust was $10,000, how much will he receive if the money has earned interest at the rate of 8%/year compounded annually? Compounded quarterly? Compounded monthly?

55. INVESTMENT PLANNING Find how much money should be deposited in a bank paying interest at the rate of 8.5%/year compounded quarterly so that at the end of 5 years, the accumulated amount will be $40,000.

56. PROMISSORY NOTES An individual purchased a 4-year, $10,000 promissory note with an interest rate of 8.5%/year compounded semiannually. How much did the note cost?

57. FINANCING A COLLEGE EDUCATION The parents of a child have just come into a large inheritance and wish to establish a trust fund for her college education. If they estimate that they will need $100,000 in 13 years, how much should they set aside in the trust now if they can invest the money at $8\frac{1}{2}$%/year compounded (a) annually, (b) semiannually, and (c) quarterly?

58. INVESTMENTS Anthony invested a sum of money 5 years ago in a savings account that has since paid interest at the rate of 8%/year compounded quarterly. His investment is now worth $22,289.22. How much did he originally invest?

59. RATE COMPARISONS In the last 5 years, Bendix Mutual Fund grew at the rate of 10.4%/year compounded quarterly. Over the same period, Acme Mutual Fund grew at the rate of 10.6%/year compounded semiannually. Which mutual fund has a better rate of return?

60. **RATE COMPARISONS** Fleet Street Savings Bank pays interest at the rate of 4.25%/year compounded weekly in a savings account, whereas Washington Bank pays interest at the rate of 4.125%/year compounded daily (assume a 365-day year). Which bank offers a better rate of interest?

61. **LOAN CONSOLIDATION** The proprietors of The Coachmen Inn secured two loans from Union Bank: one for $8000 due in 3 years and one for $15,000 due in 6 years, both at an interest rate of 10%/year compounded semiannually. The bank has agreed to allow the two loans to be consolidated into one loan payable in 5 years at the same interest rate. What amount will the proprietors of the inn be required to pay the bank at the end of 5 years?
Hint: Find the present value of the first two loans.

62. **EFFECTIVE RATE OF INTEREST** Find the effective rate of interest corresponding to a nominal rate of 9%/year compounded annually, semiannually, quarterly, and monthly.

63. **ZERO COUPON BONDS** Juan is contemplating buying a zero coupon bond that matures in 10 years and has a face value of $10,000. If the bond yields a return of 5.25%/year compounded annually, how much should Juan pay for the bond?

64. **REVENUE GROWTH OF A HOME THEATER BUSINESS** Maxwell started a home theater business in 2008. The revenue of his company for that year was $240,000. The revenue grew by 20% in 2009 and by 30% in 2010. Maxwell projected that the revenue growth for his company in the next 3 years will be at least 25%/year. How much did Maxwell expect his minimum revenue to be for 2013?

65. **ONLINE RETAIL SALES** Online retail sales stood at $141.4 billion for the year 2004. For the next 2 years, they grew by 24.3% and 14.0% per year, respectively. For the next 3 years, online retail sales were projected to grow at 30.5%, 17.6%, and 10.5% per year, respectively. What were the projected online sales for 2009?
Source: Jupiter Research.

66. **PURCHASING POWER** The U.S. inflation rates for 2003 through 2006 are 1.6%, 2.3%, 2.7%, and 3.4%, respectively. What was the purchasing power of a dollar at the beginning of 2007 compared to that at the beginning of 2003?
Source: U.S. Census Bureau.

67. **INVESTMENT OPTIONS** Investment A offers a 10% return compounded semiannually, and Investment B offers a 9.75% return compounded continuously. Which investment has a higher rate of return over a 4-year period?

68. **EFFECT OF INFLATION ON SALARIES** Leonard's current annual salary is $45,000. Ten years from now, how much will he need to earn to retain his present purchasing power if the rate of inflation over that period is 3%/year compounded continuously?

69. **SAVING FOR COLLEGE** Having received a large inheritance, Jing-mei's parents wish to establish a trust for her college education. If 7 years from now they need an estimated $70,000, how much should they set aside in trust now, if they invest the money at 10.5%/year compounded quarterly? Continuously?

70. **PENSIONS** Maria, who is now 50 years old, is employed by a firm that guarantees her a pension of $40,000/year at age 65. What is the present value of her first year's pension if the inflation rate over the next 15 years is 6%/year compounded continuously? 8%/year compounded continuously? 12%/year compounded continuously?

71. **REAL ESTATE INVESTMENTS** An investor purchased a piece of waterfront property. Because of the development of a marina in the vicinity, the market value of the property is expected to increase according to the rule

$$V(t) = 80,000e^{\sqrt{t}/2}$$

where $V(t)$ is measured in dollars and t is the time (in years) from the present. If the rate of appreciation is expected to be 9%/year compounded continuously for the next 8 years, find an expression for the present value $P(t)$ of the property's market price valid for the next 8 years. What is $P(t)$ expected to be in 4 years?

72. The simple interest formula $A = P(1 + rt)$ [Equation (1b)] can be written in the form $A = Prt + P$, which is the slope-intercept form of a straight line with slope Pr and A-intercept P.
  a. Describe the family of straight lines obtained by keeping the value of r fixed and allowing the value of P to vary. Interpret your results.
  b. Describe the family of straight lines obtained by keeping the value of P fixed and allowing the value of r to vary. Interpret your results.

73. **EFFECTIVE RATE OF INTEREST** Suppose an initial investment of $\$P$ grows to an accumulated amount of $\$A$ in t years. Show that the effective rate (annual effective yield) is

$$r_{\text{eff}} = (A/P)^{1/t} - 1$$

Use the formula given in Exercise 73 to solve Exercises 74–78.

74. **EFFECTIVE RATE OF INTEREST** Martha invested $40,000 in a boutique 5 years ago. Her investment is worth $70,000 today. What is the effective rate (annual effective yield) of her investment?

75. **HOUSING APPRECIATION** Georgia purchased a house in January 2000 for $200,000. In January 2006, she sold the house and made a net profit of $56,000. Find the effective annual rate of return on her investment over the 6-year period.

76. **COMMON STOCK TRANSACTION** Steven purchased 1000 shares of a certain stock for $25,250 (including commissions). He sold the shares 2 years later and received $32,100 after deducting commissions. Find the effective annual rate of return on his investment over the 2-year period.

77. ZERO COUPON BONDS Nina purchased a zero coupon bond for $6724.53. The bond matures in 7 years and has a face value of $10,000. Find the effective annual rate of interest for the bond.

Hint: Assume that the purchase price of the bond is the initial investment and that the face value of the bond is the accumulated amount.

78. MONEY MARKET MUTUAL FUNDS Carlos invested $5000 in a money market mutual fund that pays interest on a daily basis. The balance in his account at the end of 8 months (245 days) was $5170.42. Find the effective rate at which Carlos's account earned interest over this period (assume a 365-day year).

In Exercises 79–82, determine whether the statement is true or false. If it is true, explain why it is true. If it is false, give an example to show why it is false.

79. When simple interest is used, the accumulated amount is a linear function of t.

80. If interest is compounded annually, then the accumulated amount after t years is the same as the accumulated amount under simple interest over t years.

81. If interest is compounded annually, then the effective rate is the same as the nominal rate.

82. Susan's salary increased from $50,000/year to $60,000/year over a 4-year period. Therefore, Susan received annual increases of 5% over that period.

4.1 Solutions to Self-Check Exercises

1. Using Formula (7) with $A = 20{,}000$, $r = 0.054$, $m = 12$, and $t = 3$ so that $i = \frac{0.054}{12} = 0.0045$ and $n = (3)(12) = 36$, we find the required present value to be

$$P = 20{,}000(1 + 0.0045)^{-36} \approx 17{,}015.01$$

or $17,015.01.

2. The accumulated amount of Paul's current investment is found by using Formula (3) with $P = 100{,}000$, $r = 0.046$, and $m = 365$. Thus, $i = \frac{0.046}{365}$ and $n = 365$, so the required accumulated amount is given by

$$A_1 = 100{,}000\left(1 + \frac{0.046}{365}\right)^{365} \approx 104{,}707.14$$

or $104,707.14. Next, we compute the accumulated amount of Paul's reinvestment. Using (3) with $P = 100{,}000$, $r = 0.032$, and $m = 365$ so that $i = \frac{0.032}{365}$ and $n = 365$, we find the required accumulated amount in this case to be

$$A_2 = 100{,}000\left(1 + \frac{0.032}{365}\right)^{365} \approx 103{,}251.61$$

or approximately $103,251.61. Therefore, Paul can expect to experience a net decrease in yearly income of approximately $104{,}707.14 - 103{,}251.61$, or $1455.53.

USING TECHNOLOGY

Finding the Accumulated Amount of an Investment, the Effective Rate of Interest, and the Present Value of an Investment

Graphing Utility

Some graphing utilities have built-in routines for solving problems involving the mathematics of finance. For example, the TI-83/84 TVM SOLVER function incorporates several functions that can be used to solve the problems that are encountered in Sections 4.1–4.3. To access the TVM SOLVER on the TI-83, press 2nd, press FINANCE, and then select 1: TVM Solver. To access the TVM Solver on the TI-83 plus and the TI-84, press APPS, press 1: Finance, and then select 1: TVM Solver. Step-by-step procedures for using these functions can be found on CourseMate.

EXAMPLE 1 Finding the Accumulated Amount of an Investment   Find the accumulated amount after 10 years if $5000 is invested at a rate of 10% per year compounded monthly.

Note: Boldfaced words/characters enclosed in a box (for example, Enter) indicate that an action (click, select, or press) is required. Words/characters printed blue (for example, Chart sub-type:) indicate words/characters appearing on the screen.

(continued)

```
N=120
I%=10
PV=-5000
PMT=0
■ FV=13535.20745
P/Y=12
C/Y=12
PMT:END BEGIN
```

FIGURE **T1**
The TI-83/84 screen showing the future value (FV) of an investment

Solution   Using the TI-83/84 TVM SOLVER with the following inputs,

$$N = 120 \quad \text{(10)(12)}$$
$$I\% = 10$$
$$PV = -5000 \quad \text{Recall that an investment is an outflow.}$$
$$PMT = 0$$
$$FV = 0$$
$$P/Y = 12 \quad \text{The number of payments each year}$$
$$C/Y = 12 \quad \text{The number of conversion periods each year}$$
$$PMT\text{:END BEGIN}$$

we obtain the display shown in Figure T1. We conclude that the required accumulated amount is $13,535.21. ■

```
▶ Eff(10,4)
         10.38128906
```

FIGURE **T2**
The TI-83/84 screen showing the effective rate of interest (Eff)

EXAMPLE 2 Finding the Effective Rate of Interest Find the effective rate of interest corresponding to a nominal rate of 10% per year compounded quarterly.

Solution   Here we use the **Eff** function of the TI-83/84 calculator to obtain the result shown in Figure T2. The required effective rate is approximately 10.38% per year. ■

EXAMPLE 3 Finding the Present Value of an Investment Find the present value of $20,000 due in 5 years if the interest rate is 7.5% per year compounded daily.

Solution   Using the TI-83/84 TVM SOLVER with the following inputs,

$$N = 1825 \quad \text{(5)(365)}$$
$$I\% = 7.5$$
$$PV = 0$$
$$PMT = 0$$
$$FV = 20000$$
$$P/Y = 365 \quad \text{The number of payments each year}$$
$$C/Y = 365 \quad \text{The number of conversions each year}$$
$$PMT\text{:END BEGIN}$$

```
N=1825
I%=7.5
■ PV=-13746.3151
PMT=0
FV=20000
P/Y=365
C/Y=365
PMT:END BEGIN
```

FIGURE **T3**
The TI-83/84 screen showing the present value (PV) of an investment

we obtain the display shown in Figure T3. We see that the required present value is approximately $13,746.32. Note that PV is negative because an investment is an outflow (money is paid out). ■

Excel

Excel has many built-in functions for solving problems involving the mathematics of finance. Here we illustrate the use of the FV (future value), EFFECT (effective rate), and PV (present value) functions to solve problems of the type that we have encountered in Section 4.1.

EXAMPLE 4 Finding the Accumulated Amount of an Investment Find the accumulated amount after 10 years if $5000 is invested at a rate of 10% per year compounded monthly.

Solution   Here we are computing the future value of a lump-sum investment, so we use the FV (future value) function. Click ⎡**Financial**⎤ from the Function Library on the Formulas tab and select ⎡**FV**⎤. The Function Arguments dialog box will appear (see Figure T4). In our example, the mouse cursor is in the edit box headed by Type, so a definition of that term appears near the bottom of the box. Figure T4 shows the entries for each edit box in our example.

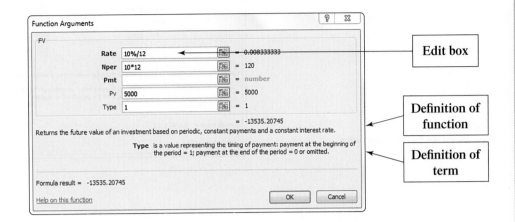

FIGURE **T4**
Excel's dialog box for computing the future value (FV) of an investment

Note that the entry for Nper is given by the total number of periods for which the investment earns interest. The Pmt box is left blank, since no money is added to the original investment. The Pv entry is 5000. The entry for Type is a 1 because the lump-sum payment is made at the beginning of the investment period. The answer, $-\$13,535.21$, is shown at the bottom of the dialog box. It is negative because an investment is considered to be an outflow of money (money is paid out). (Click OK , and the answer will also appear on your spreadsheet.)

EXAMPLE 5 Finding the Effective Rate of Interest Find the effective rate of interest corresponding to a nominal rate of 10% per year compounded quarterly.

Solution   Here we use the EFFECT function to compute the effective rate of interest. Accessing this function from the Financial function library subgroup and making the required entries, we obtain the Function Arguments dialog box shown in Figure T5. The required effective rate is approximately 10.38% per year.

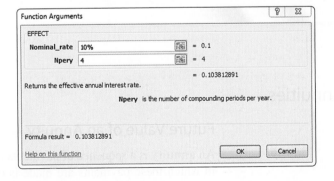

FIGURE **T5**
Excel's dialog box for the effective rate of interest function (EFFECT)

EXAMPLE 6 Finding the Present Value of an Investment Find the present value of $20,000 due in 5 years if the interest rate is 7.5% per year compounded daily.

Solution   We use the PV function to compute the present value of a lump-sum investment. Accessing this function as above and making the required entries, we obtain the PV dialog box shown in Figure T6 (see next page). Once again, the Pmt edit box is left blank, since no additional money is added to the original investment. The Fv entry is 20000. The answer is negative because an investment is considered to be an outflow of money (money is paid out). We deduce that the required amount is $13,746.32.

(continued)

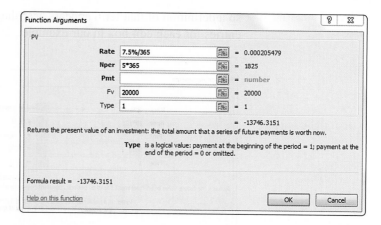

FIGURE **T6**
Excel dialog box for the present value
function (PV)

TECHNOLOGY EXERCISES

1. Find the accumulated amount A if $5000 is invested at the interest rate of $5\frac{3}{8}\%$/year compounded monthly for 3 years.

2. Find the accumulated amount A if $2850 is invested at the interest rate of $6\frac{5}{8}\%$/year compounded monthly for 4 years.

3. Find the accumulated amount A if $327.35 is invested at the interest rate of $5\frac{1}{3}\%$/year compounded daily for 7 years.

4. Find the accumulated amount A if $327.35 is invested at the interest rate of $6\frac{7}{8}\%$/year compounded daily for 8 years.

5. Find the effective rate corresponding to $8\frac{2}{3}\%$/year compounded quarterly.

6. Find the effective rate corresponding to $10\frac{5}{8}\%$/year compounded monthly.

7. Find the effective rate corresponding to $9\frac{3}{4}\%$/year compounded monthly.

8. Find the effective rate corresponding to $4\frac{3}{8}\%$/year compounded quarterly.

9. Find the present value of $38,000 due in 3 years at $8\frac{1}{4}\%$/year compounded quarterly.

10. Find the present value of $150,000 due in 5 years at $9\frac{3}{8}\%$/year compounded monthly.

11. Find the present value of $67,456 due in 3 years at $7\frac{7}{8}\%$/year compounded monthly.

12. Find the present value of $111,000 due in 5 years at $11\frac{5}{8}\%$/year compounded monthly.

4.2   Annuities

Future Value of an Annuity

An **annuity** is a sequence of payments made at regular time intervals. The time period in which these payments are made is called the **term** of the annuity. Depending on whether the term is given by a *fixed time interval*, a time interval that begins at a definite date but extends indefinitely, or one that is not fixed in advance, an annuity is called an **annuity certain,** a *perpetuity*, or a *contingent annuity*, respectively. In general, the payments in an annuity need not be equal, but in many important applications they are equal. In this section, we assume that annuity payments are equal. Examples of annuities are regular deposits to a savings account, monthly home mortgage payments, and monthly insurance payments.

Annuities are also classified by payment dates. An annuity in which the payments are made at the *end* of each payment period is called an **ordinary annuity,** whereas an annuity in which the payments are made at the beginning of each period is called an *annuity due.* Furthermore, an annuity in which the payment period coincides with the interest conversion period is called a **simple annuity,** whereas an annuity in which the payment period differs from the interest conversion period is called a *complex annuity.*

In this section, we consider ordinary annuities that are certain and simple, with periodic payments that are equal in size. In other words, we study annuities that are subject to the following conditions:

1. The terms are given by fixed time intervals.
2. The periodic payments are equal in size.
3. The payments are made at the *end* of the payment periods.
4. The payment periods coincide with the interest conversion periods.

To find a formula for the accumulated amount S of an annuity, suppose a sum of $100 is paid into an account at the end of each quarter over a period of 3 years. Furthermore, suppose the account earns interest on the deposit at the rate of 8% per year, compounded quarterly. Then the first payment of $100 made at the end of the first quarter earns interest at the rate of 8% per year compounded four times a year (or $8/4 = 2\%$ per quarter) over the remaining 11 quarters and therefore, by the compound interest formula, has an accumulated amount of

$$100\left(1 + \frac{0.08}{4}\right)^{11} \quad \text{or} \quad 100(1 + 0.02)^{11}$$

dollars at the end of the term of the annuity (Figure 2).

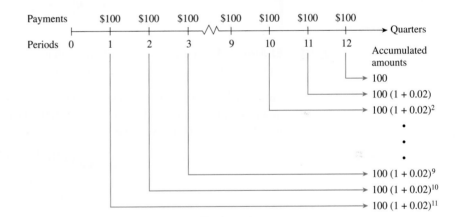

FIGURE 2
The sum of the accumulated amounts is the amount of the annuity.

The second payment of $100 made at the end of the second quarter earns interest at the same rate over the remaining 10 quarters and therefore has an accumulated amount of

$$100(1 + 0.02)^{10}$$

dollars at the end of the term of the annuity, and so on. The last payment earns no interest because it is due at the end of the term. The amount of the annuity is obtained by adding all the terms in Figure 2. Thus,

$$S = 100 + 100(1 + 0.02) + 100(1 + 0.02)^2 + \cdots + 100(1 + 0.02)^{11}$$

The sum on the right is the sum of the first n terms of a *geometric progression* with first term 100 and common ratio $(1 + 0.02)$. We show in Section 4.4 that the sum S can be written in the more compact form

$$S = 100\left[\frac{(1 + 0.02)^{12} - 1}{0.02}\right]$$
$$\approx 1341.21$$

or approximately $1341.21.

To find a general formula for the accumulated amount S of an annuity, suppose that a sum of $\$R$ is paid into an account at the end of each period for n periods and that the account earns interest at the rate of i per period. Then, proceeding as we did with the numerical example, we obtain

$$S = R + R(1 + i) + R(1 + i)^2 + \cdots + R(1 + i)^{n-1}$$

$$= R\left[\frac{(1 + i)^n - 1}{i}\right] \tag{9}$$

The expression inside the brackets is commonly denoted by $s_{\overline{n}|i}$ (read "s angle n at i") and is called the **compound-amount factor.** Extensive tables have been constructed that give values of $s_{\overline{n}|i}$ for different values of i and n (see, for example, Table 1 on the CourseMate site). In terms of the compound-amount factor,

$$S = Rs_{\overline{n}|i} \tag{10}$$

The quantity S in Equations (9) and (10) is realizable at some future date and is accordingly called the future value of an annuity.

Future Value of an Annuity

The **future value S of an annuity** of n payments of R dollars each, paid at the end of each investment period into an account that earns interest at the rate of i per period, is

$$S = R\left[\frac{(1 + i)^n - 1}{i}\right]$$

VIDEO ▶ **EXAMPLE 1** Find the amount of an ordinary annuity consisting of 12 monthly payments of $\$100$ that earn interest at 12% per year compounded monthly.

Solution Since i is the interest rate per *period* and since interest is compounded monthly in this case, we have $i = \frac{0.12}{12} = 0.01$. Using Formula (9) with $R = 100$, $n = 12$, and $i = 0.01$, we have

$$S = 100\left[\frac{(1.01)^{12} - 1}{0.01}\right]$$

$$\approx 1268.25 \quad \text{Use a calculator.}$$

or $\$1268.25$. The same result is obtained by observing that

$$S = 100s_{\overline{12}|0.01}$$
$$= 100(12.6825)$$
$$= 1268.25 \quad \text{Use Table 1 from CourseMate.}$$ ■

Explore & Discuss

Future Value S of an Annuity Due

1. Consider an annuity satisfying conditions 1, 2, and 4 on page 209 but with condition 3 replaced by the condition that payments are made at the *beginning* of the payment periods. By using an argument similar to that used to establish Formula (9), show that the future value S of an annuity due of n payments of R dollars each, paid at the beginning of each investment into an account that earns interest at the rate of i per period, is

$$S = R(1 + i)\left[\frac{(1 + i)^n - 1}{i}\right]$$

2. Use the result of part 1 to see how large your nest egg will be at age 65 if you start saving $\$4000$ annually at age 30, assuming a 10% average annual return; if you start saving at 35; if you start saving at 40. [Moral of the story: It is never too early to start saving!]

Refer to the preceding Explore & Discuss problem.

1. Show that if $R = 4000$ and $i = 0.1$, then $S = 44,000[(1.1)^n - 1]$. Using a graphing utility, plot the graph of $f(x) = 44,000[(1.1)^x - 1]$, using the viewing window $[0, 40] \times [0, 1,200,000]$.

2. Verify the results of part 1 by evaluating $f(35)$, $f(30)$, and $f(25)$ using the EVAL function.

Present Value of an Annuity

In certain instances, you may want to determine the current value P of a sequence of equal periodic payments that will be made over a certain period of time. After each payment is made, the new balance continues to earn interest at some nominal rate. The amount P is referred to as the present value of an annuity.

To derive a formula for determining the present value P of an annuity, we may argue as follows. The amount P invested now and earning interest at the rate of i per period will have an accumulated value of $P(1 + i)^n$ at the end of n periods. But this must be equal to the future value of the annuity S given by Formula (9). Therefore, equating the two expressions, we have

$$P(1 + i)^n = R\left[\frac{(1 + i)^n - 1}{i}\right]$$

Multiplying both sides of this equation by $(1 + i)^{-n}$ gives

$$P = R(1 + i)^{-n}\left[\frac{(1 + i)^n - 1}{i}\right]$$

$$= R\left[\frac{(1 + i)^n(1 + i)^{-n} - (1 + i)^{-n}}{i}\right]$$

$$= R\left[\frac{1 - (1 + i)^{-n}}{i}\right] \qquad (1 + i)^n(1 + i)^{-n} = 1$$

$$= Ra_{\overline{n}|i}$$

where the factor $a_{\overline{n}|i}$ (read "a angle n at i") represents the expression inside the brackets. Extensive tables have also been constructed giving values of $a_{\overline{n}|i}$ for different values of i and n (see Table 1 on the CourseMate site).

Present Value of an Annuity
The **present value P of an annuity** consisting of n payments of R dollars each, paid at the end of each investment period into an account that earns interest at the rate of i per period, is

$$P = R\left[\frac{1 - (1 + i)^{-n}}{i}\right] \qquad (11)$$

EXAMPLE 2 Find the present value of an ordinary annuity consisting of 24 monthly payments of $100 each and earning interest at 9% per year compounded monthly.

Solution   Here, $R = 100$, $i = \frac{r}{m} = \frac{0.09}{12} = 0.0075$, and $n = 24$, so by Formula (11), we have

$$P = 100\left[\frac{1 - (1.0075)^{-24}}{0.0075}\right]$$

$$\approx 2188.91$$

or $2188.91. The same result may be obtained by using Table 1 from the CourseMate site. Thus,

$$P = 100a_{\overline{24}|0.0075}$$
$$= 100(21.8891)$$
$$= 2188.91$$

APPLIED EXAMPLE 3 Saving for a College Education As a savings program toward Alberto's college education, his parents decide to deposit $100 at the end of every month into a bank account paying interest at the rate of 6% per year compounded monthly. If the savings program began when Alberto was 6 years old, how much money would have accumulated by the time he turns 18?

Solution   By the time the child turns 18, the parents would have made 144 deposits into the account. Thus, $n = 144$. Furthermore, we have $R = 100$, $r = 0.06$, and $m = 12$, so $i = \frac{0.06}{12} = 0.005$. Using Formula (9), we find that the amount of money that would have accumulated is given by

$$S = 100\left[\frac{(1.005)^{144} - 1}{0.005}\right]$$

$$\approx 21{,}015$$

or $21,015.

APPLIED EXAMPLE 4 Financing a Car After making a down payment of $6000 for an automobile, Murphy paid $600 per month for 36 months with interest charged at 6% per year compounded monthly on the unpaid balance. What was the original cost of the car? What portion of Murphy's total car payments went toward interest charges?

Solution   The loan taken up by Murphy is given by the present value of the annuity

$$P = 600\left[\frac{1 - (1.005)^{-36}}{0.005}\right] = 600a_{\overline{36}|0.005}$$

$$\approx 19{,}723$$

or $19,723. Therefore, the original cost of the automobile is $25,723 ($19,723 plus the $6000 down payment). The interest charges paid by Murphy are given by $(36)(600) - 19{,}723 = 1877$, or $1877.

One important application of annuities arises in the area of tax planning. During the 1980s, Congress created many tax-sheltered retirement savings plans, such as Individual Retirement Accounts (IRAs), Keogh plans, and Simplified Employee Pension (SEP) plans. These plans are examples of annuities in which the individual is allowed to make contributions (which are often tax deductible) to an investment account. The amount of the contribution is limited by congressional legislation. The taxes on the contributions and/or the interest accumulated in these accounts are deferred until the money is withdrawn—ideally during retirement, when tax brackets should be lower. In the interim period, the individual has the benefit of tax-free growth on his or her investment.

Suppose, for example, you are eligible to make a fully deductible contribution to an IRA and you are in a marginal tax bracket of 28%. Additionally, suppose you receive a year-end bonus of $2000 from your employer and have the option of depositing the $2000 into either an IRA or a regular savings account, where both accounts earn interest at an effective annual rate of 8% per year. If you choose to invest your bonus in a regular savings account, you will first have to pay taxes on the $2000, leaving $1440 to invest. At the end of 1 year, you will also have to pay taxes on the interest earned, leaving you with

$$\underset{\text{amount}}{\text{Accumulated}} - \underset{\text{interest}}{\text{Tax on}} = \underset{\text{amount}}{\text{Net}}$$

$$1555.20 \quad - \quad 32.26 \quad = 1522.94$$

or $1522.94.

On the other hand, if you put the money into the IRA, the entire sum will earn interest, and at the end of 1 year you will have (1.08)($2000), or $2160, in your account. Of course, you will still have to pay taxes on this money when you withdraw it, but you will have gained the advantage of tax-free growth of the larger principal over the years. The disadvantage of this option is that if you withdraw the money before you reach the age of $59\frac{1}{2}$, you will be liable for taxes on both your contributions and the interest earned, *and* you will also have to pay a 10% penalty.

Note   In practice, the size of the contributions an individual might make to the various retirement plans might vary from year to year. Also, he or she might make the contributions at different payment periods. To simplify our discussion, we will consider examples in which fixed payments are made at regular intervals.   ■

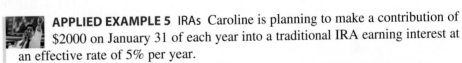 **APPLIED EXAMPLE 5** IRAs Caroline is planning to make a contribution of $2000 on January 31 of each year into a traditional IRA earning interest at an effective rate of 5% per year.

a. After she makes her 25th payment on January 31 of the year following her retirement at age 65, how much will she have in her IRA?

b. Suppose that Caroline withdraws all of her money from her traditional IRA after she makes her 25th payment in the year following her retirement at age 65 and that her investment is subjected to a tax of 28% at that time. How much money will she end up with after taxes?

Solution

a. The amount of money Caroline will have after her 25th payment into her account is found by using Formula (9) with $R = 2000$, $r = 0.05$, $m = 1$, and $t = 25$, so that $i = \frac{r}{m} = 0.05$ and $n = mt = 25$. The required amount is given by

$$S = 2000\left[\frac{(1.05)^{25} - 1}{0.05}\right]$$

$$\approx 95{,}454.20$$

or $95,454.20.

b. If she withdraws the entire amount from her account, she will end up with

$$(1 - 0.28)(95{,}454.20) \approx 68{,}727.02$$

that is, she will have approximately $68,727.02 after paying taxes.   ■

After-tax-deferred annuities are another type of investment vehicle that allows an individual to build assets for retirement, college funds, or other future needs. The advantage gained in this type of investment is that the tax on the accumulated interest is deferred to a later date. Note that in this type of investment, the contributions themselves are not tax deductible. At first glance, the advantage thus gained may seem to be relatively inconsequential, but its true effect is illustrated by the next example.

 APPLIED EXAMPLE 6 Investment Analysis Both Clark and Colby are salaried individuals, 45 years of age, who are saving for their retirement 20 years from now. Both Clark and Colby are also in the 28% marginal tax bracket. Clark makes a $1000 contribution annually on December 31 into a savings account earning an effective rate of 8% per year. At the same time, Colby makes a $1000 annual payment to an insurance company for an after-tax-deferred annuity. The annuity also earns interest at an effective rate of 8% per year. (Assume that both men remain in the same tax bracket throughout this period, and disregard state income taxes.)

a. Calculate how much each man will have in his investment account at the end of 20 years.

b. Compute the interest earned on each account.

c. Show that even if the interest on Colby's investment were subjected to a tax of 28% upon withdrawal of his investment at the end of 20 years, the net accumulated amount of his investment would still be greater than that of Clark's.

Solution

a. Because Clark is in the 28% marginal tax bracket, the net yield for his investment is (0.72)(8), or 5.76%, per year.

Using Formula (9) with $R = 1000$, $r = 0.0576$, $m = 1$, and $t = 20$, so that $i = 0.0576$ and $n = mt = 20$, we see that Clark's investment will be worth

$$S = 1000\left[\frac{(1 + 0.0576)^{20} - 1}{0.0576}\right]$$

$$\approx 35{,}850.49$$

or $35,850.49 at his retirement.

Colby has a tax-sheltered investment with an effective yield of 8% per year. Using Formula (9) with $R = 1000$, $r = 0.08$, $m = 1$, and $t = 20$, so that $i = 0.08$ and $n = mt = 20$, we see that Colby's investment will be worth

$$S = 1000\left[\frac{(1 + 0.08)^{20} - 1}{0.08}\right]$$

$$\approx 45{,}761.96$$

or $45,761.96 at his retirement.

b. Each man will have paid 20(1000), or $20,000, into his account. Therefore, the total interest earned in Clark's account will be (35,850.49 − 20,000), or $15,850.49, whereas the total interest earned in Colby's account will be (45,761.96 − 20,000), or $25,761.96.

c. From part (b) we see that the total interest earned in Colby's account will be $25,761.96. If it were taxed at 28%, he would still end up with (0.72)(25,761.96), or $18,548.61. This is larger than the total interest of $15,850.49 earned by Clark. ∎

In 1997, another type of tax-sheltered retirement savings plan was created by Congress: the Roth IRA. In contrast to traditional IRAs, contributions to Roth IRAs are not tax-deferrable. However, direct contributions to Roth IRAs (but not rollovers) may be withdrawn tax-free at any time. Also, holders of a Roth IRA are not required to take minimum distributions after age $70\frac{1}{2}$.

APPLIED EXAMPLE 7 Roth IRAs Refer to Example 5. Suppose that Caroline decides to invest her money in a Roth IRA instead of a traditional IRA. Also, suppose that she is in the 28% tax bracket and remains in that bracket for the next 25 years until her retirement at age 65. If she pays taxes on $2000 and then invests the remaining $1440 into a Roth IRA earning interest at a rate of 5%/year, compounded annually, how much will she have in her Roth IRA after her 25th payment on January 31 of the year following her retirement? (Disregard state and city taxes.) How does this compare with the amount of money she would have if she had stayed with a traditional IRA and withdrawn all of her money from that account at that time? (See Example 5b.)

Solution We use Formula (9) with $R = 1440$, $r = 0.05$, and $n = 25$, obtaining

$$S = 1440\left[\frac{(1.05)^{25} - 1}{0.05}\right]$$

$$\approx 68{,}727.02$$

that is, she will have approximately $68,727.02 in her account. This is the same as the amount she ended up with in her traditional IRA after paying taxes (see Example 5b).

4.2 Self-Check Exercises

1. Phyliss opened an IRA on January 31, 1996, with a contribution of $2000. She plans to make a contribution of $2000 thereafter on January 31 of each year until her retirement in the year 2015 (20 payments). If the account earns interest at the rate of 8%/year compounded yearly, how much will Phyliss have in her account when she retires?

2. Denver Wildcatting Company has an immediate need for a loan. In an agreement worked out with its banker, Denver assigns its royalty income of $4800/month for the next 3 years from certain oil properties to the bank, with the first payment due at the end of the first month. If the bank charges interest at the rate of 9%/year compounded monthly, what is the amount of the loan negotiated between the parties?

Solutions to Self-Check Exercises 4.2 can be found on page 217.

4.2 Concept Questions

1. In an ordinary annuity, is the term fixed or variable? Are the periodic payments all of the same size, or do they vary in size? Are the payments made at the beginning or the end of the payment period? Do the payment periods coincide with the interest conversion periods?

2. What is the difference between an ordinary annuity and an annuity due?

3. What is the future value of an annuity? Give an example.

4. What is the present value of an annuity? Give an example.

4.2 Exercises

In Exercises 1–8, find the amount (future value) of each ordinary annuity.

1. $1000/year for 10 years at 10%/year compounded annually

2. $1500/semiannual period for 8 years at 9%/year compounded semiannually

3. $1800/quarter for 6 years at 8%/year compounded quarterly

4. $500/semiannual period for 12 years at 11%/year compounded semiannually

5. $600/quarter for 9 years at 12%/year compounded quarterly

6. $150/month for 15 years at 10%/year compounded monthly

7. $200/month for $20\frac{1}{4}$ years at 9%/year compounded monthly

8. $100/week for $7\frac{1}{2}$ years at 7.5%/year compounded weekly

In Exercises 9–14, find the present value of each ordinary annuity.

9. $5000/year for 8 years at 6%/year compounded annually

10. $1200/semiannual period for 6 years at 10%/year compounded semiannually

11. $4000/year for 5 years at 9%/year compounded yearly

12. $3000/semiannual period for 6 years at 11%/year compounded semiannually

13. $800/quarter for 7 years at 12%/year compounded quarterly

14. $150/month for 10 years at 8%/year compounded monthly

15. **IRAs** If a merchant deposits $1500 at the end of each tax year in an IRA paying interest at the rate of 8%/year compounded annually, how much will she have in her account at the end of 20 years?

16. **SAVINGS ACCOUNTS** If Jackson deposits $100 at the end of each month in a savings account earning interest at the rate of 8%/year compounded monthly, how much will he have on deposit in his savings account at the end of 6 years, assuming that he makes no withdrawals during that period?

17. **SAVINGS ACCOUNTS** Linda has joined a Christmas Fund Club at her bank. At the end of every month, December through October inclusive, she will make a deposit of $40 in her fund. If the money earns interest at the rate of 7%/year compounded monthly, how much will she have in her account on December 1 of the following year?

18. **KEOGH ACCOUNTS** Robin, who is self-employed, contributes $5000/year into a Keogh account. How much will he have in the account after 25 years if the account earns interest at the rate of 8.5%/year compounded yearly?

19. **RETIREMENT PLANNING** As a fringe benefit for the past 12 years, Colin's employer has contributed $100 at the end of each month into an employee retirement account for Colin that pays interest at the rate of 7%/year compounded monthly. Colin has also contributed $2000 at the end of each of the last 8 years into an IRA that pays interest at the rate of 9%/year compounded yearly. How much does Colin have in his retirement fund at this time?

20. **SAVINGS ACCOUNTS** The Pirerras are planning to go to Europe 3 years from now and have agreed to set aside $150/month for their trip. If they deposit this money at the end of each month into a savings account paying interest at the rate of 8%/year compounded monthly, how much money will be in their travel fund at the end of the third year?

21. **INVESTMENT ANALYSIS** Karen has been depositing $150 at the end of each month in a tax-free retirement account since she was 25. Matt, who is the same age as Karen, started depositing $250 at the end of each month in a tax-free retirement account when he was 35. Assuming that both accounts have been and will be earning interest at the rate of 5%/year compounded monthly, who will end up with the larger retirement account at the age of 65?

22. **INVESTMENT ANALYSIS** Luis has $150,000 in his retirement account at his present company. Because he is assuming a position with another company, Luis is planning to "roll over" his assets to a new account. Luis also plans to put $3000/quarter into the new account until his retirement 20 years from now. If the new account earns interest at the rate of 8%/year compounded quarterly, how much will Luis have in his account at the time of his retirement?
Hint: Use the compound interest formula and the annuity formula.

23. **AUTO LEASING** The Betzes have leased an auto for 2 years at $450/month. If money is worth 9%/year compounded monthly, what is the equivalent cash payment (present value) of this annuity?

24. **AUTO FINANCING** Lupé made a down payment of $8000 toward the purchase of a new car. To pay the balance of the purchase price, she has secured a loan from her bank at the rate of 12%/year compounded monthly. Under the terms of her finance agreement, she is required to make payments of $420/month for 36 months. What is the cash price of the car?

25. **INSTALLMENT PLANS** Mike's Sporting Goods sells elliptical trainers under two payment plans: cash or installment. Under the installment plan, the customer pays $22/month over 3 years with interest charged on the balance at a rate of 18%/year compounded monthly. Find the cash price for an elliptical trainer if it is equivalent to the price paid by a customer using the installment plan.

26. **LOTTERY PAYOUTS** A state lottery commission pays the winner of the Million Dollar lottery 20 installments of $50,000/year. The commission makes the first payment of $50,000 immediately and the other $n = 19$ payments at the end of each of the next 19 years. Determine how much money the commission should have in the bank initially to guarantee the payments, assuming that the balance on deposit with the bank earns interest at the rate of 8%/year compounded yearly.
Hint: Find the present value of an annuity.

27. **PURCHASING A HOME** The Johnsons have accumulated a nest egg of $40,000 that they intend to use as a down payment toward the purchase of a new house. They have decided to invest a minimum of $2400/month in monthly payments (to take advantage of tax deductions) toward the purchase of their house. However, because of other financial obligations, their monthly payments should not exceed $3000. If local mortgage rates are 7.5%/year compounded monthly for a conventional 30-year mortgage, what is the price range of houses that they should consider?

28. **PURCHASING A HOME** Refer to Exercise 27. If local mortgage rates fell to 7%/year compounded monthly, how would this affect the price range of houses that the Johnsons should consider?

29. **PURCHASING A HOME** Refer to Exercise 27. If the Johnsons decide to secure a 15-year mortgage instead of a 30-year mortgage, what is the price range of houses they should consider when the local mortgage rate for this type of loan is 7%?

30. **SAVINGS PLAN** Lauren plans to deposit $5000 into a bank account at the beginning of next month and $200/month into the same account at the end of that month and at the end of each subsequent month for the next 5 years. If her bank pays interest at the rate of 6%/year compounded monthly, how much will Lauren have in her account at the end of 5 years? (Assume that she makes no withdrawals during the 5-year period.)

31. **FINANCIAL PLANNING** Joe plans to deposit $200 at the end of each month into a bank account for a period of 2 years, after which he plans to deposit $300 at the end of each month into the same account for another 3 years. If the bank pays interest at the rate of 6%/year compounded monthly, how much will Joe have in his account by the end of 5 years? (Assume that no withdrawals are made during the 5-year period.)

32. **INVESTMENT ANALYSIS** From age 25 to age 40, Jessica deposited $200 at the end of each month into a tax-free retirement account. She made no withdrawals or further contributions until age 65. Alex made deposits of $300 into his tax-free retirement account from age 40 to age 65. If both accounts earned interest at the rate of 5%/year compounded monthly, who ends up with a bigger nest egg upon reaching the age of 65?
Hint: Use both the annuity formula and the compound interest formula.

33. **ROTH IRAS** Suppose that Jacob deposits $3000/year for 10 years into a Roth IRA earning interest at the rate of 6%/year, compounded annually. During the next 10 years, he makes no withdrawals and no further contributions, but the account continues to earn interest at the same rate. How much will Jacob have in his retirement account at the end of the 20-year period?

34. **RETIREMENT PLANNING** Suppose that Ramos contributes $5000/year into a traditional IRA earning interest at the rate of 4%/year compounded annually, every year after age 35 until his retirement at age 65. At the same time, his wife Vanessa deposits $3600/year into a Roth IRA earning interest at the same rate as that of Ramos and also for a period of 30 years. Suppose that both Ramos and Vanessa wish to withdraw all of the money in their IRAs at the time of their retirement, and that Ramos's investment is subjected to a tax of 30% at that time.
(a) After all due taxes are paid, who will have the larger amount?
(b) How much larger will that amount be?

In Exercises 35 and 36, determine whether the statement is true or false. If it is true, explain why it is true. If it is false, give an example to show why it is false.

35. The future value of an annuity can be found by adding together all the payments that are paid into the account.

36. If the future value of an annuity consisting of n payments of R dollars each—paid at the end of each investment period into an account that earns interest at the rate of i per period—is S dollars, then

$$R = \frac{iS}{(1 + i)^n - 1}$$

4.2 Solutions to Self-Check Exercises

1. The amount Phyliss will have in her account when she retires may be found by using Formula (9) with $R = 2000$, $r = 0.08$, $m = 1$, and $t = 20$, so that $i = r = 0.08$ and $n = mt = 20$. Thus,

$$S = 2000 \left[\frac{(1.08)^{20} - 1}{0.08} \right]$$

$$\approx 91{,}523.93$$

or $91,523.93.

2. We want to find the present value of an ordinary annuity consisting of 36 monthly payments of $4800 each and earning interest at 9%/year compounded monthly. Using Formula (11) with $R = 4800$, $m = 12$, and $t = 3$, so that $i = \frac{r}{m} = \frac{0.09}{12} = 0.0075$ and $n = (12)(3) = 36$, we find

$$P = 4800 \left[\frac{1 - (1.0075)^{-36}}{0.0075} \right] \approx 150{,}944.67$$

or $150,944.67, the amount of the loan negotiated.

USING TECHNOLOGY

Finding the Amount of an Annuity

Graphing Utility

As was mentioned in Using Technology, Section 4.1, the TI-83/84 can facilitate the solution of problems in finance. We continue to exploit its versatility in this section.

EXAMPLE 1 Finding the Future Value of an Annuity Find the amount of an ordinary annuity of 36 quarterly payments of $220 each that earn interest at the rate of 10% per year compounded quarterly.

(*continued*)

Solution   We use the TI-83/84 TVM SOLVER with the following inputs:

$$N = 36$$
$$I\% = 10$$
$$PV = 0$$
$$PMT = -220 \qquad \text{Recall that a payment is an outflow.}$$
$$FV = 0$$
$$P/Y = 4 \qquad \text{The number of payments each year}$$
$$C/Y = 4 \qquad \text{The number of conversion periods each year}$$
$$\text{PMT:END BEGIN}$$

The result is displayed in Figure T1. We deduce that the desired amount is $12,606.31.

```
N=36
I%=10
PV=0
PMT=-220
FV=12606.31078
P/Y=4
C/Y=4
PMT:END BEGIN
```

FIGURE **T1**
The TI-83/84 screen showing the future
value (FV) of an annuity

EXAMPLE 2 Finding the Present Value of an Annuity Find the present value of an
ordinary annuity consisting of 48 monthly payments of $300 each and earning inter-
est at the rate of 9% per year compounded monthly.

Solution   We use the TI-83/84 TVM SOLVER with the following inputs:

$$N = 48$$
$$I\% = 9$$
$$PV = 0$$
$$PMT = -300 \qquad \text{A payment is an outflow.}$$
$$FV = 0$$
$$P/Y = 12 \qquad \text{The number of payments each year}$$
$$C/Y = 12 \qquad \text{The number of conversion periods each year}$$
$$\text{PMT:END BEGIN}$$

The output is displayed in Figure T2. We see that the required present value of the
annuity is $12,055.43.

```
N=48
I%=9
PV=12055.43457
PMT=-300
FV=0
P/Y=12
C/Y=12
PMT:END BEGIN
```

FIGURE **T2**
The TI-83/84 screen showing the present
value (PV) of an ordinary annuity

Excel

Now we show how Excel can be used to solve financial problems involving annuities.

EXAMPLE 3 Finding the Future Value of an Annuity Find the amount of an ordinary
annuity of 36 quarterly payments of $220 each that earn interest at the rate of 10%
per year compounded quarterly.

Solution   Here, we are computing the future value of a series of equal payments,
so we use the FV (future value) function. As before, we choose this function from

---

Note: Words/characters printed blue (for example, Chart sub-type:) indicate words/characters that appear on the screen.

the Financial function library to obtain the Function Arguments dialog box. After making each of the required entries, we obtain the dialog box shown in Figure T3.

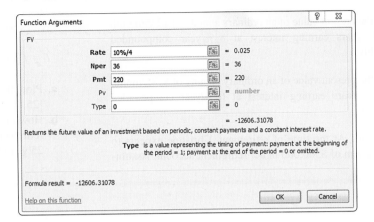

FIGURE **T3**
Excel's dialog box for the future value (FV) of an annuity

Note that a 0 is entered in the Type edit box because payments are made at the end of each payment period. Once again, the answer is negative because cash is paid out. We deduce that the desired amount is $12,606.31.

EXAMPLE 4 Finding the Present Value of an Annuity Find the present value of an ordinary annuity consisting of 48 monthly payments of $300 each and earning interest at the rate of 9% per year compounded monthly.

Solution Here, we use the PV function to compute the present value of an annuity. Accessing the PV (present value) function from the Financial function library and making the required entries, we obtain the PV dialog box shown in Figure T4. We see that the required present value of the annuity is $12,055.43.

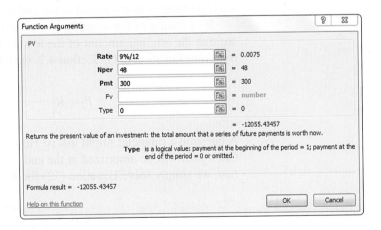

FIGURE **T4**
Excel's dialog box for computing the present value (PV) of an annuity

TECHNOLOGY EXERCISES

1. Find the amount of an ordinary annuity of 20 payments of $2500/quarter at $7\frac{1}{4}$%/year compounded quarterly.

2. Find the amount of an ordinary annuity of 24 payments of $1790/quarter at $8\frac{3}{4}$%/year compounded quarterly.

3. Find the amount of an ordinary annuity of $120/month for 5 years at $6\frac{3}{8}$%/year compounded monthly.

4. Find the amount of an ordinary annuity of $225/month for 6 years at $7\frac{5}{8}$%/year compounded monthly.

5. Find the present value of an ordinary annuity of $4500/semiannual period for 5 years earning interest at 9%/year compounded semiannually.

(continued)

6. Find the present value of an ordinary annuity of $2100/quarter for 7 years earning interest at $7\frac{1}{8}\%$/year compounded quarterly.

7. Find the present value of an ordinary annuity of $245/month for 6 years earning interest at $8\frac{3}{8}\%$/year compounded monthly.

8. Find the present value of an ordinary annuity of $185/month for 12 years earning interest at $6\frac{5}{8}\%$/year compounded monthly.

9. ANNUITIES At the time of retirement, Christine expects to have a sum of $500,000 in her retirement account. Assum-

ing that the account pays interest at the rate of 5%/year compounded continuously, her accountant pointed out to her that if she made withdrawals amounting to x dollars per year ($x > 25,000$), then the time required to deplete her savings would be T years, where

$$T = f(x) = 20 \ln\left(\frac{x}{x - 25,000}\right) \qquad (x > 25,000)$$

a. Plot the graph of f, using the viewing window $[25,000, 50,000] \times [0, 100]$.
b. How much should Christine plan to withdraw from her retirement account each year if she wants it to last for 25 years?

4.3   Amortization and Sinking Funds

Amortization of Loans

The annuity formulas derived in Section 4.2 may be used to answer questions involving the amortization of certain types of installment loans. For example, in a typical housing loan, the mortgagor makes periodic payments toward reducing his or her indebtedness to the lender, who charges interest at a fixed rate on the unpaid portion of the debt. In practice, the borrower is required to repay the lender in periodic installments, usually of the same size and over a fixed term, so that the loan (principal plus interest charges) is amortized at the end of the term.

By thinking of the monthly loan repayments R as the payments in an annuity, we see that the original amount of the loan is given by P, the present value of the annuity. From Equation (11), Section 4.2, we have

$$P = R\left[\frac{1 - (1 + i)^{-n}}{i}\right] = Ra_{\overline{n}|i} \qquad (12)$$

A question a financier might ask is: How much should the monthly installment be so that a loan will be amortized at the end of the term of the loan? To answer this question, we simply solve Equation (12) for R in terms of P, obtaining

$$R = \frac{Pi}{1 - (1 + i)^{-n}} = \frac{P}{a_{\overline{n}|i}}$$

Amortization Formula

The periodic payment R on a loan of P dollars to be amortized over n periods with interest charged at the rate of i per period is

$$R = \frac{Pi}{1 - (1 + i)^{-n}} \qquad (13)$$

VIDEO

APPLIED EXAMPLE 1 Amortization Schedule A sum of $50,000 is to be repaid over a 5-year period through equal installments made at the end of each year. If an interest rate of 8% per year is charged on the unpaid balance and interest calculations are made at the end of each year, determine the size of each installment so that the loan (principal plus interest charges) is amortized at the end of 5 years. Verify the result by displaying the amortization schedule.

Solution Substituting $P = 50,000$, $i = r = 0.08$ (here, $m = 1$), and $n = 5$ into Formula (13), we obtain

$$R = \frac{(50,000)(0.08)}{1 - (1.08)^{-5}} \approx 12,522.82$$

giving the required yearly installment as $12,522.82.

The amortization schedule is presented in Table 3. The outstanding principal at the end of 5 years is, of course, zero. (The figure of $0.01 in Table 3 is the result of round-off errors.) Observe that initially the larger portion of the repayment goes toward payment of interest charges, but as time goes by, more and more of the payment goes toward repayment of the principal.

TABLE 3

An Amortization Schedule

End of Period	Interest Charged	Repayment Made	Payment Toward Principal	Outstanding Principal
0	—	—	—	$50,000.00
1	$4,000.00	$12,522.82	$ 8,522.82	41,477.18
2	3,318.17	12,522.82	9,204.65	32,272.53
3	2,581.80	12,522.82	9,941.02	22,331.51
4	1,786.52	12,522.82	10,736.30	11,595.21
5	927.62	12,522.82	11,595.20	0.01

Financing a Home

APPLIED EXAMPLE 2 Home Mortgage Payments The Blakelys borrowed $120,000 from a bank to help finance the purchase of a house. The bank charges interest at a rate of 5.4% per year on the unpaid balance, with interest computations made at the end of each month. The Blakelys have agreed to repay the loan in equal monthly installments over 30 years. How much should each payment be if the loan is to be amortized at the end of the term?

Solution Here, $P = 120,000$, $i = \frac{r}{m} = \frac{0.054}{12} = 0.0045$, and $n = (30)(12) = 360$. Using Formula (13), we find that the size of each monthly installment required is given by

$$R = \frac{(120,000)(0.0045)}{1 - (1.0045)^{-360}}$$

$$\approx 673.84$$

or $673.84.

APPLIED EXAMPLE 3 Home Equity Teresa and Raul purchased a house 10 years ago for $200,000. They made a down payment of 20% of the purchase price and secured a 30-year conventional home mortgage at 6% per year compounded monthly on the unpaid balance. The house is now worth $380,000. How much equity do Teresa and Raul have in their house now (after making 120 monthly payments)?

Solution   Since the down payment was 20%, we know that they secured a loan of 80% of $200,000, or $160,000. Furthermore, using Formula (13) with $P = 160,000$, $i = \frac{r}{m} = \frac{0.06}{12} = 0.005$ and $n = (30)(12) = 360$, we determine that their monthly installment is

$$R = \frac{(160,000)(0.005)}{1 - (1.005)^{-360}}$$
$$\approx 959.28$$

or $959.28.

After 120 monthly payments have been made, the outstanding principal is given by the sum of the present values of the remaining installments (that is, $360 - 120 = 240$ installments). But this sum is just the present value of an annuity with $n = 240$, $R = 959.28$, and $i = 0.005$. Using Formula (11), we find

$$P = 959.28\left[\frac{1 - (1 + 0.005)^{-240}}{0.005}\right]$$
$$\approx 133,897.04$$

or approximately $133,897. Therefore, Teresa and Raul have an equity of $380,000 - 133,897$, that is, $246,103.

Explore & Discuss and Exploring with Technology

1. Consider the amortization Formula (13):

$$R = \frac{Pi}{1 - (1 + i)^{-n}}$$

Suppose you know the values of R, P, and n and you wish to determine i. Explain why you can accomplish this task by finding the point of intersection of the graphs of the functions

$$y_1 = R \quad \text{and} \quad y_2 = \frac{Pi}{1 - (1 + i)^{-n}}$$

2. Thalia knows that her monthly repayment on her 30-year conventional home loan of $150,000 is $1100.65 per month. Help Thalia determine the interest rate for her loan by verifying or executing the following steps:
   a. Plot the graphs of

$$y_1 = 1100.65 \quad \text{and} \quad y_2 = \frac{150,000x}{1 - (1 + x)^{-360}}$$

using the viewing window $[0, 0.01] \times [0, 1200]$.
   b. Use the ISECT (intersection) function of the graphing utility to find the point of intersection of the graphs of part (a). Explain why this gives the value of i.
   c. Compute r from the relationship $r = 12i$.

Explore & Discuss and Exploring with Technology

1. Suppose you secure a home mortgage loan of $\$P$ with an interest rate of r per year to be amortized over t years through monthly installments of $\$R$. Show that after N installments, your outstanding principal is given by

$$B(N) = P\left[\frac{(1 + i)^n - (1 + i)^N}{(1 + i)^n - 1}\right] \qquad (0 \le N \le n)$$

Hint: $B(N) = R\left[\dfrac{1 - (1 + i)^{-n+N}}{i}\right]$. To see this, study Example 3, page 222. Replace R using Formula (13).

2. Refer to Example 3, page 222. Using the result of part 1, show that Teresa and Raul's outstanding balance after making N payments is

$$E(N) = 160{,}000\left[\frac{1.005^{360} - 1.005^N}{1.005^{360} - 1}\right] \qquad (0 \le N \le 360)$$

3. Using a graphing utility, plot the graph of

$$E(x) = 160{,}000\left[\frac{1.005^{360} - 1.005^x}{1.005^{360} - 1}\right]$$

using the viewing window $[0, 360] \times [0, 160{,}000]$.

4. Referring to the graph in part 3, observe that the outstanding principal drops off slowly in the early years and accelerates quickly to zero toward the end of the loan. Can you explain why?

5. How long does it take Teresa and Raul to repay half of the loan of $160,000?
   Hint: See the previous Explore & Discuss and Exploring with Technology box.

APPLIED EXAMPLE 4 Home Affordability The Jacksons have determined that, after making a down payment, they could afford at most $2000 for a monthly house payment. The bank charges interest at the rate of 6% per year on the unpaid balance, with interest computations made at the end of each month. If the loan is to be amortized in equal monthly installments over 30 years, what is the maximum amount that the Jacksons can borrow from the bank?

Solution   Here, $i = \frac{r}{m} = \frac{0.06}{12} = 0.005$, $n = (30)(12) = 360$, and $R = 2000$; we are required to find P. From Formula (12), we have

$$P = R\left[\frac{1 - (1 + i)^{-n}}{i}\right]$$

Substituting the numerical values for R, n, and i into this expression for P, we obtain

$$P = 2000\left[\frac{1 - (1.005)^{-360}}{0.005}\right] \approx 333{,}583.23$$

Therefore, the Jacksons can borrow at most $333,583.   ◼

An adjustable-rate mortgage (ARM) is a home loan in which the interest rate is changed periodically based on a financial index. For example, a 5/1 ARM is one that has an initial rate for the first 5 years and thereafter is adjusted every year for the remaining term of the loan. Similarly, a 7/1 ARM is one that has an initial rate for the first 7 years and thereafter is adjusted every year for the remaining term of the loan.

During the housing boom of 2000–2006, lenders aggressively promoted another type of loan—the interest-only mortgage loan—to help prospective buyers qualify for larger mortgages. With an interest-only loan, the homeowner pays only the interest on the mortgage for a fixed term, usually 5 to 7 years. At the end of that period, the borrower usually has an option to convert the loan to one that will be amortized.

APPLIED EXAMPLE 5 Home Affordability Refer to Example 4. Suppose that the bank has also offered the Jacksons (a) a 7/1 ARM with a term of 30 years and an interest rate of 5.70%/year compounded monthly for the first 7 years and (b) an interest-only loan for a term of 30 years and an interest rate of 5.94%/year for the first 7 years. If the Jacksons limit their monthly payment to $2000/month, what is the maximum amount they can borrow with each of these mortgages?

Solution

a. Here, $i = \frac{r}{m} = \frac{0.057}{12} = 0.00475$, $n = (30)(12) = 360$, and $R = 2000$, and we want to find P. From Formula (12), we have

$$P = R\left[\frac{1 - (1 + i)^{-n}}{i}\right]$$

$$= 2000\left[\frac{1 - (1.00475)^{-360}}{0.00475}\right]$$

$$\approx 344,589.68$$

Therefore, if the Jacksons choose the 7/1 ARM, they can borrow at most $344,590.

b. If P denotes the maximum amount that the Jacksons can borrow, then the interest per year on their loan is $0.0594P$ dollars. But this is equal to their payments for the year, which are $(12)(2000)$ dollars. So we have

$$0.0594P = (12)(2000)$$

or

$$P = \frac{(12)(2000)}{0.0594} \approx 404,040.40$$

So if the Jacksons choose the 7-year interest-only loan, they can borrow at most $404,040.

APPLIED EXAMPLE 6 Adjustable Rate Mortgages Five years ago, the Campbells secured a 5/1 ARM to help finance the purchase of their home. The amount of the original loan was $350,000 for a term of 30 years, with interest at the rate of 5.76% per year, compounded monthly for the first 5 years. The Campbells' mortgage is due to reset next month, and the new interest rate will be 6.96% per year, compounded monthly.

a. What was the Campbells' monthly mortgage payment for the first 5 years?
b. What will the Campbells' new monthly mortgage payment be (after the reset)? By how much will the monthly payment increase?

Solution

a. First, we find the Campbells' monthly payment on the original loan amount. Using Formula (13) with $P = 350{,}000$, $i = \frac{r}{m} = \frac{0.0576}{12} = 0.0048$, and $n = mt = (12)(30) = 360$, we find that the monthly payment was

$$R = \frac{350{,}000(0.0048)}{1 - (1 + 0.0048)^{-360}} \approx 2044.729$$

or \$2044.73 for the first 5 years.

b. To find the amount of the Campbells' new mortgage payment, we first need to find their outstanding principal. This is given by the present value of their remaining mortgage payments. Using Formula (11), with $R = 2044.729$, $i = \frac{r}{m} = \frac{0.0576}{12} = 0.0048$, and $n = mt = 360 - 5(12) = 300$, we find that their outstanding principal is

$$P = 2044.729 \left[\frac{1 - (1 + 0.0048)^{-300}}{0.0048} \right] \approx 324{,}709.194$$

or \$324,709.19.

   Next, we compute the amount of their new mortgage payment for the remaining term (300 months). Using Formula (13) with $P = 324{,}709.194$, $i = \frac{r}{m} = \frac{0.0696}{12} = 0.0058$, and $n = mt = 300$, we find that the monthly payment is

$$R = \frac{324{,}709.194(0.0058)}{1 - (1 + 0.0058)^{-300}} \approx 2286.698$$

or \$2286.70—an increase of \$241.97.

Sinking Funds

Sinking funds are another important application of the annuity formulas. Simply stated, a sinking fund is an account that is set up for a specific purpose at some future date. For example, an individual might establish a sinking fund for the purpose of discharging a debt at a future date. A corporation might establish a sinking fund in order to accumulate sufficient capital to replace equipment that is expected to be obsolete at some future date.

   By thinking of the amount to be accumulated by a specific date in the future as the future value of an annuity [Formula (9), Section 4.2], we can answer questions about a large class of sinking fund problems.

APPLIED EXAMPLE 7 Sinking Fund The proprietor of Carson Hardware has decided to set up a sinking fund for the purpose of purchasing a truck in 2 years' time. It is expected that the truck will cost \$30,000. If the fund earns 10% interest per year compounded quarterly, determine the size of each (equal) quarterly installment the proprietor should pay into the fund. Verify the result by displaying the schedule.

Solution   The problem at hand is to find the size of each quarterly payment R of an annuity, given that its future value is $S = 30{,}000$, the interest earned per conversion period is $i = \frac{r}{m} = \frac{0.1}{4} = 0.025$, and the number of payments is $n = (2)(4) = 8$. The formula for an annuity,

$$S = R \left[\frac{(1 + i)^n - 1}{i} \right]$$

when solved for R yields

$$R = \frac{iS}{(1 + i)^n - 1} \qquad (14)$$

or, equivalently,

$$R = \frac{S}{s_{\overline{n}|i}}$$

Substituting the appropriate numerical values for i, S, and n into Equation (14), we obtain the desired quarterly payment

$$R = \frac{(0.025)(30,000)}{(1.025)^8 - 1} \approx 3434.02$$

or \$3434.02. Table 4 shows the required schedule.

TABLE 4

A Sinking Fund Schedule

End of Period	Deposit Made	Interest Earned	Addition to Fund	Accumulated Amount in Fund
1	\$3,434.02	0	\$3,434.02	\$ 3,434.02
2	3,434.02	\$ 85.85	3,519.87	6,953.89
3	3,434.02	173.85	3,607.87	10,561.76
4	3,434.02	264.04	3,698.06	14,259.82
5	3,434.02	356.50	3,790.52	18,050.34
6	3,434.02	451.26	3,885.28	21,935.62
7	3,434.02	548.39	3,982.41	25,918.03
8	3,434.02	647.95	4,081.97	30,000.00

The formula derived in this last example is restated as follows.

Sinking Fund Payment

The periodic payment R required to accumulate a sum of S dollars over n periods with interest charged at the rate of i per period is

$$R = \frac{iS}{(1 + i)^n - 1} \qquad (15)$$

APPLIED EXAMPLE 8 Retirement Planning Jason is the owner of a computer consulting firm. He is currently planning to retire in 25 years and wishes to withdraw \$8000/month from his retirement account for 25 years starting at that time. How much must he contribute each month into a retirement account earning interest at the rate of 6%/year compounded monthly to meet his retirement goal?

Solution In this case, we work backwards. We first calculate the amount of the sinking fund that Jason needs to accumulate to fund his retirement needs. Using Formula (11) with $R = 8000$, $i = \frac{r}{m} = \frac{0.06}{12} = 0.005$, and $n = 300$, we have

$$P = 8000 \left[\frac{1 - (1 + 0.005)^{-300}}{0.005} \right]$$

$$\approx 1,241,654.91$$

Next, we calculate how much Jason should deposit into his retirement account each month to accumulate $1,241,654.91 in 25 years. Here, we use Formula (15) with $S = 1,241,654.91$, $i = \frac{r}{m} = \frac{0.06}{12} = 0.005$, and $n = 300$. We have

$$R = 1,241,654.91 \left[\frac{0.005}{(1 + 0.005)^{300} - 1} \right]$$

$$\approx 1791.73$$

or approximately $1791.73/month. ∎

Here is a summary of the formulas developed thus far in this chapter:

1. Simple and compound interest; annuities

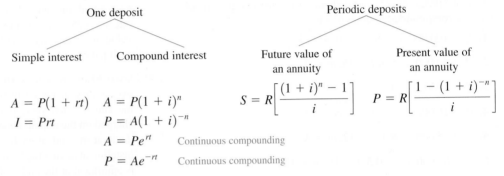

One deposit

Simple interest   Compound interest

$A = P(1 + rt)$ $A = P(1 + i)^n$
$I = Prt$ $P = A(1 + i)^{-n}$
            $A = Pe^{rt}$ Continuous compounding
            $P = Ae^{-rt}$ Continuous compounding

Periodic deposits

Future value of an annuity   Present value of an annuity

$$S = R \left[\frac{(1 + i)^n - 1}{i} \right] \qquad P = R \left[\frac{1 - (1 + i)^{-n}}{i} \right]$$

2. Effective rate of interest

$$r_{\text{eff}} = \left(1 + \frac{r}{m} \right)^m - 1$$

3. Amortization

$$R = \frac{Pi}{1 - (1 + i)^{-n}} \qquad \text{Periodic payment}$$

$$P = R \left[\frac{1 - (1 + i)^{-n}}{i} \right] \qquad \text{Amount amortized}$$

4. Sinking fund

$$R = \frac{iS}{(1 + i)^n - 1} \qquad \text{Periodic payment taken out}$$

4.3 Self-Check Exercises

1. The Mendozas wish to borrow $300,000 from a bank to help finance the purchase of a house. Their banker has offered the following plans for their consideration. In Plan I, the Mendozas have 30 years to repay the loan in monthly installments with interest on the unpaid balance charged at 6.09%/year compounded monthly. In Plan II, the loan is to be repaid in monthly installments over 15 years with interest on the unpaid balance charged at 5.76%/year compounded monthly.
   a. Find the monthly repayment for each plan.
   b. What is the difference in total payments made under each plan?

2. Harris, a self-employed individual who is 46 years old, is setting up a defined-benefit retirement plan. If he wishes to have $250,000 in this retirement account by age 65, what is the size of each yearly installment he will be required to make into a savings account earning interest at $8\frac{1}{4}$%/year?

Solutions to Self-Check Exercises 4.3 can be found on *page 231.*

4.3   Concept Questions

1. Write the amortization formula.
   a. If P and i are fixed and n is allowed to increase, what will happen to R?
   b. Interpret the result of part (a).

2. Using the formula for computing a sinking fund payment, show that if the number of payments into a sinking fund increases, then the size of the periodic payment into the sinking fund decreases.

4.3   Exercises

In Exercises 1–8, find the periodic payment R required to amortize a loan of P dollars over t years with interest charged at the rate of $r\%$/year compounded m times a year.

1. $P = 100,000, r = 8, t = 10, m = 1$

2. $P = 40,000, r = 3, t = 15, m = 2$

3. $P = 5000, r = 4, t = 3, m = 4$

4. $P = 16,000, r = 9, t = 4, m = 12$

5. $P = 25,000, r = 3, t = 12, m = 4$

6. $P = 80,000, r = 10.5, t = 15, m = 12$

7. $P = 80,000, r = 10.5, t = 20, m = 12$

8. $P = 100,000, r = 10.5, t = 25, m = 12$

In Exercises 9–14, find the periodic payment R required to accumulate a sum of S dollars over t years with interest earned at the rate of $r\%$/year compounded m times a year.

9. $S = 20,000, r = 4, t = 6, m = 2$

10. $S = 40,000, r = 4, t = 9, m = 4$

11. $S = 100,000, r = 4.5, t = 20, m = 6$

12. $S = 120,000, r = 4.5, t = 30, m = 6$

13. $S = 250,000, r = 10.5, t = 25, m = 12$

14. $S = 350,000, r = 7.5, t = 10, m = 12$

15. Suppose payments were made at the end of each quarter into an ordinary annuity earning interest at the rate of 10%/year compounded quarterly. If the future value of the annuity after 5 years is $60,000, what was the size of each payment?

16. Suppose payments were made at the end of each month into an ordinary annuity earning interest at the rate of 9%/year compounded monthly. If the future value of the annuity after 10 years is $60,000, what was the size of each payment?

17. Suppose payments will be made for $6\frac{1}{2}$ years at the end of each semiannual period into an ordinary annuity earning interest at the rate of 7.5%/year compounded semiannually. If the present value of the annuity is $35,000, what should be the size of each payment?

18. Suppose payments will be made for $9\frac{1}{4}$ years at the end of each month into an ordinary annuity earning interest at the rate of 6.25%/year compounded monthly. If the present value of the annuity is $42,000, what should be the size of each payment?

19. **LOAN AMORTIZATION** A sum of $100,000 is to be repaid over a 10-year period through equal installments made at the end of each year. If an interest rate of 10%/year is charged on the unpaid balance and interest calculations are made at the end of each year, determine the size of each installment so that the loan (principal plus interest charges) is amortized at the end of 10 years.

20. **LOAN AMORTIZATION** What monthly payment is required to amortize a loan of $30,000 over 10 years if interest at the rate of 12%/year is charged on the unpaid balance and interest calculations are made at the end of each month?

21. **HOME MORTGAGES** Complete the following table, which shows the monthly payments on a $100,000, 30-year mortgage at the interest rates shown. Use this information to answer the following questions.

Amount of Mortgage, $	Interest Rate, %	Monthly Payment, $
100,000	5	536.82
100,000	6	. . .
100,000	7	. . .
100,000	8	. . .
100,000	9	. . .
100,000	10	877.57

   a. What is the difference in monthly payments between a $100,000, 30-year mortgage secured at 7%/year and one secured at 10%/year?
   b. Use the table to calculate the monthly mortgage payments on a $150,000 mortgage at 10%/year over 30 years and a $50,000 mortgage at 10%/year over 30 years.

22. **FINANCING A HOME** The Flemings secured a bank loan of $288,000 to help finance the purchase of a house. The bank charges interest at a rate of 9%/year on the unpaid balance, and interest computations are made at the end of each month. The Flemings have agreed to repay the loan in equal monthly installments over 25 years. What should be the size of each repayment if the loan is to be amortized at the end of the term?

23. **FINANCING A CAR** The price of a new car is $20,000. Assume that an individual makes a down payment of 25% toward the purchase of the car and secures financing for the balance at the rate of 6%/year compounded monthly.
    a. What monthly payment will she be required to make if the car is financed over a period of 36 months? Over a period of 48 months?
    b. What will the interest charges be if she elects the 36-month plan? The 48-month plan?

24. **FINANCIAL ANALYSIS** A group of private investors purchased a condominium complex for $2 million. They made an initial down payment of 10% and obtained financing for the balance. If the loan is to be amortized over 15 years at an interest rate of 6.6%/year compounded quarterly, find the required quarterly payment.

25. **FINANCING A HOME** The Taylors have purchased a $270,000 house. They made an initial down payment of $30,000 and secured a mortgage with interest charged at the rate of 6%/year on the unpaid balance. Interest computations are made at the end of each month. If the loan is to be amortized over 30 years, what monthly payment will the Taylors be required to make? What is their equity (disregarding appreciation) after 5 years? After 10 years? After 20 years?

26. **FINANCIAL PLANNING** Jessica wants to accumulate $10,000 by the end of 5 years in a special bank account, which she had opened for this purpose. To achieve this goal, Jessica plans to deposit a fixed sum of money into the account at the end of each month over the 5-year period. If the bank pays interest at the rate of 5%/year compounded monthly, how much does she have to deposit each month into her account?

27. **SINKING FUNDS** A city has $2.5 million worth of school bonds that are due in 20 years and has established a sinking fund to retire this debt. If the fund earns interest at the rate of 7%/year compounded annually, what amount must be deposited annually in this fund?

28. **TRUST FUNDS** Carl is the beneficiary of a $20,000 trust fund set up for him by his grandparents. Under the terms of the trust, he is to receive the money over a 5-year period in equal installments at the end of each year. If the fund earns interest at the rate of 5%/year compounded annually, what amount will he receive each year?

29. **SINKING FUNDS** Lowell Corporation wishes to establish a sinking fund to retire a $200,000 debt that is due in 10 years. If the investment will earn interest at the rate of 9%/year compounded quarterly, find the amount of the quarterly deposit that must be made in order to accumulate the required sum.

30. **SINKING FUNDS** The management of Gibraltar Brokerage Services anticipates a capital expenditure of $20,000 in 3 years for the purchase of new computers and has decided to set up a sinking fund to finance this purchase. If the fund earns interest at the rate of 5%/year compounded quarterly, determine the size of each (equal) quarterly installment that should be deposited in the fund.

31. **RETIREMENT ACCOUNTS** Andrea, a self-employed individual, wishes to accumulate a retirement fund of $250,000. How much should she deposit each month into her retirement account, which pays interest at the rate of 6.6%/year compounded monthly, to reach her goal upon retirement 25 years from now?

32. **STUDENT LOANS** Joe secured a loan of $12,000 3 years ago from a bank for use toward his college expenses. The bank charged interest at the rate of 4%/year compounded monthly on his loan. Now that he has graduated from college, Joe wishes to repay the loan by amortizing it through monthly payments over 10 years at the same interest rate. Find the size of the monthly payments he will be required to make.

33. **RETIREMENT ACCOUNTS** Robin wishes to accumulate a sum of $450,000 in a retirement account by the time of her retirement 30 years from now. If she wishes to do this through monthly payments into the account that earn interest at the rate of 6%/year compounded monthly, what should be the size of each payment?

34. **FINANCING COLLEGE EXPENSES** Yumi's grandparents presented her with a gift of $20,000 when she was 10 years old to be used for her college education. Over the next 7 years, until she turned 17, Yumi's parents had invested this money in a tax-free account that had yielded interest at the rate of 5.5%/year compounded monthly. Upon turning 17, Yumi now plans to withdraw her funds in equal annual installments over the next 4 years, starting at age 18. If the college fund is expected to earn interest at the rate of 6%/year, compounded annually, what will be the size of each installment?

35. **IRAs** Martin has deposited $375 in his IRA at the end of each quarter for the past 20 years. His investment has earned interest at the rate of 6%/year compounded quarterly over this period. Now, at age 60, he is considering retirement. What quarterly payment will he receive over the next 15 years? (Assume that the money continues to earn interest at the same rate and that payments are made at the end of each quarter.) If he continues working and makes quarterly payments of the same amount in his IRA until age 65, what quarterly payment will he receive from his fund upon retirement over the following 10 years?

36. **RETIREMENT PLANNING** Jennifer is the owner of a video game and entertainment software retail store. She is currently planning to retire in 30 years and wishes to withdraw $10,000/month for 20 years from her retirement account starting at that time. How much must she contribute each month for 30 years into a retirement account earning interest at the rate of 5%/year compounded monthly to meet her retirement goal?

37. **EFFECT OF DELAYING RETIREMENT ON RETIREMENT FUNDS** Refer to Example 8. Suppose that Jason delays his retirement plans and decides to continue working and contributing to his retirement fund for an additional 5 years. By delaying his retirement, he will need to withdraw only $8000/month for 20 years. In this case, how much must he contribute each month for 30 years into a retirement account earning

interest at the rate of 6%/year compounded monthly to meet his retirement goal?

38. FINANCING A CAR Darla purchased a new car during a special sales promotion by the manufacturer. She secured a loan from the manufacturer in the amount of $16,000 at a rate of 7.9%/year compounded monthly. Her bank is now charging 11.5%/year compounded monthly for new car loans. Assuming that each loan would be amortized by 36 equal monthly installments, determine the amount of interest she would have paid at the end of 3 years for each loan. How much less will she have paid in interest payments over the life of the loan by borrowing from the manufacturer instead of her bank?

39. AUTO FINANCING Dan is contemplating trading in his car for a new one. He can afford a monthly payment of at most $400. If the prevailing interest rate is 7.2%/year compounded monthly for a 48-month loan, what is the most expensive car that Dan can afford, assuming that he will receive $8000 for the trade-in?

40. AUTO FINANCING Paula is considering the purchase of a new car. She has narrowed her search to two cars that are equally appealing to her. Car A costs $28,000, and Car B costs $28,200. The manufacturer of Car A is offering 0% financing for 48 months with zero down, while the manufacturer of Car B is offering a rebate of $2000 at the time of purchase plus financing at the rate of 3%/year compounded monthly over 48 months with zero down. If Paula has decided to buy the car with the lower net cost to her, which car should she purchase?

41. FINANCING A HOME The Sandersons are planning to refinance their home. The outstanding principal on their original loan is $100,000 and was to be amortized in 240 equal monthly installments at an interest rate of 10%/year compounded monthly. The new loan they expect to secure is to be amortized over the same period at an interest rate of 7.8%/year compounded monthly. How much less can they expect to pay over the life of the loan in interest payments by refinancing the loan at this time?

42. INVESTMENT ANALYSIS Since he was 22 years old, Ben has been depositing $200 at the end of each month into a tax-free retirement account earning interest at the rate of 6.5%/year compounded monthly. Larry, who is the same age as Ben, decided to open a tax-free retirement account 5 years after Ben opened his. If Larry's account earns interest at the same rate as Ben's, determine how much Larry should deposit each month into his account so that both men will have the same amount of money in their accounts at age 65.

43. PERSONAL LOANS Two years ago, Paul borrowed $10,000 from his sister Gerri to start a business. Paul agreed to pay Gerri interest for the loan at the rate of 6%/year, compounded continuously. Paul will now begin repaying the amount he owes by amortizing the loan (plus the interest that has accrued over the past 2 years) through monthly payments over the next 5 years at an interest rate of 5%/year compounded monthly. Find the size of the monthly payments Paul will be required to make.

44. REFINANCING A HOME Josh purchased a condominium 5 years ago for $180,000. He made a down payment of 20% and financed the balance with a 30-year conventional mortgage to be amortized through monthly payments with an interest rate of 7%/year compounded monthly on the unpaid balance. The condominium is now appraised at $250,000. Josh plans to start his own business and wishes to tap into the equity that he has in the condominium. If Josh can secure a new mortgage to refinance his condominium based on a loan of 80% of the appraised value, how much cash can Josh muster for his business? (Disregard taxes.)

45. FINANCING A HOME Eight years ago, Kim secured a bank loan of $180,000 to help finance the purchase of a house. The mortgage was for a term of 30 years, with an interest rate of 9.5%/year compounded monthly on the unpaid balance to be amortized through monthly payments. What is the outstanding principal on Kim's house now?

46. BALLOON PAYMENT MORTGAGES Olivia plans to secure a 5-year balloon mortgage of $200,000 toward the purchase of a condominium. Her monthly payment for the 5 years is calculated on the basis of a 30-year conventional mortgage at the rate of 6%/year compounded monthly. At the end of the 5 years, Olivia is required to pay the balance owed (the "balloon" payment). What will be her monthly payment, and what will be her balloon payment?

47. BALLOON PAYMENT MORTGAGES Emilio is securing a 7-year Fannie Mae "balloon" mortgage for $280,000 to finance the purchase of his first home. The monthly payments are based on a 30-year amortization. If the prevailing interest rate is 7.5%/year compounded monthly, what will be Emilio's monthly payment? What will be his "balloon" payment at the end of 7 years?

48. FINANCING A HOME Sarah secured a bank loan of $200,000 for the purchase of a house. The mortgage is to be amortized through monthly payments for a term of 15 years, with an interest rate of 6%/year compounded monthly on the unpaid balance. She plans to sell her house in 5 years. How much will Sarah still owe on her house?

49. HOME REFINANCING Four years ago, Emily secured a bank loan of $200,000 to help finance the purchase of an apartment in Boston. The term of the mortgage is 30 years, and the interest rate is 9.5%/year compounded monthly. Because the interest rate for a conventional 30-year home mortgage has now dropped to 6.75%/year compounded monthly, Emily is thinking of refinancing her property.
a. What is Emily's current monthly mortgage payment?
b. What is Emily's current outstanding principal?
c. If Emily decides to refinance her property by securing a 30-year home mortgage loan in the amount of the current outstanding principal at the prevailing interest rate of 6.75%/year compounded monthly, what will be her monthly mortgage payment?
d. How much less would Emily's monthly mortgage payment be if she refinances?

50. HOME REFINANCING Five years ago, Diane secured a bank loan of $300,000 to help finance the purchase of a loft in the San Francisco Bay area. The term of the mortgage was 30 years, and the interest rate was 9%/year compounded monthly on the unpaid balance. Because the interest rate for a conventional 30-year home mortgage has now dropped to 7%/year compounded monthly, Diane is thinking of refinancing her property.
  a. What is Diane's current monthly mortgage payment?
  b. What is Diane's current outstanding principal?
  c. If Diane decides to refinance her property by securing a 30-year home mortgage loan in the amount of the current outstanding principal at the prevailing interest rate of 7%/year compounded monthly, what will be her monthly mortgage payment?
  d. How much less would Diane's monthly mortgage payment be if she refinances?

51. ADJUSTABLE-RATE MORTGAGES Three years ago, Samantha secured an adjustable-rate mortgage (ARM) loan to help finance the purchase of a house. The amount of the original loan was $150,000 for a term of 30 years, with interest at the rate of 7.5%/year compounded monthly. Currently the interest rate is 7%/year compounded monthly, and Samantha's monthly payments are due to be recalculated. What will be her new monthly payment?
  Hint: Calculate her current outstanding principal. Then, to amortize the loan in the next 27 years, determine the monthly payment based on the current interest rate.

52. ADJUSTABLE-RATE MORTGAGES George secured an adjustable-rate mortgage (ARM) loan to help finance the purchase of his home 5 years ago. The amount of the loan was $300,000 for a term of 30 years, with interest at the rate of 8%/year compounded monthly. Currently, the interest rate for his ARM is 6.5%/year compounded monthly, and George's monthly payments are due to be reset. What will be the new monthly payment?

53. FINANCING A HOME After making a down payment of $25,000, the Meyers need to secure a loan of $280,000 to purchase a certain house. Their bank's current rate for 25-year home loans is 5.5%/year compounded monthly. The

owner has offered to finance the loan at 4.9%/year compounded monthly. Assuming that both loans would be amortized over a 25-year period by 300 equal monthly installments, determine the difference in the amount of interest the Meyers would pay by choosing the seller's financing rather than their bank's.

54. REFINANCING A HOME The Martinezes are planning to refinance their home. The outstanding balance on their original loan is $150,000. Their finance company has offered them two options:

Option A: A fixed-rate mortgage at an interest rate of 7.5%/year compounded monthly, payable over a 30-year period in 360 equal monthly installments.

Option B: A fixed-rate mortgage at an interest rate of 7.25%/year compounded monthly, payable over a 15-year period in 180 equal monthly installments.

  a. Find the monthly payment required to amortize each of these loans over the life of the loan.
  b. How much interest would the Martinezes save if they chose the 15-year mortgage instead of the 30-year mortgage?

55. HOME AFFORDABILITY Suppose that the Carlsons have decided that they can afford a maximum of $3000/month for a monthly house payment. The bank has offered them (a) a 5/1 ARM for a term of 30 years with interest at the rate of 5.40%/year compounded monthly for the first 5 years and (b) an interest-only loan for a term of 30 years at the rate of 5.62%/year for the first 5 years. What is the maximum amount that they can borrow with each of these mortgages if they keep to their budget?

56. COMPARING MORTGAGES Refer to Example 5. Suppose that the Jacksons choose the 7/1 ARM and borrow the maximum amount of $344,589.68.
  a. By how much will the principal of their loan be reduced at the end of 7 years?
  Hint: See Example 3.
  b. If they choose the 7-year interest-only mortgage, by how much will the principal of their loan be reduced after 7 years?

4.3 Solutions to Self-Check Exercises

1. a. We use Formula (13) in each instance. Under Plan I,

$$P = 300{,}000 \qquad i = \frac{r}{m} = \frac{0.0609}{12} = 0.005075$$

$$n = (30)(12) = 360$$

Therefore, the size of each monthly repayment under Plan I is

$$R = \frac{300{,}000(0.005075)}{1 - (1.005075)^{-360}}$$

$$\approx 1816.05$$

or $1816.05.

Under Plan II,

$$P = 300{,}000 \qquad i = \frac{r}{m} = \frac{0.0576}{12} = 0.0048$$

$$n = (15)(12) = 180$$

Therefore, the size of each monthly repayment under Plan II is

$$R = \frac{300{,}000(0.0048)}{1 - (1.0048)^{-180}}$$

$$\approx 2492.84$$

or $2492.84.

b. Under Plan I, the total amount of repayments will be

$(360)(1816.05) = 653,778$ Number of payments
· the size of each installment

or \$653,778. Under Plan II, the total amount of repayments will be

$(180)(2492.84) = 448,711.20$

or \$448,711.20. Therefore, the difference in payments is

$653,778 - 448,711.20 = 205,066.80$

or \$205,066.80.

2. We use Formula (15) with

$S = 250,000$

$i = r = 0.0825$ Since $m = 1$

$n = 20$

giving the required size of each installment as

$$R = \frac{(0.0825)(250,000)}{(1.0825)^{20} - 1}$$

$$\approx 5313.59$$

or \$5313.59.

USING TECHNOLOGY

Amortizing a Loan

Graphing Utility

Here we use the TI-83/84 **TVM SOLVER** function to help us solve problems involving amortization and sinking funds.

APPLIED EXAMPLE 1 Finding the Payment to Amortize a Loan The Wongs are considering obtaining a preapproved 30-year loan of \$120,000 to help finance the purchase of a house. The mortgage company charges interest at the rate of 8% per year on the unpaid balance, with interest computations made at the end of each month. What will be the monthly installments if the loan is amortized?

Solution   We use the TI-83/84 **TVM SOLVER** with the following inputs:

$N = 360$ (30)(12)

$I\% = 8$

$PV = 120000$

$PMT = 0$

$FV = 0$

$P/Y = 12$ The number of payments each year

$C/Y = 12$ The number of conversion periods each year

PMT:END BEGIN

From the output shown in Figure T1, we see that the required payment is \$880.52.

```
N=360
I%=8
PV=120000
■ PMT=-880.51748...
  FV=0
  P/Y=12
  C/Y=12
  PMT:END BEGIN
```

FIGURE **T1**
The TI-83/84 screen showing the monthly installment, PMT

APPLIED EXAMPLE 2 Finding the Payment in a Sinking Fund Heidi wishes to establish a retirement account that will be worth \$500,000 in 20 years' time. She expects that the account will earn interest at the rate of 11% per year compounded monthly. What should be the monthly contribution into her account each month?

Solution We use the TI-83/84 TVM SOLVER with the following inputs:

$$N = 240 \quad \text{(20)(12)}$$
$$I\% = 11$$
$$PV = 0$$
$$PMT = 0$$
$$FV = 500000$$
$$P/Y = 12 \quad \text{The number of payments each year}$$
$$C/Y = 12 \quad \text{The number of conversion periods each year}$$
$$PMT\text{:END BEGIN}$$

The result is displayed in Figure T2. We see that Heidi's monthly contribution should be $577.61. (*Note:* The display for PMT is negative because it is an out-flow.)

```
N=240
I%=11
PV=0
■ PMT=-577.60862...
FV=500000
P/Y=12
C/Y=12
PMT:END BEGIN
```

FIGURE T2
The TI-83/84 screen showing the monthly payment, PMT

Excel

Here we use Excel to help us solve problems involving amortization and sinking funds.

 APPLIED EXAMPLE 3 Finding the Payment to Amortize a Loan The Wongs are considering a preapproved 30-year loan of $120,000 to help finance the purchase of a house. The mortgage company charges interest at the rate of 8% per year on the unpaid balance, with interest computations made at the end of each month. What will be the monthly installments if the loan is amortized at the end of the term?

Solution We use the PMT function to solve this problem. Accessing this function from the Financial function library subgroup and making the required entries, we obtain the Function Arguments dialog box shown in Figure T3. We see that the desired result is $880.52. (Recall that cash you pay out is represented by a negative number.)

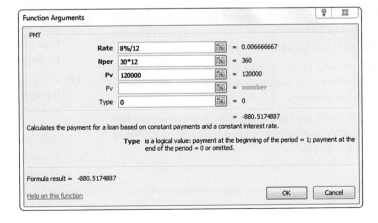

FIGURE T3
Excel's dialog box giving the payment function, PMT

APPLIED EXAMPLE 4 Finding the Payment in a Sinking Fund Heidi wishes to establish a retirement account that will be worth $500,000 in 20 years' time. She expects that the account will earn interest at the rate of 11% per year compounded monthly. What should be the monthly contribution into her account each month?

Note: Words/characters printed blue (for example, Chart sub-type:) indicate words/characters on the screen.

(continued)

Solution   As in Example 3, we use the PMT function, but this time, we are given the future value of the investment. Accessing the PMT function as before and making the required entries, we obtain the Function Arguments dialog box shown in Figure T4. We see that Heidi's monthly contribution should be $577.61. (Note that the value for PMT is negative because it is an outflow.)

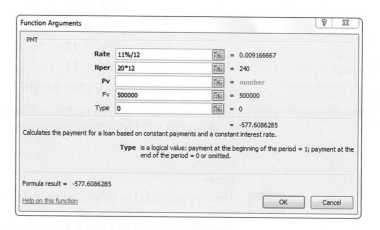

FIGURE **T4**
Excel's dialog box giving the payment function, PMT

TECHNOLOGY EXERCISES

1. Find the periodic payment required to amortize a loan of $55,000 over 120 months with interest charged at the rate of $6\frac{5}{8}$%/year compounded monthly.

2. Find the periodic payment required to amortize a loan of $178,000 over 180 months with interest charged at the rate of $7\frac{1}{8}$%/year compounded monthly.

3. Find the periodic payment required to amortize a loan of $227,000 over 360 months with interest charged at the rate of $8\frac{1}{8}$%/year compounded monthly.

4. Find the periodic payment required to amortize a loan of $150,000 over 360 months with interest charged at the rate of $7\frac{3}{8}$%/year compounded monthly.

5. Find the periodic payment required to accumulate $25,000 over 12 quarters with interest earned at the rate of $4\frac{3}{8}$%/year compounded quarterly.

6. Find the periodic payment required to accumulate $50,000 over 36 quarters with interest earned at the rate of $3\frac{7}{8}$%/year compounded quarterly.

7. Find the periodic payment required to accumulate $137,000 over 120 months with interest earned at the rate of $4\frac{3}{4}$%/year compounded monthly.

8. Find the periodic payment required to accumulate $144,000 over 120 months with interest earned at the rate of $4\frac{5}{8}$%/year compounded monthly.

9. A loan of $120,000 is to be repaid over a 10-year period through equal installments made at the end of each year. If an interest rate of 8.5%/year is charged on the unpaid balance and interest calculations are made at the end of each year, determine the size of each installment such that the loan is amortized at the end of 10 years. Verify the result by displaying the amortization schedule.

10. A loan of $265,000 is to be repaid over an 8-year period through equal installments made at the end of each year. If an interest rate of 7.4%/year is charged on the unpaid balance and interest calculations are made at the end of each year, determine the size of each installment so that the loan is amortized at the end of 8 years. Verify the result by displaying the amortization schedule.

4.4   Arithmetic and Geometric Progressions

Arithmetic Progressions

An **arithmetic progression** is a sequence of numbers in which each term after the first is obtained by adding a constant d to the preceding term. The constant d is called the **common difference.** For example, the sequence

$$2, 5, 8, 11, \ldots$$

is an arithmetic progression with common difference equal to 3.

Observe that an arithmetic progression is completely determined if the first term and the common difference are known. In fact, if

$$a_1, a_2, a_3, \ldots, a_n, \ldots$$

is an arithmetic progression with the first term given by a and common difference given by d, then by definition we have

$$a_1 = a$$
$$a_2 = a_1 + d = a + d$$
$$a_3 = a_2 + d = (a + d) + d = a + 2d$$
$$a_4 = a_3 + d = (a + 2d) + d = a + 3d$$
$$\vdots$$
$$a_n = a_{n-1} + d = a + (n - 2)d + d = a + (n - 1)d$$

Thus, we have the following formula for finding the nth term of an arithmetic progression with first term a and common difference:

---

nth Term of an Arithmetic Progression

The nth term of an arithmetic progression with first term a and common difference d is given by

$$a_n = a + (n - 1)d \qquad \textbf{(16)}$$

VIDEO ▶ **EXAMPLE 1** Find the twelfth term of the arithmetic progression

$$2, 7, 12, 17, 22, \ldots$$

Solution   The first term of the arithmetic progression is $a_1 = a = 2$, and the common difference is $d = 5$; so upon setting $n = 12$ in Equation (16), we find

$$a_{12} = 2 + (12 - 1)5 = 57 \qquad \blacksquare$$

EXAMPLE 2 Write the first five terms of an arithmetic progression whose third and eleventh terms are 21 and 85, respectively.

Solution   Using Equation (16), we obtain

$$a_3 = a + 2d = 21$$
$$a_{11} = a + 10d = 85$$

Subtracting the first equation from the second gives $8d = 64$, or $d = 8$. Substituting this value of d into the first equation yields $a + 16 = 21$, or $a = 5$. Thus, the required arithmetic progression is given by the sequence

$$5, 13, 21, 29, 37, \ldots \qquad \blacksquare$$

Let S_n denote the sum of the first n terms of an arithmetic progression with first term $a_1 = a$ and common difference d. Then

$$S_n = a + (a + d) + (a + 2d) + \cdots + [a + (n - 1)d] \qquad \textbf{(17)}$$

Rewriting the expression for S_n with the terms in reverse order gives

$$S_n = [a + (n - 1)d] + [a + (n - 2)d] + \cdots + (a + d) + a \qquad \textbf{(18)}$$

Adding Equations (17) and (18), we obtain

$$2S_n = [2a + (n-1)d] + [2a + (n-1)d]$$
$$+ \cdots + [2a + (n-1)d]$$
$$= n[2a + (n-1)d]$$
$$S_n = \frac{n}{2}[2a + (n-1)d]$$

> **Sum of Terms in an Arithmetic Progression**
>
> The sum of the first n terms of an arithmetic progression with first term a and common difference d is given by
>
> $$S_n = \frac{n}{2}[2a + (n-1)d] \qquad \textbf{(19)}$$

EXAMPLE 3 Find the sum of the first 20 terms of the arithmetic progression of Example 1.

Solution   Letting $a = 2$, $d = 5$, and $n = 20$ in Equation (19), we obtain

$$S_{20} = \frac{20}{2}[2 \cdot 2 + 19 \cdot 5] = 990$$

APPLIED EXAMPLE 4 Company Sales Madison Electric Company had sales of \$200,000 in its first year of operation. If the sales increased by \$30,000 per year thereafter, find Madison's sales in the fifth year and its total sales over the first 5 years of operation.

Solution   Madison's yearly sales follow an arithmetic progression, with the first term given by $a = 200{,}000$ and the common difference given by $d = 30{,}000$. The sales in the fifth year are found by using Equation (16) with $n = 5$. Thus,

$$a_5 = 200{,}000 + (5-1)30{,}000 = 320{,}000$$

or \$320,000.

Madison's total sales over the first 5 years of operation are found by using Equation (19) with $n = 5$. Thus,

$$S_5 = \frac{5}{2}[2(200{,}000) + (5-1)30{,}000]$$
$$= 1{,}300{,}000$$

or \$1,300,000.

Geometric Progressions

A **geometric progression** is a sequence of numbers in which each term after the first is obtained by multiplying the preceding term by a constant r. The constant r is called the **common ratio.**

A geometric progression is completely determined if the first term and the common ratio are known. Thus, if

$$a_1, a_2, a_3, \ldots, a_n, \ldots$$

is a geometric progression with the first term given by a and common ratio given by r, then by definition we have

$$a_1 = a$$
$$a_2 = a_1 r = ar$$
$$a_3 = a_2 r = ar^2$$
$$a_4 = a_3 r = ar^3$$
$$\vdots$$
$$a_n = a_{n-1} r = ar^{n-1}$$

This gives the following:

nth Term of a Geometric Progression

The nth term of a geometric progression with first term a and common ratio r is given by

$$a_n = ar^{n-1} \tag{20}$$

EXAMPLE 5 Find the eighth term of a geometric progression whose first five terms are 162, 54, 18, 6, and 2.

Solution   The common ratio is found by taking the ratio of any term other than the first to the preceding term. Taking the ratio of the fourth term to the third term, for example, gives $r = \frac{6}{18} = \frac{1}{3}$. To find the eighth term of the geometric progression, use Equation (20) with $a = 162$, $r = \frac{1}{3}$, and $n = 8$, obtaining

$$a_8 = 162 \left(\frac{1}{3} \right)^7$$
$$= \frac{2}{27}$$

EXAMPLE 6 Find the tenth term of a geometric progression with positive terms and third term equal to 16 and seventh term equal to 1.

Solution   Using Equation (20) with $n = 3$ and $n = 7$, respectively, yields

$$a_3 = ar^2 = 16$$
$$a_7 = ar^6 = 1$$

Dividing a_7 by a_3 gives

$$\frac{ar^6}{ar^2} = \frac{1}{16}$$

from which we obtain $r^4 = \frac{1}{16}$, or $r = \frac{1}{2}$. Substituting this value of r into the expression for a_3, we obtain

$$a \left(\frac{1}{2} \right)^2 = 16 \quad \text{or} \quad a = 64$$

Finally, using Equation (20) once again with $a = 64$, $r = \frac{1}{2}$, and $n = 10$ gives

$$a_{10} = 64 \left(\frac{1}{2} \right)^9 = \frac{1}{8}$$

To find the sum of the first n terms of a geometric progression with the first term $a_1 = a$ and common ratio r, denote the required sum by S_n. Then

$$S_n = a + ar + ar^2 + \cdots + ar^{n-2} + ar^{n-1} \qquad \text{(21)}$$

Upon multiplying (21) by r, we obtain

$$rS_n = ar + ar^2 + ar^3 + \cdots + ar^{n-1} + ar^n \qquad \text{(22)}$$

Subtracting Equation (22) from (21) gives

$$S_n - rS_n = a - ar^n$$
$$(1 - r)S_n = a(1 - r^n)$$

If $r \neq 1$, we may divide both sides of the last equation by $(1 - r)$, obtaining

$$S_n = \frac{a(1 - r^n)}{1 - r}$$

If $r = 1$, then (21) gives

$$S_n = a + a + a + \cdots + a \qquad n \text{ terms}$$
$$= na$$

Thus,

$$S_n = \begin{cases} \dfrac{a(1 - r^n)}{1 - r} & \text{if } r \neq 1 \\ na & \text{if } r = 1 \end{cases}$$

Sum of Terms in a Geometric Progression

The sum of the first n terms of a geometric progression with first term a and common ratio r is given by

$$S_n = \begin{cases} \dfrac{a(1 - r^n)}{1 - r} & \text{if } r \neq 1 \\ na & \text{if } r = 1 \end{cases} \qquad \text{(23)}$$

EXAMPLE 7 Find the sum of the first six terms of the following geometric progression:

$$3, 6, 12, 24, \ldots$$

Solution Here, $a = 3$, $r = \frac{6}{3} = 2$, and $n = 6$, so Equation (23) gives

$$S_6 = \frac{3(1 - 2^6)}{1 - 2} = 189$$

VIDEO

APPLIED EXAMPLE 8 Company Sales Michaelson Land Development Company had sales of $1 million in its first year of operation. If sales increased by 10% per year thereafter, find Michaelson's sales in the fifth year and its total sales over the first 5 years of operation.

Solution Michaelson's yearly sales follow a geometric progression, with the first term given by $a = 1{,}000{,}000$ and the common ratio given by $r = 1.1$. The sales in the fifth year are found by using Formula (20) with $n = 5$. Thus,

$$a_5 = 1{,}000{,}000(1.1)^4 = 1{,}464{,}100$$

or $1,464,100.

Michaelson's total sales over the first 5 years of operation are found by using Equation (23) with $n = 5$. Thus,

$$S_5 = \frac{1{,}000{,}000[1 - (1.1)^5]}{1 - 1.1}$$

$$= 6{,}105{,}100$$

or $6,105,100.   ■

Double Declining–Balance Method of Depreciation

In Section 2.5, we discussed the straight-line, or linear, method of depreciating an asset. Linear depreciation assumes that the asset depreciates at a constant rate. For certain assets (such as machines) whose market values drop rapidly in the early years of usage and thereafter less rapidly, another method of depreciation called the **double declining–balance method** is often used. In practice, a business firm normally employs the double declining–balance method for depreciating such assets for a certain number of years and then switches over to the linear method.

To derive an expression for the book value of an asset being depreciated by the double declining–balance method, let C (in dollars) denote the original cost of the asset and let the asset be depreciated over N years. When this method is used, the amount depreciated each year is $\frac{2}{N}$ times the value of the asset at the beginning of that year. Thus, the amount by which the asset is depreciated in its first year of use is given by $\frac{2C}{N}$, so if $V(1)$ denotes the book value of the asset at the end of the first year, then

$$V(1) = C - \frac{2C}{N} = C\left(1 - \frac{2}{N}\right)$$

Next, if $V(2)$ denotes the book value of the asset at the end of the second year, then a similar argument leads to

$$V(2) = C\left(1 - \frac{2}{N}\right) - C\left(1 - \frac{2}{N}\right)\frac{2}{N}$$

$$= C\left(1 - \frac{2}{N}\right)\left(1 - \frac{2}{N}\right)$$

$$= C\left(1 - \frac{2}{N}\right)^2$$

Continuing, we find that if $V(n)$ denotes the book value of the asset at the end of n years, then the terms C, $V(1)$, $V(2)$, . . . , $V(N)$ form a geometric progression with first term C and common ratio $(1 - \frac{2}{N})$. Consequently, the nth term, $V(n)$, is given by

$$V(n) = C\left(1 - \frac{2}{N}\right)^n \qquad (1 \le n \le N) \tag{24}$$

Also, if $D(n)$ denotes the amount by which the asset has been depreciated by the end of the nth year, then

$$D(n) = C - C\left(1 - \frac{2}{N}\right)^n$$

$$= C\left[1 - \left(1 - \frac{2}{N}\right)^n\right] \tag{25}$$

APPLIED EXAMPLE 9 Depreciation of Equipment A tractor purchased at a cost of $60,000 is to be depreciated by the double declining–balance method over 10 years. What is the book value of the tractor at the end of 5 years? By what amount has the tractor been depreciated by the end of the fifth year?

Solution We have $C = 60,000$ and $N = 10$. Thus, using Equation (24) with $n = 5$ gives the book value of the tractor at the end of 5 years as

$$V(5) = 60,000\left(1 - \frac{2}{10}\right)^5$$

$$= 60,000\left(\frac{4}{5}\right)^5 = 19,660.80$$

or $19,660.80.

The amount by which the tractor has been depreciated by the end of the fifth year is given by

$$60,000 - 19,660.80 = 40,339.20$$

or $40,339.20. You may verify the last result by using Equation (25) directly.

Exploring with
TECHNOLOGY

A tractor purchased at a cost of $60,000 is to be depreciated over 10 years with a residual value of $0. When the double declining–balance method is used, its value at the end of n years is $V_1(n) = 60,000(0.8)^n$ dollars. When straight-line depreciation is used, its value at the end of n years is $V_2(n) = 60,000 - 6000n$. Use a graphing utility to sketch the graphs of V_1 and V_2 in the viewing window $[0, 10] \times [0, 70,000]$. Comment on the relative merits of each method of depreciation.

4.4 Self-Check Exercises

1. Find the sum of the first five terms of the geometric progression with first term -24 and common ratio $-\frac{1}{2}$.

2. Office equipment purchased for $75,000 is to be depreciated by the double declining-balance method over 5 years. Find the book value at the end of 3 years.

3. Derive the formula for the future value of an annuity [Formula (9), Section 4.2].

Solutions to Self-Check Exercises 4.4 can be found on page 242.

4.4 Concept Questions

1. Suppose an arithmetic progression has first term a and common difference d.
   a. What is the formula for the nth term of this progression?
   b. What is the formula for the sum of the first n terms of this progression?

2. Suppose a geometric progression has first term a and common ratio r.
   a. What is the formula for the nth term of this progression?
   b. What is the formula for the sum of the first n terms of this progression?

4.4 Exercises

In Exercises 1–4, find the *n*th term of the arithmetic progression that has the given values of *a*, *d*, and *n*.

1. $a = 6, d = 3, n = 9$ **2.** $a = -5, d = 3, n = 7$

3. $a = -15, d = \dfrac{3}{2}, n = 8$ **4.** $a = 1.2, d = 0.4, n = 98$

5. Find the first five terms of the arithmetic progression whose fourth and eleventh terms are 30 and 107, respectively.

6. Find the first five terms of the arithmetic progression whose seventh and twenty-third terms are -5 and -29, respectively.

7. Find the seventh term of the arithmetic progression $x, x + y, x + 2y, \ldots$.

8. Find the eleventh term of the arithmetic progression $a + b, 2a, 3a - b, \ldots$.

9. Find the sum of the first 15 terms of the arithmetic progression 4, 11, 18, ...

10. Find the sum of the first 20 terms of the arithmetic progression 5, -1, -7, ...

11. Find the sum of the odd integers between 14 and 58.

12. Find the sum of the even integers between 21 and 99.

13. Find $f(1) + f(2) + f(3) + \cdots + f(22)$, given that $f(x) = 3x - 4$.

14. Find $g(1) + g(2) + g(3) + \cdots + g(50)$, given that $g(x) = 12 - 4x$.

15. Show that Equation (19) can be written as

$$S_n = \frac{n}{2}(a + a_n)$$

where a_n represents the last term of an arithmetic progression. Use this formula to find:

a. The sum of the first 11 terms of the arithmetic progression whose first and eleventh terms are 3 and 47, respectively.

b. The sum of the first 20 terms of the arithmetic progression whose first and twentieth terms are 5 and -33, respectively.

16. SALES GROWTH Moderne Furniture Company had sales of $1,500,000 during its first year of operation. If the sales increased by $160,000/year thereafter, find Moderne's sales in the fifth year and its total sales over the first 5 years of operation.

17. EXERCISE PROGRAM As part of her fitness program, Karen has taken up jogging. If she jogs 1 mi the first day and increases her daily run by $\frac{1}{4}$ mi every week, when will she reach her goal of 10 mi/day?

18. COST OF DRILLING A 100-ft oil well is to be drilled. The cost of drilling the first foot is $10.00, and the cost of drilling each additional foot is $4.50 more than that of the preceding foot. Find the cost of drilling the entire 100 ft.

19. CONSUMER DECISIONS Kunwoo wishes to go from the airport to his hotel, which is 25 mi away. The taxi rate is $2.00 for the first mile and $1.20 for each additional mile. The airport limousine also goes to his hotel and charges a flat rate of $15.00. How much money will he save by taking the airport limousine?

20. SALARY COMPARISONS Markeeta, a recent college graduate, received two job offers. Company *A* offered her an initial salary of $48,800 with guaranteed annual increases of $2000/year for the first 5 years. Company *B* offered an initial salary of $50,400 with guaranteed annual increases of $1500/year for the first 5 years.

a. Which company is offering a higher salary for the fifth year of employment?

b. Which company is offering more money for the first 5 years of employment?

21. SUM-OF-THE-YEARS'-DIGITS METHOD OF DEPRECIATION One of the methods that the Internal Revenue Service allows for computing depreciation of certain business property is the sum-of-the-years'-digits method. If a property valued at *C* dollars has an estimated useful life of *N* years and a salvage value of *S* dollars, then the amount of depreciation D_n allowed during the *n*th year is given by

$$D_n = (C - S)\frac{N - (n - 1)}{S_N} \qquad (0 \le n \le N)$$

where S_N is the sum of the first *N* positive integers. Thus,

$$S_N = 1 + 2 + \cdots + N = \frac{N(N + 1)}{2}$$

a. Verify that the sum of the arithmetic progression $S_N = 1 + 2 + \cdots + N$ is given by

$$\frac{N(N + 1)}{2}$$

b. If office furniture worth $6000 is to be depreciated by this method over $N = 10$ years and the salvage value of the furniture is $500, find the depreciation for the third year by computing D_3.

22. SUM-OF-THE-YEARS'-DIGITS METHOD OF DEPRECIATION Refer to Example 1, Section 2.5. The amount of depreciation allowed for a network server, which has an estimated useful life of 5 years, an initial value of $10,000, and a salvage value of $3000, was $1400/year using the straight-line method of depreciation. Determine the amount of depreciation that would be allowed for the first year if the network server were depreciated using the sum-of-the-years'-digits method described in Exercise 21. Which method would result in a larger depreciation of the asset in its first year of use?

In Exercises 23–28, determine which of the sequences are geometric progressions. For each geometric progression, find the seventh term and the sum of the first seven terms.

23. 4, 8, 16, 32, . . .

24. $1, -\dfrac{1}{2}, \dfrac{1}{4}, -\dfrac{1}{8}, \ldots$

25. $\dfrac{1}{2}, -\dfrac{3}{8}, \dfrac{1}{4}, -\dfrac{9}{64}, \ldots$

26. 0.004, 0.04, 0.4, 4, . . .

27. 243, 81, 27, 9, . . .

28. −1, 1, 3, 5, . . .

29. Find the twentieth term and sum of the first 20 terms of the geometric progression −3, 3, −3, 3, ...

30. Find the twenty-third term in a geometric progression having the first term $a = 0.1$ and ratio $r = 2$.

31. POPULATION GROWTH It has been projected that the population of a certain city in the southwest will increase by 8% during each of the next 5 years. If the current population is 200,000, what is the expected population after 5 years?

32. SALES GROWTH Metro Cable TV had sales of $2,500,000 in its first year of operation. If thereafter the sales increased by 12% of the previous year, find the sales of the company in the fifth year and the total sales over the first 5 years of operation.

33. COLAs Suppose the cost-of-living index had increased by 3% during each of the past 6 years and that a member of the EUW Union had been guaranteed an annual increase equal to 2% above the increase in the cost-of-living index over that period. What would be the present salary of a union member whose salary 6 years ago was $42,000?

34. SAVINGS PLANS The parents of a 9-year-old boy have agreed to deposit $10 in their son's bank account on his 10th birthday and to double the size of their deposit every year thereafter until his 18th birthday.
a. How much will they have to deposit on his 18th birthday?
b. How much will they have deposited by his 18th birthday?

35. SALARY COMPARISONS The starting salary of a Stenton Printing Co. employee is $48,000. The employee has the option of taking an annual raise of 8%/year or a fixed annual raise of $4000/year for each of the following three years. Which option would be more profitable to him considering his total earnings over the 4-year period?

36. BACTERIA GROWTH A culture of a certain bacteria is known to double in number every 3 hr. If the culture has an initial count of 20, what will be the population of the culture at the end of 24 hr?

37. TRUST FUNDS Sarah is the recipient of a trust fund that she will receive over a period of 6 years. Under the terms of the trust, she is to receive $10,000 the first year and each succeeding annual payment is to be increased by 15%.
a. How much will she receive during the sixth year?
b. What is the total amount of the six payments she will receive?

In Exercises 38–40, find the book value of office equipment purchased at a cost C at the end of the nth year if it is to be depreciated by the double declining-balance method over 10 years.

38. $C = \$20{,}000, n = 4$

39. $C = \$150{,}000, n = 8$

40. $C = \$80{,}000, n = 7$

41. DOUBLE DECLINING–BALANCE METHOD OF DEPRECIATION Restaurant equipment purchased at a cost of $150,000 is to be depreciated by the double declining–balance method over 10 years. What is the book value of the equipment at the end of 6 years? By what amount has the equipment been depreciated at the end of the sixth year?

42. DOUBLE DECLINING–BALANCE METHOD OF DEPRECIATION Refer to Exercise 22. Recall that a network server with an estimated useful life of 5 years, an initial value of $10,000, and a salvage value of $3000 was to be depreciated. At the end of the first year, the amount of depreciation allowed was $1400 using the straight-line method and $2333 using the sum-of-the-years'-digits method. Determine the amount of depreciation that would be allowed for the first year if the network server were depreciated by the double declining–balance method. Which of these three methods would result in the largest depreciation of the network server at the end of its first year of use?

In Exercises 43 and 44, determine whether the statement is true or false. If it is true, explain why it is true. If it is false, give an example to show why it is false.

43. If $a_1, a_2, a_3, \ldots, a_n$ and $b_1, b_2, b_3, \ldots, b_n$ are arithmetic progressions, then $a_1 + b_1, a_2 + b_2, a_3 + b_3, \ldots, a_n + b_n$ is also an arithmetic progression.

44. If $a_1, a_2, a_3, \ldots, a_n$ and $b_1, b_2, b_3, \ldots, b_n$ are geometric progressions, then $a_1 b_1, a_2 b_2, a_3 b_3, \ldots, a_n b_n$ is also a geometric progression.

4.4 Solutions to Self-Check Exercises

1. Use Equation (23) with $a = -24$, $n = 5$ and $r = -\frac{1}{2}$, obtaining

$$S_5 = \frac{-24\left[1 - \left(-\frac{1}{2}\right)^5\right]}{1 - \left(-\frac{1}{2}\right)}$$

$$= \frac{-24\left(1 + \frac{1}{32}\right)}{\frac{3}{2}} = -\frac{33}{2}$$

2. Use Equation (24) with $C = 75{,}000$, $N = 5$, and $n = 3$, giving the book value of the office equipment at the end of 3 years as

$$V(3) = 75{,}000\left(1 - \frac{2}{5}\right)^3 = 16{,}200$$

or $16,200.

3. We have

$$S = R + R(1 + i) + R(1 + i)^2 + \cdots + R(1 + i)^{n-1}$$

The sum on the right is easily seen to be the sum of the first n terms of a geometric progression with first term R and

common ratio $(1 + i)$, so by virtue of Equation (23), we obtain

$$S = R\left[\frac{1 - (1 + i)^n}{1 - (1 + i)}\right] = R\left[\frac{(1 + i)^n - 1}{i}\right]$$

CHAPTER 4   Summary of Principal Formulas and Terms

FORMULAS

1. Simple interest (accumulated amount)	$A = P(1 + rt)$
2. Compound interest	
a. Accumulated amount	$A = P(1 + i)^n$
b. Present value	$P = A(1 + i)^{-n}$
c. Interest rate per conversion period	$i = \dfrac{r}{m}$
d. Number of conversion periods	$n = mt$
3. Continuous compound interest	
a. Accumulated amount	$A = Pe^{rt}$
b. Present value	$P = Ae^{-rt}$
4. Effective rate of interest	$r_{\text{eff}} = \left(1 + \dfrac{r}{m}\right)^m - 1$
5. Annuities	
a. Future value	$S = R\left[\dfrac{(1 + i)^n - 1}{i}\right]$
b. Present value	$P = R\left[\dfrac{1 - (1 + i)^{-n}}{i}\right]$
6. Amortization payment	$R = \dfrac{Pi}{1 - (1 + i)^{-n}}$
7. Amount amortized	$P = R\left[\dfrac{1 - (1 + i)^{-n}}{i}\right]$
8. Sinking fund payment	$R = \dfrac{iS}{(1 + i)^n - 1}$

TERMS

simple interest (190)	effective rate (196)	ordinary annuity (208)
accumulated amount (190)	present value (198)	simple annuity (209)
compound interest (191)	future value (198)	future value of an annuity (210)
nominal rate (stated rate) (191)	annuity (208)	present value of an annuity (211)
conversion period (192)	annuity certain (208)	sinking fund (225)

CHAPTER 4   Concept Review Questions

Fill in the blanks.

1. a. Simple interest is computed on the _____ principal only. The formula for the accumulated amount using simple interest is $A =$ _____.

   b. In calculations using compound interest, earned interest is periodically added to the principal and thereafter itself earns _____. The formula for the accumulated amount using compound interest is $A =$ _____. Solving this equation for P gives the present value formula using compound interest as $P =$ _____.

2. The effective rate of interest is the _____ interest rate that would produce the same accumulated amount in _____ year as the _____ rate compounded _____ times a year. The formula for calculating the effective rate is $r_{\text{eff}} = $ _____.

3. A sequence of payments made at regular time intervals is called a/an _____; if the payments are made at the end of each payment period, then it is called a/an _____ _____; if the payment period coincides with the interest conversion period, then it is called a/an _____ _____.

4. The formula for the future value of an annuity is $S = $ _____. The formula for the present value of an annuity is $P = $ _____.

5. The periodic payment R on a loan of P dollars to be amortized over n periods with interest charged at the rate of i per period is $R = $ _____.

6. A sinking fund is an account that is set up for a specific purpose at some _____ date. The periodic payment R required to accumulate a sum of S dollars over n periods with interest charged at the rate of i per period is $R = $ _____.

7. An arithmetic progression is a sequence of numbers in which each term after the first is obtained by adding a/an _____ _____ to the preceding term. The nth term of an arithmetic progression is $a_n = $ _____. The sum of the first n terms of an arithmetic progression is $S_n = $ _____.

8. A geometric progression is a sequence of numbers in which each term after the first is obtained by multiplying the preceding term by a/an _____ _____. The nth term of a geometric progression is $a_n = $ _____. If $r \neq 1$, the sum of the first n terms of a geometric progression is $S_n = $ _____.

CHAPTER 4   Review Exercises

1. Find the accumulated amount after 4 years if $5000 is invested at 10%/year compounded (a) annually, (b) semiannually, (c) quarterly, and (d) monthly.

2. Find the accumulated amount after 8 years if $12,000 is invested at 6.5%/year compounded (a) annually, (b) semiannually, (c) quarterly, and (d) monthly.

3. Find the effective rate of interest corresponding to a nominal rate of 12%/year compounded (a) annually, (b) semiannually, (c) quarterly, and (d) monthly.

4. Find the effective rate of interest corresponding to a nominal rate of 11.5%/year compounded (a) annually, (b) semiannually, (c) quarterly, and (d) monthly.

5. Find the present value of $41,413 due in 5 years at an interest rate of 6.5%/year compounded quarterly.

6. Find the present value of $64,540 due in 6 years at an interest rate of 8%/year compounded monthly.

7. Find the amount (future value) of an ordinary annuity of $150/quarter for 7 years at 8%/year compounded quarterly.

8. Find the future value of an ordinary annuity of $120/month for 10 years at 9%/year compounded monthly.

9. Find the present value of an ordinary annuity of 36 payments of $250 each made monthly and earning interest at 9%/year compounded monthly.

10. Find the present value of an ordinary annuity of 60 payments of $5000 each made quarterly and earning interest at 8%/year compounded quarterly.

11. Find the payment R needed to amortize a loan of $22,000 at 8.5%/year compounded monthly with 36 monthly installments over a period of 3 years.

12. Find the payment R needed to amortize a loan of $10,000 at 9.2%/year compounded monthly with 36 monthly installments over a period of 3 years.

13. Find the payment R needed to accumulate $18,000 with 48 monthly installments over a period of 4 years at an interest rate of 6%/year compounded monthly.

14. Find the payment R needed to accumulate $15,000 with 60 monthly installments over a period of 5 years at an interest rate of 7.2%/year compounded monthly.

15. Find the effective rate of interest corresponding to a nominal rate of 7.2%/year compounded monthly.

16. Find the effective rate of interest corresponding to a nominal rate of 9.6%/year compounded monthly.

17. Find the present value of $119,346 due in 4 years at an interest rate of 10%/year compounded continuously.

18. **COMPANY SALES** JCN Media had sales of $1,750,000 in the first year of operation. If the sales increased by 14%/year thereafter, find the company's sales in the fourth year and the total sales over the first 4 years of operation.

19. **CDs** The manager of a money market fund has invested $4.2 million in certificates of deposit that pay interest at the rate of 5.4%/year compounded quarterly over a period of 5 years. How much will the investment be worth at the end of 5 years?

20. **SAVINGS ACCOUNTS** Emily deposited $2000 into a bank account 5 years ago. The bank paid interest at the rate of 8%/year compounded weekly. What is Emily's account worth today?

21. **SAVINGS ACCOUNTS** Kim invested a sum of money 4 years ago in a savings account that has since paid interest at the

rate of 6.5%/year compounded monthly. Her investment is now worth $19,440.31. How much did she originally invest?

22. **SAVINGS ACCOUNTS** Andrew withdrew $5986.09 from a savings account, which he closed this morning. The account had earned interest at the rate of 6%/year compounded continuously during the 3-year period that the money was on deposit. How much did Andrew originally deposit into the account?

23. **MUTUAL FUNDS** Juan invested $24,000 in a mutual fund 5 years ago. Today his investment is worth $34,616. Find the effective annual rate of return on his investment over the 5-year period.

24. **COLLEGE SAVINGS PROGRAM** The Blakes have decided to start a monthly savings program to provide for their son's college education. How much should they deposit at the end of each month in a savings account earning interest at the rate of 8%/year compounded monthly so that, at the end of the tenth year, the accumulated amount will be $40,000?

25. **RETIREMENT ACCOUNTS** Mai Lee has contributed $200 at the end of each month into her company's employee retirement account for the past 10 years. Her employer has matched her contribution each month. If the account has earned interest at the rate of 8%/year compounded monthly over the 10-year period, determine how much Mai Lee now has in her retirement account.

26. **AUTOMOBILE LEASING** Maria has leased an auto for 4 years at $300/month. If money is worth 5%/year compounded monthly, what is the equivalent cash payment (present value) of this annuity? (Assume that the payments are made at the end of each month.)

27. **INSTALLMENT FINANCING** Peggy made a down payment of $400 toward the purchase of new furniture. To pay the balance of the purchase price, she has secured a loan from her bank at 12%/year compounded monthly. Under the terms of her finance agreement, she is required to make payments of $75.32 at the end of each month for 24 months. What was the purchase price of the furniture?

28. **HOME FINANCING** The Turners have purchased a house for $150,000. They made an initial down payment of $30,000 and secured a mortgage with interest charged at the rate of 9%/year on the unpaid balance. (Interest computations are made at the end of each month.) Assume that the loan is amortized over 30 years.

a. What monthly payment will the Turners be required to make?

b. What will be their total interest payment?

c. What will be their equity (disregard depreciation) after 10 years?

29. **HOME FINANCING** Refer to Exercise 28. If the loan is amortized over 15 years:

a. What monthly payment will the Turners be required to make?

b. What will be their total interest payment?

c. What will be their equity (disregard depreciation) after 10 years?

30. **SINKING FUNDS** The management of a corporation anticipates a capital expenditure of $500,000 in 5 years for the purpose of purchasing replacement machinery. To finance this purchase, a sinking fund that earns interest at the rate of 10%/year compounded quarterly will be set up. Determine the amount of each (equal) quarterly installment that should be deposited in the fund. (Assume that the payments are made at the end of each quarter.)

31. **SINKING FUNDS** The management of a condominium association anticipates a capital expenditure of $120,000 in 2 years for the purpose of painting the exterior of the condominium. To pay for this maintenance, a sinking fund will be set up that will earn interest at the rate of 5.8%/year compounded monthly. Determine the amount of each (equal) monthly installment the association will be required to deposit into the fund at the end of each month for the next 2 years.

32. **CREDIT CARD PAYMENTS** The outstanding balance on Bill's credit card account is $3200. The bank issuing the credit card is charging 18.6%/year compounded monthly. If Bill decides to pay off this balance in equal monthly installments at the end of each month for the next 18 months, how much will be his monthly payment? What is the effective rate of interest the bank is charging Bill?

33. **FINANCIAL PLANNING** Matt's parents have agreed to contribute $250/month toward the rent for his apartment in his junior year in college. The plan is for Matt's parents to deposit a lump sum in Matt's bank account on August 1 and then have Matt withdraw $250 on the first of each month starting on September 1 and ending on May 1 the following year. If the bank pays interest on the balance at the rate of 5%/year compounded monthly, how much should Matt's parents deposit into his account?

CHAPTER 4   **Before Moving On . . .**

1. Find the accumulated amount at the end of 3 years if $2000 is deposited in an account paying interest at the rate of 8%/year compounded monthly.

2. Find the effective rate of interest corresponding to a nominal rate of 6%/year compounded daily. (Use a 365-day year.)

3. Find the future value of an ordinary annuity of $800/week for 10 years at 6%/year compounded weekly.

4. Find the monthly payment required to amortize a loan of $100,000 over 10 years with interest earned at the rate of 8%/year compounded monthly.

5. Find the weekly payment required to accumulate a sum of $15,000 over 6 years with interest earned at the rate of 10%/year compounded weekly.

6. a. Find the sum of the first ten terms of the arithmetic progression 3, 7, 11, 15, 19, ...

b. Find the sum of the first eight terms of the geometric progression $\frac{1}{2}$, 1, 2, 4, 8, ...

5

SYSTEMS OF LINEAR EQUATIONS AND MATRICES

Checkers Rent-A-Car is planning to expand its fleet of cars next quarter. How should the company use its budget of $12 million to meet the expected additional demand for compact and full-size cars? In Example 5, page 316, we will see how we can find the solution to this problem by solving a system of equations.

THE LINEAR EQUATIONS in two variables that we studied in Chapter 2 are readily extended to cases involving more than two variables. For example, a linear equation in three variables represents a plane in three-dimensional space. In this chapter, we see how some real-world problems can be formulated in terms of systems of linear equations, and we develop two methods for solving these equations.

In addition, we see how *matrices* (rectangular arrays of numbers) can be used to write systems of linear equations in compact form. We then go on to consider some real-life applications of matrices.

5.1   Systems of Linear Equations: An Introduction

Systems of Equations

Recall that in Section 2.5, we had to solve two simultaneous linear equations to find the *break-even point* and the *equilibrium point*. These are two examples of real-world problems that call for the solution of a **system of linear equations** in two or more variables. In this chapter, we take up a more systematic study of such systems.

We begin by considering a system of two linear equations in two variables. Recall that such a system may be written in the general form

$$ax + by = h$$
$$cx + dy = k \tag{1}$$

where a, b, c, d, h, and k are real constants and neither a and b nor c and d are both zero.

Now let's study the nature of the **solution of a system of linear equations** in more detail. Recall that the graph of each equation in System (1) is a straight line in the plane, so geometrically, the solution to the system is the point(s) of intersection of the two straight lines L_1 and L_2, represented by the first and second equations of the system.

Given two lines L_1 and L_2, *one and only one* of the following may occur:

a. L_1 and L_2 intersect at exactly one point.
b. L_1 and L_2 are parallel and coincident.
c. L_1 and L_2 are parallel and distinct.

(See Figure 1.) In the first case, the system has a unique solution corresponding to the single point of intersection of the two lines. In the second case, the system has infinitely many solutions corresponding to the points lying on the same line. Finally, in the third case, the system has no solution because the two lines do not intersect.

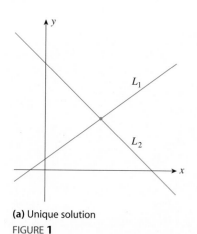

(a) Unique solution

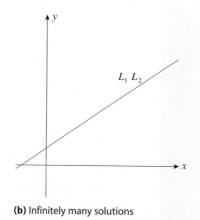

(b) Infinitely many solutions

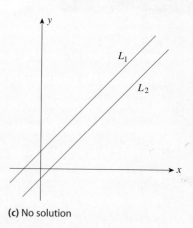

(c) No solution

FIGURE **1**

Explore & Discuss

Generalize the discussion on this page to the case in which there are three straight lines in the plane defined by three linear equations. What if there are n lines defined by n equations?

Let's illustrate each of these possibilities by considering some specific examples.

1. A system of equations with exactly one solution Consider the system

$$2x - y = 1$$
$$3x + 2y = 12$$

Solving the first equation for y in terms of x, we obtain the equation

$$y = 2x - 1$$

Substituting this expression for y into the second equation yields

$$3x + 2(2x - 1) = 12$$
$$3x + 4x - 2 = 12$$
$$7x = 14$$
$$x = 2$$

Finally, substituting this value of x into the expression for y obtained earlier gives

$$y = 2(2) - 1 = 3$$

Therefore, the unique solution of the system is given by $x = 2$ and $y = 3$. Geometrically, the two lines represented by the two linear equations that make up the system intersect at the point $(2, 3)$ (Figure 2).

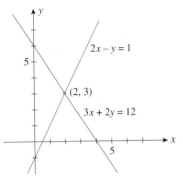

FIGURE **2**
A system of equations with one solution

Note We can check our result by substituting the values $x = 2$ and $y = 3$ into the equations. Thus,

$$2(2) - (3) = 1 \quad \checkmark$$
$$3(2) + 2(3) = 12 \quad \checkmark$$

From the geometric point of view, we have just verified that the point $(2, 3)$ lies on both lines.

2. A system of equations with infinitely many solutions Consider the system

$$2x - y = 1$$
$$6x - 3y = 3$$

Solving the first equation for y in terms of x, we obtain the equation

$$y = 2x - 1$$

Substituting this expression for y into the second equation gives

$$6x - 3(2x - 1) = 3$$
$$6x - 6x + 3 = 3$$
$$0 = 0$$

which is a true statement. This result follows from the fact that the second equation is equivalent to the first. (To see this, just multiply both sides of the first equation by 3.) Our computations have revealed that the system of two equations is equivalent to the single equation $2x - y = 1$. Thus, any ordered pair of numbers (x, y) satisfying the equation $2x - y = 1$ (or $y = 2x - 1$) constitutes a solution to the system.

In particular, by assigning the value t to x, where t is any real number, we find that $y = 2t - 1$, so the ordered pair $(t, 2t - 1)$ is a solution of the system. The variable t is called a **parameter.** For example, setting $t = 0$ gives the point $(0, -1)$ as a solution of the system, and setting $t = 1$ gives the point $(1, 1)$ as another solution. Since t represents any real number, there are infinitely many solutions of the sys-

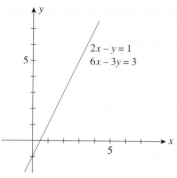

FIGURE 3
A system of equations with infinitely many solutions; each point on the line is a solution.

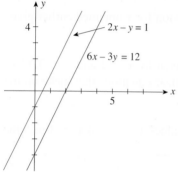

FIGURE 4
A system of equations with no solution

tem. Geometrically, the two equations in the system represent the same line, and all solutions of the system are points lying on the line (Figure 3). Such a system is said to be **dependent**.

3. **A system of equations that has no solution** Consider the system

$$2x - y = 1$$
$$6x - 3y = 12$$

The first equation is equivalent to $y = 2x - 1$. Substituting this expression for y into the second equation gives

$$6x - 3(2x - 1) = 12$$
$$6x - 6x + 3 = 12$$
$$0 = 9$$

which is clearly impossible. Thus, there is no solution to the system of equations. To interpret this situation geometrically, cast both equations in the slope-intercept form, obtaining

$$y = 2x - 1$$
$$y = 2x - 4$$

We see at once that the lines represented by these equations are parallel (each has slope 2) and distinct, since the first has y-intercept -1 and the second has y-intercept -4 (Figure 4). Systems with no solutions, such as this one, are said to be **inconsistent**.

Explore & Discuss

1. Consider a system composed of two linear equations in two variables. Can the system have exactly two solutions? Exactly three solutions? Exactly a finite number of solutions?

2. Suppose at least one of the equations in a system composed of two equations in two variables is nonlinear. Can the system have no solution? Exactly one solution? Exactly two solutions? Exactly a finite number of solutions? Infinitely many solutions? Illustrate each answer with a sketch.

Note   We have used the method of substitution in solving each of these systems. If you are familiar with the method of elimination, you might want to re-solve each of these systems using this method. We will study the method of elimination in detail in Section 5.2.

In Section 2.5, we presented some real-world applications of systems involving two linear equations in two variables. Here is an example involving a system of three linear equations in three variables.

VIDEO **APPLIED EXAMPLE 1** Manufacturing: Production Scheduling Ace Novelty wishes to produce three types of souvenirs: Types A, B, and C. To manufacture a Type A souvenir requires 2 minutes on Machine I, 1 minute on Machine II, and 2 minutes on Machine III. A Type B souvenir requires 1 minute on Machine I, 3 minutes on Machine II, and 1 minute on Machine III. A Type C souvenir requires 1 minute on Machine I and 2 minutes each on Machines II and III. There are 3 hours available on Machine I, 5 hours available on Machine II, and 4 hours available on Machine III for processing the order. How many souvenirs of each type should Ace Novelty make in order to use all of the available time? Formulate but do not solve the problem. (We will solve this problem in Example 7, Section 5.2.)

Solution   The given information may be tabulated as follows:

	Type A	Type B	Type C	Time Available (min)
Machine I	2	1	1	180
Machine II	1	3	2	300
Machine III	2	1	2	240

We have to determine the number of each of *three* types of souvenirs to be made. So let x, y, and z denote the respective numbers of Type A, Type B, and Type C souvenirs to be made. The total amount of time that Machine I is used is given by $2x + y + z$ minutes and must equal 180 minutes. This leads to the equation

$$2x + y + z = 180 \quad \text{Time spent on Machine I}$$

Similar considerations on the use of Machines II and III lead to the following equations:

$$x + 3y + 2z = 300 \quad \text{Time spent on Machine II}$$
$$2x + y + 2z = 240 \quad \text{Time spent on Machine III}$$

Since the variables x, y, and z must satisfy simultaneously the three conditions represented by the three equations, the solution to the problem is found by solving the following system of linear equations:

$$2x + y + z = 180$$
$$x + 3y + 2z = 300$$
$$2x + y + 2z = 240$$

Solutions of Systems of Equations

We will complete the solution of the problem posed in Example 1 later on (page 265). For the moment, let's look at a geometric interpretation of a system of linear equations, such as the system in Example 1, to gain some insight into the nature of the solution.

A linear system composed of three linear equations in three variables x, y, and z has the general form

$$a_1x + b_1y + c_1z = d_1$$
$$a_2x + b_2y + c_2z = d_2 \tag{2}$$
$$a_3x + b_3y + c_3z = d_3$$

Just as a linear equation in two variables represents a straight line in the plane, it can be shown that a linear equation $ax + by + cz = d$ (a, b, and c not all equal to zero) in three variables represents a plane in three-dimensional space. Thus, each equation in System (2) represents a *plane* in three-dimensional space, and the *solution(s) of the system* is precisely the point(s) of intersection of the three planes defined by the three linear equations that make up the system. As before, the system has one and only one solution, infinitely many solutions, or no solution, depending on whether and how the planes intersect one another. Figure 5 illustrates each of these possibilities.

In Figure 5a, the three planes intersect at a point corresponding to the situation in which System (2) has a unique solution. Figure 5b depicts a situation in which there are infinitely many solutions to the system. Here, the three planes intersect along a line, and the solutions are represented by the infinitely many points lying on this line. In Figure 5c, the three planes are parallel and distinct, so there is no point common to all three planes; System (2) has no solution in this case.

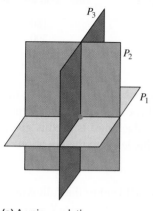

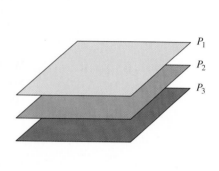

(a) A unique solution
FIGURE **5**

(b) Infinitely many solutions

(c) No solution

Note   The situations depicted in Figure 5 are by no means exhaustive. You may consider various other orientations of the three planes that would illustrate the three possible outcomes in solving a system of linear equations involving three variables.   ▪

Linear Equations in *n* Variables

A linear equation in *n* variables, $x_1, x_2, \ldots, x_n$ is an equation of the form

$$a_1x_1 + a_2x_2 + \cdots + a_nx_n = c$$

where $a_1, a_2, \ldots, a_n$ (not all zero) and c are constants.

For example, the equation

$$3x_1 + 2x_2 - 4x_3 + 6x_4 = 8$$

is a linear equation in the four variables, x_1, x_2, x_3, and x_4.

When the number of variables involved in a linear equation exceeds three, we no longer have the geometric interpretation we had for the lower-dimensional spaces. Nevertheless, the algebraic concepts of the lower-dimensional spaces generalize to higher dimensions. For this reason, a linear equation in *n* variables, $a_1x_1 + a_2x_2 + \cdots + a_nx_n = c$, where $a_1, a_2, \ldots, a_n$ are not all zero, is referred to as an *n-dimensional hyperplane*. We may interpret the solution(s) to a system comprising a finite number of such linear equations to be the *point(s) of intersection* of the hyperplanes defined by the equations that make up the system. As in the case of systems involving two or three variables, it can be shown that only three possibilities exist regarding the nature of the solution of such a system: (1) a unique solution, (2) infinitely many solutions, or (3) no solution.

Explore & Discuss

Refer to the Note above.

Using the orientation of three planes, illustrate the outcomes in solving a system of three linear equations in three variables that result in no solution or infinitely many solutions.

5.1 Self-Check Exercises

DEO **1.** Determine whether the system of linear equations

$$2x - 3y = 12$$
$$x + 2y = 6$$

has (a) a unique solution, (b) infinitely many solutions, or (c) no solution. Find all solutions whenever they exist. Make a sketch of the set of lines described by the system.

2. A farmer has 200 acres of land suitable for cultivating Crops A, B, and C. The cost per acre of cultivating Crops A, B, and C is \$40, \$60, and \$80, respectively. The farmer has

\$12,600 available for cultivation. Each acre of Crop A requires 20 labor-hours, each acre of Crop B requires 25 labor-hours, and each acre of Crop C requires 40 labor-hours. The farmer has a maximum of 5950 labor-hours available. If she wishes to use all of her cultivatable land, the entire budget, and all the labor available, how many acres of each crop should she plant? Formulate but do not solve the problem.

Solutions to Self-Check Exercises 5.1 can be found on page 255.

5.1 Concept Questions

1. Suppose you are given a system of two linear equations in two variables.
  a. What can you say about the solution(s) of the system of equations?
  b. Give a geometric interpretation of your answers to the question in part (a).

2. Suppose you are given a system of two linear equations in two variables.
  a. Explain what it means for the system to be dependent.
  b. Explain what it means for the system to be inconsistent.

5.1 Exercises

In Exercises 1–16, determine whether each system of linear equations has (a) one and only one solution, (b) infinitely many solutions, or (c) no solution. Find all solutions whenever they exist.

1. $x - 3y = -1$
    $4x + 3y = 11$

2. $2x - 4y = 5$
    $3x + 2y = 6$

3. $x + 4y = 7$
    $\frac{1}{2}x + 2y = 5$

4. $3x - 4y = 7$
    $9x - 12y = 14$

5. $x + 2y = 7$
    $2x - y = 4$

6. $\frac{3}{2}x - 2y = 4$
    $x + \frac{1}{3}y = 2$

7. $2x - 5y = 10$
    $6x - 15y = 30$

8. $5x - 6y = 8$
    $10x - 12y = 16$

9. $4x - 5y = 14$
    $2x + 3y = -4$

10. $\frac{5}{4}x - \frac{2}{3}y = 3$
    $\frac{1}{4}x + \frac{5}{3}y = 6$

11. $2x - 3y = 6$
    $6x - 9y = 12$

12. $\frac{2}{3}x + y = 5$
    $\frac{1}{2}x + \frac{3}{4}y = \frac{15}{4}$

13. $-3x + 5y = 1$
    $2x - 4y = -1$

14. $-10x + 15y = -3$
    $4x - 6y = -3$

15. $3x - 6y = 2$
    $-\frac{3}{2}x + 3y = -1$

16. $\frac{3}{2}x - \frac{1}{2}y = 1$
    $-x + \frac{1}{3}y = -\frac{2}{3}$

17. Determine the value of k for which the system of linear equations

$$2x - y = 3$$
$$4x + ky = 4$$

has no solution.

18. Determine the value of k for which the system of linear equations

$$3x + 4y = 12$$
$$x + ky = 4$$

has infinitely many solutions. Then find all solutions corresponding to this value of k.

In Exercises 19–31, formulate but do not solve the problem. You will be asked to solve these problems in the next section.

19. AGRICULTURE The Johnson Farm has 500 acres of land allotted for cultivating corn and wheat. The cost of cultivating corn and wheat (including seeds and labor) is $42 and $30 per acre, respectively. Jacob Johnson has $18,600 available for cultivating these crops. If he wishes to use all the allotted land and his entire budget for cultivating these two crops, how many acres of each crop should he plant?

20. **INVESTMENTS** Michael Perez has a total of $2000 on deposit with two savings institutions. One pays interest at the rate of 6%/year, whereas the other pays interest at the rate of 8%/year. If Michael earned a total of $144 in interest during a single year, how much does he have on deposit in each institution?

21. **MIXTURES** The Coffee Shoppe sells a coffee blend made from two coffees, one costing $5/lb and the other costing $6/lb. If the blended coffee sells for $5.60/lb, find how much of each coffee is used to obtain the desired blend. Assume that the weight of the blended coffee is 100 lb.

22. **INVESTMENTS** Kelly Fisher has a total of $30,000 invested in two municipal bonds that have yields of 8% and 10% interest per year, respectively. If the interest Kelly receives from the bonds in a year is $2640, how much does she have invested in each bond?

23. **RIDERSHIP** The total number of passengers riding a certain city bus during the morning shift is 1000. If the child's fare is $0.50, the adult fare is $1.50, and the total revenue from the fares in the morning shift is $1300, how many children and how many adults rode the bus during the morning shift?

24. **REAL ESTATE** Cantwell Associates, a real estate developer, is planning to build a new apartment complex consisting of one-bedroom units and two- and three-bedroom townhouses. A total of 192 units is planned, and the number of family units (two- and three-bedroom townhouses) will equal the number of one-bedroom units. If the number of one-bedroom units will be 3 times the number of three-bedroom units, find how many units of each type will be in the complex.

25. **INVESTMENT PLANNING** The annual returns on Sid Carrington's three investments amounted to $21,600: 6% on a savings account, 8% on mutual funds, and 12% on bonds. The amount of Sid's investment in bonds was twice the amount of his investment in the savings account, and the interest earned from his investment in bonds was equal to the dividends he received from his investment in mutual funds. Find how much money he placed in each type of investment.

26. **INVESTMENT CLUB** A private investment club has $200,000 earmarked for investment in stocks. To arrive at an acceptable overall level of risk, the stocks that management is considering have been classified into three categories: high-risk, medium-risk, and low-risk. Management estimates that high-risk stocks will have a rate of return of 15%/year; medium-risk stocks, 10%/year; and low-risk stocks, 6%/year. The members have decided that the investment in low-risk stocks should be equal to the sum of the investments in the stocks of the other two categories. Determine how much the club should invest in each type of

stock if the investment goal is to have a return of $20,000/year on the total investment. (Assume that all the money available for investment is invested.)

27. **MIXTURE PROBLEM—FERTILIZER** Lawnco produces three grades of commercial fertilizers. A 100-lb bag of grade A fertilizer contains 18 lb of nitrogen, 4 lb of phosphate, and 5 lb of potassium. A 100-lb bag of grade B fertilizer contains 20 lb of nitrogen and 4 lb each of phosphate and potassium. A 100-lb bag of grade C fertilizer contains 24 lb of nitrogen, 3 lb of phosphate, and 6 lb of potassium. How many 100-lb bags of each of the three grades of fertilizers should Lawnco produce if 26,400 lb of nitrogen, 4900 lb of phosphate, and 6200 lb of potassium are available and all the nutrients are used?

28. **BOX-OFFICE RECEIPTS** A theater has a seating capacity of 900 and charges $4 for children, $6 for students, and $8 for adults. At a certain screening with full attendance, there were half as many adults as children and students combined. The receipts totaled $5600. How many children attended the show?

29. **MANAGEMENT DECISIONS** The management of Hartman Rent-A-Car has allocated $2.25 million to buy a fleet of new automobiles consisting of compact, intermediate-size, and full-size cars. Compacts cost $18,000 each, intermediate-size cars cost $27,000 each, and full-size cars cost $36,000 each. If Hartman purchases twice as many compacts as intermediate-size cars and the total number of cars to be purchased is 100, determine how many cars of each type will be purchased. (Assume that the entire budget will be used.)

30. **INVESTMENT CLUBS** The management of a private investment club has a fund of $200,000 earmarked for investment in stocks. To arrive at an acceptable overall level of risk, the stocks that management is considering have been classified into three categories: high-risk, medium-risk, and low-risk. Management estimates that high-risk stocks will have a rate of return of 15%/year; medium-risk stocks, 10%/year; and low-risk stocks, 6%/year. The investment in low-risk stocks is to be twice the sum of the investments in stocks of the other two categories. If the investment goal is to have an average rate of return of 9%/year on the total investment, determine how much the club should invest in each type of stock. (Assume that all the money available for investment is invested.)

31. **DIET PLANNING** A dietitian wishes to plan a meal around three foods. The percentages of the daily requirements of proteins, carbohydrates, and iron contained in each ounce of the three foods are summarized in the following table:

	Food I	Food II	Food III
Proteins (%)	10	6	8
Carbohydrates (%)	10	12	6
Iron (%)	5	4	12

Determine how many ounces of each food the dietitian should include in the meal to meet exactly the daily requirement of proteins, carbohydrates, and iron (100% of each).

In Exercises 32–34, determine whether the statement is true or false. If it is true, explain why it is true. If it is false, give an example to show why it is false.

32. A system composed of two linear equations must have at least one solution if the straight lines represented by these equations are nonparallel.

33. Suppose the straight lines represented by a system of three linear equations in two variables are parallel to each other. Then the system has no solution, or it has infinitely many solutions.

34. If at least two of the three lines represented by a system composed of three linear equations in two variables are parallel, then the system has no solution.

5.1 Solutions to Self-Check Exercises

1. Solving the first equation for y in terms of x, we obtain

$$y = \frac{2}{3}x - 4$$

Next, substituting this result into the second equation of the system, we find

$$x + 2\left(\frac{2}{3}x - 4\right) = 6$$

$$x + \frac{4}{3}x - 8 = 6$$

$$\frac{7}{3}x = 14$$

$$x = 6$$

Substituting this value of x into the expression for y obtained earlier, we have

$$y = \frac{2}{3}(6) - 4 = 0$$

Therefore, the system has the unique solution $x = 6$ and $y = 0$. Both lines are shown in the accompanying figure.

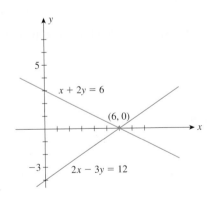

2. Let x, y, and z denote the number of acres of Crop A, Crop B, and Crop C, respectively, to be cultivated. Then the condition that all the cultivatable land be used translates into the equation

$$x + y + z = 200$$

Next, the total cost incurred in cultivating all three crops is $40x + 60y + 80z$ dollars, and since the entire budget is to be expended, we have

$$40x + 60y + 80z = 12,600$$

Finally, the amount of labor required to cultivate all three crops is $20x + 25y + 40z$ hours, and since all the available labor is to be used, we have

$$20x + 25y + 40z = 5950$$

Thus, the solution is found by solving the following system of linear equations:

$$
\begin{aligned}
x + \quad y + \quad z &= \quad 200 \\
40x + 60y + 80z &= 12,600 \\
20x + 25y + 40z &= \quad 5,950
\end{aligned}
$$

5.2 Systems of Linear Equations: Unique Solutions

The Gauss–Jordan Method

The method of substitution used in Section 5.1 is well suited to solving a system of linear equations when the number of linear equations and variables is small. But for large systems, the steps involved in the procedure become difficult to manage.

The **Gauss–Jordan elimination method** is a suitable technique for solving systems of linear equations of any size. One advantage of this technique is its adaptability to the computer. This method involves a sequence of operations on a system of linear equations to obtain at each stage an **equivalent system**—that is, a system having the same solution as the original system. The reduction is complete when the original system has been transformed so that it is in a certain standard form from which the solution can be easily read.

The operations of the Gauss–Jordan elimination method are as follows:

1. Interchange any two equations.
2. Replace an equation by a nonzero constant multiple of itself.
3. Replace an equation by the sum of that equation and a constant multiple of any other equation.

To illustrate the Gauss–Jordan elimination method for solving systems of linear equations, let's apply it to the solution of the following system:

$$2x + 4y = 8$$
$$3x - 2y = 4$$

We begin by working with the first, or x, column. First, we transform the system into an equivalent system in which the coefficient of x in the first equation is 1:

$$2x + 4y = 8$$
$$3x - 2y = 4 \tag{3a}$$

$$x + 2y = 4$$
$$3x - 2y = 4 \tag{3b}$$

Multiply the first equation in System (3a) by $\frac{1}{2}$ (operation 2).

Next, we eliminate x from the second equation:

$$x + 2y = 4$$
$$-8y = -8 \tag{3c}$$

Replace the second equation in System (3b) by the sum of $-3 \times$ the first equation and the second equation (operation 3):

$$-3x - 6y = -12$$
$$\underline{3x - 2y = 4}$$
$$-8y = -8$$

Then we obtain the following equivalent system, in which the coefficient of y in the second equation is 1:

$$x + 2y = 4$$
$$y = 1 \tag{3d}$$

Multiply the second equation in System (3c) by $-\frac{1}{8}$ (operation 2).

Next, we eliminate y in the first equation:

$$x = 2$$
$$y = 1$$

Replace the first equation in System (3d) by the sum of $-2 \times$ the second equation and the first equation (operation 3):

$$x + 2y = 4$$
$$\underline{-2y = -2}$$
$$x = 2$$

This system is now in standard form, and we can read off the solution to System (3a) as $x = 2$ and $y = 1$. We can also express this solution as $(2, 1)$ and interpret it geometrically as the point of intersection of the two lines represented by the two linear equations that make up the given system of equations.

Let's consider another example, involving a system of three linear equations and three variables.

EXAMPLE 1 Solve the following system of equations:

$$2x + 4y + 6z = 22$$
$$3x + 8y + 5z = 27$$
$$-x + y + 2z = 2$$

Solution First, we transform this system into an equivalent system in which the coefficient of x in the first equation is 1:

$$2x + 4y + 6z = 22$$
$$3x + 8y + 5z = 27 \qquad \textbf{(4a)}$$
$$-x + y + 2z = 2$$

$$x + 2y + 3z = 11$$
$$3x + 8y + 5z = 27 \qquad \text{Multiply the first equation in System} \qquad \textbf{(4b)}$$
$$-x + y + 2z = 2 \qquad \text{(4a) by } \tfrac{1}{2}.$$

Next, we eliminate the variable x from all equations except the first:

$$x + 2y + 3z = 11 \qquad \text{Replace the second equation in System}$$
$$2y - 4z = -6 \qquad \text{(4b) by the sum of } -3 \times \text{ the first equa-} \qquad \textbf{(4c)}$$
$$-x + y + 2z = 2 \qquad \text{tion and the second equation:}$$

$$\begin{array}{r} -3x - 6y - 9z = -33 \\ 3x + 8y + 5z = 27 \\ \hline 2y - 4z = -6 \end{array}$$

$$x + 2y + 3z = 11 \qquad \text{Replace the third equation in System}$$
$$2y - 4z = -6 \qquad \text{(4c) by the sum of the first equation} \qquad \textbf{(4d)}$$
$$3y + 5z = 13 \qquad \text{and the third equation:}$$

$$\begin{array}{r} x + 2y + 3z = 11 \\ -x + y + 2z = 2 \\ \hline 3y + 5z = 13 \end{array}$$

Then we transform System (4d) into yet another equivalent system, in which the coefficient of y in the second equation is 1:

$$x + 2y + 3z = 11$$
$$y - 2z = -3 \qquad \text{Multiply the second equation in} \qquad \textbf{(4e)}$$
$$3y + 5z = 13 \qquad \text{System (4d) by } \tfrac{1}{2}.$$

We now eliminate y from all equations except the second, using operation 3 of the elimination method:

$$x + 7z = 17 \qquad \text{Replace the first equation in System}$$
$$y - 2z = -3 \qquad \text{(4e) by the sum of the first equation} \qquad \textbf{(4f)}$$
$$3y + 5z = 13 \qquad \text{and } (-2) \times \text{ the second equation:}$$

$$\begin{array}{r} x + 2y + 3z = 11 \\ -2y + 4z = 6 \\ \hline x + 7z = 17 \end{array}$$

$$x + 7z = 17 \qquad \text{Replace the third equation in System}$$
$$y - 2z = -3 \qquad \text{(4f) by the sum of } (-3) \times \text{ the sec-} \qquad \textbf{(4g)}$$
$$11z = 22 \qquad \text{ond equation and the third equation:}$$

$$\begin{array}{r} -3y + 6z = 9 \\ 3y + 5z = 13 \\ \hline 11z = 22 \end{array}$$

Multiplying the third equation by $\frac{1}{11}$ in (4g) leads to the system

$$
\begin{aligned}
x \qquad\quad + 7z &= 17 \\
y - 2z &= -3 \\
z &= 2
\end{aligned}
$$

Eliminating z from all equations except the third (try it!) then leads to the system

$$
\begin{aligned}
x \qquad\qquad &= 3 \\
y \qquad\quad &= 1 \qquad\qquad\qquad \textbf{(4h)}\\
z &= 2
\end{aligned}
$$

In its final form, the solution to the given system of equations can be easily read off! We have $x = 3$, $y = 1$, and $z = 2$. Geometrically, the point $(3, 1, 2)$ is the intersection of the three planes described by the three equations comprising the given system.   ∎

Augmented Matrices

Observe from the preceding example that the variables x, y, and z play no significant role in each step of the reduction process, except as a reminder of the position of each coefficient in the system. With the aid of **matrices,** which are rectangular arrays of numbers, we can eliminate writing the variables at each step of the reduction and thus save ourselves a great deal of work. For example, the system

$$
\begin{aligned}
2x + 4y + 6z &= 22 \\
3x + 8y + 5z &= 27 \qquad\qquad\qquad \textbf{(5)}\\
-x + y + 2z &= 2
\end{aligned}
$$

may be represented by the matrix

$$
\left[\begin{array}{ccc|c}
2 & 4 & 6 & 22 \\
3 & 8 & 5 & 27 \\
-1 & 1 & 2 & 2
\end{array}\right] \qquad\qquad\qquad \textbf{(6)}
$$

The augmented matrix representing System (5)

The submatrix consisting of the first three columns of Matrix (6) is called the **coefficient matrix** of System (5). The matrix itself, Matrix (6), is referred to as the **augmented matrix** of System (5), since it is obtained by joining the matrix of coefficients to the column (matrix) of constants. The vertical line separates the column of constants from the matrix of coefficients.

The next example shows how much work you can save by using matrices instead of the standard representation of the systems of linear equations.

EXAMPLE 2 Write the augmented matrix corresponding to each equivalent system given in Systems (4a) through (4h).

Solution   The required sequence of augmented matrices follows.

Equivalent System	Augmented Matrix

a.
$$
\begin{aligned}
2x + 4y + 6z &= 22 \\
3x + 8y + 5z &= 27 \\
-x + y + 2z &= 2
\end{aligned}
\qquad
\left[\begin{array}{ccc|c}
2 & 4 & 6 & 22 \\
3 & 8 & 5 & 27 \\
-1 & 1 & 2 & 2
\end{array}\right] \quad \textbf{(7a)}
$$

b.
$$
\begin{aligned}
x + 2y + 3z &= 11 \\
3x + 8y + 5z &= 27 \\
-x + y + 2z &= 2
\end{aligned}
\qquad
\left[\begin{array}{ccc|c}
1 & 2 & 3 & 11 \\
3 & 8 & 5 & 27 \\
-1 & 1 & 2 & 2
\end{array}\right] \quad \textbf{(7b)}
$$

c. $x + 2y + 3z = 11$
    $2y - 4z = -6$
    $-x + y + 2z = 2$

$$\begin{bmatrix} 1 & 2 & 3 & | & 11 \\ 0 & 2 & -4 & | & -6 \\ -1 & 1 & 2 & | & 2 \end{bmatrix}$$ (7c)

d. $x + 2y + 3z = 11$
    $2y - 4z = -6$
    $3y + 5z = 13$

$$\begin{bmatrix} 1 & 2 & 3 & | & 11 \\ 0 & 2 & -4 & | & -6 \\ 0 & 3 & 5 & | & 13 \end{bmatrix}$$ (7d)

e. $x + 2y + 3z = 11$
    $y - 2z = -3$
    $3y + 5z = 13$

$$\begin{bmatrix} 1 & 2 & 3 & | & 11 \\ 0 & 1 & -2 & | & -3 \\ 0 & 3 & 5 & | & 13 \end{bmatrix}$$ (7e)

f. $x + 7z = 17$
    $y - 2z = -3$
    $3y + 5z = 13$

$$\begin{bmatrix} 1 & 0 & 7 & | & 17 \\ 0 & 1 & -2 & | & -3 \\ 0 & 3 & 5 & | & 13 \end{bmatrix}$$ (7f)

g. $x + 7z = 17$
    $y - 2z = -3$
    $11z = 22$

$$\begin{bmatrix} 1 & 0 & 7 & | & 17 \\ 0 & 1 & -2 & | & -3 \\ 0 & 0 & 11 & | & 22 \end{bmatrix}$$ (7g)

h. $x = 3$
    $y = 1$
    $z = 2$

$$\begin{bmatrix} 1 & 0 & 0 & | & 3 \\ 0 & 1 & 0 & | & 1 \\ 0 & 0 & 1 & | & 2 \end{bmatrix}$$ (7h)   ∎

The augmented matrix in (7h) is an example of a matrix in row-reduced form. In general, an augmented matrix with m rows and n columns (called an $m \times n$ matrix) is in **row-reduced form** if it satisfies the following conditions.

Row-Reduced Form of a Matrix

1. Each row consisting entirely of zeros lies below all rows having nonzero entries.

2. The first nonzero entry in each (nonzero) row is 1 (called a **leading 1**).

3. In any two successive (nonzero) rows, the leading 1 in the lower row lies to the right of the leading 1 in the upper row.

4. If a column in the coefficient matrix contains a leading 1, then the other entries in that column are zeros.

VIDEO ▶   **EXAMPLE 3** Determine which of the following matrices are in row-reduced form. If a matrix is not in row-reduced form, state the condition that is violated.

a. $$\begin{bmatrix} 1 & 0 & 0 & | & 0 \\ 0 & 1 & 0 & | & 0 \\ 0 & 0 & 1 & | & 3 \end{bmatrix}$$ b. $$\begin{bmatrix} 1 & 0 & 0 & | & 4 \\ 0 & 1 & 0 & | & 3 \\ 0 & 0 & 0 & | & 0 \end{bmatrix}$$ c. $$\begin{bmatrix} 1 & 2 & 0 & | & 0 \\ 0 & 0 & 1 & | & 0 \\ 0 & 0 & 0 & | & 1 \end{bmatrix}$$

d. $$\begin{bmatrix} 0 & 1 & 2 & | & -2 \\ 1 & 0 & 0 & | & 3 \\ 0 & 0 & 1 & | & 2 \end{bmatrix}$$ e. $$\begin{bmatrix} 1 & 2 & 0 & | & 0 \\ 0 & 0 & 1 & | & 3 \\ 0 & 0 & 2 & | & 1 \end{bmatrix}$$ f. $$\begin{bmatrix} 1 & 0 & | & 4 \\ 0 & 3 & | & 0 \\ 0 & 0 & | & 0 \end{bmatrix}$$

g. $$\begin{bmatrix} 0 & 0 & 0 & | & 0 \\ 1 & 0 & 0 & | & 3 \\ 0 & 1 & 0 & | & 2 \end{bmatrix}$$

Solution   The matrices in parts (a)–(c) are in row-reduced form.

d. This matrix is not in row-reduced form. Conditions 3 and 4 are violated: The leading 1 in row 2 lies to the left of the leading 1 in row 1. Also, column 3 contains a leading 1 in row 3 and a nonzero element above it.

e. This matrix is not in row-reduced form. Conditions 2 and 4 are violated: The first nonzero entry in row 3 is a 2, not a 1. Also, column 3 contains a leading 1 and has a nonzero entry below it.

f. This matrix is not in row-reduced form. Condition 2 is violated: The first nonzero entry in row 2 is not a leading 1.

g. This matrix is not in row-reduced form. Condition 1 is violated: Row 1 consists of all zeros and does not lie below the nonzero rows.   ■

The foregoing discussion suggests the following adaptation of the Gauss–Jordan elimination method in solving systems of linear equations using matrices. First, the three operations on the equations of a system (see page 256) translate into the following **row operations** on the corresponding augmented matrices.

> **Row Operations**
> **1.** Interchange any two rows.
> **2.** Replace any row by a nonzero constant multiple of itself.
> **3.** Replace any row by the sum of that row and a constant multiple of any other row.

We obtained the augmented matrices in Example 2 by using the same operations that we used on the equivalent system of equations in Example 1.

To help us describe the Gauss–Jordan elimination method using matrices, let's introduce some terminology. We begin by defining what is meant by a **unit column**.

> **Unit Column**
> A column in a coefficient matrix is called a **unit column** if one of the entries in the column is a 1 and the other entries are zeros.

For example, in the coefficient matrix of (7d), only the first column is in unit form; in the coefficient matrix of (7h), all three columns are in unit form. Now, the sequence of row operations that transforms the augmented matrix (7a) into the equivalent matrix (7d) in which the first column

$$
\begin{array}{r}
2 \\
3 \\
-1
\end{array}
$$

of (7a) is transformed into the unit column

$$
\begin{array}{r}
1 \\
0 \\
0
\end{array}
$$

is called **pivoting** the matrix about the element (number) 2. Similarly, we have pivoted about the element 2 in the second column of (7d), shown circled,

$$
\begin{array}{c}
2 \\
⓶ \\
3
\end{array}
$$

to obtain the augmented matrix (7g), in which the second column

$$
\begin{array}{c}
0 \\
1 \\
0
\end{array}
$$

is a unit column. Finally, pivoting about the element 11 in column 3 of (7g)

$$
\begin{array}{c}
7 \\
-2 \\
⑪
\end{array}
$$

leads to the augmented matrix (7h), in which the third column

$$
\begin{array}{c}
0 \\
0 \\
1
\end{array}
$$

is a unit column. Observe that in the final augmented matrix, all three columns to the left of the vertical line are in unit form. The element about which a matrix is pivoted is called the *pivot element*.

Before looking at the next example, let's introduce the following notation for the three types of row operations.

Notation for Row Operations

Letting R_i denote the ith row of a matrix, we write:

Operation 1 $R_i \leftrightarrow R_j$ to mean: Interchange row i with row j.

Operation 2 cR_i to mean: Replace row i with c times row i.

Operation 3 $R_i + aR_j$ to mean: Replace row i with the sum of row i and a times row j.

EXAMPLE 4 Pivot the matrix about the circled element.

$$
\begin{bmatrix} ③ & 5 & | & 9 \\ 2 & 3 & | & 5 \end{bmatrix}
$$

Solution We need a **1** in row 1 where the pivot element (the circled **3**) is. One way of doing this is to replace row 1 by $\frac{1}{3}$ times R_1. In other words, we use operation 2. Thus,

$$
\begin{bmatrix} 3 & 5 & | & 9 \\ 2 & 3 & | & 5 \end{bmatrix} \xrightarrow{\frac{1}{3}R_1} \begin{bmatrix} 1 & \frac{5}{3} & | & 3 \\ 2 & 3 & | & 5 \end{bmatrix}
$$

Next, we need to replace row 2 by a row with a **0** in the position that is currently occupied by the number **2**. This can be accomplished by replacing row 2 by the sum of row 2 and -2 times row 1. In other words, we use operation 3. Thus,

$$
\begin{bmatrix} 1 & \frac{5}{3} & | & 3 \\ 2 & 3 & | & 5 \end{bmatrix} \xrightarrow{R_2 - 2R_1} \begin{bmatrix} 1 & \frac{5}{3} & | & 3 \\ 0 & -\frac{1}{3} & | & -1 \end{bmatrix}
$$

Putting these two steps together, we can write the required operations as follows:

$$\begin{bmatrix} 3 & 5 & | & 9 \\ 2 & 3 & | & 5 \end{bmatrix} \xrightarrow{\frac{1}{3}R_1} \begin{bmatrix} 1 & \frac{5}{3} & | & 3 \\ 2 & 3 & | & 5 \end{bmatrix} \xrightarrow{R_2 - 2R_1} \begin{bmatrix} 1 & \frac{5}{3} & | & 3 \\ 0 & -\frac{1}{3} & | & -1 \end{bmatrix}$$

The first column, which originally contained the entry 3, is now in unit form, with a 1 where the pivot element used to be, and we are done.

Alternative Solution   In the first solution, we used operation 2 to obtain a 1 where the pivot element was originally. Alternatively, we can use operation 3 as follows:

$$\begin{bmatrix} 3 & 5 & | & 9 \\ 2 & 3 & | & 5 \end{bmatrix} \xrightarrow{R_1 - R_2} \begin{bmatrix} 1 & 2 & | & 4 \\ 2 & 3 & | & 5 \end{bmatrix} \xrightarrow{R_2 - 2R_1} \begin{bmatrix} 1 & 2 & | & 4 \\ 0 & -1 & | & -3 \end{bmatrix}$$ ∎

Note   In Example 4, the two matrices

$$\begin{bmatrix} 1 & \frac{5}{3} & | & 3 \\ 0 & -\frac{1}{3} & | & -1 \end{bmatrix} \quad \text{and} \quad \begin{bmatrix} 1 & 2 & | & 4 \\ 0 & -1 & | & -3 \end{bmatrix}$$

look quite different, but they are in fact equivalent. You can verify this by observing that they represent the systems of equations

$$x + \frac{5}{3}y = 3 \qquad\qquad x + 2y = 4$$

$$\text{and}$$

$$-\frac{1}{3}y = -1 \qquad\qquad -y = -3$$

respectively, and both have the same solution: $x = -2$ and $y = 3$. Example 4 also shows that we can sometimes avoid working with fractions by using an appropriate row operation. ∎

A summary of the Gauss–Jordan method follows.

The Gauss–Jordan Elimination Method

1. Write the augmented matrix corresponding to the linear system.
2. Interchange rows (operation 1), if necessary, to obtain an augmented matrix in which the first entry in the first row is nonzero. Then pivot the matrix about this entry.
3. Interchange the second row with any row below it, if necessary, to obtain an augmented matrix in which the second entry in the second row is nonzero. Pivot the matrix about this entry.
4. Continue until the final matrix is in row-reduced form.

⚠ Before writing the augmented matrix, be sure to write all equations with the variables on the left and constant terms on the right of the equal sign. Also, make sure that the variables are in the same order in all equations.

VIDEO   **EXAMPLE 5** Solve the system of linear equations given by

$$3x - 2y + 8z = 9$$
$$-2x + 2y + z = 3$$
$$x + 2y - 3z = 8$$

(8)

Solution Using the Gauss–Jordan elimination method, we obtain the following sequence of equivalent augmented matrices:

$$\left[\begin{array}{ccc|c} ③ & -2 & 8 & 9 \\ -2 & 2 & 1 & 3 \\ 1 & 2 & -3 & 8 \end{array}\right] \xrightarrow{R_1 + R_2} \left[\begin{array}{ccc|c} 1 & 0 & 9 & 12 \\ -2 & 2 & 1 & 3 \\ 1 & 2 & -3 & 8 \end{array}\right]$$

$$\xrightarrow[R_3 - R_1]{R_2 + 2R_1} \left[\begin{array}{ccc|c} 1 & 0 & 9 & 12 \\ 0 & 2 & 19 & 27 \\ 0 & 2 & -12 & -4 \end{array}\right]$$

$$\xrightarrow{R_2 \leftrightarrow R_3} \left[\begin{array}{ccc|c} 1 & 0 & 9 & 12 \\ 0 & ② & -12 & -4 \\ 0 & 2 & 19 & 27 \end{array}\right]$$

$$\xrightarrow{\frac{1}{2}R_2} \left[\begin{array}{ccc|c} 1 & 0 & 9 & 12 \\ 0 & 1 & -6 & -2 \\ 0 & 2 & 19 & 27 \end{array}\right]$$

$$\xrightarrow{R_3 - 2R_2} \left[\begin{array}{ccc|c} 1 & 0 & 9 & 12 \\ 0 & 1 & -6 & -2 \\ 0 & 0 & ㉛ & 31 \end{array}\right]$$

$$\xrightarrow{\frac{1}{31}R_3} \left[\begin{array}{ccc|c} 1 & 0 & 9 & 12 \\ 0 & 1 & -6 & -2 \\ 0 & 0 & 1 & 1 \end{array}\right]$$

$$\xrightarrow[R_2 + 6R_3]{R_1 - 9R_3} \left[\begin{array}{ccc|c} 1 & 0 & 0 & 3 \\ 0 & 1 & 0 & 4 \\ 0 & 0 & 1 & 1 \end{array}\right]$$

The solution to System (8) is given by $x = 3$, $y = 4$, and $z = 1$. This may be verified by substitution into System (8) as follows:

$$3(3) - 2(4) + 8(1) = 9 \quad ✓$$
$$-2(3) + 2(4) + 1 \quad\;\; = 3 \quad ✓$$
$$3 + 2(4) - 3(1) = 8 \quad ✓$$

When you are searching for an element to serve as a pivot, it is important to keep in mind that you may work only with the row containing the potential pivot or any row *below* it. To see what can go wrong if this caution is not heeded, consider the following augmented matrix for some linear system:

$$\left[\begin{array}{ccc|c} 1 & 1 & 2 & 3 \\ 0 & 0 & 3 & 1 \\ 0 & 2 & 1 & -2 \end{array}\right]$$

Observe that column 1 is in unit form. The next step in the Gauss–Jordan elimination procedure calls for obtaining a nonzero element in the second position of row 2. If you use row 1 (which is *above* the row under consideration) to help you obtain the pivot, you might proceed as follows:

$$\begin{bmatrix} 1 & 1 & 2 & | & 3 \\ 0 & 0 & 3 & | & 1 \\ 0 & 2 & 1 & | & -2 \end{bmatrix} \xrightarrow{R_2 \leftrightarrow R_1} \begin{bmatrix} 0 & 0 & 3 & | & 1 \\ 1 & 1 & 2 & | & 3 \\ 0 & 2 & 1 & | & -2 \end{bmatrix}$$

As you can see, not only have we obtained a nonzero element to serve as the next pivot, but it is already a 1, thus obviating the next step. This seems like a good move. But beware—we have undone some of our earlier work: Column 1 is no longer a unit column in which a 1 appears first. The correct move in this case is to interchange row 2 with row 3 in the first augmented matrix.

Explore & Discuss

1. Can the phrase "a nonzero constant multiple of itself" in a type 2 row operation be replaced by "a constant multiple of itself"? Explain.

2. Can a row of an augmented matrix be replaced by a row obtained by adding a constant to every element in that row without changing the solution of the system of linear equations? Explain.

The next example illustrates how to handle a situation in which the first entry in row 1 of the augmented matrix is zero.

EXAMPLE 6 Solve the system of linear equations given by

$$\begin{aligned} 2y + 3z &= 7 \\ 3x + 6y - 12z &= -3 \\ 5x - 2y + 2z &= -7 \end{aligned}$$

Solution Using the Gauss–Jordan elimination method, we obtain the following sequence of equivalent augmented matrices:

$$\begin{bmatrix} 0 & 2 & 3 & | & 7 \\ 3 & 6 & -12 & | & -3 \\ 5 & -2 & 2 & | & -7 \end{bmatrix} \xrightarrow{R_1 \leftrightarrow R_2} \begin{bmatrix} ③ & 6 & -12 & | & -3 \\ 0 & 2 & 3 & | & 7 \\ 5 & -2 & 2 & | & -7 \end{bmatrix}$$

$$\xrightarrow{\frac{1}{3}R_1} \begin{bmatrix} 1 & 2 & -4 & | & -1 \\ 0 & 2 & 3 & | & 7 \\ 5 & -2 & 2 & | & -7 \end{bmatrix}$$

$$\xrightarrow{R_3 - 5R_1} \begin{bmatrix} 1 & 2 & -4 & | & -1 \\ 0 & ② & 3 & | & 7 \\ 0 & -12 & 22 & | & -2 \end{bmatrix}$$

$$\xrightarrow{\frac{1}{2}R_2} \begin{bmatrix} 1 & 2 & -4 & | & -1 \\ 0 & 1 & \frac{3}{2} & | & \frac{7}{2} \\ 0 & -12 & 22 & | & -2 \end{bmatrix}$$

$$\xrightarrow[\;R_3 + 12R_2\;]{R_1 - 2R_2}\begin{bmatrix} 1 & 0 & -7 & | & -8 \\ 0 & 1 & \frac{3}{2} & | & \frac{7}{2} \\ 0 & 0 & 40 & | & 40 \end{bmatrix}$$

$$\xrightarrow{\frac{1}{40}R_3}\begin{bmatrix} 1 & 0 & -7 & | & -8 \\ 0 & 1 & \frac{3}{2} & | & \frac{7}{2} \\ 0 & 0 & 1 & | & 1 \end{bmatrix}$$

$$\xrightarrow[\;R_2 - \frac{3}{2}R_3\;]{R_1 + 7R_3}\begin{bmatrix} 1 & 0 & 0 & | & -1 \\ 0 & 1 & 0 & | & 2 \\ 0 & 0 & 1 & | & 1 \end{bmatrix}$$

The solution to the system is given by $x = -1$, $y = 2$, and $z = 1$; this may be verified by substituting these values into each equation of the system. ∎

 APPLIED EXAMPLE 7 Manufacturing: Production Scheduling Complete the solution to Example 1 in Section 5.1, page 250.

Solution To complete the solution of the problem posed in Example 1, recall that the mathematical formulation of the problem led to the following system of linear equations:

$$\begin{aligned} 2x + y + z &= 180 \\ x + 3y + 2z &= 300 \\ 2x + y + 2z &= 240 \end{aligned}$$

where x, y, and z denote the respective numbers of Type A, Type B, and Type C souvenirs to be made.

Solving the foregoing system of linear equations by the Gauss–Jordan elimination method, we obtain the following sequence of equivalent augmented matrices:

$$\begin{bmatrix} 2 & 1 & 1 & | & 180 \\ 1 & 3 & 2 & | & 300 \\ 2 & 1 & 2 & | & 240 \end{bmatrix} \xrightarrow{R_1 \leftrightarrow R_2} \begin{bmatrix} 1 & 3 & 2 & | & 300 \\ 2 & 1 & 1 & | & 180 \\ 2 & 1 & 2 & | & 240 \end{bmatrix}$$

$$\xrightarrow[\;R_3 - 2R_1\;]{R_2 - 2R_1} \begin{bmatrix} 1 & 3 & 2 & | & 300 \\ 0 & -5 & -3 & | & -420 \\ 0 & -5 & -2 & | & -360 \end{bmatrix}$$

$$\xrightarrow{-\frac{1}{5}R_2} \begin{bmatrix} 1 & 3 & 2 & | & 300 \\ 0 & 1 & \frac{3}{5} & | & 84 \\ 0 & -5 & -2 & | & -360 \end{bmatrix}$$

$$\xrightarrow[\;R_3 + 5R_2\;]{R_1 - 3R_2} \begin{bmatrix} 1 & 0 & \frac{1}{5} & | & 48 \\ 0 & 1 & \frac{3}{5} & | & 84 \\ 0 & 0 & 1 & | & 60 \end{bmatrix}$$

$$\xrightarrow[\;R_2 - \frac{3}{5}R_3\;]{R_1 - \frac{1}{5}R_3} \begin{bmatrix} 1 & 0 & 0 & | & 36 \\ 0 & 1 & 0 & | & 48 \\ 0 & 0 & 1 & | & 60 \end{bmatrix}$$

Thus, $x = 36$, $y = 48$, and $z = 60$; that is, Ace Novelty should make 36 Type A souvenirs, 48 Type B souvenirs, and 60 Type C souvenirs to use all available machine time. ∎

5.2 Self-Check Exercises

1. Solve the system of linear equations

$$2x + 3y + z = 6$$
$$x - 2y + 3z = -3$$
$$3x + 2y - 4z = 12$$

using the Gauss–Jordan elimination method.

2. A farmer has 200 acres of land suitable for cultivating Crops A, B, and C. The cost per acre of cultivating Crop A, Crop B, and Crop C is $40, $60, and $80, respectively. The

farmer has $12,600 available for land cultivation. Each acre of Crop A requires 20 labor-hours, each acre of Crop B requires 25 labor-hours, and each acre of Crop C requires 40 labor-hours. The farmer has a maximum of 5950 labor-hours available. If she wishes to use all of her cultivatable land, the entire budget, and all the labor available, how many acres of each crop should she plant?

Solutions to Self-Check Exercises 5.2 can be found on page 269.

5.2 Concept Questions

1. **a.** Explain what it means for two systems of linear equations to be equivalent to each other.
   b. Give the meaning of the following notation used for row operations in the Gauss–Jordan elimination method:

   i. $R_i \leftrightarrow R_j$ **ii.** cR_i **iii.** $R_i + aR_j$

2. **a.** What is an augmented matrix? A coefficient matrix? A unit column?
   b. Explain what is meant by a pivot operation.

3. Suppose that a matrix is in row-reduced form.
   a. What is the position of a row consisting entirely of zeros relative to the nonzero rows?
   b. What is the first nonzero entry in each row?
   c. What is the position of the leading 1s in successive nonzero rows?
   d. If a column contains a leading 1, then what is the value of the other entries in that column?

5.2 Exercises

In Exercises 1–4, write the augmented matrix corresponding to each system of equations.

1. $2x - 3y = 7$
$3x + y = 4$

2. $3x + 7y - 8z = 5$
$x \quad + 3z = -2$
$4x - 3y \quad = 7$

3. $\quad -y + 2z = 5$
$2x + 2y - 8z = 4$
$3y + 4z = 0$

4. $3x_1 + 2x_2 \quad = 0$
$x_1 - x_2 + 2x_3 = 4$
$2x_2 - 3x_3 = 5$

In Exercises 5–8, write the system of equations corresponding to each augmented matrix.

5. $\begin{bmatrix} 3 & 2 & | & -4 \\ 1 & -1 & | & 5 \end{bmatrix}$

6. $\begin{bmatrix} 0 & 3 & 2 & | & 4 \\ 1 & -1 & -2 & | & -3 \\ 4 & 0 & 3 & | & 2 \end{bmatrix}$

7. $\begin{bmatrix} 1 & 3 & 2 & | & 4 \\ 2 & 0 & 0 & | & 5 \\ 3 & -3 & 2 & | & 6 \end{bmatrix}$

8. $\begin{bmatrix} 2 & 3 & 1 & | & 6 \\ 4 & 3 & 2 & | & 5 \\ 0 & 0 & 0 & | & 0 \end{bmatrix}$

In Exercises 9–18, indicate whether the matrix is in row-reduced form.

9. $\begin{bmatrix} 1 & 0 & | & 3 \\ 0 & 1 & | & -2 \end{bmatrix}$

10. $\begin{bmatrix} 1 & 1 & | & 3 \\ 0 & 0 & | & 0 \end{bmatrix}$

11. $\begin{bmatrix} 0 & 1 & | & 3 \\ 1 & 0 & | & 5 \end{bmatrix}$

12. $\begin{bmatrix} 0 & 1 & | & 3 \\ 0 & 0 & | & 5 \end{bmatrix}$

13. $\begin{bmatrix} 1 & 0 & 0 & | & 3 \\ 0 & 1 & 0 & | & 4 \\ 0 & 0 & 1 & | & 5 \end{bmatrix}$

14. $\begin{bmatrix} 1 & 0 & 0 & | & -1 \\ 0 & 1 & 0 & | & -2 \\ 0 & 0 & 2 & | & -3 \end{bmatrix}$

15. $\begin{bmatrix} 1 & 0 & 1 & | & 3 \\ 0 & 1 & 0 & | & 4 \\ 0 & 0 & -1 & | & 6 \end{bmatrix}$

16. $\begin{bmatrix} 1 & 0 & | & -10 \\ 0 & 1 & | & 2 \\ 0 & 0 & | & 0 \end{bmatrix}$

17. $\begin{bmatrix} 0 & 0 & 0 & | & 0 \\ 0 & 1 & 2 & | & 4 \\ 0 & 0 & 0 & | & 0 \end{bmatrix}$

18. $\begin{bmatrix} 1 & 0 & 0 & | & 3 \\ 0 & 1 & 0 & | & 6 \\ 0 & 0 & 0 & | & 4 \\ 0 & 0 & 1 & | & 5 \end{bmatrix}$

In Exercises 19–26, pivot the system about the circled element.

19. $\begin{bmatrix} ② & 4 & | & 8 \\ 3 & 1 & | & 2 \end{bmatrix}$
20. $\begin{bmatrix} 3 & 2 & | & 6 \\ ④ & 2 & | & 5 \end{bmatrix}$

21. $\begin{bmatrix} -1 & 2 & | & 3 \\ 6 & 8 & | & 2 \end{bmatrix}$
22. $\begin{bmatrix} ① & 3 & | & 4 \\ 2 & 4 & | & 6 \end{bmatrix}$

23. $\begin{bmatrix} ② & 4 & 6 & | & 12 \\ 2 & 3 & 1 & | & 5 \\ 3 & -1 & 2 & | & 4 \end{bmatrix}$
24. $\begin{bmatrix} 1 & 3 & 2 & | & 4 \\ ② & 4 & 8 & | & 6 \\ -1 & 2 & 3 & | & 4 \end{bmatrix}$

25. $\begin{bmatrix} 0 & 1 & 3 & | & 4 \\ 2 & 4 & ① & | & 3 \\ 5 & 6 & 2 & | & -4 \end{bmatrix}$
26. $\begin{bmatrix} 1 & 2 & 3 & | & 5 \\ 0 & -3 & 3 & | & 2 \\ 0 & 4 & -1 & | & 3 \end{bmatrix}$

In Exercises 27–30, fill in the missing entries by performing the indicated row operations to obtain the row-reduced matrices.

27. $\begin{bmatrix} 3 & 9 & | & 6 \\ 2 & 1 & | & 4 \end{bmatrix} \xrightarrow{\frac{1}{3}R_1} \begin{bmatrix} \cdot & \cdot & | & \cdot \\ 2 & 1 & | & 4 \end{bmatrix} \xrightarrow{R_2 - 2R_1}$

$\begin{bmatrix} 1 & 3 & | & 2 \\ \cdot & \cdot & | & \cdot \end{bmatrix} \xrightarrow{-\frac{1}{5}R_2} \begin{bmatrix} 1 & 3 & | & 2 \\ \cdot & \cdot & | & \cdot \end{bmatrix} \xrightarrow{R_1 - 3R_2} \begin{bmatrix} 1 & 0 & | & 2 \\ 0 & 1 & | & 0 \end{bmatrix}$

28. $\begin{bmatrix} 1 & 2 & | & 1 \\ 2 & 3 & | & -1 \end{bmatrix} \xrightarrow{R_2 - 2R_1} \begin{bmatrix} 1 & 2 & | & 1 \\ \cdot & \cdot & | & \cdot \end{bmatrix} \xrightarrow{-R_2}$

$\begin{bmatrix} 1 & 2 & | & 1 \\ \cdot & \cdot & | & \cdot \end{bmatrix} \xrightarrow{R_1 - 2R_2} \begin{bmatrix} 1 & 0 & | & -5 \\ 0 & 1 & | & 3 \end{bmatrix}$

29. $\begin{bmatrix} 1 & 3 & 1 & | & 3 \\ 3 & 8 & 3 & | & 7 \\ 2 & -3 & 1 & | & -10 \end{bmatrix} \xrightarrow[R_3 - 2R_1]{R_2 - 3R_1} \begin{bmatrix} 1 & 3 & 1 & | & 3 \\ \cdot & \cdot & \cdot & | & \cdot \\ \cdot & \cdot & \cdot & | & \cdot \end{bmatrix} \xrightarrow{-R_2}$

$\begin{bmatrix} 1 & 3 & 1 & | & 3 \\ \cdot & \cdot & \cdot & | & \cdot \\ 0 & -9 & -1 & | & -16 \end{bmatrix} \xrightarrow[R_3 + 9R_2]{R_1 - 3R_2}$

$\begin{bmatrix} \cdot & \cdot & \cdot & | & \cdot \\ 0 & 1 & 0 & | & 2 \\ \cdot & \cdot & \cdot & | & \cdot \end{bmatrix} \xrightarrow[-R_3]{R_1 + R_3} \begin{bmatrix} 1 & 0 & 0 & | & -1 \\ 0 & 1 & 0 & | & 2 \\ 0 & 0 & 1 & | & -2 \end{bmatrix}$

30. $\begin{bmatrix} 0 & 1 & 3 & | & -4 \\ 1 & 2 & 1 & | & 7 \\ 1 & -2 & 0 & | & 1 \end{bmatrix} \xrightarrow{R_1 \leftrightarrow R_2} \begin{bmatrix} \cdot & \cdot & \cdot & | & \cdot \\ \cdot & \cdot & \cdot & | & \cdot \\ 1 & -2 & 0 & | & 1 \end{bmatrix}$

$\xrightarrow{R_3 - R_1} \begin{bmatrix} 1 & 2 & 1 & | & 7 \\ 0 & 1 & 3 & | & -4 \\ \cdot & \cdot & \cdot & | & \cdot \end{bmatrix} \xrightarrow[R_3 + 4R_2]{R_1 - 2R_2} \begin{bmatrix} \cdot & \cdot & \cdot & | & \cdot \\ 0 & 1 & 3 & | & -4 \\ \cdot & \cdot & \cdot & | & \cdot \end{bmatrix}$

$\xrightarrow{\frac{1}{11}R_3} \begin{bmatrix} 1 & 0 & -5 & | & 15 \\ 0 & 1 & 3 & | & -4 \\ \cdot & \cdot & \cdot & | & \cdot \end{bmatrix} \xrightarrow[R_2 - 3R_3]{R_1 + 5R_3} \begin{bmatrix} 1 & 0 & 0 & | & 5 \\ 0 & 1 & 0 & | & 2 \\ 0 & 0 & 1 & | & -2 \end{bmatrix}$

31. Write a system of linear equations for the augmented matrix of Exercise 27. Using the results of Exercise 27, determine the solution of the system.

32. Repeat Exercise 31 for the augmented matrix of Exercise 28.

33. Repeat Exercise 31 for the augmented matrix of Exercise 29.

34. Repeat Exercise 31 for the augmented matrix of Exercise 30.

In Exercises 35–56, solve the system of linear equations using the Gauss–Jordan elimination method.

35. $x + y = 3$
$2x - y = 3$

36. $x - 2y = -3$
$2x + 3y = 8$

37. $x - 2y = 8$
$3x + 4y = 4$

38. $3x + y = 1$
$-7x - 2y = -1$

39. $2x - 3y = -8$
$4x + y = -2$

40. $5x + 3y = 9$
$-2x + y = -8$

41. $6x + 8y = 15$
$2x - 4y = -5$

42. $2x + 10y = 1$
$-4x + 6y = 11$

43. $3x - 2y = 1$
$2x + 4y = 2$

44. $x - \frac{1}{2}y = \frac{7}{6}$
$-\frac{1}{2}x + 4y = \frac{2}{3}$

45. $x + y + z = 0$
$2x - y + z = 1$
$x + y - 2z = 2$

46. $2x + y - 2z = 4$
$x + 3y - z = -3$
$3x + 4y - z = 7$

47. $2x + 2y + z = 9$
$x + z = 4$
$4y - 3z = 17$

48. $2x + 3y - 2z = 10$
$3x - 2y + 2z = 0$
$4x - y + 3z = -1$

49. $-x_2 + x_3 = 2$
$4x_1 - 3x_2 + 2x_3 = 16$
$3x_1 + 2x_2 + x_3 = 11$

50. $2x + 4y - 6z = 38$
$x + 2y + 3z = 7$
$3x - 4y + 4z = -19$

51. $x_1 - 2x_2 + x_3 = 6$
$2x_1 + x_2 - 3x_3 = -3$
$x_1 - 3x_2 + 3x_3 = 10$

52. $2x + 3y - 6z = -11$
$x - 2y + 3z = 9$
$3x + y = 7$

53. $2x + 3z = -1$
$3x - 2y + z = 9$
$x + y + 4z = 4$

54. $2x_1 - x_2 + 3x_3 = -4$
$x_1 - 2x_2 + x_3 = -1$
$x_1 - 5x_2 + 2x_3 = -3$

55. $x_1 - x_2 + 3x_3 = 14$
$x_1 + x_2 + x_3 = 6$
$-2x_1 - x_2 + x_3 = -4$

56. $2x_1 - x_2 - x_3 = 0$
$3x_1 + 2x_2 + x_3 = 7$
$x_1 + 2x_2 + 2x_3 = 5$

The problems in Exercises 57–69 correspond to those in Exercises 19–31, Section 5.1. Use the results of your previous work to help you solve these problems.

57. AGRICULTURE The Johnson Farm has 500 acres of land allotted for cultivating corn and wheat. The cost of cultivating corn and wheat (including seeds and labor) is $42 and $30 per acre, respectively. Jacob Johnson has $18,600 available for cultivating these crops. If he wishes to use all the allotted land and his entire budget for cultivating these two crops, how many acres of each crop should he plant?

58. **INVESTMENTS** Michael Perez has a total of $2000 on deposit with two savings institutions. One pays interest at the rate of 6%/year, whereas the other pays interest at the rate of 8%/year. If Michael earned a total of $144 in interest during a single year, how much does he have on deposit in each institution?

59. **MIXTURES** The Coffee Shoppe sells a coffee blend made from two coffees, one costing $5/lb and the other costing $6/lb. If the blended coffee sells for $5.60/lb, find how much of each coffee is used to obtain the desired blend. Assume that the weight of the blended coffee is 100 lb.

60. **INVESTMENTS** Kelly Fisher has a total of $30,000 invested in two municipal bonds that have yields of 8% and 10% interest per year, respectively. If the interest Kelly receives from the bonds in a year is $2640, how much does she have invested in each bond?

61. **RIDERSHIP** The total number of passengers riding a certain city bus during the morning shift is 1000. If the child's fare is $0.50, the adult fare is $1.50, and the total revenue from the fares in the morning shift is $1300, how many children and how many adults rode the bus during the morning shift?

62. **REAL ESTATE** Cantwell Associates, a real estate developer, is planning to build a new apartment complex consisting of one-bedroom units and two- and three-bedroom townhouses. A total of 192 units is planned, and the number of family units (two- and three-bedroom townhouses) will equal the number of one-bedroom units. If the number of one-bedroom units will be 3 times the number of three-bedroom units, find how many units of each type will be in the complex.

63. **INVESTMENT PLANNING** The annual returns on Sid Carrington's three investments amounted to $21,600: 6% on a savings account, 8% on mutual funds, and 12% on bonds. The amount of Sid's investment in bonds was twice the amount of his investment in the savings account, and the interest earned from his investment in bonds was equal to the dividends he received from his investment in mutual funds. Find how much money he placed in each type of investment.

64. **INVESTMENT CLUB** A private investment club has $200,000 earmarked for investment in stocks. To arrive at an acceptable overall level of risk, the stocks that management is considering have been classified into three categories: high-risk, medium-risk, and low-risk. Management estimates that high-risk stocks will have a rate of return of 15%/year; medium-risk stocks, 10%/year; and low-risk stocks, 6%/year. The members have decided that the investment in low-risk stocks should be equal to the sum of the investments in the stocks of the other two categories. Determine how much the club should invest in each type of stock if the investment goal is to have a return of $20,000/year on the total investment. (Assume that all the money available for investment is invested.)

65. **MIXTURE PROBLEM—FERTILIZER** Lawnco produces three grades of commercial fertilizers. A 100-lb bag of grade A fertilizer contains 18 lb of nitrogen, 4 lb of phosphate, and 5 lb of potassium. A 100-lb bag of grade B fertilizer contains 20 lb of nitrogen and 4 lb each of phosphate and potassium. A 100-lb bag of grade C fertilizer contains 24 lb of nitrogen, 3 lb of phosphate, and 6 lb of potassium. How many 100-lb bags of each of the three grades of fertilizers should Lawnco produce if 26,400 lb of nitrogen, 4900 lb of phosphate, and 6200 lb of potassium are available and all the nutrients are used?

66. **BOX-OFFICE RECEIPTS** A theater has a seating capacity of 900 and charges $4 for children, $6 for students, and $8 for adults. At a certain screening with full attendance, there were half as many adults as children and students combined. The receipts totaled $5600. How many children attended the show?

67. **MANAGEMENT DECISIONS** The management of Hartman Rent-A-Car has allocated $2.25 million to buy a fleet of new automobiles consisting of compact, intermediate-size, and full-size cars. Compacts cost $18,000 each, intermediate-size cars cost $27,000 each, and full-size cars cost $36,000 each. If Hartman purchases twice as many compacts as intermediate-size cars and the total number of cars to be purchased is 100, determine how many cars of each type will be purchased. (Assume that the entire budget will be used.)

68. **INVESTMENT CLUBS** The management of a private investment club has a fund of $200,000 earmarked for investment in stocks. To arrive at an acceptable overall level of risk, the stocks that management is considering have been classified into three categories: high-risk, medium-risk, and low-risk. Management estimates that high-risk stocks will have a rate of return of 15%/year; medium-risk stocks, 10%/year; and low-risk stocks, 6%/year. The investment in low-risk stocks is to be twice the sum of the investments in stocks of the other two categories. If the investment goal is to have an average rate of return of 9%/year on the total investment, determine how much the club should invest in each type of stock. (Assume that all of the money available for investment is invested.)

69. **DIET PLANNING** A dietitian wishes to plan a meal around three foods. The percentages of the daily requirements of proteins, carbohydrates, and iron contained in each ounce of the three foods are summarized in the following table:

	Food I	Food II	Food III
Proteins (%)	10	6	8
Carbohydrates (%)	10	12	6
Iron (%)	5	4	12

Determine how many ounces of each food the dietitian should include in the meal to meet exactly the daily requirement of proteins, carbohydrates, and iron (100% of each).

70. **INVESTMENTS** Mr. and Mrs. Garcia have a total of $100,000 to be invested in stocks, bonds, and a money market account. The stocks have a rate of return of 12%/year, while the bonds and the money market account pay 8%/year and 4%/year, respectively. The Garcias have stipulated that the amount invested in the money market account should be equal to the sum of 20% of the amount invested in stocks and 10% of the amount invested in bonds. How should the Garcias allocate their resources if they require an annual income of $10,000 from their investments?

71. **BOX-OFFICE RECEIPTS** For the opening night at the Opera House, a total of 1000 tickets were sold. Front orchestra seats cost $80 apiece, rear orchestra seats cost $60 apiece, and front balcony seats cost $50 apiece. The combined number of tickets sold for the front orchestra and rear orchestra exceeded twice the number of front balcony tickets sold by 400. The total receipts for the performance were $62,800. Determine how many tickets of each type were sold.

72. **PRODUCTION SCHEDULING** A manufacturer of women's blouses makes three types of blouses: sleeveless, short-sleeve, and long-sleeve. The time (in minutes) required by each department to produce a dozen blouses of each type is shown in the following table:

	Sleeveless	Short-Sleeve	Long-Sleeve
Cutting	9	12	15
Sewing	22	24	28
Packaging	6	8	8

The cutting, sewing, and packaging departments have available a maximum of 80, 160, and 48 labor-hours, respectively, per day. How many dozens of each type of blouse can be produced each day if the plant is operated at full capacity?

73. **BUSINESS TRAVEL EXPENSES** An executive of Trident Communications recently traveled to London, Paris, and Rome. He paid $180, $230, and $160 per night for lodging in London, Paris, and Rome, respectively, and his hotel bills totaled $2660. He spent $110, $120, and $90 per day for his meals in London, Paris, and Rome, respectively, and his expenses for meals totaled $1520. If he spent as many days in London as he did in Paris and Rome combined, how many days did he stay in each city?

74. **VACATION COSTS** Joan and Dick spent 2 weeks (14 nights) touring four cities on the East Coast—Boston, New York, Philadelphia, and Washington. They paid $240, $400, $160, and $200 per night for lodging in each city, respectively, and their total hotel bill came to $4040. The number of days they spent in New York was the same as the total number of days they spent in Boston and Washington, and the couple spent 3 times as many days in New York as they did in Philadelphia. How many days did Joan and Dick stay in each city?

In Exercises 75 and 76, determine whether the statement is true or false. If it is true, explain why it is true. If it is false, give an example to show why it is false.

75. An equivalent system of linear equations can be obtained from a system of equations by replacing one of its equations by any constant multiple of itself.

76. If the augmented matrix corresponding to a system of three linear equations in three variables has a row of the form $[0 \ 0 \ 0 \ | \ a]$, where a is a nonzero number, then the system has no solution.

5.2 Solutions to Self-Check Exercises

1. We obtain the following sequence of equivalent augmented matrices:

$$\begin{bmatrix} 2 & 3 & 1 & | & 6 \\ 1 & -2 & 3 & | & -3 \\ 3 & 2 & -4 & | & 12 \end{bmatrix} \xrightarrow{R_1 \leftrightarrow R_2} \begin{bmatrix} ① & -2 & 3 & | & -3 \\ 2 & 3 & 1 & | & 6 \\ 3 & 2 & -4 & | & 12 \end{bmatrix}$$

$$\xrightarrow[R_3 - 3R_1]{R_2 - 2R_1} \begin{bmatrix} 1 & -2 & 3 & | & -3 \\ 0 & 7 & -5 & | & 12 \\ 0 & 8 & -13 & | & 21 \end{bmatrix} \xrightarrow{R_2 \leftrightarrow R_3}$$

$$\begin{bmatrix} 1 & -2 & 3 & | & -3 \\ 0 & ⑧ & -13 & | & 21 \\ 0 & 7 & -5 & | & 12 \end{bmatrix} \xrightarrow{R_2 - R_3} \begin{bmatrix} 1 & -2 & 3 & | & -3 \\ 0 & 1 & -8 & | & 9 \\ 0 & 7 & -5 & | & 12 \end{bmatrix}$$

$$\xrightarrow[R_3 - 7R_2]{R_1 + 2R_2} \begin{bmatrix} 1 & 0 & -13 & | & 15 \\ 0 & 1 & -8 & | & 9 \\ 0 & 0 & 51 & | & -51 \end{bmatrix} \xrightarrow{\frac{1}{51}R_3} \begin{bmatrix} 1 & 0 & -13 & | & 15 \\ 0 & 1 & -8 & | & 9 \\ 0 & 0 & ① & | & -1 \end{bmatrix}$$

$$\xrightarrow[R_2 + 8R_3]{R_1 + 13R_3} \begin{bmatrix} 1 & 0 & 0 & | & 2 \\ 0 & 1 & 0 & | & 1 \\ 0 & 0 & 1 & | & -1 \end{bmatrix}$$

The solution to the system is $x = 2$, $y = 1$, and $z = -1$.

2. Referring to the solution of Exercise 2, Self-Check Exercises 5.1, we see that the problem reduces to solving the following system of linear equations:

$$\begin{aligned} x + \quad y + \quad z &= \quad 200 \\ 40x + 60y + 80z &= 12,600 \\ 20x + 25y + 40z &= 5,950 \end{aligned}$$

Using the Gauss–Jordan elimination method, we have

$$\begin{bmatrix} ① & 1 & 1 & | & 200 \\ 40 & 60 & 80 & | & 12,600 \\ 20 & 25 & 40 & | & 5,950 \end{bmatrix} \xrightarrow[R_3 - 20R_1]{R_2 - 40R_1} \begin{bmatrix} 1 & 1 & 1 & | & 200 \\ 0 & ⑳ & 40 & | & 4600 \\ 0 & 5 & 20 & | & 1950 \end{bmatrix}$$

$$\xrightarrow{\frac{1}{20}R_2} \begin{bmatrix} 1 & 1 & 1 & | & 200 \\ 0 & 1 & 2 & | & 230 \\ 0 & 5 & 20 & | & 1950 \end{bmatrix} \xrightarrow[R_3 - 5R_2]{R_1 - R_2} \begin{bmatrix} 1 & 0 & -1 & | & -30 \\ 0 & 1 & 2 & | & 230 \\ 0 & 0 & ⑩ & | & 800 \end{bmatrix}$$

$$\xrightarrow{\frac{1}{10}R_3} \begin{bmatrix} 1 & 0 & -1 & | & -30 \\ 0 & 1 & 2 & | & 230 \\ 0 & 0 & 1 & | & 80 \end{bmatrix} \xrightarrow[R_2 - 2R_3]{R_1 + R_3} \begin{bmatrix} 1 & 0 & 0 & | & 50 \\ 0 & 1 & 0 & | & 70 \\ 0 & 0 & 1 & | & 80 \end{bmatrix}$$

From the last augmented matrix in reduced form, we see that $x = 50$, $y = 70$, and $z = 80$. Therefore, the farmer should plant 50 acres of Crop A, 70 acres of Crop B, and 80 acres of Crop C.

USING TECHNOLOGY

Systems of Linear Equations: Unique Solutions

Solving a System of Linear Equations Using the Gauss–Jordan Method

The three matrix operations can be performed on a matrix by using a graphing utility. The commands are summarized in the following table.

	Calculator Function		
Operation	**TI-83/84**	**TI-86**	
$R_i \leftrightarrow R_j$	**rowSwap([A], i, j)**	**rSwap(A, i, j)**	or equivalent
cR_i	***row(c, [A], i)**	**multR(c, A, i)**	or equivalent
$R_i + aR_j$	***row+(a, [A], j, i)**	**mRAdd(a, A, j, i)**	or equivalent

When a row operation is performed on a matrix, the result is stored as an answer in the calculator. If another operation is performed on this matrix, then the matrix is erased. Should a mistake be made in the operation, the previous matrix may be lost. For this reason, you should store the results of each operation. We do this by pressing STO, followed by the name of a matrix, and then ENTER. We use this process in the following example.

EXAMPLE 1 Use a graphing utility to solve the following system of linear equations by the Gauss–Jordan method (see Example 5 in Section 5.2):

$$3x - 2y + 8z = 9$$
$$-2x + 2y + z = 3$$
$$x + 2y - 3z = 8$$

Solution Using the Gauss–Jordan method, we obtain the following sequence of equivalent matrices.

$$\begin{bmatrix} 3 & -2 & 8 & | & 9 \\ -2 & 2 & 1 & | & 3 \\ 1 & 2 & -3 & | & 8 \end{bmatrix} \xrightarrow{\text{*row+ (1, [A], 2, 1) } \blacktriangleright B}$$

$$\begin{bmatrix} 1 & 0 & 9 & | & 12 \\ -2 & 2 & 1 & | & 3 \\ 1 & 2 & -3 & | & 8 \end{bmatrix} \xrightarrow{\text{*row+ (2, [B], 1, 2) } \blacktriangleright C}$$

$$\begin{bmatrix} 1 & 0 & 9 & | & 12 \\ 0 & 2 & 19 & | & 27 \\ 1 & 2 & -3 & | & 8 \end{bmatrix} \xrightarrow{\text{*row+ (-1, [C], 1, 3) } \blacktriangleright B}$$

$$\begin{bmatrix} 1 & 0 & 9 & | & 12 \\ 0 & 2 & 19 & | & 27 \\ 0 & 2 & -12 & | & -4 \end{bmatrix} \xrightarrow{\text{*row}(\frac{1}{2}, [B], 2) \blacktriangleright C}$$

$$\begin{bmatrix} 1 & 0 & 9 & | & 12 \\ 0 & 1 & 9.5 & | & 13.5 \\ 0 & 2 & -12 & | & -4 \end{bmatrix} \xrightarrow{\text{*row+ (-2, [C], 2, 3) } \blacktriangleright B}$$

$$\begin{bmatrix} 1 & 0 & 9 & | & 12 \\ 0 & 1 & 9.5 & | & 13.5 \\ 0 & 0 & -31 & | & -31 \end{bmatrix} \xrightarrow{\text{*row}(-\frac{1}{31}, [B], 3) \blacktriangleright C}$$

$$\begin{bmatrix} 1 & 0 & 9 & | & 12 \\ 0 & 1 & 9.5 & | & 13.5 \\ 0 & 0 & 1 & | & 1 \end{bmatrix} \xrightarrow{\text{*row+ } (-9, [C], 3, 1) \blacktriangleright B}$$

$$\begin{bmatrix} 1 & 0 & 0 & | & 3 \\ 0 & 1 & 9.5 & | & 13.5 \\ 0 & 0 & 1 & | & 1 \end{bmatrix} \xrightarrow{\text{*row+ } (-9.5, [B], 3, 2) \blacktriangleright C} \begin{bmatrix} 1 & 0 & 0 & | & 3 \\ 0 & 1 & 0 & | & 4 \\ 0 & 0 & 1 & | & 1 \end{bmatrix}$$

The last matrix is in row-reduced form, and we see that the solution of the system is $x = 3$, $y = 4$, and $z = 1$.

Using rref (TI-83/84 and TI-86) to Solve a System of Linear Equations

The operation **rref** (or equivalent function in your utility, if there is one) will transform an augmented matrix into one that is in row-reduced form. For example, using **rref**, we find

$$\begin{bmatrix} 3 & -2 & 8 & | & 9 \\ -2 & 2 & 1 & | & 3 \\ 1 & 2 & -3 & | & 8 \end{bmatrix} \xrightarrow{\text{rref}} \begin{bmatrix} 1 & 0 & 0 & | & 3 \\ 0 & 1 & 0 & | & 4 \\ 0 & 0 & 1 & | & 1 \end{bmatrix}$$

as obtained earlier!

Using SIMULT (TI-86) to Solve a System of Equations

The operation **SIMULT** (or equivalent operation on your utility, if there is one) of a graphing utility can be used to solve a system of n linear equations in n variables, where n is an integer between 2 and 30, inclusive.

EXAMPLE 2 Use the **SIMULT** operation to solve the system of Example 1.

Solution Call for the **SIMULT** operation. Since the system under consideration has three equations in three variables, enter $n = 3$. Next, enter a1, 1 = 3, a1, 2 = −2, a1, 3 = 8, b1 = 9, a2, 1 = −2, . . . , b3 = 8. Select <SOLVE>, and the display

$$x1 = 3$$
$$x2 = 4$$
$$x3 = 1$$

appears on the screen, giving $x = 3$, $y = 4$, and $z = 1$ as the required solution.

TECHNOLOGY EXERCISES

Use a graphing utility to solve the system of equations (a) by the Gauss–Jordan method, (b) using the rref operation, and (c) using SIMULT.

1. $x_1 - 2x_2 + 2x_3 - 3x_4 = -7$
$3x_1 + 2x_2 - x_3 + 5x_4 = 22$
$2x_1 - 3x_2 + 4x_3 - x_4 = -3$
$3x_1 - 2x_2 - x_3 + 2x_4 = 12$

2. $2x_1 - x_2 + 3x_3 - 2x_4 = -2$
$x_1 - 2x_2 + x_3 - 3x_4 = 2$
$x_1 - 5x_2 + 2x_3 + 3x_4 = -6$
$-3x_1 + 3x_2 - 4x_3 - 4x_4 = 9$

3. $2x_1 + x_2 + 3x_3 - x_4 = 9$
$-x_1 - 2x_2 - 3x_4 = -1$
$x_1 - 3x_3 + x_4 = 10$
$x_1 - x_2 - x_3 - x_4 = 8$

4. $x_1 - 2x_2 - 2x_3 + x_4 = 1$
$2x_1 - x_2 + 2x_3 + 3x_4 = -2$
$-x_1 - 5x_2 + 7x_3 - 2x_4 = 3$
$3x_1 - 4x_2 + 3x_3 + 4x_4 = -4$

(continued)

5. $\begin{aligned} 2x_1 - 2x_2 + 3x_3 - x_4 + 2x_5 &= 16 \\ 3x_1 + x_2 - 2x_3 + x_4 - 3x_5 &= -11 \\ x_1 + 3x_2 - 4x_3 + 3x_4 - x_5 &= -13 \\ 2x_1 - x_2 + 3x_3 - 2x_4 + 2x_5 &= 15 \\ 3x_1 + 4x_2 - 3x_3 + 5x_4 - x_5 &= -10 \end{aligned}$

6. $\begin{aligned} 2.1x_1 - 3.2x_2 + 6.4x_3 + 7x_4 - 3.2x_5 &= 54.3 \\ 4.1x_1 + 2.2x_2 - 3.1x_3 - 4.2x_4 + 3.3x_5 &= -20.81 \\ 3.4x_1 - 6.2x_2 + 4.7x_3 + 2.1x_4 - 5.3x_5 &= 24.7 \\ 4.1x_1 + 7.3x_2 + 5.2x_3 + 6.1x_4 - 8.2x_5 &= 29.25 \\ 2.8x_1 + 5.2x_2 + 3.1x_3 + 5.4x_4 + 3.8x_5 &= 43.72 \end{aligned}$

5.3   Systems of Linear Equations: Underdetermined and Overdetermined Systems

In this section, we continue our study of systems of linear equations. More specifically, we look at systems that have infinitely many solutions and those that have no solution. We also study systems of linear equations in which the number of variables is not equal to the number of equations in the system.

Solution(s) of Linear Equations

Our first two examples illustrate the situation in which a system of linear equations has infinitely many solutions.

EXAMPLE 1 A System of Equations with an Infinite Number of Solutions Solve the system of linear equations given by

$$\begin{aligned} x + 2y &= 4 \\ 3x + 6y &= 12 \end{aligned} \qquad (9)$$

Solution   Using the Gauss-Jordan elimination method, we obtain the following system of equivalent matrices:

$$\begin{bmatrix} ①　2 & | & 4 \\ 3 & 6 & | & 12 \end{bmatrix} \xrightarrow{R_2 - 3R_1} \begin{bmatrix} 1 & 2 & | & 4 \\ 0 & 0 & | & 0 \end{bmatrix}$$

The last augmented matrix is in row-reduced form. Interpreting it as a system of linear equations, we see that the given System (9) is equivalent to the single equation

$$x + 2y = 4 \quad \text{or} \quad x = 4 - 2y$$

If we assign a particular value to y—say, $y = 0$—we obtain $x = 4$, giving the solution $(4, 0)$ to System (9). By setting $y = 1$, we obtain the solution $(2, 1)$. In general, if we set $y = t$, where t represents some real number (called a parameter), we obtain the solution given by $(4 - 2t, t)$. Since the parameter t may be any real number, we see that System (9) has infinitely many solutions. Geometrically, the solutions of System (9) lie on the line on the plane with equation $x + 2y = 4$. The two equations in the system have the same graph (straight line), which you can verify graphically.   ■

EXAMPLE 2 A System of Equations with an Infinite Number of Solutions Solve the system of linear equations given by

$$\begin{aligned} x + 2y - 3z &= -2 \\ 3x - y - 2z &= 1 \\ 2x + 3y - 5z &= -3 \end{aligned} \qquad (10)$$

Solution Using the Gauss–Jordan elimination method, we obtain the following sequence of equivalent augmented matrices:

$$\begin{bmatrix} ① & 2 & -3 & | & -2 \\ 3 & -1 & -2 & | & 1 \\ 2 & 3 & -5 & | & -3 \end{bmatrix} \xrightarrow[R_3 - 2R_1]{R_2 - 3R_1} \begin{bmatrix} 1 & 2 & -3 & | & -2 \\ 0 & ⑦ & 7 & | & 7 \\ 0 & -1 & 1 & | & 1 \end{bmatrix} \xrightarrow{-\frac{1}{7}R_2}$$

$$\begin{bmatrix} 1 & 2 & -3 & | & -2 \\ 0 & 1 & -1 & | & -1 \\ 0 & -1 & 1 & | & 1 \end{bmatrix} \xrightarrow[R_3 + R_2]{R_1 - 2R_2} \begin{bmatrix} 1 & 0 & -1 & | & 0 \\ 0 & 1 & -1 & | & -1 \\ 0 & 0 & 0 & | & 0 \end{bmatrix}$$

The last augmented matrix is in row-reduced form. Interpreting it as a system of linear equations gives

$$\begin{aligned} x - z &= 0 \\ y - z &= -1 \end{aligned}$$

a system of two equations in the three variables x, y, and z.

Let's now single out one variable—say, z—and solve for x and y in terms of it. We obtain

$$\begin{aligned} x &= z \\ y &= z - 1 \end{aligned}$$

If we set $z = t$, where t is a parameter, then System (10) has infinitely many solutions given by $(t, t - 1, t)$. For example, letting $t = 0$ gives the solution $(0, -1, 0)$, and letting $t = 1$ gives the solution $(1, 0, 1)$. Geometrically, the solutions of System (10) lie on the straight line in three-dimensional space given by the intersection of the three planes determined by the three equations in the system. ■

Note In Example 2, we chose the parameter to be z because it is more convenient to solve for x and y (both the x- and y-columns are in unit form) in terms of z. ■

The next example shows what happens in the elimination procedure when the system does not have a solution.

VIDEO **EXAMPLE 3** A System of Equations That Has No Solution Solve the system of linear equations given by

$$\begin{aligned} x + y + z &= 1 \\ 3x - y - z &= 4 \\ x + 5y + 5z &= -1 \end{aligned}$$ **(11)**

Solution Using the Gauss–Jordan elimination method, we obtain the following sequence of equivalent augmented matrices:

$$\begin{bmatrix} ① & 1 & 1 & | & 1 \\ 3 & -1 & -1 & | & 4 \\ 1 & 5 & 5 & | & -1 \end{bmatrix} \xrightarrow[R_3 - R_1]{R_2 - 3R_1} \begin{bmatrix} 1 & 1 & 1 & | & 1 \\ 0 & -4 & -4 & | & 1 \\ 0 & 4 & 4 & | & -2 \end{bmatrix}$$

$$\xrightarrow{R_3 + R_2} \begin{bmatrix} 1 & 1 & 1 & | & 1 \\ 0 & -4 & -4 & | & 1 \\ 0 & 0 & 0 & | & -1 \end{bmatrix}$$

Observe that row 3 in the last matrix reads $0x + 0y + 0z = -1$—that is, $0 = -1$! We therefore conclude that System (11) is inconsistent and has no solution. Geometrically, we have a situation in which two of the planes intersect in a straight line but the third plane is parallel to this line of intersection of the two planes and does not intersect it. Consequently, there is no point of intersection of the three planes. ■

Example 3 illustrates the following more general result of using the Gauss–Jordan elimination procedure.

> ### Systems with No Solution
>
> If there is a row in an augmented matrix containing all zeros to the left of the vertical line and a nonzero entry to the right of the line, then the corresponding system of equations has no solution.

It may have dawned on you that in all the previous examples, we have dealt only with systems involving exactly the same number of linear equations as there are variables. However, systems in which the number of equations is different from the number of variables also occur in practice. Indeed, we will consider such systems in Examples 4 and 5.

The following theorem provides us with some preliminary information on a system of linear equations.

> ### THEOREM 1
>
> **a.** If the number of equations is greater than or equal to the number of variables in a linear system, then one of the following is true:
> > **i.** The system has no solution.
> > **ii.** The system has exactly one solution.
> > **iii.** The system has infinitely many solutions.
> **b.** If there are fewer equations than variables in a linear system, then the system either has no solution or has infinitely many solutions.

Note   Theorem 1 may be used to tell us, before we even begin to solve a problem, what the nature of the solution may be.   ∎

Although we will not prove this theorem, you should recall that we have illustrated geometrically part (a) for the case in which there are exactly as many equations (three) as there are variables. To show the validity of part (b), let us once again consider the case in which a system has three variables. Now, if there is only one equation in the system, then it is clear that there are infinitely many solutions corresponding geometrically to all the points lying on the plane represented by the equation.

Next, if there are two equations in the system, then *only* the following possibilities exist:

1. The two planes are parallel and distinct.
2. The two planes intersect in a straight line.
3. The two planes are coincident (the two equations define the same plane) (Figure 6).

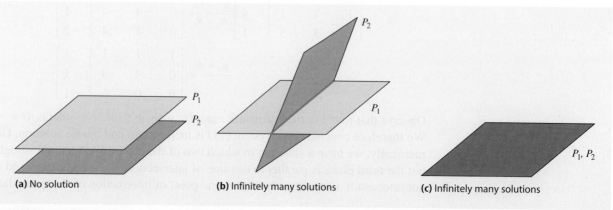

(a) No solution   (b) Infinitely many solutions   (c) Infinitely many solutions

FIGURE **6**

Thus, either there is no solution or there are infinitely many solutions corresponding to the points lying on a line of intersection of the two planes or on a single plane determined by the two equations. In the case in which two planes intersect in a straight line, the solutions will involve one parameter, and in the case in which the two planes are coincident, the solutions will involve two parameters.

Explore & Discuss

Give a geometric interpretation of Theorem 1 for a linear system composed of equations involving two variables. Specifically, illustrate what can happen if there are three linear equations in the system (the case involving two linear equations has already been discussed in Section 5.1). What if there are four linear equations? What if there is only one linear equation in the system?

EXAMPLE 4 A System with More Equations Than Variables Solve the following system of linear equations:

$$x + 2y = 4$$
$$x - 2y = 0$$
$$4x + 3y = 12$$

Solution   We obtain the following sequence of equivalent augmented matrices:

$$\begin{bmatrix} \boxed{1} & 2 & | & 4 \\ 1 & -2 & | & 0 \\ 4 & 3 & | & 12 \end{bmatrix} \xrightarrow[R_3 - 4R_1]{R_2 - R_1} \begin{bmatrix} 1 & 2 & | & 4 \\ 0 & \boxed{-4} & | & -4 \\ 0 & -5 & | & -4 \end{bmatrix} \xrightarrow{-\frac{1}{4}R_2}$$

$$\begin{bmatrix} 1 & 2 & | & 4 \\ 0 & 1 & | & 1 \\ 0 & -5 & | & -4 \end{bmatrix} \xrightarrow[R_3 + 5R_2]{R_1 - 2R_2} \begin{bmatrix} 1 & 0 & | & 2 \\ 0 & 1 & | & 1 \\ 0 & 0 & | & 1 \end{bmatrix}$$

The last row of the row-reduced augmented matrix implies that $0 = 1$, which is impossible, so we conclude that the given system has no solution. Geometrically, the three lines defined by the three equations in the system do not intersect at a point. (To see this for yourself, draw the graphs of these equations.) ■

EXAMPLE 5 A System with More Variables Than Equations Solve the following system of linear equations:

$$x + 2y - 3z + w = -2$$
$$3x - y - 2z - 4w = 1$$
$$2x + 3y - 5z + w = -3$$

Solution   First, observe that the given system consists of three equations in four variables, so by Theorem 1b, either the system has no solution or it has infinitely many solutions. To solve it, we use the Gauss–Jordan method and obtain the following sequence of equivalent augmented matrices:

$$\begin{bmatrix} \boxed{1} & 2 & -3 & 1 & | & -2 \\ 3 & -1 & -2 & -4 & | & 1 \\ 2 & 3 & -5 & 1 & | & -3 \end{bmatrix} \xrightarrow[R_3 - 2R_1]{R_2 - 3R_1} \begin{bmatrix} 1 & 2 & -3 & 1 & | & -2 \\ 0 & \boxed{-7} & 7 & -7 & | & 7 \\ 0 & -1 & 1 & -1 & | & 1 \end{bmatrix} \xrightarrow{-\frac{1}{7}R_2}$$

$$\begin{bmatrix} 1 & 2 & -3 & 1 & | & -2 \\ 0 & 1 & -1 & 1 & | & -1 \\ 0 & -1 & 1 & -1 & | & 1 \end{bmatrix} \xrightarrow[R_3 + R_2]{R_1 - 2R_2} \begin{bmatrix} 1 & 0 & -1 & -1 & | & 0 \\ 0 & 1 & -1 & 1 & | & -1 \\ 0 & 0 & 0 & 0 & | & 0 \end{bmatrix}$$

The last augmented matrix is in row-reduced form. Observe that the given system is equivalent to the system

$$x - z - w = 0$$
$$y - z + w = -1$$

of two equations in four variables. Thus, we may solve for two of the variables in terms of the other two. Letting $z = s$ and $w = t$ (where s and t are any real numbers), we find that

$$x = s + t$$
$$y = s - t - 1$$
$$z = s$$
$$w = t$$

The solutions may be written in the form $(s + t, s - t - 1, s, t)$. Geometrically, the three equations in the system represent three hyperplanes in four-dimensional space (since there are four variables), and their "points" of intersection lie in a two-dimensional subspace of four-space (since there are two parameters).   ■

Note   In Example 5, we assigned parameters to z and w rather than to x and y because x and y are readily solved in terms of z and w.   ■

The following example illustrates a situation in which a system of linear equations has infinitely many solutions.

VIDEO ▶   **APPLIED EXAMPLE 6**   Traffic Control Figure 7 shows the flow of down-town traffic in a certain city during the rush hours on a typical weekday. The arrows indicate the direction of traffic flow on each one-way road, and the average number of vehicles per hour entering and leaving each intersection appears beside each road. 5th Avenue and 6th Avenue can each handle up to 2000 vehicles per hour without causing congestion, whereas the maximum capacity of both 4th Street and 5th Street is 1000 vehicles per hour. The flow of traffic is controlled by traffic lights installed at each of the four intersections.

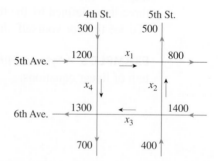

FIGURE **7**

a. Write a general expression involving the rates of flow—x_1, x_2, x_3, x_4—and suggest two possible traffic-flow patterns that will ensure no traffic congestion.
b. Suppose that the part of 4th Street between 5th Avenue and 6th Avenue is to be resurfaced and that traffic flow between the two junctions must therefore be reduced to at most 300 vehicles per hour. Find two possible traffic-flow patterns that will result in a smooth flow of traffic.

Solution

a. To avoid congestion, all traffic entering an intersection must also leave that intersection. Applying this condition to each of the four intersections in a clockwise direction beginning with the 5th Avenue and 4th Street intersection, we obtain the following equations:

$$1500 = x_1 + x_4$$
$$1300 = x_1 + x_2$$
$$1800 = x_2 + x_3$$
$$2000 = x_3 + x_4$$

This system of four linear equations in the four variables x_1, x_2, x_3, x_4 may be rewritten in the more standard form

$$x_1 \qquad\qquad + x_4 = 1500$$
$$x_1 + x_2 \qquad\qquad = 1300$$
$$x_2 + x_3 \qquad = 1800$$
$$x_3 + x_4 = 2000$$

Using the Gauss–Jordan elimination method to solve the system, we obtain

$$\left[\begin{array}{cccc|c} 1 & 0 & 0 & 1 & 1500 \\ 1 & 1 & 0 & 0 & 1300 \\ 0 & 1 & 1 & 0 & 1800 \\ 0 & 0 & 1 & 1 & 2000 \end{array}\right] \xrightarrow{R_2 - R_1} \left[\begin{array}{cccc|c} 1 & 0 & 0 & 1 & 1500 \\ 0 & 1 & 0 & -1 & -200 \\ 0 & 1 & 1 & 0 & 1800 \\ 0 & 0 & 1 & 1 & 2000 \end{array}\right]$$

$$\xrightarrow{R_3 - R_2} \left[\begin{array}{cccc|c} 1 & 0 & 0 & 1 & 1500 \\ 0 & 1 & 0 & -1 & -200 \\ 0 & 0 & 1 & 1 & 2000 \\ 0 & 0 & 1 & 1 & 2000 \end{array}\right]$$

$$\xrightarrow{R_4 - R_3} \left[\begin{array}{cccc|c} 1 & 0 & 0 & 1 & 1500 \\ 0 & 1 & 0 & -1 & -200 \\ 0 & 0 & 1 & 1 & 2000 \\ 0 & 0 & 0 & 0 & 0 \end{array}\right]$$

The last augmented matrix is in row-reduced form and is equivalent to a system of three linear equations in the four variables x_1, x_2, x_3, x_4. Thus, we may express three of the variables—say, x_1, x_2, x_3—in terms of the fourth, x_4. Setting $x_4 = t$ (t a parameter), we may write the infinitely many solutions of the system as

$$x_1 = 1500 - t$$
$$x_2 = -200 + t$$
$$x_3 = 2000 - t$$
$$x_4 = t$$

Observe that for a meaningful solution, we must have $200 \le t \le 1000$, since x_1, x_2, x_3, and x_4 must all be nonnegative and the maximum capacity of a street is 1000. For example, picking $t = 300$ gives the flow pattern

$$x_1 = 1200 \qquad x_2 = 100 \qquad x_3 = 1700 \qquad x_4 = 300$$

Selecting $t = 500$ gives the flow pattern

$$x_1 = 1000 \qquad x_2 = 300 \qquad x_3 = 1500 \qquad x_4 = 500$$

b. In this case, x_4 must not exceed 300. Again, using the results of part (a), we find, upon setting $x_4 = t = 300$, the flow pattern

$$x_1 = 1200 \qquad x_2 = 100 \qquad x_3 = 1700 \qquad x_4 = 300$$

obtained earlier. Picking $t = 250$ gives the flow pattern

$$x_1 = 1250 \qquad x_2 = 50 \qquad x_3 = 1750 \qquad x_4 = 250$$

5.3 Self-Check Exercises

1. The following augmented matrix in row-reduced form is equivalent to the augmented matrix of a certain system of linear equations. Use this result to solve the system of equations.

$$\begin{bmatrix} 1 & 0 & -1 & | & 3 \\ 0 & 1 & 5 & | & -2 \\ 0 & 0 & 0 & | & 0 \end{bmatrix}$$

2. Solve the system of linear equations

$$2x - 3y + z = 6$$
$$x + 2y + 4z = -4$$
$$x - 5y - 3z = 10$$

using the Gauss–Jordan elimination method.

3. Solve the system of linear equations

$$x - 2y + 3z = 9$$
$$2x + 3y - z = 4$$
$$x + 5y - 4z = 2$$

using the Gauss–Jordan elimination method.

Solutions to Self-Check Exercises 5.3 can be found on page 280.

5.3 Concept Questions

1. **a.** If a system of linear equations has the same number of equations or more equations than variables, what can you say about the nature of its solution(s)?
   b. If a system of linear equations has fewer equations than variables, what can you say about the nature of its solution(s)?

2. A system consists of three linear equations in four variables. Can the system have a unique solution?

5.3 Exercises

In Exercises 1–12, given that the augmented matrix in row-reduced form is equivalent to the augmented matrix of a system of linear equations, (a) determine whether the system has a solution and (b) find the solution or solutions to the system, if they exist.

1. $\begin{bmatrix} 1 & 0 & 0 & | & 3 \\ 0 & 1 & 0 & | & -1 \\ 0 & 0 & 1 & | & 2 \end{bmatrix}$

2. $\begin{bmatrix} 1 & 0 & 0 & | & 3 \\ 0 & 1 & 0 & | & -2 \\ 0 & 0 & 1 & | & 1 \end{bmatrix}$

3. $\begin{bmatrix} 1 & 0 & | & 2 \\ 0 & 1 & | & 5 \\ 0 & 0 & | & 0 \end{bmatrix}$

4. $\begin{bmatrix} 1 & 0 & 0 & | & 3 \\ 0 & 1 & 0 & | & 1 \\ 0 & 0 & 0 & | & 0 \end{bmatrix}$

5. $\begin{bmatrix} 1 & 0 & 1 & | & 4 \\ 0 & 1 & 0 & | & -2 \end{bmatrix}$

6. $\begin{bmatrix} 1 & 0 & 0 & 0 & | & 3 \\ 0 & 1 & 1 & 0 & | & -1 \\ 0 & 0 & 0 & 1 & | & 2 \end{bmatrix}$

7. $\begin{bmatrix} 1 & 0 & 0 & 0 & | & 2 \\ 0 & 1 & 0 & 0 & | & 1 \\ 0 & 0 & 1 & 0 & | & 3 \\ 0 & 0 & 0 & 0 & | & 1 \end{bmatrix}$

8. $\begin{bmatrix} 1 & 0 & 0 & | & 4 \\ 0 & 1 & 0 & | & -1 \\ 0 & 0 & 1 & | & 3 \\ 0 & 0 & 0 & | & 1 \end{bmatrix}$

9. $\begin{bmatrix} 1 & 0 & 0 & 0 & | & 4 \\ 0 & 1 & 0 & 0 & | & -1 \\ 0 & 0 & 1 & 1 & | & 3 \\ 0 & 0 & 0 & 0 & | & 0 \end{bmatrix}$

10. $\begin{bmatrix} 0 & 1 & 0 & 1 & | & 3 \\ 0 & 0 & 1 & -2 & | & 4 \\ 0 & 0 & 0 & 0 & | & 0 \\ 0 & 0 & 0 & 0 & | & 0 \end{bmatrix}$

11. $\begin{bmatrix} 1 & 0 & 3 & 0 & | & 2 \\ 0 & 1 & -1 & 0 & | & 1 \\ 0 & 0 & 0 & 0 & | & 0 \\ 0 & 0 & 0 & 0 & | & 0 \end{bmatrix}$

12. $\begin{bmatrix} 1 & 0 & 3 & -1 & | & 4 \\ 0 & 1 & -2 & 3 & | & 2 \\ 0 & 0 & 0 & 0 & | & 0 \\ 0 & 0 & 0 & 0 & | & 0 \end{bmatrix}$

In Exercises 13–34, solve the system of linear equations, using the Gauss–Jordan elimination method.

13. $2x - y = 3$
    $x + 2y = 4$
    $2x + 3y = 7$

14. $x + 2y = 3$
    $2x - 3y = -8$
    $x - 4y = -9$

15. $3x - 2y = -3$
    $2x + y = 3$
    $x - 2y = -5$

16. $2x + 3y = 2$
    $x + 3y = -2$
    $x - y = 3$

17. $3x - 2y = 5$
    $-x + 3y = -4$
    $2x - 4y = 6$

18. $4x + 6y = 8$
    $3x - 2y = -7$
    $x + 3y = 5$

19. $\begin{aligned} x - 2y &= 2 \\ 7x - 14y &= 14 \\ 3x - 6y &= 6 \end{aligned}$

20. $\begin{aligned} 3x - y + 2z &= 5 \\ x - y + 2z &= 1 \\ 5x - 2y + 4z &= 12 \end{aligned}$

21. $\begin{aligned} x + 2y + z &= -2 \\ -2x - 3y - z &= 1 \\ 2x + 4y + 2z &= -4 \end{aligned}$

22. $\begin{aligned} 3y + 2z &= 4 \\ 2x - y - 3z &= 3 \\ 2x + 2y - z &= 7 \end{aligned}$

23. $\begin{aligned} 3x + 2y &= 4 \\ -\tfrac{3}{2}x - y &= -2 \\ 6x + 4y &= 8 \end{aligned}$

24. $\begin{aligned} 2x_1 - x_2 + x_3 &= -4 \\ 3x_1 - \tfrac{3}{2}x_2 + \tfrac{3}{2}x_3 &= -6 \\ -6x_1 + 3x_2 - 3x_3 &= 12 \end{aligned}$

25. $\begin{aligned} x + y - 2z &= -3 \\ 2x - y + 3z &= 7 \\ x - 2y + 5z &= 0 \end{aligned}$

26. $\begin{aligned} 2x_1 + 6x_2 - 5x_3 &= 5 \\ x_1 + 3x_2 + x_3 + 7x_4 &= -1 \\ 3x_1 + 9x_2 - x_3 + 13x_4 &= 1 \end{aligned}$

27. $\begin{aligned} x - 2y + 3z &= 4 \\ 2x + 3y - z &= 2 \\ x + 2y - 3z &= -6 \end{aligned}$

28. $\begin{aligned} x_1 - 2x_2 + x_3 &= -3 \\ 2x_1 + x_2 - 2x_3 &= 2 \\ x_1 + 3x_2 - 3x_3 &= 5 \end{aligned}$

29. $\begin{aligned} 4x + y - z &= 4 \\ 8x + 2y - 2z &= 8 \end{aligned}$

30. $\begin{aligned} x_1 + 2x_2 + 4x_3 &= 2 \\ x_1 + x_2 + 2x_3 &= 1 \end{aligned}$

31. $\begin{aligned} 2x + y - 3z &= 1 \\ x - y + 2z &= 1 \\ 5x - 2y + 3z &= 6 \end{aligned}$

32. $\begin{aligned} 3x - 9y + 6z &= -12 \\ x - 3y + 2z &= -4 \\ 2x - 6y + 4z &= 8 \end{aligned}$

33. $\begin{aligned} x + 2y - z &= -4 \\ 2x + y + z &= 7 \\ x + 3y + 2z &= 7 \\ x - 3y + z &= 9 \end{aligned}$

34. $\begin{aligned} 3x - 2y + z &= 4 \\ x + 3y - 4z &= -3 \\ 2x - 3y + 5z &= 7 \\ x - 8y + 9z &= 10 \end{aligned}$

35. MANAGEMENT DECISIONS The management of Hartman Rent-A-Car has allocated $1,512,000 to purchase 60 new automobiles to add to the existing fleet of rental cars. The company will choose from compact, mid-sized, and full-sized cars costing $18,000, $28,800, and $39,600 each, respectively. Find formulas giving the options available to the company. Give two specific options. (*Note:* Your answers will *not* be unique.)

36. NUTRITION A dietitian wishes to plan a meal around three foods. The meal is to include 8800 units of vitamin A, 3380 units of vitamin C, and 1020 units of calcium. The number of units of the vitamins and calcium in each ounce of the foods is summarized in the following table:

	Food I	Food II	Food III
Vitamin A	400	1200	800
Vitamin C	110	570	340
Calcium	90	30	60

Determine the amount of each food the dietitian should include in the meal in order to meet the vitamin and calcium requirements.

37. NUTRITION Refer to Exercise 36. In planning for another meal, the dietitian changes the requirement of vitamin C from 3380 units to 2160 units. All other requirements remain the same. Show that such a meal cannot be planned around the same foods.

38. MANUFACTURING PRODUCTION SCHEDULE Ace Novelty manufactures Giant Pandas, Saint Bernards, and Big Birds. Each Giant Panda requires 1.5 yd^2 of plush, 30 ft^3 of stuffing, and 5 pieces of trim; each Saint Bernard requires 2 yd^2 of plush, 35 ft^3 of stuffing, and 8 pieces of trim; and each Big Bird requires 2.5 yd^2 of plush, 25 ft^3 of stuffing, and 15 pieces of trim. If 4700 yd^2 of plush, 65,000 ft^3 of stuffing, and 23,400 pieces of trim are available, how many of each of the stuffed animals should the company manufacture if all the material is to be used? Give two specific options.

39. INVESTMENTS Mr. and Mrs. Garcia have a total of $100,000 to be invested in stocks, bonds, and a money market account. The stocks have a rate of return of 12%/year, while the bonds and the money market account pay 8%/year and 4%/year, respectively. The Garcias have stipulated that the amount invested in stocks should be equal to the sum of the amount invested in bonds and 3 times the amount invested in the money market account. How should the Garcias allocate their resources if they require an annual income of $10,000 from their investments? Give two specific options.

40. TRAFFIC CONTROL The accompanying figure shows the flow of traffic near a city's Civic Center during the rush hours on a typical weekday. Each road can handle a maximum of 1000 cars/hour without causing congestion. The flow of traffic is controlled by traffic lights at each of the five intersections.

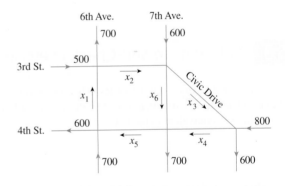

a. Set up a system of linear equations describing the traffic flow.

b. Solve the system devised in part (a), and suggest two possible traffic-flow patterns that will ensure no traffic congestion.

c. Suppose 7th Avenue between 3rd and 4th Streets is soon to be closed for road repairs. Find one possible traffic-flow pattern that will result in a smooth flow of traffic.

41. TRAFFIC CONTROL The accompanying figure shows the flow of downtown traffic during the rush hours on a typical weekday. Each avenue can handle up to 1500 vehicles/hour without causing congestion, whereas the maximum capacity of each street is 1000 vehicles/hour. The flow of traffic is controlled by traffic lights at each of the six intersections.

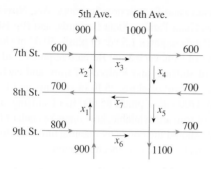

a. Set up a system of linear equations describing the traffic flow.

b. Solve the system devised in part (a), and suggest two possible traffic-flow patterns that will ensure no traffic congestion.

c. Suppose the traffic flow along 9th Street between 5th and 6th Avenues, x_6, is restricted because of sewer construction. What is the minimum permissible traffic flow along this road that will not result in traffic congestion?

42. Determine the value of k such that the following system of linear equations has a unique solution, and then find the solution:

$$2x + 3y = 2$$
$$x + 4y = 6$$
$$5x + ky = 2$$

43. Determine the value of k such that the following system of linear equations has infinitely many solutions, and then find the solutions:

$$3x - 2y + 4z = 12$$
$$-9x + 6y - 12z = k$$

44. Solve the system:

$$\frac{1}{x} + \frac{1}{y} + \frac{1}{z} = -1$$
$$\frac{2}{x} + \frac{3}{y} + \frac{2}{z} = 3$$
$$\frac{2}{x} + \frac{1}{y} + \frac{2}{z} = -7$$

In Exercises 45 and 46, determine whether the statement is true or false. If it is true, explain why it is true. If it is false, give an example to show why it is false.

45. A system of linear equations having fewer equations than variables has no solution, a unique solution, or infinitely many solutions.

46. A system of linear equations having more equations than variables has no solution, a unique solution, or infinitely many solutions.

5.3   Solutions to Self-Check Exercises

1. Let x, y, and z denote the variables. Then the given row-reduced augmented matrix tells us that the system of linear equations is equivalent to the two equations

$$x \quad - z = 3$$
$$y + 5z = -2$$

Letting $z = t$, where t is a parameter, we find the infinitely many solutions given by

$$x = t + 3$$
$$y = -5t - 2$$
$$z = t$$

2. We obtain the following sequence of equivalent augmented matrices:

$$\begin{bmatrix} 2 & -3 & 1 & 6 \\ 1 & 2 & 4 & -4 \\ 1 & -5 & -3 & 10 \end{bmatrix} \xrightarrow{R_1 \leftrightarrow R_2}$$

$$\begin{bmatrix} ① & 2 & 4 & -4 \\ 2 & -3 & 1 & 6 \\ 1 & -5 & -3 & 10 \end{bmatrix} \xrightarrow[R_3 - R_1]{R_2 - 2R_1}$$

$$\begin{bmatrix} 1 & 2 & 4 & -4 \\ 0 & ⑦ & -7 & 14 \\ 0 & -7 & -7 & 14 \end{bmatrix} \xrightarrow{-\frac{1}{7}R_2}$$

$$\begin{bmatrix} 1 & 2 & 4 & -4 \\ 0 & 1 & 1 & -2 \\ 0 & -7 & -7 & 14 \end{bmatrix} \xrightarrow[R_3 + 7R_2]{R_1 - 2R_2} \begin{bmatrix} 1 & 0 & 2 & 0 \\ 0 & 1 & 1 & -2 \\ 0 & 0 & 0 & 0 \end{bmatrix}$$

The last augmented matrix, which is in row-reduced form, tells us that the given system of linear equations is equivalent to the following system of two equations:

$$x \quad + 2z = 0$$
$$y + z = -2$$

Letting $z = t$, where t is a parameter, we see that the infinitely many solutions are given by

$$x = -2t$$
$$y = -t - 2$$
$$z = t$$

3. We obtain the following sequence of equivalent augmented matrices:

$$\begin{bmatrix} ① & -2 & 3 & | & 9 \\ 2 & 3 & -1 & | & 4 \\ 1 & 5 & -4 & | & 2 \end{bmatrix} \xrightarrow[R_3 - R_1]{R_2 - 2R_1}$$

$$\begin{bmatrix} 1 & -2 & 3 & | & 9 \\ 0 & 7 & -7 & | & -14 \\ 0 & 7 & -7 & | & -7 \end{bmatrix} \xrightarrow{R_3 - R_2} \begin{bmatrix} 1 & -2 & 3 & | & 9 \\ 0 & 7 & -7 & | & -14 \\ 0 & 0 & 0 & | & 7 \end{bmatrix}$$

Since the last row of the final augmented matrix is equivalent to the equation $0 = 7$, a contradiction, we conclude that the given system has no solution.

USING TECHNOLOGY

Systems of Linear Equations: Underdetermined and Overdetermined Systems

We can use the row operations of a graphing utility to solve a system of m linear equations in n unknowns by the Gauss–Jordan method, as we did in the previous technology section. We can also use the **rref** or equivalent operation to obtain the row-reduced form without going through all the steps of the Gauss–Jordan method. The **SIMULT** function, however, cannot be used to solve a system where the number of equations and the number of variables are not the same.

EXAMPLE 1 Solve the system

$$\begin{aligned} x_1 - 2x_2 + 4x_3 &= 2 \\ 2x_1 + x_2 - 2x_3 &= -1 \\ 3x_1 - x_2 + 2x_3 &= 1 \\ 2x_1 + 6x_2 - 12x_3 &= -6 \end{aligned}$$

Solution   First, we enter the augmented matrix A into the calculator as

$$A = \begin{bmatrix} 1 & -2 & 4 & | & 2 \\ 2 & 1 & -2 & | & -1 \\ 3 & -1 & 2 & | & 1 \\ 2 & 6 & -12 & | & -6 \end{bmatrix}$$

Then using the **rref** or equivalent operation, we obtain the equivalent matrix

$$\begin{bmatrix} 1 & 0 & 0 & | & 0 \\ 0 & 1 & -2 & | & -1 \\ 0 & 0 & 0 & | & 0 \\ 0 & 0 & 0 & | & 0 \end{bmatrix}$$

in reduced form. Thus, the given system is equivalent to

$$\begin{aligned} x_1 \qquad\quad &= 0 \\ x_2 - 2x_3 &= -1 \end{aligned}$$

If we let $x_3 = t$, where t is a parameter, then we find that the solutions are $(0, 2t - 1, t)$.

(continued)

TECHNOLOGY EXERCISES

Use a graphing utility to solve the system of equations using the rref or equivalent operation.

1.
$$\begin{aligned}
2x_1 - x_2 - x_3 &= 0 \\
3x_1 - 2x_2 - x_3 &= -1 \\
-x_1 + 2x_2 - x_3 &= 3 \\
2x_2 - 2x_3 &= 4
\end{aligned}$$

2.
$$\begin{aligned}
3x_1 + x_2 - 4x_3 &= 5 \\
2x_1 - 3x_2 + 2x_3 &= -4 \\
-x_1 - 2x_2 + 4x_3 &= 6 \\
4x_1 + 3x_2 - 5x_3 &= 9
\end{aligned}$$

3.
$$\begin{aligned}
2x_1 + 3x_2 + 2x_3 + x_4 &= -1 \\
x_1 - x_2 + x_3 - 2x_4 &= -8 \\
5x_1 + 6x_2 - 2x_3 + 2x_4 &= 11 \\
x_1 + 3x_2 + 8x_3 + x_4 &= -14
\end{aligned}$$

4.
$$\begin{aligned}
x_1 - x_2 + 3x_3 - 6x_4 &= 2 \\
x_1 + x_2 + x_3 - 2x_4 &= 2 \\
-2x_1 - x_2 + x_3 + 2x_4 &= 0
\end{aligned}$$

5.
$$\begin{aligned}
x_1 + x_2 - x_3 - x_4 &= -1 \\
x_1 - x_2 + x_3 + 4x_4 &= -6 \\
3x_1 + x_2 - x_3 + 2x_4 &= -4 \\
5x_1 + x_2 - 3x_3 + x_4 &= -9
\end{aligned}$$

6.
$$\begin{aligned}
1.2x_1 - 2.3x_2 + 4.2x_3 + 5.4x_4 - 1.6x_5 &= 4.2 \\
2.3x_1 + 1.4x_2 - 3.1x_3 + 3.3x_4 - 2.4x_5 &= 6.3 \\
1.7x_1 + 2.6x_2 - 4.3x_3 + 7.2x_4 - 1.8x_5 &= 7.8 \\
2.6x_1 - 4.2x_2 + 8.3x_3 - 1.6x_4 + 2.5x_5 &= 6.4
\end{aligned}$$

5.4 Matrices

Using Matrices to Represent Data

Many practical problems are solved by using arithmetic operations on the data associated with the problems. By properly organizing the data into *blocks* of numbers, we can then carry out these arithmetic operations in an orderly and efficient manner. In particular, this systematic approach enables us to use the computer to full advantage.

Let's begin by considering how the monthly output data of a manufacturer may be organized. The Acrosonic Company manufactures four different loudspeaker systems at three separate locations. The company's May output is described in Table 1.

TABLE 1

	Model A	Model B	Model C	Model D
Location I	320	280	460	280
Location II	480	360	580	0
Location III	540	420	200	880

Now, if we agree to preserve the relative location of each entry in Table 1, we can summarize the set of data as follows:

$$\begin{bmatrix} 320 & 280 & 460 & 280 \\ 480 & 360 & 580 & 0 \\ 540 & 420 & 200 & 880 \end{bmatrix}$$

A matrix summarizing the data in Table 1

The array of numbers displayed here is an example of a matrix. Observe that the numbers in row 1 give the output of models A, B, C, and D of Acrosonic loudspeaker systems manufactured at Location I; similarly, the numbers in rows 2 and 3 give the respective outputs of these loudspeaker systems at Locations II and III. The numbers in each column of the matrix give the outputs of a particular model of loudspeaker system manufactured at each of the company's three manufacturing locations.

More generally, a matrix is a rectangular array of real numbers. For example, each of the following arrays is a matrix:

$$A = \begin{bmatrix} 3 & 0 & -1 \\ 2 & 1 & 4 \end{bmatrix} \quad B = \begin{bmatrix} 3 & 2 \\ 0 & 1 \\ -1 & 4 \end{bmatrix} \quad C = \begin{bmatrix} 1 \\ 2 \\ 4 \\ 0 \end{bmatrix} \quad D = \begin{bmatrix} 1 & 3 & 0 & 1 \end{bmatrix}$$

The real numbers that make up the array are called the **entries,** or *elements*, of the matrix. The entries in a row in the array are referred to as a **row** of the matrix, whereas the entries in a column in the array are referred to as a **column** of the matrix. Matrix A, for example, has two rows and three columns, which may be identified as follows:

$$\begin{array}{c} \quad\quad \text{Column 1} \quad \text{Column 2} \quad \text{Column 3} \\ \begin{array}{c} \text{Row 1} \\ \text{Row 2} \end{array} \begin{bmatrix} 3 & 0 & -1 \\ 2 & 1 & 4 \end{bmatrix} \end{array}$$

A 2×3 matrix

The **size,** or *dimension*, **of a matrix** is described in terms of the number of rows and columns of the matrix. For example, matrix A has two rows and three columns and is said to have size 2 by 3, denoted 2×3. In general, a matrix having m rows and n columns is said to have size $m \times n$.

> **Matrix**
>
> A **matrix** is an ordered rectangular array of numbers. A matrix with m rows and n columns has size $m \times n$. The entry in the ith row and jth column of a matrix A is denoted by a_{ij}.

A matrix of size $1 \times n$—a matrix having one row and n columns—is referred to as a **row matrix,** or *row vector*, of dimension n. For example, the matrix D is a row vector of dimension 4. Similarly, a matrix having m rows and one column is referred to as a **column matrix,** or *column vector*, of dimension m. The matrix C is a column vector of dimension 4. Finally, an $n \times n$ matrix—that is, a matrix having the same number of rows as columns—is called a **square matrix.** For example, the matrix

$$\begin{bmatrix} -3 & 8 & 6 \\ 2 & \frac{1}{4} & 4 \\ 1 & 3 & 2 \end{bmatrix}$$

A 3×3 square matrix

is a square matrix of size 3×3, or simply of size 3.

APPLIED EXAMPLE 1 Organizing Production Data Consider the matrix

$$P = \begin{bmatrix} 320 & 280 & 460 & 280 \\ 480 & 360 & 580 & 0 \\ 540 & 420 & 200 & 880 \end{bmatrix}$$

representing the output of loudspeaker systems of the Acrosonic Company discussed earlier (see Table 1).

a. What is the size of the matrix P?
b. Find p_{24} (the entry in row 2 and column 4 of the matrix P), and give an interpretation of this number.

c. Find the sum of the entries that make up row 1 of P, and interpret the result.

d. Find the sum of the entries that make up column 4 of P, and interpret the result.

Solution

a. The matrix P has three rows and four columns and hence has size 3×4.

b. The required entry lies in row 2 and column 4 and is the number 0. This means that no model D loudspeaker system was manufactured at Location II in May.

c. The required sum is given by

$$320 + 280 + 460 + 280 = 1340$$

which gives the total number of loudspeaker systems manufactured at Location I in May as 1340 units.

d. The required sum is given by

$$280 + 0 + 880 = 1160$$

giving the output of model D loudspeaker systems at all locations of the company in May as 1160 units. ∎

Equality of Matrices

Two matrices are said to be *equal* if they have the same size and their corresponding entries are equal. For example,

$$\begin{bmatrix} 2 & 3 & 1 \\ 4 & 6 & 2 \end{bmatrix} = \begin{bmatrix} (3-1) & 3 & 1 \\ 4 & (4+2) & 2 \end{bmatrix}$$

Also,

$$\begin{bmatrix} 1 & 3 & 5 \\ 2 & 4 & 3 \end{bmatrix} \neq \begin{bmatrix} 1 & 2 \\ 3 & 4 \\ 5 & 3 \end{bmatrix}$$

since the matrix on the left has size 2×3, whereas the matrix on the right has size 3×2, and

$$\begin{bmatrix} 2 & 3 \\ 4 & 6 \end{bmatrix} \neq \begin{bmatrix} 2 & 3 \\ 4 & 7 \end{bmatrix}$$

since the corresponding elements in row 2 and column 2 of the two matrices are not equal.

> **Equality of Matrices**
>
> Two matrices are equal if they have the same size and their corresponding entries are equal.

VIDEO **EXAMPLE 2** Solve the following matrix equation for x, y, and z:

$$\begin{bmatrix} 1 & x & 3 \\ 2 & y-1 & 2 \end{bmatrix} = \begin{bmatrix} 1 & 4 & z \\ 2 & 1 & 2 \end{bmatrix}$$

Solution   Since the corresponding elements of the two matrices must be equal, we find that $x = 4$, $z = 3$, and $y - 1 = 1$, or $y = 2$. ∎

Addition and Subtraction

Two matrices A and B of the *same size* can be added or subtracted to produce a matrix of the same size. This is done by adding or subtracting the corresponding entries in the two matrices. For example,

$$\begin{bmatrix} 1 & 3 & 4 \\ -1 & 2 & 0 \end{bmatrix} + \begin{bmatrix} 1 & 4 & 3 \\ 6 & 1 & -2 \end{bmatrix} = \begin{bmatrix} 1+1 & 3+4 & 4+3 \\ -1+6 & 2+1 & 0+(-2) \end{bmatrix} = \begin{bmatrix} 2 & 7 & 7 \\ 5 & 3 & -2 \end{bmatrix}$$

Adding two matrices of the same size

and

$$\begin{bmatrix} 1 & 2 \\ -1 & 3 \\ 4 & 0 \end{bmatrix} - \begin{bmatrix} 2 & -1 \\ 3 & 2 \\ -1 & 0 \end{bmatrix} = \begin{bmatrix} 1-2 & 2-(-1) \\ -1-3 & 3-2 \\ 4-(-1) & 0-0 \end{bmatrix} = \begin{bmatrix} -1 & 3 \\ -4 & 1 \\ 5 & 0 \end{bmatrix}$$

Subtracting two matrices of the same size

> **Addition and Subtraction of Matrices**
>
> If A and B are two matrices of the same size, then:
>
> 1. The *sum $A + B$* is the matrix obtained by adding the corresponding entries in the two matrices.
> 2. The *difference $A - B$* is the matrix obtained by subtracting the corresponding entries in B from those in A.

 APPLIED EXAMPLE 3 Organizing Production Data The total output of Acrosonic for June is shown in Table 2.

TABLE 2

	Model A	Model B	Model C	Model D
Location I	210	180	330	180
Location II	400	300	450	40
Location III	420	280	180	740

The output for May was given earlier, in Table 1. Find the total output of the company for May and June.

Solution As we saw earlier, the production matrix for Acrosonic in May is given by

$$A = \begin{bmatrix} 320 & 280 & 460 & 280 \\ 480 & 360 & 580 & 0 \\ 540 & 420 & 200 & 880 \end{bmatrix}$$

Next, from Table 2, we see that the production matrix for June is given by

$$B = \begin{bmatrix} 210 & 180 & 330 & 180 \\ 400 & 300 & 450 & 40 \\ 420 & 280 & 180 & 740 \end{bmatrix}$$

Finally, the total output of Acrosonic for May and June is given by the matrix

$$A + B = \begin{bmatrix} 320 & 280 & 460 & 280 \\ 480 & 360 & 580 & 0 \\ 540 & 420 & 200 & 880 \end{bmatrix} + \begin{bmatrix} 210 & 180 & 330 & 180 \\ 400 & 300 & 450 & 40 \\ 420 & 280 & 180 & 740 \end{bmatrix}$$

$$= \begin{bmatrix} 530 & 460 & 790 & 460 \\ 880 & 660 & 1030 & 40 \\ 960 & 700 & 380 & 1620 \end{bmatrix}$$

The following laws hold for matrix addition.

Laws for Matrix Addition

If A, B, and C are matrices of the same size, then

1. $A + B = B + A$ Commutative law
2. $(A + B) + C = A + (B + C)$ Associative law

The *commutative law* for matrix addition states that the order in which matrix addition is performed is immaterial. The *associative law* states that, when adding three matrices together, we may first add A and B and then add the resulting sum to C. Equivalently, we can add A to the sum of B and C.

EXAMPLE 4 Let

$$A = \begin{bmatrix} 2 & 1 \\ 3 & -2 \\ 1 & 0 \end{bmatrix} \qquad B = \begin{bmatrix} -1 & 2 \\ 3 & 0 \\ 2 & 4 \end{bmatrix} \qquad C = \begin{bmatrix} 1 & 1 \\ 2 & 3 \\ 0 & -1 \end{bmatrix}$$

a. Show that $A + B = B + A$.
b. Show that $(A + B) + C = A + (B + C)$.

Solution

a. $A + B = \begin{bmatrix} 2 & 1 \\ 3 & -2 \\ 1 & 0 \end{bmatrix} + \begin{bmatrix} -1 & 2 \\ 3 & 0 \\ 2 & 4 \end{bmatrix} = \begin{bmatrix} 2 + (-1) & 1 + 2 \\ 3 + 3 & -2 + 0 \\ 1 + 2 & 0 + 4 \end{bmatrix} = \begin{bmatrix} 1 & 3 \\ 6 & -2 \\ 3 & 4 \end{bmatrix}$

On the other hand,

$$B + A = \begin{bmatrix} -1 & 2 \\ 3 & 0 \\ 2 & 4 \end{bmatrix} + \begin{bmatrix} 2 & 1 \\ 3 & -2 \\ 1 & 0 \end{bmatrix} = \begin{bmatrix} -1 + 2 & 2 + 1 \\ 3 + 3 & 0 + -2 \\ 2 + 1 & 4 + 0 \end{bmatrix} = \begin{bmatrix} 1 & 3 \\ 6 & -2 \\ 3 & 4 \end{bmatrix}$$

so $A + B = B + A$, as was to be shown.
b. Using the results of part (a), we have

$$(A + B) + C = \begin{bmatrix} 1 & 3 \\ 6 & -2 \\ 3 & 4 \end{bmatrix} + \begin{bmatrix} 1 & 1 \\ 2 & 3 \\ 0 & -1 \end{bmatrix} = \begin{bmatrix} 2 & 4 \\ 8 & 1 \\ 3 & 3 \end{bmatrix}$$

Next,

$$B + C = \begin{bmatrix} -1 & 2 \\ 3 & 0 \\ 2 & 4 \end{bmatrix} + \begin{bmatrix} 1 & 1 \\ 2 & 3 \\ 0 & -1 \end{bmatrix} = \begin{bmatrix} 0 & 3 \\ 5 & 3 \\ 2 & 3 \end{bmatrix}$$

so

$$A + (B + C) = \begin{bmatrix} 2 & 1 \\ 3 & -2 \\ 1 & 0 \end{bmatrix} + \begin{bmatrix} 0 & 3 \\ 5 & 3 \\ 2 & 3 \end{bmatrix} = \begin{bmatrix} 2 & 4 \\ 8 & 1 \\ 3 & 3 \end{bmatrix}$$

This shows that $(A + B) + C = A + (B + C)$. ∎

A *zero matrix* is one in which all entries are zero. A zero matrix O has the property that

$$A + O = O + A = A$$

for any matrix A having the same size as that of O. For example, the zero matrix of size 3×2 is

$$O = \begin{bmatrix} 0 & 0 \\ 0 & 0 \\ 0 & 0 \end{bmatrix}$$

If A is any 3×2 matrix, then

$$A + O = \begin{bmatrix} a_{11} & a_{12} \\ a_{21} & a_{22} \\ a_{31} & a_{32} \end{bmatrix} + \begin{bmatrix} 0 & 0 \\ 0 & 0 \\ 0 & 0 \end{bmatrix} = \begin{bmatrix} a_{11} & a_{12} \\ a_{21} & a_{22} \\ a_{31} & a_{32} \end{bmatrix} = A$$

where a_{ij} denotes the entry in the ith row and jth column of the matrix A.

The matrix that is obtained by interchanging the rows and columns of a given matrix A is called the *transpose* of A and is denoted A^T. For example, if

$$A = \begin{bmatrix} 1 & 2 & 3 \\ 4 & 5 & 6 \\ 7 & 8 & 9 \end{bmatrix}$$

then

$$A^T = \begin{bmatrix} 1 & 4 & 7 \\ 2 & 5 & 8 \\ 3 & 6 & 9 \end{bmatrix}$$

Transpose of a Matrix

If A is an $m \times n$ matrix with elements a_{ij}, then the **transpose** of A is the $n \times m$ matrix A^T with elements a_{ji}.

Scalar Multiplication

A matrix A may be multiplied by a real number, called a **scalar** in the context of matrix algebra. The scalar product, denoted by cA, is a matrix obtained by multiplying each entry of A by c. For example, the scalar product of the matrix

$$A = \begin{bmatrix} 3 & -1 & 2 \\ 0 & 1 & 4 \end{bmatrix}$$

and the scalar 3 is the matrix

$$3A = 3\begin{bmatrix} 3 & -1 & 2 \\ 0 & 1 & 4 \end{bmatrix} = \begin{bmatrix} 9 & -3 & 6 \\ 0 & 3 & 12 \end{bmatrix}$$

> **Scalar Product**
>
> If A is a matrix and c is a real number, then the **scalar product** cA is the matrix obtained by multiplying each entry of A by c.

VIDEO ▶ **EXAMPLE 5** Given

$$A = \begin{bmatrix} 3 & 4 \\ -1 & 2 \end{bmatrix} \quad \text{and} \quad B = \begin{bmatrix} 3 & 2 \\ -1 & 2 \end{bmatrix}$$

find the matrix X satisfying the *matrix equation* $2X + B = 3A$.

Solution From the given equation $2X + B = 3A$, we find that

$$2X = 3A - B$$
$$= 3\begin{bmatrix} 3 & 4 \\ -1 & 2 \end{bmatrix} - \begin{bmatrix} 3 & 2 \\ -1 & 2 \end{bmatrix}$$
$$= \begin{bmatrix} 9 & 12 \\ -3 & 6 \end{bmatrix} - \begin{bmatrix} 3 & 2 \\ -1 & 2 \end{bmatrix} = \begin{bmatrix} 6 & 10 \\ -2 & 4 \end{bmatrix}$$
$$X = \frac{1}{2}\begin{bmatrix} 6 & 10 \\ -2 & 4 \end{bmatrix} = \begin{bmatrix} 3 & 5 \\ -1 & 2 \end{bmatrix}$$

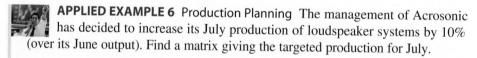 **APPLIED EXAMPLE 6** Production Planning The management of Acrosonic has decided to increase its July production of loudspeaker systems by 10% (over its June output). Find a matrix giving the targeted production for July.

Solution From the results of Example 3, we see that Acrosonic's total output for June may be represented by the matrix

$$B = \begin{bmatrix} 210 & 180 & 330 & 180 \\ 400 & 300 & 450 & 40 \\ 420 & 280 & 180 & 740 \end{bmatrix}$$

The required matrix is given by

$$(1.1)B = 1.1\begin{bmatrix} 210 & 180 & 330 & 180 \\ 400 & 300 & 450 & 40 \\ 420 & 280 & 180 & 740 \end{bmatrix}$$

$$= \begin{bmatrix} 231 & 198 & 363 & 198 \\ 440 & 330 & 495 & 44 \\ 462 & 308 & 198 & 814 \end{bmatrix}$$

and is interpreted in the usual manner.

5.4 Self-Check Exercises

1. Perform the indicated operations:

$$\begin{bmatrix} 1 & 3 & 2 \\ -1 & 4 & 7 \end{bmatrix} - 3\begin{bmatrix} 2 & 1 & 0 \\ 1 & 3 & 4 \end{bmatrix}$$

2. Solve the following matrix equation for x, y, and z:

$$\begin{bmatrix} x & 3 \\ z & 2 \end{bmatrix} + \begin{bmatrix} 2 - y & z \\ 2 - z & -x \end{bmatrix} = \begin{bmatrix} 3 & 7 \\ 2 & 0 \end{bmatrix}$$

3. Jack owns two gas stations, one downtown and the other in the Wilshire district. Over two consecutive days, his gas stations recorded gasoline sales represented by the following matrices:

		Regular	Regular plus	Premium
$A =$	Downtown	1200	750	650
	Wilshire	1100	850	600

and

		Regular	Regular plus	Premium
$B =$	Downtown	1250	825	550
	Wilshire	1150	750	750

Find a matrix representing the total sales of the two gas stations over the 2-day period.

Solutions to Self-Check Exercises 5.4 can be found on page 292.

5.4 Concept Questions

1. Define (a) a matrix, (b) the size of a matrix, (c) a row matrix, (d) a column matrix, and (e) a square matrix.

2. When are two matrices equal? Give an example of two matrices that are equal.

3. Construct a 3×3 matrix A having the property that $A = A^T$. What special characteristic does A have?

5.4 Exercises

In Exercises 1–6, refer to the following matrices:

$$A = \begin{bmatrix} 2 & -3 & 9 & -4 \\ -11 & 2 & 6 & 7 \\ 6 & 0 & 2 & 9 \\ 5 & 1 & 5 & -8 \end{bmatrix}$$

$$B = \begin{bmatrix} 3 & -1 & 2 \\ 0 & 1 & 4 \\ 3 & 2 & 1 \\ -1 & 0 & 8 \end{bmatrix}$$

$$C = \begin{bmatrix} 1 & 0 & 3 & 4 & 5 \end{bmatrix}$$

$$D = \begin{bmatrix} 1 \\ 3 \\ -2 \\ 0 \end{bmatrix}$$

1. What is the size of A? Of B? Of C? Of D?

2. Find a_{14}, a_{21}, a_{31}, and a_{43}.

3. Find b_{13}, b_{31}, and b_{43}.

4. Identify the row matrix. What is its transpose?

5. Identify the column matrix. What is its transpose?

6. Identify the square matrix. What is its transpose?

In Exercises 7–12, refer to the following matrices:

$$A = \begin{bmatrix} -1 & 2 \\ 3 & -2 \\ 4 & 0 \end{bmatrix} \quad B = \begin{bmatrix} 2 & 4 \\ 3 & 1 \\ -2 & 2 \end{bmatrix}$$

$$C = \begin{bmatrix} 3 & -1 & 0 \\ 2 & -2 & 3 \\ 4 & 6 & 2 \end{bmatrix} \quad D = \begin{bmatrix} 2 & -2 & 4 \\ 3 & 6 & 2 \\ -2 & 3 & 1 \end{bmatrix}$$

7. What is the size of A? Of B? Of C? Of D?

8. Explain why the matrix $A + C$ does *not* exist.

9. Compute $A + B$.

10. Compute $2A - 3B$.

11. Compute $C - D$.

12. Compute $4D - 2C$.

In Exercises 13–20, perform the indicated operations.

13. $\begin{bmatrix} 6 & 3 & 8 \\ 4 & 5 & 6 \end{bmatrix} - \begin{bmatrix} 1 & -2 & -1 \\ 2 & -5 & -7 \end{bmatrix}$

14. $\begin{bmatrix} 2 & -3 & 4 & -1 \\ 3 & 1 & 0 & 0 \end{bmatrix} + \begin{bmatrix} 4 & 3 & -2 & -4 \\ 6 & 2 & 0 & -3 \end{bmatrix}$

15. $\begin{bmatrix} 1 & 4 & -5 \\ 3 & -8 & 6 \end{bmatrix} + \begin{bmatrix} 4 & 0 & -2 \\ 3 & 6 & 5 \end{bmatrix} - \begin{bmatrix} 2 & 8 & 9 \\ -11 & 2 & -5 \end{bmatrix}$

16. $3\begin{bmatrix} 1 & 1 & -3 \\ 3 & 2 & 3 \\ 7 & -1 & 6 \end{bmatrix} + 4\begin{bmatrix} -2 & -1 & 8 \\ 4 & 2 & 2 \\ 3 & 6 & 3 \end{bmatrix}$

17. $\begin{bmatrix} 1.2 & 4.5 & -4.2 \\ 8.2 & 6.3 & -3.2 \end{bmatrix} - \begin{bmatrix} 3.1 & 1.5 & -3.6 \\ 2.2 & -3.3 & -4.4 \end{bmatrix}$

18. $\begin{bmatrix} 0.06 & 0.12 \\ 0.43 & 1.11 \\ 1.55 & -0.43 \end{bmatrix} - \begin{bmatrix} 0.77 & -0.75 \\ 0.22 & -0.65 \\ 1.09 & -0.57 \end{bmatrix}$

19. $\dfrac{1}{2}\begin{bmatrix} 1 & 0 & 0 & -4 \\ 3 & 0 & -1 & 6 \\ -2 & 1 & -4 & 2 \end{bmatrix} + \dfrac{4}{3}\begin{bmatrix} 3 & 0 & -1 & 4 \\ -2 & 1 & -6 & 2 \\ 8 & 2 & 0 & -2 \end{bmatrix}$
$- \dfrac{1}{3}\begin{bmatrix} 3 & -9 & -1 & 0 \\ 6 & 2 & 0 & -6 \\ 0 & 1 & -3 & 1 \end{bmatrix}$

20. $0.5\begin{bmatrix} 1 & 3 & 5 \\ 5 & 2 & -1 \\ -2 & 0 & 1 \end{bmatrix} - 0.2\begin{bmatrix} 2 & 3 & 4 \\ -1 & 1 & -4 \\ 3 & 5 & -5 \end{bmatrix}$
$+ 0.6\begin{bmatrix} 3 & 4 & -1 \\ 4 & 5 & 1 \\ 1 & 0 & 0 \end{bmatrix}$

In Exercises 21–24, solve for u, x, y, and z in the given matrix equation.

21. $\begin{bmatrix} 2x - 2 & 3 & 2 \\ 2 & 4 & y - 2 \\ 2z & -3 & 2 \end{bmatrix} = \begin{bmatrix} 3 & u & 2 \\ 2 & 4 & 5 \\ 4 & -3 & 2 \end{bmatrix}$

22. $\begin{bmatrix} x & -2 \\ 3 & y \end{bmatrix} + \begin{bmatrix} -2 & z \\ -1 & 2 \end{bmatrix} = \begin{bmatrix} 4 & -2 \\ 2u & 4 \end{bmatrix}$

23. $\begin{bmatrix} 1 & x \\ 2y & -3 \end{bmatrix} - 4\begin{bmatrix} 2 & -2 \\ 0 & 3 \end{bmatrix} = \begin{bmatrix} 3z & 10 \\ 4 & -u \end{bmatrix}$

24. $\begin{bmatrix} 1 & 2 \\ 3 & 4 \\ x & -1 \end{bmatrix} - 3\begin{bmatrix} y - 1 & 2 \\ 1 & 2 \\ 4 & 2z + 1 \end{bmatrix} = 2\begin{bmatrix} -4 & -u \\ 0 & -1 \\ 4 & 4 \end{bmatrix}$

In Exercises 25 and 26, let

$$A = \begin{bmatrix} 2 & -4 & 3 \\ 4 & 2 & 1 \end{bmatrix} \qquad B = \begin{bmatrix} 4 & 3 & 2 \\ 1 & 0 & 4 \end{bmatrix}$$

$$C = \begin{bmatrix} 1 & 0 & 2 \\ 3 & -2 & 1 \end{bmatrix}$$

25. Verify by direct computation the validity of the commutative law for matrix addition.

26. Verify by direct computation the validity of the associative law for matrix addition.

In Exercises 27–30, let

$$A = \begin{bmatrix} 3 & 1 \\ 2 & 4 \\ -4 & 0 \end{bmatrix} \quad \text{and} \quad B = \begin{bmatrix} 1 & 2 \\ -1 & 0 \\ 3 & 2 \end{bmatrix}$$

Verify each equation by direct computation.

27. $(3 + 5)A = 3A + 5A$ **28.** $2(4A) = (2 \cdot 4)A = 8A$

29. $4(A + B) = 4A + 4B$ **30.** $2(A - 3B) = 2A - 6B$

In Exercises 31–34, find the transpose of each matrix.

31. $\begin{bmatrix} 3 & 2 & -1 & 5 \end{bmatrix}$ **32.** $\begin{bmatrix} 4 & 2 & 0 & -1 \\ 3 & 4 & -1 & 5 \end{bmatrix}$

33. $\begin{bmatrix} 1 & -1 & 2 \\ 3 & 4 & 2 \\ 0 & 1 & 0 \end{bmatrix}$ **34.** $\begin{bmatrix} 1 & 2 & 6 & 4 \\ 2 & 3 & 2 & 5 \\ 6 & 2 & 3 & 0 \\ 4 & 5 & 0 & 2 \end{bmatrix}$

35. **CHOLESTEROL LEVELS** Mr. Cross, Mr. Jones, and Mr. Smith all suffer from coronary heart disease. As part of their treatment, they were put on special low-cholesterol diets: Cross on Diet I, Jones on Diet II, and Smith on Diet III. Progressive records of each patient's cholesterol level were kept. At the beginning of the first, second, third, and fourth months, the cholesterol levels of the three patients were:

Cross: 220, 215, 210, and 205
Jones: 220, 210, 200, and 195
Smith: 215, 205, 195, and 190

Represent this information in a 3×4 matrix.

36. **INVESTMENT PORTFOLIOS** The following table gives the number of shares of certain corporations held by Leslie and Tom in their respective IRA accounts at the beginning of the year:

	IBM	GE	Ford	Wal-Mart
Leslie	500	350	200	400
Tom	400	450	300	200

Over the year, they added more shares to their accounts, as shown in the following table:

	IBM	GE	Ford	Wal-Mart
Leslie	50	50	0	100
Tom	0	80	100	50

a. Write a matrix A giving the holdings of Leslie and Tom at the beginning of the year and a matrix B giving the shares they have added to their portfolios.
b. Find a matrix C giving their total holdings at the end of the year.

37. HOME SALES K & R Builders build three models of houses, M_1, M_2, and M_3, in three subdivisions I, II, and III located in three different areas of a city. The prices of the houses (in thousands of dollars) are given in matrix A:

$$A = \begin{array}{c} \text{I} \\ \text{II} \\ \text{III} \end{array} \begin{bmatrix} 340 & 360 & 380 \\ 410 & 430 & 440 \\ 620 & 660 & 700 \end{bmatrix} \begin{array}{ccc} M_1 & M_2 & M_3 \end{array}$$

K & R Builders has decided to raise the price of each house by 3% next year. Write a matrix B giving the new prices of the houses.

38. HOME SALES K & R Builders build three models of houses, M_1, M_2, and M_3, in three subdivisions I, II, and III located in three different areas of a city. The prices of the homes (in thousands of dollars) are given in matrix A:

$$A = \begin{array}{c} \text{I} \\ \text{II} \\ \text{III} \end{array} \begin{bmatrix} 340 & 360 & 380 \\ 410 & 430 & 440 \\ 620 & 660 & 700 \end{bmatrix} \begin{array}{ccc} M_1 & M_2 & M_3 \end{array}$$

The new price schedule for next year, reflecting a uniform percentage increase in each house, is given by matrix B:

$$B = \begin{array}{c} \text{I} \\ \text{II} \\ \text{III} \end{array} \begin{bmatrix} 357 & 378 & 399 \\ 430.5 & 451.5 & 462 \\ 651 & 693 & 735 \end{bmatrix} \begin{array}{ccc} M_1 & M_2 & M_3 \end{array}$$

What was the percentage increase in the prices of the houses?
Hint: Find r such that $(1 + 0.01r)A = B$.

39. BANKING The numbers of three types of bank accounts on January 1 at the Central Bank and its branches are represented by matrix A:

	Checking accounts	Savings accounts	Fixed-deposit accounts
Main office	2820	1470	1120
A = Westside branch	1030	520	480
Eastside branch	1170	540	460

The number and types of accounts opened during the first quarter are represented by matrix B, and the number and types of accounts closed during the same period are represented by matrix C. Thus,

$$B = \begin{bmatrix} 260 & 120 & 110 \\ 140 & 60 & 50 \\ 120 & 70 & 50 \end{bmatrix} \quad \text{and} \quad C = \begin{bmatrix} 120 & 80 & 80 \\ 70 & 30 & 40 \\ 60 & 20 & 40 \end{bmatrix}$$

a. Find matrix D, which represents the number of each type of account at the end of the first quarter at each location.
b. Because a new manufacturing plant is opening in the immediate area, it is anticipated that there will be a 10% increase in the number of accounts at each location during the second quarter. Write a matrix $E = 1.1D$ to reflect this anticipated increase.

40. BOOKSTORE INVENTORIES The Campus Bookstore's inventory of books is

Hardcover: textbooks, 5280; fiction, 1680; nonfiction, 2320; reference, 1890

Paperback: fiction, 2810; nonfiction, 1490; reference, 2070; textbooks, 1940

The College Bookstore's inventory of books is

Hardcover: textbooks, 6340; fiction, 2220; nonfiction, 1790; reference, 1980

Paperback: fiction, 3100; nonfiction, 1720; reference, 2710; textbooks, 2050

a. Represent Campus's inventory as a matrix A.
b. Represent College's inventory as a matrix B.
c. The two companies decide to merge, so now write a matrix C that represents the total inventory of the newly amalgamated company.

41. INSURANCE CLAIMS The property damage claim frequencies per 100 cars in Massachusetts in the years 2000, 2001, and 2002 were 6.88, 7.05, and 7.18, respectively. The corresponding claim frequencies in the United States were 4.13, 4.09, and 4.06, respectively. Express this information using a 2 × 3 matrix.
Sources: Registry of Motor Vehicles; Federal Highway Administration.

42. MORTALITY RATES Mortality actuarial tables in the United States were revised in 2001, the fourth time since 1858. On the basis of the new life insurance mortality rates, 1% of 60-year-old men, 2.6% of 70-year-old men, 7% of 80-year-old men, 18.8% of 90-year-old men, and 36.3% of 100-year-old men would die within a year. The corresponding rates for women are 0.8%, 1.8%, 4.4%, 12.2%, and 27.6%, respectively. Express this information using a 2 × 5 matrix.
Source: Society of Actuaries.

43. **LIFE EXPECTANCY** Figures for life expectancy at birth of Massachusetts residents in 2008 are 82.6, 80.5, and 91.2 years for white, black, and Hispanic women, respectively, and 78.0, 73.9, and 84.8 years for white, black, and Hispanic men, respectively. Express this information using a 2×3 matrix and a 3×2 matrix.

    Source: Massachusetts Department of Public Health.

44. **MARKET SHARE OF MOTORCYCLES** The market share of motorcycles in the United States in 2001 follows: Honda 27.9%, Harley-Davidson 21.9%, Yamaha 19.2%, Suzuki 11.0%, Kawasaki 9.1%, and others 10.9%. The corresponding figures for 2002 are 27.6%, 23.3%, 18.2%, 10.5%, 8.8%, and 11.6%, respectively. Express this information using a 2×6 matrix. What is the sum of all the elements in the first row? In the second row? Is this expected?

Which company gained the most market share between 2001 and 2002?

Source: Motorcycle Industry Council.

In Exercises 45–48, determine whether the statement is true or false. If it is true, explain why it is true. If it is false, give an example to show why it is false.

45. If A and B are matrices of the same size and c is a scalar, then $c(A + B) = cA + cB$.

46. If A and B are matrices of the same size, then $A - B = A + (-1)B$.

47. If A is a matrix and c is a nonzero scalar, then $(cA)^T = (1/c)A^T$.

48. If A is a matrix, then $(A^T)^T = A$.

5.4 Solutions to Self-Check Exercises

1. $\begin{bmatrix} 1 & 3 & 2 \\ -1 & 4 & 7 \end{bmatrix} - 3\begin{bmatrix} 2 & 1 & 0 \\ 1 & 3 & 4 \end{bmatrix} = \begin{bmatrix} 1 & 3 & 2 \\ -1 & 4 & 7 \end{bmatrix} - \begin{bmatrix} 6 & 3 & 0 \\ 3 & 9 & 12 \end{bmatrix}$

   $= \begin{bmatrix} -5 & 0 & 2 \\ -4 & -5 & -5 \end{bmatrix}$

2. We are given

   $\begin{bmatrix} x & 3 \\ z & 2 \end{bmatrix} + \begin{bmatrix} 2 - y & z \\ 2 - z & -x \end{bmatrix} = \begin{bmatrix} 3 & 7 \\ 2 & 0 \end{bmatrix}$

   Performing the indicated operation on the left-hand side, we obtain

   $\begin{bmatrix} 2 + x - y & 3 + z \\ 2 & 2 - x \end{bmatrix} = \begin{bmatrix} 3 & 7 \\ 2 & 0 \end{bmatrix}$

By the equality of matrices, we have

$$2 + x - y = 3$$
$$3 + z = 7$$
$$2 - x = 0$$

from which we deduce that $x = 2$, $y = 1$, and $z = 4$.

3. The required matrix is

   $A + B = \begin{bmatrix} 1200 & 750 & 650 \\ 1100 & 850 & 600 \end{bmatrix} + \begin{bmatrix} 1250 & 825 & 550 \\ 1150 & 750 & 750 \end{bmatrix}$

   $= \begin{bmatrix} 2450 & 1575 & 1200 \\ 2250 & 1600 & 1350 \end{bmatrix}$

USING TECHNOLOGY

Matrix Operations

Graphing Utility

A graphing utility can be used to perform matrix addition, matrix subtraction, and scalar multiplication. It can also be used to find the transpose of a matrix.

EXAMPLE 1 Let

$$A = \begin{bmatrix} 1.2 & 3.1 \\ -2.1 & 4.2 \\ 3.1 & 4.8 \end{bmatrix} \quad \text{and} \quad B = \begin{bmatrix} 4.1 & 3.2 \\ 1.3 & 6.4 \\ 1.7 & 0.8 \end{bmatrix}$$

Find (a) $A + B$, (b) $2.1A - 3.2B$, and (c) $(2.1A + 3.2B)^T$.

Solution We first enter the matrices A and B into the calculator.

a. Using matrix operations, we enter the expression $A + B$ and obtain

$$A + B = \begin{bmatrix} 5.3 & 6.3 \\ -0.8 & 10.6 \\ 4.8 & 5.6 \end{bmatrix}$$

b. Using matrix operations, we enter the expression $2.1A - 3.2B$ and obtain

$$2.1A - 3.2B = \begin{bmatrix} -10.6 & -3.73 \\ -8.57 & -11.66 \\ 1.07 & 7.52 \end{bmatrix}$$

c. Using matrix operations, we enter the expression $(2.1A + 3.2B)^T$ and obtain

$$(2.1A + 3.2B)^T = \begin{bmatrix} 15.64 & -0.25 & 11.95 \\ 16.75 & 29.3 & 12.64 \end{bmatrix}$$

APPLIED EXAMPLE 2 John operates three gas stations at three locations, I, II, and III. Over two consecutive days, his gas stations recorded the following fuel sales (in gallons):

Day 1

	Regular	Regular Plus	Premium	Diesel
Location I	1400	1200	1100	200
Location II	1600	900	1200	300
Location III	1200	1500	800	500

Day 2

	Regular	Regular Plus	Premium	Diesel
Location I	1000	900	800	150
Location II	1800	1200	1100	250
Location III	800	1000	700	400

Find a matrix representing the total fuel sales at John's gas stations.

Solution The fuel sales can be represented by the matrix A (day 1) and matrix B (day 2):

$$A = \begin{bmatrix} 1400 & 1200 & 1100 & 200 \\ 1600 & 900 & 1200 & 300 \\ 1200 & 1500 & 800 & 500 \end{bmatrix} \quad \text{and} \quad B = \begin{bmatrix} 1000 & 900 & 800 & 150 \\ 1800 & 1200 & 1100 & 250 \\ 800 & 1000 & 700 & 400 \end{bmatrix}$$

We enter the matrices A and B into the calculator. Using matrix operations, we enter the expression $A + B$ and obtain

$$A + B = \begin{bmatrix} 2400 & 2100 & 1900 & 350 \\ 3400 & 2100 & 2300 & 550 \\ 2000 & 2500 & 1500 & 900 \end{bmatrix}$$

Excel

First, we show how basic operations on matrices can be carried out by using Excel.

EXAMPLE 3 Given the following matrices,

$$A = \begin{bmatrix} 1.2 & 3.1 \\ -2.1 & 4.2 \\ 3.1 & 4.8 \end{bmatrix} \quad \text{and} \quad B = \begin{bmatrix} 4.1 & 3.2 \\ 1.3 & 6.4 \\ 1.7 & 0.8 \end{bmatrix}$$

a. Compute $A + B$.   **b.** Compute $2.1A - 3.2B$.

(continued)

Solution

a. *First, represent the matrices A and B in a spreadsheet.* Enter the elements of each matrix in a block of cells as shown in Figure T1.

	A	B	C	D	E
1		A			B
2	1.2	3.1		4.1	3.2
3	-2.1	4.2		1.3	6.4
4	3.1	4.8		1.7	0.8

FIGURE T1
The elements of matrix A and matrix B in a spreadsheet

Second, compute the sum of matrix A and matrix B. Highlight the cells that will contain matrix $A + B$, type =, highlight the cells in matrix A, type +, highlight the cells in matrix B, and press $\boxed{\textbf{Ctrl-Shift-Enter}}$. The resulting matrix $A + B$ is shown in Figure T2.

	A	B
8		A + B
9	5.3	6.3
10	-0.8	10.6
11	4.8	5.6

FIGURE T2
The matrix $A + B$

b. *Highlight the cells that will contain matrix* $2.1A - 3.2B$. *Type* = 2.1*, highlight matrix A, type −3.2*, highlight the cells in matrix B, and press $\boxed{\textbf{Ctrl-Shift-Enter}}$. The resulting matrix $2.1A - 3.2B$ is shown in Figure T3.

	A	B
13		2.1A - 3.2B
14	-10.6	-3.73
15	-8.57	-11.66
16	1.07	7.52

FIGURE T3
The matrix $2.1A - 3.2B$

APPLIED EXAMPLE 4 John operates three gas stations at three locations, I, II, and III. Over two consecutive days, his gas stations recorded the following fuel sales (in gallons):

	Day 1			
	Regular	**Regular Plus**	**Premium**	**Diesel**
Location I	1400	1200	1100	200
Location II	1600	900	1200	300
Location III	1200	1500	800	500

	Day 2			
	Regular	**Regular Plus**	**Premium**	**Diesel**
Location I	1000	900	800	150
Location II	1800	1200	1100	250
Location III	800	1000	700	400

Find a matrix representing the total fuel sales at John's gas stations.

Note: Boldfaced words/characters enclosed in a box (for example, $\boxed{\textbf{Enter}}$) indicate that an action (click, select, or press) is required. Words/characters printed blue (for example, Chart sub-type:) indicate words/characters that appear on the screen. Words/characters printed in a monospace font (for example, = (−2/3) *A2+2) indicate words/characters that need to be typed and entered.

Solution The fuel sales can be represented by the matrices A (day 1) and B (day 2):

$$A = \begin{bmatrix} 1400 & 1200 & 1100 & 200 \\ 1600 & 900 & 1200 & 300 \\ 1200 & 1500 & 800 & 500 \end{bmatrix} \quad \text{and} \quad B = \begin{bmatrix} 1000 & 900 & 800 & 150 \\ 1800 & 1200 & 1100 & 250 \\ 800 & 1000 & 700 & 400 \end{bmatrix}$$

We first enter the elements of the matrices A and B onto a spreadsheet. Next, we highlight the cells that will contain the matrix $A + B$, type =, highlight A, type +, highlight B, and then press **Ctrl-Shift-Enter**. The resulting matrix $A + B$ is shown in Figure T4.

	A	B	C	D
23		A + B		
24	2400	2100	1900	350
25	3400	2100	2300	550
26	2000	2500	1500	900

FIGURE **T4**
The matrix $A + B$

TECHNOLOGY EXERCISES

Refer to the following matrices and perform the indicated operations.

$$A = \begin{bmatrix} 1.2 & 3.1 & -5.4 & 2.7 \\ 4.1 & 3.2 & 4.2 & -3.1 \\ 1.7 & 2.8 & -5.2 & 8.4 \end{bmatrix}$$

$$B = \begin{bmatrix} 6.2 & -3.2 & 1.4 & -1.2 \\ 3.1 & 2.7 & -1.2 & 1.7 \\ 1.2 & -1.4 & -1.7 & 2.8 \end{bmatrix}$$

1. $12.5A$
2. $-8.4B$
3. $A - B$
4. $B - A$
5. $1.3A + 2.4B$
6. $2.1A - 1.7B$
7. $3(A + B)$
8. $1.3(4.1A - 2.3B)$

5.5 Multiplication of Matrices

Matrix Product

In Section 5.4, we saw how matrices of the same size may be added or subtracted and how a matrix may be multiplied by a scalar (real number), an operation referred to as scalar multiplication. In this section we see how, with certain restrictions, one matrix may be multiplied by another matrix.

To define matrix multiplication, let's consider the following problem. On a certain day, Al's Service Station sold 1600 gallons of regular, 1000 gallons of regular plus, and 800 gallons of premium gasoline. If the price of gasoline on this day was $3.59 for regular, $3.79 for regular plus, and $3.95 for premium gasoline, find the total revenue realized by Al's for that day.

The day's sale of gasoline may be represented by the matrix

$$A = \begin{bmatrix} 1600 & 1000 & 800 \end{bmatrix} \quad \text{Row matrix } (1 \times 3)$$

Next, we let the unit selling price of regular, regular plus, and premium gasoline be the entries in the matrix

$$B = \begin{bmatrix} 3.59 \\ 3.79 \\ 3.95 \end{bmatrix} \quad \text{Column matrix } (3 \times 1)$$

The first entry in matrix A gives the number of gallons of regular gasoline sold, and the first entry in matrix B gives the selling price for each gallon of regular gasoline, so their product $(1600)(3.59)$ gives the revenue realized from the sale of regular gasoline for the day. A similar interpretation of the second and third entries in the two matrices suggests that we multiply the corresponding entries to obtain the respective revenues realized from the sale of regular, regular plus, and premium gasoline. Finally, the total revenue realized by Al's from the sale of gasoline is given by adding these products to obtain

$$(1600)(3.59) + (1000)(3.79) + (800)(3.95) = 12{,}694$$

or $12,694.

This example suggests that if we have a row matrix of size $1 \times n$,

$$A = \begin{bmatrix} a_1 & a_2 & a_3 & \cdots & a_n \end{bmatrix}$$

and a column matrix of size $n \times 1$,

$$B = \begin{bmatrix} b_1 \\ b_2 \\ b_3 \\ \vdots \\ b_n \end{bmatrix}$$

then we may define the **matrix product** of A and B, written AB, by

$$AB = \begin{bmatrix} a_1 & a_2 & a_3 & \cdots & a_n \end{bmatrix} \begin{bmatrix} b_1 \\ b_2 \\ b_3 \\ \vdots \\ b_n \end{bmatrix} = a_1b_1 + a_2b_2 + a_3b_3 + \cdots + a_nb_n \qquad \textbf{(12)}$$

EXAMPLE 1 Let

$$A = \begin{bmatrix} 1 & -2 & 3 & 5 \end{bmatrix} \quad \text{and} \quad B = \begin{bmatrix} 2 \\ 3 \\ 0 \\ -1 \end{bmatrix}$$

Then

$$AB = \begin{bmatrix} 1 & -2 & 3 & 5 \end{bmatrix} \begin{bmatrix} 2 \\ 3 \\ 0 \\ -1 \end{bmatrix} = (1)(2) + (-2)(3) + (3)(0) + (5)(-1) = -9$$

 APPLIED EXAMPLE 2 Stock Transactions　Judy's stock holdings are given by the matrix

$$\begin{array}{ccc} \text{GM} & \text{IBM} & \text{BAC} \end{array}$$
$$A = \begin{bmatrix} 700 & 400 & 200 \end{bmatrix}$$

At the close of trading on a certain day, the prices (in dollars per share) of these stocks are

$$B = \begin{array}{c} \text{GM} \\ \text{IBM} \\ \text{BAC} \end{array} \begin{bmatrix} 50 \\ 120 \\ 42 \end{bmatrix}$$

What is the total value of Judy's holdings as of that day?

Solution   Judy's holdings are worth

$$AB = \begin{bmatrix} 700 & 400 & 200 \end{bmatrix} \begin{bmatrix} 50 \\ 120 \\ 42 \end{bmatrix} = (700)(50) + (400)(120) + (200)(42)$$

$$= 91,400$$

or $91,400.

Returning once again to the matrix product AB in Equation (12), observe that the number of columns of the row matrix A is *equal* to the number of rows of the column matrix B. Observe further that the product matrix AB has size 1×1 (a real number may be thought of as a 1×1 matrix). Schematically,

Size of A Size of B

$1 \times n$ $n \times 1$

(1×1)

Size of AB

More generally, if A is a matrix of size $m \times n$ and B is a matrix of size $n \times p$ (the number of columns of A equals the numbers of rows of B), then the *matrix product* of A and B, AB, is defined and is a matrix of size $m \times p$. Schematically,

Size of A Size of B

$m \times n$ $n \times p$

$(m \times p)$

Size of AB

Next, let's illustrate the mechanics of matrix multiplication by computing the product of a 2×3 matrix A and a 3×4 matrix B. Suppose

$$A = \begin{bmatrix} a_{11} & a_{12} & a_{13} \\ a_{21} & a_{22} & a_{23} \end{bmatrix}$$

$$B = \begin{bmatrix} b_{11} & b_{12} & b_{13} & b_{14} \\ b_{21} & b_{22} & b_{23} & b_{24} \\ b_{31} & b_{32} & b_{33} & b_{34} \end{bmatrix}$$

From the schematic

Same

Size of A 2×3 3×4 Size of B

(2×4)

Size of AB

we see that the matrix product $C = AB$ is defined (since the number of columns of A equals the number of rows of B) and has size 2×4. Thus,

$$C = \begin{bmatrix} c_{11} & c_{12} & c_{13} & c_{14} \\ c_{21} & c_{22} & c_{23} & c_{24} \end{bmatrix}$$

The entries of C are computed as follows: The entry c_{11} (the entry in the *first* row, *first* column of C) is the product of the row matrix composed of the entries from the *first* row of A and the column matrix composed of the *first* column of B. Thus,

$$c_{11} = \begin{bmatrix} a_{11} & a_{12} & a_{13} \end{bmatrix} \begin{bmatrix} b_{11} \\ b_{21} \\ b_{31} \end{bmatrix} = a_{11}b_{11} + a_{12}b_{21} + a_{13}b_{31}$$

The entry c_{12} (the entry in the *first* row, *second* column of C) is the product of the row matrix composed of the *first* row of A and the column matrix composed of the *second* column of B. Thus,

$$c_{12} = \begin{bmatrix} a_{11} & a_{12} & a_{13} \end{bmatrix} \begin{bmatrix} b_{12} \\ b_{22} \\ b_{32} \end{bmatrix} = a_{11}b_{12} + a_{12}b_{22} + a_{13}b_{32}$$

The other entries in C are computed in a similar manner.

VIDEO   **EXAMPLE 3** Let

$$A = \begin{bmatrix} 3 & 1 & 4 \\ -1 & 2 & 3 \end{bmatrix} \quad \text{and} \quad B = \begin{bmatrix} 1 & 3 & -3 \\ 4 & -1 & 2 \\ 2 & 4 & 1 \end{bmatrix}$$

Compute AB.

Solution   The size of matrix A is 2×3, and the size of matrix B is 3×3. Since the number of columns of matrix A is equal to the number of rows of matrix B, the matrix product $C = AB$ is defined. Furthermore, the size of matrix C is 2×3. Thus,

$$\begin{bmatrix} 3 & 1 & 4 \\ -1 & 2 & 3 \end{bmatrix} \begin{bmatrix} 1 & 3 & -3 \\ 4 & -1 & 2 \\ 2 & 4 & 1 \end{bmatrix} = \begin{bmatrix} c_{11} & c_{12} & c_{13} \\ c_{21} & c_{22} & c_{23} \end{bmatrix}$$

It remains now to determine the entries $c_{11}, c_{12}, c_{13}, c_{21}, c_{22}$, and c_{23}. We have

$$c_{11} = \begin{bmatrix} 3 & 1 & 4 \end{bmatrix} \begin{bmatrix} 1 \\ 4 \\ 2 \end{bmatrix} = (3)(1) + (1)(4) + (4)(2) = 15$$

$$c_{12} = \begin{bmatrix} 3 & 1 & 4 \end{bmatrix} \begin{bmatrix} 3 \\ -1 \\ 4 \end{bmatrix} = (3)(3) + (1)(-1) + (4)(4) = 24$$

$$c_{13} = \begin{bmatrix} 3 & 1 & 4 \end{bmatrix} \begin{bmatrix} -3 \\ 2 \\ 1 \end{bmatrix} = (3)(-3) + (1)(2) + (4)(1) = -3$$

$$c_{21} = \begin{bmatrix} -1 & 2 & 3 \end{bmatrix} \begin{bmatrix} 1 \\ 4 \\ 2 \end{bmatrix} = (-1)(1) + (2)(4) + (3)(2) = 13$$

$$c_{22} = \begin{bmatrix} -1 & 2 & 3 \end{bmatrix} \begin{bmatrix} 3 \\ -1 \\ 4 \end{bmatrix} = (-1)(3) + (2)(-1) + (3)(4) = 7$$

$$c_{23} = \begin{bmatrix} -1 & 2 & 3 \end{bmatrix} \begin{bmatrix} -3 \\ 2 \\ 1 \end{bmatrix} = (-1)(-3) + (2)(2) + (3)(1) = 10$$

so the required product AB is given by

$$AB = \begin{bmatrix} 15 & 24 & -3 \\ 13 & 7 & 10 \end{bmatrix}$$

EXAMPLE 4 Let

$$A = \begin{bmatrix} 3 & 2 & 1 \\ -1 & 2 & 3 \\ 3 & 1 & 4 \end{bmatrix} \quad \text{and} \quad B = \begin{bmatrix} 1 & 3 & 4 \\ 2 & 4 & 1 \\ -1 & 2 & 3 \end{bmatrix}$$

Then

$$AB = \begin{bmatrix} 3 \cdot 1 & + 2 \cdot 2 + 1 \cdot (-1) & 3 \cdot 3 & + 2 \cdot 4 + 1 \cdot 2 & 3 \cdot 4 & + 2 \cdot 1 + 1 \cdot 3 \\ (-1) \cdot 1 + 2 \cdot 2 + 3 \cdot (-1) & (-1) \cdot 3 + 2 \cdot 4 + 3 \cdot 2 & (-1) \cdot 4 + 2 \cdot 1 + 3 \cdot 3 \\ 3 \cdot 1 & + 1 \cdot 2 + 4 \cdot (-1) & 3 \cdot 3 & + 1 \cdot 4 + 4 \cdot 2 & 3 \cdot 4 & + 1 \cdot 1 + 4 \cdot 3 \end{bmatrix}$$

$$= \begin{bmatrix} 6 & 19 & 17 \\ 0 & 11 & 7 \\ 1 & 21 & 25 \end{bmatrix}$$

$$BA = \begin{bmatrix} 1 \cdot 3 & + 3 \cdot (-1) + 4 \cdot 3 & 1 \cdot 2 & + 3 \cdot 2 + 4 \cdot 1 & 1 \cdot 1 & + 3 \cdot 3 + 4 \cdot 4 \\ 2 \cdot 3 & + 4 \cdot (-1) + 1 \cdot 3 & 2 \cdot 2 & + 4 \cdot 2 + 1 \cdot 1 & 2 \cdot 1 & + 4 \cdot 3 + 1 \cdot 4 \\ (-1) \cdot 3 + 2 \cdot (-1) + 3 \cdot 3 & (-1) \cdot 2 + 2 \cdot 2 + 3 \cdot 1 & (-1) \cdot 1 + 2 \cdot 3 + 3 \cdot 4 \end{bmatrix}$$

$$= \begin{bmatrix} 12 & 12 & 26 \\ 5 & 13 & 18 \\ 4 & 5 & 17 \end{bmatrix}$$

The preceding example shows that, in general, $AB \neq BA$ for two square matrices A and B. However, the following laws are valid for matrix multiplication.

Laws for Matrix Multiplication

If the products and sums are defined for the matrices A, B, and C, then

1. $(AB)C = A(BC)$ Associative law

2. $A(B + C) = AB + AC$ Distributive law

The square matrix of size n having 1s along the main diagonal and 0s elsewhere is called the identity matrix of size n.

Identity Matrix

The **identity matrix** of size n is given by

$$I_n = \begin{bmatrix} 1 & 0 & \cdots & & & 0 \\ 0 & 1 & & \cdots & & 0 \\ & & \cdot & & & \\ \cdot & \cdot & & \cdot & \cdot & \cdot \\ & & & & \cdot & \\ 0 & 0 & \cdots & & & 1 \end{bmatrix} \quad n \text{ rows}$$

n columns

The identity matrix has the properties that $I_n A = A$ for every $n \times r$ matrix A and $B I_n = B$ for every $s \times n$ matrix B. In particular, if A is a square matrix of size n, then

$$I_n A = A I_n = A$$

EXAMPLE 5 Let

$$A = \begin{bmatrix} 1 & 3 & 1 \\ -4 & 3 & 2 \\ 1 & 0 & 1 \end{bmatrix}$$

Then

$$I_3 A = \begin{bmatrix} 1 & 0 & 0 \\ 0 & 1 & 0 \\ 0 & 0 & 1 \end{bmatrix} \begin{bmatrix} 1 & 3 & 1 \\ -4 & 3 & 2 \\ 1 & 0 & 1 \end{bmatrix} = \begin{bmatrix} 1 & 3 & 1 \\ -4 & 3 & 2 \\ 1 & 0 & 1 \end{bmatrix} = A$$

$$AI_3 = \begin{bmatrix} 1 & 3 & 1 \\ -4 & 3 & 2 \\ 1 & 0 & 1 \end{bmatrix} \begin{bmatrix} 1 & 0 & 0 \\ 0 & 1 & 0 \\ 0 & 0 & 1 \end{bmatrix} = \begin{bmatrix} 1 & 3 & 1 \\ -4 & 3 & 2 \\ 1 & 0 & 1 \end{bmatrix} = A$$

so $I_3 A = AI_3 = A$, confirming our result for this special case.   ∎

APPLIED EXAMPLE 6 Production Planning Ace Novelty received an order from Magic World Amusement Park for 900 Giant Pandas, 1200 Saint Bernards, and 2000 Big Birds. Ace's management decided that 500 Giant Pandas, 800 Saint Bernards, and 1300 Big Birds could be manufactured in their Los Angeles plant, and the balance of the order could be filled by their Seattle plant. Each Panda requires 1.5 square yards of plush, 30 cubic feet of stuffing, and 5 pieces of trim; each Saint Bernard requires 2 square yards of plush, 35 cubic feet of stuffing, and 8 pieces of trim; and each Big Bird requires 2.5 square yards of plush, 25 cubic feet of stuffing, and 15 pieces of trim. The plush costs $4.50 per square yard, the stuffing costs 10 cents per cubic foot, and the trim costs 25 cents per unit.

a. Find how much of each type of material must be purchased for each plant.
b. What is the total cost of materials incurred by each plant and the total cost of materials incurred by Ace Novelty in filling the order?

Solution   The quantities of each type of stuffed animal to be produced at each plant location may be expressed as a 2×3 *production matrix P*. Thus,

$$P = \begin{matrix} & \text{Pandas} & \text{St. Bernards} & \text{Birds} \\ \text{L.A.} & & & \\ \text{Seattle} & \end{matrix} \begin{bmatrix} 500 & 800 & 1300 \\ 400 & 400 & 700 \end{bmatrix}$$

Similarly, we may represent the amount and type of material required to manufacture each type of animal by a 3×3 *activity matrix A*. Thus,

$$A = \begin{matrix} & \text{Plush} & \text{Stuffing} & \text{Trim} \\ \text{Pandas} & & & \\ \text{St. Bernards} & & & \\ \text{Birds} & \end{matrix} \begin{bmatrix} 1.5 & 30 & 5 \\ 2 & 35 & 8 \\ 2.5 & 25 & 15 \end{bmatrix}$$

Finally, the unit cost for each type of material may be represented by the 3×1 *cost matrix C*.

$$C = \begin{matrix} \text{Plush} \\ \text{Stuffing} \\ \text{Trim} \end{matrix} \begin{bmatrix} 4.50 \\ 0.10 \\ 0.25 \end{bmatrix}$$

a. The amount of each type of material required for each plant is given by the matrix *PA*. Thus,

$$PA = \begin{bmatrix} 500 & 800 & 1300 \\ 400 & 400 & 700 \end{bmatrix} \begin{bmatrix} 1.5 & 30 & 5 \\ 2 & 35 & 8 \\ 2.5 & 25 & 15 \end{bmatrix}$$

$$= \begin{matrix} \text{L.A.} \\ \text{Seattle} \end{matrix} \begin{bmatrix} \overset{\text{Plush}}{5600} & \overset{\text{Stuffing}}{75{,}500} & \overset{\text{Trim}}{28{,}400} \\ 3150 & 43{,}500 & 15{,}700 \end{bmatrix}$$

b. The total cost of materials for each plant is given by the matrix PAC:

$$PAC = \begin{bmatrix} 5600 & 75{,}500 & 28{,}400 \\ 3150 & 43{,}500 & 15{,}700 \end{bmatrix} \begin{bmatrix} 4.50 \\ 0.10 \\ 0.25 \end{bmatrix}$$

$$= \begin{matrix} \text{L.A.} \\ \text{Seattle} \end{matrix} \begin{bmatrix} 39{,}850 \\ 22{,}450 \end{bmatrix}$$

or \$39,850 for the L.A. plant and \$22,450 for the Seattle plant. Thus, the total cost of materials incurred by Ace Novelty is \$62,300.

Matrix Representation

Example 7 shows how a system of linear equations may be written in a compact form with the help of matrices. (We will use this matrix equation representation in Section 5.6.)

EXAMPLE 7 Write the following system of linear equations in matrix form.

$$\begin{aligned} 2x - 4y + z &= 6 \\ -3x + 6y - 5z &= -1 \\ x - 3y + 7z &= 0 \end{aligned}$$

Solution Let's write

$$A = \begin{bmatrix} 2 & -4 & 1 \\ -3 & 6 & -5 \\ 1 & -3 & 7 \end{bmatrix} \quad X = \begin{bmatrix} x \\ y \\ z \end{bmatrix} \quad B = \begin{bmatrix} 6 \\ -1 \\ 0 \end{bmatrix}$$

Note that A is just the 3×3 matrix of coefficients of the system, X is the 3×1 column matrix of unknowns (variables), and B is the 3×1 column matrix of constants. We now show that the required matrix representation of the system of linear equations is

$$AX = B$$

To see this, observe that

$$AX = \begin{bmatrix} 2 & -4 & 1 \\ -3 & 6 & -5 \\ 1 & -3 & 7 \end{bmatrix} \begin{bmatrix} x \\ y \\ z \end{bmatrix} = \begin{bmatrix} 2x - 4y + z \\ -3x + 6y - 5z \\ x - 3y + 7z \end{bmatrix}$$

Equating this 3×1 matrix with matrix B now gives

$$\begin{bmatrix} 2x - 4y + z \\ -3x + 6y - 5z \\ x - 3y + 7z \end{bmatrix} = \begin{bmatrix} 6 \\ -1 \\ 0 \end{bmatrix}$$

which, by matrix equality, is easily seen to be equivalent to the given system of linear equations.

5.5 Self-Check Exercises

1. Compute

$$\begin{bmatrix} 1 & 3 & 0 \\ 2 & 4 & -1 \end{bmatrix} \begin{bmatrix} 3 & 1 & 4 \\ 2 & 0 & 3 \\ 1 & 2 & -1 \end{bmatrix}$$

2. Write the following system of linear equations in matrix form:

$$\begin{aligned} y - 2z &= 1 \\ 2x - y + 3z &= 0 \\ x \phantom{{}- y} + 4z &= 7 \end{aligned}$$

3. On June 1, the stock holdings of Ash and Joan Robinson were given by the matrix

$$\begin{array}{cccc} & \text{AT\&T} & \text{TWX} & \text{IBM} & \text{GM} \end{array}$$
$$A = \begin{array}{c} \text{Ash} \\ \text{Joan} \end{array} \begin{bmatrix} 2000 & 1000 & 500 & 5000 \\ 1000 & 2500 & 2000 & 0 \end{bmatrix}$$

and the closing prices of AT&T, TWX, IBM, and GM were $54, $113, $112, and $70 per share, respectively. Use matrix multiplication to determine the separate values of Ash's and Joan's stock holdings as of that date.

Solutions to Self-Check Exercises 5.5 can be found on page 307.

5.5 Concept Questions

1. What is the difference between scalar multiplication and matrix multiplication? Give examples of each operation.

2. a. Suppose A and B are matrices whose products AB and BA are both defined. What can you say about the sizes of A and B?

b. If A, B, and C are matrices such that $A(B + C)$ is defined, what can you say about the relationship between the number of columns of A and the number of rows of C? Explain.

5.5 Exercises

In Exercises 1–4, the sizes of matrices A and B are given. Find the size of AB and BA whenever they are defined.

1. A is of size 2×3, and B is of size 3×5.

2. A is of size 3×4, and B is of size 4×3.

3. A is of size 1×7, and B is of size 7×1.

4. A is of size 4×4, and B is of size 4×4.

5. Let A be a matrix of size $m \times n$ and B be a matrix of size $s \times t$. Find conditions on m, n, s, and t such that both matrix products AB and BA are defined.

6. Find condition(s) on the size of a matrix A such that A^2 (that is, AA) is defined.

In Exercises 7–24, compute the indicated products.

7. $\begin{bmatrix} 1 & 2 \\ 3 & 0 \end{bmatrix} \begin{bmatrix} 1 \\ -1 \end{bmatrix}$

8. $\begin{bmatrix} -1 & 3 \\ 5 & 0 \end{bmatrix} \begin{bmatrix} 7 \\ 2 \end{bmatrix}$

9. $\begin{bmatrix} 4 & 1 & 2 \\ -1 & 2 & 4 \end{bmatrix} \begin{bmatrix} 4 \\ 1 \\ -2 \end{bmatrix}$

10. $\begin{bmatrix} 3 & 2 & -1 \\ 4 & -1 & 0 \\ -5 & 2 & 1 \end{bmatrix} \begin{bmatrix} 3 \\ -2 \\ 0 \end{bmatrix}$

11. $\begin{bmatrix} -1 & 2 \\ 3 & 1 \end{bmatrix} \begin{bmatrix} 2 & 4 \\ 3 & 1 \end{bmatrix}$

12. $\begin{bmatrix} 1 & 3 \\ -1 & 2 \end{bmatrix} \begin{bmatrix} 1 & 3 & 0 \\ 3 & 0 & 2 \end{bmatrix}$

13. $\begin{bmatrix} 2 & 1 & 2 \\ 3 & 2 & 4 \end{bmatrix} \begin{bmatrix} -1 & 2 \\ 4 & 3 \\ 0 & 1 \end{bmatrix}$

14. $\begin{bmatrix} -1 & 2 \\ 4 & 3 \\ 0 & 1 \end{bmatrix} \begin{bmatrix} 2 & 1 & 2 \\ 3 & 2 & 4 \end{bmatrix}$

15. $\begin{bmatrix} 0.1 & 0.9 \\ 0.2 & 0.8 \end{bmatrix} \begin{bmatrix} 1.2 & 0.4 \\ 0.5 & 2.1 \end{bmatrix}$

16. $\begin{bmatrix} 1.2 & 0.3 \\ 0.4 & 0.5 \end{bmatrix} \begin{bmatrix} 0.2 & 0.6 \\ 0.4 & -0.5 \end{bmatrix}$

17. $\begin{bmatrix} 6 & -3 & 0 \\ -2 & 1 & -8 \\ 4 & -4 & 9 \end{bmatrix} \begin{bmatrix} 1 & 0 & 0 \\ 0 & 1 & 0 \\ 0 & 0 & 1 \end{bmatrix}$

18. $\begin{bmatrix} 2 & 4 \\ -1 & -5 \\ 3 & -1 \end{bmatrix} \begin{bmatrix} 2 & -2 & 4 \\ 1 & 3 & -1 \end{bmatrix}$

19. $\begin{bmatrix} 3 & 0 & -2 & 1 \\ 1 & 2 & 0 & -1 \end{bmatrix} \begin{bmatrix} 2 & 1 & -2 \\ -1 & 2 & 0 \\ 0 & 0 & 1 \\ -1 & -2 & 2 \end{bmatrix}$

20. $\begin{bmatrix} 2 & 1 & -3 & 0 \\ 4 & -2 & -1 & 1 \\ -1 & 2 & 0 & 1 \end{bmatrix} \begin{bmatrix} 2 & -1 \\ 1 & 4 \\ 3 & -3 \\ 0 & -5 \end{bmatrix}$

21. $4 \begin{bmatrix} 1 & -2 & 0 \\ 2 & -1 & 1 \\ 3 & 0 & -1 \end{bmatrix} \begin{bmatrix} 1 & 3 & 1 \\ 1 & 4 & 0 \\ 0 & 1 & -2 \end{bmatrix}$

22. $3\begin{bmatrix} 2 & -1 & 0 \\ 2 & 1 & 2 \\ 1 & 0 & -1 \end{bmatrix}\begin{bmatrix} 2 & 3 & 1 \\ 3 & -3 & 0 \\ 0 & 1 & -1 \end{bmatrix}$

23. $\begin{bmatrix} 1 & 0 \\ 0 & 1 \end{bmatrix}\begin{bmatrix} 4 & -3 & 2 \\ 7 & 1 & -5 \end{bmatrix}\begin{bmatrix} 1 & 0 & 0 \\ 0 & 1 & 0 \\ 0 & 0 & 1 \end{bmatrix}$

24. $2\begin{bmatrix} 3 & 2 & -1 \\ 0 & 1 & 3 \\ 2 & 0 & 3 \end{bmatrix}\begin{bmatrix} 1 & 0 & 0 \\ 0 & 1 & 0 \\ 0 & 0 & 1 \end{bmatrix}\begin{bmatrix} 1 & 2 & 0 \\ 0 & -1 & -2 \\ 1 & 3 & 1 \end{bmatrix}$

In Exercises 25 and 26, let

$$A = \begin{bmatrix} 1 & 0 & -2 \\ 1 & -3 & 2 \\ -2 & 1 & 1 \end{bmatrix} \quad B = \begin{bmatrix} 3 & 1 & 0 \\ 2 & 2 & 0 \\ 1 & -3 & -1 \end{bmatrix}$$

$$C = \begin{bmatrix} 2 & -1 & 0 \\ 1 & -1 & 2 \\ 3 & -2 & 1 \end{bmatrix}$$

25. Verify the validity of the associative law for matrix multiplication.

26. Verify the validity of the distributive law for matrix multiplication.

27. Let

$$A = \begin{bmatrix} 1 & 2 \\ 3 & 4 \end{bmatrix} \quad \text{and} \quad B = \begin{bmatrix} 2 & 1 \\ 4 & 3 \end{bmatrix}$$

Compute AB and BA, and hence deduce that matrix multiplication is, in general, not commutative.

28. Let

$$A = \begin{bmatrix} 0 & 3 & 0 \\ 1 & 0 & 1 \\ 0 & 2 & 0 \end{bmatrix} \quad B = \begin{bmatrix} 2 & 4 & 5 \\ 3 & -1 & -6 \\ 4 & 3 & 4 \end{bmatrix}$$

$$C = \begin{bmatrix} 4 & 5 & 6 \\ 3 & -1 & -6 \\ 2 & 2 & 3 \end{bmatrix}$$

a. Compute AB.
b. Compute AC.
c. Using the results of parts (a) and (b), conclude that $AB = AC$ does *not* imply that $B = C$.

29. Let

$$A = \begin{bmatrix} 3 & 0 \\ 8 & 0 \end{bmatrix} \quad \text{and} \quad B = \begin{bmatrix} 0 & 0 \\ 4 & 5 \end{bmatrix}$$

Show that $AB = 0$, thereby demonstrating that for matrix multiplication, the equation $AB = 0$ does not imply that one or both of the matrices A and B must be the zero matrix.

30. Let

$$A = \begin{bmatrix} 2 & 2 \\ -2 & -2 \end{bmatrix}$$

Show that $A^2 = 0$. Compare this with the equation $a^2 = 0$, where a is a real number.

31. Find the matrix A such that

$$A\begin{bmatrix} 1 & 0 \\ -1 & 3 \end{bmatrix} = \begin{bmatrix} -1 & -3 \\ 3 & 6 \end{bmatrix}$$

Hint: Let $A = \begin{bmatrix} a & b \\ c & d \end{bmatrix}$.

32. Find the matrix A such that

$$\begin{bmatrix} 1 & 0 \\ -1 & 3 \end{bmatrix}A = \begin{bmatrix} -1 & -3 \\ 3 & 6 \end{bmatrix}$$

Hint: Let $A = \begin{bmatrix} a & b \\ c & d \end{bmatrix}$.

33. A square matrix is called an *upper triangular matrix* if all its entries below the main diagonal are zero. For example, the matrix

$$A = \begin{bmatrix} a & b \\ 0 & d \end{bmatrix}$$

is a 2×2 *upper triangular matrix*.
a. Show that the sum and the product of two upper triangular matrices of size two are upper triangular matrices.
b. If A and B are two upper triangular matrices of size two, then is it true that $AB = BA$, in general?

34. Let

$$A = \begin{bmatrix} 3 & 1 \\ 0 & 2 \end{bmatrix} \quad \text{and} \quad B = \begin{bmatrix} 4 & -2 \\ 2 & 1 \end{bmatrix}$$

a. Compute $(A + B)^2$.
b. Compute $A^2 + 2AB + B^2$.
c. From the results of parts (a) and (b), show that in general, $(A + B)^2 \neq A^2 + 2AB + B^2$.

35. Let

$$A = \begin{bmatrix} 2 & 4 \\ 5 & -6 \end{bmatrix} \quad \text{and} \quad B = \begin{bmatrix} 4 & 8 \\ -7 & 3 \end{bmatrix}$$

a. Find A^T and show that $(A^T)^T = A$.
b. Show that $(A + B)^T = A^T + B^T$.
c. Show that $(AB)^T = B^T A^T$.

36. Let

$$A = \begin{bmatrix} 1 & 3 \\ -2 & -1 \end{bmatrix} \quad \text{and} \quad B = \begin{bmatrix} 3 & -4 \\ 2 & -2 \end{bmatrix}$$

a. Find A^T and show that $(A^T)^T = A$.
b. Show that $(A + B)^T = A^T + B^T$.
c. Show that $(AB)^T = B^T A^T$.

(*continued*)

In Exercises 37–42, write the given system of linear equations in matrix form.

37. $2x - 3y = 7$
$3x - 4y = 8$

38. $2x \quad\;\;\, = 7$
$3x - 2y = 12$

39. $2x - 3y + 4z = 6$
$2y - 3z = 7$
$x - y + 2z = 4$

40. $x - 2y + 3z = -1$
$3x + 4y - 2z = 1$
$2x - 3y + 7z = 6$

41. $-x_1 + x_2 + x_3 = 0$
$2x_1 - x_2 - x_3 = 2$
$-3x_1 + 2x_2 + 4x_3 = 4$

42. $3x_1 - 5x_2 + 4x_3 = 10$
$4x_1 + 2x_2 - 3x_3 = -12$
$-x_1 \quad\;\;\, + x_3 = -2$

43. INVESTMENTS William's and Michael's stock holdings are given by the matrix

$$A = \begin{array}{c} \\ \text{William} \\ \text{Michael} \end{array} \begin{array}{cccc} \text{BAC} & \text{GM} & \text{IBM} & \text{TRW} \\ \begin{bmatrix} 200 & 300 & 100 & 200 \\ 100 & 200 & 400 & 0 \end{bmatrix} \end{array}$$

At the close of trading on a certain day, the prices (in dollars per share) of the stocks are given by the matrix

$$B = \begin{array}{c} \text{BAC} \\ \text{GM} \\ \text{IBM} \\ \text{TRW} \end{array} \begin{bmatrix} 54 \\ 48 \\ 98 \\ 82 \end{bmatrix}$$

a. Find AB.
b. Explain the meaning of the entries in the matrix AB.

44. FOREIGN EXCHANGE Ethan has just returned to the United States from a Southeast Asian trip and wishes to exchange the various foreign currencies that he has accumulated for U.S. dollars. He has 1200 Thai bahts, 80,000 Indonesian rupiahs, 42 Malaysian ringgits, and 36 Singapore dollars. Suppose the foreign exchange rates are U.S. $0.033 for one baht, U.S. $0.0001 for one rupiah, U.S. $0.3320 for one Malaysian ringgit, and U.S. $0.8075 for one Singapore dollar.

a. Write a row matrix A giving the value of the various currencies that Ethan holds. (*Note:* The answer is *not* unique.)
b. Write a column matrix B giving the exchange rates for the various currencies.
c. If Ethan exchanges all of his foreign currencies for U.S. dollars, how many dollars will he have?

45. FOREIGN EXCHANGE Kaitlin and her friend Emma have returned to the United States from a tour of four cities: Oslo, Stockholm, Copenhagen, and Saint Petersburg. They now wish to exchange the various foreign currencies that they have accumulated for U.S. dollars. Kaitlin has 82 Norwegian kroner, 68 Swedish kronor, 62 Danish kroner, and 1200 Russian rubles. Emma has 64 Norwegian kroner, 74 Swedish kronor, 44 Danish kroner, and 1600 Russian rubles. Suppose the exchange rates are U.S. $0.1805 for one Norwegian krone, U.S. $0.1582 for one Swedish krone, U.S. $0.1901 for one Danish krone, and U.S. $0.0356 for one Russian ruble.

a. Write a 2×4 matrix A giving the values of the various foreign currencies held by Kaitlin and Emma. (*Note:* The answer is *not* unique.)
b. Write a column matrix B giving the exchange rate for the various currencies.
c. If both Kaitlin and Emma exchange all their foreign currencies for U.S. dollars, how many dollars will each have?

46. REAL ESTATE Bond Brothers, a real estate developer, builds houses in three states. The projected number of units of each model to be built in each state is given by the matrix

$$A = \begin{array}{c} \\ \text{N.Y.} \\ \text{Conn.} \\ \text{Mass.} \end{array} \begin{array}{cccc} \;\;\text{I} & \;\;\text{II} & \;\text{III} & \;\text{IV} \\ \begin{bmatrix} 60 & 80 & 120 & 40 \\ 20 & 30 & 60 & 10 \\ 10 & 15 & 30 & 5 \end{bmatrix} \end{array}$$

The profits to be realized are $20,000, $22,000, $25,000, and $30,000, respectively, for each Model I, II, III, and IV house sold.

a. Write a column matrix B representing the profit for each type of house.
b. Find the total profit Bond Brothers expects to earn in each state if all the houses are sold.

47. REAL ESTATE Refer to Exercise 46. Let $B = \begin{bmatrix} 1 & 1 & 1 \end{bmatrix}$ and $C = \begin{bmatrix} 1 & 1 & 1 & 1 \end{bmatrix}$.

a. Compute BA, and explain what the entries of the matrix represent.
b. Compute AC^T, and give an interpretation of the matrix.

48. CHARITIES The amount of money raised by Charity I, Charity II, and Charity III (in millions of dollars) in each of the years 2009, 2010, and 2011 is represented by the matrix A:

$$A = \begin{array}{c} \\ 2009 \\ 2010 \\ 2011 \end{array} \begin{array}{ccc} \;\text{I} & \;\;\text{II} & \;\text{III} \\ \begin{bmatrix} 18.2 & 28.2 & 40.5 \\ 19.6 & 28.6 & 42.6 \\ 20.8 & 30.4 & 46.4 \end{bmatrix} \end{array}$$

On average, Charity I puts 78% toward program cost, Charity II puts 88% toward program cost, and Charity III puts 80% toward program cost. Write a 3×1 matrix B reflecting the percentage put toward program cost by the charities. Then use matrix multiplication to find the total amount of money put toward program cost in each of the 3 years by the charities under consideration.

49. BOX-OFFICE RECEIPTS The Cinema Center consists of four theaters: Cinemas I, II, III, and IV. The admission price for one feature at the Center is $4 for children, $6 for students, and $8 for adults. The attendance for the Sunday matinee is given by the matrix

$$A = \begin{array}{c} \\ \text{Cinema I} \\ \text{Cinema II} \\ \text{Cinema III} \\ \text{Cinema IV} \end{array} \begin{array}{ccc} \text{Children} & \text{Students} & \text{Adults} \\ \begin{bmatrix} 225 & 110 & 50 \\ 75 & 180 & 225 \\ 280 & 85 & 110 \\ 0 & 250 & 225 \end{bmatrix} \end{array}$$

Write a column vector B representing the admission prices. Then compute AB, the column vector showing the gross receipts for each theater. Finally, find the total revenue collected at the Cinema Center for admission that Sunday afternoon.

50. **Box-Office Receipts** Refer to Exercise 49.
    a. Find a 1×4 matrix B such that the entries in BA give the total number of children, the total number of students, and the total number of adults who attended the Sunday matinee.
    b. Find a 1×3 matrix C such that the entries in AC^T give the total number of people (children, students, and adults) who attended Cinema I, Cinema II, Cinema III, and Cinema IV.

51. **Politics: Voter Affiliation** Matrix A gives the percentage of eligible voters in the city of Newton, classified according to party affiliation and age group.

$$A = \begin{array}{c} \text{Under 30} \\ \text{30 to 50} \\ \text{Over 50} \end{array} \begin{array}{ccc} \text{Dem.} & \text{Rep.} & \text{Ind.} \\ \begin{bmatrix} 0.50 & 0.30 & 0.20 \\ 0.45 & 0.40 & 0.15 \\ 0.40 & 0.50 & 0.10 \end{bmatrix} \end{array}$$

The population of eligible voters in the city by age group is given by the matrix B:

$$\begin{array}{ccc} \text{Under 30} & \text{30 to 50} & \text{Over 50} \end{array}$$
$$B = \begin{bmatrix} 30{,}000 & 40{,}000 & 20{,}000 \end{bmatrix}$$

Find a matrix giving the total number of eligible voters in the city who will vote Democratic, Republican, and Independent.

52. **401(k) Retirement Plans** Three network consultants, Alan, Maria, and Steven, each received a year-end bonus of $10,000, which they decided to invest in a 401(k) retirement plan sponsored by their employer. Under this plan, employees are allowed to place their investments in three funds: an equity index fund (I), a growth fund (II), and a global equity fund (III). The allocations of the investments (in dollars) of the three employees at the beginning of the year are summarized in the matrix

$$A = \begin{array}{c} \text{Alan} \\ \text{Maria} \\ \text{Steven} \end{array} \begin{array}{ccc} \text{I} & \text{II} & \text{III} \\ \begin{bmatrix} 4000 & 3000 & 3000 \\ 2000 & 5000 & 3000 \\ 2000 & 3000 & 5000 \end{bmatrix} \end{array}$$

The returns of the three funds after 1 year are given in the matrix

$$B = \begin{array}{c} \text{I} \\ \text{II} \\ \text{III} \end{array} \begin{bmatrix} 0.18 \\ 0.24 \\ 0.12 \end{bmatrix}$$

Which employee realized the best return on his or her investment for the year in question? The worst return?

53. **College Admissions** A university admissions committee anticipates an enrollment of 8000 students in its freshman class next year. To satisfy admission quotas, incoming students have been categorized according to their sex and place of residence. The number of students in each category is given by the matrix

$$A = \begin{array}{c} \text{In-state} \\ \text{Out-of-state} \\ \text{Foreign} \end{array} \begin{array}{cc} \text{Male} & \text{Female} \\ \begin{bmatrix} 2700 & 3000 \\ 800 & 700 \\ 500 & 300 \end{bmatrix} \end{array}$$

By using data accumulated in previous years, the admissions committee has determined that these students will elect to enter the College of Letters and Science, the College of Fine Arts, the School of Business Administration, and the School of Engineering according to the percentages that appear in the following matrix:

$$B = \begin{array}{c} \text{Male} \\ \text{Female} \end{array} \begin{array}{cccc} \text{L. \& S.} & \text{Fine Arts} & \text{Bus. Ad.} & \text{Eng.} \\ \begin{bmatrix} 0.25 & 0.20 & 0.30 & 0.25 \\ 0.30 & 0.35 & 0.25 & 0.10 \end{bmatrix} \end{array}$$

Find the matrix AB that shows the number of in-state, out-of-state, and foreign students expected to enter each discipline.

54. **Production Planning** Refer to Example 6 in this section. Suppose Ace Novelty received an order from another amusement park for 1200 Pink Panthers, 1800 Giant Pandas, and 1400 Big Birds. The quantity of each type of stuffed animal to be produced at each plant is shown in the following production matrix:

$$P = \begin{array}{c} \text{L.A.} \\ \text{Seattle} \end{array} \begin{array}{ccc} \text{Panthers} & \text{Pandas} & \text{Birds} \\ \begin{bmatrix} 700 & 1000 & 800 \\ 500 & 800 & 600 \end{bmatrix} \end{array}$$

Each Panther requires 1.3 yd^2 of plush, 20 ft^3 of stuffing, and 12 pieces of trim. Assume the materials required to produce the other two stuffed animals and the unit cost for each type of material are as given in Example 6.
    a. How much of each type of material must be purchased for each plant?
    b. What is the total cost of materials that will be incurred at each plant?
    c. What is the total cost of materials incurred by Ace Novelty in filling the order?

55. **Computing Phone Bills** Cindy regularly makes long-distance phone calls to three foreign cities: London, Tokyo, and Hong Kong. The matrices A and B give the lengths (in minutes) of her calls during peak and nonpeak hours, respectively, to each of these three cities during the month of June.

$$\begin{array}{cccc} & \text{London} & \text{Tokyo} & \text{Hong Kong} \\ A = & \begin{bmatrix} 80 & 60 & 40 \end{bmatrix} \end{array}$$

and

$$\begin{array}{cccc} & \text{London} & \text{Tokyo} & \text{Hong Kong} \\ B = & \begin{bmatrix} 300 & 150 & 250 \end{bmatrix} \end{array}$$

The costs for the calls (in dollars per minute) for the peak and nonpeak periods in the month in question are given, respectively, by the matrices

$$C = \begin{matrix} \text{London} \\ \text{Tokyo} \\ \text{Hong Kong} \end{matrix} \begin{bmatrix} 0.34 \\ 0.42 \\ 0.48 \end{bmatrix} \quad \text{and} \quad D = \begin{matrix} \text{London} \\ \text{Tokyo} \\ \text{Hong Kong} \end{matrix} \begin{bmatrix} 0.24 \\ 0.31 \\ 0.35 \end{bmatrix}$$

Compute the matrix $AC + BD$, and explain what it represents.

56. **PRODUCTION PLANNING** The total output of loudspeaker systems of the Acrosonic Company at their three production facilities for May and June is given by the matrices A and B, respectively, where

$$A = \begin{matrix} \text{Location I} \\ \text{Location II} \\ \text{Location III} \end{matrix} \begin{matrix} \text{Model} & \text{Model} & \text{Model} & \text{Model} \\ A & B & C & D \end{matrix} \begin{bmatrix} 320 & 280 & 460 & 280 \\ 480 & 360 & 580 & 0 \\ 540 & 420 & 200 & 880 \end{bmatrix}$$

$$B = \begin{matrix} \text{Location I} \\ \text{Location II} \\ \text{Location III} \end{matrix} \begin{matrix} \text{Model} & \text{Model} & \text{Model} & \text{Model} \\ A & B & C & D \end{matrix} \begin{bmatrix} 210 & 180 & 330 & 180 \\ 400 & 300 & 450 & 40 \\ 420 & 280 & 180 & 740 \end{bmatrix}$$

The unit production costs and selling prices for these loudspeakers are given by matrices C and D, respectively, where

$$C = \begin{matrix} \text{Model A} \\ \text{Model B} \\ \text{Model C} \\ \text{Model D} \end{matrix} \begin{bmatrix} 120 \\ 180 \\ 260 \\ 500 \end{bmatrix} \quad \text{and} \quad D = \begin{matrix} \text{Model A} \\ \text{Model B} \\ \text{Model C} \\ \text{Model D} \end{matrix} \begin{bmatrix} 160 \\ 250 \\ 350 \\ 700 \end{bmatrix}$$

Compute the following matrices, and explain the meaning of the entries in each matrix.

a. AC b. AD c. BC d. BD e. $(A + B)C$
f. $(A + B)D$ g. $A(D - C)$
h. $B(D - C)$ i. $(A + B)(D - C)$

57. **DIET PLANNING** A dietitian plans a meal around three foods. The number of units of vitamin A, vitamin C, and calcium in each ounce of these foods is represented by the matrix M, where

$$M = \begin{matrix} \text{Vitamin A} \\ \text{Vitamin C} \\ \text{Calcium} \end{matrix} \begin{matrix} \text{Food I} & \text{Food II} & \text{Food III} \end{matrix} \begin{bmatrix} 400 & 1200 & 800 \\ 110 & 570 & 340 \\ 90 & 30 & 60 \end{bmatrix}$$

The matrices A and B represent the amount of each food (in ounces) consumed by a girl at two different meals, where

$$A = \begin{matrix} \text{Food I} & \text{Food II} & \text{Food III} \end{matrix} \begin{bmatrix} 7 & 1 & 6 \end{bmatrix}$$

$$B = \begin{matrix} \text{Food I} & \text{Food II} & \text{Food III} \end{matrix} \begin{bmatrix} 9 & 3 & 2 \end{bmatrix}$$

Calculate the following matrices, and explain the meaning of the entries in each matrix.
a. MA^T b. MB^T c. $M(A + B)^T$

58. **PRODUCTION PLANNING** Hartman Lumber Company has two branches in the city. The sales of four of its products for the last year (in thousands of dollars) are represented by the matrix

$$B = \begin{matrix} \text{Branch I} \\ \text{Branch II} \end{matrix} \begin{matrix} \text{Product} \\ A \quad B \quad C \quad D \end{matrix} \begin{bmatrix} 5 & 2 & 8 & 10 \\ 3 & 4 & 6 & 8 \end{bmatrix}$$

For the present year, management has projected that the sales of the four products in Branch I will be 10% more than the corresponding sales for last year and the sales of the four products in Branch II will be 15% more than the corresponding sales for last year.
a. Show that the sales of the four products in the two branches for the current year are given by the matrix AB, where

$$A = \begin{bmatrix} 1.1 & 0 \\ 0 & 1.15 \end{bmatrix}$$

Compute AB.
b. Hartman has m branches nationwide, and the sales of n of its products (in thousands of dollars) last year are represented by the matrix

$$B = \begin{matrix} \text{Branch 1} \\ \text{Branch 2} \\ \vdots \\ \text{Branch } m \end{matrix} \begin{matrix} \text{Product} \\ 1 \quad 2 \quad 3 \quad \cdots \quad n \end{matrix} \begin{bmatrix} a_{11} & a_{12} & a_{13} & \cdots & a_{1n} \\ a_{21} & a_{22} & a_{23} & \cdots & a_{2n} \\ \vdots & \vdots & \vdots & & \vdots \\ a_{m1} & a_{m2} & a_{m3} & \cdots & a_{mn} \end{bmatrix}$$

Also, management has projected that the sales of the n products in Branch 1, Branch 2, . . . , Branch m will be $r_1\%$, $r_2\%$, . . . , $r_m\%$, respectively, more than the corresponding sales for last year. Write the matrix A such that AB gives the sales of the n products in the m branches for the current year.

In Exercises 59–62, determine whether the statement is true or false. If it is true, explain why it is true. If it is false, give an example to show why it is false.

59. If A and B are matrices such that AB and BA are both defined, then A and B must be square matrices of the same size.

60. If A and B are matrices such that AB is defined and if c is a scalar, then $(cA)B = A(cB) = cAB$.

61. If A, B, and C are matrices and $A(B + C)$ is defined, then B must have the same size as C, and the number of columns of A must be equal to the number of rows of B.

62. If A is a 2 × 4 matrix and B is a matrix such that ABA is defined, then the size of B must be 4 × 2.

5.5 Solutions to Self-Check Exercises

1. We compute

$$\begin{bmatrix} 1 & 3 & 0 \\ 2 & 4 & -1 \end{bmatrix} \begin{bmatrix} 3 & 1 & 4 \\ 2 & 0 & 3 \\ 1 & 2 & -1 \end{bmatrix}$$

$$= \begin{bmatrix} 1(3) + 3(2) + 0(1) & 1(1) + 3(0) + 0(2) & 1(4) + 3(3) + 0(-1) \\ 2(3) + 4(2) - 1(1) & 2(1) + 4(0) - 1(2) & 2(4) + 4(3) - 1(-1) \end{bmatrix}$$

$$= \begin{bmatrix} 9 & 1 & 13 \\ 13 & 0 & 21 \end{bmatrix}$$

2. Let

$$A = \begin{bmatrix} 0 & 1 & -2 \\ 2 & -1 & 3 \\ 1 & 0 & 4 \end{bmatrix} \quad X = \begin{bmatrix} x \\ y \\ z \end{bmatrix} \quad B = \begin{bmatrix} 1 \\ 0 \\ 7 \end{bmatrix}$$

Then the given system may be written as the matrix equation

$$AX = B$$

3. Write

$$B = \begin{bmatrix} 54 \\ 113 \\ 112 \\ 70 \end{bmatrix} \begin{matrix} \text{AT\&T} \\ \text{TWX} \\ \text{IBM} \\ \text{GM} \end{matrix}$$

and compute the following:

$$AB = \begin{matrix} \text{Ash} \\ \text{Joan} \end{matrix} \begin{bmatrix} 2000 & 1000 & 500 & 5000 \\ 1000 & 2500 & 2000 & 0 \end{bmatrix} \begin{bmatrix} 54 \\ 113 \\ 112 \\ 70 \end{bmatrix}$$

$$= \begin{bmatrix} 627{,}000 \\ 560{,}500 \end{bmatrix} \begin{matrix} \text{Ash} \\ \text{Joan} \end{matrix}$$

We conclude that Ash's stock holdings were worth $627,000 and Joan's stock holdings were worth $560,500 on June 1.

USING TECHNOLOGY

Matrix Multiplication

Graphing Utility

A graphing utility can be used to perform matrix multiplication.

EXAMPLE 1 Let

$$A = \begin{bmatrix} 1.2 & 3.1 & -1.4 \\ 2.7 & 4.2 & 3.4 \end{bmatrix}$$

$$B = \begin{bmatrix} 0.8 & 1.2 & 3.7 \\ 6.2 & -0.4 & 3.3 \end{bmatrix}$$

$$C = \begin{bmatrix} 1.2 & 2.1 & 1.3 \\ 4.2 & -1.2 & 0.6 \\ 1.4 & 3.2 & 0.7 \end{bmatrix}$$

Find (a) AC and (b) $(1.1A + 2.3B)C$.

Solution First, we enter the matrices A, B, and C into the calculator.

a. Using matrix operations, we enter the expression $A*C$. We obtain the matrix

$$\begin{bmatrix} 12.5 & -5.68 & 2.44 \\ 25.64 & 11.51 & 8.41 \end{bmatrix}$$

(You might need to scroll the display on the screen to obtain the complete matrix.)

b. Using matrix operations, we enter the expression $(1.1A + 2.3B)C$. We obtain the matrix

$$\begin{bmatrix} 39.464 & 21.536 & 12.689 \\ 52.078 & 67.999 & 32.55 \end{bmatrix}$$

(continued)

Excel

We use the MMULT function in Excel to perform matrix multiplication.

EXAMPLE 2 Let

$$A = \begin{bmatrix} 1.2 & 3.1 & -1.4 \\ 2.7 & 4.2 & 3.4 \end{bmatrix} \quad B = \begin{bmatrix} 0.8 & 1.2 & 3.7 \\ 6.2 & -0.4 & 3.3 \end{bmatrix} \quad C = \begin{bmatrix} 1.2 & 2.1 & 1.3 \\ 4.2 & -1.2 & 0.6 \\ 1.4 & 3.2 & 0.7 \end{bmatrix}$$

Find (a) AC and (b) $(1.1A + 2.3B)C$.

Solution

a. *First, enter the matrices A, B, and C onto a spreadsheet* (Figure T1).

	A	B	C	D	E	F	G
1	A					B	
2	1.2	3.1	-1.4		0.8	1.2	3.7
3	2.7	4.2	3.4		6.2	-0.4	3.3
4							
5	C						
6	1.2	2.1	1.3				
7	4.2	-1.2	0.6				
8	1.4	3.2	0.7				

FIGURE **T1**
Spreadsheet showing the matrices A, B, and C

Second, compute AC. Highlight the cells that will contain the matrix product AC, which has size 2×3. Type =MMULT (, highlight the cells in matrix A, type , , highlight the cells in matrix C, type), and press **Ctrl-Shift-Enter** . The matrix product AC shown in Figure T2 will appear on your spreadsheet.

	A	B	C
10		AC	
11	12.5	-5.68	2.44
12	25.64	11.51	8.41

FIGURE **T2**
The matrix product AC

b. *Compute* $(1.1A + 2.3B)C$. Highlight the cells that will contain the matrix product $(1.1A + 2.3B)C$. Next, type =MMULT (1.1*, highlight the cells in matrix A, type +2.3*, highlight the cells in matrix B, type , , highlight the cells in matrix C, type), and then press **Ctrl-Shift-Enter** . The matrix product shown in Figure T3 will appear on your spreadsheet.

	A	B	C
13		(1.1A + 2.3B)C	
14	39.464	21.536	12.689
15	52.078	67.999	32.55

FIGURE **T3**
The matrix product (1.1A + 2.3B)C

Note: Boldfaced words/characters in a box (for example, **Enter**) indicate that an action (click, select, or press) is required. Words/characters printed blue (for example, Chart sub-type:) indicate words/characters that appear on the screen. Words/characters printed in a monospace font (for example, =(−2/3)*A2+2) indicate words/characters that need to be typed and entered.

TECHNOLOGY EXERCISES

In Exercises 1–8, refer to the following matrices, and perform the indicated operations. Round your answers to two decimal places.

$$A = \begin{bmatrix} 1.2 & 3.1 & -1.2 & 4.3 \\ 7.2 & 6.3 & 1.8 & -2.1 \\ 0.8 & 3.2 & -1.3 & 2.8 \end{bmatrix}$$

$$B = \begin{bmatrix} 0.7 & 0.3 & 1.2 & -0.8 \\ 1.2 & 1.7 & 3.5 & 4.2 \\ -3.3 & -1.2 & 4.2 & 3.2 \end{bmatrix}$$

$$C = \begin{bmatrix} 0.8 & 7.1 & 6.2 \\ 3.3 & -1.2 & 4.8 \\ 1.3 & 2.8 & -1.5 \\ 2.1 & 3.2 & -8.4 \end{bmatrix}$$

1. AC

2. CB

3. $(A + B)C$

4. $(2A + 3B)C$

5. $(2A - 3.1B)C$

6. $C(2.1A + 3.2B)$

7. $(4.1A + 2.7B)1.6C$

8. $2.5C(1.8A - 4.3B)$

In Exercises 9–12, refer to the following matrices, and perform the indicated operations. Round your answers to two decimal places.

$$A = \begin{bmatrix} 2 & 5 & -4 & 2 & 8 \\ 6 & 7 & 2 & 9 & 6 \\ 4 & 5 & 4 & 4 & 4 \\ 9 & 6 & 8 & 3 & 2 \end{bmatrix}$$

$$B = \begin{bmatrix} 2 & 6 & 7 & 5 \\ 3 & 4 & 6 & 2 \\ -5 & 8 & 4 & 3 \\ 8 & 6 & 9 & 5 \\ 4 & 7 & 8 & 8 \end{bmatrix}$$

$$C = \begin{bmatrix} 6.2 & 7.3 & -4.0 & 7.1 & 9.3 \\ 4.8 & 6.5 & 8.4 & -6.3 & 8.4 \\ 5.4 & 3.2 & 6.3 & 9.1 & -2.8 \\ 8.2 & 7.3 & 6.5 & 4.1 & 9.8 \\ 10.3 & 6.8 & 4.8 & -9.1 & 20.4 \end{bmatrix}$$

$$D = \begin{bmatrix} 4.6 & 3.9 & 8.4 & 6.1 & 9.8 \\ 2.4 & -6.8 & 7.9 & 11.4 & 2.9 \\ 7.1 & 9.4 & 6.3 & 5.7 & 4.2 \\ 3.4 & 6.1 & 5.3 & 8.4 & 6.3 \\ 7.1 & -4.2 & 3.9 & -6.4 & 7.1 \end{bmatrix}$$

9. Find AB and BA.

10. Find CD and DC. Is $CD = DC$?

11. Find $AC + AD$.

12. Find:

   a. AC

   b. AD

   c. $A(C + D)$

   d. Is $A(C + D) = AC + AD$?

5.6   The Inverse of a Square Matrix

The Inverse of a Square Matrix

In this section, we discuss a procedure for finding the inverse of a matrix, and we show how the inverse can be used to help us solve a system of linear equations.

Recall that if a is a nonzero real number, then there exists a unique real number a^{-1} (that is, $\frac{1}{a}$) such that

$$a^{-1}a = \left(\frac{1}{a}\right)(a) = 1$$

The use of the (multiplicative) inverse of a real number enables us to solve algebraic equations of the form

$$ax = b \tag{13}$$

Multiplying both sides of (13) by a^{-1}, we have

$$a^{-1}(ax) = a^{-1}b$$

$$\left(\frac{1}{a}\right)(ax) = \frac{1}{a}(b)$$

$$x = \frac{b}{a}$$

For example, since the inverse of 2 is $2^{-1} = \frac{1}{2}$, we can solve the equation

$$2x = 5$$

by multiplying both sides of the equation by $2^{-1} = \frac{1}{2}$, giving

$$2^{-1}(2x) = 2^{-1} \cdot 5$$

$$x = \frac{5}{2}$$

We can use a similar procedure to solve the matrix equation

$$AX = B$$

where A, X, and B are matrices of the proper sizes. To do this, we need the matrix equivalent of the inverse of a real number. Such a matrix, whenever it exists, is called the **inverse of a matrix**.

Inverse of a Matrix

Let A be a square matrix of size n. A square matrix A^{-1} of size n such that

$$A^{-1}A = AA^{-1} = I_n$$

is called the inverse of A.

Let's show that the matrix

$$A = \begin{bmatrix} 1 & 2 \\ 3 & 4 \end{bmatrix}$$

has the matrix

$$A^{-1} = \begin{bmatrix} -2 & 1 \\ \frac{3}{2} & -\frac{1}{2} \end{bmatrix}$$

as its inverse. Since

$$AA^{-1} = \begin{bmatrix} 1 & 2 \\ 3 & 4 \end{bmatrix}\begin{bmatrix} -2 & 1 \\ \frac{3}{2} & -\frac{1}{2} \end{bmatrix} = \begin{bmatrix} 1 & 0 \\ 0 & 1 \end{bmatrix} = I$$

$$A^{-1}A = \begin{bmatrix} -2 & 1 \\ \frac{3}{2} & -\frac{1}{2} \end{bmatrix}\begin{bmatrix} 1 & 2 \\ 3 & 4 \end{bmatrix} = \begin{bmatrix} 1 & 0 \\ 0 & 1 \end{bmatrix} = I$$

we see that A^{-1} is the inverse of A, as asserted.

Explore & Discuss

In defining the inverse of a matrix A, why is it necessary to require that A be a square matrix?

Not every square matrix has an inverse. A square matrix that has an inverse is said to be **nonsingular.** A matrix that does not have an inverse is said to be **singular.** An example of a singular matrix is given by

$$B = \begin{bmatrix} 0 & 1 \\ 0 & 0 \end{bmatrix}$$

If B had an inverse given by

$$B^{-1} = \begin{bmatrix} a & b \\ c & d \end{bmatrix}$$

where a, b, c, and d are some appropriate numbers, then by the definition of an inverse, we would have $BB^{-1} = I$; that is,

$$\begin{bmatrix} 0 & 1 \\ 0 & 0 \end{bmatrix}\begin{bmatrix} a & b \\ c & d \end{bmatrix} = \begin{bmatrix} 1 & 0 \\ 0 & 1 \end{bmatrix}$$

$$\begin{bmatrix} c & d \\ 0 & 0 \end{bmatrix} = \begin{bmatrix} 1 & 0 \\ 0 & 1 \end{bmatrix}$$

which implies that $0 = 1$—an impossibility! This contradiction shows that B does not have an inverse.

A Method for Finding the Inverse of a Square Matrix

The methods of Section 5.5 can be used to find the inverse of a nonsingular matrix. To discover such an algorithm, let's find the inverse of the matrix

$$A = \begin{bmatrix} 1 & 2 \\ -1 & 3 \end{bmatrix}$$

Suppose A^{-1} exists and is given by

$$A^{-1} = \begin{bmatrix} a & b \\ c & d \end{bmatrix}$$

where a, b, c, and d are to be determined. By the definition of an inverse, we have $AA^{-1} = I$; that is,

$$\begin{bmatrix} 1 & 2 \\ -1 & 3 \end{bmatrix}\begin{bmatrix} a & b \\ c & d \end{bmatrix} = \begin{bmatrix} 1 & 0 \\ 0 & 1 \end{bmatrix}$$

which simplifies to

$$\begin{bmatrix} a + 2c & b + 2d \\ -a + 3c & -b + 3d \end{bmatrix} = \begin{bmatrix} 1 & 0 \\ 0 & 1 \end{bmatrix}$$

But this matrix equation is equivalent to the two systems of linear equations

$$\left.\begin{array}{r} a + 2c = 1 \\ -a + 3c = 0 \end{array}\right\} \quad \text{and} \quad \left.\begin{array}{r} b + 2d = 0 \\ -b + 3d = 1 \end{array}\right\}$$

with augmented matrices given by

$$\left[\begin{array}{cc|c} 1 & 2 & 1 \\ -1 & 3 & 0 \end{array}\right] \quad \text{and} \quad \left[\begin{array}{cc|c} 1 & 2 & 0 \\ -1 & 3 & 1 \end{array}\right]$$

Note that the matrices of coefficients of the two systems are identical. This suggests that we solve the two systems of simultaneous linear equations by writing the following augmented matrix, which we obtain by joining the coefficient matrix and the two columns of constants:

$$\left[\begin{array}{cc|cc} 1 & 2 & 1 & 0 \\ -1 & 3 & 0 & 1 \end{array}\right]$$

Using the Gauss–Jordan elimination method, we obtain the following sequence of equivalent matrices:

$$\left[\begin{array}{cc|cc} 1 & 2 & 1 & 0 \\ -1 & 3 & 0 & 1 \end{array}\right] \xrightarrow{R_2 + R_1} \left[\begin{array}{cc|cc} 1 & 2 & 1 & 0 \\ 0 & 5 & 1 & 1 \end{array}\right] \xrightarrow{\frac{1}{5}R_2}$$

$$\left[\begin{array}{cc|cc} 1 & 2 & 1 & 0 \\ 0 & 1 & \frac{1}{5} & \frac{1}{5} \end{array}\right] \xrightarrow{R_1 - 2R_2} \left[\begin{array}{cc|cc} 1 & 0 & \frac{3}{5} & -\frac{2}{5} \\ 0 & 1 & \frac{1}{5} & \frac{1}{5} \end{array}\right]$$

Thus, $a = \frac{3}{5}$, $b = -\frac{2}{5}$, $c = \frac{1}{5}$, and $d = \frac{1}{5}$, giving

$$A^{-1} = \begin{bmatrix} \frac{3}{5} & -\frac{2}{5} \\ \frac{1}{5} & \frac{1}{5} \end{bmatrix}$$

The following computations verify that A^{-1} is indeed the inverse of A:

$$\begin{bmatrix} 1 & 2 \\ -1 & 3 \end{bmatrix}\begin{bmatrix} \frac{3}{5} & -\frac{2}{5} \\ \frac{1}{5} & \frac{1}{5} \end{bmatrix} = \begin{bmatrix} 1 & 0 \\ 0 & 1 \end{bmatrix} = \begin{bmatrix} \frac{3}{5} & -\frac{2}{5} \\ \frac{1}{5} & \frac{1}{5} \end{bmatrix}\begin{bmatrix} 1 & 2 \\ -1 & 3 \end{bmatrix}$$

The preceding example suggests a general algorithm for computing the inverse of a square matrix of size n when it exists.

> **Finding the Inverse of a Matrix**
>
> Given the $n \times n$ matrix A:
>
> **1.** Adjoin the $n \times n$ identity matrix I to obtain the augmented matrix
>
> $$[A \mid I]$$
>
> **2.** Use a sequence of row operations to reduce $[A \mid I]$ to the form
>
> $$[I \mid B]$$
>
> if possible.
>
> Then the matrix B is the inverse of A.

Note　Although matrix multiplication is not generally commutative, it is possible to prove that if A has an inverse and $AB = I$, then $BA = I$ also. Hence, to verify that B is the inverse of A, it suffices to show that $AB = I$. ■

EXAMPLE 1　Find the inverse of the matrix

$$A = \begin{bmatrix} 2 & 1 & 1 \\ 3 & 2 & 1 \\ 2 & 1 & 2 \end{bmatrix}$$

Solution　We form the augmented matrix

$$\left[\begin{array}{ccc|ccc} 2 & 1 & 1 & 1 & 0 & 0 \\ 3 & 2 & 1 & 0 & 1 & 0 \\ 2 & 1 & 2 & 0 & 0 & 1 \end{array}\right]$$

and use the Gauss–Jordan elimination method to reduce it to the form $[I \mid B]$:

$$\begin{bmatrix} 2 & 1 & 1 & | & 1 & 0 & 0 \\ 3 & 2 & 1 & | & 0 & 1 & 0 \\ 2 & 1 & 2 & | & 0 & 0 & 1 \end{bmatrix} \xrightarrow{R_1 - R_2} \begin{bmatrix} -1 & -1 & 0 & | & 1 & -1 & 0 \\ 3 & 2 & 1 & | & 0 & 1 & 0 \\ 2 & 1 & 2 & | & 0 & 0 & 1 \end{bmatrix}$$

$$\xrightarrow[\substack{R_2 + 3R_1 \\ R_3 + 2R_1}]{-R_1} \begin{bmatrix} 1 & 1 & 0 & | & -1 & 1 & 0 \\ 0 & -1 & 1 & | & 3 & -2 & 0 \\ 0 & -1 & 2 & | & 2 & -2 & 1 \end{bmatrix}$$

$$\xrightarrow[\substack{-R_2 \\ R_3 - R_2}]{R_1 + R_2} \begin{bmatrix} 1 & 0 & 1 & | & 2 & -1 & 0 \\ 0 & 1 & -1 & | & -3 & 2 & 0 \\ 0 & 0 & 1 & | & -1 & 0 & 1 \end{bmatrix}$$

$$\xrightarrow[\substack{R_2 + R_3}]{R_1 - R_3} \begin{bmatrix} 1 & 0 & 0 & | & 3 & -1 & -1 \\ 0 & 1 & 0 & | & -4 & 2 & 1 \\ 0 & 0 & 1 & | & -1 & 0 & 1 \end{bmatrix}$$

The inverse of A is the matrix

$$A^{-1} = \begin{bmatrix} 3 & -1 & -1 \\ -4 & 2 & 1 \\ -1 & 0 & 1 \end{bmatrix}$$

We leave it to you to verify these results. ∎

Example 2 illustrates what happens to the reduction process when a matrix A does *not* have an inverse.

VIDEO   **EXAMPLE 2** Find the inverse of the matrix

$$A = \begin{bmatrix} 1 & 2 & 3 \\ 2 & 1 & 2 \\ 3 & 3 & 5 \end{bmatrix}$$

Solution   We form the augmented matrix

$$\begin{bmatrix} 1 & 2 & 3 & | & 1 & 0 & 0 \\ 2 & 1 & 2 & | & 0 & 1 & 0 \\ 3 & 3 & 5 & | & 0 & 0 & 1 \end{bmatrix}$$

Explore & Discuss

Explain in terms of solutions to systems of linear equations why the final augmented matrix in Example 2 implies that A has no inverse. *Hint:* See the discussion on page 311.

and use the Gauss–Jordan elimination method:

$$\begin{bmatrix} 1 & 2 & 3 & | & 1 & 0 & 0 \\ 2 & 1 & 2 & | & 0 & 1 & 0 \\ 3 & 3 & 5 & | & 0 & 0 & 1 \end{bmatrix} \xrightarrow[\substack{R_3 - 3R_1}]{R_2 - 2R_1} \begin{bmatrix} 1 & 2 & 3 & | & 1 & 0 & 0 \\ 0 & -3 & -4 & | & -2 & 1 & 0 \\ 0 & -3 & -4 & | & -3 & 0 & 1 \end{bmatrix}$$

$$\xrightarrow[\substack{R_3 - R_2}]{-R_2} \begin{bmatrix} 1 & 2 & 3 & | & 1 & 0 & 0 \\ 0 & 3 & 4 & | & 2 & -1 & 0 \\ 0 & 0 & 0 & | & -1 & -1 & 1 \end{bmatrix}$$

Since the entries in the last row of the 3×3 submatrix that comprises the left-hand side of the augmented matrix just obtained are all equal to zero, the latter cannot be reduced to the form $[I \mid B]$. Accordingly, we draw the conclusion that A is singular—that is, it does not have an inverse. ∎

More generally, we have the following criterion for determining when the inverse of a matrix does not exist.

> **Matrices That Have No Inverses**
> If there is a row to the left of the vertical line in the augmented matrix containing all zeros, then the matrix does not have an inverse.

A Formula for the Inverse of a 2 × 2 Matrix

Before turning to some applications, we show an alternative method that employs a formula for finding the inverse of a 2 × 2 matrix. This method will prove useful in many situations; we will see an application in Example 5. The derivation of this formula is left as an exercise (Exercise 52).

> **Formula for the Inverse of a 2 × 2 Matrix**
> Let
> $$A = \begin{bmatrix} a & b \\ c & d \end{bmatrix}$$
> Suppose $D = ad - bc$ is not equal to zero. Then A^{-1} exists and is given by
> $$A^{-1} = \frac{1}{D} \begin{bmatrix} d & -b \\ -c & a \end{bmatrix} \tag{14}$$

Note   As an aid to memorizing the formula, note that D is the product of the elements along the main diagonal minus the product of the elements along the other diagonal:

$$\begin{bmatrix} a & b \\ c & d \end{bmatrix} \qquad D = ad - bc$$

Main diagonal

Next, the matrix

$$\begin{bmatrix} d & -b \\ -c & a \end{bmatrix}$$

is obtained by interchanging a and d and reversing the signs of b and c. Finally, A^{-1} is obtained by dividing this matrix by D.

Explore & Discuss

Suppose A is a square matrix with the property that one of its rows is a nonzero constant multiple of another row. What can you say about the existence or nonexistence of A^{-1}? Explain your answer.

EXAMPLE 3   Find the inverse of

$$A = \begin{bmatrix} 1 & 2 \\ 3 & 4 \end{bmatrix}$$

Solution   We first compute $D = (1)(4) - (2)(3) = 4 - 6 = -2$. Next, we rewrite the given matrix obtaining

$$\begin{bmatrix} 4 & -2 \\ -3 & 1 \end{bmatrix}$$

Finally, dividing this matrix by D, we obtain

$$A^{-1} = -\frac{1}{2}\begin{bmatrix} 4 & -2 \\ -3 & 1 \end{bmatrix} = \begin{bmatrix} -2 & 1 \\ \frac{3}{2} & -\frac{1}{2} \end{bmatrix}$$

Solving Systems of Equations with Inverses

We now show how the inverse of a matrix may be used to solve certain systems of linear equations in which the number of equations in the system is equal to the number of variables. For simplicity, let's illustrate the process for a system of three linear equations in three variables:

$$
\begin{aligned}
a_{11}x_1 + a_{12}x_2 + a_{13}x_3 &= b_1 \\
a_{21}x_1 + a_{22}x_2 + a_{23}x_3 &= b_2 \\
a_{31}x_1 + a_{32}x_2 + a_{33}x_3 &= b_3
\end{aligned}
\tag{15}
$$

Let's write

$$
A = \begin{bmatrix} a_{11} & a_{12} & a_{13} \\ a_{21} & a_{22} & a_{23} \\ a_{31} & a_{32} & a_{33} \end{bmatrix} \qquad X = \begin{bmatrix} x_1 \\ x_2 \\ x_3 \end{bmatrix} \qquad B = \begin{bmatrix} b_1 \\ b_2 \\ b_3 \end{bmatrix}
$$

You should verify that System (15) of linear equations may be written in the form of the matrix equation

$$
AX = B \tag{16}
$$

If A is nonsingular, then the method of this section may be used to compute A^{-1}. Next, multiplying both sides of Equation (16) by A^{-1} (on the left), we obtain

$$
A^{-1}AX = A^{-1}B \quad \text{or} \quad IX = A^{-1}B \quad \text{or} \quad X = A^{-1}B
$$

the desired solution to the problem.

In the case of a system of n equations with n unknowns, we have the following more general result.

Using Inverses to Solve Systems of Equations

If $AX = B$ is a linear system of n equations in n unknowns and if A^{-1} exists, then

$$
X = A^{-1}B
$$

is the unique solution of the system.

The use of inverses to solve systems of equations is particularly advantageous when we are required to solve more than one system of equations, $AX = B$, involving the same coefficient matrix, A, and different matrices of constants, B. As you will see in Examples 4 and 5, we need to compute A^{-1} just once in each case.

EXAMPLE 4 Solve the following systems of linear equations:

a. $\begin{aligned} 2x + y + z &= 1 \\ 3x + 2y + z &= 2 \\ 2x + y + 2z &= -1 \end{aligned}$ **b.** $\begin{aligned} 2x + y + z &= 2 \\ 3x + 2y + z &= -3 \\ 2x + y + 2z &= 1 \end{aligned}$

Solution   We may write the given systems of equations in the form

$$
AX = B \quad \text{and} \quad AX = C
$$

respectively, where

$$
A = \begin{bmatrix} 2 & 1 & 1 \\ 3 & 2 & 1 \\ 2 & 1 & 2 \end{bmatrix} \qquad X = \begin{bmatrix} x \\ y \\ z \end{bmatrix} \qquad B = \begin{bmatrix} 1 \\ 2 \\ -1 \end{bmatrix} \qquad C = \begin{bmatrix} 2 \\ -3 \\ 1 \end{bmatrix}
$$

The inverse of the matrix A,

$$A^{-1} = \begin{bmatrix} 3 & -1 & -1 \\ -4 & 2 & 1 \\ -1 & 0 & 1 \end{bmatrix}$$

was found in Example 1. Using this result, we find that the solution of the first system (a) is

$$X = A^{-1}B = \begin{bmatrix} 3 & -1 & -1 \\ -4 & 2 & 1 \\ -1 & 0 & 1 \end{bmatrix}\begin{bmatrix} 1 \\ 2 \\ -1 \end{bmatrix}$$

$$= \begin{bmatrix} (3)(1) + (-1)(2) + (-1)(-1) \\ (-4)(1) + (2)(2) + (1)(-1) \\ (-1)(1) + (0)(2) + (1)(-1) \end{bmatrix} = \begin{bmatrix} 2 \\ -1 \\ -2 \end{bmatrix}$$

or $x = 2$, $y = -1$, and $z = -2$.

The solution of the second system (b) is

$$X = A^{-1}C = \begin{bmatrix} 3 & -1 & -1 \\ -4 & 2 & 1 \\ -1 & 0 & 1 \end{bmatrix}\begin{bmatrix} 2 \\ -3 \\ 1 \end{bmatrix} = \begin{bmatrix} 8 \\ -13 \\ -1 \end{bmatrix}$$

or $x = 8$, $y = -13$, and $z = -1$.   ∎

APPLIED EXAMPLE 5 Capital Expenditure Planning The management of Checkers Rent-A-Car plans to expand its fleet of rental cars for the next quarter by purchasing compact and full-size cars. The average cost of a compact car is $15,000, and the average cost of a full-size car is $36,000.

a. If a total of 800 cars is to be purchased with a budget of $18 million, how many cars of each size will be acquired?
b. If the predicted demand calls for a total purchase of 1000 cars with a budget of $21 million, how many cars of each type will be acquired?

Solution Let x and y denote the number of compact and full-size cars, respectively, to be purchased. Furthermore, let n denote the total number of cars to be acquired, and let b denote the amount of money budgeted for the purchase of these cars. Then

$$x + y = n$$
$$15{,}000x + 36{,}000y = b$$

This system of two equations in two variables may be written in the matrix form

$$AX = B$$

where

$$A = \begin{bmatrix} 1 & 1 \\ 15{,}000 & 36{,}000 \end{bmatrix} \qquad X = \begin{bmatrix} x \\ y \end{bmatrix} \qquad B = \begin{bmatrix} n \\ b \end{bmatrix}$$

Therefore,

$$X = A^{-1}B$$

Since A is a 2×2 matrix, its inverse may be found by using Formula (14). We find $D = (1)(36{,}000) - (1)(15{,}000) = 21{,}000$, so

$$A^{-1} = \frac{1}{21{,}000}\begin{bmatrix} 36{,}000 & -1 \\ -15{,}000 & 1 \end{bmatrix} = \begin{bmatrix} \frac{36{,}000}{21{,}000} & -\frac{1}{21{,}000} \\ -\frac{15{,}000}{21{,}000} & \frac{1}{21{,}000} \end{bmatrix}$$

Thus,

$$X = \begin{bmatrix} \frac{12}{7} & -\frac{1}{21,000} \\ -\frac{5}{7} & \frac{1}{21,000} \end{bmatrix} \begin{bmatrix} n \\ b \end{bmatrix}$$

a. Here, $n = 800$ and $b = 18,000,000$, so

$$X = A^{-1}B = \begin{bmatrix} \frac{12}{7} & -\frac{1}{21,000} \\ -\frac{5}{7} & \frac{1}{21,000} \end{bmatrix} \begin{bmatrix} 800 \\ 18,000,000 \end{bmatrix} \approx \begin{bmatrix} 514.3 \\ 285.7 \end{bmatrix}$$

Therefore, 514 compact cars and 286 full-size cars will be acquired in this case.

b. Here, $n = 1000$ and $b = 21,000,000$, so

$$X = A^{-1}B = \begin{bmatrix} \frac{12}{7} & -\frac{1}{21,000} \\ -\frac{5}{7} & \frac{1}{21,000} \end{bmatrix} \begin{bmatrix} 1000 \\ 21,000,000 \end{bmatrix} \approx \begin{bmatrix} 714.3 \\ 285.7 \end{bmatrix}$$

Therefore, 714 compact cars and 286 full-size cars will be purchased in this case.
■

5.6 Self-Check Exercises

1. Find the inverse of the matrix

$$A = \begin{bmatrix} 2 & 1 & -1 \\ 1 & 1 & -1 \\ -1 & -2 & 3 \end{bmatrix}$$

if it exists.

2. Solve the system of linear equations

$$2x + y - z = b_1$$
$$x + y - z = b_2$$
$$-x - 2y + 3z = b_3$$

where (a) $b_1 = 5$, $b_2 = 4$, $b_3 = -8$ and (b) $b_1 = 2$, $b_2 = 0$, $b_3 = 5$, by finding the inverse of the coefficient matrix.

3. Grand Canyon Tours offers air and ground scenic tours of the Grand Canyon. Tickets for the $7\frac{1}{2}$-hour tour cost $169 for an adult and $129 for a child, and each tour group is limited to 19 people. On three recent fully booked tours, total receipts were $2931 for the first tour, $3011 for the second tour, and $2771 for the third tour. Determine how many adults and how many children were in each tour.

Solutions to Self-Check Exercises 5.6 can be found on page 320.

5.6 Concept Questions

1. What is the inverse of a matrix A?

2. Explain how you would find the inverse of a nonsingular matrix.

3. Give the formula for the inverse of the 2 × 2 matrix

$$A = \begin{bmatrix} a & b \\ c & d \end{bmatrix}$$

4. Explain how the inverse of a matrix can be used to solve a system of n linear equations in n unknowns. Can the method work for a system of m linear equations in n unknowns with $m \neq n$? Explain.

5.6 Exercises

In Exercises 1–4, show that the matrices are inverses of each other by showing that their product is the identity matrix *I*.

1. $\begin{bmatrix} 1 & -3 \\ 1 & -2 \end{bmatrix}$ and $\begin{bmatrix} -2 & 3 \\ -1 & 1 \end{bmatrix}$

2. $\begin{bmatrix} 4 & 5 \\ 2 & 3 \end{bmatrix}$ and $\begin{bmatrix} \frac{3}{2} & -\frac{5}{2} \\ -1 & 2 \end{bmatrix}$

3. $\begin{bmatrix} 3 & 2 & 3 \\ 2 & 2 & 1 \\ 2 & 1 & 1 \end{bmatrix}$ and $\begin{bmatrix} -\frac{1}{3} & -\frac{1}{3} & \frac{4}{3} \\ 0 & 1 & -1 \\ \frac{2}{3} & -\frac{1}{3} & -\frac{2}{3} \end{bmatrix}$

4. $\begin{bmatrix} 2 & 4 & -2 \\ -4 & -6 & 1 \\ 3 & 5 & -1 \end{bmatrix}$ and $\begin{bmatrix} \frac{1}{2} & -3 & -4 \\ -\frac{1}{2} & 2 & 3 \\ -1 & 1 & 2 \end{bmatrix}$

In Exercises 5–16, find the inverse of the matrix, if it exists. Verify your answer.

5. $\begin{bmatrix} 2 & 5 \\ 1 & 3 \end{bmatrix}$

6. $\begin{bmatrix} 2 & 3 \\ 3 & 5 \end{bmatrix}$

7. $\begin{bmatrix} 3 & -3 \\ -2 & 2 \end{bmatrix}$

8. $\begin{bmatrix} 4 & 2 \\ 6 & 3 \end{bmatrix}$

9. $\begin{bmatrix} 2 & -3 & -4 \\ 0 & 0 & -1 \\ 1 & -2 & 1 \end{bmatrix}$

10. $\begin{bmatrix} 1 & -1 & 3 \\ 2 & 1 & 2 \\ -2 & -2 & 1 \end{bmatrix}$

11. $\begin{bmatrix} 4 & 2 & 2 \\ -1 & -3 & 4 \\ 3 & -1 & 6 \end{bmatrix}$

12. $\begin{bmatrix} 1 & 2 & 0 \\ -3 & 4 & -2 \\ -5 & 0 & -2 \end{bmatrix}$

13. $\begin{bmatrix} 1 & 4 & -1 \\ 2 & 3 & -2 \\ -1 & 2 & 3 \end{bmatrix}$

14. $\begin{bmatrix} 3 & -2 & 7 \\ -2 & 1 & 4 \\ 6 & -5 & 8 \end{bmatrix}$

15. $\begin{bmatrix} 1 & 1 & -1 & 1 \\ 2 & 1 & 1 & 0 \\ 2 & 1 & 0 & 1 \\ 2 & -1 & -1 & 3 \end{bmatrix}$

16. $\begin{bmatrix} 1 & 1 & 2 & 3 \\ 2 & 3 & 0 & -1 \\ 0 & 2 & -1 & 1 \\ 1 & 2 & 1 & 1 \end{bmatrix}$

In Exercises 17–24, (a) write a matrix equation that is equivalent to the system of linear equations, and (b) solve the system using the inverses found in Exercises 5–16.

17. $2x + 5y = 3$
$x + 3y = 2$
(See Exercise 5.)

18. $2x + 3y = 5$
$3x + 5y = 8$
(See Exercise 6.)

19. $2x - 3y - 4z = 4$
$\qquad\qquad -z = 3$
$x - 2y + z = -8$
(See Exercise 9.)

20. $x_1 - x_2 + 3x_3 = 2$
$2x_1 + x_2 + 2x_3 = 2$
$-2x_1 - 2x_2 + x_3 = 3$
(See Exercise 10.)

21. $x + 4y - z = 3$
$2x + 3y - 2z = 1$
$-x + 2y + 3z = 7$
(See Exercise 13.)

22. $3x_1 - 2x_2 + 7x_3 = 6$
$-2x_1 + x_2 + 4x_3 = 4$
$6x_1 - 5x_2 + 8x_3 = 4$
(See Exercise 14.)

23. $x_1 + x_2 - x_3 + x_4 = 6$
$2x_1 + x_2 + x_3 \qquad = 4$
$2x_1 + x_2 \qquad + x_4 = 7$
$2x_1 - x_2 - x_3 + 3x_4 = 9$
(See Exercise 15.)

24. $x_1 + x_2 + 2x_3 + 3x_4 = 4$
$2x_1 + 3x_2 \qquad - x_4 = 11$
$\qquad 2x_2 - x_3 + x_4 = 7$
$x_1 + 2x_2 + x_3 + x_4 = 6$
(See Exercise 16.)

In Exercises 25–32, (a) write each system of equations as a matrix equation, and (b) solve the system of equations by using the inverse of the coefficient matrix.

25.
$$x + 2y = b_1$$
$$2x - y = b_2$$
where (i) $b_1 = 14, b_2 = 5$
and (ii) $b_1 = 4, b_2 = -1$

26.
$$3x - 2y = b_1$$
$$4x + 3y = b_2$$
where (i) $b_1 = -6, b_2 = 10$
and (ii) $b_1 = 3, b_2 = -2$

27.
$$x + 2y + z = b_1$$
$$x + y + z = b_2$$
$$3x + y + z = b_3$$
where (i) $b_1 = 7, b_2 = 4, b_3 = 2$
and (ii) $b_1 = 5, b_2 = -3, b_3 = -1$

28.
$$x_1 + x_2 + x_3 = b_1$$
$$x_1 - x_2 + x_3 = b_2$$
$$x_1 - 2x_2 - x_3 = b_3$$
where (i) $b_1 = 5, b_2 = -3, b_3 = -1$
and (ii) $b_1 = 1, b_2 = 4, b_3 = -2$

29.
$$3x + 2y - z = b_1$$
$$2x - 3y + z = b_2$$
$$x - y - z = b_3$$
where (i) $b_1 = 2, b_2 = -2, b_3 = 4$
and (ii) $b_1 = 8, b_2 = -3, b_3 = 6$

30.
$$2x_1 + x_2 + x_3 = b_1$$
$$x_1 - 3x_2 + 4x_3 = b_2$$
$$-x_1 \qquad + x_3 = b_3$$
where (i) $b_1 = 1, b_2 = 4, b_3 = -3$
and (ii) $b_1 = 2, b_2 = -5, b_3 = 0$

31.
$$x_1 + x_2 + x_3 + x_4 = b_1$$
$$x_1 - x_2 - x_3 + x_4 = b_2$$
$$x_2 + 2x_3 + 2x_4 = b_3$$
$$x_1 + 2x_2 + x_3 - 2x_4 = b_4$$
where (i) $b_1 = 1, b_2 = -1, b_3 = 4, b_4 = 0$
and (ii) $b_1 = 2, b_2 = 8, b_3 = 4, b_4 = -1$

32.
$$x_1 + x_2 + 2x_3 + x_4 = b_1$$
$$4x_1 + 5x_2 + 9x_3 + x_4 = b_2$$
$$3x_1 + 4x_2 + 7x_3 + x_4 = b_3$$
$$2x_1 + 3x_2 + 4x_3 + 2x_4 = b_4$$
where (i) $b_1 = 3, b_2 = 6, b_3 = 5, b_4 = 7$
and (ii) $b_1 = 1, b_2 = -1, b_3 = 0, b_4 = -4$

33. Let
$$A = \begin{bmatrix} 2 & 3 \\ -4 & -5 \end{bmatrix}$$
a. Find A^{-1}.
b. Show that $(A^{-1})^{-1} = A$.

34. Let

$$A = \begin{bmatrix} 6 & -4 \\ -4 & 3 \end{bmatrix} \quad \text{and} \quad B = \begin{bmatrix} 3 & -5 \\ 4 & -7 \end{bmatrix}$$

a. Find AB, A^{-1}, and B^{-1}.
b. Show that $(AB)^{-1} = B^{-1}A^{-1}$.

35. Let

$$A = \begin{bmatrix} 2 & -5 \\ 1 & -3 \end{bmatrix} \quad B = \begin{bmatrix} 4 & 3 \\ 1 & 1 \end{bmatrix} \quad C = \begin{bmatrix} 2 & 3 \\ -2 & 1 \end{bmatrix}$$

a. Find ABC, A^{-1}, B^{-1}, and C^{-1}.
b. Show that $(ABC)^{-1} = C^{-1}B^{-1}A^{-1}$.

36. Find the matrix A if

$$\begin{bmatrix} 2 & 1 \\ -1 & 3 \end{bmatrix} A = \begin{bmatrix} 3 & 2 \\ 1 & 4 \end{bmatrix}$$

37. Find the matrix A if

$$A \begin{bmatrix} 1 & 2 \\ 3 & -1 \end{bmatrix} = \begin{bmatrix} 2 & 1 \\ 3 & -2 \end{bmatrix}$$

38. TICKET REVENUES Rainbow Harbor Cruises charges $16/adult and $8/child for a round-trip ticket. The records show that, on a certain weekend, 1000 people took the cruise on Saturday, and 800 people took the cruise on Sunday. The total receipts for Saturday were $12,800, and the total receipts for Sunday were $9,600. Determine how many adults and children took the cruise on Saturday and on Sunday.

39. PRICING BelAir Publishing publishes a deluxe leather edition and a standard edition of its daily organizer. The company's marketing department estimates that x copies of the deluxe edition and y copies of the standard edition will be demanded per month when the unit prices are p dollars and q dollars, respectively, where x, y, p, and q are related by the following system of linear equations:

$$5x + y = 1000(70 - p)$$
$$x + 3y = 1000(40 - q)$$

Find the monthly demand for the deluxe edition and the standard edition when the unit prices are set according to the following schedules:
a. $p = 50$ and $q = 25$
b. $p = 45$ and $q = 25$
c. $p = 45$ and $q = 20$

40. NUTRITION/DIET PLANNING Bob, a nutritionist who works for the University Medical Center, has been asked to prepare special diets for two patients, Susan and Tom. Bob has decided that Susan's meals should contain at least 400 mg of calcium, 20 mg of iron, and 50 mg of vitamin C, whereas Tom's meals should contain at least 350 mg of calcium, 15 mg of iron, and 40 mg of vitamin C. Bob has also decided that the meals are to be prepared from three basic foods: Food A, Food B, and Food C. The special nutritional contents of these foods are summarized in the accompanying table. Find how many ounces of each type of food should be used in a meal so that the minimum requirements of calcium, iron, and vitamin C are met for each patient's meals.

	Contents (mg/oz)		
	Calcium	Iron	Vitamin C
Food A	30	1	2
Food B	25	1	5
Food C	20	2	4

41. AGRICULTURE Jackson Farms has allotted a certain amount of land for cultivating soybeans, corn, and wheat. Cultivating 1 acre of soybeans requires 2 labor-hours, and cultivating 1 acre of corn or wheat requires 6 labor-hours. The cost of seeds for 1 acre of soybeans is $12, the cost for 1 acre of corn is $20, and the cost for 1 acre of wheat is $8. If all resources are to be used, how many acres of each crop should be cultivated if the following hold?
a. 1000 acres of land are allotted, 4400 labor-hours are available, and $13,200 is available for seeds.
b. 1200 acres of land are allotted, 5200 labor-hours are available, and $16,400 is available for seeds.

42. MIXTURE PROBLEM—FERTILIZER Lawnco produces three grades of commercial fertilizers. A 100-lb bag of grade A fertilizer contains 18 lb of nitrogen, 4 lb of phosphate, and 5 lb of potassium. A 100-lb bag of grade B fertilizer contains 20 lb of nitrogen and 4 lb each of phosphate and potassium. A 100-lb bag of grade C fertilizer contains 24 lb of nitrogen, 3 lb of phosphate, and 6 lb of potassium. How many 100-lb bags of each of the three grades of fertilizers should Lawnco produce if
a. 26,400 lb of nitrogen, 4900 lb of phosphate, and 6200 lb of potassium are available and all the nutrients are used?
b. 21,800 lb of nitrogen, 4200 lb of phosphate, and 5300 lb of potassium are available and all the nutrients are used?

43. INVESTMENT CLUBS A private investment club has a certain amount of money earmarked for investment in stocks. To arrive at an acceptable overall level of risk, the stocks that management is considering have been classified into three categories: high-risk, medium-risk, and low-risk. Management estimates that high-risk stocks will have a rate of return of 15%/year; medium-risk stocks, 10%/year; and low-risk stocks, 6%/year. The members have decided that the investment in low-risk stocks should be equal to the sum of the investments in the stocks of the other two categories. Determine how much the club should invest in each type of stock in each of the following scenarios. (In all cases, assume that the entire sum available for investment is invested.)
a. The club has $200,000 to invest, and the investment goal is to have a return of $20,000/year on the total investment.
b. The club has $220,000 to invest, and the investment goal is to have a return of $22,000/year on the total investment.
c. The club has $240,000 to invest, and the investment goal is to have a return of $22,000/year on the total investment.

44. RESEARCH FUNDING The Carver Foundation funds three nonprofit organizations engaged in alternative-energy research activities. From past data, the proportion of funds spent by each organization in research on solar energy, energy from harnessing the wind, and energy from the motion of ocean tides is given in the accompanying table.

| | Proportion of Money Spent | | |
	Solar	Wind	Tides
Organization I	0.6	0.3	0.1
Organization II	0.4	0.3	0.3
Organization III	0.2	0.6	0.2

Find the amount awarded to each organization if the total amount spent by all three organizations on solar, wind, and tidal research is

a. $9.2 million, $9.6 million, and $5.2 million, respectively.
b. $8.2 million, $7.2 million, and $3.6 million, respectively.

45. Find the value(s) of k such that

$$A = \begin{bmatrix} 1 & 2 \\ k & 3 \end{bmatrix}$$

has an inverse. What is the inverse of A?
Hint: Use Formula (14).

46. Find the value(s) of k such that

$$A = \begin{bmatrix} 1 & 0 & 1 \\ -2 & 1 & k \\ -1 & 2 & k^2 \end{bmatrix}$$

has an inverse.
Hint: Find the value(s) of k such that the augmented matrix $[A \mid I]$ can be reduced to the form $[I \mid B]$.

47. Find conditions on a and d such that the matrix

$$A = \begin{bmatrix} a & 0 \\ 0 & d \end{bmatrix}$$

has an inverse. A square matrix is said to be a *diagonal matrix* if all the entries not lying on the main diagonal are

zero. Discuss the existence of the inverse matrix of a diagonal matrix of size $n \times n$.

48. Find conditions a, b, and d such that the 2×2 upper triangular matrix

$$A = \begin{bmatrix} a & b \\ 0 & d \end{bmatrix}$$

has an inverse. A square matrix is said to be an *upper triangular matrix* if all its entries below the main diagonal are zero. Discuss the existence of the inverse of an upper triangular matrix of size $n \times n$.

In Exercises 49–51, determine whether the statement is true or false. If it is true, explain why it is true. If it is false, give an example to show why it is false.

49. If A is a square matrix with inverse A^{-1} and c is a nonzero real number, then

$$(cA)^{-1} = \left(\frac{1}{c}\right)A^{-1}$$

50. The matrix

$$A = \begin{bmatrix} a & b \\ c & d \end{bmatrix}$$

has an inverse if and only if $ad - bc = 0$.

51. If A^{-1} does not exist, then the system $AX = B$ of n linear equations in n unknowns does not have a unique solution.

52. Let

$$A = \begin{bmatrix} a & b \\ c & d \end{bmatrix}$$

a. Find A^{-1} if it exists.
b. Find a necessary condition for A to be nonsingular.
c. Verify that $AA^{-1} = A^{-1}A = I$.

5.6 Solutions to Self-Check Exercises

1. We form the augmented matrix

$$\begin{bmatrix} 2 & 1 & -1 & | & 1 & 0 & 0 \\ 1 & 1 & -1 & | & 0 & 1 & 0 \\ -1 & -2 & 3 & | & 0 & 0 & 1 \end{bmatrix}$$

and row-reduce as follows:

$$\begin{bmatrix} 2 & 1 & -1 & | & 1 & 0 & 0 \\ 1 & 1 & -1 & | & 0 & 1 & 0 \\ -1 & -2 & 3 & | & 0 & 0 & 1 \end{bmatrix} \xrightarrow{R_1 \leftrightarrow R_2}$$

$$\begin{bmatrix} 1 & 1 & -1 & | & 0 & 1 & 0 \\ 2 & 1 & -1 & | & 1 & 0 & 0 \\ -1 & -2 & 3 & | & 0 & 0 & 1 \end{bmatrix} \xrightarrow[R_3 + R_1]{R_2 - 2R_1}$$

$$\begin{bmatrix} 1 & 1 & -1 & | & 0 & 1 & 0 \\ 0 & -1 & 1 & | & 1 & -2 & 0 \\ 0 & -1 & 2 & | & 0 & 1 & 1 \end{bmatrix} \xrightarrow[R_3 - R_2]{R_1 + R_2}$$

$$\begin{bmatrix} 1 & 0 & 0 & | & 1 & -1 & 0 \\ 0 & 1 & -1 & | & -1 & 2 & 0 \\ 0 & 0 & 1 & | & -1 & 3 & 1 \end{bmatrix} \xrightarrow{R_2 + R_3}$$

$$\begin{bmatrix} 1 & 0 & 0 & | & 1 & -1 & 0 \\ 0 & 1 & 0 & | & -2 & 5 & 1 \\ 0 & 0 & 1 & | & -1 & 3 & 1 \end{bmatrix}$$

From the preceding results, we see that

$$A^{-1} = \begin{bmatrix} 1 & -1 & 0 \\ -2 & 5 & 1 \\ -1 & 3 & 1 \end{bmatrix}$$

2. a. We write the systems of linear equations in the matrix form

$$AX = B_1$$

where

$$A = \begin{bmatrix} 2 & 1 & -1 \\ 1 & 1 & -1 \\ -1 & -2 & 3 \end{bmatrix} \quad X = \begin{bmatrix} x \\ y \\ z \end{bmatrix} \quad B_1 = \begin{bmatrix} 5 \\ 4 \\ -8 \end{bmatrix}$$

Now, using the results of Exercise 1, we have

$$X = \begin{bmatrix} x \\ y \\ z \end{bmatrix} = A^{-1}B_1 = \begin{bmatrix} 1 & -1 & 0 \\ -2 & 5 & 1 \\ -1 & 3 & 1 \end{bmatrix}\begin{bmatrix} 5 \\ 4 \\ -8 \end{bmatrix} = \begin{bmatrix} 1 \\ 2 \\ -1 \end{bmatrix}$$

Therefore, $x = 1$, $y = 2$, and $z = -1$.

b. Here, A and X are as in part (a), but

$$B_2 = \begin{bmatrix} 2 \\ 0 \\ 5 \end{bmatrix}$$

Therefore,

$$X = \begin{bmatrix} x \\ y \\ z \end{bmatrix} = A^{-1}B_2 = \begin{bmatrix} 1 & -1 & 0 \\ -2 & 5 & 1 \\ -1 & 3 & 1 \end{bmatrix}\begin{bmatrix} 2 \\ 0 \\ 5 \end{bmatrix} = \begin{bmatrix} 2 \\ 1 \\ 3 \end{bmatrix}$$

or $x = 2$, $y = 1$, and $z = 3$.

3. Let x denote the number of adults, and let y denote the number of children on a tour. Since the tours are filled to capacity, we have

$$x + y = 19$$

Next, since the total receipts for the first tour were $2931, we have

$$169x + 129y = 2931$$

Therefore, the number of adults and the number of children in the first tour are found by solving the system of linear equations

$$\begin{align} x + y &= 19 \\ 169x + 129y &= 2931 \end{align} \quad \text{(a)}$$

Similarly, we see that the number of adults and the number of children in the second and third tours are found by solving the systems

$$\begin{align} x + y &= 19 \\ 169x + 129y &= 3011 \end{align} \quad \text{(b)}$$

$$\begin{align} x + y &= 19 \\ 169x + 129y &= 2771 \end{align} \quad \text{(c)}$$

These systems may be written in the form

$$AX = B_1 \qquad AX = B_2 \qquad AX = B_3$$

where

$$A = \begin{bmatrix} 1 & 1 \\ 169 & 129 \end{bmatrix} \quad X = \begin{bmatrix} x \\ y \end{bmatrix}$$

$$B_1 = \begin{bmatrix} 19 \\ 2931 \end{bmatrix} \quad B_2 = \begin{bmatrix} 19 \\ 3011 \end{bmatrix} \quad B_3 = \begin{bmatrix} 19 \\ 2771 \end{bmatrix}$$

To solve these systems, we first find A^{-1}. Using Formula (14) with $D = (1)(129) - (1)(169) = -40$, we obtain

$$A^{-1} = -\frac{1}{40}\begin{bmatrix} 129 & -1 \\ -169 & 1 \end{bmatrix} = \begin{bmatrix} -\frac{129}{40} & \frac{1}{40} \\ \frac{169}{40} & -\frac{1}{40} \end{bmatrix}$$

Then, solving each system, we find

$$X = \begin{bmatrix} x \\ y \end{bmatrix} = A^{-1}B_1$$

$$= \begin{bmatrix} -\frac{129}{40} & \frac{1}{40} \\ \frac{169}{40} & -\frac{1}{40} \end{bmatrix}\begin{bmatrix} 19 \\ 2931 \end{bmatrix} = \begin{bmatrix} 12 \\ 7 \end{bmatrix} \quad \text{(a)}$$

$$X = \begin{bmatrix} x \\ y \end{bmatrix} = A^{-1}B_2$$

$$= \begin{bmatrix} -\frac{129}{40} & \frac{1}{40} \\ \frac{169}{40} & -\frac{1}{40} \end{bmatrix}\begin{bmatrix} 19 \\ 3011 \end{bmatrix}$$

$$= \begin{bmatrix} 14 \\ 5 \end{bmatrix} \quad \text{(b)}$$

$$X = \begin{bmatrix} x \\ y \end{bmatrix} = A^{-1}B_3$$

$$= \begin{bmatrix} -\frac{129}{40} & \frac{1}{40} \\ \frac{169}{40} & -\frac{1}{40} \end{bmatrix}\begin{bmatrix} 19 \\ 2771 \end{bmatrix} = \begin{bmatrix} 8 \\ 11 \end{bmatrix} \quad \text{(c)}$$

We conclude that there were:
a. 12 adults and 7 children on the first tour.
b. 14 adults and 5 children on the second tour.
c. 8 adults and 11 children on the third tour.

USING TECHNOLOGY

Finding the Inverse of a Square Matrix

Graphing Utility

A graphing utility can be used to find the inverse of a square matrix.

(continued)

EXAMPLE 1 Use a graphing utility to find the inverse of

$$\begin{bmatrix} 1 & 3 & 5 \\ -2 & 2 & 4 \\ 5 & 1 & 3 \end{bmatrix}$$

Solution   We first enter the given matrix as

$$A = \begin{bmatrix} 1 & 3 & 5 \\ -2 & 2 & 4 \\ 5 & 1 & 3 \end{bmatrix}$$

Then, recalling the matrix A and using the $\boxed{x^{-1}}$ key, we find

$$A^{-1} = \begin{bmatrix} 0.1 & -0.2 & 0.1 \\ 1.3 & -1.1 & -0.7 \\ -0.6 & 0.7 & 0.4 \end{bmatrix}$$

EXAMPLE 2 Use a graphing utility to solve the system

$$
\begin{aligned}
x + 3y + 5z &= 4 \\
-2x + 2y + 4z &= 3 \\
5x + y + 3z &= 2
\end{aligned}
$$

by using the inverse of the coefficient matrix.

Solution   The given system can be written in the matrix form $AX = B$, where

$$A = \begin{bmatrix} 1 & 3 & 5 \\ -2 & 2 & 4 \\ 5 & 1 & 3 \end{bmatrix} \qquad X = \begin{bmatrix} x \\ y \\ z \end{bmatrix} \qquad B = \begin{bmatrix} 4 \\ 3 \\ 2 \end{bmatrix}$$

The solution is $X = A^{-1}B$. Entering the matrices A and B in the graphing utility and using the matrix multiplication capability of the utility gives the output shown in Figure T1—that is, $x = 0$, $y = 0.5$, and $z = 0.5$.

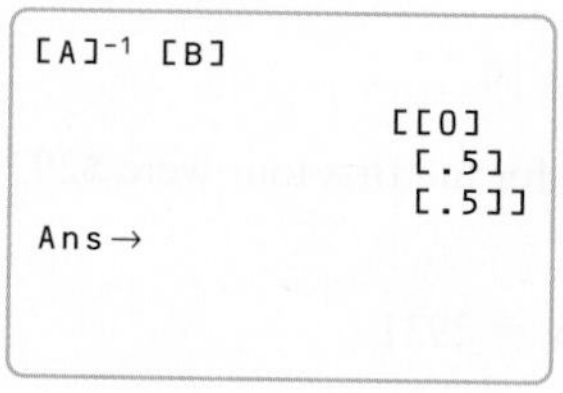

```
[A]⁻¹ [B]
              [[0]
               [.5]
               [.5]]
Ans→
```

FIGURE **T1**
The TI-83/84 screen showing $A^{-1}B$

Excel

We use the function **MINVERSE** to find the inverse of a square matrix using Excel.

EXAMPLE 3   Find the inverse of

$$A = \begin{bmatrix} 1 & 3 & 5 \\ -2 & 2 & 4 \\ 5 & 1 & 3 \end{bmatrix}$$

Note: Boldfaced words/characters enclosed in a box (for example, $\boxed{\text{Enter}}$) indicate that an action (click, select, or press) is required. Words/characters printed blue (for example, Chart sub-type:) indicate words/characters on the screen. Words/characters printed in a monospace font (for example, =(−2/3)*A2+2) indicate words/characters that need to be typed and entered.

Solution

1. Enter the elements of matrix A onto a spreadsheet (Figure T2).
2. Compute the inverse of the matrix A: Highlight the cells that will contain the inverse matrix A^{-1}, type = MINVERSE (, highlight the cells containing matrix A, type), and press Ctrl-Shift-Enter. The desired matrix will appear in your spreadsheet (Figure T2).

	A	B	C
1		Matrix A	
2	1	3	5
3	-2	2	4
4	5	1	3
5			
6		Matrix A^{-1}	
7	0.1	-0.2	0.1
8	1.3	-1.1	-0.7
9	-0.6	0.7	0.4

FIGURE **T2**
Matrix A and its inverse, matrix A^{-1}

EXAMPLE 4 Solve the system

$$\begin{aligned} x + 3y + 5z &= 4 \\ -2x + 2y + 4z &= 3 \\ 5x + y + 3z &= 2 \end{aligned}$$

by using the inverse of the coefficient matrix.

Solution The given system can be written in the matrix form $AX = B$, where

$$A = \begin{bmatrix} 1 & 3 & 5 \\ -2 & 2 & 4 \\ 5 & 1 & 3 \end{bmatrix} \qquad X = \begin{bmatrix} x \\ y \\ z \end{bmatrix} \qquad B = \begin{bmatrix} 4 \\ 3 \\ 2 \end{bmatrix}$$

The solution is $X = A^{-1}B$.

1. *Enter the matrix B on a spreadsheet.*
2. *Compute $A^{-1}B$.* Highlight the cells that will contain the matrix X, and then type =MMULT (, highlight the cells in the matrix A^{-1}, type , , highlight the cells in the matrix B, type), and press Ctrl-Shift-Enter. (*Note*: The matrix A^{-1} was found in Example 3.) The matrix X shown in Figure T3 will appear on your spreadsheet. Thus, $x = 0$, $y = 0.5$, and $z = 0.5$.

	A
12	Matrix X
13	5.55112E-17
14	0.5
15	0.5

FIGURE **T3**
Matrix X gives the solution to the problem.

TECHNOLOGY EXERCISES

In Exercises 1–6, find the inverse of the matrix. Round your answers to two decimal places.

1. $\begin{bmatrix} 1.2 & 3.1 & -2.1 \\ 3.4 & 2.6 & 7.3 \\ -1.2 & 3.4 & -1.3 \end{bmatrix}$

2. $\begin{bmatrix} 4.2 & 3.7 & 4.6 \\ 2.1 & -1.3 & -2.3 \\ 1.8 & 7.6 & -2.3 \end{bmatrix}$

3. $\begin{bmatrix} 1.1 & 2.3 & 3.1 & 4.2 \\ 1.6 & 3.2 & 1.8 & 2.9 \\ 4.2 & 1.6 & 1.4 & 3.2 \\ 1.6 & 2.1 & 2.8 & 7.2 \end{bmatrix}$

4. $\begin{bmatrix} 2.1 & 3.2 & -1.4 & -3.2 \\ 6.2 & 7.3 & 8.4 & 1.6 \\ 2.3 & 7.1 & 2.4 & -1.3 \\ -2.1 & 3.1 & 4.6 & 3.7 \end{bmatrix}$

5. $\begin{bmatrix} 2 & -1 & 3 & 2 & 4 \\ 3 & 2 & -1 & 4 & 1 \\ 3 & 2 & 6 & 4 & -1 \\ 2 & 1 & -1 & 4 & 2 \\ 3 & 4 & 2 & 5 & 6 \end{bmatrix}$

(*continued*)

6. $\begin{bmatrix} 1 & 4 & 2 & 3 & 1.4 \\ 6 & 2.4 & 5 & 1.2 & 3 \\ 4 & 1 & 2 & 3 & 1.2 \\ -1 & 2 & -3 & 4 & 2 \\ 1.1 & 2.2 & 3 & 5.1 & 4 \end{bmatrix}$

In Exercises 7–10, solve the system of linear equations by first writing the system in the form $AX = B$ and then solving the resulting system by using A^{-1}. Round your answers to two decimal places.

7. $\begin{aligned} 2x - 3y + 4z &= 2.4 \\ 3x + 2y - 7z &= -8.1 \\ x + 4y - 2z &= 10.2 \end{aligned}$

8. $\begin{aligned} 3.2x - 4.7y + 3.2z &= 7.1 \\ 2.1x + 2.6y + 6.2z &= 8.2 \\ 5.1x - 3.1y - 2.6z &= -6.5 \end{aligned}$

9. $\begin{aligned} 3x_1 - 2x_2 + 4x_3 - 8x_4 &= 8 \\ 2x_1 + 3x_2 - 2x_3 + 6x_4 &= 4 \\ 3x_1 + 2x_2 - 6x_3 - 7x_4 &= -2 \\ 4x_1 - 7x_2 + 4x_3 + 6x_4 &= 22 \end{aligned}$

10. $\begin{aligned} 1.2x_1 + 2.1x_2 - 3.2x_3 + 4.6x_4 &= 6.2 \\ 3.1x_1 - 1.2x_2 + 4.1x_3 - 3.6x_4 &= -2.2 \\ 1.8x_1 + 3.1x_2 - 2.4x_3 + 8.1x_4 &= 6.2 \\ 2.6x_1 - 2.4x_2 + 3.6x_3 - 4.6x_4 &= 3.6 \end{aligned}$

CHAPTER 5   Summary of Principal Formulas and Terms

FORMULAS

1. Laws for matrix addition	
a. Commutative law	$A + B = B + A$
b. Associative law	$(A + B) + C = A + (B + C)$
2. Laws for matrix multiplication	
a. Associative law	$(AB)C = A(BC)$
b. Distributive law	$A(B + C) = AB + AC$
3. Inverse of a 2 × 2 matrix	If $\quad A = \begin{bmatrix} a & b \\ c & d \end{bmatrix}$ and $\quad D = ad - bc \neq 0$ then $\quad A^{-1} = \dfrac{1}{D} \begin{bmatrix} d & -b \\ -c & a \end{bmatrix}$
4. Solution of system $AX = B$ (A nonsingular)	$X = A^{-1}B$

TERMS

system of linear equations (248)

solution of a system of linear equations (248)

parameter (249)

dependent system (250)

inconsistent system (250)

Gauss–Jordan elimination method (256)

equivalent system (256)

coefficient matrix (258)

augmented matrix (258)

row-reduced form of a matrix (259)

row operations (260)

unit column (260)

pivoting (261)

size of a matrix (283)

matrix (283)

row matrix (283)

column matrix (283)

square matrix (283)

transpose of a matrix (287)

scalar (288)

scalar product (288)

matrix product (296)

identity matrix (299)

inverse of a matrix (310)

nonsingular matrix (311)

singular matrix (311)

CHAPTER 5 Concept Review Questions

Fill in the blanks.

1. a. Two lines in the plane can intersect at (a) exactly _____ point, (b) infinitely _____ points, or (c) _____ point.
  b. A system of two linear equations in two variables can have (a) exactly _____ solution, (b) infinitely _____ solutions, or (c) _____ solution.

2. To find the point(s) of intersection of two lines, we solve the system of _____ describing the two lines.

3. The row operations used in the Gauss–Jordan elimination method are denoted by _____, _____, and _____. The use of each of these operations does not alter the _____ of the system of linear equations.

4. a. A system of linear equations with fewer equations than variables cannot have a/an _____ solution.
  b. A system of linear equations with at least as many equations as variables may have _____ solution, _____ _____ solutions, or a _____ solution.

5. Two matrices are equal provided they have the same _____ and their corresponding _____ are equal.

6. Two matrices may be added (subtracted) if they both have the same _____. To add or subtract two matrices, we add or subtract their _____ entries.

7. The transpose of a/an _____ matrix with elements a_{ij} is the matrix of size _____ with entries _____.

8. The scalar product of a matrix A by the scalar c is the matrix _____ obtained by multiplying each entry of A by _____.

9. a. For the product AB of two matrices A and B to be defined, the number of _____ of A must be equal to the number of _____ of B.
  b. If A is an $m \times n$ matrix and B is an $n \times p$ matrix, then the size of AB is _____.

10. a. If the products and sums are defined for the matrices A, B, and C, then the associative law states that $(AB)C =$ _____; the distributive law states that $A(B + C) =$ _____.
  b. If I is an identity matrix of size n, then $IA = A$ if A is any matrix of size _____.

11. A matrix A is nonsingular if there exists a matrix A^{-1} such that _____ = _____ = I. If A^{-1} does not exist, then A is said to be _____.

12. A system of n linear equations in n variables written in the form $AX = B$ has a unique solution given by $X =$ _____ if A has an inverse.

CHAPTER 5 Review Exercises

In Exercises 1–4, perform the operations if possible.

1. $\begin{bmatrix} 1 & 2 \\ -1 & 3 \\ 2 & 1 \end{bmatrix} + \begin{bmatrix} 1 & 0 \\ 0 & 1 \\ 1 & 2 \end{bmatrix}$

2. $\begin{bmatrix} -1 & 2 \\ 3 & 4 \end{bmatrix} - \begin{bmatrix} 1 & 2 \\ 5 & -2 \end{bmatrix}$

3. $\begin{bmatrix} -3 & 2 & 1 \end{bmatrix} \begin{bmatrix} 2 & 1 \\ -1 & 0 \\ 2 & 1 \end{bmatrix}$

4. $\begin{bmatrix} 1 & 3 & 2 \\ -1 & 2 & 3 \end{bmatrix} \begin{bmatrix} 1 \\ 4 \\ 2 \end{bmatrix}$

In Exercises 5–8, find the values of the variables.

5. $\begin{bmatrix} 1 & x \\ y & 3 \end{bmatrix} = \begin{bmatrix} z & 2 \\ 3 & w \end{bmatrix}$

6. $\begin{bmatrix} 3 & x \\ y & 3 \end{bmatrix} \begin{bmatrix} 1 \\ 2 \end{bmatrix} = \begin{bmatrix} 7 \\ 4 \end{bmatrix}$

7. $\begin{bmatrix} 3 & a + 3 \\ -1 & b \\ c + 1 & d \end{bmatrix} = \begin{bmatrix} 3 & 6 \\ e + 2 & 4 \\ -1 & 2 \end{bmatrix}$

8. $\begin{bmatrix} x & 3 & 1 \\ 0 & y & 2 \end{bmatrix} \begin{bmatrix} 1 & 1 \\ 3 & z \\ 4 & 2 \end{bmatrix} = \begin{bmatrix} 12 & 4 \\ 2 & 2 \end{bmatrix}$

In Exercises 9–16, compute the expressions if possible, given that

$$A = \begin{bmatrix} 1 & 3 & 1 \\ -2 & 1 & 3 \\ 4 & 0 & 2 \end{bmatrix}$$

$$B = \begin{bmatrix} 2 & 1 & 3 \\ -2 & -1 & -1 \\ 1 & 4 & 2 \end{bmatrix}$$

$$C = \begin{bmatrix} 3 & -1 & 2 \\ 1 & 6 & 4 \\ 2 & 1 & 3 \end{bmatrix}$$

9. $2A + 3B$

10. $3A - 2B$

11. $2(3A)$

12. $2(3A - 4B)$

13. $A(B - C)$

14. $AB + AC$

15. $A(BC)$

16. $\frac{1}{2}(CA - CB)$

In Exercises 17–24, solve the system of linear equations using the Gauss–Jordan elimination method.

17. $2x - 3y = 5$
$3x + 4y = -1$

18. $3x + 2y = 3$
$2x - 4y = -14$

19. $x - y + 2z = 5$
$3x + 2y + z = 10$
$2x - 3y - 2z = -10$

20. $3x - 2y + 4z = 16$
$2x + y - 2z = -1$
$x + 4y - 8z = -18$

21. $3x - 2y + 4z = 11$
$2x - 4y + 5z = 4$
$x + 2y - z = 10$

22. $x - 2y + 3z + 4w = 17$
$2x + y - 2z - 3w = -9$
$3x - y + 2z - 4w = 0$
$4x + 2y - 3z + w = -2$

23. $3x - 2y + z = 4$
$x + 3y - 4z = -3$
$2x - 3y + 5z = 7$
$x - 8y + 9z = 10$

24. $2x - 3y + z = 10$
$3x + 2y - 2z = -2$
$x - 3y - 4z = -7$
$4x + y - z = 4$

In Exercises 25–32, find the inverse of the matrix (if it exists).

25. $A = \begin{bmatrix} 3 & 1 \\ 1 & 2 \end{bmatrix}$

26. $A = \begin{bmatrix} 2 & 4 \\ 1 & 6 \end{bmatrix}$

27. $A = \begin{bmatrix} 3 & 4 \\ 2 & 2 \end{bmatrix}$

28. $A = \begin{bmatrix} 2 & 4 \\ 1 & -2 \end{bmatrix}$

29. $A = \begin{bmatrix} 2 & 3 & 1 \\ 1 & -1 & 2 \\ 1 & 2 & 1 \end{bmatrix}$

30. $A = \begin{bmatrix} 1 & 2 & 4 \\ 2 & 1 & 3 \\ -1 & 0 & 2 \end{bmatrix}$

31. $A = \begin{bmatrix} 1 & 2 & 4 \\ 3 & 1 & 2 \\ 1 & 0 & -6 \end{bmatrix}$

32. $A = \begin{bmatrix} 2 & 1 & -3 \\ 1 & 2 & -4 \\ 3 & 1 & -2 \end{bmatrix}$

In Exercises 33–36, compute the value of the expressions if possible, given that

$$A = \begin{bmatrix} 1 & 2 \\ -1 & 2 \end{bmatrix} \quad B = \begin{bmatrix} 3 & 1 \\ 4 & 2 \end{bmatrix} \quad C = \begin{bmatrix} 1 & 1 \\ -1 & 2 \end{bmatrix}$$

33. $(A^{-1}B)^{-1}$

34. $(ABC)^{-1}$

35. $(2A - C)^{-1}$

36. $(A + B)^{-1}$

In Exercises 37–40, write each system of linear equations in the form $AX = C$. Find A^{-1} and use the result to solve the system.

37. $2x + 3y = -8$
$x - 2y = 3$

38. $x - 3y = -1$
$2x + 4y = 8$

39. $x - 2y + 4z = 13$
$2x + 3y - 2z = 0$
$x + 4y - 6z = -15$

40. $2x - 3y + 4z = 17$
$x + 2y - 4z = -7$
$3x - y + 2z = 14$

41. GASOLINE SALES Gloria Newburg operates three self-service gasoline stations in different parts of town. On a certain day, station A sold 600 gal of premium, 800 gal of super, 1000 gal of regular gasoline, and 700 gal of diesel fuel; station B sold 700 gal of premium, 600 gal of super, 1200 gal of regular gasoline, and 400 gal of diesel fuel; station C sold 900 gal of premium, 700 gal of super, 1400 gal of regular gasoline, and 800 gal of diesel fuel. Assume that the price of gasoline was \$3.70/gal for premium, \$3.48/gal for super, and \$3.30/gal for regular and that diesel fuel sold for \$3.60/gal. Use matrix algebra to find the total revenue at each station.

42. COMMON STOCK TRANSACTIONS Jack Worthington bought 10,000 shares of Stock X, 20,000 shares of Stock Y, and 30,000 shares of Stock Z at a unit price of \$20, \$30, and \$50 per share, respectively. Six months later, the closing prices of Stocks X, Y, and Z were \$22, \$35, and \$51 per share, respectively. Jack made no other stock transactions during the period in question. Compare the value of Jack's stock holdings at the time of purchase and 6 months later.

43. INVESTMENTS William's and Michael's stock holdings are given in the following table:

	BAC	GM	IBM	TRW
William	800	1200	400	1500
Michael	600	1400	600	2000

The prices (in dollars per share) of the stocks of BAC, GM, IBM, and TRW at the close of the stock market on a certain day are \$50.26, \$31.00, \$103.07, and \$38.67, respectively.

a. Write a 2×4 matrix A giving the stock holdings of William and Michael.

b. Write a 4×1 matrix B giving the closing prices of the stocks of BAC, GM, IBM, and TRW.

c. Use matrix multiplication to find the total value of the stock holdings of William and Michael at the market close.

44. INVESTMENT PORTFOLIOS The following table gives the number of shares of certain corporations held by Olivia and Max in their stock portfolios at the beginning of September and at the beginning of October:

September	IBM	Google	Boeing	GM
Olivia	800	500	1200	1500
Max	500	600	2000	800

October	IBM	Google	Boeing	GM
Olivia	900	600	1000	1200
Max	700	500	2100	900

a. Write matrices A and B giving the stock portfolios of Olivia and Max at the beginning of September and at the beginning of October, respectively.

b. Find a matrix C reflecting the change in the stock portfolios of Olivia and Max between the beginning of September and the beginning of October.

45. **MACHINE SCHEDULING** Desmond Jewelry wishes to produce three types of pendants: Type A, Type B, and Type C. To manufacture a Type A pendant requires 2 min on Machines I and II and 3 min on Machine III. A Type B pendant requires 2 min on Machine I, 3 min on Machine II, and 4 min on Machine III. A Type C pendant requires 3 min on Machine I, 4 min on Machine II, and 3 min on Machine III. There are $3\frac{1}{2}$ hr available on Machine I, $4\frac{1}{2}$ hr available on Machine II, and 5 hr available on Machine III. How many pendants of each type should Desmond make to use all the available time?

46. **PETROLEUM PRODUCTION** Wildcat Oil Company has two refineries, one located in Houston and the other in Tulsa. The Houston refinery ships 60% of its petroleum to a Chicago distributor and 40% of its petroleum to a Los Angeles distributor. The Tulsa refinery ships 30% of its petroleum to the Chicago distributor and 70% of its petroleum to the Los Angeles distributor. Assume that, over the year, the Chicago distributor received 240,000 gal of petroleum and the Los Angeles distributor received 460,000 gal of petroleum. Find the amount of petroleum produced at each of Wildcat's refineries.

CHAPTER 5　Before Moving On ...

1. Solve the following system of linear equations, using the Gauss–Jordan elimination method:

$$2x + y - z = -1$$
$$x + 3y + 2z = 2$$
$$3x + 3y - 3z = -5$$

2. Find the solution(s), if it exists, of the system of linear equations whose augmented matrix in reduced form follows.

   a. $\begin{bmatrix} 1 & 0 & 0 & | & 2 \\ 0 & 1 & 0 & | & -3 \\ 0 & 0 & 1 & | & 1 \end{bmatrix}$　　**b.** $\begin{bmatrix} 1 & 0 & 0 & | & 3 \\ 0 & 1 & 0 & | & 0 \\ 0 & 0 & 0 & | & 1 \end{bmatrix}$

   c. $\begin{bmatrix} 1 & 0 & 0 & | & 2 \\ 0 & 1 & 3 & | & 1 \\ 0 & 0 & 0 & | & 0 \end{bmatrix}$　　**d.** $\begin{bmatrix} 1 & 0 & 0 & 0 & | & 0 \\ 0 & 1 & 0 & 0 & | & 0 \\ 0 & 0 & 1 & 0 & | & 0 \\ 0 & 0 & 0 & 1 & | & 0 \end{bmatrix}$

   e. $\begin{bmatrix} 1 & 0 & -1 & | & 2 \\ 0 & 1 & 2 & | & 3 \end{bmatrix}$

3. Solve each system of linear equations using the Gauss–Jordan elimination method.

   a. $x + 2y = 3$　　　**b.** $x - 2y + 4z = 2$
   $3x - y = -5$　　　$3x + y - 2z = 1$
   $4x + y = -2$

4. Let

$$A = \begin{bmatrix} 1 & -2 & 4 \\ 3 & 0 & 1 \end{bmatrix} \qquad B = \begin{bmatrix} 1 & -1 & 2 \\ 3 & 1 & -1 \\ 2 & 1 & 0 \end{bmatrix}$$

$$C = \begin{bmatrix} 2 & -2 \\ 1 & 1 \\ 3 & 4 \end{bmatrix}$$

   Find (a) AB, (b) $(A + C^T)B$, and (c) $C^T B - AB^T$.

5. Find A^{-1} if

$$A = \begin{bmatrix} 2 & 1 & 2 \\ 0 & -1 & 3 \\ 1 & 1 & 0 \end{bmatrix}$$

6. Solve the system

$$2x + z = 4$$
$$2x + y - z = -1$$
$$3x + y - z = 0$$

   by first writing it in the matrix form $AX = B$ and then finding A^{-1}.

6

LINEAR PROGRAMMING

Aya73aya/Shutterstock.com

How many souvenirs should Ace Novelty make to maximize its profit? The company produces two types of souvenirs, each of which requires a certain amount of time on each of two different machines. Each machine can be operated for only a certain number of hours per day. In Example 1, page 338, we show how this production problem can be formulated as a linear programming problem, and in Example 1, page 349, we solve this linear programming problem.

ANY PRACTICAL PROBLEMS involve maximizing or minimizing a function subject to certain constraints. For example, we may wish to maximize a profit function subject to certain limitations on the amount of material and labor available. Maximization or minimization problems that can be formulated in terms of a *linear* objective function and constraints in the form of linear inequalities are called *linear programming problems*. In this chapter, we first look at linear programming problems involving two variables. These problems are amenable to geometric analysis, and the method of solution introduced here will shed much light on the basic nature of a linear programming problem. Solving linear programming problems involving more than two variables requires algebraic techniques. One such technique, the *simplex method,* was developed by George Dantzig in the late 1940s and remains in wide use to this day.

6.1   Graphing Systems of Linear Inequalities in Two Variables

Graphing Linear Inequalities

In Chapter 2, we saw that a linear equation in two variables x and y

$$ax + by + c = 0 \qquad \text{\footnotesize a, b not both equal to zero}$$

has a *solution set* that may be exhibited graphically as points on a straight line in the xy-plane. We now show that there is also a simple graphical representation for **linear inequalities** in two variables:

$$ax + by + c < 0 \qquad ax + by + c \le 0$$
$$ax + by + c > 0 \qquad ax + by + c \ge 0$$

Before turning to a general procedure for graphing such inequalities, let's consider a specific example. Suppose we wish to graph

$$2x + 3y < 6 \tag{1}$$

We first graph the equation $2x + 3y = 6$, which is obtained by replacing the given inequality "$<$" with an equality "$=$" (Figure 1).

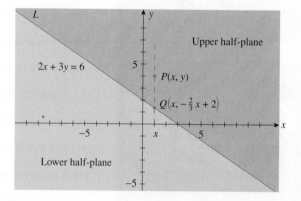

FIGURE 1
A straight line divides the xy-plane into two half-planes.

Observe that this line divides the xy-plane into two half-planes: an upper half-plane and a lower half-plane. Let's show that the upper half-plane is the graph of the linear inequality

$$2x + 3y > 6 \tag{2}$$

whereas the lower half-plane is the graph of the linear inequality

$$2x + 3y < 6 \tag{3}$$

To see this, let's write Inequalities (2) and (3) in the equivalent forms

$$y > -\frac{2}{3}x + 2 \tag{4}$$

and

$$y < -\frac{2}{3}x + 2 \tag{5}$$

The equation of the line itself is

$$y = -\frac{2}{3}x + 2 \tag{6}$$

Now pick any point $P(x, y)$ lying above the line L. Let Q be the point lying on L and directly below P (see Figure 1). Since Q lies on L, its coordinates must satisfy Equation (6). In other words, Q has representation $Q(x, -\frac{2}{3}x + 2)$. Comparing the y-coordinates of P and Q and recalling that P lies above Q, so that its y-coordinate must be larger than that of Q, we have

$$y > -\frac{2}{3}x + 2$$

But this inequality is just Inequality (4) or, equivalently, Inequality (2). Similarly, we can show that every point lying below L must satisfy Inequality (5) and therefore Inequality (3).

This analysis shows that the lower half-plane provides a solution to our problem (Figure 2). (By convention, we draw the line as a dashed line to show that the points on L do not belong to the solution set.) Observe that the two half-planes in question are disjoint; that is, they do not have any points in common.

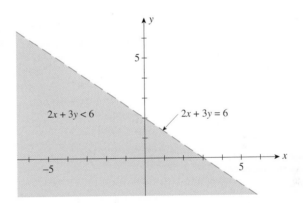

FIGURE 2
The set of points lying below the dashed line satisfies the given inequality.

Alternatively, there is a simpler method for determining the half-plane that provides the solution to the problem. To determine the required half-plane, let's pick *any* point lying in one of the half-planes. For simplicity, pick the origin $(0, 0)$, which lies in the lower half-plane. Substituting $x = 0$ and $y = 0$ (the coordinates of this point) into the given Inequality (1), we find

$$2(0) + 3(0) < 6$$

or $0 < 6$, which is certainly true. This tells us that the required half-plane is the one containing the test point—namely, the lower half-plane.

Next, let's see what happens if we choose the point $(2, 3)$, which lies in the upper half-plane. Substituting $x = 2$ and $y = 3$ into the given inequality, we find

$$2(2) + 3(3) < 6$$

or $13 < 6$, which is false. This tells us that the upper half-plane is *not* the required half-plane, as expected. Note, too, that no point $P(x, y)$ lying on the line constitutes a solution to our problem, given the *strict* inequality $<$.

This discussion suggests the following procedure for graphing a linear inequality in two variables.

Procedure for Graphing Linear Inequalities

1. Draw the graph of the equation obtained for the given inequality by replacing the inequality sign with an equal sign. Use a dashed or dotted line if the problem involves a strict inequality, $<$ or $>$. Otherwise, use a solid line to indicate that the line itself constitutes part of the solution.
2. Pick a test point (a, b) lying in one of the half-planes determined by the line sketched in Step 1 and substitute the numbers a and b for the values of x and y in the given inequality. For simplicity, use the origin whenever possible.
3. If the inequality is satisfied, the graph of the solution to the inequality is the half-plane containing the test point. Otherwise, the solution is the half-plane not containing the test point.

VIDEO   **EXAMPLE 1** Determine the solution set for the inequality $2x + 3y \geq 6$.

Solution Replacing the inequality $\geq$ with an equality $=$, we obtain the equation $2x + 3y = 6$, whose graph is the straight line shown in Figure 3.

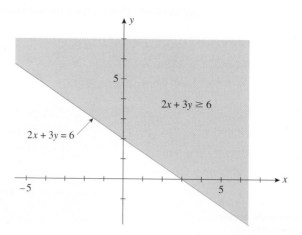

FIGURE 3
The set of points lying on the line and in the upper half-plane satisfies the given inequality.

Instead of a dashed line as before, we use a solid line to show that all points on the line are also solutions to the inequality. Picking the origin as our test point, we find $2(0) + 3(0) \geq 6$, or $0 \geq 6$, which is false. So we conclude that the solution set is made up of the half-plane that does not contain the origin, including (in this case) the line given by $2x + 3y = 6$. ■

EXAMPLE 2 Graph $x \leq -1$.

Solution The graph of $x = -1$ is the vertical line shown in Figure 4. Picking the origin $(0, 0)$ as a test point, we find $0 \leq -1$, which is false. Therefore, the required solution is the *left* half-plane, which does not contain the origin. ■

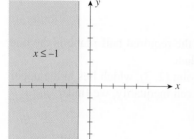

FIGURE 4
The set of points lying on the line $x = -1$ and in the left half-plane satisfies the given inequality.

EXAMPLE 3 Graph $x - 2y > 0$.

Solution We first graph the equation $x - 2y = 0$, or $y = \frac{1}{2}x$ (Figure 5). Since the origin lies on the line, we may not use it as a test point. (Why?) Let's pick $(1, 2)$ as a test point. Substituting $x = 1$ and $y = 2$ into the given inequality, we find $1 - 2(2) > 0$, or $-3 > 0$, which is false. Therefore, the required solution is the half-plane that does not contain the test point—namely, the lower half-plane.

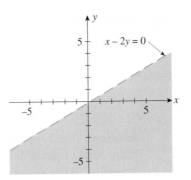

FIGURE **5**
The set of points in the lower half-plane
satisfies $x - 2y > 0$.

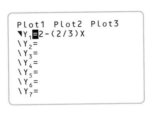

Exploring with
TECHNOLOGY

A graphing utility can be used to plot the graph of a linear inequality. For example, to plot the solution set for Example 1, first rewrite the equation $2x + 3y = 6$ in the form $y = 2 - \frac{2}{3}x$. Next, enter this expression for Y_1 in the calculator, and move the cursor to the left of Y_1. Then press **ENTER** repeatedly, and select the icon that indicates the shading option desired (see Figure a). The required graph follows (see Figure b).

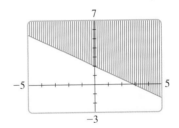

```
Plot1  Plot2  Plot3
▼Y₁■2-(2/3)X
 \Y₂=
 \Y₃=
 \Y₄=
 \Y₅=
 \Y₆=
 \Y₇=
```

FIGURE **a**
TI 83/84 screen

FIGURE **b**
Graph of the inequality $2x + 3y \geq 6$

Graphing Systems of Linear Inequalities

By the **solution set of a system of linear inequalities** in the two variables x and y, we mean the set of all points (x, y) satisfying each inequality of the system. The graphical solution of such a system may be obtained by graphing the solution set for each inequality independently and then determining the region in common with each solution set.

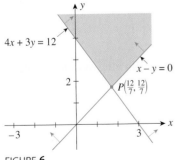

FIGURE **6**
The set of points in the shaded area satisfies the system

$$4x + 3y \geq 12$$
$$x - y \leq 0$$

EXAMPLE 4 Determine the solution set for the system

$$4x + 3y \geq 12$$
$$x - y \leq 0$$

Solution   Proceeding as in the previous examples, you should have no difficulty locating the half-planes determined by each of the linear inequalities that make up the system. These half-planes are shown in Figure 6. The intersection of the two half-planes is the shaded region. A point in this region is an element of the solution set for the given system. The point P, the intersection of the two straight lines determined by the equations, is found by solving the simultaneous equations

$$4x + 3y = 12$$
$$x - y = 0$$

VIDEO ➤   **EXAMPLE 5** Sketch the solution set for the system

$$x \geq 0$$
$$y \geq 0$$
$$x + y - 6 \leq 0$$
$$2x + y - 8 \leq 0$$

Solution   The first inequality in the system defines the right half-plane—all points to the right of the y-axis plus all points lying on the y-axis itself. The second inequality in the system defines the upper half-plane, including the x-axis. The half-planes defined by the third and fourth inequalities are indicated by arrows in Figure 7. Thus, the required region—the intersection of the four half-planes defined by the four inequalities in the given system of linear inequalities—is the shaded region. The point P is found by solving the simultaneous equations $x + y - 6 = 0$ and $2x + y - 8 = 0$.

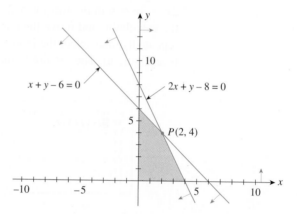

FIGURE 7
The set of points in the shaded region, including the x- and y-axes, satisfies the given inequalities.

The solution set found in Example 5 is an example of a bounded set. Observe that the set can be enclosed by a circle. For example, if you draw a circle of radius 10 with center at the origin, you will see that the set lies entirely inside the circle. On the other hand, the solution set of Example 4 cannot be enclosed by a circle and is said to be unbounded.

> **Bounded and Unbounded Solution Sets**
>
> The solution set of a system of linear inequalities is **bounded** if it can be enclosed by a circle. Otherwise, it is **unbounded**.

EXAMPLE 6 Determine the graphical solution set for the following system of linear inequalities:

$$2x + \ y \geq 50$$
$$x + 2y \geq 40$$
$$x \geq 0$$
$$y \geq 0$$

Solution   The required solution set is the unbounded region shown in Figure 8.

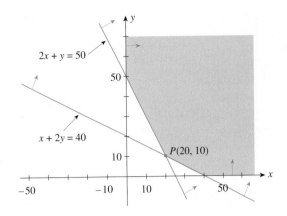

FIGURE **8**
The solution set is an unbounded region.

6.1 Self-Check Exercises

1. Determine graphically the solution set for the following system of inequalities:

$$x + 2y \leq 10$$
$$5x + 3y \leq 30$$
$$x \geq 0, y \geq 0$$

2. Determine graphically the solution set for the following system of inequalities:

$$5x + 3y \geq 30$$
$$x - 3y \leq 0$$
$$x \geq 2$$

Solutions to Self-Check Exercises 6.1 can be found on page 337.

6.1 Concept Questions

1. a. What is the difference, geometrically, between the solution set of $ax + by < c$ and the solution set of $ax + by \leq c$?
   b. Describe the set that is obtained by intersecting the solution set of $ax + by \leq c$ with the solution set of $ax + by \geq c$.

2. a. What is the solution set of a system of linear inequalities?
   b. How do you find the solution of a system of linear inequalities graphically?

6.1 Exercises

In Exercises 1–10, find the graphical solution of each inequality.

1. $4x - 8 < 0$

2. $3y + 2 > 0$

3. $x - y \leq 0$

4. $3x + 4y \leq -2$

5. $x \leq -3$

6. $y \geq -1$

7. $2x + y \leq 4$

8. $-3x + 6y \geq 12$

9. $4x - 3y \leq -24$

10. $5x - 3y \geq 15$

In Exercises 11–18, write a system of linear inequalities that describes the shaded region.

11.

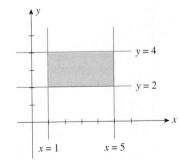

12.

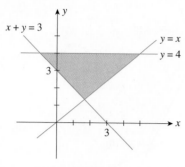

13.

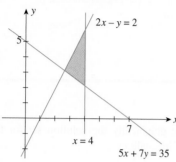

14.

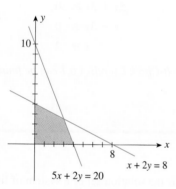

15.

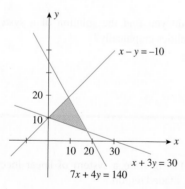

16.

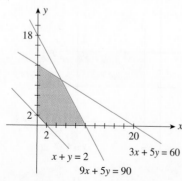

17.

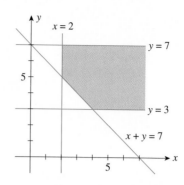

18.

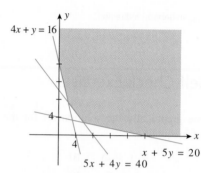

In Exercises 19–36, determine graphically the solution set for each system of inequalities and indicate whether the solution set is bounded or unbounded.

19. $2x + 4y > 16$
$-x + 3y \geq 7$

20. $3x - 2y > -13$
$-x + 2y > \quad 5$

21. $x - y \leq 0$
$2x + 3y \geq 10$

22. $x + y \geq -2$
$3x - y \leq \quad 6$

23. $x + 2y \geq \quad 3$
$2x + 4y \leq -2$

24. $2x - y \geq \quad 4$
$4x - 2y < -2$

25. $x + y \leq 6$
$0 \leq x \leq 3$
$y \geq 0$

26. $4x - 3y \leq 12$
$5x + 2y \leq 10$
$x \geq 0, y \geq 0$

27. $3x - 6y \leq 12$
$-x + 2y \leq \quad 4$
$x \geq 0, y \geq 0$

28. $x + \quad y \geq 20$
$x + 2y \geq 40$
$x \geq 0, y \geq 0$

29. $3x - 7y \geq -24$
$x + 3y \geq \quad 8$
$x \geq 0, y \geq 0$

30. $3x + 4y \geq \quad 12$
$2x - \quad y \geq -2$
$0 \leq y \leq \quad 3$
$x \geq \quad 0$

31. $x + 2y \geq \quad 3$
$5x - 4y \leq 16$
$0 \leq y \leq \quad 2$
$x \geq \quad 0$

32. $x + y \leq \quad 4$
$2x + y \leq \quad 6$
$2x - y \geq -1$
$x \geq 0, y \geq 0$

33. $6x + 5y \leq 30$
$3x + \quad y \geq \quad 6$
$x + \quad y \geq \quad 4$
$x \geq 0, y \geq 0$

34. $6x + \quad 7y \leq 84$
$12x - 11y \leq 18$
$6x - \quad 7y \leq 28$
$x \geq 0, y \geq 0$

35.
$$x - y \geq -6$$
$$x - 2y \leq -2$$
$$x + 2y \geq 6$$
$$x - 2y \geq -14$$
$$x \geq 0, y \geq 0$$

36.
$$x - 3y \geq -18$$
$$3x - 2y \geq 2$$
$$x - 3y \leq -4$$
$$3x - 2y \leq 16$$
$$x \geq 0, y \geq 0$$

In Exercises 37–40, determine whether the statement is true or false. If it is true, explain why it is true. If it is false, give an example to show why it is false.

37. The solution set of a linear inequality involving two variables is either a half-plane or a straight line.

38. The solution set of the inequality $ax + by + c \leq 0$ is either a left half-plane or a lower half-plane.

39. The solution set of a system of linear inequalities in two variables is bounded if it can be enclosed by a rectangle.

40. The solution set of the system

$$ax + by \leq e$$
$$cx + dy \leq f$$
$$x \geq 0, y \geq 0$$

where a, b, c, d, e, and f are positive real numbers, is a bounded set.

6.1 Solutions to Self-Check Exercises

1. The required solution set is shown in the following figure:

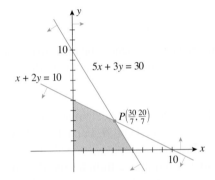

The point P is found by solving the system of equations

$$x + 2y = 10$$
$$5x + 3y = 30$$

Solving the first equation for x in terms of y gives

$$x = 10 - 2y$$

Substituting this value of x into the second equation of the system gives

$$5(10 - 2y) + 3y = 30$$
$$50 - 10y + 3y = 30$$
$$-7y = -20$$

so $y = \frac{20}{7}$. Substituting this value of y into the expression for x found earlier, we obtain

$$x = 10 - 2\left(\frac{20}{7}\right) = \frac{30}{7}$$

giving the point of intersection as $\left(\frac{30}{7}, \frac{20}{7}\right)$.

2. The required solution set is shown in the following figure:

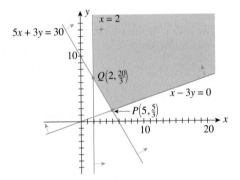

To find the coordinates of P, we solve the system

$$5x + 3y = 30$$
$$x - 3y = 0$$

Solving the second equation for x in terms of y and substituting this value of x in the first equation gives

$$5(3y) + 3y = 30$$

or $y = \frac{5}{3}$. Substituting this value of y into the second equation gives $x = 5$. Next, the coordinates of Q are found by solving the system

$$5x + 3y = 30$$
$$x = 2$$

yielding $x = 2$ and $y = \frac{20}{3}$.

6.2 Linear Programming Problems

In many business and economic problems, we are asked to optimize (maximize or minimize) a function subject to a system of equalities or inequalities. The function to be optimized is called the **objective function.** Profit functions and cost functions are examples of objective functions. The system of equalities or inequalities to which the objective function is subjected reflects the constraints (for example, limitations on resources such as materials and labor) imposed on the solution(s) to the problem. Problems of this nature are called **mathematical programming problems.** In particular, problems in which both the objective function and the constraints are expressed as linear equations or inequalities are called linear programming problems.

> **Linear Programming Problem**
>
> A **linear programming problem** consists of a linear objective function to be maximized or minimized subject to certain constraints in the form of linear equations or inequalities.

A Maximization Problem

As an example of a linear programming problem in which the objective function is to be maximized, let's consider the following simplified version of a production problem involving two variables.

VIDEO

APPLIED EXAMPLE 1 A Production Problem Ace Novelty wishes to produce two types of souvenirs: Type *A* and Type *B*. Each Type *A* souvenir will result in a profit of $1, and each Type *B* souvenir will result in a profit of $1.20. To manufacture a Type *A* souvenir requires 2 minutes on Machine I and 1 minute on Machine II. A Type *B* souvenir requires 1 minute on Machine I and 3 minutes on Machine II. There are 3 hours available on Machine I and 5 hours available on Machine II. How many souvenirs of each type should Ace make to maximize its profit?

Solution As a first step toward the mathematical formulation of this problem, we tabulate the given information (see Table 1).

TABLE 1

	Type *A*	Type *B*	Time Available
Machine I	2 min	1 min	180 min
Machine II	1 min	3 min	300 min
Profit/Unit	$1	$1.20	

Let *x* be the number of Type *A* souvenirs and *y* the number of Type *B* souvenirs to be made. Then, the total profit *P* (in dollars) is given by

$$P = x + 1.2y$$

which is the objective function to be maximized.

The total amount of time that Machine I is used is given by $2x + y$ minutes and must not exceed 180 minutes. Thus, we have the inequality

$$2x + y \leq 180$$

Similarly, the total amount of time that Machine II is used is $x + 3y$ minutes and cannot exceed 300 minutes, so we are led to the inequality

$$x + 3y \le 300$$

Finally, neither x nor y can be negative, so

$$x \ge 0$$
$$y \ge 0$$

To summarize, the problem at hand is one of maximizing the objective function $P = x + 1.2y$ subject to the system of inequalities

$$2x + y \le 180$$
$$x + 3y \le 300$$
$$x \ge 0$$
$$y \ge 0$$

The solution to this problem will be completed in Example 1, Section 6.3. ■

Minimization Problems

In the following linear programming problem, the objective function is to be minimized.

VIDEO▶

APPLIED EXAMPLE 2 A Nutrition Problem A nutritionist advises an individual who is suffering from iron and vitamin B deficiency to take at least 2400 milligrams (mg) of iron, 2100 mg of vitamin B_1 (thiamine), and 1500 mg of vitamin B_2 (riboflavin) over a period of time. Two vitamin pills are suitable, Brand A and Brand B. Each Brand A pill costs 6 cents and contains 40 mg of iron, 10 mg of vitamin B_1, and 5 mg of vitamin B_2. Each Brand B pill costs 8 cents and contains 10 mg of iron and 15 mg each of vitamins B_1 and B_2 (Table 2). What combination of pills should the individual purchase to meet the minimum iron and vitamin requirements at the lowest cost?

TABLE 2

	Brand A	**Brand B**	**Minimum Requirement**
Iron	40 mg	10 mg	2400 mg
Vitamin B_1	10 mg	15 mg	2100 mg
Vitamin B_2	5 mg	15 mg	1500 mg
Cost/Pill	6¢	8¢	

Solution   Let x be the number of Brand A pills to be purchased, and let y be the number of Brand B pills. The cost C (in cents) is given by

$$C = 6x + 8y$$

and is the objective function to be minimized.

The amount of iron contained in x Brand A pills and y Brand B pills is given by $40x + 10y$ mg, and this must be greater than or equal to 2400 mg. This translates into the inequality

$$40x + 10y \ge 2400$$

Similar considerations involving the minimum requirements of vitamins B_1 and B_2 lead to the inequalities

$$10x + 15y \geq 2100$$
$$5x + 15y \geq 1500$$

respectively. Thus, the problem here is to minimize $C = 6x + 8y$ subject to

$$40x + 10y \geq 2400$$
$$10x + 15y \geq 2100$$
$$5x + 15y \geq 1500$$
$$x \geq 0, y \geq 0$$

The solution to this problem will be completed in Example 2, Section 6.3.   ■

APPLIED EXAMPLE 3   A Transportation Problem Curtis-Roe Aviation Industries has two plants, I and II, that produce the Zephyr jet engines used in their light commercial airplanes. There are 100 units of the engines in Plant I and 110 units in Plant II. The engines are shipped to two of Curtis-Roe's main assembly plants, A and B. The shipping costs (in dollars) per engine from Plants I and II to the Main Assembly Plants A and B are as follows:

From	To Assembly Plant	
	A	B
Plant I	100	60
Plant II	120	70

In a certain month, Assembly Plant A needs 80 engines, whereas Assembly Plant B needs 70 engines. Find how many engines should be shipped from each plant to each main assembly plant if shipping costs are to be kept to a minimum.

Solution   Let x denote the number of engines shipped from Plant I to Assembly Plant A, and let y denote the number of engines shipped from Plant I to Assembly Plant B. Since the requirements of Assembly Plants A and B are 80 and 70 engines, respectively, the number of engines shipped from Plant II to Assembly Plants A and B are $(80 - x)$ and $(70 - y)$, respectively. These numbers may be displayed in a schematic. With the aid of the accompanying schematic (Figure 9) and the shipping cost schedule, we find that the total shipping cost incurred by Curtis-Roe is given by

$$C = 100x + 60y + 120(80 - x) + 70(70 - y)$$
$$= 14,500 - 20x - 10y$$

Next, the production constraints on Plants I and II lead to the inequalities

$$x + y \leq 100$$
$$(80 - x) + (70 - y) \leq 110$$

The last inequality simplifies to

$$x + y \geq 40$$

Also, the requirements of the two main assembly plants lead to the inequalities

$$x \geq 0 \qquad y \geq 0 \qquad 80 - x \geq 0 \qquad 70 - y \geq 0$$

The last two may be written as $x \leq 80$ and $y \leq 70$.

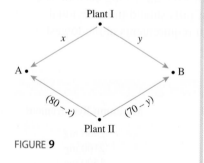

Plant I

x y

A •   • B

$(80 - x)$ $(70 - y)$

Plant II

FIGURE **9**

Summarizing, we have the following linear programming problem: Minimize the objective (cost) function $C = 14{,}500 - 20x - 10y$ subject to the constraints

$$x + y \geq 40$$
$$x + y \leq 100$$
$$x \leq 80$$
$$y \leq 70$$

where $x \geq 0$ and $y \geq 0$.

You will be asked to complete the solution to this problem in Exercise 47, Section 6.3.

APPLIED EXAMPLE 4 A Warehouse Problem Acrosonic manufactures its Brentwood loudspeaker systems in two separate locations, Plant I and Plant II. The output at Plant I is at most 400 per month, whereas the output at Plant II is at most 600 per month. These loudspeaker systems are shipped to three warehouses that serve as distribution centers for the company. For the warehouses to meet their orders, the minimum monthly requirements of Warehouses A, B, and C are 200, 300, and 400 systems, respectively. Shipping costs from Plant I to Warehouses A, B, and C are $20, $8, and $10 per loudspeaker system, respectively, and shipping costs from Plant II to each of these warehouses are $12, $22, and $18, respectively. What should the shipping schedule be if Acrosonic wishes to meet the requirements of the distribution centers and at the same time keep its shipping costs to a minimum?

Solution The respective shipping costs (in dollars) per loudspeaker system may be tabulated as in Table 3. Letting x_1 denote the number of loudspeaker systems shipped from Plant I to Warehouse A, x_2 the number shipped from Plant I to Warehouse B, and so on leads to Table 4.

TABLE 3

Plant	Warehouse		
	A	B	C
I	20	8	10
II	12	22	18

TABLE 4

Plant	Warehouse			Max. Prod.
	A	B	C	
I	x_1	x_2	x_3	400
II	x_4	x_5	x_6	600
Min. Req.	200	300	400	

From Tables 3 and 4, we see that the cost of shipping x_1 loudspeaker systems from Plant I to Warehouse A is $20x_1$, the cost of shipping x_2 loudspeaker systems from Plant I to Warehouse B is $8x_2$, and so on. Thus, the total monthly shipping cost (in dollars) incurred by Acrosonic is given by

$$C = 20x_1 + 8x_2 + 10x_3 + 12x_4 + 22x_5 + 18x_6$$

Next, the production constraints on Plants I and II lead to the inequalities

$$x_1 + x_2 + x_3 \leq 400$$
$$x_4 + x_5 + x_6 \leq 600$$

(see Table 4). Also, the minimum requirements of each of the three warehouses lead to the three inequalities

$$x_1 + x_4 \geq 200$$
$$x_2 + x_5 \geq 300$$
$$x_3 + x_6 \geq 400$$

Summarizing, we have the following linear programming problem:

$$\text{Minimize} \quad C = 20x_1 + 8x_2 + 10x_3 + 12x_4 + 22x_5 + 18x_6$$

$$\text{subject to} \quad x_1 + x_2 + x_3 \leq 400$$

$$x_4 + x_5 + x_6 \leq 600$$

$$x_1 + x_4 \geq 200$$

$$x_2 + x_5 \geq 300$$

$$x_3 + x_6 \geq 400$$

$$x_1 \geq 0, x_2 \geq 0, \ldots, x_6 \geq 0$$

The solution to this problem will be completed in Example 5, Section 6.5.   ■

6.2   Self-Check Exercise

Gino Balduzzi, proprietor of Luigi's Pizza Palace, allocates $9000 a month for advertising in two newspapers, the *City Tribune* and the *Daily News*. The *City Tribune* charges $300 for a certain advertisement, whereas the *Daily News* charges $100 for the same ad. Gino has stipulated that the ad is to appear in at least 15 but no more than 30 editions of the *Daily News* per month. The *City Tribune* has a daily circulation of 50,000, and the *Daily News* has a circulation of 20,000. Under these condi-

tions, determine how many ads Gino should place in each newspaper to reach the largest number of readers. Formulate but do not solve the problem. (The solution to this problem can be found in Exercise 3 of Solutions to Self-Check Exercises 6.3.)

The solution to Self-Check Exercise 6.2 can be found on page 346.

6.2   Concept Questions

1. What is a linear programming problem?

2. Suppose you are asked to formulate a linear programming problem in two variables x and y. How would you express the fact that x and y are nonnegative? Why are these conditions often required in practical problems?

3. What is the difference between a maximization linear programming problem and a minimization linear programming problem?

6.2   Exercises

Formulate but do not solve each of the following exercises as a linear programming problem. You will be asked to solve these problems later.

1. **MANUFACTURING—PRODUCTION SCHEDULING** A company manufactures two products, A and B, on two machines, I and II. It has been determined that the company will realize a profit of $3 on each unit of Product A and a profit of $4 on each unit of Product B. To manufacture a unit of Product A requires 6 min on Machine I and 5 min on Machine II. To manufacture a unit of Product B requires 9 min on Machine I and 4 min on Machine II. There are 5 hr of machine time available on Machine I and 3 hr of machine time available on Machine II in each work shift. How many units of each product should be produced in each shift to maximize the company's profit?

2. **MANUFACTURING—PRODUCTION SCHEDULING** National Business Machines manufactures two models of fax machines: A and B. Each model A costs $100 to make, and each model B costs $150. The profits are $30 for each model A and $40 for each model B fax machine. If the total number of fax machines demanded per month does not exceed 2500 and the company has earmarked no more than $600,000/month for manufacturing costs, how many units of each model should National make each month to maximize its monthly profit?

3. **MANUFACTURING—PRODUCTION SCHEDULING** Kane Manufacturing has a division that produces two models of fireplace grates, model A and model B. To produce each model A grate requires 3 lb of cast iron and 6 min of labor. To pro-

duce each model B grate requires 4 lb of cast iron and 3 min of labor. The profit for each model A grate is $2.00, and the profit for each model B grate is $1.50. If 1000 lb of cast iron and 20 hr of labor are available for the production of grates per day, how many grates of each model should the division produce per day to maximize Kane's profits?

4. **MANUFACTURING—PRODUCTION SCHEDULING** Refer to Exercise 3. Because of a backlog of orders on model A grates, the manager of Kane Manufacturing has decided to produce at least 150 of these grates a day. Operating under this additional constraint, how many grates of each model should Kane produce to maximize profit?

5. **MANUFACTURING—PRODUCTION SCHEDULING** A division of the Winston Furniture Company manufactures dining tables and chairs. Each table requires 40 board feet of wood and 3 labor-hours. Each chair requires 16 board feet of wood and 4 labor-hours. The profit for each table is $45, and the profit for each chair is $20. In a certain week, the company has 3200 board feet of wood available and 520 labor-hours. How many tables and chairs should Winston manufacture to maximize its profits?

6. **MANUFACTURING—PRODUCTION SCHEDULING** Refer to Exercise 5. If the profit for each table is $50 and the profit for each chair is $18, how many tables and chairs should Winston manufacture to maximize its profits?

7. **FINANCE—ALLOCATION OF FUNDS** Madison Finance has a total of $20 million earmarked for homeowner loans and auto loans. On the average, homeowner loans have a 10% annual rate of return, whereas auto loans yield a 12% annual rate of return. Management has also stipulated that the total amount of homeowner loans should be greater than or equal to 4 times the total amount of automobile loans. Determine the total amount of loans of each type Madison should extend to each category to maximize its returns.

8. **INVESTMENTS—ASSET ALLOCATION** A financier plans to invest up to $500,000 in two projects. Project A yields a return of 10% on the investment whereas Project B yields a return of 15% on the investment. Because the investment in Project B is riskier than the investment in Project A, the financier has decided that the investment in Project B should not exceed 40% of the total investment. How much should she invest in each project to maximize the return on her investment?

9. **MANUFACTURING—PRODUCTION SCHEDULING** Acoustical Company manufactures a CD storage cabinet that can be bought fully assembled or as a kit. Each cabinet is processed in the fabrication department and the assembly department. If the fabrication department manufactures only fully assembled cabinets, it can produce 200 units/day; and if it manufactures only kits, it can produce 200 units/day. If the assembly department produces only fully assembled cabinets, it can produce 100 units/day; but if it produces only kits, then it can produce 300 units/day. Each fully assembled cabinet contributes $50 to the profits of the company, whereas each kit contributes $40 to its profits. How many fully assembled units and how many kits should the company produce per day to maximize its profits?

10. **AGRICULTURE—CROP PLANNING** A farmer plans to plant two crops, A and B. The cost of cultivating Crop A is $40/acre, whereas the cost of cultivating Crop B is $60/acre. The farmer has a maximum of $7400 available for land cultivation. Each acre of Crop A requires 20 labor-hours, and each acre of Crop B requires 25 labor-hours. The farmer has a maximum of 3300 labor-hours available. If she expects to make a profit of $150/acre on Crop A, and $200/acre on Crop B, how many acres of each crop should she plant to maximize her profit?

11. **MINING—PRODUCTION** Perth Mining Company operates two mines for the purpose of extracting gold and silver. The Saddle Mine costs $14,000/day to operate, and it yields 50 oz of gold and 3000 oz of silver each day. The Horseshoe Mine costs $16,000/day to operate, and it yields 75 oz of gold and 1000 oz of silver each day. Company management has set a target of at least 650 oz of gold and 18,000 oz of silver. How many days should each mine be operated so that the target can be met at a minimum cost?

12. **TRANSPORTATION** Deluxe River Cruises operates a fleet of river vessels. The fleet has two types of vessels: A type A vessel has 60 deluxe cabins and 160 standard cabins, whereas a type B vessel has 80 deluxe cabins and 120 standard cabins. Under a charter agreement with Odyssey Travel Agency, Deluxe River Cruises is to provide Odyssey with a minimum of 360 deluxe and 680 standard cabins for their 15-day cruise in May. It costs $44,000 to operate a type A vessel and $54,000 to operate a type B vessel for that period. How many of each type vessel should be used to keep the operating costs to a minimum?

13. **WATER SUPPLY** The water-supply manager for a Midwestern city needs to supply the city with at least 10 million gallons of potable (drinkable) water per day. The supply may be drawn from the local reservoir or from a pipeline to an adjacent town. The local reservoir has a maximum daily yield of 5 million gallons of potable water, and the pipeline has a maximum daily yield of 10 million gallons. By contract, the pipeline is required to supply a minimum of 6 million gallons/day. If the cost for 1 million gallons of reservoir water is $300 and that for pipeline water is $500, how much water should the manager get from each source to minimize daily water costs for the city?

14. **MANUFACTURING—PRODUCTION SCHEDULING** Ace Novelty manufactures Giant Pandas and Saint Bernards. Each Panda requires 1.5 yd^2 of plush, 30 ft^3 of stuffing, and 5 pieces of trim; each Saint Bernard requires 2 yd^2 of plush, 35 ft^3 of stuffing, and 8 pieces of trim. The profit for each Panda is $10 and the profit for each Saint Bernard is $15. If 3600 yd^2 of plush, 66,000 ft^3 of stuffing and 13,600 pieces of trim are available, how many of each of the stuffed animals should the company manufacture to maximize profit?

15. **NUTRITION—DIET PLANNING** A nutritionist at the Medical Center has been asked to prepare a special diet for certain patients. She has decided that the meals should contain a minimum of 400 mg of calcium, 10 mg of iron, and 40 mg of vitamin C. She has further decided that the meals are to be prepared from Foods A and B. Each ounce of Food A contains 30 mg of calcium, 1 mg of iron, 2 mg of vitamin C, and 2 mg of cholesterol. Each ounce of Food B contains 25 mg of calcium, 0.5 mg of iron, 5 mg of vitamin C, and 5 mg of cholesterol. Find how many ounces of each type of food should be used in a meal so that the cholesterol content is minimized and the minimum requirements of calcium, iron, and vitamin C are met.

16. **SOCIAL PROGRAMS PLANNING** AntiFam, a hunger-relief organization, has earmarked between $2 million and $2.5 million (inclusive) for aid to two African countries, Country A and Country B. Country A is to receive between $1 million and $1.5 million (inclusive), and Country B is to receive at least $0.75 million. It has been estimated that each dollar spent in Country A will yield an effective return of $0.60, whereas a dollar spent in Country B will yield an effective return of $0.80. How should the aid be allocated if the money is to be utilized most effectively according to these criteria?

Hint: If x and y denote the amount of money to be given to Country A and Country B, respectively, then the objective function to be maximized is $P = 0.6x + 0.8y$.

17. **ADVERTISING** Everest Deluxe World Travel has decided to advertise in the Sunday editions of two major newspapers in town. These advertisements are directed at three groups of potential customers. Each advertisement in Newspaper I is seen by 70,000 Group A customers, 40,000 Group B customers, and 20,000 Group C customers. Each advertisement in Newspaper II is seen by 10,000 Group A, 20,000 Group B, and 40,000 Group C customers. Each advertisement in Newspaper I costs $1000, and each advertisement in Newspaper II costs $800. Everest would like their advertisements to be read by at least 2 million people from Group A, 1.4 million people from Group B, and 1 million people from Group C. How many advertisements should Everest place in each newspaper to achieve its advertisement goals at a minimum cost?

18. **MANUFACTURING—SHIPPING COSTS** TMA manufactures 37-in. high-definition LCD televisions in two separate locations: Location I and Location II. The output at Location I is at most 6000 televisions/month, whereas the output at Location II is at most 5000 televisions/month. TMA is the main supplier of televisions to Pulsar Corporation, its holding company, which has priority in having all its requirements met. In a certain month, Pulsar placed orders for 3000 and 4000 televisions to be shipped to two of its factories located in City A and City B, respectively. The shipping costs (in dollars) per television from the two TMA plants to the two Pulsar factories are as follows:

From TMA	To Pulsar Factories City A	City B
Location I	$6	$4
Location II	$8	$10

Find a shipping schedule that meets the requirements of both companies while keeping costs to a minimum.

19. **INVESTMENTS—ASSET ALLOCATION** A financier plans to invest up to $2 million in three projects. She estimates that Project A will yield a return of 10% on her investment, Project B will yield a return of 15% on her investment, and Project C will yield a return of 20% on her investment. Because of the risks associated with the investments, she decided to put not more than 20% of her total investment in Project C. She also decided that her investments in Projects B and C should not exceed 60% of her total investment. Finally, she decided that her investment in Project A should be at least 60% of her investments in Projects B and C. How much should the financier invest in each project if she wishes to maximize the total returns on her investments?

20. **INVESTMENTS—ASSET ALLOCATION** Ashley has earmarked at most $250,000 for investment in three mutual funds: a money market fund, an international equity fund, and a growth-and-income fund. The money market fund has a rate of return of 6%/year, the international equity fund has a rate of return of 10%/year, and the growth-and-income fund has a rate of return of 15%/year. Ashley has stipulated that no more than 25% of her total portfolio should be in the growth-and-income fund and that no more than 50% of her total portfolio should be in the international equity fund. To maximize the return on her investment, how much should Ashley invest in each type of fund?

21. **MANUFACTURING—PRODUCTION SCHEDULING** A company manufactures Products A, B, and C. Each product is processed in three departments: I, II, and III. The total available labor-hours per week for Departments I, II, and III are 900, 1080, and 840, respectively. The time requirements (in hours per unit) and profit per unit for each product are as follows:

	Product A	Product B	Product C
Dept. I	2	1	2
Dept. II	3	1	2
Dept. III	2	2	1
Profit	$18	$12	$15

How many units of each product should the company produce to maximize its profit?

22. **ADVERTISING** As part of a campaign to promote its annual clearance sale, the Excelsior Company decided to buy television advertising time on Station KAOS. Excelsior's advertising budget is $102,000. Morning time costs $3000/minute, afternoon time costs $1000/minute, and evening (prime) time costs $12,000/minute. Because of previous commitments, KAOS cannot offer Excelsior more than 6 min of prime time or more than a total of 25 min of advertising time over the 2 weeks in which the commercials are to be run. KAOS estimates that morning commercials are seen by 200,000 people, afternoon commercials are seen by 100,000 people, and evening com-

mercials are seen by 600,000 people. How much morning, afternoon, and evening advertising time should Excelsior buy to maximize exposure of its commercials?

23. **MANUFACTURING—PRODUCTION SCHEDULING** Custom Office Furniture Company is introducing a new line of executive desks made from a specially selected grade of walnut. Initially, three different models—A, B, and C—are to be marketed. Each model A desk requires $1\frac{1}{4}$ hr for fabrication, 1 hr for assembly, and 1 hr for finishing; each model B desk requires $1\frac{1}{2}$ hr for fabrication, 1 hr for assembly, and 1 hr for finishing; each model C desk requires $1\frac{1}{2}$ hr, $\frac{3}{4}$ hr, and $\frac{1}{2}$ hr for fabrication, assembly, and finishing, respectively. The profit on each model A desk is $26, the profit on each model B desk is $28, and the profit on each model C desk is $24. The total time available in the fabrication department, the assembly department, and the finishing department in the first month of production is 310 hr, 205 hr, and 190 hr, respectively. To maximize Custom's profit, how many desks of each model should be made in the month?

24. **MANUFACTURING—SHIPPING COSTS** Acrosonic of Example 4 also manufactures a model G loudspeaker system in plants I and II. The output at Plant I is at most 800 systems/month whereas the output at Plant II is at most 600/month. These loudspeaker systems are also shipped to three warehouses— A, B, and C—whose minimum monthly requirements are 500, 400, and 400, respectively. Shipping costs from Plant I to Warehouse A, Warehouse B, and Warehouse C are $16, $20, and $22 per system, respectively, and shipping costs from Plant II to each of these warehouses are $18, $16, and $14 per system, respectively. What shipping schedule will enable Acrosonic to meet the warehouses' requirements and at the same time keep its shipping costs to a minimum?

25. **MANUFACTURING—SHIPPING COSTS** Steinwelt Piano manufactures upright and console pianos in two plants, Plant I and Plant II. The output of Plant I is at most 300/month, whereas the output of Plant II is at most 250/month. These pianos are shipped to three warehouses, which serve as distribution centers for the company. To fill current and projected future orders, Warehouse A requires at least 200 pianos/month, Warehouse B requires at least 150 pianos/month, and Warehouse C requires at least 200 pianos/month. The shipping cost of each piano from Plant I to Warehouse A, Warehouse B, and Warehouse C is $60, $60, and $80, respectively, and the shipping cost of each piano from Plant II to Warehouse A, Warehouse B, and Warehouse C is $80, $70, and $50, respectively. What shipping schedule will enable Steinwelt to meet the warehouses' requirements while keeping shipping costs to a minimum?

26. **MANUFACTURING—PREFABRICATED HOUSING PRODUCTION** Boise Lumber has decided to enter the lucrative prefabricated housing business. Initially, it plans to offer three models: standard, deluxe, and luxury. Each house is prefabricated and partially assembled in the factory, and the final assembly is completed on site. The dollar amount of building material required, the amount of labor required in the factory for prefabrication and partial assembly, the amount of on-site labor required, and the profit per unit are as follows:

	Standard Model	Deluxe Model	Luxury Model
Material	$6,000	$8,000	$10,000
Factory Labor (hr)	240	220	200
On-site Labor (hr)	180	210	300
Profit	$3,400	$4,000	$5,000

For the first year's production, a sum of $8.2 million is budgeted for the building material; the number of labor-hours available for work in the factory (for prefabrication and partial assembly) is not to exceed 218,000 hr; and the amount of labor for on-site work is to be less than or equal to 237,000 labor-hours. Determine how many houses of each type Boise should produce (market research has confirmed that there should be no problems with sales) to maximize its profit from this new venture.

27. **PRODUCTION—JUICE PRODUCTS** CalJuice Company has decided to introduce three fruit juices made from blending two or more concentrates. These juices will be packaged in 2-qt (64-oz) cartons. One carton of pineapple–orange juice requires 8 oz each of pineapple and orange juice concentrates. One carton of orange–banana juice requires 12 oz of orange juice concentrate and 4 oz of banana pulp concentrate. Finally, one carton of pineapple–orange–banana juice requires 4 oz of pineapple juice concentrate, 8 oz of orange juice concentrate, and 4 oz of banana pulp. The company has decided to allot 16,000 oz of pineapple juice concentrate, 24,000 oz of orange juice concentrate, and 5000 oz of banana pulp concentrate for the initial production run. The company has also stipulated that the production of pineapple–orange–banana juice should not exceed 800 cartons. Its profit on one carton of pineapple–orange juice is $1.00, its profit on one carton of orange–banana juice is $0.80, and its profit on one carton of pineapple–orange–banana juice is $0.90. To realize a maximum profit, how many cartons of each blend should the company produce?

28. **MANUFACTURING—COLD FORMULA PRODUCTION** Beyer Pharmaceutical produces three kinds of cold formulas: Formula I, Formula II, and Formula III. It takes 2.5 hr to produce 1000 bottles of Formula I, 3 hr to produce 1000 bottles of Formula II, and 4 hr to produce 1000 bottles of Formula III. The profits for each 1000 bottles of Formula I, Formula II, and Formula III are $180, $200, and $300, respectively. For a certain production run, there are enough ingredients on hand to make at most 9000 bottles of Formula I, 12,000 bottles of Formula II, and 6000 bottles of Formula III. Furthermore, the time for the production run is limited to a maximum of 70 hr. How many bottles of each formula should be produced in this production run so that the profit is maximized?

In Exercises 29 and 30, determine whether the statement is true or false. If it is true, explain why it is true. If it is false, give an example to show why it is false.

29. The problem

$$\begin{aligned} \text{Maximize} \quad & P = xy \\ \text{subject to} \quad & 2x + 3y \le 12 \\ & 2x + y \le 8 \\ & x \ge 0, y \ge 0 \end{aligned}$$

is a linear programming problem.

30. The problem

$$\begin{aligned} \text{Minimize} \quad & C = 2x + 3y \\ \text{subject to} \quad & 2x + 3y \le 6 \\ & x - y = 0 \\ & x \ge 0, y \ge 0 \end{aligned}$$

is a linear programming problem.

6.2 Solution to Self-Check Exercise

Let x denote the number of ads to be placed in the *City Tribune*, and let y denote the number to be placed in the *Daily News*. The total cost for placing x ads in the *City Tribune* and y ads in the *Daily News* is $300x + 100y$ dollars, and since the monthly budget is $9000, we must have

$$300x + 100y \le 9000$$

Next, the condition that the ad must appear in at least 15 but no more than 30 editions of the *Daily News* translates into the inequalities

$$y \ge 15$$
$$y \le 30$$

Finally, the objective function to be maximized is

$$P = 50,000x + 20,000y$$

To summarize, we have the following linear programming problem:

$$\begin{aligned} \text{Maximize} \quad & P = 50,000x + 20,000y \\ \text{subject to} \quad & 300x + 100y \le 9000 \\ & y \ge 15 \\ & y \le 30 \\ & x \ge 0, y \ge 0 \end{aligned}$$

6.3 Graphical Solution of Linear Programming Problems

The Graphical Method

Linear programming problems in two variables have relatively simple geometric interpretations. For example, the system of linear constraints associated with a two-dimensional linear programming problem, unless it is inconsistent, defines a planar region or a line segment whose boundary is composed of straight-line segments and/or half-lines. Such problems are therefore amenable to graphical analysis.

Consider the following two-dimensional linear programming problem:

$$\begin{aligned} \text{Maximize} \quad & P = 3x + 2y \\ \text{subject to} \quad & 2x + 3y \le 12 \\ & 2x + y \le 8 \\ & x \ge 0, y \ge 0 \end{aligned} \tag{7}$$

The system of linear inequalities in (7) defines the planar region S shown in Figure 10. Each point in S is a candidate for the solution of the problem at hand and is referred to as a **feasible solution.** The set S itself is referred to as a **feasible set.** Our goal is to find, from among all the points in the set S, the point(s) that optimize(s) the objective function P. Such a feasible solution is called an **optimal solution** and constitutes the solution to the linear programming problem under consideration.

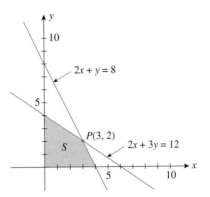

FIGURE **10**
Each point in the feasible set *S* is a candidate for the optimal solution.

As was noted earlier, each point $P(x, y)$ in S is a candidate for the optimal solution to the problem at hand. For example, the point $(1, 3)$ is easily seen to lie in S and is therefore in the running. The value of the objective function P at the point $(1, 3)$ is given by $P = 3(1) + 2(3) = 9$. Now, if we could compute the value of P corresponding to each point in S, then the point(s) in S that gave the largest value to P would constitute the solution set sought. Unfortunately, in most problems, the number of candidates either is too large or, as in this problem, is infinite. Therefore, this method is at best unwieldy and at worst impractical.

Let's turn the question around. Instead of asking for the value of the objective function P at a feasible point, let's assign a value to the objective function P and ask whether there are feasible points that would correspond to the given value of P. Toward this end, suppose we assign a value of 6 to P. Then the objective function P becomes $3x + 2y = 6$, a linear equation in x and y; thus, it has a graph that is a straight line L_1 in the plane. In Figure 11, we have drawn the graph of this straight line superimposed on the feasible set S.

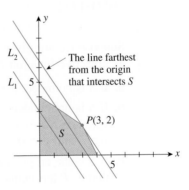

FIGURE **11**
A family of parallel lines that intersect the feasible set *S*

It is clear that each point on the straight-line segment given by the intersection of the straight line L_1 and the feasible set S corresponds to the given value, 6, of P. For this reason, the line L_1 is called an **isoprofit line.** Let's repeat the process, this time assigning a value of 10 to P. We obtain the equation $3x + 2y = 10$ and the line L_2 (see Figure 11), which suggests that there are feasible points that correspond to a larger value of P. Observe that the line L_2 is parallel to the line L_1 because both lines have slope equal to $-\frac{3}{2}$, which is easily seen by casting the corresponding equations in the slope-intercept form.

In general, by assigning different values to the objective function, we obtain a family of parallel lines, each with slope equal to $-\frac{3}{2}$. Furthermore, a line corresponding to a larger value of P lies farther away from the origin than one with a smaller value of P. The implication is clear. To obtain the optimal solution(s) to the problem at hand, find the straight line, from this family of straight lines, that is farthest from the origin and still intersects the feasible set S. The required line is the one that passes through the point $P(3, 2)$ (see Figure 11), so the solution to the problem is given by $x = 3$, $y = 2$, resulting in a maximum value of $P = 3(3) + 2(2) = 13$.

That the optimal solution to this problem was found to occur at a vertex of the feasible set S is no accident. In fact, the result is a consequence of the following basic theorem on linear programming, which we state without proof.

THEOREM 1

Linear Programming

If a linear programming problem has a solution, then it must occur at a vertex, or corner point, of the feasible set S associated with the problem.

Furthermore, if the objective function P is optimized at two adjacent vertices of S, then it is optimized at every point on the line segment joining these vertices, in which case there are infinitely many solutions to the problem.

Theorem 1 tells us that our search for the solution(s) to a linear programming problem may be restricted to the examination of the set of vertices of the feasible set S associated with the problem. Since a feasible set S has finitely many vertices, the theorem suggests that the solution(s) to the linear programming problem may be found by inspecting the values of the objective function P at these vertices.

Although Theorem 1 sheds some light on the nature of the solution of a linear programming problem, it does not tell us when a linear programming problem has a solution. The following theorem states conditions that guarantee when a solution exists.

THEOREM 2

Existence of a Solution

Suppose we are given a linear programming problem with a feasible set S and an objective function $P = ax + by$.

a. If S is bounded, then P has both a maximum and a minimum value on S.

b. If S is unbounded and both a and b are nonnegative, then P has a minimum value on S provided that the constraints defining S include the inequalities $x \geq 0$ and $y \geq 0$.

c. If S is the empty set, then the linear programming problem has no solution; that is, P has neither a maximum nor a minimum value.

The **method of corners,** a simple procedure for solving linear programming problems based on Theorem 1, follows.

The Method of Corners

1. Graph the feasible set.

2. Find the coordinates of all corner points (vertices) of the feasible set.

3. Evaluate the objective function at each corner point.

4. Find the vertex that renders the objective function a maximum (minimum). If there is only one such vertex, then this vertex constitutes a unique solution to the problem. If the objective function is maximized (minimized) at two adjacent corner points of S, there are infinitely many optimal solutions given by the points on the line segment determined by these two vertices.

APPLIED EXAMPLE 1 Maximizing Profit We are now in a position to complete the solution to the production problem posed in Example 1, Section 6.2. Recall that the mathematical formulation led to the following linear programming problem:

$$\text{Maximize } P = x + 1.2y$$
$$\text{subject to } 2x + y \le 180$$
$$x + 3y \le 300$$
$$x \ge 0, y \ge 0$$

Solution The feasible set S for the problem is shown in Figure 12.

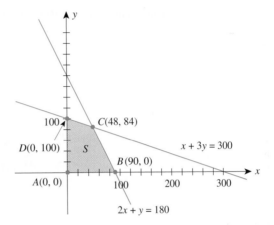

FIGURE **12**
The corner point that yields the maximum profit is C(48, 84).

The vertices of the feasible set are $A(0, 0)$, $B(90, 0)$, $C(48, 84)$, and $D(0, 100)$. The values of P at these vertices may be tabulated as follows:

Vertex	$P = x + 1.2y$
$A(0, 0)$	0
$B(90, 0)$	90
$C(48, 84)$	148.8
$D(0, 100)$	120

From the table, we see that the maximum of $P = x + 1.2y$ occurs at the vertex $(48, 84)$ and has a value of 148.8. Recalling what the symbols x, y, and P represent, we conclude that Ace Novelty would maximize its profit ($148.80) by producing 48 Type A souvenirs and 84 Type B souvenirs. ◼

Explore & Discuss

Consider the linear programming problem

$$\text{Maximize } P = 4x + 3y$$
$$\text{subject to } 2x + y \le 10$$
$$2x + 3y \le 18$$
$$x \ge 0, y \ge 0$$

1. Sketch the feasible set S for the linear programming problem.
2. Draw the isoprofit lines superimposed on S corresponding to $P = 12, 16, 20$, and 24, and show that these lines are parallel to each other.
3. Show that the solution to the linear programming problem is $x = 3$ and $y = 4$. Is this result the same as that found by using the method of corners?

VIDEO **APPLIED EXAMPLE 2** A Nutrition Problem Complete the solution of the nutrition problem posed in Example 2, Section 6.2.

Solution   Recall that the mathematical formulation of the problem led to the following linear programming problem in two variables:

$$\text{Minimize}\quad C = 6x + 8y$$
$$\text{subject to}\quad 40x + 10y \geq 2400$$
$$10x + 15y \geq 2100$$
$$5x + 15y \geq 1500$$
$$x \geq 0, y \geq 0$$

The feasible set S defined by the system of constraints is shown in Figure 13.

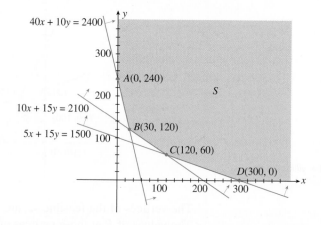

FIGURE 13
The corner point that yields the minimum cost is $B(30, 120)$.

The vertices of the feasible set S are $A(0, 240)$, $B(30, 120)$, $C(120, 60)$, and $D(300, 0)$. The values of the objective function C at these vertices are given in the following table:

Vertex	$C = 6x + 8y$
$A(0, 240)$	1920
$B(30, 120)$	1140
$C(120, 60)$	1200
$D(300, 0)$	1800

From the table, we can see that the minimum for the objective function $C = 6x + 8y$ occurs at the vertex $B(30, 120)$ and has a value of 1140. Thus, the individual should purchase 30 Brand A pills and 120 Brand B pills at a minimum cost of \$11.40.

EXAMPLE 3 A Linear Programming Problem with Multiple Solutions Find the maximum and minimum of $P = 2x + 3y$ subject to the following system of linear inequalities:

$$2x + 3y \leq 30$$
$$-x + y \leq 5$$
$$x + y \geq 5$$
$$x \leq 10$$
$$x \geq 0, y \geq 0$$

Solution   The feasible set S is shown in Figure 14. The vertices of the feasible set S are $A(5, 0)$, $B(10, 0)$, $C(10, \frac{10}{3})$, $D(3, 8)$, and $E(0, 5)$. The values of the objective function P at these vertices are given in the following table:

Vertex	$P = 2x + 3y$
$A(5, 0)$	10
$B(10, 0)$	20
$C(10, \frac{10}{3})$	30
$D(3, 8)$	30
$E(0, 5)$	15

From the table, we see that the maximum for the objective function $P = 2x + 3y$ occurs at the vertices $C(10, \frac{10}{3})$ and $D(3, 8)$. This tells us that every point on the line segment joining the points $C(10, \frac{10}{3})$ and $D(3, 8)$ maximizes P. The value of P at each of these points is 30. From the table, it is also clear that P is minimized at the point $(5, 0)$, where it attains a value of 10.

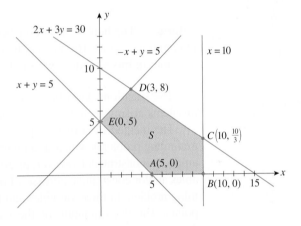

FIGURE 14
Every point lying on the line segment joining C and D maximizes P.

Explore & Discuss

Consider the linear programming problem

$$\text{Maximize} \quad P = 2x + 3y$$
$$\text{subject to} \quad 2x + y \leq 10$$
$$2x + 3y \leq 18$$
$$x \geq 0, y \geq 0$$

1. Sketch the feasible set S for the linear programming problem.
2. Draw the isoprofit lines superimposed on S corresponding to $P = 6, 8, 12$, and 18, and show that these lines are parallel to each other.
3. Show that there are infinitely many solutions to the problem. Is this result as predicted by the method of corners?

We close this section by examining two situations in which a linear programming problem has no solution.

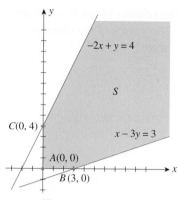

FIGURE **15**
This maximization problem has no solution because the feasible set is unbounded.

EXAMPLE 4 An Unbounded Linear Programming Problem with No Solution
Solve the following linear programming problem:

$$\text{Maximize} \quad P = x + 2y$$
$$\text{subject to} \quad -2x + y \le 4$$
$$x - 3y \le 3$$
$$x \ge 0, y \ge 0$$

Solution   The feasible set S for this problem is shown in Figure 15. Since the set S is unbounded (both x and y can take on arbitrarily large positive values), we see that we can make P as large as we please by choosing x and y large enough. This problem has no solution. The problem is said to be unbounded.   ■

EXAMPLE 5 An Infeasible Linear Programming Problem Solve the following linear programming problem:

$$\text{Maximize} \quad P = x + 2y$$
$$\text{subject to} \quad x + 2y \le 4$$
$$2x + 3y \ge 12$$
$$x \ge 0, y \ge 0$$

Solution   The half-planes described by the constraints (inequalities) have no points in common (Figure 16). Hence, there are no feasible points, and the problem has no solution. In this situation, we say that the problem is **infeasible,** or **inconsistent.**   ■

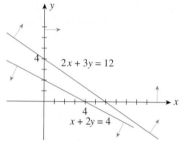

FIGURE **16**
This problem is inconsistent because there is no point that satisfies all of the given inequalities.

The situations described in Examples 4 and 5 are unlikely to occur in well-posed problems arising from practical applications of linear programming.

The method of corners is particularly effective in solving two-variable linear programming problems with a small number of constraints, as the preceding examples have amply demonstrated. However, its effectiveness decreases rapidly as the number of variables and/or constraints increases. For example, it may be shown that a linear programming problem in three variables and five constraints may have up to ten feasible corner points. The determination of the feasible corner points calls for the solution of ten 3×3 systems of linear equations and then the verification—by the substitution of each of these solutions into the system of constraints—to see whether it is, in fact, a feasible point. When the number of variables and constraints goes up to five and ten, respectively (still a very small system from the standpoint of applications in economics), the number of vertices to be found and checked for feasible corner points increases dramatically to 252, and each of these vertices is found by solving a 5×5 linear system! For this reason, the method of corners is seldom used to solve linear programming problems; its redeeming value lies in the fact that much insight is gained into the nature of the solutions of linear programming problems through its use in solving two-variable problems.

6.3   Self-Check Exercises

1. Use the method of corners to solve the following linear programming problem:

$$\text{Maximize} \quad P = 4x + 5y$$
$$\text{subject to} \quad x + 2y \le 10$$
$$5x + 3y \le 30$$
$$x \ge 0, y \ge 0$$

2. Use the method of corners to solve the following linear programming problem:

$$\text{Minimize} \quad C = 5x + 3y$$
$$\text{subject to} \quad 5x + 3y \ge 30$$
$$x - 3y \le 0$$
$$x \ge 2$$

3. Gino Balduzzi, proprietor of Luigi's Pizza Palace, allocates $9000 a month for advertising in two newspapers, the *City Tribune* and the *Daily News*. The *City Tribune* charges $300 for a certain advertisement, whereas the *Daily News* charges $100 for the same ad. Gino has stipulated that the ad is to appear in at least 15 but no more than 30 editions of the *Daily News* per month. The *City Tribune* has a daily circulation of 50,000, and the *Daily News* has a circulation of 20,000. Under these conditions, determine how many ads Gino should place in each newspaper to reach the largest number of readers.

Solutions to Self-Check Exercises 6.3 can be found on page 358.

6.3 Concept Questions

1. a. What is the feasible set associated with a linear programming problem?
   b. What is a feasible solution of a linear programming problem?
   c. What is an optimal solution of a linear programming problem?

2. Describe the method of corners.

6.3 Exercises

In Exercises 1–6, find the maximum and/or minimum value(s) of the objective function on the feasible set S.

1. $Z = 2x + 3y$

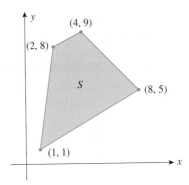

2. $Z = 3x - y$

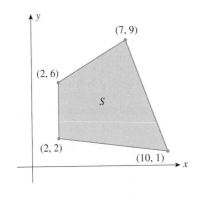

3. $Z = 2x + 3y$

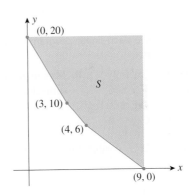

4. $Z = 7x + 9y$

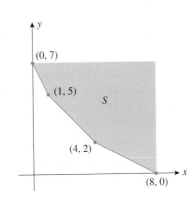

5. $Z = x + 4y$

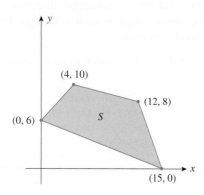

6. $Z = 3x + 2y$

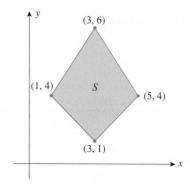

In Exercises 7–28, solve each linear programming problem by the method of corners.

7. Maximize $P = 3x + 2y$
subject to $\quad x + y \leq 6$
$\quad\quad\quad\quad x \leq 3$
$\quad\quad\quad\quad x \geq 0, y \geq 0$

8. Maximize $P = x + 2y$
subject to $\quad x + y \leq 4$
$\quad\quad\quad 2x + y \leq 5$
$\quad\quad\quad x \geq 0, y \geq 0$

9. Maximize $P = 2x + y$ subject to the constraints of Exercise 8.

10. Maximize $P = 4x + 2y$
subject to $\quad x + y \leq 8$
$\quad\quad\quad 2x + y \leq 10$
$\quad\quad\quad x \geq 0, y \geq 0$

11. Maximize $P = x + 8y$ subject to the constraints of Exercise 10.

12. Maximize $P = 3x - 4y$
subject to $\quad x + 3y \leq 15$
$\quad\quad\quad 4x + y \leq 16$
$\quad\quad\quad x \geq 0, y \geq 0$

13. Maximize $P = x + 3y$
subject to $\quad 2x + y \leq 6$
$\quad\quad\quad x + y \leq 4$
$\quad\quad\quad\quad x \leq 1$
$\quad\quad\quad x \geq 0, y \geq 0$

14. Maximize $P = 2x + 5y$
subject to $\quad 2x + y \leq 16$
$\quad\quad\quad 2x + 3y \leq 24$
$\quad\quad\quad\quad y \leq 6$
$\quad\quad\quad x \geq 0, y \geq 0$

15. Minimize $C = 2x + 5y$
subject to $\quad x + y \geq 3$
$\quad\quad\quad x + 2y \geq 4$
$\quad\quad\quad x \geq 0, y \geq 0$

16. Minimize $C = 2x + 4y$ subject to the constraints of Exercise 15.

17. Minimize $C = 3x + 6y$
subject to $\quad x + 2y \geq 40$
$\quad\quad\quad x + y \geq 30$
$\quad\quad\quad x \geq 0, y \geq 0$

18. Minimize $C = 3x + y$ subject to the constraints of Exercise 17.

19. Minimize $C = 2x + 10y$
subject to $\quad 5x + 2y \geq 40$
$\quad\quad\quad x + 2y \geq 20$
$\quad\quad\quad y \geq 3, x \geq 0$

20. Minimize $C = 2x + 5y$
subject to $\quad 4x + y \geq 40$
$\quad\quad\quad 2x + y \geq 30$
$\quad\quad\quad x + 3y \geq 30$
$\quad\quad\quad x \geq 0, y \geq 0$

21. Minimize $C = 10x + 15y$
subject to $\quad x + y \leq 10$
$\quad\quad\quad 3x + y \geq 12$
$\quad\quad\quad -2x + 3y \geq 3$
$\quad\quad\quad x \geq 0, y \geq 0$

22. Maximize $P = 2x + 5y$ subject to the constraints of Exercise 21.

23. Maximize $P = 3x + 4y$
subject to $\quad x + 2y \leq 50$
$\quad\quad\quad 5x + 4y \leq 145$
$\quad\quad\quad 2x + y \geq 25$
$\quad\quad\quad y \geq 5, x \geq 0$

24. Maximize $P = 4x - 3y$ subject to the constraints of Exercise 23.

25. Maximize $P = 2x + 3y$
subject to $\quad x + y \leq 48$
$\quad\quad\quad x + 3y \geq 60$
$\quad\quad\quad 9x + 5y \leq 320$
$\quad\quad\quad x \geq 10, y \geq 0$

26. Minimize $C = 5x + 3y$ subject to the constraints of Exercise 25.

27. Find the maximum and minimum of $P = 8x + 5y$ subject to

$$5x + 2y \geq 63$$
$$x + y \geq 18$$
$$3x + 2y \leq 51$$
$$x \geq 0, y \geq 0$$

28. Find the maximum and minimum of $P = 4x + 3y$ subject to

$$3x + 5y \geq 20$$
$$3x + y \leq 16$$
$$-2x + y \leq 1$$
$$x \geq 0, y \geq 0$$

The problems in Exercises 29–46 correspond to those in Exercises 1–18, Section 6.2. Use the results of your previous work to help you solve these problems.

29. **MANUFACTURING—PRODUCTION SCHEDULING** A company manufactures two products, A and B, on two machines, I and II. It has been determined that the company will realize a profit of $3/unit of Product A and a profit of $4/unit of Product B. To manufacture a unit of Product A requires 6 min on Machine I and 5 min on Machine II. To manufacture a unit of Product B requires 9 min on Machine I and 4 min on Machine II. There are 5 hr of machine time available on Machine I and 3 hr of machine time available on Machine II in each work shift. How many units of each product should be produced in each shift to maximize the company's profit? What is the optimal profit?

30. **MANUFACTURING—PRODUCTION SCHEDULING** National Business Machines manufactures two models of fax machines: A and B. Each model A costs $100 to make, and each model B costs $150. The profits are $30 for each model A and $40 for each model B fax machine. If the total number of fax machines demanded per month does not exceed 2500 and the company has earmarked no more than $600,000/month for manufacturing costs, how many units of each model should National make each month to maximize its monthly profit? What is the optimal profit?

31. **MANUFACTURING—PRODUCTION SCHEDULING** Kane Manufacturing has a division that produces two models of fireplace grates, model A and model B. To produce each model A grate requires 3 lb of cast iron and 6 min of labor. To produce each model B grate requires 4 lb of cast iron and 3 min of labor. The profit for each model A grate is $2.00, and the profit for each model B grate is $1.50. If 1000 lb of cast iron and 20 labor-hours are available for the production of fireplace grates per day, how many grates of each model should the division produce to maximize Kane's profit? What is the optimal profit?

32. **MANUFACTURING—PRODUCTION SCHEDULING** Refer to Exercise 31. Because of a backlog of orders for model A grates, Kane's manager had decided to produce at least 150 of these models a day. Operating under this additional constraint, how many grates of each model should Kane produce to maximize profit? What is the optimal profit?

33. **MANUFACTURING—PRODUCTION SCHEDULING** A division of the Winston Furniture Company manufactures dining tables and chairs. Each table requires 40 board feet of wood and 3 labor-hours. Each chair requires 16 board feet of wood and 4 labor-hours. The profit for each table is $45, and the profit for each chair is $20. In a certain week, the company has 3200 board feet of wood available and 520 labor-hours available. How many tables and chairs should Winston manufacture to maximize its profit? What is the maximum profit?

34. **MANUFACTURING—PRODUCTION SCHEDULING** Refer to Exercise 33. If the profit for each table is $50 and the profit for each chair is $18, how many tables and chairs should Win-

ston manufacture to maximize its profit? What is the maximum profit?

35. **FINANCE—ALLOCATION OF FUNDS** Madison Finance has a total of $20 million earmarked for homeowner loans and auto loans. On the average, homeowner loans have a 10% annual rate of return, whereas auto loans yield a 12% annual rate of return. Management has also stipulated that the total amount of homeowner loans should be greater than or equal to 4 times the total amount of automobile loans. Determine the total amount of loans of each type that Madison should extend to each category to maximize its returns. What are the optimal returns?

36. **INVESTMENTS—ASSET ALLOCATION** A financier plans to invest up to $500,000 in two projects. Project A yields a return of 10% on the investment, whereas Project B yields a return of 15% on the investment. Because the investment in Project B is riskier than the investment in Project A, the financier has decided that the investment in Project B should not exceed 40% of the total investment. How much should she invest in each project to maximize the return on her investment? What is the maximum return?

37. **MANUFACTURING—PRODUCTION SCHEDULING** Acoustical manufactures a CD storage cabinet that can be bought fully assembled or as a kit. Each cabinet is processed in the fabrications department and the assembly department. If the fabrication department manufactures only fully assembled cabinets, then it can produce 200 units/day; and if it manufactures only kits, it can produce 200 units/day. If the assembly department produces only fully assembled cabinets, then it can produce 100 units/day; but if it produces only kits, then it can produce 300 units/day. Each fully assembled cabinet contributes $50 to the profits of the company whereas each kit contributes $40 to its profits. How many fully assembled units and how many kits should the company produce per day to maximize its profit? What is the optimal profit?

38. **AGRICULTURE—CROP PLANNING** A farmer plans to plant two crops, A and B. The cost of cultivating Crop A is $40/acre whereas the cost of cultivating Crop B is $60/acre. The farmer has a maximum of $7400 available for land cultivation. Each acre of Crop A requires 20 labor-hours, and each acre of Crop B requires 25 labor-hours. The farmer has a maximum of 3300 labor-hours available. If she expects to make a profit of $150/acre on Crop A and $200/acre on Crop B, how many acres of each crop should she plant to maximize her profit? What is the optimal profit?

39. **MINING—PRODUCTION** Perth Mining Company operates two mines for the purpose of extracting gold and silver. The Saddle Mine costs $14,000/day to operate, and it yields 50 oz of gold and 3000 oz of silver each day. The Horseshoe Mine costs $16,000/day to operate, and it yields 75 oz of gold and 1000 oz of silver each day. Company management has set a target of at least 650 oz of gold and 18,000 oz of silver. How many days should each mine be operated so that the target can be met at a minimum cost? What is the minimum cost?

40. **TRANSPORTATION** Deluxe River Cruises operates a fleet of river vessels. The fleet has two types of vessels: A type A vessel has 60 deluxe cabins and 160 standard cabins, whereas a type B vessel has 80 deluxe cabins and 120 standard cabins. Under a charter agreement with Odyssey Travel Agency, Deluxe River Cruises is to provide Odyssey with a minimum of 360 deluxe and 680 standard cabins for their 15-day cruise in May. It costs $44,000 to operate a type A vessel and $54,000 to operate a type B vessel for that period. How many of each type vessel should be used to keep the operating costs to a minimum? What is the minimum cost?

41. **WATER SUPPLY** The water-supply manager for a Midwestern city needs to supply the city with at least 10 million gallons of potable (drinkable) water per day. The supply may be drawn from the local reservoir or from a pipeline to an adjacent town. The local reservoir has a maximum daily yield of 5 million gal of potable water, and the pipeline has a maximum daily yield of 10 million gallons. By contract, the pipeline is required to supply a minimum of 6 million gallons/day. If the cost for 1 million gallons of reservoir water is $300 and that for pipeline water is $500, how much water should the manager get from each source to minimize daily water costs for the city? What is the minimum daily cost?

42. **MANUFACTURING—PRODUCTION SCHEDULING** Ace Novelty manufactures Giant Pandas and Saint Bernards. Each Panda requires 1.5 yd^2 of plush, 30 ft^3 of stuffing, and 5 pieces of trim; each Saint Bernard requires 2 yd^2 of plush, 35 ft^3 of stuffing, and 8 pieces of trim. The profit for each Panda is $10, and the profit for each Saint Bernard is $15. If 3600 yd^2 of plush, 66,000 ft^3 of stuffing, and 13,600 pieces of trim are available, how many of each of the stuffed animals should the company manufacture to maximize profit? What is the maximum profit?

43. **NUTRITION—DIET PLANNING** A nutritionist at the Medical Center has been asked to prepare a special diet for certain patients. She has decided that the meals should contain a minimum of 400 mg of calcium, 10 mg of iron, and 40 mg of vitamin C. She has further decided that the meals are to be prepared from Foods A and B. Each ounce of Food A contains 30 mg of calcium, 1 mg of iron, 2 mg of vitamin C, and 2 mg of cholesterol. Each ounce of Food B contains 25 mg of calcium, 0.5 mg of iron, 5 mg of vitamin C, and 5 mg of cholesterol. Find how many ounces of each type of food should be used in a meal so that the cholesterol content is minimized and the minimum requirements of calcium, iron, and vitamin C are met.

44. **SOCIAL PROGRAMS PLANNING** AntiFam, a hunger-relief organization, has earmarked between $2 and $2.5 million (inclusive) for aid to two African countries, Country A and Country B. Country A is to receive between $1 million and $1.5 million (inclusive), and Country B is to receive at least $0.75 million. It has been estimated that each dollar spent in Country A will yield an effective return of $0.60, whereas a dollar spent in Country B will yield an effective return of $0.80. How should the aid be allocated if the money is to be utilized most effectively according to these criteria?
    Hint: If x and y denote the amount of money to be given to Country A and Country B, respectively, then the objective function to be maximized is $P = 0.6x + 0.8y$.

45. **ADVERTISING** Everest Deluxe World Travel has decided to advertise in the Sunday editions of two major newspapers in town. These advertisements are directed at three groups of potential customers. Each advertisement in Newspaper I is seen by 70,000 Group A customers, 40,000 Group B customers, and 20,000 Group C customers. Each advertisement in Newspaper II is seen by 10,000 Group A, 20,000 Group B, and 40,000 Group C customers. Each advertisement in Newspaper I costs $1000, and each advertisement in Newspaper II costs $800. Everest would like their advertisements to be read by at least 2 million people from Group A, 1.4 million people from Group B, and 1 million people from Group C. How many advertisements should Everest place in each newspaper to achieve its advertising goals at a minimum cost? What is the minimum cost?
    Hint: Use different scales for drawing the feasible set.

46. **MANUFACTURING—SHIPPING COSTS** TMA manufactures 37-in. high definition LCD televisions in two separate locations, Locations I and II. The output at Location I is at most 6000 televisions/month, whereas the output at Location II is at most 5000 televisions/month. TMA is the main supplier of televisions to the Pulsar Corporation, its holding company, which has priority in having all its requirements met. In a certain month, Pulsar placed orders for 3000 and 4000 televisions to be shipped to two of its factories located in City A and City B, respectively. The shipping costs (in dollars) per television from the two TMA plants to the two Pulsar factories are as follows:

From TMA	To Pulsar Factories City A	City B
Location I	$6	$4
Location II	$8	$10

Find a shipping schedule that meets the requirements of both companies while keeping costs to a minimum.

47. Complete the solution to Example 3, Section 6.2.

48. **MANUFACTURING—PRODUCTION SCHEDULING** Bata Aerobics manufactures two models of steppers used for aerobic exercises. Manufacturing each luxury model requires 10 lb of plastic and 10 min of labor. Manufacturing each standard model requires 16 lb of plastic and 8 min of labor. The profit for each luxury model is $40, and the profit for each standard model is $30. If 6000 lb of plastic and 60 labor-hours are available for the production of the steppers per day, how many steppers of each model should Bata produce each day to maximize its profit? What is the optimal profit?

49. **INVESTMENT PLANNING** Patricia has at most $30,000 to invest in securities in the form of corporate stocks. She has narrowed her choices to two groups of stocks: growth stocks that she assumes will yield a 15% return (dividends and capital appreciation) within a year and speculative stocks that she assumes will yield a 25% return (mainly in capital appreciation) within a year. Determine how much she should invest in each group of stocks to maximize the return on her investments within a year if she has decided to invest at least 3 times as much in growth stocks as in speculative stocks. What is the maximum return?

50. VETERINARY SCIENCE A veterinarian has been asked to prepare a diet for a group of dogs to be used in a nutrition study at the School of Animal Science. It has been stipulated that each serving should be no larger than 8 oz and must contain at least 29 units of Nutrient I and 20 units of Nutrient II. The vet has decided that the diet may be prepared from two brands of dog food: Brand *A* and Brand *B*. Each ounce of Brand *A* contains 3 units of Nutrient I and 4 units of Nutrient II. Each ounce of Brand *B* contains 5 units of Nutrient I and 2 units of Nutrient II. Brand *A* costs 3 cents/ounce, and Brand *B* costs 4 cents/ounce. Determine how many ounces of each brand of dog food should be used per serving to meet the given requirements at a minimum cost.

51. MARKET RESEARCH Trendex, a telephone survey company, has been hired to conduct a television-viewing poll among urban and suburban families in the Los Angeles area. The client has stipulated that a maximum of 1500 families is to be interviewed. At least 500 urban families must be interviewed, and at least half of the total number of families interviewed must be from the suburban area. For this service, Trendex will be paid $6000 plus $8 for each completed interview. From previous experience, Trendex has determined that it will incur an expense of $4.40 for each successful interview with an urban family and $5 for each successful interview with a suburban family. How many urban and suburban families should Trendex interview to maximize its profit? What is the optimal profit?

In Exercises 52–55, determine whether the statement is true or false. If it is true, explain why it is true. If it is false, give an example to show why it is false.

52. An optimal solution of a linear programming problem is a feasible solution, but a feasible solution of a linear programming problem need not be an optimal solution.

53. An optimal solution of a linear programming problem can occur inside the feasible set of the problem.

54. If a maximization problem has no solution, then the feasible set associated with the linear programming problem must be unbounded.

55. Suppose you are given the following linear programming problem: Maximize $P = ax + by$ on the unbounded feasible set *S* shown in the accompanying figure.

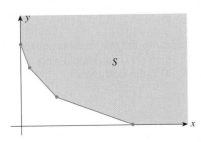

a. If $a > 0$ or $b > 0$, then the linear programming problem has no optimal solution.
b. If $a \le 0$ and $b \le 0$, then the linear programming problem has at least one optimal solution.

56. Suppose you are given the following linear programming problem: Maximize $P = ax + by$, where $a > 0$ and $b > 0$, on the feasible set *S* shown in the accompanying figure.

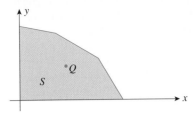

Explain, without using Theorem 1, why the optimal solution of the linear programming problem cannot occur at the point *Q*.

57. Suppose you are given the following linear programming problem: Maximize $P = ax + by$, where $a > 0$ and $b > 0$, on the feasible set *S* shown in the accompanying figure.

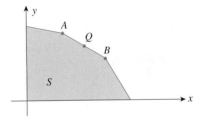

Explain, without using Theorem 1, why the optimal solution of the linear programming problem cannot occur at the point *Q* unless the problem has infinitely many solutions lying along the line segment joining the vertices *A* and *B*.
Hint: Let $A(x_1, y_1)$ and $B(x_2, y_2)$. Then let *Q* be the point $(\bar{x}, \bar{y})$, where $\bar{x} = x_1 + (x_2 - x_1)t$ and $\bar{y} = y_1 + (y_2 - y_1)t$ with $0 < t < 1$. Study the value of *P* at and near *Q*.

58. Consider the linear programming problem

$$\text{Maximize} \quad P = 2x + 7y$$
$$\text{subject to} \quad 2x + y \ge 8$$
$$x + y \ge 6$$
$$x \ge 0, y \ge 0$$

a. Sketch the feasible set *S*.
b. Find the corner points of *S*.
c. Find the values of *P* at the corner points of *S* found in part (b).
d. Show that the linear programming problem has no (optimal) solution. Does this contradict Theorem 1?

59. Consider the linear programming problem

$$\text{Minimize} \quad C = -2x + 5y$$
$$\text{subject to} \quad x + y \le 3$$
$$2x + y \le 4$$
$$5x + 8y \ge 40$$
$$x \ge 0, y \ge 0$$

a. Sketch the feasible set.
b. Find the solution(s) of the linear programming problem, if it exists.

6.3   Solutions to Self-Check Exercises

1. The feasible set S for the problem was graphed in the solution to Exercise 1, Self-Check Exercises 6.1. It is reproduced in the following figure.

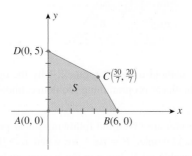

The values of the objective function P at the vertices of S are summarized in the following table:

Vertex	$P = 4x + 5y$
$A(0, 0)$	0
$B(6, 0)$	24
$C(\frac{30}{7}, \frac{20}{7})$	$\frac{220}{7} = 31\frac{3}{7}$
$D(0, 5)$	25

From the table, we see that the maximum for the objective function P is attained at the vertex $C(\frac{30}{7}, \frac{20}{7})$. Therefore, the solution to the problem is $x = \frac{30}{7}$, $y = \frac{20}{7}$, and $P = 31\frac{3}{7}$.

2. The feasible set S for the problem was graphed in the solution to Exercise 2, Self-Check Exercises 6.1. It is reproduced in the following figure.

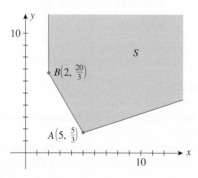

Evaluating the objective function $C = 5x + 3y$ at each corner point, we obtain the following table:

Vertex	$C = 5x + 3y$
$A(5, \frac{5}{3})$	30
$B(2, \frac{20}{3})$	30

We conclude that (i) the objective function is minimized at every point on the line segment joining the points $(5, \frac{5}{3})$ and $(2, \frac{20}{3})$, and (ii) the minimum value of C is 30.

3. Refer to Self-Check Exercise 6.2. The problem is to maximize $P = 50,000x + 20,000y$ subject to

$$300x + 100y \leq 9000$$
$$y \geq 15$$
$$y \leq 30$$
$$x \geq 0, y \geq 0$$

The feasible set S for the problem is shown in the following figure:

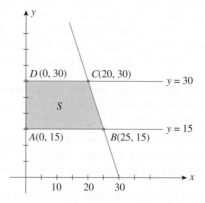

Evaluating the objective function $P = 50,000x + 20,000y$ at each vertex of S, we obtain the following table:

Vertex	$P = 50,000x + 20,000y$
$A(0, 15)$	300,000
$B(25, 15)$	1,550,000
$C(20, 30)$	1,600,000
$D(0, 30)$	600,000

From the table, we see that P is maximized when $x = 20$ and $y = 30$. Therefore, Gino should place 20 ads in the *City Tribune* and 30 in the *Daily News*.

6.4   The Simplex Method: Standard Maximization Problems

The Simplex Method

As was mentioned earlier, the method of corners is not suitable for solving linear programming problems when the number of variables or constraints is large. Its major shortcoming is that a knowledge of all the corner points of the feasible set S associated with the problem is required. What we need is a method of solution that is based on a judicious selection of the corner points of the feasible set S, thereby reducing the number of points to be inspected. One such technique, called the *simplex method*, was developed in the late 1940s by George Dantzig and is based on the Gauss–Jordan elimination method. The simplex method is readily adaptable to the computer, which makes it ideally suitable for solving linear programming problems involving large numbers of variables and constraints.

Basically, the simplex method is an iterative procedure; that is, it is repeated over and over again. Beginning at some initial feasible solution (a corner point of the feasible set S, usually the origin), each iteration brings us to another corner point of S, usually with an improved (but certainly no worse) value of the objective function. The iteration is terminated when the optimal solution is reached (if it exists).

In this section, we describe the simplex method for solving a large class of problems that are referred to as standard maximization problems.

Before stating a formal procedure for solving standard linear programming problems based on the simplex method, let's consider the following analysis of a two-variable problem. The ensuing discussion will clarify the general procedure and at the same time enhance our understanding of the simplex method by examining the motivation that led to the steps of the procedure.

A Standard Linear Programming Problem

A **standard maximization problem** is one in which

1. The objective function is to be maximized.
2. All the variables involved in the problem are nonnegative.
3. All other linear constraints may be written so that the expression involving the variables is less than or equal to a nonnegative constant.

Consider the linear programming problem presented at the beginning of Section 6.3:

$$\text{Maximize} \quad P = 3x + 2y \qquad \textbf{(8)}$$
$$\text{subject to} \quad 2x + 3y \le 12$$
$$2x + \ y \le \ 8 \qquad \textbf{(9)}$$
$$x \ge 0, y \ge 0$$

You can easily verify that this is a standard maximization problem. The feasible set S associated with this problem is reproduced in Figure 17, where we have labeled the four feasible corner points $A(0, 0)$, $B(4, 0)$, $C(3, 2)$, and $D(0, 4)$. Recall that the optimal solution to the problem occurs at the corner point $C(3, 2)$.

To solve this problem using the simplex method, we first replace the system of inequality constraints (9) with a system of equality constraints. This may be accomplished by using nonnegative variables called **slack variables.** Let's begin by considering the inequality

$$2x + 3y \le 12$$

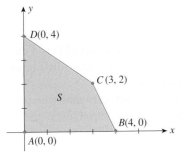

FIGURE **17**
The optimal solution occurs at $C(3, 2)$.

Observe that the left-hand side of this equation is always less than or equal to the right-hand side. Therefore, by adding a nonnegative variable u to the left-hand side to compensate for this difference, we obtain the equality

$$2x + 3y + u = 12$$

For example, if $x = 1$ and $y = 1$, then $u = 7$, because

$$2(1) + 3(1) + 7 = 12$$

(You can see by referring to Figure 1 that the point $(1, 1)$ is a feasible point of S.) If $x = 2$ and $y = 1$, then $u = 5$, because

$$2(2) + 3(1) + 5 = 12$$

(So the point $(2, 1)$ is also a feasible point of S.) The variable u is a slack variable.

Similarly, the inequality $2x + y \leq 8$ is converted into the equation $2x + y + v = 8$ through the introduction of the slack variable v. System (9) of linear inequalities may now be viewed as the system of linear equations

$$2x + 3y + u = 12$$
$$2x + y + v = 8$$

where x, y, u, and v are all nonnegative.

Finally, rewriting the objective function (8) in the form $-3x - 2y + P = 0$, where the coefficient of P is $+1$, we are led to the following system of linear equations:

$$
\begin{aligned}
2x + 3y + u & = 12 \\
2x + y + v & = 8 \\
-3x - 2y + P & = 0
\end{aligned}
\tag{10}
$$

Since System (10) consists of three linear equations in the five variables x, y, u, v, and P, we may solve for three of the variables in terms of the other two. Thus, there are infinitely many solutions to this system expressible in terms of two parameters. Our linear programming problem is now seen to be equivalent to the following: From among all the solutions of System (10) for which x, y, u, and v are nonnegative (such solutions are called **feasible solutions**), determine the solution(s) that maximizes P.

The augmented matrix associated with System (10) is

Nonbasic variables ⟶ | Basic variables | Column of constants

$$
\begin{array}{ccccc|c}
x & y & u & v & P & \\
2 & 3 & 1 & 0 & 0 & 12 \\
2 & 1 & 0 & 1 & 0 & 8 \\
-3 & -2 & 0 & 0 & 1 & 0
\end{array}
\tag{11}
$$

Observe that each of the u-, v-, and P-columns of the augmented matrix (11) is a unit column (see page 261). The variables associated with unit columns are called **basic variables**; all other variables are called **nonbasic variables**.

Now, the configuration of the augmented matrix (11) suggests that we solve for the basic variables u, v, and P in terms of the nonbasic variables x and y, obtaining

$$
\begin{aligned}
u & = 12 - 2x - 3y \\
v & = 8 - 2x - y \\
P & = 3x + 2y
\end{aligned}
\tag{12}
$$

Of the infinitely many feasible solutions that are obtainable by assigning arbitrary nonnegative values to the parameters x and y, a particular solution is obtained by letting $x = 0$ and $y = 0$. In fact, this solution is given by

$$x = 0 \qquad y = 0 \qquad u = 12 \qquad v = 8 \qquad P = 0$$

Such a solution, obtained by setting the nonbasic variables equal to zero, is called a **basic solution** of the system. This particular solution corresponds to the corner point $A(0, 0)$ of the feasible set associated with the linear programming problem (see Figure 17). Observe that $P = 0$ at this point.

Now, if the value of P cannot be increased, we have found the optimal solution to the problem at hand. To determine whether the value of P can in fact be improved, let's turn our attention to the objective function in Equation (8). Since the coefficients of both x and y are positive, the value of P can be improved by increasing x and/or y—that is, by moving away from the origin. Note that we arrive at the same conclusion by observing that the last row of the augmented matrix (11) contains entries that are *negative*. (Compare the original objective function, $P = 3x + 2y$, with the rewritten objective function, $-3x - 2y + P = 0$.)

Continuing our quest for an optimal solution, our next task is to determine whether it is more profitable to increase the value of x or that of y (increasing x and y simultaneously is more difficult). Since the coefficient of x is greater than that of y, a unit increase in the x-direction will result in a greater increase in the value of the objective function P than will a unit increase in the y-direction. Therefore, we should increase the value of x while holding y constant. How much can x be increased while holding $y = 0$? Upon setting $y = 0$ in the first two equations of System (12), we see that

$$\begin{aligned} u &= 12 - 2x \\ v &= 8 - 2x \end{aligned} \tag{13}$$

Since u must be nonnegative, the first equation of System (13) implies that x cannot exceed $\frac{12}{2}$, or 6. The second equation of System (13) and the nonnegativity of v imply that x cannot exceed $\frac{8}{2}$, or 4. Thus, we conclude that x can be increased by at most 4.

Now, if we set $y = 0$ and $x = 4$ in System (12), we obtain the solution

$$x = 4 \qquad y = 0 \qquad u = 4 \qquad v = 0 \qquad P = 12$$

which is a basic solution to System (10), this time with y and v as nonbasic variables. (Recall that the nonbasic variables are precisely the variables that are set equal to zero.)

Let's see how this basic solution may be found by working with the augmented matrix of the system. Since x is to replace v as a basic variable, our aim is to find an augmented matrix that is equivalent to the matrix (11) and has a configuration in which the x-column is in the unit form

$$\begin{bmatrix} 0 \\ 1 \\ 0 \end{bmatrix}$$

replacing what is presently the form of the v-column in augmented matrix (11). This may be accomplished by pivoting about the circled number 2.

$$\begin{array}{ccccccc} & x & y & u & v & P & \text{Const.} \\ \begin{bmatrix} & 2 & 3 & 1 & 0 & 0 & 12 \\ & ② & 1 & 0 & 1 & 0 & 8 \\ & -3 & -2 & 0 & 0 & 1 & 0 \end{bmatrix} \end{array} \xrightarrow{\frac{1}{2}R_2} \begin{array}{ccccccc} & x & y & u & v & P & \text{Const.} \\ \begin{bmatrix} 2 & 3 & 1 & 0 & 0 & 12 \\ ① & \frac{1}{2} & 0 & \frac{1}{2} & 0 & 4 \\ -3 & -2 & 0 & 0 & 1 & 0 \end{bmatrix} \end{array} \tag{14}$$

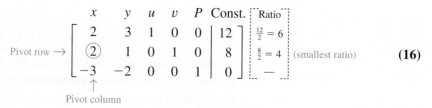

$$\begin{array}{c}\\ \xrightarrow[R_3 + 3R_2]{R_1 - 2R_2}\end{array} \begin{array}{ccccc} x & y & u & v & P \quad \text{Const.} \\ \begin{bmatrix} 0 & 2 & 1 & -1 & 0 & 4 \\ 1 & \frac{1}{2} & 0 & \frac{1}{2} & 0 & 4 \\ 0 & -\frac{1}{2} & 0 & \frac{3}{2} & 1 & 12 \end{bmatrix} \end{array} \qquad \text{(15)}$$

Using System (15), we now solve for the basic variables x, u, and P in terms of the nonbasic variables y and v, obtaining

$$x = 4 - \frac{1}{2}y - \frac{1}{2}v$$

$$u = 4 - 2y + v$$

$$P = 12 + \frac{1}{2}y - \frac{3}{2}v$$

Setting the nonbasic variables y and v equal to zero gives

$$x = 4 \qquad y = 0 \qquad u = 4 \qquad v = 0 \qquad P = 12$$

as before.

We have now completed one iteration of the simplex procedure, and our search has brought us from the feasible corner point $A(0, 0)$, where $P = 0$, to the feasible corner point $B(4, 0)$, where P attained a value of 12, which is certainly an improvement! (See Figure 18.)

Before going on, let's introduce the following terminology. In what follows, refer to the augmented matrix (16), which is reproduced from the first augmented matrix in (14):

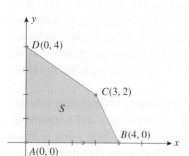

FIGURE 18
One iteration has taken us from $A(0, 0)$, where $P = 0$, to $B(4, 0)$, where $P = 12$.

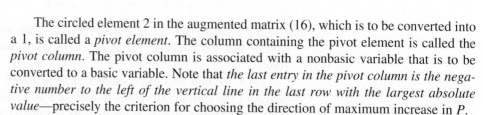

(the negative number in the last row to the left
of the vertical line with the largest absolute value)

The circled element 2 in the augmented matrix (16), which is to be converted into a 1, is called a *pivot element*. The column containing the pivot element is called the *pivot column*. The pivot column is associated with a nonbasic variable that is to be converted to a basic variable. Note that *the last entry in the pivot column is the negative number to the left of the vertical line in the last row with the largest absolute value*—precisely the criterion for choosing the direction of maximum increase in P.

The row containing the pivot element is called the *pivot row*. The pivot row can also be found by dividing each positive number in the pivot column into the corresponding number in the last column (the column of constants). *The pivot row is the one with the smallest ratio*. In augmented matrix (16), the pivot row is the second row because the ratio $\frac{8}{2}$, or 4, is less than the ratio $\frac{12}{2}$, or 6. (Compare this with the earlier analysis pertaining to the determination of the largest permissible increase in the value of x.) Then pivoting about the pivot element, we obtain the second tableau in (14).

The following is a summary of the procedure for selecting the pivot element.

Selecting the Pivot Element

1. *Select the pivot column:* Locate the most negative entry to the left of the vertical line in the last row. The column containing this entry is the **pivot column.** (If there is more than one such column, choose any one.)

> **2.** *Select the pivot row*: Divide each positive entry in the pivot column into its corresponding entry in the column of constants. The **pivot row** is the row corresponding to the smallest ratio thus obtained. (If there is more than one such entry, choose any one.)
>
> **3.** The **pivot element** is the element common to both the pivot column and the pivot row.

Continuing with the solution to our problem, we observe that the last row of the augmented matrix (15) contains a negative number—namely, $-\frac{1}{2}$. This indicates that P is not maximized at the feasible corner point $B(4, 0)$, so another iteration is required. Without once again going into a detailed analysis, we proceed immediately to the selection of a pivot element. In accordance with the rules, we perform the necessary row operations as follows:

$$
\begin{array}{c} \text{Pivot} \rightarrow \\ \text{row} \end{array}
\begin{array}{ccccc} x & y & u & v & P \end{array}
\left[\begin{array}{ccccc|c}
0 & ② & 1 & -1 & 0 & 4 \\
1 & \frac{1}{2} & 0 & \frac{1}{2} & 0 & 4 \\
0 & -\frac{1}{2} & 0 & \frac{3}{2} & 1 & 12
\end{array}\right]
\begin{array}{c} \text{Ratio} \\ \frac{4}{2} = 2 \\ \frac{4}{1/2} = 8 \end{array}
$$

$$
\begin{array}{c} \uparrow \\ \text{Pivot} \\ \text{column} \end{array}
$$

$$
\xrightarrow{\frac{1}{2}R_1}
\begin{array}{ccccc} x & y & u & v & P \end{array}
\left[\begin{array}{ccccc|c}
0 & ① & \frac{1}{2} & -\frac{1}{2} & 0 & 2 \\
1 & \frac{1}{2} & 0 & \frac{1}{2} & 0 & 4 \\
0 & -\frac{1}{2} & 0 & \frac{3}{2} & 1 & 12
\end{array}\right]
$$

$$
\begin{array}{c} R_2 - \frac{1}{2}R_1 \\ \xrightarrow{\hspace{1cm}} \\ R_3 + \frac{1}{2}R_1 \end{array}
\begin{array}{ccccc} x & y & u & v & P \end{array}
\left[\begin{array}{ccccc|c}
0 & 1 & \frac{1}{2} & -\frac{1}{2} & 0 & 2 \\
1 & 0 & -\frac{1}{4} & \frac{3}{4} & 0 & 3 \\
0 & 0 & \frac{1}{4} & \frac{5}{4} & 1 & 13
\end{array}\right]
$$

Interpreting the last augmented matrix in the usual fashion, we find the basic solution $x = 3$, $y = 2$, and $P = 13$. Since there are no negative entries in the last row, the solution is optimal, and P cannot be increased further. The optimal solution is the feasible corner point $C(3, 2)$ (Figure 19). Observe that this agrees with the solution we found using the method of corners in Section 6.3.

Having seen how the simplex method works, let's list the steps involved in the procedure. The first step is to set up the initial **simplex tableau.**

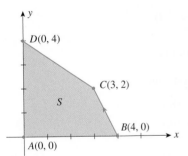

FIGURE 19
The next iteration has taken us from $B(4, 0)$, where $P = 12$, to $C(3, 2)$, where $P = 13$.

> ### Setting Up the Initial Simplex Tableau
>
> **1.** Transform the system of linear inequalities into a system of linear equations by introducing slack variables.
>
> **2.** Rewrite the objective function
>
> $$P = c_1x_1 + c_2x_2 + \cdots + c_nx_n$$
>
> in the form
>
> $$-c_1x_1 - c_2x_2 - \cdots - c_nx_n + P = 0$$
>
> where all the variables are on the left and the coefficient of P is $+1$. Write this equation below the equations of Step 1.
>
> **3.** Write the tableau associated with this system of linear equations.

EXAMPLE 1 Set up the initial simplex tableau for the linear programming problem posed in Example 1, Section 6.2.

Solution The problem at hand is to maximize

$$P = x + 1.2y$$

or, equivalently,

$$P = x + \frac{6}{5}y$$

subject to

$$2x + y \leq 180$$
$$x + 3y \leq 300 \qquad \qquad \textbf{(17)}$$
$$x \geq 0, y \geq 0$$

This is a standard maximization problem and may be solved by the simplex method. Since System (17) has two linear inequalities (other than $x \geq 0$, $y \geq 0$), we introduce the two slack variables u and v to convert it to a system of linear equations:

$$2x + y + u \qquad = 180$$
$$x + 3y \qquad + v = 300$$

Next, by rewriting the objective function in the form

$$-x - \frac{6}{5}y + P = 0$$

where the coefficient of P is $+1$, and placing it below the system of equations, we obtain the system of linear equations

$$2x + y + u \qquad \qquad = 180$$
$$x + 3y \qquad + v \qquad = 300$$
$$-x - \frac{6}{5}y \qquad \qquad + P = 0$$

The initial simplex tableau associated with this system is

x	y	u	v	P	Constant
2	1	1	0	0	180
1	3	0	1	0	300
-1	$-\frac{6}{5}$	0	0	1	0

Before completing the solution to the problem posed in Example 1, let's summarize the main steps of the **simplex method.**

The Simplex Method

1. *Set up the initial simplex tableau.*
2. *Determine whether the optimal solution has been reached by examining all entries in the last row to the left of the vertical line.*
   a. If all the entries are nonnegative, the optimal solution has been reached. Proceed to Step 4.
   b. If there are one or more negative entries, the optimal solution has not been reached. Proceed to Step 3.

3. *Perform the pivot operation.* Locate the pivot element and convert it to a 1 by dividing all the elements in the pivot row by the pivot element. Using row operations, convert the pivot column into a unit column by adding suitable multiples of the pivot row to each of the other rows as required. Return to Step 2.

4. *Determine the optimal solution(s).* The value of the variable heading each unit column is given by the entry lying in the column of constants in the row containing the 1. The variables heading columns not in unit form are assigned the value zero.

EXAMPLE 2 Complete the solution to the problem discussed in Example 1.

Solution   The first step in our procedure, setting up the initial simplex tableau, was completed in Example 1. We continue with Step 2.

Step 2   *Determine whether the optimal solution has been reached.* First, refer to the initial simplex tableau:

x	y	u	v	P	Constant
2	1	1	0	0	180
1	3	0	1	0	300
-1	$-\frac{6}{5}$	0	0	1	0

(18)

Since there are negative entries in the last row of the initial simplex tableau, the initial solution is not optimal. We proceed to Step 3.

Step 3   *Perform the following iterations.* First, locate the pivot element:

a. Since the entry $-\frac{6}{5}$ is the most negative entry to the left of the vertical line in the last row of the initial simplex tableau, the second column in the tableau is the pivot column.

b. Divide each positive number of the pivot column into the corresponding entry in the column of constants, and compare the ratios thus obtained. We see that the ratio $\frac{300}{3}$ is less than the ratio $\frac{180}{1}$, so row 2 is the pivot row.

c. The entry 3 lying in the pivot column and the pivot row is the pivot element.

	x	y	u	v	P	Constant	Ratio
	2	1	1	0	0	180	$\frac{180}{1} = 180$
Pivot row $\rightarrow$	1	③	0	1	0	300	$\frac{300}{3} = 100$
	-1	$-\frac{6}{5}$	0	0	1	0	

Pivot column

Next, we convert this pivot element into a 1 by multiplying all the entries in the pivot row by $\frac{1}{3}$. Then, using elementary row operations, we complete the conversion of the pivot column into a unit column. The details of the iteration follow:

$\xrightarrow{\frac{1}{3}R_2}$

x	y	u	v	P	Constant
2	1	1	0	0	180
$\frac{1}{3}$	①	0	$\frac{1}{3}$	0	100
-1	$-\frac{6}{5}$	0	0	1	0

	x	y	u	v	P	Constant
	$\frac{5}{3}$	0	1	$-\frac{1}{3}$	0	80
	$\frac{1}{3}$	1	0	$\frac{1}{3}$	0	100
	$-\frac{3}{5}$	0	0	$\frac{2}{5}$	1	120

$$\xrightarrow[R_3 + \frac{6}{5}R_2]{R_1 - R_2}$$

(19)

This completes one iteration. The last row of the simplex tableau contains a negative number, so an optimal solution has not been reached. Therefore, we repeat the iterative step once again, as follows:

	x	y	u	v	P	Constant	Ratio
Pivot row →	$\left(\frac{5}{3}\right)$	0	1	$-\frac{1}{3}$	0	80	$\frac{80}{5/3} = 48$
	$\frac{1}{3}$	1	0	$\frac{1}{3}$	0	100	$\frac{100}{1/3} = 300$
	$-\frac{3}{5}$	0	0	$\frac{2}{5}$	1	120	

↑
Pivot column

	x	y	u	v	P	Constant
	$\left(1\right)$	0	$\frac{3}{5}$	$-\frac{1}{5}$	0	48
	$\frac{1}{3}$	1	0	$\frac{1}{3}$	0	100
	$-\frac{3}{5}$	0	0	$\frac{2}{5}$	1	120

$$\xrightarrow{\frac{3}{5}R_1}$$

	x	y	u	v	P	Constant
	1	0	$\frac{3}{5}$	$-\frac{1}{5}$	0	48
	0	1	$-\frac{1}{5}$	$\frac{2}{5}$	0	84
	0	0	$\frac{9}{25}$	$\frac{7}{25}$	1	$148\frac{4}{5}$

$$\xrightarrow[R_3 + \frac{3}{5}R_1]{R_2 - \frac{1}{3}R_1}$$

(20)

The last row of the simplex tableau (20) does not contain any negative numbers, so we conclude that the optimal solution has been reached.

Step 4   *Determine the optimal solution.* Locate the basic variables in the final tableau. In this case, the basic variables (those heading unit columns) are x, y, and P. The value assigned to the basic variable x is the number 48, which is the entry lying in the column of constants and in row 1 (the row that contains the 1).

x	y	u	v	P	Constant
$\left(1\right)$	0	$\frac{3}{5}$	$-\frac{1}{5}$	0	48 ←
0	$\left(1\right)$	$-\frac{1}{5}$	$\frac{2}{5}$	0	84 ←
0	0	$\frac{9}{25}$	$\frac{7}{25}$	$\left(1\right)$	$148\frac{4}{5}$ ←

Similarly, we conclude that $y = 84$ and $P = 148.8$. Next, we note that the variables u and v are nonbasic and are accordingly assigned the values $u = 0$ and $v = 0$. These results agree with those obtained in Example 1, Section 6.3. ■

VIDEO **EXAMPLE 3**

$$\begin{aligned}
\text{Maximize} \quad & P = 2x + 2y + z \\
\text{subject to} \quad & 2x + y + 2z \le 14 \\
& 2x + 4y + z \le 26 \\
& x + 2y + 3z \le 28 \\
& x \ge 0, y \ge 0, z \ge 0
\end{aligned}$$

Solution   Introducing the slack variables u, v, and w and rewriting the objective function in the standard form gives the system of linear equations

$$
\begin{aligned}
2x + y + 2z + u & & & = 14 \\
2x + 4y + z & + v & & = 26 \\
x + 2y + 3z & & + w & = 28 \\
-2x - 2y - z & & & + P = 0
\end{aligned}
$$

The initial simplex tableau is given by

x	y	z	u	v	w	P	Constant
2	1	2	1	0	0	0	14
2	4	1	0	1	0	0	26
1	2	3	0	0	1	0	28
-2	-2	-1	0	0	0	1	0

Since the most negative entry in the last row (-2) occurs twice, we may choose either the x- or the y-column as the pivot column. Choosing the x-column as the pivot column and proceeding with the first iteration, we obtain the following sequence of tableaus:

	x	y	z	u	v	w	P	Constant		Ratio
Pivot row $\rightarrow$	②	1	2	1	0	0	0	14		$\frac{14}{2} = 7$
	2	4	1	0	1	0	0	26		$\frac{26}{2} = 13$
	1	2	3	0	0	1	0	28		$\frac{28}{1} = 28$
	-2	-2	-1	0	0	0	1	0		

Pivot column $\uparrow$

	x	y	z	u	v	w	P	Constant
$\tfrac{1}{2}R_1 \longrightarrow$	①	$\frac{1}{2}$	1	$\frac{1}{2}$	0	0	0	7
	2	4	1	0	1	0	0	26
	1	2	3	0	0	1	0	28
	-2	-2	-1	0	0	0	1	0

	x	y	z	u	v	w	P	Constant
$R_2 - 2R_1$	1	$\frac{1}{2}$	1	$\frac{1}{2}$	0	0	0	7
$R_3 - R_1 \longrightarrow$	0	3	-1	-1	1	0	0	12
$R_4 + 2R_1$	0	$\frac{3}{2}$	2	$-\frac{1}{2}$	0	1	0	21
	0	-1	1	1	0	0	1	14

Since there is a negative number in the last row of the simplex tableau, we perform another iteration, as follows:

	x	y	z	u	v	w	P	Constant		Ratio
	1	$\frac{1}{2}$	1	$\frac{1}{2}$	0	0	0	7		$\frac{7}{1/2} = 14$
Pivot row $\rightarrow$	0	③	-1	-1	1	0	0	12		$\frac{12}{3} = 4$
	0	$\frac{3}{2}$	2	$-\frac{1}{2}$	0	1	0	21		$\frac{21}{3/2} = 14$
	0	-1	1	1	0	0	1	14		

Pivot column $\uparrow$

	x	y	z	u	v	w	P	Constant
	1	$\frac{1}{2}$	1	$\frac{1}{2}$	0	0	0	7
$\tfrac{1}{3}R_2 \longrightarrow$	0	①	$-\frac{1}{3}$	$-\frac{1}{3}$	$\frac{1}{3}$	0	0	4
	0	$\frac{3}{2}$	2	$-\frac{1}{2}$	0	1	0	21
	0	-1	1	1	0	0	1	14

	x	y	z	u	v	w	P	Constant
$R_1 - \frac{1}{2}R_2$	1	0	$\frac{7}{6}$	$\frac{2}{3}$	$-\frac{1}{6}$	0	0	5
$\xrightarrow{}$	0	1	$-\frac{1}{3}$	$-\frac{1}{3}$	$\frac{1}{3}$	0	0	4
$R_3 - \frac{3}{2}R_2$ $R_4 + R_2$	0	0	$\frac{5}{2}$	0	$-\frac{1}{2}$	1	0	15
	0	0	$\frac{2}{3}$	$\frac{2}{3}$	$\frac{1}{3}$	0	1	18

All entries in the last row are nonnegative, so we have reached the optimal solution. We conclude that $x = 5$, $y = 4$, $z = 0$, $u = 0$, $v = 0$, $w = 15$, and $P = 18$. ■

Explore & Discuss

Consider the linear programming problem

$$\text{Maximize} \quad P = x + 2y$$
$$\text{subject to} \quad -2x + y \le 4$$
$$x - 3y \le 3$$
$$x \ge 0, y \ge 0$$

1. Sketch the feasible set S for the linear programming problem and explain why the problem has an unbounded solution.

2. Use the simplex method to solve the problem as follows:
   a. Perform one iteration on the initial simplex tableau. Interpret your result. Indicate the point on S corresponding to this (nonoptimal) solution.
   b. Show that the simplex procedure breaks down when you attempt to perform another iteration by demonstrating that there is no pivot element.
   c. Describe what happens if you violate the rule for finding the pivot element by allowing the ratios to be negative and proceeding with the iteration.

The following example is constructed to illustrate the geometry associated with the simplex method when used to solve a problem in three-dimensional space. We sketch the feasible set for the problem and show the path dictated by the simplex method in arriving at the optimal solution for the problem. The use of a calculator will help in the arithmetic operations if you wish to verify the steps.

EXAMPLE 4 Geometric Illustration of Simplex Method in 3-Space

$$\text{Maximize} \quad P = 20x + 12y + 18z$$
$$\text{subject to} \quad 3x + y + 2z \le 9$$
$$2x + 3y + z \le 8$$
$$x + 2y + 3z \le 7$$
$$x \ge 0, y \ge 0, z \ge 0$$

Solution Introducing the slack variables u, v, and w and rewriting the objective function in standard form give the following system of linear equations:

$$3x + y + 2z + u \qquad\qquad = 9$$
$$2x + 3y + z \quad + v \qquad\quad = 8$$
$$x + 2y + 3z \qquad\quad + w \quad = 7$$
$$-20x - 12y - 18z \qquad\qquad\quad + P = 0$$

The initial simplex tableau is given by

x	y	z	u	v	w	P	Constant
3	1	2	1	0	0	0	9
2	3	1	0	1	0	0	8
1	2	3	0	0	1	0	7
-20	-12	-18	0	0	0	1	0

Since the most negative entry in the last row (-20) occurs in the x-column, we choose the x-column as the pivot column. Proceeding with the first iteration, we obtain the following sequence of tableaus:

	x	y	z	u	v	w	P	Constant	Ratio
Pivot row →	③	1	2	1	0	0	0	9	$\frac{9}{3} = 3$
	2	3	1	0	1	0	0	8	$\frac{8}{2} = 4$
	1	2	3	0	0	1	0	7	$\frac{7}{1} = 7$
	-20	-12	-18	0	0	0	1	0	

Pivot column

	x	y	z	u	v	w	P	Constant
$\frac{1}{3}R_1$ →	①	$\frac{1}{3}$	$\frac{2}{3}$	$\frac{1}{3}$	0	0	0	3
	2	3	1	0	1	0	0	8
	1	2	3	0	0	1	0	7
	-20	-12	-18	0	0	0	1	0

	x	y	z	u	v	w	P	Constant	Ratio
	1	$\frac{1}{3}$	$\frac{2}{3}$	$\frac{1}{3}$	0	0	0	3	9
$R_2 - 2R_1$ →	0	$\frac{7}{3}$	$-\frac{1}{3}$	$-\frac{2}{3}$	1	0	0	2	$\frac{6}{7}$
$R_3 - R_1$	0	$\frac{5}{3}$	$\frac{7}{3}$	$-\frac{1}{3}$	0	1	0	4	$\frac{12}{5}$
$R_4 + 20R_1$ Pivot row	0	$-\frac{16}{3}$	$-\frac{14}{3}$	$\frac{20}{3}$	0	0	1	60	

Pivot column

Interpreting this tableau, we see that $x = 3$, $y = 0$, $z = 0$, and $P = 60$. Thus, after one iteration, we are at the point B $(3, 0, 0)$ with $P = 60$. (See Figure 20 on the following page.)

Since the most negative entry in the last row is $-\frac{16}{3}$, we choose the y-column as the pivot column. Proceeding with this iteration, we obtain

	x	y	z	u	v	w	P	Constant
	1	$\frac{1}{3}$	$\frac{2}{3}$	$\frac{1}{3}$	0	0	0	3
$\frac{3}{7}R_2$ →	0	①	$-\frac{1}{7}$	$-\frac{2}{7}$	$\frac{3}{7}$	0	0	$\frac{6}{7}$
	0	$\frac{5}{3}$	$\frac{7}{3}$	$-\frac{1}{3}$	0	1	0	4
	0	$-\frac{16}{3}$	$-\frac{14}{3}$	$\frac{20}{3}$	0	0	1	60

		x	y	z	u	v	w	P	Constant		Ratio
$R_1 - \frac{1}{3}R_2$		1	0	$\frac{5}{7}$	$\frac{3}{7}$	$-\frac{1}{7}$	0	0	$\frac{19}{7}$		$\frac{19}{5}$
$\xrightarrow{}$		0	1	$-\frac{1}{7}$	$-\frac{2}{7}$	$\frac{3}{7}$	0	0	$\frac{6}{7}$		—
$R_3 - \frac{5}{3}R_2$		0	0	$\boxed{\frac{18}{7}}$	$\frac{1}{7}$	$-\frac{5}{7}$	1	0	$\frac{18}{7}$		1
$R_4 + \frac{16}{3}R_2$	Pivot row	0	0	$-\frac{38}{7}$	$\frac{36}{7}$	$\frac{16}{7}$	0	1	$64\frac{4}{7}$		

Pivot column

Interpreting this tableau, we see that $x = \frac{19}{7}, y = \frac{6}{7}, z = 0$ and $P = 64\frac{4}{7}$. Thus, the second iteration brings us to the point $C(\frac{19}{7}, \frac{6}{7}, 0)$. (See Figure 20.)

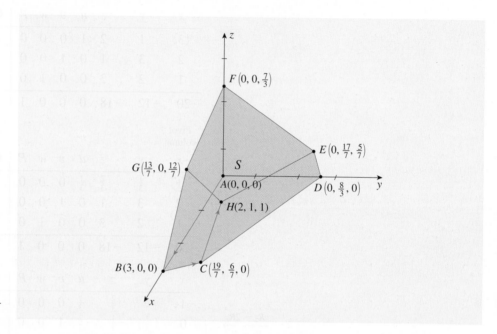

FIGURE 20
The simplex method brings us from the point A to the point H, at which the objective function is maximized.

Since there is a negative number in the last row of the simplex tableau, we perform yet another iteration, as follows:

	x	y	z	u	v	w	P	Constant
$\xrightarrow{\frac{7}{18}R_3}$	1	0	$\frac{5}{7}$	$\frac{3}{7}$	$-\frac{1}{7}$	0	0	$\frac{19}{7}$
	0	1	$-\frac{1}{7}$	$-\frac{2}{7}$	$\frac{3}{7}$	0	0	$\frac{6}{7}$
	0	0	$\boxed{1}$	$\frac{1}{18}$	$-\frac{5}{18}$	$\frac{7}{18}$	0	1
	0	0	$-\frac{38}{7}$	$\frac{36}{7}$	$\frac{16}{7}$	0	1	$64\frac{4}{7}$

	x	y	z	u	v	w	P	Constant
$R_1 - \frac{5}{7}R_3$	1	0	0	$\frac{7}{18}$	$\frac{1}{18}$	$-\frac{5}{18}$	0	2
$\xrightarrow{}$	0	1	0	$-\frac{5}{18}$	$\frac{7}{18}$	$\frac{1}{18}$	0	1
$R_2 + \frac{1}{7}R_3$	0	0	1	$\frac{1}{18}$	$-\frac{5}{18}$	$\frac{7}{18}$	0	1
$R_4 + \frac{38}{7}R_3$	0	0	0	$\frac{49}{9}$	$\frac{7}{9}$	$\frac{19}{9}$	1	70

All entries in the last row are nonnegative, so we have reached the optimal solution (corresponding to the point $H(2, 1, 1)$). We conclude that $x = 2, y = 1, z = 1, u = 0, v = 0, w = 0$, and $P = 70$.

The feasible set S for the problem is the hexahedron shown in Figure 21. It is the intersection of the half-spaces determined by the planes P_1, P_2, and P_3 with equations $3x + y + 2z = 9$, $2x + 3y + z = 8$, $x + 2y + 3z = 7$, respectively, and the coordinate planes $x = 0$, $y = 0$, and $z = 0$. The portion of the figure showing the feasible set S is shown in Figure 20. Observe that the first iteration of the simplex method brings us from $A(0, 0, 0)$ with $P = 0$ to $B(3, 0, 0)$ with $P = 60$. The second iteration brings us from $B(3, 0, 0)$ to $C\left(\frac{19}{7}, \frac{6}{7}, 0\right)$ with $P = 64\frac{4}{7}$, and the third iteration brings us from $C\left(\frac{19}{7}, \frac{6}{7}, 0\right)$ to the point $H(2, 1, 1)$ with an optimal value of 70 for P.

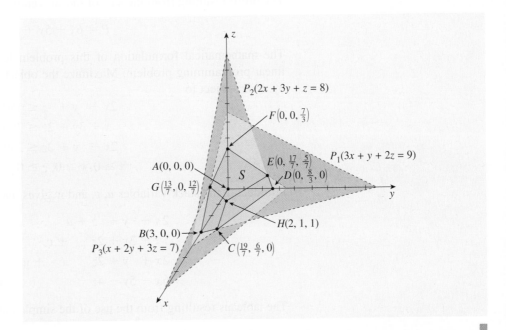

FIGURE 21
The feasible set S is obtained from the intersection of the half-spaces determined by P_1, P_2, and P_3 with the coordinate planes $x = 0$, $y = 0$, and $z = 0$.

VIDEO▶

APPLIED EXAMPLE 5 Production Planning Ace Novelty Company has determined that the profit for each Type A, Type B, and Type C souvenir that it plans to produce is $6, $5, and $4, respectively. To manufacture a Type A souvenir requires 2 minutes on Machine I, 1 minute on Machine II, and 2 minutes on Machine III. A Type B souvenir requires 1 minute on Machine I, 3 minutes on Machine II, and 1 minute on Machine III. A Type C souvenir requires 1 minute on Machine I and 2 minutes on each of Machines II and III. Each day, there are 3 hours available on Machine I, 5 hours available on Machine II, and 4 hours available on Machine III for manufacturing these souvenirs. How many souvenirs of each type should Ace Novelty make per day to maximize its profit? (Compare with Example 1, Section 5.1.)

Solution   The given information is tabulated as follows:

	Type A	Type B	Type C	Time Available (min)
Machine I	2	1	1	180
Machine II	1	3	2	300
Machine III	2	1	2	240
Profit per Unit	$6	$5	$4	

Let x, y, and z denote the respective numbers of Type A, Type B, and Type C souvenirs to be made. The total amount of time that Machine I is used is given by $2x + y + z$ minutes and must not exceed 180 minutes. Thus, we have the inequality

$$2x + y + z \leq 180$$

Similar considerations on the use of Machines II and III lead to the inequalities

$$x + 3y + 2z \leq 300$$
$$2x + y + 2z \leq 240$$

The profit resulting from the sale of the souvenirs produced is given by

$$P = 6x + 5y + 4z$$

The mathematical formulation of this problem leads to the following standard linear programming problem: Maximize the objective (profit) function $P = 6x + 5y + 4z$ subject to

$$2x + y + z \leq 180$$
$$x + 3y + 2z \leq 300$$
$$2x + y + 2z \leq 240$$
$$x \geq 0, y \geq 0, z \geq 0$$

Introducing the slack variables u, v, and w gives the system of linear equations

$$
\begin{aligned}
2x + y + z + u & & & = 180 \\
x + 3y + 2z & + v & & = 300 \\
2x + y + 2z & & + w & = 240 \\
-6x - 5y - 4z & & & + P = 0
\end{aligned}
$$

The tableaus resulting from the use of the simplex algorithm are as follows:

	x	y	z	u	v	w	P	Constant	Ratio
Pivot row →	②	1	1	1	0	0	0	180	$\frac{180}{2} = 90$
	1	3	2	0	1	0	0	300	$\frac{300}{1} = 300$
	2	1	2	0	0	1	0	240	$\frac{240}{2} = 120$
	-6	-5	-4	0	0	0	1	0	

↑ Pivot column

$\frac{1}{2}R_1 \longrightarrow$

	x	y	z	u	v	w	P	Constant
	①	$\frac{1}{2}$	$\frac{1}{2}$	$\frac{1}{2}$	0	0	0	90
	1	3	2	0	1	0	0	300
	2	1	2	0	0	1	0	240
	-6	-5	-4	0	0	0	1	0

$R_2 - R_1$
$R_3 - 2R_1$
$R_4 + 6R_1$ Pivot row

	x	y	z	u	v	w	P	Constant	Ratio
	1	$\frac{1}{2}$	$\frac{1}{2}$	$\frac{1}{2}$	0	0	0	90	$\frac{90}{1/2} = 180$
	0	$\frac{5}{2}$	$\frac{3}{2}$	$-\frac{1}{2}$	1	0	0	210	$\frac{210}{5/2} = 84$
	0	0	1	-1	0	1	0	60	
	0	-2	-1	3	0	0	1	540	

↑ Pivot column

	x	y	z	u	v	w	P	Constant
	1	$\frac{1}{2}$	$\frac{1}{2}$	$\frac{1}{2}$	0	0	0	90
$\xrightarrow{\frac{2}{5}R_2}$	0	①	$\frac{3}{5}$	$-\frac{1}{5}$	$\frac{2}{5}$	0	0	84
	0	0	1	-1	0	1	0	60
	0	-2	-1	3	0	0	1	540

	x	y	z	u	v	w	P	Constant
	1	0	$\frac{1}{5}$	$\frac{3}{5}$	$-\frac{1}{5}$	0	0	48
$\xrightarrow[R_4 + 2R_2]{R_1 - \frac{1}{2}R_2}$	0	1	$\frac{3}{5}$	$-\frac{1}{5}$	$\frac{2}{5}$	0	0	84
	0	0	1	-1	0	1	0	60
	0	0	$\frac{1}{5}$	$\frac{13}{5}$	$\frac{4}{5}$	0	1	708

From the final simplex tableau, we read off the solution

$$x = 48 \qquad y = 84 \qquad z = 0 \qquad u = 0 \qquad v = 0 \qquad w = 60 \qquad P = 708$$

Thus, to maximize its profit, Ace Novelty should produce 48 Type A souvenirs, 84 Type B souvenirs, and no Type C souvenirs. The resulting profit is \$708 per day. The value of the slack variable $w = 60$ tells us that 1 hour of the available time on Machine III is left unused. ▪

Interpreting Our Results Let's compare the results obtained here with those obtained in Example 7, Section 5.2. Recall that to use all available machine time on each of the three machines, Ace Novelty had to produce 36 Type A, 48 Type B, and 60 Type C souvenirs. This would have resulted in a profit of \$696. Example 5 shows how, through the optimal use of equipment, a company can boost its profit while reducing machine wear!

Problems with Multiple Solutions and Problems with No Solutions

As we saw in Section 6.3, a linear programming problem may have infinitely many solutions. We also saw that a linear programming problem may have no solution. How do we spot each of these phenomena when using the simplex method to solve a problem?

A linear programming problem will have infinitely many solutions if and only if the last row to the left of the vertical line of the final simplex tableau has a zero in a column that is not a unit column. Also, a linear programming problem will have *no* solution if the simplex method breaks down at some stage. For example, if at some stage there are no nonnegative ratios in our computation, then the linear programming problem has no solution (see Exercise 50).

Explore & Discuss

Consider the linear programming problem

$$\text{Maximize} \quad P = 4x + 6y$$
$$\text{subject to} \quad 2x + y \le 10$$
$$2x + 3y \le 18$$
$$x \ge 0, y \ge 0$$

1. Sketch the feasible set for the linear programming problem.

2. Use the method of corners to show that there are infinitely many optimal solutions. What are they?

(continued)

3. Use the simplex method to solve the problem as follows.

a. Perform one iteration on the initial simplex tableau and conclude that you have arrived at an optimal solution. What is the value of P, and where is it attained? Compare this result with that obtained in Step 2.

b. Observe that the tableau obtained in part (a) indicates that there are infinitely many solutions (see the comment above on multiple solutions). Now perform another iteration on the simplex tableau using the x-column as the pivot column. Interpret the final tableau.

6.4 Self-Check Exercises

1. Solve the following linear programming problem by the simplex method:

$$\text{Maximize} \quad P = 2x + 3y + 6z$$
$$\text{subject to} \quad 2x + 3y + z \le 10$$
$$x + y + 2z \le 8$$
$$2y + 3z \le 6$$
$$x \ge 0, y \ge 0, z \ge 0$$

2. The LaCrosse Iron Works makes two models of cast-iron fireplace grates: model A and model B. Producing one model A grate requires 20 lb of cast iron and 20 min of labor, whereas producing one model B grate requires 30 lb of cast iron and 15 min of labor. The profit for a model A grate is \$6, and the profit for a model B grate is \$8. There are 7200 lb of cast iron and 100 labor-hours available each week. Because of a surplus from the previous week, the proprietor has decided to make no more than 150 units of model A grates this week. Determine how many of each model he should make to maximize his profit.

Solutions to Self-Check Exercises 6.4 can be found on page 379.

6.4 Concept Questions

1. Give the three characteristics of a standard maximization linear programming problem.

2. a. When the initial simplex tableau is set up, how is the system of linear inequalities transformed into a system of linear equations? How is the objective function $P = c_1x_1 + c_2x_2 + \cdots + c_nx_n$ rewritten?

b. If you are given a simplex tableau, how do you determine whether the optimal solution has been reached?

3. In the simplex method, how is a pivot column selected? A pivot row? A pivot element?

6.4 Exercises

In Exercises 1–6, (a) write the linear programming problem as a standard maximization problem if it is not already in that form, and (b) write the initial simplex tableau.

1. Maximize $P = 2x + 4y$ subject to the constraints

$$x + 4y \le 12$$
$$x + 3y \le 10$$
$$x \ge 0, y \ge 0$$

2. Maximize $P = 3x + 5y$ subject to the constraints

$$x + 3y \le 12$$
$$-2x - 3y \ge -18$$
$$x \ge 0, y \ge 0$$

3. Maximize $P = 2x + 3y$ subject to the constraints

$$x + y \le 10$$
$$-x - 2y \ge -12$$
$$2x + y \le 12$$
$$x \ge 0, y \ge 0$$

4. Maximize $P = 2x + 5y$ subject to the constraints

$$3x + 8y \le 1$$
$$4x - 5y \le 4$$
$$2x + 7y \le 6$$
$$x \ge 0, y \ge 0$$

5. Maximize $P = x + 3y + 4z$ subject to the constraints

$$x + 2y + z \leq 40$$
$$-x - y - z \geq -30$$
$$x \geq 0, y \geq 0, z \geq 0$$

6. Maximize $P = 4x + 5y + 6z$ subject to the constraints

$$2x + 3y + z \leq 900$$
$$3x + y + z \leq 350$$
$$4x + 2y + z \leq 400$$
$$x \geq 0, y \geq 0, z \geq 0$$

In Exercises 7–16, determine whether the given simplex tableau is in final form. If so, find the solution to the associated standard linear programming problem. If not, find the pivot element to be used in the next iteration of the simplex method.

7.

x	y	u	v	P	Constant
0	1	$\frac{5}{7}$	$-\frac{1}{7}$	0	$\frac{20}{7}$
1	0	$-\frac{3}{7}$	$\frac{2}{7}$	0	$\frac{30}{7}$
0	0	$\frac{13}{7}$	$\frac{3}{7}$	1	$\frac{220}{7}$

8.

x	y	u	v	P	Constant
1	1	1	0	0	6
1	0	-1	1	0	2
3	0	5	0	1	30

9.

x	y	u	v	P	Constant
0	$\frac{1}{2}$	1	$-\frac{1}{2}$	0	2
1	$\frac{1}{2}$	0	$\frac{1}{2}$	0	4
0	$-\frac{1}{2}$	0	$\frac{3}{2}$	1	12

10.

x	y	z	u	v	P	Constant
3	0	5	1	1	0	28
2	1	3	0	1	0	16
2	0	8	0	3	1	48

11.

x	y	z	u	v	w	P	Constant
1	$-\frac{1}{3}$	0	$\frac{1}{3}$	0	$-\frac{2}{3}$	0	$\frac{1}{3}$
0	2	0	0	1	1	0	6
0	$\frac{2}{3}$	1	$\frac{1}{3}$	0	$\frac{1}{3}$	0	$\frac{13}{3}$
0	4	0	1	0	2	1	17

12.

x	y	z	u	v	w	P	Constant
$\frac{1}{2}$	0	$\frac{1}{4}$	1	$-\frac{1}{4}$	0	0	$\frac{19}{2}$
$\frac{1}{2}$	1	$\frac{3}{4}$	0	$\frac{1}{4}$	0	0	$\frac{21}{2}$
2	0	3	0	0	1	0	30
-1	0	$-\frac{1}{2}$	6	$\frac{3}{2}$	0	1	63

13.

x	y	z	s	t	u	v	P	Constant
$\frac{5}{2}$	3	0	1	0	0	-4	0	46
1	0	0	0	1	0	0	0	9
0	1	0	0	0	1	0	0	12
0	0	1	0	0	0	1	0	6
-180	-200	0	0	0	0	300	1	1800

14.

x	y	z	s	t	u	v	P	Constant
1	0	0	$\frac{2}{5}$	0	$-\frac{6}{5}$	$-\frac{8}{5}$	0	4
0	0	0	$-\frac{2}{5}$	1	$\frac{6}{5}$	$\frac{8}{5}$	0	5
0	1	0	0	0	1	0	0	12
0	0	1	0	0	0	1	0	6
0	0	0	72	0	-16	12	1	4920

15.

x	y	z	u	v	P	Constant
1	0	$\frac{3}{5}$	0	$\frac{1}{5}$	0	30
0	1	$-\frac{19}{5}$	1	$-\frac{3}{5}$	0	10
0	0	$\frac{26}{5}$	0	0	1	60

16.

x	y	z	u	v	w	P	Constant
0	$\frac{1}{2}$	0	1	$-\frac{1}{2}$	0	0	2
1	$\frac{1}{2}$	1	0	$\frac{1}{2}$	0	0	13
2	$\frac{1}{2}$	0	0	$-\frac{3}{2}$	1	0	4
-1	3	0	0	1	0	1	26

In Exercises 17–31, solve each linear programming problem by the simplex method.

17. Maximize $P = 3x + 4y$
subject to $\quad x + y \leq 4$
$$2x + y \leq 5$$
$$x \geq 0, y \geq 0$$

18. Maximize $P = 5x + 3y$
subject to $\quad x + y \leq 80$
$$3x \quad\quad \leq 90$$
$$x \geq 0, y \geq 0$$

19. Maximize $P = 10x + 12y$
subject to $\quad x + 2y \leq 12$
$$3x + 2y \leq 24$$
$$x \geq 0, y \geq 0$$

20. Maximize $P = 5x + 4y$
subject to $\quad 3x + 5y \leq 78$
$$4x + y \leq 36$$
$$x \geq 0, y \geq 0$$

21. Maximize $P = 4x + 6y$
subject to $\quad 3x + y \leq 24$
$$2x + y \leq 18$$
$$x + 3y \leq 24$$
$$x \geq 0, y \geq 0$$

22. Maximize　$P = 15x + 12y$

subject to　$\quad x + \ y \le 12$

$\qquad\qquad 3x + \ y \le 30$

$\qquad\qquad 10x + 7y \le 70$

$\qquad\qquad x \ge 0, y \ge 0$

23. Maximize　$P = 3x + 4y + 5z$

subject to　$\quad x + \ y + \ z \le \ 8$

$\qquad\qquad 3x + 2y + 4z \le 24$

$\qquad\qquad x \ge 0, y \ge 0, z \ge 0$

24. Maximize　$P = 3x + 3y + 4z$

subject to　$\quad x + \ y + 3z \le 15$

$\qquad\qquad 4x + 4y + 3z \le 65$

$\qquad\qquad x \ge 0, y \ge 0, z \ge 0$

25. Maximize　$P = 3x + 4y + z$

subject to　$\quad 3x + 10y + 5z \le 120$

$\qquad\qquad 5x + \ 2y + 8z \le \ \ 6$

$\qquad\qquad 8x + 10y + 3z \le 105$

$\qquad\qquad x \ge 0, y \ge 0, z \ge 0$

26. Maximize　$P = x + 2y - z$

subject to　$\quad 2x + \ y + \ z \le 14$

$\qquad\qquad 4x + 2y + 3z \le 28$

$\qquad\qquad 2x + 5y + 5z \le 30$

$\qquad\qquad x \ge 0, y \ge 0, z \ge 0$

27. Maximize　$P = 4x + 6y + 5z$

subject to　$\quad x + \ y + \ z \le 20$

$\qquad\qquad 2x + 4y + 3z \le 42$

$\qquad\qquad 2x \qquad + 3z \le 30$

$\qquad\qquad x \ge 0, y \ge 0, z \ge 0$

28. Maximize　$P = x + 4y - 2z$

subject to　$\quad 3x + y - z \le 80$

$\qquad\qquad 2x + y - z \le 40$

$\qquad\qquad -x + y + z \le 80$

$\qquad\qquad x \ge 0, y \ge 0, z \ge 0$

29. Maximize　$P = 12x + 10y + 5z$

subject to　$\quad 2x + \ y + z \le 10$

$\qquad\qquad 3x + 5y + z \le 45$

$\qquad\qquad 2x + 5y + z \le 40$

$\qquad\qquad x \ge 0, y \ge 0, z \ge 0$

30. Maximize　$P = 2x + 6y + 6z$

subject to　$\quad 2x + \ y + 3z \le \ \ 10$

$\qquad\qquad 4x + \ y + 2z \le \ \ 56$

$\qquad\qquad 6x + 4y + 3z \le 126$

$\qquad\qquad 2x + \ y + \ z \le \ \ 32$

$\qquad\qquad x \ge 0, y \ge 0, z \ge 0$

31. Maximize　$P = 24x + 16y + 23z$

subject to　$\quad 2x + \ y + 2z \le 7$

$\qquad\qquad 2x + 3y + \ z \le 8$

$\qquad\qquad x + 2y + 3z \le 7$

$\qquad\qquad x \ge 0, y \ge 0, z \ge 0$

32. Rework Example 3 using the y-column as the pivot column in the first iteration of the simplex method.

33. Show that the linear programming problem

Maximize　$P = 2x + 2y - 4z$

subject to　$3x + 3y - 2z \le 100$

$\qquad\qquad 5x + 5y + 3z \le 150$

$\qquad\qquad x \ge 0, y \ge 0, z \ge 0$

has optimal solutions $x = 30$, $y = 0$, $z = 0$, $P = 60$ and $x = 0$, $y = 30$, $z = 0$, $P = 60$.

34. MANUFACTURING—PRODUCTION SCHEDULING A company manufactures two products, A and B, on two machines, I and II. It has been determined that the company will realize a profit of \$3/unit on Product A and a profit of \$4/unit on Product B. To manufacture 1 unit of Product A requires 6 min on Machine I and 5 min on Machine II. To manufacture 1 unit of Product B requires 9 min on Machine I and 4 min on Machine II. There are 5 hr of machine time available on Machine I and 3 hr of machine time available on Machine II in each work shift. How many units of each product should be produced in each shift to maximize the company's profit? What is the largest profit the company can realize? Is there any time left unused on the machines?

35. MANUFACTURING—PRODUCTION SCHEDULING National Business Machines Corporation manufactures two models of fax machines: A and B. Each model A costs \$100 to make, and each model B costs \$150. The profits are \$30 for each model A and \$40 for each model B fax machine. If the total number of fax machines demanded each month does not exceed 2500 and the company has earmarked no more than \$600,000/month for manufacturing costs, find how many units of each model National should make each month to maximize its monthly profit. What is the largest monthly profit the company can make?

36. MANUFACTURING—PRODUCTION SCHEDULING Kane Manufacturing has a division that produces two models of hibachis, model A and model B. To produce each model A hibachi requires 3 lb of cast iron and 6 min of labor. To produce each model B hibachi requires 4 lb of cast iron and 3 min of labor. The profit for each model A hibachi is \$2, and the profit for each model B hibachi is \$1.50. If 1000 lb of cast iron and 20 labor-hours are available for the production of hibachis each day, how many hibachis of each model should the division produce to maximize Kane's profit? What is the largest profit the company can realize? Is there any raw material left over?

37. **AGRICULTURE—CROP PLANNING** A farmer has 150 acres of land suitable for cultivating Crops A and B. The cost of cultivating Crop A is $40/acre, whereas the cost of cultivating Crop B is $60/acre. The farmer has a maximum of $7400 available for land cultivation. Each acre of Crop A requires 20 labor-hours, and each acre of Crop B requires 25 labor-hours. The farmer has a maximum of 3300 labor-hours available. If he expects to make a profit of $150/acre on Crop A and $200/acre on Crop B, how many acres of each crop should he plant to maximize his profit? What is the largest profit the farmer can realize? Are there any resources left over?

38. **INVESTMENTS—ASSET ALLOCATION** A financier plans to invest up to $500,000 in two projects. Project A yields a return of 10% on the investment, whereas Project B yields a return of 15% on the investment. Because the investment in Project B is riskier than the investment in Project A, the financier has decided that the investment in Project B should not exceed 40% of the total investment. How much should she invest in each project to maximize the return on her investment? What is the maximum return?

39. **INVESTMENTS—ASSET ALLOCATION** Ashley has earmarked at most $250,000 for investment in three mutual funds: a money market fund, an international equity fund, and a growth-and-income fund. The money market fund has a rate of return of 6%/year, the international equity fund has a rate of return of 10%/year, and the growth-and-income fund has a rate of return of 15%/year. Ashley has stipulated that no more than 25% of her total portfolio should be in the growth-and-income fund and that no more than 50% of her total portfolio should be in the international equity fund. To maximize the return on her investment, how much should Ashley invest in each type of fund? What is the maximum return?

40. **MANUFACTURING—PRODUCTION SCHEDULING** A division of the Winston Furniture Company manufactures dining tables and chairs. Each table requires 40 board feet of wood and 3 labor-hours. Each chair requires 16 board feet of wood and 4 labor-hours. The profit for each table is $45, and the profit for each chair is $20. In a certain week, the company has 3200 board feet of wood available and 520 labor-hours available. How many tables and chairs should Winston manufacture to maximize its profit? What is the maximum profit?

41. **MANUFACTURING—PRODUCTION SCHEDULING** A company manufactures Products A, B, and C. Each product is processed in three departments: I, II, and III. The total available labor-hours per week for Departments I, II, and III are 900, 1080, and 840, respectively. The time requirements (in hours per unit) and profit per unit for each product are as follows:

	Product A	Product B	Product C
Dept. I	2	1	2
Dept. II	3	1	2
Dept. III	2	2	1
Profit	$18	$12	$15

How many units of each product should the company produce to maximize its profit? What is the largest profit the company can realize? Are there any resources left over?

42. **MANUFACTURING—PRODUCTION SCHEDULING** Ace Novelty manufactures Giant Pandas and Saint Bernards. Each Panda requires 1.5 yd^2 of plush, 30 ft^3 of stuffing, and 5 pieces of trim; each Saint Bernard requires 2 yd^2 of plush, 35 ft^3 of stuffing, and 8 pieces of trim. The profit for each Panda is $10, and the profit for each Saint Bernard is $15. If 3600 yd^2 of plush, 66,000 ft^3 of stuffing, and 13,600 pieces of trim are available, how many of each of the stuffed animals should the company manufacture to maximize its profit? What is the maximum profit?

43. **ADVERTISING—TELEVISION COMMERCIALS** As part of a campaign to promote its annual clearance sale, Excelsior Company decided to buy television advertising time on Station KAOS. Excelsior's television advertising budget is $102,000. Morning time costs $3000/min, afternoon time costs $1000/min, and evening (prime) time costs $12,000/min. Because of previous commitments, KAOS cannot offer Excelsior more than 6 min of prime time or more than a total of 25 min of advertising time over the 2 weeks in which the commercials are to be run. KAOS estimates that morning commercials are seen by 200,000 people, afternoon commercials are seen by 100,000 people, and evening commercials are seen by 600,000 people. How much morning, afternoon, and evening advertising time should Excelsior buy to maximize exposure of its commercials?

44. **INVESTMENTS—ASSET ALLOCATION** Sharon has a total of $200,000 to invest in three types of mutual funds: growth, balanced, and income funds. Growth funds have a rate of return of 12%/year, balanced funds have a rate of return of 10%/year, and income funds have a return of 6%/year. The growth, balanced, and income mutual funds are assigned risk factors of 0.1, 0.06, and 0.02, respectively. Sharon has decided that at least 50% of her total portfolio is to be in income funds and at least 25% in balanced funds. She has also decided that the average risk factor for her investment should not exceed 0.05. How much should Sharon invest in each type of fund to realize a maximum return on her investment? What is the maximum return?

Hint: The constraint for the average risk factor for the investment is given by $0.1x + 0.06y + 0.02z \leq 0.05(x + y + z)$.

45. **MANUFACTURING—PRODUCTION CONTROL** Custom Office Furniture is introducing a new line of executive desks made from a specially selected grade of walnut. Initially, three models—A, B, and C—are to be marketed. Each model A desk requires $1\frac{1}{4}$ hr for fabrication, 1 hr for assembly, and 1 hr for finishing; each model B desk requires $1\frac{1}{2}$ hr for fabrication, 1 hr for assembly, and 1 hr for finishing; each model C desk requires $1\frac{1}{2}$ hr, $\frac{3}{4}$ hr, and $\frac{1}{2}$ hr for fabrication, assembly, and finishing, respectively. The profit on each model A desk is $26, the profit on each model B desk is $28, and the profit on each model C desk is $24. The total time available in the fabrication department, the assembly department, and the finishing department in the first month of production is 310 hr, 205 hr, and 190 hr, respectively. To max-

imize Custom's profit, how many desks of each model should be made in the month? What is the largest profit the company can realize? Are there any resources left over?

46. **MANUFACTURING—PREFABRICATED HOUSING PRODUCTION** Boise Lumber has decided to enter the lucrative prefabricated housing business. Initially, it plans to offer three models: standard, deluxe, and luxury. Each house is prefabricated and partially assembled in the factory, and the final assembly is completed on site. The dollar amount of building material required, the amount of labor required in the factory for prefabrication and partial assembly, the amount of on-site labor required, and the profit per unit are as follows:

	Standard Model	Deluxe Model	Luxury Model
Material	$6,000	$8,000	$10,000
Factory Labor (hr)	240	220	200
On-site Labor (hr)	180	210	300
Profit	$3,400	$4,000	$5,000

For the first year's production, a sum of $8,200,000 is budgeted for the building material; the number of labor-hours available for work in the factory (for prefabrication and partial assembly) is not to exceed 218,000 hr; and the amount of labor for on-site work is to be less than or equal to 237,000 labor-hours. Determine how many houses of each type Boise should produce to maximize its profit from this new venture. (Market research has confirmed that there should be no problems with sales.)

47. **MANUFACTURING—COLD FORMULA PRODUCTION** Beyer Pharmaceutical produces three kinds of cold formulas: I, II, and III. It takes 2.5 hr to produce 1000 bottles of Formula I, 3 hr to produce 1000 bottles of Formula II, and 4 hr to produce 1000 bottles of Formula III. The profits for each 1000 bottles of Formula I, Formula II, and Formula III are $180, $200, and $300, respectively. Suppose that for a certain production run, there are enough ingredients on hand to make at most 9000 bottles of Formula I, 12,000 bottles of Formula II, and 6000 bottles of Formula III. Furthermore, suppose the time for the production run is limited to a maximum of 70 hr. How many bottles of each formula should be produced in this production run so that the profit is maximized? What is the maximum profit realizable by the company? Are there any resources left over?

48. **PRODUCTION—JUICE PRODUCTS** CalJuice Company has decided to introduce three fruit juices made from blending two or more concentrates. These juices will be packaged in 2-qt (64-oz) cartons. One carton of pineapple–orange juice requires 8 oz each of pineapple and orange juice concentrates. One carton of orange–banana juice requires 12 oz of orange juice concentrate and 4 oz of banana pulp concentrate. Finally, one carton of pineapple–orange–banana juice requires 4 oz of pineapple juice concentrate, 8 oz of orange juice concentrate, and 4 oz of banana pulp. The company has decided to allot 16,000 oz of pineapple juice concentrate, 24,000 oz of orange juice concentrate, and 5000 oz of banana pulp concentrate for the initial production run. The company has also

stipulated that the production of pineapple–orange–banana juice should not exceed 800 cartons. Its profit on one carton of pineapple–orange juice is $1.00, its profit on one carton of orange–banana juice is $0.80, and its profit on one carton of pineapple–orange–banana juice is $0.90. To realize a maximum profit, how many cartons of each blend should the company produce? What is the largest profit it can realize? Are there any concentrates left over?

49. **INVESTMENTS—ASSET ALLOCATION** A financier plans to invest up to $2 million in three projects. She estimates that Project A will yield a return of 10% on her investment, Project B will yield a return of 15% on her investment, and Project C will yield a return of 20% on her investment. Because of the risks associated with the investments, she decided to put not more than 20% of her total investment in Project C. She also decided that her investments in Projects B and C should not exceed 60% of her total investment. Finally, she decided that her investment in Project A should be at least 60% of her investments in Projects B and C. How much should the financier invest in each project if she wishes to maximize the total returns on her investments? What is the maximum amount she can expect to make from her investments?

50. Consider the linear programming problem

$$\text{Maximize} \quad P = 3x + 2y$$
$$\text{subject to} \quad x - y \leq 3$$
$$x \leq 2$$
$$x \geq 0, y \geq 0$$

a. Sketch the feasible set for the linear programming problem.
b. Show that the linear programming problem is unbounded.
c. Solve the linear programming problem using the simplex method. How does the method break down?
d. Explain why the result in part (c) implies that no solution exists for the linear programming problem.

In Exercises 51–54, determine whether the statement is true or false. If it is true, explain why it is true. If it is false, give an example to show why it is false.

51. If at least one of the coefficients $a_1, a_2, \ldots, a_n$ of the objective function $P = a_1x_1 + a_2x_2 + \cdots + a_nx_n$ is positive, then $(0, 0, \ldots, 0)$ cannot be the optimal solution of the standard (maximization) linear programming problem.

52. Choosing the pivot row by requiring that the ratio associated with that row be the smallest ensures that the iteration will not take us from a feasible point to a nonfeasible point.

53. Choosing the pivot column by requiring that it be the column associated with the most negative entry to the left of the vertical line in the last row of the simplex tableau ensures that the iteration will result in the greatest increase or, at worse, no decrease in the objective function.

54. If, at any stage of an iteration of the simplex method, it is not possible to compute the ratios (division by zero) or the ratios are negative, then we can conclude that the standard linear programming problem may have no solution.

6.4 Solutions to Self-Check Exercises

1. Introducing the slack variables u, v, and w, we obtain the system of linear equations

$$
\begin{aligned}
2x + 3y + z + u &&&= 10 \\
x + y + 2z &+ v &&= 8 \\
2y + 3z &&+ w &= 6 \\
-2x - 3y - 6z &&&+ P = 0
\end{aligned}
$$

The initial simplex tableau and the successive tableaus resulting from the use of the simplex procedure follow:

	x	y	z	u	v	w	P	Constant		Ratio	
	2	3	1	1	0	0	0	10		$\frac{10}{1}=10$	
	1	1	2	0	1	0	0	8		$\frac{8}{2}=4$	$\frac{1}{3}R_3$
Pivot row →	0	2	③	0	0	1	0	6		$\frac{6}{3}=2$	→
	-2	-3	-6	0	0	0	1	0			

Pivot column

x	y	z	u	v	w	P	Constant		
2	3	1	1	0	0	0	10		
1	1	2	0	1	0	0	8	$R_1 - R_3$	
0	$\frac{2}{3}$	①	0	0	$\frac{1}{3}$	0	2	$R_2 - 2R_3$	
-2	-3	-6	0	0	0	1	0	$R_4 + 6R_3$	

	x	y	z	u	v	w	P	Constant		Ratio	
	2	$\frac{7}{3}$	0	1	0	$-\frac{1}{3}$	0	8		$\frac{8}{2}=4$	
Pivot row →	①	$-\frac{1}{3}$	0	0	1	$-\frac{2}{3}$	0	4		$\frac{4}{1}=4$	$R_1 - 2R_2$
	0	$\frac{2}{3}$	1	0	0	$\frac{1}{3}$	0	2		—	$R_4 + 2R_2$
	-2	1	0	0	0	2	1	12			

Pivot column

x	y	z	u	v	w	P	Constant
0	3	0	1	-2	1	0	0
1	$-\frac{1}{3}$	0	0	1	$-\frac{2}{3}$	0	4
0	$\frac{2}{3}$	1	0	0	$\frac{1}{3}$	0	2
0	$\frac{1}{3}$	0	0	2	$\frac{2}{3}$	1	20

All entries in the last row are nonnegative, and the tableau is final. We conclude that $x = 4$, $y = 0$, $z = 2$, and $P = 20$.

2. Let x denote the number of model A grates, and let y denote the number of model B grates to be made this week. Then the profit function to be maximized is given by

$$P = 6x + 8y$$

The limitations on the availability of material and labor may be expressed by the linear inequalities

$$
\begin{aligned}
20x + 30y &\le 7200 &\quad\text{or}\quad& 2x + 3y \le 720 \\
20x + 15y &\le 6000 &\quad\text{or}\quad& 4x + 3y \le 1200
\end{aligned}
$$

Finally, the condition that no more than 150 units of model A grates be made this week may be expressed by the linear inequality

$$x \le 150$$

Thus, we are led to the following linear programming problem:

$$
\begin{aligned}
\text{Maximize} \quad & P = 6x + 8y \\
\text{subject to} \quad & 2x + 3y \le 720 \\
& 4x + 3y \le 1200 \\
& x \le 150 \\
& x \ge 0, y \ge 0
\end{aligned}
$$

To solve this problem, we introduce slack variables u, v, and w and use the simplex method, obtaining the following sequence of simplex tableaus:

	x	y	u	v	w	P	Constant		Ratio	
Pivot row →	2	③	1	0	0	0	720		$\frac{720}{3}=240$	
	4	3	0	1	0	0	1200		$\frac{1200}{3}=400$	
	1	0	0	0	1	0	150		—	
	-6	-8	0	0	0	1	0			

Pivot column

	x	y	u	v	w	P	Constant
$\frac{1}{3}R_1$ →	$\frac{2}{3}$	①	$\frac{1}{3}$	0	0	0	240
	4	3	0	1	0	0	1200
	1	0	0	0	1	0	150
	-6	-8	0	0	0	1	0

	x	y	u	v	w	P	Constant		Ratio	
$R_2 - 3R_1$ →	$\frac{2}{3}$	1	$\frac{1}{3}$	0	0	0	240		$\frac{240}{2/3}=360$	
$R_4 + 8R_1$	2	0	-1	1	0	0	480		$\frac{480}{2}=240$	
Pivot row →	①	0	0	0	1	0	150		$\frac{150}{1}=150$	
	$-\frac{2}{3}$	0	$\frac{8}{3}$	0	0	1	1920			

Pivot column

	x	y	u	v	w	P	Constant
$R_1 - \frac{2}{3}R_3$ →	0	1	$\frac{1}{3}$	0	$-\frac{2}{3}$	0	140
$R_2 - 2R_3$	0	0	-1	1	-2	0	180
$R_4 + \frac{2}{3}R_3$	1	0	0	0	1	0	150
	0	0	$\frac{8}{3}$	0	$\frac{2}{3}$	1	2020

The last tableau is final, and we see that $x = 150$, $y = 140$, and $P = 2020$. Therefore, LaCrosse should make 150 model A grates and 140 model B grates this week. The profit will be $2020.

USING TECHNOLOGY

The Simplex Method: Solving Maximization Problems

Graphing Utility

A graphing utility can be used to solve a linear programming problem by the simplex method, as illustrated in Example 1.

EXAMPLE 1 (Refer to Example 5, Section 6.4.) The problem reduces to the following linear programming problem:

$$\text{Maximize}\quad P = 6x + 5y + 4z$$
$$\text{subject to}\quad 2x + y + z \le 180$$
$$x + 3y + 2z \le 300$$
$$2x + y + 2z \le 240$$
$$x \ge 0, y \ge 0, z \ge 0$$

With u, v, and w as slack variables, we are led to the following sequence of simplex tableaus, where the first tableau is entered as the matrix A:

	x	y	z	u	v	w	P	Constant		Ratio
Pivot row →	②	1	1	1	0	0	0	180		$\frac{180}{2} = 90$
	1	3	2	0	1	0	0	300		$\frac{300}{1} = 300$
	2	1	2	0	0	1	0	240		$\frac{240}{2} = 120$
	−6	−5	−4	0	0	0	1	0		

$\xrightarrow{\ *\mathbf{row}(\frac{1}{2}, A, 1)\ \blacktriangleright\ B\ }$

↑ Pivot column

	x	y	z	u	v	w	P	Constant
	①	0.5	0.5	0.5	0	0	0	90
	1	3	2	0	1	0	0	300
	2	1	2	0	0	1	0	240
	−6	−5	−4	0	0	0	1	0

$\xrightarrow{\ *\mathbf{row}+(-1, B, 1, 2)\ \blacktriangleright\ C\ }$
$\xrightarrow{\ *\mathbf{row}+(-2, C, 1, 3)\ \blacktriangleright\ B\ }$
$*\mathbf{row}+(6, B, 1, 4)\ \ \blacktriangleright\ C$

	x	y	z	u	v	w	P	Constant		Ratio
	1	0.5	0.5	0.5	0	0	0	90		$\frac{90}{0.5} = 180$
Pivot row →	0	②.5	1.5	−0.5	1	0	0	210		$\frac{210}{2.5} = 84$
	0	0	1	−1	0	1	0	60		
	0	−2	−1	3	0	0	1	540		

$\xrightarrow{\ *\mathbf{row}(\frac{1}{2.5}, C, 2)\ \blacktriangleright\ B\ }$

↑ Pivot column

	x	y	z	u	v	w	P	Constant
	1	0.5	0.5	0.5	0	0	0	90
	0	①	0.6	−0.2	0.4	0	0	84
	0	0	1	−1	0	1	0	60
	0	−2	−1	3	0	0	1	540

$\xrightarrow{\ *\mathbf{row}+(-0.5, B, 2, 1)\ \blacktriangleright\ C\ }$
$*\mathbf{row}+(2, C, 2, 4)\ \ \ \blacktriangleright\ B$

x	y	z	u	v	w	P	Constant
1	0	0.2	0.6	-0.2	0	0	48
0	1	0.6	-0.2	0.4	0	0	84
0	0	1	-1	0	1	0	60
0	0	0.2	2.6	0.8	0	1	708

The final simplex tableau is the same as the one obtained earlier. We see that $x = 48$, $y = 84$, $z = 0$, and $P = 708$. Hence, Ace Novelty should produce 48 Type A souvenirs, 84 Type B souvenirs, and no Type C souvenirs, resulting in a profit of \$708 per day. ∎

Excel

Solver is an Excel add-in that is used to solve linear programming problems. When you start the Excel program, check the *Tools* menu for the *Solver* command. If it is not there, you will need to install it. (Check your manual for installation instructions.)

EXAMPLE 2 Solve the following linear programming problem:

$$\begin{aligned}
\text{Maximize} \quad & P = 6x + 5y + 4z \\
\text{subject to} \quad & 2x + y + z \le 180 \\
& x + 3y + 2z \le 300 \\
& 2x + y + 2z \le 240 \\
& x \ge 0, y \ge 0, z \ge 0
\end{aligned}$$

Solution

1. *Enter the data for the linear programming problem onto a spreadsheet.* Enter the labels shown in column A and the variables with which we are working under Decision Variables in cells B4:B6, as shown in Figure T1. This optional step will help us to organize our work.

	A	B	C	D	E	F	G	H	I
1	Maximization Problem								
2							Formulas for indicated cells		
3	Decision Variables						C8: = 6*C4 + 5*C5 + 4*C6		
4		x	0				C11: = 2*C4 + C5 + C6		
5		y	0				C12: = C4 + 3*C5 + 2*C6		
6		z	0				C13: = 2*C4 + C5 + 2*C6		
7									
8	Objective Function		0						
9									
10	Constraints								
11			0	<=	180				
12			0	<=	300				
13			0	<=	240				

FIGURE **T1**
Setting up the spreadsheet for Solver

For the moment, the cells that will contain the values of the variables (C4:C6) are left blank. In C8, type the formula for the objective function: =6*C4+5*C5+ 4*C6. In C11, type the formula for the left-hand side of the first constraint: =2*C4+C5+C6. In C12, type the formula for the left-hand side of the second constraint: `=C4+3*C5+2*C6`. In C13, type the formula for the left-hand side of the third constraint: `=2*C4+C5+2*C6`. Zeros will then appear in cell C8 and cells C11:C13. In cells D11:D13, type <= to indicate that each constraint is of the form ≤. Finally, in cells E11:E13, type the right-hand value of each constraint—in this case, 180, 300, and 240, respectively. Note that we need not enter the nonnegativity constraints $x \geq 0$, $y \geq 0$, and $z \geq 0$. The resulting spreadsheet is shown in Figure T1, where the formulas that were entered for the objective function and the constraints are shown in the comment box.

2. *Use Solver to solve the problem.* Click the **Data** tab, and then click **Solver** in the **Analysis** group. The Solver Parameters dialog box will appear.

   a. The pointer will be in the Set Objective: box (refer to Figure T2). Highlight the cell on your spreadsheet containing the formula for the objective function—in this case, C8.

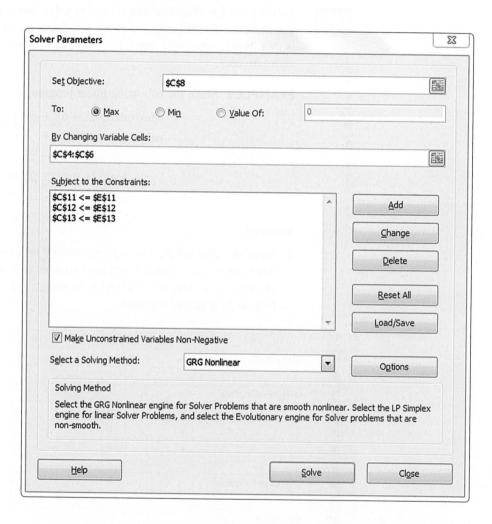

FIGURE **T2**
The completed Solver Parameters dialog box

Then, next to To:, select **Max** . Select the **By Changing Variable Cells:** box, and highlight the cells in your spreadsheet that will contain the values of the variables—in this case, C4:C6. Select the **Subject to the Constraints:** box, and then click **Add** . The Add Constraint dialog box will appear (Figure T3).

Note: Boldfaced words/characters enclosed in a box (for example, **Enter**) indicate that an action (click, select, or press) is required. Words/characters printed blue (for example, Chart sub-type:) indicate words/characters that appear on the screen. Words/characters printed in a monospace font (for example, `=(−2/3)*A2+2`) indicate words/characters that need to be typed and entered.

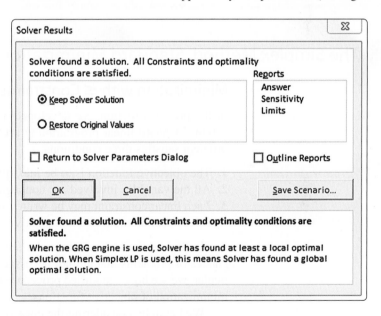

FIGURE **T3**
The Add Constraint dialog box

b. The pointer will appear in the Cell Reference: box. Highlight the cells on your spreadsheet that contain the formula for the left-hand side of the first constraint—in this case, C11. Next, select the symbol for the appropriate constraint—in this case, <= . Select the **Constraint:** box, and highlight the value of the right-hand side of the first constraint on your spreadsheet—in this case, 180. Click **Add** , and then follow the same procedure to enter the second and third constraints. Click **OK** . The resulting Solver Parameters dialog box shown in Figure T2 will appear.

c. In the Solver Parameters dialog box, go to the Select a Solving Method: box, and select **Simplex LP** .

d. Next, click **Solve** . A Solver Results dialog box will then appear (see Figure T4), and at the same time the answers will appear on your spreadsheet (see Figure T5).

FIGURE **T4**
The Solver Results dialog box

FIGURE **T5**
Completed spreadsheet after using Solver

	A	B	C	D	E
1	Maximization Problem				
2					
3	Decision Variables				
4		x	48		
5		y	84		
6		z	0		
7					
8	Objective Function		708		
9					
10	Constraints				
11			180	<=	180
12			300	<=	300
13			240	<=	240

(continued)

3. *Read off your answers.* From the spreadsheet, we see that the objective function attains a maximum value of 708 (cell C8) when $x = 48$, $y = 84$, and $z = 0$ (cells C4:C6). ∎

TECHNOLOGY EXERCISES

Solve the linear programming problems.

1. Maximize $P = 2x + 3y + 4z + 2w$
   subject to $x + 2y + 3z + 2w \leq 6$
                  $2x + 4y + z - w \leq 4$
                  $3x + 2y - 2z + 3w \leq 12$
                  $x \geq 0, y \geq 0, z \geq 0, w \geq 0$

2. Maximize $P = 3x + 2y + 2z + w$
   subject to $2x + y - z + 2w \leq 8$
                  $2x - y + 2z + 3w \leq 20$
                  $x + y + z + 2w \leq 8$
                  $4x - 2y + z + 3w \leq 24$
                  $x \geq 0, y \geq 0, z \geq 0, w \geq 0$

3. Maximize $P = x + y + 2z + 3w$
   subject to $3x + 6y + 4z + 2w \leq 12$
                  $x + 4y + 8z + 4w \leq 16$
                  $2x + y + 4z + w \leq 10$
                  $x \geq 0, y \geq 0, z \geq 0, w \geq 0$

4. Maximize $P = 2x + 4y + 3z + 5w$
   subject to $x - 2y + 3z + 4w \leq 8$
                  $2x + 2y + 4z + 6w \leq 12$
                  $3x + 2y + z + 5w \leq 10$
                  $2x + 8y - 2z + 6w \leq 24$
                  $x \geq 0, y \geq 0, z \geq 0, w \geq 0$

6.5   The Simplex Method: Standard Minimization Problems

Minimization with ≤ Constraints

In the previous section, we developed a procedure, called the simplex method, for solving standard linear programming problems. Recall that a standard maximization problem satisfies three conditions:

1. The objective function is to be maximized.
2. All the variables involved are nonnegative.
3. Each linear constraint may be written so that the expression involving the variables is less than or equal to a nonnegative constant.

In this section, we see how the simplex method may be used to solve certain classes of problems that are not necessarily standard maximization problems. In particular, we see how a modified procedure may be used to solve problems involving the minimization of objective functions.

We begin by considering the class of linear programming problems that calls for the minimization of objective functions but otherwise satisfies conditions 2 and 3 for standard maximization problems. The method that is used to solve these problems is illustrated in the following example.

VIDEO   **EXAMPLE 1**

$$\text{Minimize} \quad C = -2x - 3y$$
$$\text{subject to} \quad 5x + 4y \leq 32$$
$$x + 2y \leq 10$$
$$x \geq 0, y \geq 0$$

Solution   This problem involves the minimization of the objective function and is accordingly *not* a standard maximization problem. Note, however, that all other conditions for a standard maximization problem hold true. To solve a problem of this type, we observe that minimizing the objective function C is equivalent to maximizing the objective function $P = -C$. Thus, the solution to this problem may be found by solving the following associated standard maximization problem: Maximize $P = 2x + 3y$ subject to the given constraints. Using the simplex method with u and v as slack variables, we obtain the following sequence of simplex tableaus:

	x	y	u	v	P	Constant		Ratio
	5	4	1	0	0	32		$\frac{32}{4} = 8$
Pivot row $\rightarrow$	1	②	0	1	0	10		$\frac{10}{2} = 5$
	-2	-3	0	0	1	0		

$\uparrow$
Pivot column

	x	y	u	v	P	Constant
	5	4	1	0	0	32
$\frac{1}{2}R_2 \longrightarrow$	$\frac{1}{2}$	①	0	$\frac{1}{2}$	0	5
	-2	-3	0	0	1	0

	x	y	u	v	P	Constant		Ratio
Pivot row $\llcorner\rightarrow$	③	0	1	-2	0	12		$\frac{12}{3} = 4$
$R_1 - 4R_2$ $\xrightarrow{\quad}$ $R_3 + 3R_2$	$\frac{1}{2}$	1	0	$\frac{1}{2}$	0	5		$\frac{5}{1/2} = 10$
	$-\frac{1}{2}$	0	0	$\frac{3}{2}$	1	15		

$\uparrow$
Pivot column

	x	y	u	v	P	Constant
$\frac{1}{3}R_1 \longrightarrow$	①	0	$\frac{1}{3}$	$-\frac{2}{3}$	0	4
	$\frac{1}{2}$	1	0	$\frac{1}{2}$	0	5
	$-\frac{1}{2}$	0	0	$\frac{3}{2}$	1	15

	x	y	u	v	P	Constant
$R_2 - \frac{1}{2}R_1$ $\xrightarrow{\quad}$ $R_3 + \frac{1}{2}R_1$	1	0	$\frac{1}{3}$	$-\frac{2}{3}$	0	4
	0	1	$-\frac{1}{6}$	$\frac{5}{6}$	0	3
	0	0	$\frac{1}{6}$	$\frac{7}{6}$	1	17

Explore & Discuss

Refer to Example 1.

1. Sketch the feasible set S for the linear programming problem.

2. Solve the problem using the method of corners.

3. Indicate on S the corner points corresponding to each iteration of the simplex procedure and trace the path leading to the optimal solution.

The last tableau is in final form. The solution to the standard maximization problem associated with the given linear programming problem is $x = 4$, $y = 3$, and $P = 17$, so the required solution is given by $x = 4$, $y = 3$, and $C = -17$. You may verify that the solution is correct by using the method of corners. ∎

The Dual Problem

Another special class of linear programming problems we encounter in practical applications is characterized by the following conditions:

1. The objective function is to be *minimized*.
2. All the variables involved are nonnegative.
3. All other linear constraints may be written so that the expression involving the variables is *greater than or equal to* a constant.

Such problems are called **standard minimization problems.**

A method for solving this type of problem is based on the following observation. Each maximization linear programming problem is associated with a minimization problem, and vice versa. For the purpose of identification, the given problem is called the **primal problem**; the problem related to it is called the **dual problem.**

The following example illustrates the technique for constructing the dual of a given linear programming problem.

EXAMPLE 2   Write the dual problem associated with the following problem:

$$\left.\begin{array}{l} \text{Minimize the objective function} \quad C = 4x + 2y \\ \quad \text{subject to} \quad 5x + y \geq 5 \\ \qquad\qquad\qquad\quad 5x + 3y \geq 10 \\ \qquad\qquad\qquad\quad x \geq 0, y \geq 0 \end{array}\right\} \begin{array}{l} \text{Primal} \\ \text{problem} \end{array}$$

Solution   We first write down the following tableau for the given primal problem:

x	y	Constant
5	1	5
5	3	10
4	2	

Next, we interchange the columns and rows of the foregoing tableau and head the two columns of the resulting array with the two variables u and v, obtaining the tableau:

u	v	Constant
5	5	4
1	3	2
5	10	

Interpreting the last tableau as if it were part of the initial simplex tableau for a standard maximization problem—with the exception that the signs of the coefficients pertaining to the objective function are not reversed—we construct the required dual problem as follows:

$$\left.\begin{array}{l} \text{Maximize the objective function} \quad P = 5u + 10v \\ \text{subject to} \quad 5u + 5v \le 4 \\ \qquad\qquad\quad u + 3v \le 2 \\ \qquad\qquad\quad u \ge 0, v \ge 0 \end{array}\right\} \begin{array}{l} \text{Dual} \\ \text{problem} \end{array}$$

Since both the primal problem and the dual problem in Example 2 involve two variables, they can be solved graphically. In fact, the feasible sets of the primal problem and the dual problem are shown in Figure 22. From the following tables of values associated with the two problems, we see that the solution to the primal problem is $x = \frac{1}{2}$, $y = \frac{5}{2}$, and $C = 7$, and the solution for the dual problem is $u = \frac{1}{5}$, $v = \frac{3}{5}$, and $P = 7$.

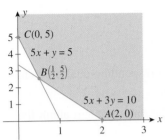

(a) Feasible set for the primal problem

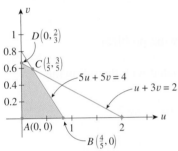

(b) Feasible set for the dual problem
FIGURE **22**

Table for the primal problem

Vertex	$C = 4x + 2y$
$A(2,0)$	8
$B(\frac{1}{2},\frac{5}{2})$	7
$C(0, 5)$	10

Table for the dual problem

Vertex	$P = 5u + 10v$
$A(0,0)$	0
$B(\frac{4}{5},0)$	4
$C(\frac{1}{5},\frac{3}{5})$	7
$D(0,\frac{2}{3})$	$\frac{20}{3}$

Observe that the *objective functions of both the primal problem and the dual problem attain the same optimal value.*

Next let's see how we can find these solutions using the simplex method. Since the dual problem is a standard maximization problem, we can use the method of Section 6.4. Proceeding, we obtain the following sequence of tableaus. Note that we use the letters x and y (the variables used in the primal problem) to denote the slack variables.

	u	v	x	y	P	Constant	Ratio	
	5	5	1	0	0	4	$\frac{4}{5}$	$\frac{1}{3}R_2$
Pivot row →	1	③	0	1	0	2	$\frac{2}{3}$	→
	−5	−10	0	0	1	0		

↑ Pivot column

	u	v	x	y	P	Constant
	5	5	1	0	0	4
	$\frac{1}{3}$	1	0	$\frac{1}{3}$	0	$\frac{2}{3}$
	−5	−10	0	0	1	0

$\xrightarrow[R_3 + 10R_2]{R_1 - 5R_2}$

	u	v	x	y	P	Constant	Ratio	
	$\left(\frac{10}{3}\right)$	0	1	$-\frac{5}{3}$	0	$\frac{2}{3}$	$\frac{1}{5}$	$\frac{3}{10}R_1$
	$\frac{1}{3}$	1	0	$\frac{1}{3}$	0	$\frac{2}{3}$	2	→
	$-\frac{5}{3}$	0	0	$\frac{10}{3}$	1	$\frac{20}{3}$		

	u	v	x	y	P	Constant
	1	0	$\frac{3}{10}$	$-\frac{1}{2}$	0	$\frac{1}{5}$
	$\frac{1}{3}$	1	0	$\frac{1}{3}$	0	$\frac{2}{3}$
	$-\frac{5}{3}$	0	0	$\frac{10}{3}$	1	$\frac{20}{3}$

$\xrightarrow[R_3 + \frac{5}{3}R_1]{R_2 - \frac{1}{3}R_1}$

	u	v	x	y	P	Constant
	1	0	$\frac{3}{10}$	$-\frac{1}{2}$	0	$\frac{1}{5}$
	0	1	$-\frac{1}{10}$	$\frac{1}{2}$	0	$\frac{3}{5}$
	0	0	$\frac{1}{2}$	$\frac{5}{2}$	1	7

Solution for the primal problem

Interpreting the final tableau in the usual manner, we see that the solution of the dual problem is $u = \frac{1}{5}, v = \frac{3}{5}$, and $P = 7$, as obtained earlier by using the graphical method. Next, observe that the solution of the primal problem $x = \frac{1}{2}$ and $y = \frac{5}{2}$ appears under the respective slack variables in the last row of the final tableau, as indicated. That we can always find the solution to both the optimal and dual problems in this manner is the content of the following theorem. The theorem, attributed to John von Neumann (1903–1957), is stated without proof.

THEOREM 3

The Fundamental Theorem of Duality

A primal problem has a solution if and only if the corresponding dual problem has a solution. Furthermore, if a solution exists, then:

a. The objective functions of both the primal and the dual problem attain the same optimal value.

b. The optimal solution to the primal problem appears under the slack variables in the last row of the final simplex tableau associated with the dual problem.

Armed with this theorem, we will solve the following problem.

EXAMPLE 3 Write and solve the dual problem associated with the following problem:

$$
\left.
\begin{aligned}
\text{Minimize the objective function } C &= 6x + 8y \\
\text{subject to}\quad 40x + 10y &\geq 2400 \\
10x + 15y &\geq 2100 \\
5x + 15y &\geq 1500 \\
x \geq 0, y &\geq 0
\end{aligned}
\right\}
\begin{array}{l}
\text{Primal} \\
\text{problem}
\end{array}
$$

Solution   We first write down the following tableau for the given primal problem:

x	y	Constant
40	10	2400
10	15	2100
5	15	1500
6	8	

Next, we interchange the columns and rows of the foregoing tableau and head the three columns of the resulting array with the three variables u, v, and w, obtaining the tableau

u	v	w	Constant
40	10	5	6
10	15	15	8
2400	2100	1500	

Interpreting the last tableau as if it were part of the initial simplex tableau for a standard maximization problem—with the exception that the signs of the coefficients pertaining to the objective function are not reversed—we construct the required dual problem as follows:

$$
\left.
\begin{aligned}
\text{Maximize the objective function } P &= 2400u + 2100v + 1500w \\
\text{subject to}\quad 40u + 10v + 5w &\leq 6 \\
10u + 15v + 15w &\leq 8 \\
u \geq 0, v \geq 0, w &\geq 0
\end{aligned}
\right\}
\begin{array}{l}
\text{Dual} \\
\text{problem}
\end{array}
$$

Observe that the dual problem associated with the given (primal) problem is a standard maximization problem. The solution may thus be found by using the simplex algorithm. Introducing the slack variables x and y, we obtain the system of linear equations

$$
\begin{aligned}
40u + 10v + 5w + x &= 6 \\
10u + 15v + 15w + y &= 8 \\
-2400u - 2100v - 1500w + P &= 0
\end{aligned}
$$

Continuing with the simplex algorithm, we obtain the following sequence of simplex tableaus:

	u	v	w	x	y	P	Constant	
Pivot row $\rightarrow$	$\boxed{40}$	10	5	1	0	0	6	Ratio: $\frac{6}{40} = \frac{3}{20}$
	10	15	15	0	1	0	8	$\frac{8}{10} = \frac{4}{5}$
	-2400	-2100	-1500	0	0	1	0	

Pivot column (u)

	u	v	w	x	y	P	Constant
$\frac{1}{40}R_1 \rightarrow$	$\boxed{1}$	$\frac{1}{4}$	$\frac{1}{8}$	$\frac{1}{40}$	0	0	$\frac{3}{20}$
	10	15	15	0	1	0	8
	-2400	-2100	-1500	0	0	1	0

	u	v	w	x	y	P	Constant	
$\dfrac{R_2 - 10R_1}{R_3 + 2400R_1} \rightarrow$	1	$\frac{1}{4}$	$\frac{1}{8}$	$\frac{1}{40}$	0	0	$\frac{3}{20}$	Ratio: $\frac{3/20}{1/4} = \frac{3}{5}$
	0	$\boxed{\frac{25}{2}}$	$\frac{55}{4}$	$-\frac{1}{4}$	1	0	$\frac{13}{2}$	$\frac{13/2}{25/2} = \frac{13}{25}$
Pivot row	0	-1500	-1200	60	0	1	360	

Pivot column (v)

	u	v	w	x	y	P	Constant
$\frac{2}{25}R_2 \rightarrow$	1	$\frac{1}{4}$	$\frac{1}{8}$	$\frac{1}{40}$	0	0	$\frac{3}{20}$
	0	$\boxed{1}$	$\frac{11}{10}$	$-\frac{1}{50}$	$\frac{2}{25}$	0	$\frac{13}{25}$
	0	-1500	-1200	60	0	1	360

	u	v	w	x	y	P	Constant
$\dfrac{R_1 - \frac{1}{4}R_2}{R_3 + 1500R_2} \rightarrow$	1	0	$-\frac{3}{20}$	$\frac{3}{100}$	$-\frac{1}{50}$	0	$\frac{1}{50}$
	0	1	$\frac{11}{10}$	$-\frac{1}{50}$	$\frac{2}{25}$	0	$\frac{13}{25}$
	0	0	450	30	120	1	1140

Solution for the primal problem (x, y)

The last tableau is final. The fundamental theorem of duality tells us that the solution to the primal problem is $x = 30$ and $y = 120$ with a minimum value for C of 1140. Observe that the solution to the dual (maximization) problem may be read from the simplex tableau in the usual manner: $u = \frac{1}{50}$, $v = \frac{13}{25}$, $w = 0$, and $P = 1140$. Note that the maximum value of P is equal to the minimum value of C, as guaranteed by the fundamental theorem of duality. The solution to the primal problem agrees with the solution of the same problem that we solved in Section 6.3, Example 2, using the method of corners. ■

Notes

1. The dual of a standard minimization problem is always a standard maximization problem provided that the coefficients of the objective function in the primal problem are all nonnegative. Such problems can always be solved by applying the simplex method to solve the dual problem.
2. Standard minimization problems in which the coefficients of the objective function are not all nonnegative do not necessarily have a dual problem that is a standard maximization problem. ■

VIDEO **EXAMPLE 4**

$$\text{Minimize} \quad C = 3x + 2y$$
$$\text{subject to} \quad 8x + y \geq 80$$
$$8x + 5y \geq 240$$
$$x + 5y \geq 100$$
$$x \geq 0, y \geq 0$$

Solution We begin by writing the dual problem associated with the given primal problem. First, we write down the following tableau for the primal problem:

x	y	Constant
8	1	80
8	5	240
1	5	100
3	2	

Next, interchanging the columns and rows of this tableau and heading the three columns of the resulting array with the three variables, u, v, and w, we obtain the tableau

u	v	w	Constant
8	8	1	3
1	5	5	2
80	240	100	

Interpreting the last tableau as if it were part of the initial simplex tableau for a standard maximization problem—with the exception that the signs of the coefficients pertaining to the objective function are not reversed—we construct the dual problem as follows: Maximize the objective function $P = 80u + 240v + 100w$ subject to the constraints

$$8u + 8v + w \leq 3$$
$$u + 5v + 5w \leq 2$$

where $u \geq 0$, $v \geq 0$, and $w \geq 0$. Having constructed the dual problem, which is a standard maximization problem, we now solve it using the simplex method. Introducing the slack variables x and y, we obtain the system of linear equations

$$8u + 8v + w + x   = 3$$
$$u + 5v + 5w + y = 2$$
$$-80u - 240v - 100w   + P = 0$$

Continuing with the simplex algorithm, we obtain the following sequence of simplex tableaus:

	u	v	w	x	y	P	Constant		Ratio
Pivot row →	8	(8)	1	1	0	0	3		$\frac{3}{8}$
	1	5	5	0	1	0	2		$\frac{2}{5}$
	-80	-240	-100	0	0	1	0		

Pivot column (↑ under -240)

	u	v	w	x	y	P	Constant
$\frac{1}{8}R_1$ →	1	(1)	$\frac{1}{8}$	$\frac{1}{8}$	0	0	$\frac{3}{8}$
	1	5	5	0	1	0	2
	-80	-240	-100	0	0	1	0

	u	v	w	x	y	P	Constant		Ratio
	1	1	$\frac{1}{8}$	$\frac{1}{8}$	0	0	$\frac{3}{8}$		3
$\dfrac{R_2 - 5R_1}{R_3 + 240R_1}$ →	-4	0	$(\frac{35}{8})$	$-\frac{5}{8}$	1	0	$\frac{1}{8}$		$\frac{1}{35}$
Pivot row	160	0	-70	30	0	1	90		

Pivot column (↑ under -70)

	u	v	w	x	y	P	Constant
$\frac{8}{35}R_2$ →	1	1	$\frac{1}{8}$	$\frac{1}{8}$	0	0	$\frac{3}{8}$
	$-\frac{32}{35}$	0	(1)	$-\frac{1}{7}$	$\frac{8}{35}$	0	$\frac{1}{35}$
	160	0	-70	30	0	1	90

	u	v	w	x	y	P	Constant
	$\frac{39}{35}$	1	0	$\frac{1}{7}$	$-\frac{1}{35}$	0	$\frac{13}{35}$
$\dfrac{R_1 - \frac{1}{8}R_2}{R_3 + 70R_2}$ →	$-\frac{32}{35}$	0	1	$-\frac{1}{7}$	$\frac{8}{35}$	0	$\frac{1}{35}$
	96	0	0	20	16	1	92

Solution for the primal problem (under 20 and 16)

The last tableau is final. The fundamental theorem of duality tells us that the solution to the primal problem is $x = 20$ and $y = 16$ with a minimum value for C of 92. ∎

Our last example illustrates how the warehouse problem posed in Section 6.2 may be solved by duality.

 APPLIED EXAMPLE 5 A Warehouse Problem Complete the solution to the warehouse problem given in Section 6.2, Example 4 (page 341).

$$\text{Minimize} \quad C = 20x_1 + 8x_2 + 10x_3 + 12x_4 + 22x_5 + 18x_6 \qquad (21)$$

subject to

$$
\begin{aligned}
x_1 + x_2 + x_3 \qquad\qquad &\le 400 \\
x_4 + x_5 + x_6 &\le 600 \\
x_1 \qquad + x_4 \qquad\qquad &\ge 200 \\
x_2 \qquad + x_5 \qquad &\ge 300 \\
x_3 \qquad + x_6 &\ge 400 \\
x_1 \ge 0, x_2 \ge 0, \ldots, x_6 \ge 0 &
\end{aligned}
\qquad (22)
$$

Solution   Upon multiplying each of the first two inequalities of (22) by -1, we obtain the following equivalent system of constraints, in which each of the expressions involving the variables is greater than or equal to a constant:

$$
\begin{aligned}
-x_1 - x_2 - x_3 &\geq -400 \\
- x_4 - x_5 - x_6 &\geq -600 \\
x_1 \quad\quad\quad + x_4 \quad\quad &\geq 200 \\
x_2 \quad\quad\quad + x_5 \quad\; &\geq 300 \\
x_3 \quad\quad\quad + x_6 &\geq 400 \\
x_1 \geq 0,\ x_2 \geq 0,\ \dots,\ x_6 &\geq 0
\end{aligned}
$$

The problem may now be solved by duality. First, we write the tableau

x_1	x_2	x_3	x_4	x_5	x_6	Constant
-1	-1	-1	0	0	0	-400
0	0	0	-1	-1	-1	-600
1	0	0	1	0	0	200
0	1	0	0	1	0	300
0	0	1	0	0	1	400
20	8	10	12	22	18	

Interchanging the rows and columns of this tableau and heading the five columns of the resulting array of numbers by the variables u_1, u_2, u_3, u_4, and u_5, we obtain the tableau

u_1	u_2	u_3	u_4	u_5	Constant
-1	0	1	0	0	20
-1	0	0	1	0	8
-1	0	0	0	1	10
0	-1	1	0	0	12
0	-1	0	1	0	22
0	-1	0	0	1	18
-400	-600	200	300	400	

from which we construct the associated dual problem:

Maximize $P = -400u_1 - 600u_2 + 200u_3 + 300u_4 + 400u_5$

subject to

$$
\begin{aligned}
- u_1 \quad\quad\quad + u_3 \quad\quad\quad\quad &\leq 20 \\
- u_1 \quad\quad\quad\quad\quad + u_4 \quad\quad &\leq 8 \\
- u_1 \quad\quad\quad\quad\quad\quad\quad + u_5 &\leq 10 \\
- u_2 + u_3 \quad\quad\quad\quad &\leq 12 \\
- u_2 \quad\quad + u_4 \quad\quad &\leq 22 \\
- u_2 \quad\quad\quad\quad + u_5 &\leq 18 \\
u_1 \geq 0,\ u_2 \geq 0,\ \dots,\ u_5 &\geq 0
\end{aligned}
$$

Solving the standard maximization problem by the simplex algorithm, we obtain the following sequence of tableaus (x_1, x_2, $\dots$, x_6 are slack variables):

	u_1	u_2	u_3	u_4	u_5	x_1	x_2	x_3	x_4	x_5	x_6	P	Constant	Ratio
	−1	0	1	0	0	1	0	0	0	0	0	0	20	—
	−1	0	0	1	0	0	1	0	0	0	0	0	8	—
Pivot row →	−1	0	0	0	(1)	0	0	1	0	0	0	0	10	10
	0	−1	1	0	0	0	0	0	1	0	0	0	12	—
	0	−1	0	1	0	0	0	0	0	1	0	0	22	—
	0	−1	0	0	1	0	0	0	0	0	1	0	18	18
	400	600	−200	−300	−400	0	0	0	0	0	0	1	0	

↑ Pivot column (u_5)

	u_1	u_2	u_3	u_4	u_5	x_1	x_2	x_3	x_4	x_5	x_6	P	Constant	Ratio
	−1	0	1	0	0	1	0	0	0	0	0	0	20	—
Pivot row →	−1	0	0	(1)	0	0	1	0	0	0	0	0	8	8
	−1	0	0	0	1	0	0	1	0	0	0	0	10	—
	0	−1	1	0	0	0	0	0	1	0	0	0	12	—
	0	−1	0	1	0	0	0	0	0	1	0	0	22	22
	1	−1	0	0	0	0	0	−1	0	0	1	0	8	—
	0	600	−200	−300	0	0	0	400	0	0	0	1	4000	

$\xrightarrow[R_7 + 400R_3]{R_6 - R_3}$

↑ Pivot column (u_4)

	u_1	u_2	u_3	u_4	u_5	x_1	x_2	x_3	x_4	x_5	x_6	P	Constant	Ratio
	−1	0	1	0	0	1	0	0	0	0	0	0	20	—
	−1	0	0	1	0	0	1	0	0	0	0	0	8	—
	−1	0	0	0	1	0	0	1	0	0	0	0	10	—
	0	−1	1	0	0	0	0	0	1	0	0	0	12	—
	1	−1	0	0	0	0	−1	0	0	1	0	0	14	14
Pivot row →	(1)	−1	0	0	0	0	0	−1	0	0	1	0	8	8
	−300	600	−200	0	0	0	300	400	0	0	0	1	6400	

$\xrightarrow[R_7 + 300R_2]{R_5 - R_2}$

↑ Pivot column (u_1)

	u_1	u_2	u_3	u_4	u_5	x_1	x_2	x_3	x_4	x_5	x_6	P	Constant	Ratio
	0	−1	1	0	0	1	0	−1	0	0	1	0	28	28
	0	−1	0	1	0	0	1	−1	0	0	1	0	16	—
	0	−1	0	0	1	0	0	0	0	0	1	0	18	—
Pivot row →	0	−1	(1)	0	0	0	0	0	1	0	0	0	12	12
	0	0	0	0	0	0	−1	1	0	1	−1	0	6	—
	1	−1	0	0	0	0	0	−1	0	0	1	0	8	—
	0	300	−200	0	0	0	300	100	0	0	300	1	8800	

$R_1 + R_6$
$R_2 + R_6$
$\xrightarrow{R_3 + R_6}$
$R_5 - R_6$
$R_7 + 300R_6$

↑ Pivot column (u_3)

	u_1	u_2	u_3	u_4	u_5	x_1	x_2	x_3	x_4	x_5	x_6	P	Constant
	0	0	0	0	0	1	0	−1	−1	0	1	0	16
	0	−1	0	1	0	0	1	−1	0	0	1	0	16
	0	−1	0	0	1	0	0	0	0	0	1	0	18
	0	−1	1	0	0	0	0	0	1	0	0	0	12
	0	0	0	0	0	0	−1	1	0	1	−1	0	6
	1	−1	0	0	0	0	0	−1	0	0	1	0	8
	0	100	0	0	0	0	300	100	200	0	300	1	11,200

$\xrightarrow[R_7 + 200R_4]{R_1 - R_4}$

The last tableau is final, and we find that

$$x_1 = 0 \qquad x_2 = 300 \qquad x_3 = 100 \qquad x_4 = 200$$
$$x_5 = 0 \qquad x_6 = 300 \qquad P = 11{,}200$$

Thus, to minimize shipping costs, Acrosonic should ship 300 loudspeaker systems from Plant I to Warehouse B, 100 systems from Plant I to Warehouse C, 200 systems from Plant II to Warehouse A, and 300 systems from Plant II to Warehouse C. The company's total shipping cost is \$11,200. ∎

6.5 Self-Check Exercises

1. Write the dual problem associated with the following problem:

$$\begin{aligned}
\text{Minimize} \quad & C = 2x + 5y \\
\text{subject to} \quad & 4x + y \geq 40 \\
& 2x + y \geq 30 \\
& x + 3y \geq 30 \\
& x \geq 0, y \geq 0
\end{aligned}$$

2. Solve the primal problem posed in Exercise 1.

Solutions to Self-Check Exercises 6.5 can be found on page 396.

6.5 Concept Questions

1. Suppose you are given the linear programming problem

$$\begin{aligned}
\text{Minimize} \quad & C = -3x - 5y \\
\text{subject to} \quad & 5x + 2y \leq 30 \\
& x + 3y \leq 21 \\
& x \geq 0, y \geq 0
\end{aligned}$$

Give the associated standard maximization problem that you would use to solve this linear programming problem via the simplex method.

2. Give three characteristics of a standard minimization linear programming problem.

3. What is the primal problem associated with a standard minimization linear programming problem? The dual problem?

4. a. What does the fundamental theorem of duality tell us about the existence of a solution to a primal problem?
   b. How are the optimal values of the primal and dual problems related?
   c. Given the final simplex tableau associated with a dual problem, how would you determine the optimal solution to the associated primal problem?

6.5 Exercises

In Exercises 1–6, use the technique developed in this section to solve the minimization problem.

1. Minimize $\quad C = -2x + y$
   subject to $\quad x + 2y \leq 6$
   $\qquad\qquad 3x + 2y \leq 12$
   $\qquad\qquad x \geq 0, y \geq 0$

2. Minimize $\quad C = -2x - 3y$
   subject to $\quad 3x + 4y \leq 24$
   $\qquad\qquad 7x - 4y \leq 16$
   $\qquad\qquad x \geq 0, y \geq 0$

3. Minimize $C = -3x - 2y$ subject to the constraints of Exercise 2.

4. Minimize $\quad C = x - 2y + z$
   subject to $\quad x - 2y + 3z \leq 10$
   $\qquad\qquad 2x + y - 2z \leq 15$
   $\qquad\qquad 2x + y + 3z \leq 20$
   $\qquad\qquad x \geq 0, y \geq 0, z \geq 0$

5. Minimize $\quad C = 2x - 3y - 4z$
   subject to $\quad -x + 2y - z \leq 8$
   $\qquad\qquad x - 2y + 2z \leq 10$
   $\qquad\qquad 2x + 4y - 3z \leq 12$
   $\qquad\qquad x \geq 0, y \geq 0, z \geq 0$

6. Minimize $C = -3x - 2y - z$ subject to the constraints of Exercise 5.

In Exercises 7–10, you are given the final simplex tableau for the dual problem. Give the solution to the primal problem and the solution to the associated dual problem.

7. Problem: Minimize $C = 8x + 12y$

   subject to $x + 3y \geq 2$

   $2x + 2y \geq 3$

   $x \geq 0, y \geq 0$

Final tableau:

u	v	x	y	P	Constant
0	1	$\frac{3}{4}$	$-\frac{1}{4}$	0	3
1	0	$-\frac{1}{2}$	$\frac{1}{2}$	0	2
0	0	$\frac{5}{4}$	$\frac{1}{4}$	1	13

8. Problem: Minimize $C = 3x + 2y$

   subject to $5x + y \geq 10$

   $2x + 2y \geq 12$

   $x + 4y \geq 12$

   $x \geq 0, y \geq 0$

Final tableau:

u	v	w	x	y	P	Constant
1	0	$-\frac{3}{4}$	$\frac{1}{4}$	$-\frac{1}{4}$	0	$\frac{1}{4}$
0	1	$\frac{19}{8}$	$-\frac{1}{8}$	$\frac{5}{8}$	0	$\frac{7}{8}$
0	0	9	1	5	1	13

9. Problem: Minimize $C = 10x + 3y + 10z$

   subject to $2x + y + 5z \geq 20$

   $4x + y + z \geq 30$

   $x \geq 0, y \geq 0, z \geq 0$

Final tableau:

u	v	x	y	z	P	Constant
0	1	$\frac{1}{2}$	-1	0	0	2
1	0	$-\frac{1}{2}$	2	0	0	1
0	0	2	-9	1	0	3
0	0	5	10	0	1	80

10. Problem: Minimize $C = 2x + 3y$

   subject to $x + 4y \geq 8$

   $x + y \geq 5$

   $2x + y \geq 7$

   $x \geq 0, y \geq 0$

Final tableau:

u	v	w	x	y	P	Constant
0	1	$\frac{7}{3}$	$\frac{4}{3}$	$-\frac{1}{3}$	0	$\frac{5}{3}$
1	0	$-\frac{1}{3}$	$-\frac{1}{3}$	$\frac{1}{3}$	0	$\frac{1}{3}$
0	0	2	4	1	1	11

In Exercises 11–20, construct the dual problem associated with the primal problem. Solve the primal problem.

11. Minimize $C = 2x + 5y$

   subject to $x + 2y \geq 4$

   $3x + 2y \geq 6$

   $x \geq 0, y \geq 0$

12. Minimize $C = 3x + 2y$

   subject to $2x + 3y \geq 90$

   $3x + 2y \geq 120$

   $x \geq 0, y \geq 0$

13. Minimize $C = 6x + 4y$

   subject to $6x + y \geq 60$

   $2x + y \geq 40$

   $x + y \geq 30$

   $x \geq 0, y \geq 0$

14. Minimize $C = 10x + y$

   subject to $4x + y \geq 16$

   $x + 2y \geq 12$

   $x \geq 2$

   $x \geq 0, y \geq 0$

15. Minimize $C = 200x + 150y + 120z$

   subject to $20x + 10y + z \geq 10$

   $x + y + 2z \geq 20$

   $x \geq 0, y \geq 0, z \geq 0$

16. Minimize $C = 40x + 30y + 11z$

   subject to $2x + y + z \geq 8$

   $x + y - z \geq 6$

   $x \geq 0, y \geq 0, z \geq 0$

17. Minimize $C = 6x + 8y + 4z$

   subject to $x + 2y + 2z \geq 10$

   $2x + y + z \geq 24$

   $x + y + z \geq 16$

   $x \geq 0, y \geq 0, z \geq 0$

18. Minimize $C = 12x + 4y + 8z$

   subject to $2x + 4y + z \geq 6$

   $3x + 2y + 2z \geq 2$

   $4x + y + z \geq 2$

   $x \geq 0, y \geq 0, z \geq 0$

19. Minimize $C = 30x + 12y + 20z$

   subject to $2x + 4y + 3z \geq 6$

   $6x + z \geq 2$

   $6y + 2z \geq 4$

   $x \geq 0, y \geq 0, z \geq 0$

20. Minimize $C = 8x + 6y + 4z$

   subject to $2x + 3y + z \geq 6$

   $x + 2y - 2z \geq 4$

   $x + y + 2z \geq 2$

   $x \geq 0, y \geq 0, z \geq 0$

21. **TRANSPORTATION** Deluxe River Cruises operates a fleet of river vessels. The fleet has two types of vessels: A type A vessel has 60 deluxe cabins and 160 standard cabins, whereas a type B vessel has 80 deluxe cabins and 120 standard cabins. Under a charter agreement with Odyssey Travel Agency, Deluxe River Cruises is to provide Odyssey with a minimum of 360 deluxe and 680 standard cabins for their 15-day cruise in May. It costs $44,000 to operate a type A vessel and $54,000 to operate a type B vessel for that period. How many of each type vessel should be used to keep the operating costs to a minimum? What is the minimum cost?

22. **SHIPPING COSTS** Acrosonic manufactures a model G loudspeaker system in Plants I and II. The output at Plant I is at most 800/month, and the output at Plant II is at most 600/month. Model G loudspeaker systems are also shipped to the three warehouses—A, B, and C—whose minimum monthly requirements are 500, 400, and 400 systems, respectively. Shipping costs from Plant I to Warehouse A, Warehouse B, and Warehouse C are $16, $20, and $22 per loudspeaker system, respectively, and shipping costs from Plant II to each of these warehouses are $18, $16, and $14, respectively. What shipping schedule will enable Acrosonic to meet the requirements of the warehouses while keeping its shipping costs to a minimum? What is the minimum cost?

23. **ADVERTISING** Everest Deluxe World Travel has decided to advertise in the Sunday editions of two major newspapers in town. These advertisements are directed at three groups of potential customers. Each advertisement in Newspaper I is seen by 70,000 Group A customers, 40,000 Group B customers, and 20,000 Group C customers. Each advertisement in Newspaper II is seen by 10,000 Group A, 20,000 Group B, and 40,000 Group C customers. Each advertisement in Newspaper I costs $1000, and each advertisement in Newspaper II costs $800. Everest would like their advertisements to be read by at least 2 million people from Group A, 1.4 million people from Group B, and 1 million people from Group C. How many advertisements should Everest place in each newspaper to achieve its advertising goals at a minimum cost? What is the minimum cost?

24. **SHIPPING COSTS** Steinwelt Piano manufactures uprights and consoles in two plants, Plant I and Plant II. The output of Plant I is at most 300/month, and the output of Plant II is at most 250/month. These pianos are shipped to three warehouses that serve as distribution centers for Steinwelt. To fill current and projected future orders, Warehouse A requires a minimum of 200 pianos/month, Warehouse B requires at least 150 pianos/month, and Warehouse C requires at least 200 pianos/month. The shipping cost of each piano from Plant I to Warehouse A, Warehouse B, and Warehouse C is $60, $60, and $80, respectively, and the shipping cost of each piano from Plant II to Warehouse A, Warehouse B, and Warehouse C is $80, $70, and $50, respectively. What shipping schedule will enable Steinwelt to meet the requirements of the warehouses while keeping the shipping costs to a minimum? What is the minimum cost?

25. **NUTRITION—DIET PLANNING** The owner of the Health Juice-Bar wishes to prepare a low-calorie fruit juice with a high vitamin A and vitamin C content by blending orange juice and pink grapefruit juice. Each glass of the blended juice is to contain at least 1200 International Units (IU) of vitamin A and 200 IU of vitamin C. One ounce of orange juice contains 60 IU of vitamin A, 16 IU of vitamin C, and 14 calories; each ounce of pink grapefruit juice contains 120 IU of vitamin A, 12 IU of vitamin C, and 11 calories. How many ounces of each juice should a glass of the blend contain if it is to meet the minimum vitamin requirements while containing a minimum number of calories?

26. **PRODUCTION CONTROL** An oil company operates two refineries in a certain city. Refinery I has an output of 200, 100, and 100 barrels of low-, medium-, and high-grade oil per day, respectively. Refinery II has an output of 100, 200, and 600 barrels of low-, medium-, and high-grade oil per day, respectively. The company wishes to produce at least 1000, 1400, and 3000 barrels of low-, medium-, and high-grade oil to fill an order. If it costs $200/day to operate Refinery I and $300/day to operate Refinery II, determine how many days each refinery should be operated to meet the production requirements at minimum cost to the company. What is the minimum cost?

In Exercises 27 and 28, determine whether the statement is true or false. If it is true, explain why it is true. If it is false, give an example to show why it is false.

27. If a standard minimization linear programming problem has a unique solution, then so does the corresponding maximization problem with objective function $P = -C$, where $C = a_1x_1 + a_2x_2 + \cdots + a_nx_n$ is the objective function for the minimization problem.

28. The optimal value attained by the objective function of the primal problem may be different from that attained by the objective function of the dual problem.

6.5   Solutions to Self-Check Exercises

1. We first write down the following tableau for the given (primal) problem:

x	y	Constant
4	1	40
2	1	30
1	3	30
2	5	

Next, we interchange the columns and rows of the tableau and head the three columns of the resulting array with the three variables, u, v, and w, obtaining the tableau

u	v	w	Constant
4	2	1	2
1	1	3	5
40	30	30	

Interpreting the last tableau as if it were the initial tableau for a standard linear programming problem—with the exception that the signs of the coefficients pertaining to the objective function are not reversed—we construct the required dual problem as follows:

$$\text{Maximize} \quad P = 40u + 30v + 30w$$
$$\text{subject to} \quad 4u + 2v + w \le 2$$
$$u + v + 3w \le 5$$
$$u \ge 0, v \ge 0, w \ge 0$$

2. We introduce slack variables x and y to obtain the system of linear equations

$$4u + 2v + w + x = 2$$
$$u + v + 3w + y = 5$$
$$-40u - 30v - 30w + P = 0$$

Using the simplex algorithm, we obtain the sequence of simplex tableaus

	u	v	w	x	y	P	Constant
Pivot row →	④	2	1	1	0	0	2
	1	1	3	0	1	0	5
	−40	−30	−30	0	0	1	0

Ratio: $\frac{2}{4} = \frac{1}{2}$; $\frac{5}{1} = 5$ $\xrightarrow{\frac{1}{4}R_1}$

↑ Pivot column

u	v	w	x	y	P	Constant
①	$\frac{1}{2}$	$\frac{1}{4}$	$\frac{1}{4}$	0	0	$\frac{1}{2}$
1	1	3	0	1	0	5
−40	−30	−30	0	0	1	0

$\xrightarrow[R_3 + 40R_1]{R_2 - R_1}$

	u	v	w	x	y	P	Constant
	1	$\frac{1}{2}$	$\frac{1}{4}$	$\frac{1}{4}$	0	0	$\frac{1}{2}$
Pivot row →	0	$\frac{1}{2}$	⑪⁄₄	$-\frac{1}{4}$	1	0	$\frac{9}{2}$
	0	−10	−20	10	0	1	20

Ratio: $\frac{1/2}{1/4} = 2$; $\frac{9/2}{11/4} = \frac{18}{11}$ $\xrightarrow{\frac{4}{11}R_2}$

↑ Pivot column

u	v	w	x	y	P	Constant
1	$\frac{1}{2}$	$\frac{1}{4}$	$\frac{1}{4}$	0	0	$\frac{1}{2}$
0	$\frac{2}{11}$	①	$-\frac{1}{11}$	$\frac{4}{11}$	0	$\frac{18}{11}$
0	−10	−20	10	0	1	20

$\xrightarrow[R_3 + 20R_2]{R_1 - \frac{1}{4}R_2}$

	u	v	w	x	y	P	Constant
Pivot row →	1	⑤⁄₁₁	0	$\frac{3}{11}$	$-\frac{1}{11}$	0	$\frac{1}{11}$
	0	$\frac{2}{11}$	1	$-\frac{1}{11}$	$\frac{4}{11}$	0	$\frac{18}{11}$
	0	$-\frac{70}{11}$	0	$\frac{90}{11}$	$\frac{80}{11}$	1	$\frac{580}{11}$

Ratio: $\frac{1/11}{5/11} = \frac{1}{5}$; $\frac{18/11}{2/11} = 9$ $\xrightarrow{\frac{11}{5}R_1}$

↑ Pivot column

u	v	w	x	y	P	Constant
$\frac{11}{5}$	①	0	$\frac{3}{5}$	$-\frac{1}{5}$	0	$\frac{1}{5}$
0	$\frac{2}{11}$	1	$-\frac{1}{11}$	$\frac{4}{11}$	0	$\frac{18}{11}$
0	$-\frac{70}{11}$	0	$\frac{90}{11}$	$\frac{80}{11}$	1	$\frac{580}{11}$

$\xrightarrow[R_3 + \frac{70}{11}R_1]{R_2 - \frac{2}{11}R_1}$

u	v	w	x	y	P	Constant
$\frac{11}{5}$	1	0	$\frac{3}{5}$	$-\frac{1}{5}$	0	$\frac{1}{5}$
$-\frac{2}{5}$	0	1	$-\frac{1}{5}$	$\frac{2}{5}$	0	$\frac{8}{5}$
14	0	0	12	6	1	54

Solution for the primal problem

The last tableau is final, and the solution to the primal problem is $x = 12$ and $y = 6$ with a minimum value for C of 54.

USING TECHNOLOGY

The Simplex Method: Solving Minimization Problems

Graphing Utility

A graphing utility can be used to solve minimization problems using the simplex method.

EXAMPLE 1

$$\text{Minimize} \quad C = 2x + 3y$$
$$\text{subject to} \quad 8x + y \ge 80$$
$$3x + 2y \ge 100$$
$$x + 4y \ge 80$$
$$x \ge 0, y \ge 0$$

(continued)

Solution We begin by writing the dual problem associated with the given primal problem. From the tableau for the primal problem

x	y	Constant
8	1	80
3	2	100
1	4	80
2	3	

we obtain—upon interchanging the columns and rows of this tableau and heading the three columns of the resulting array with the variables u, v, and w—the tableau

u	v	w	Constant
8	3	1	2
1	2	4	3
80	100	80	

This tells us that the dual problem is

$$\text{Maximize} \quad P = 80u + 100v + 80w$$
$$\text{subject to} \quad 8u + 3v + w \le 2$$
$$u + 2v + 4w \le 3$$
$$u \ge 0, v \ge 0, w \ge 0$$

To solve this standard maximization problem, we proceed as follows:

	u	v	w	x	y	P	Constant	Ratio	
Pivot row →	8	③	1	1	0	0	2	$\frac{2}{3}$	$*\text{row}(\frac{1}{3}, A, 1) \blacktriangleright B$
	1	2	4	0	1	0	3	$\frac{3}{2}$	
	−80	−100	−80	0	0	1	0		

Pivot column

u	v	w	x	y	P	Constant	
2.67	①	0.33	0.33	0	0	0.67	$*\text{row}+(-2, B, 1, 2) \blacktriangleright C$
1	2	4	0	1	0	3	$*\text{row}+(100, C, 1, 3) \blacktriangleright B$
−80	−100	−80	0	0	1	0	

	u	v	w	x	y	P	Constant	Ratio	
	2.67	1	0.33	0.33	0	0	0.67	2	$*\text{row}(\frac{1}{3.33}, B, 2) \blacktriangleright C$
Pivot row →	−4.33	0	③.33	−0.67	1	0	1.67	0.5	
	186.67	0	−46.67	33.33	0	1	66.67		

Pivot column

u	v	w	x	y	P	Constant	
2.67	1	0.33	0.33	0	0	0.67	$*\text{row}+(-0.33, C, 2, 1) \blacktriangleright B$
−1.30	0	1	−0.2	0.3	0	0.5	$*\text{row}+(46.67, B, 2, 3) \blacktriangleright C$
186.67	0	−46.67	33.33	0	1	66.67	

u	v	w	x	y	P	Constant
3.1	1	0	0.4	−0.1	0	0.50
−1.3	0	1	−0.2	0.3	0	0.50
125.93	0	0.05	23.99	14.02	1	90.03

Solution for the
primal problem

From the last tableau, we see that $x = 23.99$, $y = 14.02$, and $C = 90.03$. ∎

Excel

EXAMPLE 2

$$\begin{aligned}
\text{Minimize} \quad & C = 2x + 3y \\
\text{subject to} \quad & 8x + y \geq 80 \\
& 3x + 2y \geq 100 \\
& x + 4y \geq 80 \\
& x \geq 0, y \geq 0
\end{aligned}$$

Solution We use Solver as outlined in Example 2, pages 381–384, to obtain the spreadsheet shown in Figure T1. (In this case, select **Min** next to To: instead of Max because this is a minimization problem. Also select **>=** in the Add Constraint dialog box because the inequalities in the problem are of the form ≥.) From the spreadsheet, we read off the solution: $x = 24$, $y = 14$, and $C = 90$.

	A	B	C	D	E	F	G	H	I	J
1	Minimization Problem							Formulas for indicated cells		
2								C8: = 2*C4 + 3*C5		
3	Decision Variables							C11: = 8*C4 + C5		
4		x	24					C12: = 3*C4 + 2*C5		
5		y	14					C13: = C4 + 4*C5		
6										
7										
8	Objective Function		90							
9										
10	Constraints									
11			206	>=	80					
12			100	>=	100					
13			80	>=	80					

FIGURE **T1**
Completed spreadsheet after using Solver

Note: Boldfaced words/characters enclosed in a box (for example, **Enter**) indicate that an action (click, select, or press) is required. Words/characters printed blue (for example, Chart sub-type:) indicate words/characters that appear on the screen.

(*continued*)

TECHNOLOGY EXERCISES

In Exercises 1–4, solve the linear programming problem by the simplex method.

1. Minimize $C = x + y + 3z$
subject to $2x + y + 3z \geq 6$
$x + 2y + 4z \geq 8$
$3x + y - 2z \geq 4$
$x \geq 0, y \geq 0, z \geq 0$

2. Minimize $C = 2x + 4y + z$
subject to $x + 2y + 4z \geq 7$
$3x + y - z \geq 6$
$x + 4y + 2z \geq 24$
$x \geq 0, y \geq 0, z \geq 0$

3. Minimize $C = x + 1.2y + 3.5z$
subject to $2x + 3y + 5z \geq 12$
$3x + 1.2y - 2.2z \geq 8$
$1.2x + 3y + 1.8z \geq 14$
$x \geq 0, y \geq 0, z \geq 0$

4. Minimize $C = 2.1x + 1.2y + z$
subject to $x + y - z \geq 5.2$
$x - 2.1y + 4.2z \geq 8.4$
$x \geq 0, y \geq 0, z \geq 0$

CHAPTER 6   Summary of Principal Terms

TERMS

solution set of a system of linear inequalities (333)

bounded solution set (334)

unbounded solution set (334)

objective function (338)

linear programming problem (338)

feasible solution (346)

feasible set (346)

optimal solution (346)

isoprofit line (347)

method of corners (348)

standard maximization problem (359)

slack variable (359)

basic variable (360)

nonbasic variable (360)

basic solution (361)

pivot column (362)

pivot row (363)

pivot element (363)

simplex tableau (363)

simplex method (364)

standard minimization problem (385)

primal problem (385)

dual problem (385)

CHAPTER 6   Concept Review Questions

Fill in the blanks.

1. a. The solution set of the inequality $ax + by < c$ is a/an _____ _____ that does not include the _____ with equation $ax + by = c$.

b. If $ax + by < c$ describes the lower half plane, then the inequality _____ describes the lower half plane together with the line having equation _____.

2. a. The solution set of a system of linear inequalities in the two variables x and y is the set of all _____ satisfying _____ inequality of the system.

b. The solution set of a system of linear inequalities is _____ if it can be _____ by a circle.

3. A linear programming problem consists of a linear function, called a/an _____ _____ to be _____ or _____ subject to constraints in the form of _____ equations or _____.

4. a. If a linear programming problem has a solution, then it must occur at a/an _____ _____ of the feasible set.

b. If the objective function of a linear programming problem is optimized at two adjacent vertices of the feasible set, then it is optimized at every point on the _____ segment joining these vertices.

5. In a standard maximization problem: the objective function is to be _____; all the variables involved in the problem are _____; and each linear constraint may be written so that the expression involving the variables is _____ _____ or _____ _____ a nonnegative constant.

6. In setting up the initial simplex tableau, we first transform the system of linear inequalities into a system of linear _____, using _____ _____; the objective function is rewritten so that it has the form _____ and then is placed _____ the system of linear equations obtained earlier. Finally, the initial simplex tableau is the _____ matrix associated with this system of linear equations.

7. In a standard minimization problem: the objective function is to be _____; all the variables involved in the problem are _____; and each linear constraint may be written so that the expression involving the variables is _____ _____ or _____ _____ a constant.

8. The fundamental theorem of duality states that a primal problem has a solution if and only if the corresponding _____ problem has a solution. If a solution exists, then the _____ functions of both the primal and the dual problem attain the same _____ _____.

CHAPTER 6 Review Exercises

In Exercises 1 and 2, find the optimal value(s) of the given objective function on the feasible set S.

1. $Z = 2x + 3y$

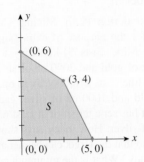

2. $Z = 4x + 3y$

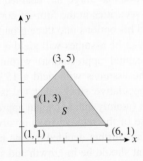

In Exercises 3–12, use the method of corners to solve the given linear programming problem.

3. Maximize $P = 3x + 5y$
   subject to $2x + 3y \le 12$
            $x + y \le 5$
            $x \ge 0, y \ge 0$

4. Maximize $P = 2x + 3y$
   subject to $2x + y \le 12$
            $x - 2y \le 1$
            $x \ge 0, y \ge 0$

5. Minimize $C = 2x + 5y$
   subject to $x + 3y \ge 15$
            $4x + y \ge 16$
            $x \ge 0, y \ge 0$

6. Minimize $C = 3x + 4y$
   subject to $2x + y \ge 4$
            $2x + 5y \ge 10$
            $x \ge 0, y \ge 0$

7. Maximize $P = 3x + 2y$
   subject to $2x + y \le 16$
            $2x + 3y \le 36$
            $4x + 5y \ge 28$
            $x \ge 0, y \ge 0$

8. Maximize $P = 6x + 2y$
   subject to $x + 2y \le 12$
            $x + y \le 8$
            $2x - 3y \ge 6$
            $x \ge 0, y \ge 0$

9. Minimize $C = 2x + 7y$
   subject to $3x + 5y \ge 45$
            $3x + 10y \ge 60$
            $x \ge 0, y \ge 0$

10. Minimize $C = 4x + y$
   subject to $6x + y \ge 18$
            $2x + y \ge 10$
            $x + 4y \ge 12$
            $x \ge 0, y \ge 0$

11. Find the maximum and minimum of $Q = x + y$ subject to
$$5x + 2y \ge 20$$
$$x + 2y \ge 8$$
$$x + 4y \le 22$$
$$x \ge 0, y \ge 0$$

12. Find the maximum and minimum of $Q = 2x + 5y$ subject to
$$x + y \ge 4$$
$$-x + y \le 6$$
$$x + 3y \le 30$$
$$x \le 12$$
$$x \ge 0, y \ge 0$$

In Exercises 13–20, use the simplex method to solve the linear programming problem.

13. Maximize $P = 3x + 4y$
   subject to $x + 3y \le 15$
            $4x + y \le 16$
            $x \ge 0, y \ge 0$

14. Maximize $P = 2x + 5y$
   subject to $2x + y \le 16$
            $2x + 3y \le 24$
            $y \le 6$
            $x \ge 0, y \ge 0$

15. Maximize $P = 2x + 3y + 5z$
   subject to $x + 2y + 3z \le 12$
            $x - 3y + 2z \le 10$
            $x \ge 0, y \ge 0, z \ge 0$

16. Maximize $P = x + 2y + 3z$
    subject to $2x + y + z \le 14$
    $3x + 2y + 4z \le 24$
    $2x + 5y - 2z \le 10$
    $x \ge 0, y \ge 0, z \ge 0$

17. Minimize $C = 3x + 2y$
    subject to $2x + 3y \ge 6$
    $2x + y \ge 4$
    $x \ge 0, y \ge 0$

18. Minimize $C = x + 2y$
    subject to $3x + y \ge 12$
    $x + 4y \ge 16$
    $x \ge 0, y \ge 0$

19. Minimize $C = 24x + 18y + 24z$
    subject to $3x + 2y + z \ge 4$
    $x + y + 3z \ge 6$
    $x \ge 0, y \ge 0, z \ge 0$

20. Minimize $C = 4x + 2y + 6z$
    subject to $x + 2y + z \ge 4$
    $2x + y + 2z \ge 2$
    $3x + 2y + z \ge 3$
    $x \ge 0, y \ge 0, z \ge 0$

In Exercises 21–23, use the method of corners to solve the linear programming problem.

21. **FINANCIAL ANALYSIS** An investor has decided to commit no more than $80,000 to the purchase of the common stocks of two companies, Company A and Company B. He has also estimated that there is a chance of at most a 1% capital loss on his investment in Company A and a chance of at most a 4% loss on his investment in Company B, and he has decided that these losses should not exceed $2000. On the other hand, he expects to make a 14% profit from his investment in Company A and a 20% profit from his investment in Company B. Determine how much he should invest in the stock of each company to maximize his investment returns.

22. **MANUFACTURING—PRODUCTION SCHEDULING** Soundex produces two models of clock radios. Model A requires 15 min of work on Assembly Line I and 10 min of work on Assembly Line II. Model B requires 10 min of work on Assembly Line I and 12 min of work on Assembly Line II. At most, 25 labor-hours of assembly time on Line I and 22 labor-hours of assembly time on Line II are available each day. It is anticipated that Soundex will realize a profit of $12 on model A and $10 on model B. How many clock radios of each model should be produced each day to maximize Soundex's profit?

23. **MANUFACTURING—PRODUCTION SCHEDULING** Kane Manufacturing has a division that produces two models of grates, model A and model B. To produce each model A grate requires 3 lb of cast iron and 6 min of labor. To produce each model B grate requires 4 lb of cast iron and 3 min of labor. The profit for each model A grate is $2.00, and the profit for each model B grate is $1.50. Available for grate production each day are 1000 lb of cast iron and 20 labor-hours. Because of a backlog of orders for model B grates, Kane's manager has decided to produce at least 180 model B grates/day. How many grates of each model should Kane produce to maximize its profits?

In Exercises 24–26, use the simplex method to solve the linear programming problem.

24. **MINING—PRODUCTION** Perth Mining Company operates two mines for the purpose of extracting gold and silver. The Saddle Mine costs $14,000/day to operate, and it yields 50 oz of gold and 3000 oz of silver each day. The Horseshoe Mine costs $16,000/day to operate and it yields 75 oz of gold and 1000 oz of silver each day. Company management has set a target of at least 650 oz of gold and 18,000 oz of silver. How many days should each mine be operated at so that the target can be met at a minimum cost to the company? What is the minimum cost?

25. **INVESTMENT ANALYSIS** Jorge has decided to invest at most $100,000 in securities in the form of corporate stocks. He has classified his options into three groups of stocks: blue-chip stocks that he assumes will yield a 10% return (dividends and capital appreciation) within a year, growth stocks that he assumes will yield a 15% return within a year, and speculative stocks that he assumes will yield a 20% return (mainly due to capital appreciation) within a year. Because of the relative risks involved in his investment, Jorge has further decided that no more than 30% of his investment should be in growth and speculative stocks and at least 50% of his investment should be in blue-chip and speculative stocks. Determine how much Jorge should invest in each group of stocks in the hope of maximizing the return on his investments.

26. **MAXIMIZING PROFIT** A company manufactures three products, A, B, and C, on two machines, I and II. It has been determined that the company will realize a profit of $4/unit of Product A, $6/unit of Product B, and $8/unit of Product C. Manufacturing a unit of Product A requires 9 min on Machine I and 6 min on Machine II; manufacturing a unit of Product B requires 12 min on Machine I and 6 min on Machine II; manufacturing a unit of Product C requires 18 min on Machine I and 10 min on Machine II. There are 6 hr of machine time available on Machine I and 4 hr of machine time available on Machine II in each work shift. How many units of each product should be produced in each shift to maximize the company's profit?

1. Find the maximum and minimum values of $Z = 3x - y$ on the following feasible set.

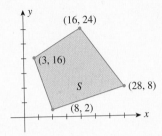

2. Use the method of corners to solve the following linear programming problem:

$$\begin{aligned} \text{Maximize} \quad & P = x + 3y \\ \text{subject to} \quad & 2x + 3y \le 11 \\ & 3x + 7y \le 24 \\ & x \ge 0, y \ge 0 \end{aligned}$$

3. Consider the following linear programming problem:

$$\begin{aligned} \text{Maximize} \quad & P = x + 2y - 3z \\ \text{subject to} \quad & 2x + y - z \le 3 \\ & x - 2y + 3z \le 1 \\ & 3x + 2y + 4z \le 17 \\ & x \ge 0, y \ge 0, z \ge 0 \end{aligned}$$

Write the initial simplex tableau for the problem and identify the pivot element to be used in the first iteration of the simplex method.

4. The following simplex tableau is in final form. Find the solution to the linear programming problem associated with this tableau.

x	y	z	u	v	w	P	Constant
0	$\frac{1}{2}$	0	1	$-\frac{1}{2}$	0	0	2
0	$\frac{1}{4}$	1	0	$\frac{5}{4}$	$-\frac{1}{2}$	0	11
1	$\frac{1}{4}$	0	0	$-\frac{3}{4}$	$\frac{1}{2}$	0	2
0	$\frac{13}{4}$	0	0	$\frac{1}{4}$	$\frac{1}{2}$	1	28

5. Using the simplex method, solve the following linear programming problem:

$$\begin{aligned} \text{Maximize} \quad & P = 5x + 2y \\ \text{subject to} \quad & 4x + 3y \le 30 \\ & 2x - 3y \le 6 \\ & x \ge 0, y \ge 0 \end{aligned}$$

Answers

CHAPTER 1

Exercises 1.1, page 6

1. Integer, rational, and real   3. Rational and real

5. Irrational and real   7. Irrational and real

9. Rational and real   11. False   13. True   15. False

17. Commutative law of addition

19. Commutative law of multiplication   21. Distributive law

23. Associative law of addition   25. Property 1 of negatives

27. Property 1 of zero properties   29. Property 2 of zero properties

31. Property 2 of quotients   33. Properties 2 and 5 of quotients

35. Property 6 of quotients and distributive law   37. False

39. False   41. False

Exercises 1.2, page 13

1. 81   3. $\frac{8}{27}$ 5. -81 7. $-\frac{81}{125}$ 9. 256

11. $243y^5$ 13. $6x - 3$ 15. $9x^2 + 3x + 1$

17. $4y^2 + 2y + 9$ 19. $1.2x^3 - 4.2x^2 + 2.5x - 8.2$

21. $6x^5$ 23. $2x^3 + 4x$ 25. $7m^2 - 9m$

27. $14a - 7b$ 29. $6x^2 + 5x - 6$

31. $6x^2 - 5xy - 6y^2$ 33. $12r^2 - rs - 6s^2$

35. $0.06x^2 - 0.06xy - 2.52y^2$ 37. $6x^3 - 3x^2y + 4xy - 2y^2$

39. $4x^2 + 12xy + 9y^2$ 41. $4u^2 - v^2$

43. $2x^2 - x + 2$ 45. $4x^2 + 6xy + 9y^2 - x + 2y + 2$

47. $2t^4 - 4t^3 + 9t^2 - 2t + 4$ 49. $-2x + 1$ 51. $-3x - 1$

53. $x^2 - 10x + 50$ 55. $22x^3 - 20x^2 - 6x$

57. $-0.000002x^3 - 0.02x^2 + 1000x - 120,000$

59. $0.7t^2 + 350t$ 61. $3.5t^2 + 2.4t + 71.2$ 63. False

65. False   67. m

Exercises 1.3, page 19

1. $2m(3m - 2)$ 3. $3ab(3b - 2a)$ 5. $5mn(2m - 3n + 4)$

7. $(3x - 5)(2x + 1)$ 9. $(2c - d)(3a + b + 4ac - 2ad)$

11. $(2m + 1)(m - 6)$ 13. $(x - 3y)(x + 2y)$ 15. Prime

17. $(2a - b)(2a + b)$ 19. $(uv - w)(uv + w)$ 21. Prime

23. Prime   25. $(x + 4)(x - 1)$ 27. $2y(3x + 2)(2x - 3)$

29. $(7r - 4)(5r + 3)$ 31. $xy(3x - 2y)(3x + 2y)$

33. $(x^2 - 4y)(x^2 + 4y)$ 35. $-8ab$

37. $(2m + 1)(4m^2 - 2m + 1)$ 39. $(2r - 3s)(4r^2 + 6rs + 9s^2)$

41. $u^2(v^2 - 2)(v^4 + 2v^2 + 4)$ 43. $(2x + 1)(x^2 + 3)$

45. $(3a + b)(x + 2y)$ 47. $(u - v)(u + v)(u^2 + v^2)$

49. $(x + y)(2x - 3y)(2x + 3y)$ 51. $(x^3 - 2)(x + 3)$

53. $(au + c)(u + 1)$ 55. $P(1 + rt)$ 57. $100x(80 - x)$

59. $kx(M - x)$ 61. $\dfrac{V_0}{273}(273 + T)$

Exercises 1.4, page 25

1. $\dfrac{4}{x}$ 3. $\frac{4}{5}$ 5. $\dfrac{2x - 1}{2x}$ 7. $\dfrac{x - 1}{x + 1}$ 9. $\dfrac{x + 3}{2x + 1}$

11. $x + y$ 13. $\frac{1}{2}x$ 15. $\frac{2}{5}x^2$ 17. $\dfrac{5x}{2}$ 19. $\frac{4}{3}$

21. $\dfrac{3(3r - 2)}{2}$ 23. $\dfrac{k + 1}{k - 2}$ 25. $\dfrac{10x + 7}{(2x + 3)(2x - 1)}$

27. $\dfrac{5x - 9}{(x - 3)(x + 2)(x - 1)}$ 29. $\dfrac{4m^3 - 3}{(2m^2 - 2m - 1)(2m^2 - 3m + 3)}$

31. $-\dfrac{x^2 - x - 3}{(x + 1)(x - 1)}$ 33. $\dfrac{2(x^2 - x + 2)}{(x + 2)(x - 2)}$

35. $\dfrac{x^3 - 2x^2 + 3x + 24}{(x + 3)(x + 2)(x - 2)(x + 1)}$ 37. $\dfrac{bx - ay}{ab(x - y)}$

39. $\dfrac{x + 1}{x - 1}$ 41. $\dfrac{x + y}{xy - 1}$ 43. $\dfrac{y - x}{x^2y^2}$ 45. $-\dfrac{1}{2x(x + h)}$

47. **a.** $\dfrac{2.2x + 2500}{x}$ **b.** $2.2x + 2500$ 49. $\dfrac{R[(1 + i)^n - 1]}{i(1 + i)^n}$

51. $\dfrac{164 + 7(t - 4.5)^2}{1 + 0.25(t - 4.5)^2}$

Exercises 1.5, page 30

1. -8 3. $\frac{1}{49}$ 5. -16 7. $\frac{7}{12}$ 9. 0.0009 11. 1

13. 1   15. $\frac{1}{32}$ 17. 1   19. $\frac{1}{27}$ 21. $\frac{1}{4}x^5$ 23. $\dfrac{3}{2x}$

25. $\dfrac{1}{a^6}$ 27. $\dfrac{8y^6}{x^6}$ 29. $\dfrac{8}{xy}$ 31. $-2x^2y^5$ 33. $\dfrac{3}{2uv^2}$

35. $864x^4$ 37. $\dfrac{1}{4x^8}$ 39. $\dfrac{4u^4}{9v^5}$ 41. $\dfrac{1}{1728x^2y^3z^2}$

43. $\dfrac{a^{10}}{b^{12}}$ 45. $9a^2b^8$ 47. $\dfrac{u^8}{16}$ 49. $\dfrac{1 - x}{1 + x}$

51. $\dfrac{1}{uv}$ 53. $\dfrac{b + a}{b - a}$ 55. False   57. False

Exercises 1.6, page 35

1. $x = 4$ 3. $y = \frac{20}{3}$ 5. $x = -\frac{2}{3}$ 7. $y = 5$

9. $p = 15$ 11. $p = 0$ 13. $k = 4$ 15. $x = 2$ 17. $x = 8$

19. $x = \frac{1}{2}$ 21. $x = \frac{1}{3}$ 23. $y = \frac{3}{2}$ 25. $x = \frac{17}{8}$

27. $q = -1$ 29. $k = -2$ 31. No solution

33. $r = \dfrac{I}{Pt}$ 35. $q = -\dfrac{P}{3} + \dfrac{1}{3}$ 37. $R = \dfrac{iS}{(1 + i)^n - 1}$

39. $x = \dfrac{Vb}{a - V}$ 41. $m = \dfrac{rB(n + 1)}{2I}$ 43. $n = \dfrac{2mI - rB}{rB}$

45. $p = \dfrac{fq}{q - f}$ 47. $t = \dfrac{I}{Pr}; \dfrac{3}{2}$ years   49. $t = \dfrac{a}{S - b}$

51. **a.** $C = \dfrac{NV - St}{N - t}$ **b.** \$115,000

53. **a.** $t = \dfrac{24c - a}{a}$ **b.** 5 years

Exercises 1.7, page 44

1. 9   3. 4   5. 4   7. 4   9. -5 11. 4

13. $\frac{2}{3}$ 15. $\frac{9}{4}$ 17. $\frac{1}{4}$ 19. $-\frac{2}{3}$ 21. 9   23. $\frac{1}{9}$

25. $\frac{3}{4}$ 27. 64   29. $x^{1/5}$ 31. x 33. $\dfrac{9}{x^6}$ 35. $\dfrac{y^{5/2}}{x^3}$

37. $x^{12/5} - 2x^{17/5}$ 39. $4p^2 - 2p$ 41. $4\sqrt{2}$ 43. $-3\sqrt[3]{2}$

45. $4xy\sqrt{y}$ 47. m^2np^4 49. $\sqrt[3]{3}$ 51. $\sqrt[6]{x}$ 53. $\dfrac{2\sqrt{3}}{3}$

55. $\dfrac{3\sqrt{x}}{2x}$ 57. $\dfrac{2\sqrt{3y}}{3}$ 59. $\dfrac{\sqrt[3]{x^2}}{x}$ 61. $\sqrt{3} - 1$

63. $-(1 + \sqrt{2})^2$ 65. $\dfrac{q(\sqrt{q} + 1)}{q - 1}$ 67. $\dfrac{y\sqrt[3]{xz^2}}{xz}$ 69. $\dfrac{4\sqrt{3}}{3}$

71. $\dfrac{\sqrt[3]{18}}{3}$ 73. $\dfrac{\sqrt{6}}{2x}$ 75. $\dfrac{\sqrt[3]{18y^2}}{3}$ 77. $\dfrac{\sqrt{a}(1 + a)}{a}$

79. $\dfrac{x + y}{x - y}$ 81. $\dfrac{\sqrt{x + 1}(3x + 2)}{2(x + 1)}$ 83. $\dfrac{3 + x^{1/3}}{6x^{1/2}(1 + x^{1/3})^2}$

85. $x = 1$ 87. $k = \frac{5}{2}$ 89. $k = \frac{1}{3}$ 91. $p = 144 - x^2$

93. True   95. True

Exercises 1.8, page 51

1. $x = -2, 3$ 3. $x = -2, 2$ 5. $x = -4, 3$

7. $t = -1, \frac{1}{2}$ 9. $x = 2$ 11. $m = 2, \frac{3}{2}$ 13. $x = -\frac{3}{2}, \frac{3}{2}$

15. $z = -2, \frac{3}{2}$ 17. $x = -4, 2$ 19. $x = 1 - \dfrac{\sqrt{6}}{2}, 1 + \dfrac{\sqrt{6}}{2}$

21. $m = -\frac{1}{2} - \frac{1}{2}\sqrt{13}, -\frac{1}{2} + \frac{1}{2}\sqrt{13}$

23. $x = -\dfrac{3}{4} - \dfrac{\sqrt{41}}{4}, -\dfrac{3}{4} + \dfrac{\sqrt{41}}{4}$ 25. $x = \pm\dfrac{\sqrt{13}}{2}$

27. $x = -\frac{3}{2}, 2$ 29. $m = 2 \pm \sqrt{3}$ 31. $x = \frac{1}{2} \pm \frac{1}{4}\sqrt{10}$

33. $x = -1 \pm \frac{1}{2}\sqrt{10}$ 35. $x = -0.93, 3.17$

37. $x = \pm\sqrt{2}, \pm\sqrt{3}$ 39. $y = \pm\sqrt{2}, \pm\sqrt{5}$

41. $x = -\frac{7}{2}, -\frac{5}{3}$ 43. $w = \frac{4}{9}, \frac{9}{4}$ 45. $x = -2, -\frac{3}{2}$

47. $x = -\frac{5}{2}, 1$ 49. $y = -2, \frac{15}{4}$ 51. $x = -8, 2$

53. $x = -\frac{16}{3}$ 55. $t = -1 \pm \frac{1}{2}\sqrt{6}$ 57. $u = -3, 2$

59. $r = 3$ 61. $s = 6$ 63. $x = \frac{22}{7}, \frac{10}{3}$

65. Two real solutions   67. No real solutions

69. One real solution   71. Two real solutions

73. 9.2 sec   75. 1.41 sec after passing the tree   77. 10,000

79. 40,000   81. 3.82 or 26.18 days

83. 12 in. $\times$ 6 in. $\times$ 2 in.   85. 30 ft and 90 ft

87. 1500 yd $\times$ 750 yd   89. 1.83 ft   91. False   93. True

Exercises 1.9, page 62

1. False   3. False

5.

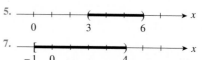

7.

9.

11. $(-\infty, 3)$ 13. $(-\infty, -5]$ 15. $(-4, 6)$

17. $(-\infty, -3) \cup (3, \infty)$ 19. $(-2, 3)$ 21. $[-3, 5]$

23. $[1, \frac{3}{2}]$ 25. $(-\infty, -3] \cup (2, \infty)$ 27. $(-\infty, 0] \cup (1, \infty)$

29. 4   31. 2   33. $5\sqrt{3}$ 35. $\pi + 1$

37. 2   39. False   41. False   43. True

45. $a - b < x < a + b$ 47. False   49. True   51. False

53. \$50.70   55. $[362, 488.7]$ 57. \$52,000

59. Between 1000 and 4000 units

61. Between 98.04% and 98.36% of the toxic pollutants

63. After 2 min   65. 4 sec   67. $|x - 0.5| \leq 0.01$

Chapter 1 Concept Review Questions, page 65

1. **a.** rational; repeating; terminating
   b. irrational; terminates; repeats

2. **a.** $b + a$; $(a + b) + c$; a; 0   **b.** ba; $(ab)c$; $1 \cdot a = a$; 1
   c. $ab + ac$

3. **a.** a; $-(ab) = a(-b)$; ab **b.** 0; 0

4. **a.** polynomial; x; degree; term; polynomial; coefficient   **b.** like

5. product; prime; $x(x + 2)(x - 1)$

6. **a.** polynomials   **b.** numerator; denominator; factors; 1; -1
   c. denominator; fractions

7. complex; $\dfrac{1 + \dfrac{1}{x}}{1 - \dfrac{1}{y}}$

8. **a.** $\underbrace{a \cdot a \cdot a \cdot \cdots \cdot a}_{n \text{ factors}}$; base; exponent; power   **b.** 1; not defined

   c. $\dfrac{1}{a^n}$

9. **a.** equation   **b.** number   **c.** $ax + b = 0$; 1

10. **a.** $a^n = b$ **b.** pairs   **c.** no   **d.** real root

11. **a.** radical; $b^{1/n}$ **b.** radical

12. **a.** $ax^2 + bx + c = 0$

   b. factoring; completing the square; $x = \dfrac{-b \pm \sqrt{b^2 - 4ac}}{2a}$

Chapter 1 Review Exercises, page 65

1. Rational and real   2. Irrational and real

3. Irrational and real   4. Whole, integer, rational, and real

5. Rational and real   6. Irrational and real   7. $\frac{27}{8}$

8. 25   9. $\frac{1}{144}$ 10. -32 11. $\frac{64}{27}$ 12. $\frac{1}{4}$ 13. $\frac{3}{5}$

14. $3\sqrt[3]{3}$ 15. $4(x^2 + y)^2$ 16. $\dfrac{a^{15}}{b^{11}}$ 17. $\dfrac{2x}{3z}$ 18. $-x^{1/2}$

19. $6xy^7$ 20. $\frac{9}{2}a^5b^8$ 21. $9x^2y^4$ 22. $\dfrac{x}{y^{1/2}}$

23. $5x^4 + 20x^3 + 12x^2 + 14x + 3$ 24. $9x^3 - 18x^2 + 17x - 12$

25. $-2x^2 + 9y^2 + 12xy + 7x + 3$ 26. $3a - b$

27. $\dfrac{180}{(t + 6)^2}$ 28. $\dfrac{15x^2 + 24x + 2}{4(3x^2 + 2)(x + 2)}$

29. $\dfrac{78x^2 - 8x - 27}{3(2x^2 - 1)(3x - 1)}$ 30. $\dfrac{2\sqrt{x + 1}(x + 2)}{x + 1}$

31. $-2\pi r^2(\pi r - 50)$ 32. $2vw(v^2 + w^2 + u^2)$ 33. $(4 - x)(4 + x)$

34. $6t(2t - 3)(t + 1)$ 35. $-2(x + 3)(x - 1)$

36. $4(3x - 5)(x - 6)$ 37. $(3a - 5b)(3a + 5b)$

38. $u^3(2uv + 3)(4u^2v^2 - 6uv + 9)$ 39. $3a^2b^2(2a^2b^2c - ac - 3)$

40. $(2x - y)(3x + y)$ 41. $\dfrac{x + 2}{x + 3}$ 42. $\dfrac{4(t^2 - 4)}{(t^2 + 4)^2}$ 43. 2

44. $\dfrac{3x(2x^3 + 2x + 1)}{(x^2 + 2)(x^3 + 1)}$ 45. $\dfrac{x}{(x + 2)(x - 3)}$

46. $\dfrac{x\sqrt{1 + 3x^2}(3x^2 - 5x + 1)}{(x - 1)^2}$ 47. $x = -\frac{3}{4}, \frac{1}{2}$

48. $x = -2, \frac{1}{3}$ 49. $x = \dfrac{3 \pm \sqrt{41}}{4}$ 50. $x = \dfrac{-5 \pm \sqrt{13}}{2}$

51. $y = \frac{1}{2}, 1$ 52. $m = -1.2871, 8.2871$ 53. $x = 0, -3, 1$

54. $x = \dfrac{\pm\sqrt{2}}{2}$ 55. $x = 14$ 56. $p = -2$ 57. $x = -\frac{5}{3}, 0$

58. $q = -\frac{6}{17}$ 59. $k = 2$ 60. $x = 1$ 61. $x = \dfrac{100C}{20 + C}$

62. $I = \dfrac{rB(n + 1)}{2m}$ 63. $[-2, \infty)$ 64. $[-1, 2]$

65. $(-\infty, -4) \cup (5, \infty)$ 66. $(-\infty, -5) \cup (5, \infty)$ 67. 4

68. 1   69. $\pi - 6$ 70. $8 - 3\sqrt{3}$ 71. $[-2, \frac{1}{2}]$ 72. $[-4, 3]$

73. $(-2, -\frac{3}{2})$ 74. $(-1, 4)$ 75. $[\frac{2}{3}, 2]$ 76. $\frac{2}{3}; \frac{3}{2}$

77. $\dfrac{1}{\sqrt{x + 1}}$ 78. $\dfrac{x}{z\sqrt[3]{xy}}$ 79. $\dfrac{x - \sqrt{x}}{2x}$ 80. $\dfrac{3(1 - 2\sqrt{x})}{1 - 4x}$

81. $x = 1 \pm \sqrt{6}$ 82. $x = -2 \pm \dfrac{\sqrt{2}}{2}$

83. \$100   84. \$400

Chapter 1 Before Moving On, page 67

1. $3(5x^2 - 9x + 4)$

2. **a.** $x^2(x - 3)(x + 2)$ **b.** $a(a + 1)(a - 2b - a^2)$

3. $\dfrac{5x^2 - 1}{(3x + 1)(x - 2)(x + 1)}$ 4. $\dfrac{1}{2xy}$ 5. $\dfrac{2s^2}{1 - 2s}$

6. $7 - 4\sqrt{3}$ 7. **a.** -4 or $\frac{3}{2}$ **b.** $\frac{1}{2}(3 + \sqrt{17})$ or $\frac{1}{2}(3 - \sqrt{17})$

8. 21   9. $[-\frac{2}{3}, \frac{3}{2}]$ 10. $[-2, -1]$

CHAPTER 2

Exercises 2.1, page 74

1. $(3, 3)$; Quadrant I   3. $(2, -2)$; Quadrant IV

5. $(-4, -6)$; Quadrant III   7. A 9. $E, F,$ and G

11. F 13–19. See the following figure.

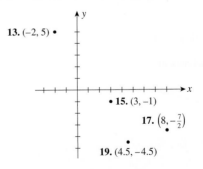

13. $(-2, 5)$ •
• 15. $(3, -1)$
17. $\left(8, -\frac{7}{2}\right)$
19. $(4.5, -4.5)$

21. $\frac{1}{2}$ 23. Not defined   25. 5

27. $\frac{5}{6}$ 29. $\dfrac{d - b}{c - a}$ $(a \neq c)$

31. **a.** 4   **b.** -8 33. Parallel

35. $a = -5$ 37. Yes

Exercises 2.2, page 82

1. (e)   3. (a)   5. (f)

7. Perpendicular   9. $y = -5$

11. $y = 2x - 10$ 13. $y = 2$

15. $y = 3x - 2$ 17. $y = x + 1$

19. $y = 3x + 5$ 21. $y = 5$

23. $y = \frac{1}{2}x$; $m = \frac{1}{2}$; $b = 0$

25. $y = \frac{2}{3}x - 3$; $m = \frac{2}{3}$; $b = -3$

27. $y = -\frac{1}{2}x + \frac{7}{2}$; $m = -\frac{1}{2}$; $b = \frac{7}{2}$

29. $y = \frac{1}{2}x + 3$ 31. $y = \frac{4}{3}x + \frac{4}{3}$

33. $y = -6$ 35. $y = b$

37. $y = \frac{2}{3}x - \frac{2}{3}$ 39. $k = 8$

41.
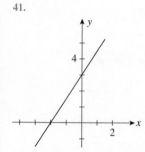

43.
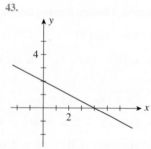

45.
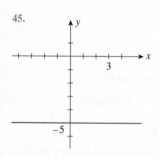

49. $y = -2x - 4$ 51. $y = \frac{1}{8}x - \frac{1}{2}$ 53. Yes

55. **a.**
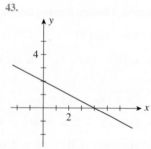

b. 1.9467; 70.082
c. The capacity utilization has been increasing by 1.9467% each year since 1990 when it stood at 70.082%.
d. In the first half of 2005

57. **a.** $y = 0.55x$ **b.** 2000   59. 84.8%

61. **a** and **b.**
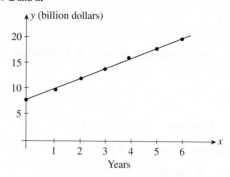

c. $y = 1.82x + 7.9$ **d.** $17 billion; same

63. **a** and **b.**
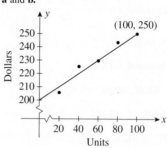

c. $y = \frac{1}{2}x + 200$ **d.** $227

65. **a** and **b.**
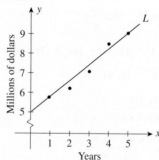

c. $y = 0.8x + 5$ **d.** $12.2 million

67. False   69. True   71. True

Using Technology Exercises 2.2, page 89

Graphing Utility

1.
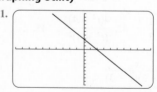

3.
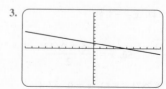

5. a.

b.

7. a.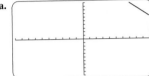

b.

9.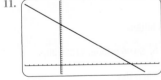

11.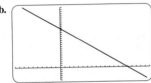

Excel

1.

$$3.2x + 2.1y - 6.72 = 0$$

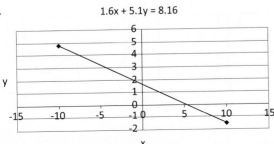

3.

$$1.6x + 5.1y = 8.16$$

5.

$$12.1x + 4.1y = 49.61$$

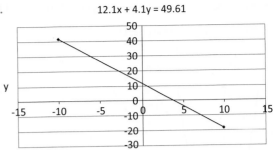

7.

$$20x + 16y = 300$$

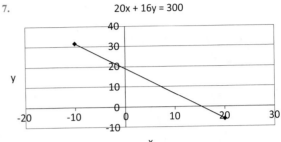

9.

$$20x + 30y = 600$$

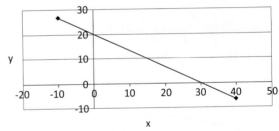

11.

$$22.4x + 16.1y - 352 = 0$$

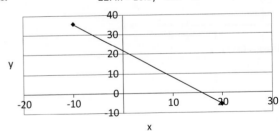

Exercises 2.3, page 97

1. $21, -9, 5a + 6, -5a + 6, 5a + 21$

3. $3, 12, 3a^2 - 6a + 3, 3a^2 + 6a + 3, 3x^2$

5. $2a + 2h + 5, -2a + 5, 2a^2 + 5, 2a - 4h + 5, 4a - 2h + 5$

7. $\dfrac{8}{15}, 0, \dfrac{2a}{a^2 - 1}, \dfrac{2(2 + a)}{a^2 + 4a + 3}, \dfrac{2(t + 1)}{t(t + 2)}$

9. $8, \dfrac{2a^2}{\sqrt{a - 1}}, \dfrac{2(x + 1)^2}{\sqrt{x}}, \dfrac{2(x - 1)^2}{\sqrt{x - 2}}$ 11. $10, 1, 1$ 13. $\frac{5}{2}, 3, 3, 9$

15. **a.** -2 **b.** (i) $x = 2$; (ii) $x = 1$ **c.** $[0, 6]$ **d.** $[-2, 6]$

17. Yes   19. Yes   21. 7 23. $(-\infty, \infty)$

25. $(-\infty, 0) \cup (0, \infty)$

27. $(-\infty, \infty)$ 29. $(-\infty, 5]$ 31. $(-\infty, -2) \cup (-2, 2) \cup (2, \infty)$

33. $[-3, \infty)$ 35. $(-\infty, -2) \cup (-2, 1]$

37. **a.** $(-\infty, \infty)$
    b. $6, 0, -4, -6, -\frac{25}{4}, -6, -4, 0$
    c.

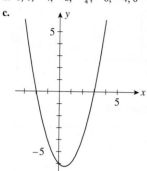

39.

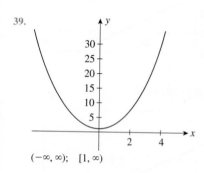

$(-\infty, \infty); \quad [1, \infty)$

41.

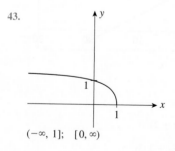

$[0, \infty); \quad [2, \infty)$

43.

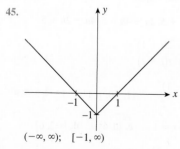

$(-\infty, 1]; \quad [0, \infty)$

45.

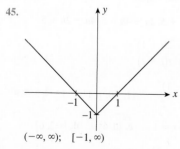

$(-\infty, \infty); \quad [-1, \infty)$

47.

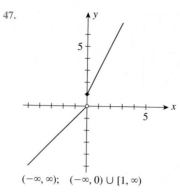

$(-\infty, \infty); \quad (-\infty, 0) \cup [1, \infty)$

49.

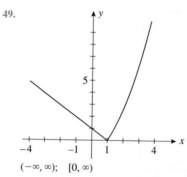

$(-\infty, \infty); \quad [0, \infty)$

51. Yes   53. No   55. Yes   57. Yes

59. 10π in.   61. 8

63. **a.** $f(t) = \begin{cases} 0.0185t + 0.58 & \text{if } 0 \le t \le 20 \\ 0.015t + 0.65 & \text{if } 20 < t \le 30 \end{cases}$

    b. 0.0185/year from 1960 through 1980; 0.015/year from 1980 through 1990
    c. 1983

65. 20; 26   67. $5.6 billion; $7.8 billion

69. **a.** $0.6 trillion; $0.6 trillion   **b.** $0.96 trillion; $1.2 trillion

71. **a.** 130 tons/day; 100 tons/day; 40 tons/day
    b.

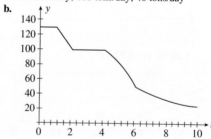

73. False   75. False

Using Technology Exercises 2.3, page 104

1. **a.**   **b.**

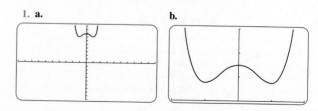

3. a.

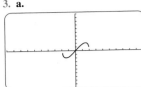

b.

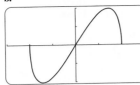

5.

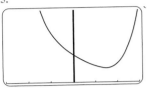

7.

9. 18.5505   **11.** 4.1616

13. a.

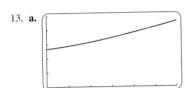

b. $9.4 billion, $13.9 billion

15. a.

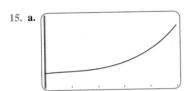

b. 44.7; 52.7; 129.2

Exercises 2.4, page 110

1. $f(x) + g(x) = x^3 + x^2 + 3$ **3.** $f(x)g(x) = x^5 - 2x^3 + 5x^2 - 10$

5. $\dfrac{f(x)}{g(x)} = \dfrac{x^3 + 5}{x^2 - 2}$ **7.** $\dfrac{f(x)g(x)}{h(x)} = \dfrac{x^5 - 2x^3 + 5x^2 - 10}{2x + 4}$

9. $f(x) + g(x) = x - 1 + \sqrt{x + 1}$

11. $f(x)g(x) = (x - 1)\sqrt{x + 1}$

13. $\dfrac{g(x)}{h(x)} = \dfrac{\sqrt{x + 1}}{2x^3 - 1}$ **15.** $\dfrac{f(x)g(x)}{h(x)} = \dfrac{(x - 1)\sqrt{x + 1}}{2x^3 - 1}$

17. $\dfrac{f(x) - h(x)}{g(x)} = \dfrac{x - 2x^3}{\sqrt{x + 1}}$

19. $f(x) + g(x) = x^2 + \sqrt{x} + 3$;

$f(x) - g(x) = x^2 - \sqrt{x} + 7$;

$f(x)g(x) = (x^2 + 5)(\sqrt{x} - 2); \dfrac{f(x)}{g(x)} = \dfrac{x^2 + 5}{\sqrt{x} - 2}$

21. $f(x) + g(x) = \dfrac{(x - 1)\sqrt{x + 3} + 1}{x - 1}$;

$f(x) - g(x) = \dfrac{(x - 1)\sqrt{x + 3} - 1}{x - 1}$;

$f(x)g(x) = \dfrac{\sqrt{x + 3}}{x - 1}; \dfrac{f(x)}{g(x)} = (x - 1)\sqrt{x + 3}$

23. $f(x) + g(x) = \dfrac{2(x^2 - 2)}{(x - 1)(x - 2)}$;

$f(x) - g(x) = \dfrac{-2x}{(x - 1)(x - 2)}$;

$f(x)g(x) = \dfrac{(x + 1)(x + 2)}{(x - 1)(x - 2)}; \dfrac{f(x)}{g(x)} = \dfrac{(x + 1)(x - 2)}{(x - 1)(x + 2)}$

25. $f(g(x)) = x^4 + 3x^2 + 3; g(f(x)) = (x^2 + x + 1)^2 + 1$

27. $f(g(x)) = \sqrt{x^2 - 1} + 1; g(f(x)) = x + 2\sqrt{x}$

29. $f(g(x)) = \dfrac{x}{x^2 + 1}; g(f(x)) = \dfrac{x^2 + 1}{x}$ **31.** 49

33. $\dfrac{\sqrt{5}}{5}$ **35.** $f(x) = 2x^3 + x^2 + 1$ and $g(x) = x^5$

37. $f(x) = x^2 - 1$ and $g(x) = \sqrt{x}$

39. $f(x) = x^2 - 1$ and $g(x) = \dfrac{1}{x}$

41. $f(x) = 3x^2 + 2$ and $g(x) = \dfrac{1}{x^{3/2}}$ **43.** $3h$

45. $-h(2a + h)$ **47.** $2(2a + h)$

49. $3a^2 + 3ah + h^2 - 1$ **51.** $-\dfrac{1}{a(a + h)}$

53. The total revenue in dollars from both restaurants at time t

55. The value in dollars of Nancy's shares of IBM at time t

57. The carbon monoxide pollution from cars in parts per million at time t

59. $C(x) = 0.6x + 12,100$

61. $0.0075t^2 + 0.13t + 0.17$; D gives the difference in year t between the deficit without the rescue package and the deficit with the rescue package.

63. a. 23; In 2002, 23% of reported serious crimes ended in the arrests or in the identification of the suspects.
b. 18; In 2007, 18% of reported serious crimes ended in the arrests or in the identification of the suspects.

65. a. $P(x) = -0.000003x^3 - 0.07x^2 + 300x - 100,000$
b. $182,375

67. a. $3.5t^2 + 2.4t + 71.2$ **b.** 71,200; 109,900

69. a. 55%; 98.2%   **b.** $444,700; $1,167,600

71. a. $s(x) = f(x) + g(x) + h(x)$

73. True   **75.** False

Exercises 2.5, page 119

1. Yes; $y = -\tfrac{2}{3}x + 2$ **3.** Yes; $y = \tfrac{1}{2}x + 2$

5. Yes; $y = \tfrac{1}{2}x + \tfrac{9}{4}$ **7.** No   **9.** No

11. a. $C(x) = 8x + 40,000$ **b.** $R(x) = 12x$
c. $P(x) = 4x - 40,000$
d. Loss of $8000; profit of $8000

13. $m = -2$; $b = 4$ **15.** $(3, 13)$

17. $(4, \frac{2}{3})$ **19.** $(-4, -6)$

21. 1000 units; $15,000   **23.** 600 units; $240

25. $900,000; $800,000

27. a. $y = 1.033x$ **b.** $1570.16

29. $C(x) = 0.6x + 12,100$; $R(x) = 1.15x$;
$P(x) = 0.55x - 12,100$

31. a. $12,000/year   **b.** $V = 60,000 - 12,000t$

c.

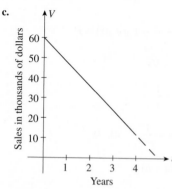

d. $24,000

33. $900,000; $800,000

35. a. $m = a/1.7$; $b = 0$ **b.** 117.65 mg

37. a. $f(t) = 6.5t + 20\ (0 \le t \le 8)$ **b.** 72 million

39. a. $F = \frac{9}{5}C + 32$ **b.** 68°F   **c.** 21.1°C

41. a.   **b.** 8000 units; $112,000

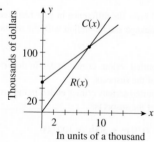

c.   **d.** $(8000, 0)$

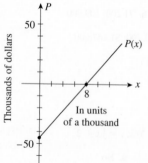

43. 9259 units; $83,331

45. a. $C_1(x) = 18,000 + 15x$
$C_2(x) = 15,000 + 20x$

b.

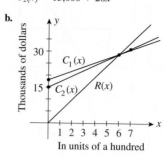

In units of a hundred

c. Machine II; Machine II; Machine I
d. ($1500); $1500; $4750

47. Middle of 2003

49. a.

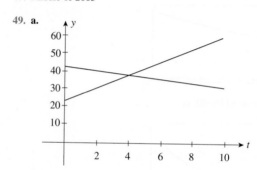

b. Feb. 2005

51. True

Using Technology Exercises 2.5, page 126

1. 2.2875   **3.** 2.880952381   **5.** 7.2851648352
7. 2.4680851064

Exercises 2.6, page 132

1. Vertex: $(-\frac{1}{2}, -\frac{25}{4})$; x-intercepts: $-3, 2$

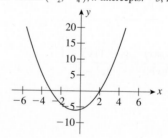

3. Vertex: $(2, 0)$; x-intercept: 2

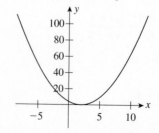

5. Vertex: $\left(\frac{5}{2}, \frac{1}{4}\right)$; x-intercepts: 2, 3

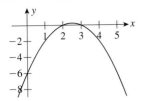

7. Vertex: $\left(\frac{5}{6}, -\frac{13}{12}\right)$; x-intercepts: 0.2324, 1.4343

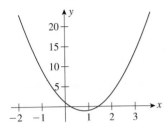

9. Vertex: $\left(\frac{3}{4}, \frac{15}{8}\right)$; no x-intercepts

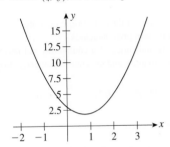

11. Vertex: $(0, -4)$; x-intercepts: ± 2

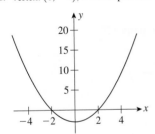

13. Vertex: $(0, 16)$; x-intercepts: ± 4

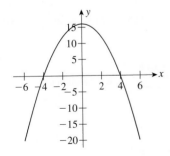

15. Vertex: $\left(\frac{8}{3}, -\frac{2}{3}\right)$; x-intercepts: $\frac{4}{3}$, 4

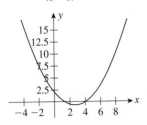

17. Vertex: $\left(-\frac{4}{3}, -\frac{10}{3}\right)$; x-intercepts: $\frac{1}{3}$, -3

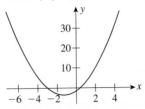

19. $(2, 0)$; $(-3, -5)$ **21.** $(-2, -2)$; $(3, 3)$

23. $(-1.1205, 0.1133)$, $(2.3205, -8.8332)$

25. a.   **b.** 5000

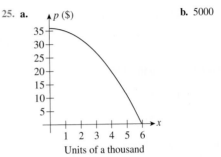

Units of a thousand

27. a.   **b.** $26

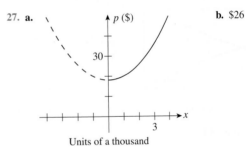

Units of a thousand

29. 2500; $67.50   **31.** 11,000; $3

33. a. 3.6 million; 9.5 million   **b.** 11.2 million

35. a.   **b.** $t = 2$; 144 ft

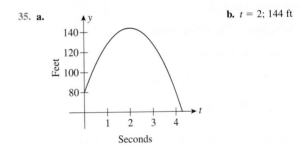

Seconds

37. 3000

39. a.   **b.** $30

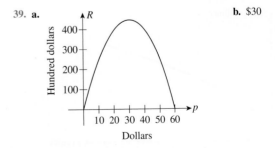

Dollars

41. 500; $32.50

43. a.

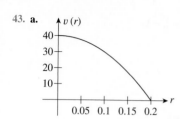

b. 0, 0.2. The velocity of blood is greatest along the central artery ($r = 0$) and smallest along the wall of the artery ($r = 0.2$). The maximum velocity is $v(0) = 40$ cm/sec, and the minimum velocity is $v(0.2) = 0$ cm/sec.

45. $\dfrac{28}{\pi + 4}$ ft by $\dfrac{28}{\pi + 4}$ ft   **47.** True   **49.** True

51. True

Using Technology Exercises 2.6, page 136

1. $(-3.0414, 0.1503)$; $(3.0414, 7.4497)$

3. $(-2.3371, 2.4117)$; $(6.0514, -2.5015)$

5. $(-1.1055, -6.5216)$; $(1.1055, -1.8784)$

7. a.

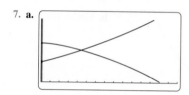

b. 438 wall clocks; $40.92

Exercises 2.7, page 144

1. Polynomial function; degree 6

3. Polynomial function; degree 6

5. Some other function

7. a. 59.7 million   **b.** 152.54 million

9. a. $2360   **b.** $4779.04

11. a. $0.72 trillion   **b.** $3.513 trillion

13. a. $R(x) = \dfrac{100x}{40 + x}$ **b.** 60%

15. a. 0.06 million terabytes   **b.** 1.122 million terabytes

17. a. 320,000   **b.** 3,923,200

19. a.

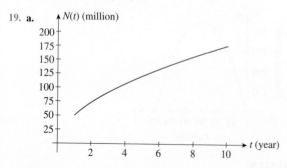

b. 176,610,000

21. 582,650; 1,605,590   **23.** $699; $130

25. $\dfrac{110}{\frac{1}{2}t + 1} - 26\left(\dfrac{1}{4}t^2 - 1\right)^2 - 52$; $32, $6.71, $3; the gap was closing.

27. $5 billion, $152 billion

29. a.

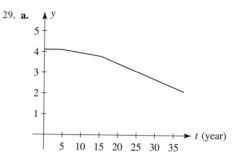

b. 3.95; 3.05   **c.** 1995 to 2000   **d.** 2010 through 2030

31. $p = \sqrt{-x^2 + 100}$; 6614

33. a. $\dfrac{b - d}{c - a}$; $\dfrac{bc - ad}{c - a}$

b. If the unit price is increased, then the equilibrium quantity decreases while the equilibrium price increases.

c. If the upper bound for the unit price of a commodity is lowered, then both the equilibrium quantity and the equilibrium price drop.

35. $f(x) = 2x + \dfrac{500}{x}$; $x > 0$ **37.** $f(x) = 0.5x^2 + \dfrac{8}{x}$; $x > 0$

39. $f(x) = (22 + x)(36 - 2x)$ bushels/acre

41. a. $P(x) = -0.0002x^2 + 3x + 50,000$ **b.** $60,800

43. False   **45.** False

Using Technology Exercises 2.7, page 150

1. a. $f(t) = 1.85t + 16.9$
b.

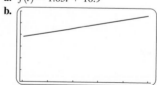

c.

t	1	2	3	4	5	6
y	18.8	20.6	22.5	24.3	26.2	28.0

d. 31.7 gallons

3. a. $f(t) = -0.221t^2 + 4.14t + 64.8$
b.

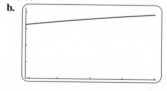

c. 77.8 million

5. a.

b. $f(t) = 2.94t^2 + 38.75t + 188.5$ **c.** $870 billion

7. **a.** $f(t) = -0.00081t^3 + 0.0206t^2 + 0.125t + 1.69$

b.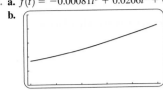

c. $1.8 trillion; $2.7 trillion; $4.2 trillion

9. **a.** $-0.0056t^3 + 0.112t^2 + 0.51t + 8$

b.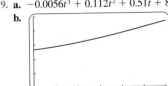

c. $8 billion, $10.4 billion, $13.9 billion

11. **a.** $f(t) = -2.4167t^3 + 24.5t^2 - 123.33t + 506$

b.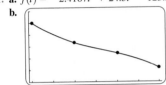

c. 506,000; 338,000; 126,000

13. **a.** $f(t) = 0.00125t^4 - 0.0051t^3 - 0.0243t^2 + 0.129t + 1.71$

b.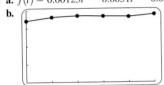

c. 1.71 mg; 1.81 mg; 1.85 mg; 1.84 mg; 1.83 mg; 1.89 mg
d. 2.13 mg/cigarette

Chapter 2 Concept Review Questions, page 153

1. ordered; abscissa (x-coordinate); ordinate (y-coordinate)

2. **a.** x; y **b.** third

3. **a.** $\dfrac{y_2 - y_1}{x_2 - x_1}$ **b.** undefined   **c.** 0   **d.** positive

4. $m_1 = m_2$; $m_1 = -\dfrac{1}{m_2}$

5. **a.** $y - y_1 = m(x - x_1)$; point-slope form
   b. $y = mx + b$; slope-intercept

6. **a.** $Ax + By + C = 0$ (A, B, not both zero)   **b.** $-\dfrac{a}{b}$

7. domain; range; B 8. domain, $f(x)$; vertical, point

9. $f(x) \pm g(x)$; $f(x)g(x)$; $\dfrac{f(x)}{g(x)}$; $A \cap B$; $A \cap B$; 0

10. $g[f(x)]$; f; $f(x)$; g

11. $ax^2 + bx + c$; parabola; upward; downward; vertex; $\dfrac{-b}{2a}$; $x = \dfrac{-b}{2a}$

12. **a.** $P(x) = a_n x^n + a_{n-1} x^{n-1} + \cdots + a_1 x + a_0$
   ($a_n \neq 0$; n, a positive integer)
   b. linear; quadratic
   c. quotient; polynomials
   d. x^r (r, a real number)

Chapter 2 Review Exercises, page 154

1. $x = -2$ 2. $y = 4$ 3. $y = -\frac{1}{10}x + \frac{19}{5}$

4. $y = -\frac{4}{5}x + \frac{12}{5}$ 5. $y = \frac{5}{2}x + 9$ 6. $y = \frac{3}{4}x + \frac{11}{2}$

7. $y = -\frac{1}{2}x - 3$ 8. $\frac{3}{5}$; $-\frac{6}{5}$ 9. $y = -\frac{3}{4}x + \frac{9}{2}$

10. $y = -\frac{3}{5}x + \frac{12}{5}$ 11. $y = -\frac{3}{2}x - 7$

12.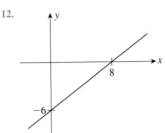

13.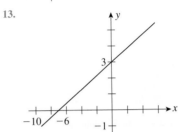

14. $(-\infty, 9]$ 15. $(-\infty, -1) \cup \left(-1, \frac{3}{2}\right) \cup \left(\frac{3}{2}, \infty\right)$

16. **a.** 0   **b.** $3a^2 + 17a + 20$ **c.** $12a^2 + 10a - 2$
   d. $3a^2 + 6ah + 3h^2 + 5a + 5h - 2$

17. **a.** From 1985 to 1990
   b. From 1990 on
   c. 1990; $3.5 billion

18. **a.** 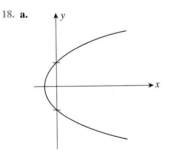   **b.** No   **c.** Yes

19.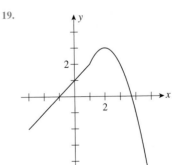

20. **a.** $\dfrac{2x + 3}{x}$ **b.** $\dfrac{1}{x(2x + 3)}$

   c. $\dfrac{1}{2x + 3}$ **d.** $\dfrac{2}{x} + 3$

21. Vertex: $\left(\frac{11}{12}, -\frac{361}{24}\right)$; x-intercepts: $-\frac{2}{3}, \frac{5}{2}$

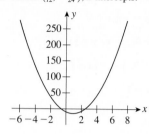

22. Vertex: $\left(\frac{1}{2}, 4\right)$; x-intercepts: $-\frac{1}{2}, \frac{3}{2}$

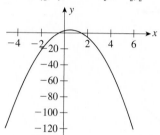

23. $(2, -3)$ 24. $\left(6, \frac{21}{2}\right)$ 25. $\left(-2, \frac{1}{3}\right)$

26. $(2500, 50{,}000)$ 27. L_2 28. L_2

29. $60{,}000$ 30. **a.** $f(x) = x + 2.4$ **b.** $\$5.4$ million

31. **a.** $C(x) = 6x + 30{,}000$ **b.** $R(x) = 10x$
    c. $P(x) = 4x - 30{,}000$ **d.** $(\$6{,}000)$; $\$2000$; $\$18{,}000$

32. **a.** $\$200{,}000$/year   **b.** $\$4{,}000{,}000$

33. $p = -0.05x + 200$

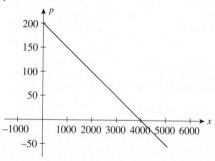

34. $p = \frac{1}{36}x + \frac{400}{9}$ 35. 117 mg

36. **a.** $y = 0.25x$ **b.** 1600   37. $\$45{,}000$

38. 400; 800   39. 990; 2240

40.

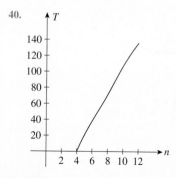

As the length of the list increases, the time taken to learn the list increases by a very large amount.

41. $648{,}000$; $901{,}900$; $1{,}345{,}200$; $1{,}762{,}800$

42. **a.** $\$16.4$ billion; $\$17.6$ billion; $\$18.3$ billion; $\$18.8$ billion; $\$19.3$ billion

b.

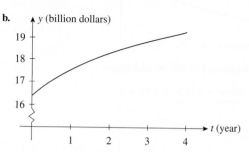

43. **a.** $f(t) = 267$; $g(t) = 2t^2 + 46t + 733$
    b. $f(t) + g(t) = 2t^2 + 46t + 1000$ **c.** 1936 tons

44. 5000; $\$20$

45. **a.** $r = f(V) = \sqrt[3]{(3V)/(4\pi)}$ **b.** $g(t) = \frac{9}{2}\pi t$
    c. $h(t) = \frac{3}{2}\sqrt[3]{t}$ **d.** 3 ft

46. **a.** 59.8%; 58.9%; 59.2%; 60.7%, 61.7%

b.

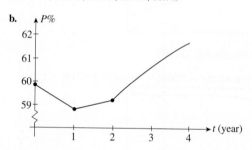

c. 60.7%

47. $x(20 - 2x)^2$

48. $100x^2 + \dfrac{1350}{x}$

Chapter 2 Before Moving On, page 156

1. $y = \frac{7}{5}x - \frac{3}{5}$ 2. $y = -\frac{1}{3}x + \frac{4}{3}$

3. **a.** 3   **b.** 2   **c.** $\frac{17}{4}$

4. **a.** $\dfrac{1}{x + 1} + x^2 + 1$ **b.** $\dfrac{x^2 + 1}{x + 1}$

   c. $\dfrac{1}{x^2 + 2}$ **d.** $\dfrac{1}{(x + 1)^2} + 1$

5. $V(x) = 108x^2 - 4x^3$

CHAPTER 3

Exercises 3.1, page 162

1. **a.** 16   **b.** 27   3. **a.** 3   **b.** $\sqrt{5}$

5. **a.** -3 **b.** 8   7. **a.** $4x^3$ **b.** $5x^{3/2}y^2$

9. **a.** $\dfrac{2}{a}$ **b.** $\frac{1}{3}b^2$ 11. **a.** $8x^9y^6$ **b.** $16x^4y^4z^6$

13. 3   15. 3   17. 3   19. $\frac{5}{4}$ 21. 1 or 2

23.

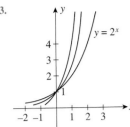

$y = 2^x$

25.

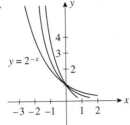

$y = 2^{-x}$

27.

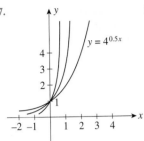

$y = 4^{0.5x}$

29.

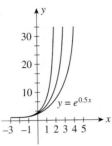

$y = e^{0.5x}$

31.

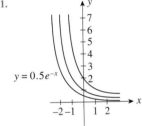

$y = 0.5e^{-x}$

33. $f(x) = 100(\frac{6}{5})^x$ **35.** 54.6

37. a. 26.30%; 24.67%; 21.71%; 19.72%

b.

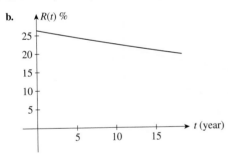

39. a.

Year	0	1	2	3	4	5
Number of Web Addresses (billions)	0.45	0.80	1.41	2.49	4.39	7.76

b.

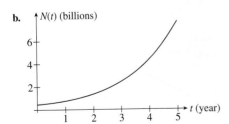

41. 34,210,000

43. a. 0.08 g/cm³   **b.** 0.12 g/cm³

45. False   **47.** True

Using Technology Exercises 3.1, page 165

1.

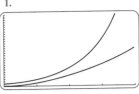

3.

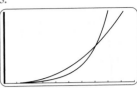

5.

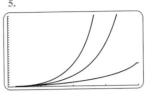

7.

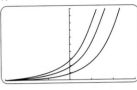

9.

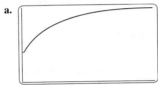

11. a.

b. 0.08 g/cm³   **c.** 0.12 g/cm³

13. a.

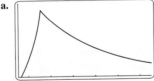

b. 20 sec   **c.** 35.1 sec

Exercises 3.2, page 172

1. $\log_2 64 = 6$ **3.** $\log_4 \frac{1}{16} = -2$ **5.** $\log_{1/3} \frac{1}{3} = 1$

7. $\log_{32} 16 = \frac{4}{5}$ **9.** $\log_{10} 0.001 = -3$ **11.** 1.0792

13. 1.2042   **15.** 1.6813   **17.** $\ln a^2 b^3$ **19.** $\ln \dfrac{3\sqrt{xy}}{\sqrt[3]{z}}$

21. $\log x + 4 \log (x + 1)$ **23.** $\frac{1}{2} \log (x + 1) - \log (x^2 + 1)$

25. $\ln x - x^2$ **27.** $-\frac{3}{2} \ln x - \frac{1}{2} \ln (1 + x^2)$

29. $x = 8$ **31.** $x = 3$ **33.** $x = 10$ **35.** $x = \frac{11}{2}$

37. $x = \frac{16}{7}$ **39.** $x = \frac{11}{3}$ **41.** $x = 3$

43.

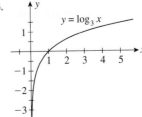

$y = \log_3 x$

45.

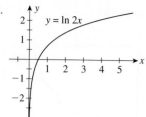

$y = \ln 2x$

47.

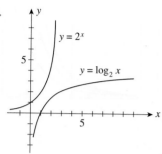

$y = 2^x$

$y = \log_2 x$

49. 5.1986   **51.** −0.0912   **53.** −8.0472   **55.** −4.9041

57. $-2 \ln \left(\dfrac{A}{B} \right)$ **59.** $f(x) = 2 + 2.8854 \ln x$ **61.** 106 mm

63. a. $10^3 I_0$ **b.** 100,000 times greater
c. 10,000,000 times greater

65. 27.4 years   **67.** 6.4 years   **69. a.** 9.1 sec   **b.** 20.3 sec

71. False   **73. a.** ln 2

Exercises 3.3, page 181

1. a. 0.02   **b.** 300
c.

t	0	10	20	100	1000
Q	300	366	448	2217	1.46×10^{11}

3. a. $Q(t) = 100e^{0.035t}$ **b.** 266 min   **c.** $Q(t) = 1000e^{0.035t}$

5. a. 54.93 years   **b.** 14.25 billion

7. 8.7 lb/in.²

9. $Q(t) = 100e^{-0.049t}$; 70.7 g

11. 13,412 years ago

13.

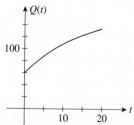

$Q(t)$

a. 60 words/min   **b.** 107 words/min   **c.** 136 words/min

15. $5.81 trillion; $8.57 trillion

17.

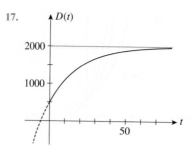

$D(t)$

a. 573 computers/month; 1177 computers/month;
1548 computers/month; 1925 computers/month
b. 2000 computers/month

19. 135.1 cm   **21.** 86%

23. 76.4 million; 85.0 million   **25.** 1080

Using Technology Exercises 3.3, page 185

1.

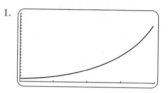

3. a.

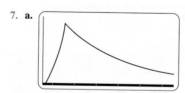

b. 666 million, 1006.6 million

5. a.

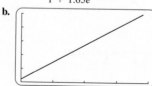

b. 325 million

7. a.

b. 0   **c.** 0.237 g/cm³
d. 0.760 g/cm³

9. a. $f(t) = \dfrac{544.61}{1 + 1.65e^{-0.1846t}}$

b.

Chapter 3 Concept Review, page 186

1. power; 0; 1; exponential

2. **a.** $(-\infty, \infty)$; $(0, \infty)$ **b.** $(0, 1)$; left; right

3. **a.** $(0, \infty)$; $(-\infty, \infty)$; $(1, 0)$ **b.** falls; rises

4. **a.** x **b.** x

5. **a.** initially; growth   **b.** decay   **c.** time; one half

6. **a.** learning curve; C
   b. logistic growth model; A, carrying capacity

Chapter 3 Review Exercises, page 187

1.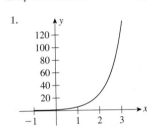

2.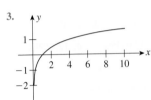

3.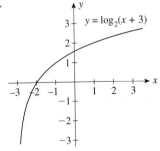

4.

5. $\log_3 81 = 4$ 6. $\log_9 3 = \frac{1}{2}$ 7. $\log_{2/3} \frac{27}{8} = -3$

8. $\log_{16} 0.125 = -\frac{3}{4}$ 9. 3.4011   10. 2.1972   11. 1.2809

12. 4.3174   13. $x + y + z$ 14. $x + 2y - z$ 15. $y + 2z$

16. $x = 3$ 17. $x = -2, 1$ 18. $x = -5$ 19. $x = -2, 3$

20. $x = \frac{15}{2}$ 21. $x = 2$ 22. $x \approx 1.1610$ 23. $x \approx -0.9464$

24. $x \approx -2.5025$ 25. $x \approx -1.2528$ 26. $x \approx 2.8332$

27. $x \approx 1.8195$ 28. $x \approx 0.2409$ 29. $x \approx 33.8672$

30. $x \approx \pm 1.8934$ 31. $x \approx 2.5903$ 32. $x = -9.1629$

33. $x \approx 8.9588$ 34. $x \approx 3.4657$ 35. $x \approx -9.1629$

36.

$y = \log_2(x + 3)$

37.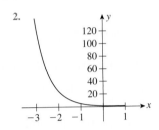

$y = \log_3 (x + 1)$

38. **a.** $Q(t) = 2000e^{0.01831t}$ **b.** 162,000   39. $k \approx 0.0004$

40.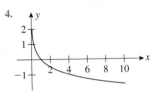

$D(t)$ **a.** 1175; 2540; 3289   **b.** 4000

41. 970
42. 12.5/1000 live births; 9.3/1000 live births; 6.9/1000 live births
43. **a.** 0 g/cm³   **b.** 0.0361 g/cm³

Chapter 3 Before Moving On, page 188

1. $\dfrac{12}{x^6}$ 2. $x = \ln 3$ 3. $x = 0$ or 8

4. -0.9589 5. 8.7 min

CHAPTER 4

Exercises 4.1, page 202

1. $80; $580   3. $836   5. $1000   7. $243\frac{1}{3}$ days

9. 10%/year   11. $1718.19   13. $4974.47

15. $27,566.93   17. $261,751.04   19. $214,986.69

21. 10.25%/year   23. 8.3%/year   25. $31,576.37

27. $30,255.95   29. $6885.64   31. 2.2 years   33. 7.7 years

35. 6.08%/year   37. 2.06 years   39. 24%/year

41. $123,000   43. 5%/year   45. 2.39%/year   47. $852.21

49. $255,256   51. $2.58 million   53. $22,163.75   55. $26,267.49

57. **a.** $34,626.88   **b.** $33,886.16   **c.** $33,506.76

59. Acme Mutual Fund   61. $23,329.48   63. $5994.86

65. $339.79 billion   67. Investment A

69. $33,885.14; $33,565.38   71. $80,000e^{(\sqrt{t}/2 - 0.09t)}$; $151,718

75. 4.2%   77. 5.83%   79. True   81. True

Using Technology Exercises 4.1, page 208

1. $5872.78   3. $475.49   5. 8.95%/year

7. 10.20%/year   9. $29,743.30   11. $53,303.25

Exercises 4.2, page 215

1. $15,937.42   3. $54,759.35   5. $37,965.57

7. $137,209.97   9. $31,048.97   11. $15,558.61

13. $15,011.29   15. $68,642.95   17. $455.70

19. $44,526.45   21. Karen   23. $9850.12   25. $608.54

27. Between $383,242 and $469,053

29. Between $307,014 and $373,768   31. $17,887.61

33. $70,814.38   35. False

Using Technology Exercises 4.2, page 219

1. $59,622.15   3. $8453.59   5. $35,607.23

7. $13,828.60

9. a.

b. $35,038.78/year

Exercises 4.3, page 228

1. $14,902.95   3. $444.24   5. $622.13

7. $798.70   9. $1491.19   11. $516.76

13. $172.95   15. $2348.83   17. $3450.87

19. $16,274.54   21. a. $212.27   b. $1316.36; $438.79

23. a. $456.33; $352.28   b. $1427.88; $1909.44

25. $1438.92; $46,670; $69,154; $140,392   27. $60,982.31

29. $3135.48   31. $328.67   33. $447.98

35. $1454.19; $2868.08   37. $1,111.63   39. $24,639.53

41. $33,835.20   43. $212.77   45. $167,341.59

47. $1957.80; $257,135.23

49. a. $1681.71   b. $194,282.85   c. $1260.11   d. $421.60

51. $1000.61   53. $29,658   55. $640,569.40

Using Technology Exercises 4.3, page 234

1. $628.02   3. $1685.47   5. $1960.96

7. $894.12   9. $18,288.92

Exercises 4.4, page 241

1. 30   3. $-\frac{9}{2}$ 5. $-3, 8, 19, 30, 41$ 7. $x + 6y$

9. 795   11. 792   13. 671

15. a. 275   b. -280 17. At the beginning of the 37th week

19. $15.80   21. b. $800   23. GP; 256; 508   25. Not a GP

27. GP; $\frac{1}{3}$; $\frac{1093}{3}$ 29. 3; 0   31. 293,866

33. $56,284   35. Annual raise of 8%/year

37. a. $20,113.57   b. $87,537.38   39. $25,165.82

41. $39,321.60; $110,678.40   43. True

Chapter 4 Concept Review Questions, page 243

1. a. original; $P(1 + rt)$ b. interest; $P(1 + i)^n$; $A(1 + i)^{-n}$

2. simple; one; nominal; m; $\left(1 + \dfrac{r}{m}\right)^m - 1$

3. annuity; ordinary annuity; simple annuity

4. $R\left[\dfrac{(1 + i)^n - 1}{i}\right]$; $R\left[\dfrac{1 - (1 + i)^{-n}}{i}\right]$

5. $\dfrac{Pi}{1 - (1 + i)^{-n}}$ 6. future; $\dfrac{iS}{(1 + i)^n - 1}$

7. constant d; $a + (n - 1)d$; $\dfrac{n}{2}[2a + (n - 1)d]$

8. constant r; ar^{n-1}; $\dfrac{a(1 - r^n)}{1 - r}$

Chapter 4 Review Exercises, page 244

1. a. $7320.50   b. $7387.28   c. $7422.53
   d. $7446.77

2. a. $19,859.95   b. $20,018.07   c. $20,100.14
   d. $20,156.03

3. a. 12%   b. 12.36%   c. 12.5509%   d. 12.6825%

4. a. 11.5%   b. 11.8306%   c. 12.0055%   d. 12.1259%

5. $30,000.29   6. $39,999.95   7. $5557.68

8. $23,221.71   9. $7861.70   10. $173,804.43

11. $694.49   12. $318.93   13. $332.73   14. $208.44

15. 7.442%   16. 10.034%   17. $80,000

18. $2,592,702; $8,612,002   19. $5,491,922   20. $2982.73

21. $15,000   22. $5000   23. 7.6%   24. $218.64

25. $73,178.41   26. $13,026.89   27. $2000

28. a. $965.55   b. $227,598   c. $42,684

29. a. $1217.12   b. $99,081.60   c. $91,367

30. $19,573.56   31. $4727.67   32. $205.09; 20.27%/year

33. $2203.83

Chapter 4 Before Moving On, page 245

1. $2540.47   2. 6.18%/year   3. $569,565.47   4. $1213.28

5. $35.13   6. a. 210   b. 127.5

CHAPTER 5

Exercises 5.1, page 253

1. Unique solution; (2, 1)   3. No solution

5. Unique solution; (3, 2)

7. Infinitely many solutions; $\left(t, \frac{2}{5}t - 2\right)$; t, a parameter

9. Unique solution; (1, -2)

11. No solution   13. Unique solution; $\left(\frac{1}{2}, \frac{1}{2}\right)$

15. Infinitely many solutions; $\left(2t + \frac{2}{3}, t\right)$ 17. $k = -2$

19. $\begin{aligned}x+\quad y&=\quad 500\\42x+30y&=18{,}600\end{aligned}$ **21.** $\begin{aligned}x+\ y&=100\\5x+6y&=560\end{aligned}$

23. $\begin{aligned}x+\quad y&=1000\\0.5x+1.5y&=1300\end{aligned}$

25. $\begin{aligned}0.06x+0.08y+0.12z&=21{,}600\\2x\quad\quad\ -\ z&=\quad 0\\0.08y-0.12z&=\quad 0\end{aligned}$

27. $\begin{aligned}18x+20y+24z&=26{,}400\\4x+\ 4y+\ 3z&=\ 4{,}900\\5x+\ 4y+\ 6z&=\ 6{,}200\end{aligned}$

29. $\begin{aligned}18{,}000x+27{,}000y+36{,}000z&=2{,}250{,}000\\x-\quad\quad 2y\quad\quad\quad&=\quad\quad 0\\x+\quad\quad y+\quad\quad z&=\quad\ 100\end{aligned}$

31. $\begin{aligned}10x+\ 6y+\ 8z&=100\\10x+12y+\ 6z&=100\\5x+\ 4y+12z&=100\end{aligned}$

33. True

Exercises 5.2, page 266

1. $\begin{bmatrix}2 & -3 & 7\\3 & 1 & 4\end{bmatrix}$

3. $\begin{bmatrix}0 & -1 & 2 & 5\\2 & 2 & -8 & 4\\0 & 3 & 4 & 0\end{bmatrix}$

5. $\begin{aligned}3x+2y&=-4\\x-\ y&=\ \ 5\end{aligned}$ **7.** $\begin{aligned}x+3y+2z&=4\\2x\quad\quad&=5\\3x-3y+2z&=6\end{aligned}$

9. Yes   **11.** No   **13.** Yes   **15.** No   **17.** No

19. $\begin{bmatrix}1 & 2 & 4\\0 & -5 & -10\end{bmatrix}$ **21.** $\begin{bmatrix}1 & -2 & -3\\0 & 20 & 20\end{bmatrix}$

23. $\begin{bmatrix}1 & 2 & 3 & 6\\0 & -1 & -5 & -7\\0 & -7 & -7 & -14\end{bmatrix}$

25. $\begin{bmatrix}-6 & -11 & 0 & -5\\2 & 4 & 1 & 3\\1 & -2 & 0 & -10\end{bmatrix}$

27. $\begin{bmatrix}3 & 9 & 6\\2 & 1 & 4\end{bmatrix}\xrightarrow{\frac{1}{3}R_1}\begin{bmatrix}1 & 3 & 2\\2 & 1 & 4\end{bmatrix}$

$\xrightarrow{R_2-2R_1}\begin{bmatrix}1 & 3 & 2\\0 & -5 & 0\end{bmatrix}\xrightarrow{-\frac{1}{5}R_2}$

$\begin{bmatrix}1 & 3 & 2\\0 & 1 & 0\end{bmatrix}\xrightarrow{R_1-3R_2}\begin{bmatrix}1 & 0 & 2\\0 & 1 & 0\end{bmatrix}$

29. $\begin{bmatrix}1 & 3 & 1 & 3\\3 & 8 & 3 & 7\\2 & -3 & 1 & -10\end{bmatrix}\xrightarrow[R_3-2R_1]{R_2-3R_1}$

$\begin{bmatrix}1 & 3 & 1 & 3\\0 & -1 & 0 & -2\\0 & -9 & -1 & -16\end{bmatrix}\xrightarrow{-R_2}$

$\begin{bmatrix}1 & 3 & 1 & 3\\0 & 1 & 0 & 2\\0 & -9 & -1 & -16\end{bmatrix}\xrightarrow[R_3+9R_2]{R_1-3R_2}$

$\begin{bmatrix}1 & 0 & 1 & -3\\0 & 1 & 0 & 2\\0 & 0 & -1 & 2\end{bmatrix}\xrightarrow[-R_3]{R_1+R_3}$

$\begin{bmatrix}1 & 0 & 0 & -1\\0 & 1 & 0 & 2\\0 & 0 & 1 & -2\end{bmatrix}$

31. $(2,0)$ **33.** $(-1,2,-2)$ **35.** $(2,1)$

37. $(4,-2)$ **39.** $(-1,2)$ **41.** $\left(\frac{1}{2},\frac{3}{2}\right)$

43. $\left(\frac{1}{2},\frac{1}{4}\right)$ **45.** $\left(\frac{7}{9},-\frac{1}{9},-\frac{2}{3}\right)$ **47.** $(19,-7,-15)$

49. $(3,0,2)$ **51.** $(1,-2,1)$

53. $(-20,-28,13)$ **55.** $(4,-1,3)$

57. 300 acres of corn, 200 acres of wheat

59. In 100 lb of blended coffee, use 40 lb of the $5/lb coffee and 60 lb of the $6/lb coffee.

61. 200 children and 800 adults

63. $40,000 in a savings account, $120,000 in mutual funds, $80,000 in bonds

65. 400 bags of grade A fertilizer; 600 bags of grade B fertilizer; 300 bags of grade C fertilizer

67. 60 compact, 30 intermediate-size, and 10 full-size cars

69. 4 oz of Food I, 2 oz of Food II, 6 oz of Food III

71. 240 front orchestra seats, 560 rear orchestra seats, 200 front balcony seats

73. 7 days in London, 4 days in Paris, and 3 days in Rome

75. False

Using Technology Exercises 5.2, page 271

1. $x_1=3;\ x_2=1;\ x_3=-1;\ x_4=2$

3. $x_1=5;\ x_2=4;\ x_3=-3;\ x_4=-4$

5. $x_1=1;\ x_2=-1;\ x_3=2;\ x_4=0;\ x_5=3$

Exercises 5.3, page 278

1. a. One solution   **b.** $(3,-1,2)$

3. a. One solution   **b.** $(2,5)$

5. a. Infinitely many solutions
   b. $(4-t,-2,t)$; t, a parameter

7. a. No solution

9. a. Infinitely many solutions
   b. $(4,-1,3-t,t)$; t, a parameter

11. a. Infinitely many solutions
   b. $(2-3s,1+s,s,t)$; s,t, parameters

13. $(2, 1)$ 15. No solution   17. $(1, -1)$

19. $(2 + 2t, t)$; t, a parameter   21. $(4 + t, -3 - t, t)$

23. $\left(\frac{4}{3} - \frac{2}{3}t, t\right)$; t, a parameter   25. No solution

27. $\left(-1, \frac{17}{7}, \frac{23}{7}\right)$

29. $\left(1 - \frac{1}{4}s + \frac{1}{4}t, s, t\right)$; s, t, parameters

31. No solution   33. $(2, -1, 4)$

35. $x = 20 + z$, $y = 40 - 2z$; 25 compact cars, 30 mid-sized cars, and 5 full-sized cars; 30 compact cars, 20 mid-sized cars, and 10 full-sized cars

39. $10,000 in money-market account, $60,000 in stocks, and $30,000 in bonds; $20,000 in money-market account, $70,000 in stocks, and $10,000 in bonds

41. **a.**
$$\begin{aligned}
x_1 &\qquad\qquad\quad + x_6 &&= 1700 \\
x_1 - x_2 &\qquad\qquad\qquad\; + x_7 &&= 700 \\
x_2 - x_3 &\qquad\qquad &&= 300 \\
-x_3 + x_4 &\qquad\qquad &&= 400 \\
-x_4 + x_5 &\;\; + x_7 &&= 700 \\
x_5 + x_6 &\qquad\qquad &&= 1800
\end{aligned}$$

b. $(1700 - s, 1000 - s + t, 700 - s + t, 1100 - s + t, 1800 - s, s, t)$; $(900, 1000, 700, 1100, 1000, 800, 800)$; $(1000, 1100, 800, 1200, 1100, 700, 800)$

c. x_6 must have at least 300 cars/hr.

43. $k = -36$; $\left(4 + \frac{2}{3}y - \frac{4}{3}z, y, z\right)$ 45. False

Using Technology Exercises 5.3, page 282

1. $(1 + t, 2 + t, t)$; t, a parameter

3. $\left(-\frac{17}{7} + \frac{6}{7}t, 3 - t, -\frac{18}{7} + \frac{1}{7}t, t\right)$; t, a parameter

5. No solution

Exercises 5.4, page 289

1. 4×4; 4×3; 1×5; 4×1 3. 2; 3; 8

5. D; $D^T = \begin{bmatrix} 1 & 3 & -2 & 0 \end{bmatrix}$ 7. 3×2; 3×2; 3×3; 3×3

9. $\begin{bmatrix} 1 & 6 \\ 6 & -1 \\ 2 & 2 \end{bmatrix}$ 11. $\begin{bmatrix} 1 & 1 & -4 \\ -1 & -8 & 1 \\ 6 & 3 & 1 \end{bmatrix}$

13. $\begin{bmatrix} 5 & 5 & 9 \\ 2 & 10 & 13 \end{bmatrix}$ 15. $\begin{bmatrix} 3 & -4 & -16 \\ 17 & -4 & 16 \end{bmatrix}$

17. $\begin{bmatrix} -1.9 & 3.0 & -0.6 \\ 6.0 & 9.6 & 1.2 \end{bmatrix}$

19. $\begin{bmatrix} \frac{7}{2} & 3 & -1 & \frac{10}{3} \\ -\frac{19}{6} & \frac{2}{3} & -\frac{17}{2} & \frac{23}{3} \\ \frac{29}{3} & \frac{17}{6} & -1 & -2 \end{bmatrix}$

21. $u = 3$, $x = \frac{5}{2}$, $y = 7$, and $z = 2$

23. $x = 2$, $y = 2$, $z = -\frac{7}{3}$, and $u = 15$

31. $\begin{bmatrix} 3 \\ 2 \\ -1 \\ 5 \end{bmatrix}$ 33. $\begin{bmatrix} 1 & 3 & 0 \\ -1 & 4 & 1 \\ 2 & 2 & 0 \end{bmatrix}$

35. $\begin{bmatrix} 220 & 215 & 210 & 205 \\ 220 & 210 & 200 & 195 \\ 215 & 205 & 195 & 190 \end{bmatrix}$ 37. $B = \begin{bmatrix} 350.2 & 370.8 & 391.4 \\ 422.3 & 442.9 & 453.2 \\ 638.6 & 679.8 & 721 \end{bmatrix}$

39. **a.** $D = \begin{bmatrix} 2960 & 1510 & 1150 \\ 1100 & 550 & 490 \\ 1230 & 590 & 470 \end{bmatrix}$

b. $E = \begin{bmatrix} 3256 & 1661 & 1265 \\ 1210 & 605 & 539 \\ 1353 & 649 & 517 \end{bmatrix}$

41.
$$\begin{array}{c} \quad\;\; 2000 \quad 2001 \quad 2002 \\ \begin{array}{c} \text{MA} \\ \text{U.S.} \end{array} \begin{bmatrix} 6.88 & 7.05 & 7.18 \\ 4.13 & 4.09 & 4.06 \end{bmatrix} \end{array}$$

43.
$$\begin{array}{c} \qquad\qquad \text{White} \;\; \text{Black} \;\; \text{Hispanic} \\ \begin{array}{c} \text{Women} \\ \text{Men} \end{array} \begin{bmatrix} 82.6 & 80.5 & 91.2 \\ 78 & 73.9 & 84.8 \end{bmatrix} \end{array}$$

$$\begin{array}{c} \qquad\qquad\;\; \text{Women} \quad \text{Men} \\ \begin{array}{c} \text{White} \\ \text{Black} \\ \text{Hispanic} \end{array} \begin{bmatrix} 82.6 & 78 \\ 80.5 & 73.9 \\ 91.2 & 84.8 \end{bmatrix} \end{array}$$

45. True   47. False

Using Technology Exercises 5.4, page 295

1. $\begin{bmatrix} 15 & 38.75 & -67.5 & 33.75 \\ 51.25 & 40 & 52.5 & -38.75 \\ 21.25 & 35 & -65 & 105 \end{bmatrix}$

3. $\begin{bmatrix} -5 & 6.3 & -6.8 & 3.9 \\ 1 & 0.5 & 5.4 & -4.8 \\ 0.5 & 4.2 & -3.5 & 5.6 \end{bmatrix}$

5. $\begin{bmatrix} 16.44 & -3.65 & -3.66 & 0.63 \\ 12.77 & 10.64 & 2.58 & 0.05 \\ 5.09 & 0.28 & -10.84 & 17.64 \end{bmatrix}$

7. $\begin{bmatrix} 22.2 & -0.3 & -12 & 4.5 \\ 21.6 & 17.7 & 9 & -4.2 \\ 8.7 & 4.2 & -20.7 & 33.6 \end{bmatrix}$

Exercises 5.5, page 302

1. 2×5; not defined   3. 1×1; 7×7

5. $n = s$; $m = t$ 7. $\begin{bmatrix} -1 \\ 3 \end{bmatrix}$

9. $\begin{bmatrix} 13 \\ -10 \end{bmatrix}$ 11. $\begin{bmatrix} 4 & -2 \\ 9 & 13 \end{bmatrix}$

13. $\begin{bmatrix} 2 & 9 \\ 5 & 16 \end{bmatrix}$ 15. $\begin{bmatrix} 0.57 & 1.93 \\ 0.64 & 1.76 \end{bmatrix}$

17. $\begin{bmatrix} 6 & -3 & 0 \\ -2 & 1 & -8 \\ 4 & -4 & 9 \end{bmatrix}$ 19. $\begin{bmatrix} 5 & 1 & -6 \\ 1 & 7 & -4 \end{bmatrix}$

21. $\begin{bmatrix} -4 & -20 & 4 \\ 4 & 12 & 0 \\ 12 & 32 & 20 \end{bmatrix}$ 23. $\begin{bmatrix} 4 & -3 & 2 \\ 7 & 1 & -5 \end{bmatrix}$

27. $AB = \begin{bmatrix} 10 & 7 \\ 22 & 15 \end{bmatrix}$; $BA = \begin{bmatrix} 5 & 8 \\ 13 & 20 \end{bmatrix}$

31. $A = \begin{bmatrix} -2 & -1 \\ 5 & 2 \end{bmatrix}$ **33. b.** No   **35. a.** $A^T = \begin{bmatrix} 2 & 5 \\ 4 & -6 \end{bmatrix}$

37. $AX = B$, where $A = \begin{bmatrix} 2 & -3 \\ 3 & -4 \end{bmatrix}$, $X = \begin{bmatrix} x \\ y \end{bmatrix}$,

and $B = \begin{bmatrix} 7 \\ 8 \end{bmatrix}$

39. $AX = B$, where $A = \begin{bmatrix} 2 & -3 & 4 \\ 0 & 2 & -3 \\ 1 & -1 & 2 \end{bmatrix}$, $X = \begin{bmatrix} x \\ y \\ z \end{bmatrix}$,

and $B = \begin{bmatrix} 6 \\ 7 \\ 4 \end{bmatrix}$

41. $AX = B$, where $A = \begin{bmatrix} -1 & 1 & 1 \\ 2 & -1 & -1 \\ -3 & 2 & 4 \end{bmatrix}$, $X = \begin{bmatrix} x_1 \\ x_2 \\ x_3 \end{bmatrix}$,

and $B = \begin{bmatrix} 0 \\ 2 \\ 4 \end{bmatrix}$

43. a. $AB = \begin{bmatrix} 51,400 \\ 54,200 \end{bmatrix}$

b. The first entry shows that William's total stock holdings are $51,400; the second shows that Michael's stock holdings are $54,200.

45. a. $A = \begin{array}{c} \text{Kaitlin} \\ \text{Emma} \end{array} \begin{bmatrix} \overset{N}{\underset{\text{kroner}}{82}} & \overset{S}{\underset{\text{kronor}}{68}} & \overset{D}{\underset{\text{kroner}}{62}} & \overset{R}{\underset{\text{rubles}}{1200}} \\ 64 & 74 & 44 & 1600 \end{bmatrix}$

b. $B = \begin{array}{c} N \\ S \\ D \\ R \end{array} \begin{bmatrix} 0.1805 \\ 0.1582 \\ 0.1901 \\ 0.0356 \end{bmatrix}$ **c.** Kaitlin: $80.06; Emma: $88.58

47. a. $[90 \ \ 125 \ \ 210 \ \ 55]$; the entries give the respective total number of Model I, II, III, and IV houses built in the three states.

b. $\begin{bmatrix} 300 \\ 120 \\ 60 \end{bmatrix}$; the entries give the respective total number of Model I, II, III, and IV houses built in all three states.

49. $B = \begin{bmatrix} 4 \\ 6 \\ 8 \end{bmatrix}$; $AB = \begin{bmatrix} 1960 \\ 3180 \\ 2510 \\ 3300 \end{bmatrix}$; $10,950

51. $BA = \begin{matrix} \text{Dem} & \text{Rep} & \text{Ind} \\ [41,000 & 35,000 & 14,000] \end{matrix}$

53. $AB = \begin{bmatrix} 1575 & 1590 & 1560 & 975 \\ 410 & 405 & 415 & 270 \\ 215 & 205 & 225 & 155 \end{bmatrix}$

55. $[277.60]$; it represents Cindy's long-distance bill for phone calls to London, Tokyo, and Hong Kong.

57. a. $\begin{bmatrix} 8800 \\ 3380 \\ 1020 \end{bmatrix}$ **b.** $\begin{bmatrix} 8800 \\ 3380 \\ 1020 \end{bmatrix}$ **c.** $\begin{bmatrix} 17,600 \\ 6,760 \\ 2,040 \end{bmatrix}$

59. False   **61.** True

Using Technology Exercises 5.5, page 309

1. $\begin{bmatrix} 18.66 & 15.2 & -12 \\ 24.48 & 41.88 & 89.82 \\ 15.39 & 7.16 & -1.25 \end{bmatrix}$

3. $\begin{bmatrix} 20.09 & 20.61 & -1.3 \\ 44.42 & 71.6 & 64.89 \\ 20.97 & 7.17 & -60.65 \end{bmatrix}$

5. $\begin{bmatrix} 32.89 & 13.63 & -57.17 \\ -12.85 & -8.37 & 256.92 \\ 13.48 & 14.29 & 181.64 \end{bmatrix}$

7. $\begin{bmatrix} 128.59 & 123.08 & -32.50 \\ 246.73 & 403.12 & 481.52 \\ 125.06 & 47.01 & -264.81 \end{bmatrix}$

9. $\begin{bmatrix} 87 & 68 & 110 & 82 \\ 119 & 176 & 221 & 143 \\ 51 & 128 & 142 & 94 \\ 28 & 174 & 174 & 112 \end{bmatrix}$

$\begin{bmatrix} 113 & 117 & 72 & 101 & 90 \\ 72 & 85 & 36 & 72 & 76 \\ 81 & 69 & 76 & 87 & 30 \\ 133 & 157 & 56 & 121 & 146 \\ 154 & 157 & 94 & 127 & 122 \end{bmatrix}$

11. $\begin{bmatrix} 170 & 18.1 & 133.1 & -106.3 & 341.3 \\ 349 & 226.5 & 324.1 & 164 & 506.4 \\ 245.2 & 157.7 & 231.5 & 125.5 & 312.9 \\ 310 & 245.2 & 291 & 274.3 & 354.2 \end{bmatrix}$

Exercises 5.6, page 317

5. $\begin{bmatrix} 3 & -5 \\ -1 & 2 \end{bmatrix}$ **7.** Does not exist

9. $\begin{bmatrix} 2 & -11 & -3 \\ 1 & -6 & -2 \\ 0 & -1 & 0 \end{bmatrix}$ **11.** Does not exist

13. $\begin{bmatrix} -\frac{13}{10} & \frac{7}{5} & \frac{1}{2} \\ \frac{2}{5} & -\frac{1}{5} & 0 \\ -\frac{7}{10} & \frac{3}{5} & \frac{1}{2} \end{bmatrix}$

15. $\begin{bmatrix} 3 & 4 & -6 & 1 \\ -2 & -3 & 5 & -1 \\ -4 & -4 & 7 & -1 \\ -4 & -5 & 8 & -1 \end{bmatrix}$

17. a. $AX = B$, where $A = \begin{bmatrix} 2 & 5 \\ 1 & 3 \end{bmatrix}$, $X = \begin{bmatrix} x \\ y \end{bmatrix}$,

and $B = \begin{bmatrix} 3 \\ 2 \end{bmatrix}$

b. $x = -1; y = 1$

19. a. $AX = B$, where $A = \begin{bmatrix} 2 & -3 & -4 \\ 0 & 0 & -1 \\ 1 & -2 & 1 \end{bmatrix}$, $X = \begin{bmatrix} x \\ y \\ z \end{bmatrix}$,

and $B = \begin{bmatrix} 4 \\ 3 \\ -8 \end{bmatrix}$

b. $x = -1$; $y = 2$; $z = -3$

21. a. $AX = B$, where $A = \begin{bmatrix} 1 & 4 & -1 \\ 2 & 3 & -2 \\ -1 & 2 & 3 \end{bmatrix}$, $X = \begin{bmatrix} x \\ y \\ z \end{bmatrix}$,

and $B = \begin{bmatrix} 3 \\ 1 \\ 7 \end{bmatrix}$

b. $x = 1$; $y = 1$; $z = 2$

23. a. $AX = B$, where $A = \begin{bmatrix} 1 & 1 & -1 & 1 \\ 2 & 1 & 1 & 0 \\ 2 & 1 & 0 & 1 \\ 2 & -1 & -1 & 3 \end{bmatrix}$, $X = \begin{bmatrix} x_1 \\ x_2 \\ x_3 \\ x_4 \end{bmatrix}$,

and $B = \begin{bmatrix} 6 \\ 4 \\ 7 \\ 9 \end{bmatrix}$

b. $x_1 = 1$; $x_2 = 2$; $x_3 = 0$; $x_4 = 3$

25. b. (i) $x = 4.8$ and $y = 4.6$ **(ii)** $x = 0.4$ and $y = 1.8$

27. b. (i) $x = -1$; $y = 3$; $z = 2$
(ii) $x = 1$; $y = 8$; $z = -12$

29. b. (i) $x = -\frac{2}{17}$; $y = -\frac{10}{17}$; $z = -\frac{60}{17}$
(ii) $x = 1$; $y = 0$; $z = -5$

31. b. (i) $x_1 = 1$; $x_2 = -4$; $x_3 = 5$; $x_4 = -1$
(ii) $x_1 = 12$; $x_2 = -24$; $x_3 = 21$; $x_4 = -7$

33. a. $A^{-1} = \begin{bmatrix} -\frac{5}{2} & -\frac{3}{2} \\ 2 & 1 \end{bmatrix}$

35. a. $ABC = \begin{bmatrix} 4 & 10 \\ 2 & 3 \end{bmatrix}$; $A^{-1} = \begin{bmatrix} 3 & -5 \\ 1 & -2 \end{bmatrix}$;

$B^{-1} = \begin{bmatrix} 1 & -3 \\ -1 & 4 \end{bmatrix}$; $C^{-1} = \begin{bmatrix} \frac{1}{8} & -\frac{3}{8} \\ \frac{1}{4} & \frac{1}{4} \end{bmatrix}$

37. $\begin{bmatrix} \frac{5}{7} & \frac{3}{7} \\ -\frac{3}{7} & \frac{8}{7} \end{bmatrix}$

39. a. 3214; 3929 **b.** 4286; 3571 **c.** 3929; 5357

41. a. 400 acres of soybeans; 300 acres of corn; 300 acres of wheat
b. 500 acres of soybeans; 400 acres of corn; 300 acres of wheat

43. a. $80,000 in high-risk stocks; $20,000 in medium-risk stocks; $100,000 in low-risk stocks
b. $88,000 in high-risk stocks; $22,000 in medium-risk stocks; $110,000 in low-risk stocks
c. $56,000 in high-risk stocks; $64,000 in medium-risk stocks; $120,000 in low-risk stocks

45. All values of k except $k = \frac{3}{2}$; $\dfrac{1}{3 - 2k} \begin{bmatrix} 3 & -2 \\ -k & 1 \end{bmatrix}$

47. A^{-1} exists provided $ad \neq 0$; every entry along the main diagonal is not equal to zero.

49. True **51.** True

Using Technology Exercises 5.6, page 323

1. $\begin{bmatrix} 0.36 & 0.04 & -0.36 \\ 0.06 & 0.05 & 0.20 \\ -0.19 & 0.10 & 0.09 \end{bmatrix}$

3. $\begin{bmatrix} 0.01 & -0.09 & 0.31 & -0.11 \\ -0.25 & 0.58 & -0.15 & -0.02 \\ 0.86 & -0.42 & 0.07 & -0.37 \\ -0.27 & 0.01 & -0.05 & 0.31 \end{bmatrix}$

5. $\begin{bmatrix} 0.30 & 0.85 & -0.10 & -0.77 & -0.11 \\ -0.21 & 0.10 & 0.01 & -0.26 & 0.21 \\ 0.03 & -0.16 & 0.12 & -0.01 & 0.03 \\ -0.14 & -0.46 & 0.13 & 0.71 & -0.05 \\ 0.10 & -0.05 & -0.10 & -0.03 & 0.11 \end{bmatrix}$

7. $x = 1.2$; $y = 3.6$; $z = 2.7$

9. $x_1 \approx 2.50$; $x_2 \approx -0.88$; $x_3 \approx 0.70$; $x_4 \approx 0.51$

Chapter 5 Concept Review Questions, page 325

1. a. one; many; no **b.** one; many; no **2.** equations

3. $R_i \leftrightarrow R_j$; cR_i; $R_i + aR_j$; solution

4. a. unique **b.** no; infinitely many; unique

5. size; entries **6.** size; corresponding

7. $m \times n$; $n \times m$; a_{ji} **8.** cA; c

9. a. columns; rows **b.** $m \times p$

10. a. $A(BC)$; $AB + AC$ **b.** $n \times r$

11. $A^{-1}A$; AA^{-1}; singular **12.** $A^{-1}B$

Chapter 5 Review Exercises, page 325

1. $\begin{bmatrix} 2 & 2 \\ -1 & 4 \\ 3 & 3 \end{bmatrix}$ **2.** $\begin{bmatrix} -2 & 0 \\ -2 & 6 \end{bmatrix}$ **3.** $[-6 \quad -2]$ **4.** $\begin{bmatrix} 17 \\ 13 \end{bmatrix}$

5. $x = 2$; $y = 3$; $z = 1$; $w = 3$ **6.** $x = 2$; $y = -2$

7. $a = 3$; $b = 4$; $c = -2$; $d = 2$; $e = -3$

8. $x = -1$; $y = -2$; $z = 1$

9. $\begin{bmatrix} 8 & 9 & 11 \\ -10 & -1 & 3 \\ 11 & 12 & 10 \end{bmatrix}$ **10.** $\begin{bmatrix} -1 & 7 & -3 \\ -2 & 5 & 11 \\ 10 & -8 & 2 \end{bmatrix}$

11. $\begin{bmatrix} 6 & 18 & 6 \\ -12 & 6 & 18 \\ 24 & 0 & 12 \end{bmatrix}$ **12.** $\begin{bmatrix} -10 & 10 & -18 \\ 4 & 14 & 26 \\ 16 & -32 & -4 \end{bmatrix}$

13. $\begin{bmatrix} -11 & -16 & -15 \\ -4 & -2 & -10 \\ -6 & 14 & 2 \end{bmatrix}$ **14.** $\begin{bmatrix} 5 & 20 & 19 \\ -2 & 20 & 8 \\ 26 & 10 & 30 \end{bmatrix}$

15. $\begin{bmatrix} -3 & 17 & 8 \\ -2 & 56 & 27 \\ 74 & 78 & 116 \end{bmatrix}$ 16. $\begin{bmatrix} \frac{3}{2} & -2 & -5 \\ \frac{11}{2} & -1 & 11 \\ \frac{7}{2} & -3 & 0 \end{bmatrix}$

17. $x = 1; y = -1$ 18. $x = -1; y = 3$

19. $x = 1; y = 2; z = 3$

20. $(2, 2t - 5, t); t$, a parameter   21. No solution

22. $x = 1; y = -1; z = 2; w = 2$

23. $x = 1; y = 0; z = 1$ 24. $x = 2; y = -1; z = 3$

25. $\begin{bmatrix} \frac{2}{5} & -\frac{1}{5} \\ -\frac{1}{5} & \frac{3}{5} \end{bmatrix}$ 26. $\begin{bmatrix} \frac{3}{4} & -\frac{1}{2} \\ -\frac{1}{8} & \frac{1}{4} \end{bmatrix}$

27. $\begin{bmatrix} -1 & 2 \\ 1 & -\frac{3}{2} \end{bmatrix}$ 28. $\begin{bmatrix} \frac{1}{4} & \frac{1}{2} \\ \frac{1}{8} & -\frac{1}{4} \end{bmatrix}$

29. $\begin{bmatrix} \frac{5}{4} & \frac{1}{4} & -\frac{7}{4} \\ -\frac{1}{4} & -\frac{1}{4} & \frac{3}{4} \\ -\frac{3}{4} & \frac{1}{4} & \frac{5}{4} \end{bmatrix}$ 30. $\begin{bmatrix} -\frac{1}{4} & \frac{1}{2} & -\frac{1}{4} \\ \frac{7}{8} & -\frac{3}{4} & -\frac{5}{8} \\ -\frac{1}{8} & \frac{1}{4} & \frac{3}{8} \end{bmatrix}$

31. $\begin{bmatrix} -\frac{1}{5} & \frac{2}{5} & 0 \\ \frac{2}{3} & -\frac{1}{3} & \frac{1}{3} \\ -\frac{1}{30} & \frac{1}{15} & -\frac{1}{6} \end{bmatrix}$ 32. $\begin{bmatrix} 0 & -\frac{1}{5} & \frac{2}{5} \\ -2 & 1 & 1 \\ -1 & \frac{1}{5} & \frac{3}{5} \end{bmatrix}$

33. $\begin{bmatrix} \frac{3}{2} & 1 \\ -\frac{7}{2} & -1 \end{bmatrix}$ 34. $\begin{bmatrix} \frac{11}{24} & -\frac{7}{8} \\ -\frac{1}{12} & \frac{1}{4} \end{bmatrix}$

35. $\begin{bmatrix} \frac{2}{5} & -\frac{3}{5} \\ \frac{1}{5} & \frac{1}{5} \end{bmatrix}$ 36. $\begin{bmatrix} \frac{4}{7} & -\frac{3}{7} \\ -\frac{3}{7} & \frac{4}{7} \end{bmatrix}$

37. $A^{-1} = \begin{bmatrix} \frac{2}{7} & \frac{3}{7} \\ \frac{1}{7} & -\frac{2}{7} \end{bmatrix}; x = -1; y = -2$

38. $A^{-1} = \begin{bmatrix} \frac{2}{5} & \frac{3}{10} \\ -\frac{1}{5} & \frac{1}{10} \end{bmatrix}; x = 2; y = 1$

39. $A^{-1} = \begin{bmatrix} 1 & -\frac{2}{5} & \frac{4}{5} \\ -1 & 1 & -1 \\ -\frac{1}{2} & \frac{3}{5} & -\frac{7}{10} \end{bmatrix}; x = 1; y = 2; z = 4$

40. $A^{-1} = \begin{bmatrix} 0 & \frac{1}{7} & \frac{2}{7} \\ -1 & -\frac{4}{7} & \frac{6}{7} \\ -\frac{1}{2} & -\frac{1}{2} & \frac{1}{2} \end{bmatrix}; x = 3; y = -1; z = 2$

41. $10,824, $10,078, and $13,266

42. $2,300,000; $2,450,000; an increase of $150,000

43. **a.** $A = \begin{bmatrix} 800 & 1200 & 400 & 1500 \\ 600 & 1400 & 600 & 2000 \end{bmatrix}$ **b.** $B = \begin{bmatrix} 50.26 \\ 31.00 \\ 103.07 \\ 38.67 \end{bmatrix}$

b. William: $176,641; Michael: $212,738

44. **a.** $A = \begin{bmatrix} \text{IBM} & \text{Google} & \text{Boeing} & \text{GM} \\ 800 & 500 & 1200 & 1500 \\ 500 & 600 & 2000 & 800 \end{bmatrix} \begin{matrix} \\ \text{Olivia} \\ \text{Max} \end{matrix}$;

$B = \begin{bmatrix} \text{IBM} & \text{Google} & \text{Boeing} & \text{GM} \\ 900 & 600 & 1000 & 1200 \\ 700 & 500 & 2100 & 900 \end{bmatrix} \begin{matrix} \\ \text{Olivia} \\ \text{Max} \end{matrix}$

b. $C = \begin{bmatrix} \text{IBM} & \text{Google} & \text{Boeing} & \text{GM} \\ 100 & 100 & -200 & -300 \\ 200 & -100 & 100 & 100 \end{bmatrix} \begin{matrix} \\ \text{Olivia} \\ \text{Max} \end{matrix}$

45. 30 of each type

46. Houston: 100,000 gallons; Tulsa: 600,000 gallons

Chapter 5 Before Moving On, page 327

1. $\left(\frac{2}{3}, -\frac{2}{3}, \frac{5}{3}\right)$

2. **a.** $(2, -3, 1)$ **b.** No solution   **c.** $(2, 1 - 3t, t), t$, a parameter
   d. $(0, 0, 0, 0)$ **e.** $(2 + t, 3 - 2t, t), t$, a parameter

3. **a.** $(-1, 2)$ **b.** $\left(\frac{4}{7}, -\frac{5}{7} + 2t, t\right), t$, a parameter

4. **a.** $\begin{bmatrix} 3 & 1 & 4 \\ 5 & -2 & 6 \end{bmatrix}$ **b.** $\begin{bmatrix} 14 & 3 & 7 \\ 14 & 5 & 1 \end{bmatrix}$ **c.** $\begin{bmatrix} 0 & 5 & 3 \\ 4 & -1 & -11 \end{bmatrix}$

5. $\begin{bmatrix} 3 & -2 & -5 \\ -3 & 2 & 6 \\ -1 & 1 & 2 \end{bmatrix}$ 6. $(1, -1, 2)$

CHAPTER 6

Exercises 6.1, page 335

1.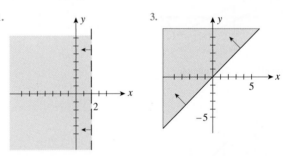

3. (graph)

5.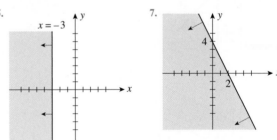

7. (graph)

9.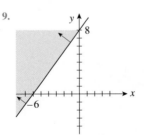

11. $1 \le x \le 5; 2 \le y \le 4$

13. $2x - y \ge 2; 5x + 7y \ge 35; x \le 4$

15. $x - y \ge -10; 7x + 4y \le 140; x + 3y \ge 30$

17. $x + y \ge 7; x \ge 2; 3 \le y \le 7$

19.

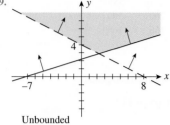

Unbounded

21.

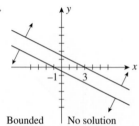

Unbounded

23.

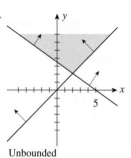

Bounded   No solution

25.

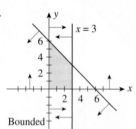

Bounded

27.

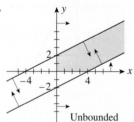

Unbounded

29.

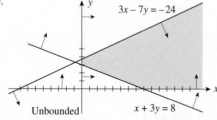

Unbounded $x + 3y = 8$

31.

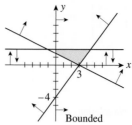

Bounded

33.

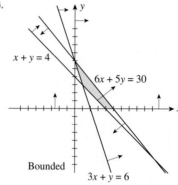

Bounded $3x + y = 6$

35.

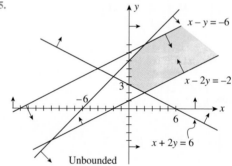

Unbounded

37. False   **39.** True

Exercises 6.2, page 342

1. Maximize $P = 3x + 4y$
subject to $6x + 9y \le 300$
$5x + 4y \le 180$
$x \ge 0, y \ge 0$

3. Maximize $P = 2x + 1.5y$
subject to $3x + 4y \le 1000$
$6x + 3y \le 1200$
$x \ge 0, y \ge 0$

5. Maximize $P = 45x + 20y$
subject to $40x + 16y \le 3200$
$3x + 4y \le 520$
$x \ge 0, y \ge 0$

7. Maximize $P = 0.1x + 0.12y$
subject to $x + y \le 20$
$x - 4y \ge 0$
$x \ge 0, y \ge 0$

9. Maximize $P = 50x + 40y$
subject to $\frac{1}{200}x + \frac{1}{200}y \le 1$
$\frac{1}{100}x + \frac{1}{300}y \le 1$
$x \ge 0, y \ge 0$

11. Minimize $C = 14{,}000x + 16{,}000y$
    subject to $50x + 75y \geq 650$
    $3000x + 1000y \geq 18{,}000$
    $x \geq 0,\ y \geq 0$

13. Minimize $C = 300x + 500y$
    subject to $x + y \geq 10$
    $x \leq 5$
    $y \leq 10$
    $y \geq 6$
    $x \geq 0$

15. Minimize $C = 2x + 5y$
    subject to $30x + 25y \geq 400$
    $x + 0.5y \geq 10$
    $2x + 5y \geq 40$
    $x \geq 0,\ y \geq 0$

17. Minimize $C = 1000x + 800y$
    subject to $70{,}000x + 10{,}000y \geq 2{,}000{,}000$
    $40{,}000x + 20{,}000y \geq 1{,}400{,}000$
    $20{,}000x + 40{,}000y \geq 1{,}000{,}000$
    $x \geq 0,\ y \geq 0$

19. Maximize $P = 0.1x + 0.15y + 0.2z$
    subject to $x + y + z \leq 2{,}000{,}000$
    $-2x - 2y + 8z \leq 0$
    $-6x + 4y + 4z \leq 0$
    $-10x + 6y + 6z \leq 0$
    $x \geq 0,\ y \geq 0,\ z \geq 0$

21. Maximize $P = 18x + 12y + 15z$
    subject to $2x + y + 2z \leq 900$
    $3x + y + 2z \leq 1080$
    $2x + 2y + z \leq 840$
    $x \geq 0,\ y \geq 0,\ z \geq 0$

23. Maximize $P = 26x + 28y + 24z$
    subject to $\frac{5}{4}x + \frac{3}{2}y + \frac{3}{2}z \leq 310$
    $x + y + \frac{3}{4}z \leq 205$
    $x + y + \frac{1}{2}z \leq 190$
    $x \geq 0,\ y \geq 0,\ z \geq 0$

25. Minimize $C = 60x_1 + 60x_2 + 80x_3 + 80x_4 + 70x_5 + 50x_6$
    subject to $x_1 + x_2 + x_3 \leq 300$
    $x_4 + x_5 + x_6 \leq 250$
    $x_1 + x_4 \geq 200$
    $x_2 + x_5 \geq 150$
    $x_3 + x_6 \geq 200$
    $x_1 \geq 0,\ x_2 \geq 0,\ \ldots,\ x_6 \geq 0$

27. Maximize $P = x + 0.8y + 0.9z$
    subject to $8x + 4z \leq 16{,}000$
    $8x + 12y + 8z \leq 24{,}000$
    $4y + 4z \leq 5000$
    $z \leq 800$
    $x \geq 0,\ y \geq 0,\ z \geq 0$

29. False

Exercises 6.3, page 353

1. Max: 35; min: 5   3. No max. value; min: 18

5. Max: 44; min: 15   7. $x = 3$; $y = 3$; $P = 15$

9. Any point (x, y) lying on the line segment joining $\left(\frac{5}{2}, 0\right)$ and $(1, 3)$; $P = 5$

11. $x = 0$; $y = 8$; $P = 64$

13. $x = 0$; $y = 4$; $P = 12$

15. $x = 4$; $y = 0$; $C = 8$

17. Any point (x, y) lying on the line segment joining $(20, 10)$ and $(40, 0)$; $C = 120$

19. $x = 14$; $y = 3$; $C = 58$

21. $x = 3$; $y = 3$; $C = 75$

23. $x = 15$; $y = 17.5$; $P = 115$

25. $x = 10$; $y = 38$; $P = 134$

27. Max: $x = 15$; $y = 3$; $P = 135$
    Min: $x = 9$; $y = 9$; $P = 117$

29. 20 Product A, 20 Product B; \$140

31. 120 model A, 160 model B; \$480

33. 40 tables; 100 chairs; \$3800

35. \$16 million in homeowner loans, \$4 million in auto loans; \$2.08 million

37. 50 fully assembled units, 150 kits; \$8500

39. Saddle Mine: 4 days; Horseshoe Mine: 6 days; \$152,000

41. Reservoir: 4 million gallons; pipeline: 6 million gallons; \$4200

43. Infinitely many solutions; 10 oz of Food A and 4 oz of Food B or 20 oz of Food A and 0 oz of Food B, etc., with a minimum value of 40 mg of cholesterol

45. 30 in Newspaper I, 10 in Newspaper II; \$38,000

47. 80 from I to A, 20 from I to B, 0 from II to A, 50 from II to B

49. \$22,500 in growth stocks and \$7500 in speculative stocks; maximum return; \$5250

51. 750 urban, 750 suburban; \$10,950

53. False   55. a. True   b. True

59. a.

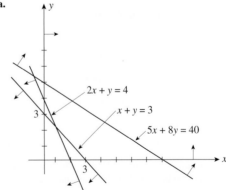

b. No solution

Exercises 6.4, page 374

1. a. It is already in standard form.
   b.

x	y	u	v	P	Constant
1	4	1	0	0	12
1	3	0	1	0	10
-2	-4	0	0	1	0

3. **a.** Maximize $P = 2x + 3y$
   subject to $x + y \le 10$
   $x + 2y \le 12$
   $2x + y \le 12$
   $x \ge 0, y \ge 0$

b.

x	y	u	v	w	P	Constant
1	1	1	0	0	0	10
1	2	0	1	0	0	12
2	1	0	0	1	0	12
−2	−3	0	0	0	1	0

5. **a.** Maximize $P = x + 3y + 4z$
   subject to $x + 2y + z \le 40$
   $x + y + z \le 30$
   $x \ge 0, y \ge 0, z \ge 0$

b.

x	y	z	u	v	P	Constant
1	2	1	1	0	0	40
1	1	1	0	1	0	30
−1	−3	−4	0	0	1	0

7. In final form; $x = \frac{30}{7}, y = \frac{20}{7}, u = 0, v = 0; P = \frac{220}{7}$

9. Not in final form; pivot element is $\frac{1}{2}$, lying in the first row, second column.

11. In final form; $x = \frac{1}{3}, y = 0, z = \frac{13}{3}, u = 0, v = 6, w = 0$; $P = 17$

13. Not in final form; pivot element is 1, lying in the third row, second column.

15. In final form; $x = 30, y = 10, z = 0, u = 0, v = 0; P = 60$;
    $x = 30, y = 0, z = 0, u = 10, v = 0; P = 60$; among others

17. $x = 0, y = 4, u = 0, v = 1; P = 16$

19. $x = 6, y = 3, u = 0, v = 0; P = 96$

21. $x = 6, y = 6, u = 0, v = 0, w = 0; P = 60$

23. $x = 0, y = 4, z = 4, u = 0, v = 0; P = 36$

25. $x = 0, y = 3, z = 0, u = 90, v = 0, w = 75; P = 12$

27. $x = 15, y = 3, z = 0, u = 2, v = 0, w = 0; P = 78$

29. $x = \frac{5}{4}, y = \frac{15}{2}, z = 0, u = 0, v = \frac{15}{4}, w = 0; P = 90$

31. $x = 2, y = 1, z = 1, u = 0, v = 0, w = 0; P = 87$

35. No model A, 2500 model B; $100,000

37. 65 acres of Crop A, 80 acres of Crop B; $25,750; yes; 5 acres of land

39. $62,500 in the money market fund, $125,000 in the international equity fund, $62,500 in the growth-and-income fund; $25,625

41. 180 units of Product A, 140 units of Product B, and 200 units of Product C; $7920; no

43. 22 min of morning advertising time, 3 min of evening advertising time

45. 80 units of model A, 80 units of model B, and 60 units of model C; maximum profit: $5760; no

47. 9000 bottles of Formula I, 7833 bottles of Formula II, 6000 bottles of Formula III; maximum profit: $4986.60; Yes, ingredients for 4167 bottles of Formula II

49. Project A: $800,000, project B: $800,000, and project C: $400,000; $280,000

51. False   53. True

Using Technology Exercises 6.4, page 384

1. $x = 1.2, y = 0, z = 1.6, w = 0; P = 8.8$

3. $x = 1.6, y = 0, z = 0, w = 3.6; P = 12.4$

Exercises 6.5, page 394

1. $x = 4, y = 0; C = -8$

3. $x = 4, y = 3; C = -18$

5. $x = 0, y = 13, z = 18, w = 14; C = -111$

7. $x = \frac{5}{4}, y = \frac{1}{4}, u = 2, v = 3; C = P = 13$

9. $x = 5, y = 10, z = 0, u = 1, v = 2; C = P = 80$

11. Maximize $P = 4u + 6v$
    subject to $u + 3v \le 2$
    $2u + 2v \le 5$; $x = 4, y = 0; C = 8$
    $u \ge 0, v \ge 0$

13. Maximize $P = 60u + 40v + 30w$
    subject to $6u + 2v + w \le 6$
    $u + v + w \le 4$; $x = 10, y = 20; C = 140$
    $u \ge 0, v \ge 0, w \ge 0$

15. Maximize $P = 10u + 20v$
    subject to $20u + v \le 200$
    $10u + v \le 150$; $x = 0, y = 0, z = 10; C = 1200$
    $u + 2v \le 120$
    $u \ge 0, v \ge 0$

17. Maximize $P = 10u + 24v + 16w$
    subject to $u + 2v + w \le 6$
    $2u + v + w \le 8$; $x = 8, y = 0, z = 8; C = 80$
    $2u + v + w \le 4$
    $u \ge 0, v \ge 0, w \ge 0$

19. Maximize $P = 6u + 2v + 4w$
    subject to $2u + 6v \le 30$
    $4u + 6w \le 12$; $x = \frac{1}{3}, y = \frac{4}{3}, z = 0; C = 26$
    $3u + v + 2w \le 20$
    $u \ge 0, v \ge 0, w \ge 0$

21. 2 type A vessels; 3 type B vessels; $250,000

23. 30 in Newspaper I; 10 in Newspaper II; $38,000

25. 8 oz of orange juice; 6 oz of pink grapefruit juice; 178 calories

27. True

Using Technology Exercises 6.5, page 400

1. $x \approx 1.333333, y \approx 3.333333, z = 0$; and $C \approx 4.66667$

3. $x = 0.9524, y = 4.2857, z = 0; C = 6.0952$

Chapter 6 Concept Review Questions, page 400

1. **a.** half-plane; line   **b.** $ax + by \le c; ax + by = c$

2. **a.** points; each   **b.** bounded; enclosed

3. objective function; maximized; minimized; linear; inequalities

4. **a.** corner point   **b.** line

5. maximized; nonnegative; less than; equal to

6. equations; slack variables; $-c_1x_1 - c_2x_2 - \cdots - c_nx_n + P = 0$; below; augmented

7. minimized; nonnegative; greater than; equal to

8. dual; objective; optimal value

Chapter 6 Review Exercises, page 401

1. Max: 18—any point (x, y) lying on the line segment joining $(0, 6)$ and $(3, 4)$; min: 0

2. Max: 27—any point (x, y) lying on the line segment joining $(3, 5)$ and $(6, 1)$; min: 7

3. $x = 0, y = 4; P = 20$

4. $x = 0, y = 12; P = 36$

5. $x = 3, y = 4; C = 26$

6. $x = 1.25, y = 1.5; C = 9.75$

7. $x = 3, y = 10; P = 29$

8. $x = 8, y = 0; P = 48$

9. $x = 20, y = 0; C = 40$

10. $x = 2, y = 6; C = 14$

11. Max: $x = 22, y = 0; Q = 22$; min: $x = 3, y = \frac{5}{2}; Q = \frac{11}{2}$

12. Max: $x = 12, y = 6; Q = 54$; min: $x = 4, y = 0; Q = 8$

13. $x = 3, y = 4, u = 0, v = 0; P = 25$

14. $x = 3, y = 6, u = 4, v = 0, w = 0; P = 36$

15. $x = \frac{56}{5}, y = \frac{2}{5}, z = 0, u = 0, v = 0; P = \frac{118}{5}$

16. $x = 0, y = \frac{11}{3}, z = \frac{25}{6}, u = \frac{37}{6}, v = 0, w = 0; P = \frac{119}{6}$

17. $x = \frac{3}{2}, y = 1; C = \frac{13}{2}$

18. $x = \frac{32}{11}, y = \frac{36}{11}; C = \frac{104}{11}$

19. $x = \frac{3}{4}, y = 0, z = \frac{7}{4}; C = 60$

20. $x = 0, y = 2, z = 0; C = 4$

21. $40,000 in each company; $13,600

22. 60 model A satellite radios; 60 model B satellite radios; $1320

23. 93 model A, 180 model B; $456

24. Saddle Mine: 4 days; Horseshoe Mine: 6 days; $152,000

25. $70,000 in blue-chip stocks; $0 in growth stocks; $30,000 in speculative stocks; maximum return: $13,000

26. 0 unit of Product A, 30 units of Product B, 0 unit of Product C; $P = \$180$

Chapter 6 Before Moving On, page 403

1. Min: $x = 3, y = 16; C = -7$
   Max: $x = 28, y = 8; P = 76$

2. Max: $x = 0, y = \frac{24}{7}; P = \frac{72}{7}$

3.

x	y	z	u	v	w	P	Constant
2	①	-1	1	0	0	0	3
1	-2	3	0	1	0	0	1
3	2	4	0	0	1	0	17
-1	-2	3	0	0	0	1	0

4. $x = 2, y = 0, z = 11, u = 2, v = 0, w = 0; P = 28$

5. $x = 6, y = 2; u = 0, v = 0; P = 34$

CHAPTER 7

Exercises 7.1, page 412

1. $\{x \mid x$ is a gold medalist in the 2010 Winter Olympic Games$\}$

3. $\{x \mid x$ is an integer greater than 2 and less than 8$\}$

5. $\{2, 3, 4, 5, 6\}$ 7. $\{-2\}$

9. **a.** True   **b.** False   11. **a.** False   **b.** False

13. True   15. **a.** True   **b.** False

17. **a** and **b.**

19. **a.** $\varnothing, \{1\}, \{2\}, \{1, 2\}$
    b. $\varnothing, \{1\}, \{2\}, \{3\}, \{1, 2\}, \{1, 3\}, \{2, 3\}, \{1, 2, 3\}$
    c. $\varnothing, \{1\}, \{2\}, \{3\}, \{4\}, \{1, 2\}, \{1, 3\}, \{1, 4\}, \{2, 3\}, \{2, 4\}, \{3, 4\}, \{1, 2, 3\}, \{1, 2, 4\}, \{1, 3, 4\}, \{2, 3, 4\}, \{1, 2, 3, 4\}$

21. $\{1, 2, 3, 4, 6, 8, 10\}$

23. $\{$Jill, John, Jack, Susan, Sharon$\}$

25. **a.**

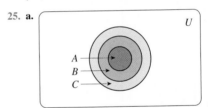

b.

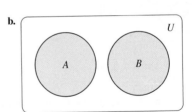

c.

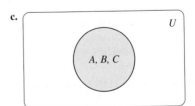

27. **a.**

b.

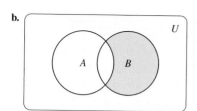

29. a.

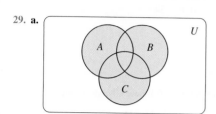

b.

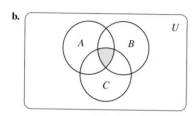

31. a.

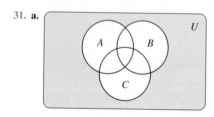

b.

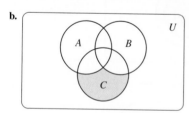

33. a. {2, 4, 6, 8, 10}   **b.** {2, 3, 4, 6, 7, 8, 10}
   c. {1, 2, 3, 4, 5, 6, 7, 8, 9, 10}

35. a. $C = \{1, 2, 4, 5, 8, 9\}$ **b.** $\varnothing$ **c.** {1, 2, 3, 4, 5, 6, 7, 8, 9, 10}

37. a. Not disjoint   **b.** Disjoint

39. a. The set of all employees at Universal Life Insurance who do not drink tea
   b. The set of all employees at Universal Life Insurance who do not drink coffee

41. a. The set of all employees at Universal Life Insurance who drink tea but not coffee
   b. The set of all employees at Universal Life Insurance who drink coffee but not tea

43. a. The set of all employees in a hospital who are not doctors
   b. The set of all employees in a hospital who are not nurses

45. a. The set of all employees in a hospital who are female doctors
   b. The set of all employees in a hospital who are both doctors and administrators

47. a. $D \cap F$ **b.** $R \cap F^c \cap L^c$

49. a. B^c **b.** $A \cap B$ **c.** $A \cap B \cap C^c$

51. a. $A \cap B \cap C$; the set of tourists who have taken the underground, a cab, and a bus over a 1-week period in London
   b. $A \cap C$; the set of tourists who have taken the underground and a bus over a 1-week period in London
   c. B^c; the set of tourists who have not taken a cab over a 1-week period in London

53.

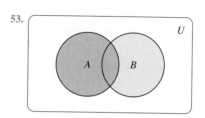

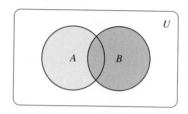

55.

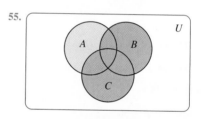

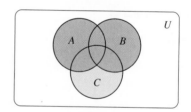

57.

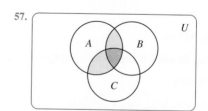

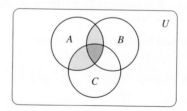

61. a. x, y, v, r, w, u **b.** v, r

63. a. s, t, y **b.** t, z, w, x, s **65.** $A \subset C$

67. False   **69.** True   **71.** True   **73.** True   **75.** True

Exercises 7.2, page 418

3. **a.** 4   **b.** 5   **c.** 7   **d.** 2

7. 19

9. **a.** 15   **b.** 30   **c.** 15   **d.** 12   **e.** 50   **f.** 20

11. **a.** 140   **b.** 100   **c.** 60

13. 12   15. 0   17. 12   19. 61

21. **a.** 106   **b.** 64   **c.** 38   **d.** 14

23. **a.** 182   **b.** 118   **c.** 56   **d.** 18   25. 30

27. **a.** 16   **b.** 31   **c.** 4   **d.** 21   **e.** 11

29. **a.** 64   **b.** 10   31. **a.** 36   **b.** 36   33. 5

35. **a.** 62   **b.** 33   **c.** 25   **d.** 38

37. **a.** 108   **b.** 15   **c.** 45   **d.** 12

39. **a.** 22   **b.** 80

41. True   43. True

Exercises 7.3, page 425

1. 12   3. 64   5. 36

7. 24   9. 60   11. 1 billion   13. 64   15. 5^{50}

17. 400   19. 9990

21. **a.** 17,576,000   **b.** 17,576,000

23. 1024; 59,049   25. 2730

27. 217   29. True

Exercises 7.4, page 436

1. 360   3. 30   5. 120

7. 60   9. n 11. 1

13. 21   15. 1   17. 84

19. $\dfrac{n(n-1)(n-2)}{6}$ 21. $\dfrac{n!}{2}$

23. Permutation   25. Combination

27. Permutation   29. Combination

31. $P(4, 4) = 24$ 33. $P(4, 4) = 24$

35. $P(9, 9) = 362{,}880$ 37. $C(12, 4) = 495$

39. 151,200   41. 2520

43. 20   45. $C(12, 3) = 220$

47. $C(100, 3) = 161{,}700$ 49. $P(6, 6) = 720$

51. $P(12, 5) = 95{,}040$

53. **a.** $P(10, 10) = 3{,}628{,}800$
b. $P(3, 3)P(4, 4)P(3, 3)P(3, 3) = 5184$

55. **a.** $P(20, 20) = 20!$
b. $P(5, 5)P[(4, 4)]^5 = 5!(4!)^5 = 955{,}514{,}880$

57. **a.** $P(12, 9) = 79{,}833{,}600$
b. $C(12, 9) = 220$ **c.** $C(12, 9) \cdot C(3, 2) = 660$

59. $2\{C(2, 2) + [C(3, 2) - C(2, 2)]\} = 6$

61. $C(3,3)[C(8, 6) + C(8, 7) + C(8, 8)] = 37$

63. **a.** $C(12, 3) = 220$ **b.** $C(11, 2) = 55$
c. $C(5, 1)C(7, 2) + C(5, 2)C(7, 1) + C(5, 3) = 185$

65. $P(7, 3) + C(7, 2)P(3, 2) = 336$

67. $C(5, 1)C(3, 1)C(6, 2)[C(4, 1) + C(3, 1)] = 1575$

69. $10C(4, 1) = 40$

71. $C(4, 1)C(13, 5) - 40 = 5108$

73. $13C(4, 3) \cdot 12C(4, 2) = 3744$

75. $C(6, 2) = 15$

77. $C(12, 6) + C(12, 7) + C(12, 8) + C(12, 9) + C(12, 10) + C(12, 11) + C(12, 12) = 2510$

79. $4! = 24$ 83. True   85. True

Using Technology Exercises 7.4, page 441

1. $1.307674368 \times 10^{12}$ 3. $2.56094948229 \times 10^{16}$

5. 674,274,182,400   7. 133,784,560

9. 4,656,960

11. 658,337,004,000

Exercises 7.5, page 446

1. $\{a, b, d, f\}; \{a\}$ 3. $\{b, c, e\}; \{a\}$ 5. No   7. S

9. $\varnothing$ 11. Yes   13. Yes   15. $E \cup G$ 17. F^c

19. $(E \cup F \cup G)^c$

21. **a.** $\{(2, 1), (3, 1), (4, 1), (5, 1), (6, 1), (3, 2), (4, 2), (5, 2), (6, 2), (4, 3), (5, 3), (6, 3), (5, 4), (6, 4), (6, 5)\}$
b. $\{(1, 2), (2, 4), (3, 6)\}$

23. $\varnothing, \{a\}, \{b\}, \{c\}, \{a, b\}, \{a, c\}, \{b, c\}, S$

25. **a.** $S = \{B, R\}$ **b.** $\varnothing, \{B\}, \{R\}, S$

27. **a.** $S = \{(H, 1), (H, 2), (H, 3), (H, 4), (H, 5), (H, 6), (T, 1), (T, 2), (T, 3), (T, 4), (T, 5), (T, 6)\}$
b. $\{(H, 1), (H, 3), (H, 5)\}$

29. **a.** No   **b.** No

31. $S = \{ddd, ddn, dnd, ndd, dnn, ndn, nnd, nnn\}$

33. **a.** $S = \{bbbb, bbbg, bbgb, bbgg, bgbb, bgbg, bggb, bggg, gbbb, gbbg, gbgb, gbgg, ggbb, ggbg, gggb, gggg\}$
b. $E = \{bbbg, bbgb, bgbb, gbbb\}$
c. $F = \{bbbg, bbgg, bgbg, bggg, gbbg, gbgg, ggbg, gggg\}$
d. $G = \{gbbg, gbgg, ggbg, gggg\}$

35. **a.** $\{ABC, ABD, ABE, ACD, ACE, ADE, BCD, BCE, BDE, CDE\}$
b. 6   **c.** 3   **d.** 6

37. **a.** E^c **b.** $E^c \cap F^c$ **c.** $E \cup F$
d. $(E \cap F^c) \cup (E^c \cap F)$

39. **a.** $\{t \mid t > 0\}$ **b.** $\{t \mid 0 < t \le 2\}$ **c.** $\{t \mid t > 2\}$

41. **a.** $S = \{0, 1, 2, 3, \ldots, 10\}$ **b.** $E = \{0, 1, 2, 3\}$
c. $F = \{5, 6, 7, 8, 9, 10\}$

43. **a.** $S = \{0, 1, 2, \ldots, 20\}$
b. $E = \{0, 1, 2, \ldots, 9\}$ **c.** $F = \{20\}$

49. False

Exercises 7.6, page 454

1. {(H, H)}, {(H, T)}, {(T, H)}, {(T, T)}

3. {(D, m)}, {(D, f)}, {(R, m)}, {(R, f)}, {(I, m)}, {(I, f)}

5. {(1, i)}, {(1, d)}, {(1, s)}, {(2, i)}, {(2, d)}, {(2, s)}, . . . , {(5, i)}, {(5, d)}, {(5, s)}

7. {(A, Rh⁺)}, {(A, Rh⁻)}, {(B, Rh⁺)}, {(B, Rh⁻)}, {(AB, Rh⁺)}, {(AB, Rh⁻)}, {(O, Rh⁺)}, {(O, Rh⁻)}

9.

Grade	A	B	C	D	F
Probability	.10	.25	.45	.15	.05

11.

Answer	Falling behind	Staying even	Increasing faster	Don't know
Probability	.40	.44	.12	.04

13. .69

15.

Number of Days	0	1	2	3	4	5	6	7
Probability	.05	.06	.09	.15	.11	.20	.17	.17

17.

Event	A	B	C	D	E
Probability	.026	.199	.570	.193	.012

19. **a.** $S = \{(0 < x \leq 200), (200 < x \leq 400),$
$(400 < x \leq 600), (600 < x \leq 800),$
$(800 < x \leq 1000), (x > 1000)\}$

b.

Cars, x	Probability
$0 < x \leq 200$	.075
$200 < x \leq 400$	.1
$400 < x \leq 600$	.175
$600 < x \leq 800$	.35
$800 < x \leq 1000$	.225
$x > 1000$	.075

21. .469

23. **a.** .856   **b.** .144   25. .46

27. **a.** $\frac{1}{4}$ **b.** $\frac{1}{2}$ **c.** $\frac{1}{13}$ 29. $\frac{3}{8}$

31. **a.** .4   **b.** .1   **c.** .1

33. **a.** .633   **b.** .276

35. **a.** .35   **b.** .33

37. .530   39. **a.** .4   **b.** .23

41. **a.** .448   **b.** .255   43. .783

45. There are six ways of obtaining a sum of 7.

47. No   49. **a.** $\frac{1}{6}$ **b.** $\frac{5}{6}$ **c.** 1   51. True

Exercises 7.7, page 463

1. $\frac{1}{2}$ 3. $\frac{1}{36}$ 5. $\frac{1}{6}$ 7. $\frac{1}{52}$

9. $\frac{3}{13}$ 11. $\frac{12}{13}$ 13. .002; .998

15. $P(a) + P(b) + P(c) \neq 1$

17. Since the five events are not mutually exclusive, Property 3 cannot be used; that is, he could win more than one purse.

19. The two events are not mutually exclusive; hence, the probability of the given event is $\frac{1}{6} + \frac{1}{6} - \frac{1}{36} = \frac{11}{36}$.

21. $E^c \cap F^c = \{e\} \neq \varnothing$

23. $P[(G \cup C)^c] \neq 1 - P(G) - P(C)$; he has not considered the case in which a customer buys both glasses and contact lenses.

25. **a.** 0   **b.** .7   **c.** .5   **d.** .3

27. **a.** $\frac{1}{2}, \frac{3}{8}$ **b.** $\frac{1}{2}, \frac{5}{8}$ **c.** $\frac{1}{8}$ **d.** $\frac{3}{4}$

29. **a.** .41   **b.** .48   31. .332

33. **a.** .24   **b.** .46   35. **a.** .16   **b.** .38   **c.** .22

37. **a.** .41   **b.** .518   39. **a.** .52   **b.** .34

41. **a.** .439   **b.** .385   43. **b.** .52   **c.** .859

45. **a.** .90   **b.** .40   **c.** .40

47. **a.** .6   **b.** .332   **c.** .232   **d.** .6

51. True   53. False

Chapter 7 Concept Review Questions, page 469

1. set; elements; set   2. equal

3. subset   4. **a.** no   **b.** all

5. union; intersection

6. complement   7. $A^c \cap B^c \cap C^c$

8. permutation; combination

9. experiment; sample; space; event

10. $\varnothing$ 11. uniform; $\dfrac{1}{n}$

Chapter 7 Review Exercises, page 469

1. {3}   2. {A, E, H, L, S, T}

3. {4, 6, 8, 10}   4. {−4}   5. Yes

6. Yes   7. Yes   8. No

9.

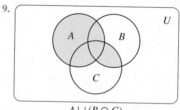

$A \cup (B \cap C)$

10.

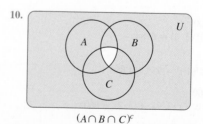

$(A \cap B \cap C)^c$

11.

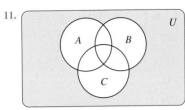

$A^c \cap B^c \cap C^c$

12.

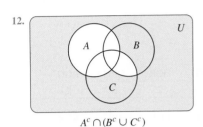

$A^c \cap (B^c \cup C^c)$

17. The set of all participants in a consumer-behavior survey who both avoided buying a product because it is not recyclable and boycotted a company's products because of its record on the environment.

18. The set of all participants in a consumer-behavior survey who avoided buying a product because it is not recyclable and/or voluntarily recycled their garbage.

19. The set of all participants in a consumer-behavior survey who both did not use cloth diapers rather than disposable diapers and voluntarily recycled their garbage.

20. The set of all participants in a consumer-behavior survey who did not boycott a company's products because of its record on the environment and/or did not voluntarily recycle their garbage.

21. 150   22. 230   23. 270

24. 30   25. 70   26. 200

27. 190   28. 181,440   29. 120   30. 8400

31. **a.** 0   **b.** .6   **c.** .6   **d.** .4   **e.** 1

32. **a.** .35   **b.** .65   **c.** .05

33. **a.** .49   **b.** .39   **c.** .48

34. $\frac{2}{7}$ 35. None

36. **a.** 446   **b.** 377   **c.** 34

37. 720   38. 20

39. **a.** 50,400   **b.** 5040

40. **a.** 60   **b.** 125   41. 80

42. **a.** 1287   **b.** 288   43. 720   44. 1050

45. **a.** 5040   **b.** 3600

46. **a.** 487,635   **b.** 550   **c.** 341,055

47. **a.** 1365   **b.** 1155

48. **a.** .019   **b.** .981

49. **a.** .54   **b.** .72

50. **a.** .757   **b.** .243

51. **a.** .56   **b.** .4   52. .5

Chapter 7 Before Moving On, page 472

1. **a.** $\{d, f, g\}$ **b.** $\{b, c, d, e, f, g\}$ **c.** $\{b, c, e\}$

2. 15   3. 200   4. $\frac{5}{12}$ 5. $\frac{4}{13}$ 6. **a.** .9   **b.** .3

CHAPTER 8

Exercises 8.1, page 478

1. $\frac{1}{32}$ 3. $\frac{31}{32}$

5. $P(E) = 13C(4, 2)/C(52, 2) \approx .059$

7. $C(26, 2)/C(52, 2) \approx .245$

9. $[C(3, 2)C(5, 2)]/C(8, 4) = 3/7$

11. $[C(5, 1)C(3, 3)]/C(8, 4) = 1/14$

13. $C(3, 2)/8 = 3/8$

15. $1/8$ 17. $C(10, 6)/2^{10} \approx .205$

19. **a.** $C(4, 2)/C(24, 2) \approx .022$
    b. $1 - C(20, 2)/C(24, 2) \approx .312$

21. **a.** $C(6, 2)/C(80, 2) \approx .005$
    b. $1 - C(74, 2)/C(80, 2) \approx .145$

23. **a.** .12; $C(98, 10)/C(100, 12) \approx .013$
    b. .15; .015

25. $[C(12, 8)C(8, 2) + C(12, 9)C(8, 1) + C(12, 10)]/C(20, 10) \approx .085$

27. **a.** $\frac{3}{5}$ **b.** $C(3, 1)/C(5, 3) = .3$ **c.** $1 - C(3, 3)/C(5, 3) = .9$

29. $\frac{1}{729}$ 31. .0001   33. .1

35. $40/C(52, 5) \approx .0000154$

37. $[4C(13, 5) - 40]/C(52, 5) \approx .00197$

39. $[13C(4, 3) \cdot 12C(4, 2)]/C(52, 5) \approx .00144$

41. **a.** .618   **b.** .059   43. .030

Exercises 8.2, page 490

1. **a.** .6   **b.** .5   3. .3   5. Not independent

7. Independent   9. **a.** .3   **b.** .8   **c.** .4   **d.** .7

11. **a.** .5   **b.** .4   **c.** .2   **d.** .35   **e.** No   **f.** No

13.

```
                          .3      C
                     A ──────<
                .4  ╱         .7 ── D
                  ╱
                 ╱
                ╲
                 ╲  .3      C
            .6   B ──────<
                           .7 ── D
```

a. .4   **b.** .3   **c.** .12   **d.** .30   **e.** Yes   **f.** Yes

15. **a.** $\frac{1}{12}$ **b.** $\frac{1}{36}$ **c.** $\frac{1}{6}$ **d.** $\frac{1}{6}$ **e.** No

17. $\frac{4}{11}$ 19. Independent   21. Not independent   23. .1875

25. **a.** $\frac{4}{9}$ **b.** $\frac{4}{9}$ 27. **a.** $\frac{1}{21}$ **b.** $\frac{1}{3}$

29. .48   31. .06   33. $\frac{1}{7}$

35. a.

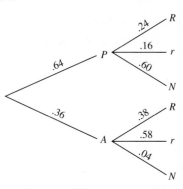

P = Professional
A = Amateur
R = Recovered within 48 hr
r = Recovered after 48 hr
N = Never recovered

b. .24   **c.** .3984

37. a. .16   **b.** .424   **c.** .1696

39. a. .092   **b.** .008

41. a. .28; .39; .18; .643; .292   **b.** Not independent

43. Not independent   **45.** .0000068   **47. a.** $\frac{7}{10}$ **b.** $\frac{1}{5}$

49. 3   **53.** 1   **57.** True   **59.** True

Exercises 8.3, page 499

1.

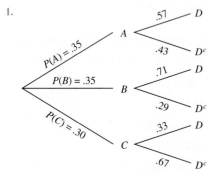

3. a. .45   **b.** .22   **5. a.** .48   **b.** .33

7. a. .08   **b.** .15   **c.** .348

9. a. $\frac{1}{12}$ **b.** $\frac{1}{4}$ **c.** $\frac{1}{18}$ **d.** $\frac{3}{14}$

11.

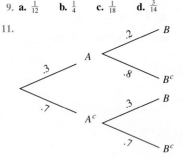

a. .27   **b.** .22   **c.** .73   **d.** .33

13. $\frac{4}{17}$ **15.** $\frac{4}{51}$

17.

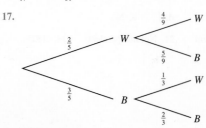

19. $\frac{9}{17}$ **21.** .423   **23. a.** $\frac{3}{4}$ **b.** $\frac{2}{9}$ **25.** .125   **27.** .407

29. a. .543   **b.** .545   **c.** .455   **31.** .377

33. a. .297   **b.** .10   **35.** .856   **37. a.** .57   **b.** .691

39. .647   **41.** .172   **43.** .301

45. a. .763   **b.** .276   **c.** .724

47. a. .4906   **b.** .62   **c.** .186   **49.** .028   **51.** .029

Exercises 8.4, page 511

1. a. See part (b).
b.

Outcome	GGG	GGR	GRG	RGG
Value	3	2	2	2

Outcome	GRR	RGR	RRG	RRR
Value	1	1	1	0

c. {GGG}

3. Any positive integer   **5.** $\frac{1}{6}$

7. Any positive integer; infinite discrete

9. $x \geq 0$; continuous

11. Any positive integer; infinite discrete

13. No. The probability assigned to a value of the random variable X cannot be negative.

15. No. The sum of the probabilities exceed 1.

17. $a = .2$

19. a. .20   **b.** .60   **c.** .30   **d.** 1   **e.** .40   **f.** 0

21.

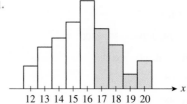

23. a.

x	1	2	3	4	5	6
$P(X = x)$	$\frac{1}{6}$	$\frac{1}{6}$	$\frac{1}{6}$	$\frac{1}{6}$	$\frac{1}{6}$	$\frac{1}{6}$

y	1	2	3	4	5	6
$P(Y = y)$	$\frac{1}{6}$	$\frac{1}{6}$	$\frac{1}{6}$	$\frac{1}{6}$	$\frac{1}{6}$	$\frac{1}{6}$

b.

$x + y$	2	3	4	5	6	7
$P(X + Y = x + y)$	$\frac{1}{36}$	$\frac{2}{36}$	$\frac{3}{36}$	$\frac{4}{36}$	$\frac{5}{36}$	$\frac{6}{36}$

$x + y$	8	9	10	11	12
$P(X + Y = x + y)$	$\frac{5}{36}$	$\frac{4}{36}$	$\frac{3}{36}$	$\frac{2}{36}$	$\frac{1}{36}$

25. a.

x	0	1	2	3	4
$P(X = x)$	.017	.067	.033	.117	.233

x	5	6	7	8	9	10
$P(X = x)$	.133	.167	.100	.050	.067	.017

b.

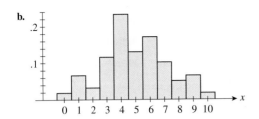

c. .217

27. a.

x	1	2	3	4	5
P(X = x)	.007	.029	.021	.079	.164

x	6	7	8	9	10
P(X = x)	.15	.20	.207	.114	.029

b. .70

29. True

Using Technology Exercises 8.4, page 516

Graphing Utility

1.

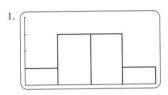

3.

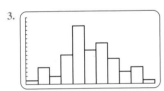

Excel

1.

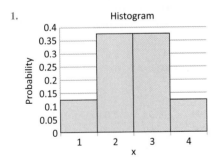

3.

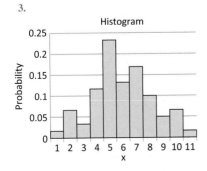

Exercises 8.5, page 525

1. a. 2.6

b.

x	0	1	2	3	4	
P(X = x)	0	.1	.4	.3	.2	; 2.6

3. 0.96   **5.** $78.50   **7.** 0.91

9. 0.12   **11.** 1.73   **13.** 4.16%

15. −39¢   **17.** $100   **19.** $118,800

21. City B **23.** Company B **25.** 2.86%

27. −5.3¢   **29.** −2.7¢   **31.** 2 to 3; 3 to 2

33. .4   **35.** $\frac{7}{12}$ **37.** $\frac{5}{14}$

41. a. Mean: 74; mode: 85; median: 80   **b.** Mode

43. 3; close   **45.** 16; 16; 16   **47.** True

Exercises 8.6, page 535

1. $\mu = 2$, Var$(X) = 1$, $\sigma = 1$

3. $\mu = 0$, Var$(X) = 1$, $\sigma = 1$

5. $\mu = 518$, Var$(X) = 1891$, $\sigma \approx 43.5$

7. Figure (a)   **9.** 1.56

11. $\mu = 4.5$, Var$(X) = 5.25$

13. a. Let X = the annual birthrate during the years 1997–2006.
b.

x	13.9	14.0	14.1	14.2	14.5	14.6	14.7
P(X = x)	.1	.2	.2	.1	.2	.1	.1

c. $\mu = 14.26$, Var$(X) = 0.0744$, $\sigma \approx 0.2728$

15. a. Mutual Fund A: $\mu = \$620$, Var$(X) = 267{,}600$;
Mutual Fund B: $\mu = \$520$, Var$(X) = 137{,}600$
b. Mutual Fund A
c. Mutual Fund B

17. 1

19. $\mu = \$439{,}600$; Var$(X) = 1{,}443{,}840{,}000$; $\sigma \approx \$37{,}998$

21. 95.3%; 0.5%   **23.** 34.95; 5.94   **25.** −25.5%; 9.23

27. 21.59%; 5.20   **29.** 9.64%; 1.64   **31.** 94.56%; 19.94

33. 1607; 182   **35.** 3.324; 0.4497

37. a. At least .75
b. At least .96

39. 7   **41.** At least $\frac{7}{16}$

43. At least $\frac{15}{16}$ **45.** True

Using Technology Exercises 8.6, page 541

1. a.

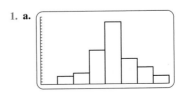

b. $\mu = 4$,
$\sigma \approx 1.40$

3. a.    b. $\mu = 17.34$,
$\sigma \approx 1.11$

5. a. Let X denote the random variable that gives the weight of a carton of sugar.

b.

x	4.96	4.97	4.98	4.99	5.00	5.01
$P(X = x)$	$\frac{3}{30}$	$\frac{4}{30}$	$\frac{4}{30}$	$\frac{1}{30}$	$\frac{1}{30}$	$\frac{5}{30}$

x	5.02	5.03	5.04	5.05	5.06
$P(X = x)$	$\frac{3}{30}$	$\frac{3}{30}$	$\frac{4}{30}$	$\frac{1}{30}$	$\frac{1}{30}$

c. $\mu \approx 5.00$; $\sigma \approx 0.03$

7. a. 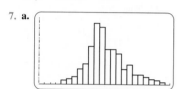   b. 65.875; 1.73

Chapter 8 Concept Review Questions, page 542

1. conditional   2. independent   3. a posteriori probability

4. random   5. finite; infinite; continuous   6. sum; .75

7. a. $\dfrac{P(E)}{P(E^C)}$ b. $\dfrac{a}{a + b}$

8. $p_1(x_1 - \mu)^2 + p_2(x_2 - \mu)^2 + \cdots + p_n(x_n - \mu)^2$; $\sqrt{\text{Var}(X)}$

Chapter 8 Review Exercises, page 543

1. .18   2. .25   3. .06   4. .49   5. .37

6. .364   7. No   8. .5   9. a. $\frac{7}{8}$ b. $\frac{7}{8}$ c. No

10. a. .284   b. .984   11. .150   12. $\frac{2}{15}$ 13. $\frac{1}{24}$

14. $\frac{1}{52}$ 15. .00018   16. .00995   17. .245   18. .510

19. .245   20. .619   21. .180   22. a. .513   b. .390

23. a. {WWW, BWW, WBW, WWB, BBW, BWB, WBB, BBB}

b.

Outcome	WWW	BWW	WBW	WWB
Value of X	0	1	1	1

Outcome	BBW	BWB	WBB	BBB
Value of X	2	2	2	3

c.

x	0	1	2	3
$P(X = x)$	$\frac{1}{35}$	$\frac{12}{35}$	$\frac{18}{35}$	$\frac{4}{35}$

d.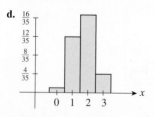

24. $100   25. a. .8   b. $\mu = 2.7$; $\sigma \approx 1.42$

26. 41.3 mph   27. $12,000   28. At least .75

29. €26.76 billion; €1.6317 billion   30. $\mu = 27.87$; $\sigma = 6.41$

Chapter 8 Before Moving On, page 544

1. .72   2. .308

3.

x	-3	-2	0	1	2	3
$P(X = x)$	.05	.1	.25	.3	.2	.1

4. a. .8   b. .92

5. 0.44; 4.0064; 2.0016

CHAPTER 9

Exercises 9.1, page 562

1. $\lim\limits_{x \to -2} f(x) = 3$ 3. $\lim\limits_{x \to 3} f(x) = 3$ 5. $\lim\limits_{x \to -2} f(x) = 3$

7. The limit does not exist.

9.

x	1.9	1.99	1.999
$f(x)$	4.61	4.9601	4.9960

x	2.001	2.01	2.1
$f(x)$	5.004	5.0401	5.41

$\lim\limits_{x \to 2} (x^2 + 1) = 5$

11.

x	-0.1	-0.01	-0.001
$f(x)$	-1	-1	-1

x	0.001	0.01	0.1
$f(x)$	1	1	1

The limit does not exist.

13.

x	0.9	0.99	0.999
$f(x)$	100	10,000	1,000,000

x	1.001	1.01	1.1
$f(x)$	1,000,000	10,000	100

The limit does not exist.

15.

x	0.9	0.99	0.999	1.001	1.01	1.1
$f(x)$	2.9	2.99	2.999	3.001	3.01	3.1

$\lim\limits_{x \to 1} \dfrac{x^2 + x - 2}{x - 1} = 3$

17.

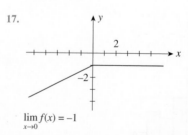

$\lim\limits_{x \to 0} f(x) = -1$

19.

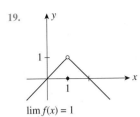

$$\lim_{x \to 1} f(x) = 1$$

21.

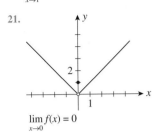

$$\lim_{x \to 0} f(x) = 0$$

23. 4   **25.** 3   **27.** -2 **29.** 2   **31.** -4 **33.** $\frac{5}{6}$

35. 2   **37.** $\sqrt{171} = 3\sqrt{19}$ **39.** 1   **41.** -1 **43.** 0

45. 2   **47.** $\frac{1}{6}$ **49.** 2   **51.** $-\frac{1}{2}$ **53.** -10

55. The limit does not exist.   **57.** $\frac{5}{3}$ **59.** $\frac{1}{2}$ **61.** $\frac{2}{3}$

63. $\lim_{x \to \infty} f(x) = \infty;\ \lim_{x \to -\infty} f(x) = \infty$ **65.** 0; 0

67. $\lim_{x \to \infty} f(x) = -\infty;\ \lim_{x \to -\infty} f(x) = -\infty$

69.

x	1	10	100	1000
$f(x)$	0.5	0.009901	0.0001	0.000001

x	-1	-10	-100	-1000
$f(x)$	0.5	0.009901	0.0001	0.000001

$$\lim_{x \to \infty} f(x) = 0 \text{ and } \lim_{x \to -\infty} f(x) = 0$$

71.

x	1	5	10	100
$f(x)$	12	360	2910	2.99×10^6

x	1000	-1	-5
$f(x)$	2.999×10^9	6	-390

x	-10	-100	-1000
$f(x)$	-3090	-3.01×10^6	-3.0×10^9

$$\lim_{x \to \infty} f(x) = \infty \text{ and } \lim_{x \to -\infty} f(x) = -\infty$$

73. 3   **75.** 3   **77.** $\lim_{x \to -\infty} f(x) = -\infty$ **79.** 0

81. a. $0.5 million; $0.75 million; $1.17 million; $2 million; $4.5 million; $9.5 million
b. The limit does not exist; as the percent of pollutant to be removed approaches 100, the cost becomes astronomical.

83. $2.20; the average cost of producing x DVDs will approach $2.20/disc in the long run.

85. a. $24 million; $60 million; $83.1 million   **b.** $120 million

87. 1000   **89.** a moles/liter/second   **91.** True

93. False   **95.** False   **97.** No

Using Technology Exercises 9.1, page 568

1. 5   **3.** 3   **5.** $\frac{2}{3}$ **7.** $e^2 \approx 7.38906$

11. a.

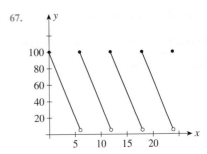

b. 25,000

13. a. 50   **c.**

Exercises 9.2, page 577

1. 3; 2; the limit does not exist.

3. The limit does not exist; 2; the limit does not exist.

5. 0; 2; the limit does not exist.

7. -2; 2; the limit does not exist.

9. True   **11.** True   **13.** False   **15.** True

17. False   **19.** True   **21.** 7   **23.** $-\frac{1}{2}$

25. The limit does not exist.   **27.** -1 **29.** 0

31. -4 **33.** The limit does not exist.

35. 4   **37.** 0; 0   **39.** $x = 0$; conditions 2 and 3

41. Continuous everywhere   **43.** $x = 0$; condition 3

45. $(-\infty, \infty)$ **47.** $(-\infty, \infty)$ **49.** $\left(-\infty, \frac{1}{2}\right)$ and $\left(\frac{1}{2}, \infty\right)$

51. $(-\infty, -2), (-2, 1),$ and $(1, \infty)$ **53.** $(-\infty, \infty)$

55. $(-\infty, \infty)$ **57.** -1 and 1   **59.** 1 and 2

61. f is discontinuous at $x = 1, 2, \ldots, 12$.

63. Michael makes progress toward solving the problem until $x = x_1$. Between $x = x_1$ and $x = x_2$, he makes no further progress. But at $x = x_2$ he suddenly achieves a breakthrough, and at $x = x_3$ he proceeds to complete the problem.

65. Conditions 2 and 3 are not satisfied at each of these points.

67.

69.

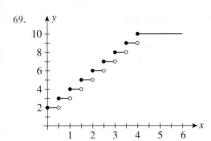

f is discontinuous at $x = \frac{1}{2}, 1, 1\frac{1}{2}, \ldots, 4$.

71. a. ∞; as the time taken to excite the tissue is made shorter and shorter, the strength of the electric current gets stronger and stronger.
  b. b; as the time taken to excite the tissue is made longer and longer, the strength of the electric current gets weaker and weaker and approaches b.

73. 3   **75. a.** Yes   **b.** No

77. a. f is a polynomial of degree 2.
  b. $f(1) = 3$ and $f(3) = -1$

79. a. f is a polynomial of degree 3.
  b. $f(-1) = -4$ and $f(1) = 4$

81. 0.59   **83.** 1.34

85. c. $\frac{1}{2}$; $\frac{7}{2}$; Joan sees the ball on its way up $\frac{1}{2}$ sec after it was thrown and again 3 sec later.

87. False   **89.** False   **91.** False   **93.** False   **95.** False

97. No   **99. c.** $\pm\dfrac{\sqrt{2}}{2}$

Using Technology Exercises 9.2, page 583

1. $x = 0, 1$ **3.** $x = 0, \frac{1}{2}$ **5.** $x = -\frac{1}{2}, 2$ **7.** $x = -2, 1$

9.
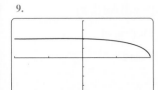

11.
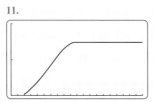

Exercises 9.3, page 596

1. 1.5 lb/month; 0.58 lb/month; 1.25 lb/month

3. 3.1%/hr; -21.2%/hr

5. a. Car A **b.** They are traveling at the same speed.
  c. Car B
  d. Both cars covered the same distance; they are again side-by-side.

7. a. P_2 **b.** P_1 **c.** Bactericide B; Bactericide A

9. 0   **11.** 2   **13.** $4x$

15. $-2x + 3$ **17.** 2; $y = 2x + 7$

19. 6; $y = 6x - 3$ **21.** $\frac{1}{9}$; $y = \frac{1}{9}x - \frac{2}{3}$

23. a. $4x$ **b.** $y = 4x - 1$
  c.

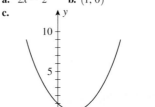

25. a. $2x - 2$ **b.** $(1, 0)$
  c.

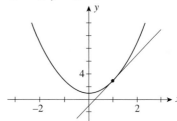

  d. 0

27. a. 6; 5.5; 5.1   **b.** 5
  c. The computations in part (a) illustrate that as h approaches zero, the average rate of change approaches the instantaneous rate of change.

29. a. 130 ft/sec; 128.2 ft/sec; 128.02 ft/sec   **b.** 128 ft/sec
  c. The computations in part (a) illustrate that as the time intervals over which the average velocity are computed become smaller and smaller, the average velocity approaches the instantaneous velocity of the car at $t = 20$.

31. a. 5 sec   **b.** 80 ft/sec   **c.** 160 ft/sec

33. a. $-\frac{1}{6}$ liter/atmosphere   **b.** $-\frac{1}{4}$ liter/atmosphere

35. a. $-\frac{2}{3}x + 7$ **b.** $333/per \$1000 spent on advertising;
                                   $-\$13,000$/per \$1000 spent on advertising

37. \$6 billion/year; \$10 billion/year

39. a. $f'(h)$ gives the instantaneous rate of change of the temperature with respect to height at a given height h, in °F per foot.
  b. Negative   **c.** -0.05°F

41. Average rate of change of the seal population over $[a, a + h]$; instantaneous rate of change of the seal population at $x = a$

43. Average rate of change of the country's industrial production over $[a, a + h]$; instantaneous rate of change of the country's industrial production at $x = a$

45. Average rate of change of atmospheric pressure with respect to altitude over $[a, a + h]$; instantaneous rate of change of atmospheric pressure with respect to altitude at $x = a$

47. a. Yes   **b.** No   **c.** No

49. a. Yes   **b.** Yes   **c.** No

51. a. No   **b.** No   **c.** No

53. 32.1, 30.939, 30.814, 30.8014, 30.8001; 30.8 ft/sec

55. False

57.

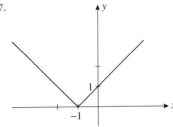

59. $a = 2, b = -1$

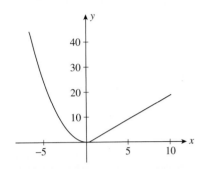

Using Technology Exercises 9.3, page 602

1. a. $y = 9x - 11$
   b.

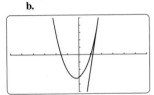

3. a. $y = \frac{1}{12}x + \frac{4}{3}$
   b.

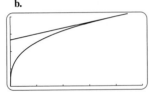

5. a. 4
   b. $y = 4x - 1$
   c.

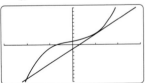

7. a. 4.02
   b. $y = 4.02x - 3.57$
   c.

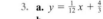

9. a.

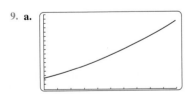

   b. 41.22¢/mi   **c.** 1.22¢/mi/year

Exercises 9.4, page 610

1. 0   **3.** $5x^4$ **5.** $3.1x^{2.1}$ **7.** $8x$ **9.** $2\pi r$ **11.** $\dfrac{2}{x^{2/3}}$

13. $\dfrac{3}{2\sqrt{x}}$ **15.** $-84x^{-13}$ **17.** $8x - 2$ **19.** $-3x^2 + 4x$

21. $0.06x - 0.4$ **23.** $4x - 4 - \dfrac{3}{x^2}$ **25.** $16x^3 - 7.5x^{3/2}$

27. $-\dfrac{5}{x^2} - \dfrac{8}{x^3}$ **29.** $-\dfrac{16}{t^5} + \dfrac{9}{t^4} - \dfrac{2}{t^2}$ **31.** $3 - \dfrac{5}{2\sqrt{x}}$

33. $-\dfrac{4}{x^3} + \dfrac{1}{x^{4/3}}$ **35. a.** 20   **b.** -4 **c.** 20   **37.** 3

39. 11   **41.** $m = 5$; $y = 5x - 4$ **43.** $m = 3$; $y = 3x - 7$

45. a. $(0, 0)$
   b.

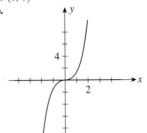

47. a. $(-2, -7), (2, 9)$
   b. $y = 12x + 17$ and $y = 12x - 15$
   c.

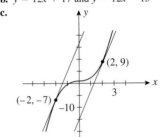

49. a. $(0, 0)$; $\left(1, -\frac{13}{12}\right)$ **b.** $(0, 0)$; $\left(2, -\frac{8}{3}\right)$; $\left(-1, -\frac{5}{12}\right)$
   c. $(0, 0)$; $\left(4, \frac{80}{3}\right)$; $\left(-3, \frac{81}{4}\right)$

51. a. $\dfrac{16\pi}{9}$ cm³/cm   **b.** $\dfrac{25\pi}{4}$ cm³/cm

53. a. 16.3 million   **b.** 14.3 million/year   **c.** 66.8 million
   d. 11.7 million/year

55. a. 49.6%; 41.1%; 36.9%; 34.1%
   b. -5.6%/decade; -3.3%/decade

57. a. 157 million   **b.** 10.4 million/year

59. a. $120 - 30t$ **b.** 120 ft/sec   **c.** 240 ft

61. a. 5%; 11.3%; 15.5%   **b.** 0.63%/year; 0.525%/year

63. a. -0.9 thousand metric tons/year; 20.3 thousand metric tons/year
   b. Yes

65. a. 15 pts/year; 12.6 pts/year; 0 pts/year   **b.** 10 pts/year

67. a. $0.000125x^{1/4}$ **b.** $0.00125/radio

69. a. $20\left(1 - \dfrac{1}{\sqrt{t}}\right)$ **b.** 50 mph; 30 mph; 33.4 mph
   c. At 6:30 A.M. the average speed is decreasing at the rate of 8.28 mph/hr; at 7 A.M., it is not changing; and at 8 A.M., it is increasing at the rate of 5.86 mph/hr.

71. 32 turtles/year; 428 turtles/year; 3260 turtles

73. a. 12%; 23.9%   **b.** 0.8%/year; 1.1%/year

75. a. The population age 65 and over, including the population of the developed countries and that of the underdeveloped/emerging countries
   b. $0.92t + 3.727$; 13 million people/year

77. True

Using Technology Exercises 9.4, page 616

1. 1   **3.** 0.4226   **5.** 0.1613

7. a.

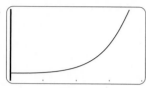

b. 3.4295 ppm/40 year; 105.4332 ppm/40 year

9. a. $f(t) = 0.611t^3 + 9.702t^2 + 32.544t + 473.5$

b.

c. At the beginning of 2000, the assets of the hedge funds were increasing at the rate of \$53.78 billion/year, and at the beginning of 2003, they were increasing at the rate of \$139.48 billion/year.

Exercises 9.5, page 625

1. $2x(2x) + (x^2 + 1)(2)$, or $6x^2 + 2$

3. $(t - 1)(2) + (2t + 1)(1)$, or $4t - 1$

5. $(3x + 2)(2x) + (x^2 - 2)(3)$, or $9x^2 + 4x - 6$

7. $(x^3 - 1)(2) + (2x + 1)(3x^2)$, or $8x^3 + 3x^2 - 2$

9. $(w^3 - w^2 + w - 1)(2w) + (w^2 + 2)(3w^2 - 2w + 1)$, or $5w^4 - 4w^3 + 9w^2 - 6w + 2$

11. $(5x^2 + 1)(x^{-1/2}) + (2x^{1/2} - 1)(10x)$, or $\dfrac{25x^2 - 10x\sqrt{x} + 1}{\sqrt{x}}$

13. $\dfrac{(x^2 - 5x + 2)(x^2 + 2)}{x^2} + \dfrac{(x^2 - 2)(2x - 5)}{x}$, or $\dfrac{3x^4 - 10x^3 + 4}{x^2}$

15. $\dfrac{-1}{(x - 2)^2}$ **17.** $\dfrac{(2x + 1)(2) - (2x - 1)(2)}{(2x + 1)^2}$, or $\dfrac{4}{(2x + 1)^2}$

19. $-\dfrac{2x}{(x^2 + 1)^2}$ **21.** $\dfrac{s^2 + 2s + 4}{(s + 1)^2}$

23. $\dfrac{(\frac{1}{2}x^{-1/2})[x^2 + 1 - 4x^{3/2}(x^{1/2} + 1)]}{(x^2 + 1)^2}$, or $\dfrac{-3x^2 - 4x^{3/2} + 1}{2\sqrt{x}(x^2 + 1)^2}$

25. $\dfrac{2x^3 + 2x^2 + 2x - 2x^3 - x^2 - 4x - 2}{(x^2 + x + 1)^2}$, or $\dfrac{x^2 - 2x - 2}{(x^2 + x + 1)^2}$

27. $\dfrac{(x - 2)(3x^2 + 2x + 1) - (x^3 + x^2 + x + 1)}{(x - 2)^2}$, or $\dfrac{2x^3 - 5x^2 - 4x - 3}{(x - 2)^2}$

29. $\dfrac{(x^2 - 4)(x^2 + 4)(2x + 8) - (x^2 + 8x - 4)(4x^3)}{(x^2 - 4)^2(x^2 + 4)^2}$, or $\dfrac{-2x^5 - 24x^4 + 16x^3 - 32x - 128}{(x^2 - 4)^2(x^2 + 4)^2}$

31. 8   **33.** −9   **35.** $2(3x^2 - x + 3)$; 10

37. $\dfrac{-3x^4 + 2x^2 - 1}{(x^4 - 2x^2 - 1)^2}$; $-\dfrac{1}{2}$ **39.** 60; $y = 60x - 102$

41. $-\dfrac{1}{2}$; $y = -\dfrac{1}{2}x + \dfrac{3}{2}$ **43.** $4x - 2$; 4   **45.** $6x^2 - 6x$; $6(2x - 1)$

47. $4t^3 - 6t^2 + 12t - 3$; $12(t^2 - t + 1)$ **49.** $12(6x - 1)$

51. $-\dfrac{6}{x^4}$ **53.** 8   **55.** $y = 7x - 5$ **57.** $(\frac{1}{3}, \frac{50}{27})$; $(1, 2)$

59. $(\frac{4}{3}, -\frac{770}{27})$; $(2, -30)$ **61.** $y = -\frac{1}{2}x + 1$; $y = 2x - \frac{3}{2}$

63. 0.125, 0.5, 2, 50; the cost of removing (essentially) all of the pollutant is prohibitively high.

65. −5000/min; −1600/min; 7000; 4000

67. a. $\dfrac{50x}{0.01x^2 + 1}$ **b.** $\dfrac{50(1 - 0.01x^2)}{(0.01x^2 + 1)^2}$

c. 6.69, 0, −3.70; the revenue is increasing at the rate of approximately \$6700/thousand watches/week when the level of sales is 8000 watches/week; the rate of change of the revenue is \$0/thousand watches/week when the level of sales is 10,000 watches/week, and the revenue is decreasing at the rate of approximately \$3700/thousand watches/week when the sales are 12,000 watches/week.

69. a. $\dfrac{180}{(t + 6)^2}$ **b.** 3.7; 2.2; 1.8; 1.1

c.

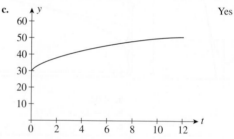

Yes

d. 50 words/min

71. Dropping at the rate of 0.0375 ppm/year; dropping at the rate of 0.006 ppm/year

73. 128 ft/sec; 32 ft/sec²

75. a. and b.

t	0	1	2	3	4	5	6	7
$N'(t)$	0	2.7	4.8	6.3	7.2	7.5	7.2	6.3
$N''(t)$				0.6	0	−0.6	−1.2

77. 8.1 million; 0.204 million/year; −0.03 million/year². At the beginning of 1998, there were 8.1 million people receiving disability benefits; the number was increasing at the rate of 0.2 million/year; the rate of the rate of change of the number of people was decreasing at the rate of 0.03 million people/year².

79. False   **81.** False

Using Technology Exercises 9.5, page 631

1. 0.8750   **3.** 0.0774

5. −0.5000   **7.** 31,312/year

9. −18   11. 15.2762   13. −0.6255   15. 0.1973

17. −68.46214; at the beginning of 1988, the rate of change of the rate of change of the rate at which banks were failing was 68 banks/year/year/year.

Exercises 9.6, page 639

1. $6(2x - 1)^2$ 3. $10x(x^2 + 2)^4$

5. $4(2x - x^2)^3(2 - 2x)$, or $8x^3(x - 2)^3(x - 1)$ 7. $\dfrac{-4}{(2x + 1)^3}$

9. $5x(x^2 - 4)^{3/2}$ 11. $\dfrac{3}{2\sqrt{3x - 2}}$ 13. $\dfrac{-2x}{3(1 - x^2)^{2/3}}$

15. $-\dfrac{6}{(2x + 3)^4}$ 17. $\dfrac{-1}{(2t - 4)^{3/2}}$ 19. $-\dfrac{3(16x^3 + 1)}{2(4x^4 + x)^{5/2}}$

21. $-2(3x^2 + 2x + 1)^{-3}(6x + 2)$ or $-4(3x + 1)(3x^2 + 2x + 1)^{-3}$

23. $3(x^2 + 1)^2(2x) - 2(x^3 + 1)(3x^2)$, or $6x(2x^2 - x + 1)$

25. $3(t^{-1} - t^{-2})^2(-t^{-2} + 2t^{-3})$

27. $\dfrac{1}{2\sqrt{x - 1}} + \dfrac{1}{2\sqrt{x + 1}}$

29. $2x^2(4)(3 - 4x)^3(-4) + (3 - 4x)^4(4x)$, or $(-12x)(4x - 1)(3 - 4x)^3$

31. $8(x - 1)^2(2x + 1)^3 + 2(x - 1)(2x + 1)^4$, or $6(x - 1)(2x - 1)(2x + 1)^3$

33. $3\left(\dfrac{x + 3}{x - 2}\right)^2\left[\dfrac{(x - 2)(1) - (x + 3)(1)}{(x - 2)^2}\right]$, or $-\dfrac{15(x + 3)^2}{(x - 2)^4}$

35. $\dfrac{3}{2}\left(\dfrac{t}{2t + 1}\right)^{1/2}\left[\dfrac{(2t + 1)(1) - t(2)}{(2t + 1)^2}\right]$, or $\dfrac{3t^{1/2}}{2(2t + 1)^{5/2}}$

37. $\dfrac{1}{2}\left(\dfrac{u + 1}{3u + 2}\right)^{-1/2}\left[\dfrac{(3u + 2)(1) - (u + 1)(3)}{(3u + 2)^2}\right]$, or

$-\dfrac{1}{2\sqrt{u + 1}\,(3u + 2)^{3/2}}$

39. $\dfrac{(x^2 - 1)^4(2x) - x^2(4)(x^2 - 1)^3(2x)}{(x^2 - 1)^8}$, or $\dfrac{(-2x)(3x^2 + 1)}{(x^2 - 1)^5}$

41. $\dfrac{2x(x^2 - 1)^3(3x^2 + 1)^2[9(x^2 - 1) - 4(3x^2 + 1)]}{(x^2 - 1)^8}$, or

$-\dfrac{2x(3x^2 + 13)(3x^2 + 1)^2}{(x^2 - 1)^5}$

43. $\dfrac{(2x + 1)^{-1/2}[(x^2 - 1) - (2x + 1)(2x)]}{(x^2 - 1)^2}$, or

$-\dfrac{3x^2 + 2x + 1}{\sqrt{2x + 1}(x^2 - 1)^2}$

45. $\dfrac{(t^2 + 1)^{1/2}(\frac{1}{2})(t + 1)^{-1/2}(1) - (t + 1)^{1/2}(\frac{1}{2})(t^2 + 1)^{-1/2}(2t)}{t^2 + 1}$, or

$-\dfrac{t^2 + 2t - 1}{2\sqrt{t + 1}(t^2 + 1)^{3/2}}$

47. $10(9x^2 + 2)(x^2 + 2)^3$

49. $\frac{4}{3}u^{1/3}$; $6x$; $8x(3x^2 - 1)^{1/3}$

51. $-\dfrac{2}{3u^{5/3}}$; $6x^2 - 1$; $-\dfrac{2(6x^2 - 1)}{3(2x^3 - x + 1)^{5/3}}$

53. $\frac{1}{2}u^{-1/2} - \frac{1}{2}u^{-3/2}$; $3x^2 - 1$; $\dfrac{(3x^2 - 1)(x^3 - x - 1)}{2(x^3 - x)^{3/2}}$

55. $2f'(2x + 1)$ 57. −12   59. 6   61. No

63. $y = -33x + 57$ 65. $y = \frac{43}{5}x - \frac{54}{5}$

67. 0.333 million/week; 0.305 million/week; 16 million; 22.7 million

69. $\dfrac{6.87775}{(5 + t)^{0.795}}$; 0.53%/year; 64.9%

71. **a.** 100, 30.0. The probability of survival at the moment of diagnosis is 100%. The probability of survival 1 year after diagnosis is approximately 30%.
  b. −129% per year, −34% per year. At the moment of diagnosis, the probability of survival is falling at the rate of approximately 129% per year; the probability of survival 1 year after diagnosis is dropping at the rate of approximately 34% per year.

73. **a.** $0.0267(0.2t^2 + 4t + 64)^{-1/3}(0.1t + 1)$ **b.** 0.0090 ppm/year

75. **a.** $0.03[3t^2(t - 7)^4 + t^3(4)(t - 7)^3]$, or $0.21t^2(t - 3)(t - 7)^3$
  b. 90.72; 0; −90.72; at 8 A.M. the level of nitrogen dioxide is increasing; at 10 A.M. the level stops increasing; at 11 A.M. the level is decreasing.

77.

$$300\left[\dfrac{(t + 25)\frac{1}{2}(\frac{1}{2}t^2 + 2t + 25)^{-1/2}(t + 2) - (\frac{1}{2}t^2 + 2t + 25)^{1/2}(1)}{(t + 25)^2}\right],$$

or $\dfrac{3450t}{(t + 25)^2\sqrt{\frac{1}{2}t^2 + 2t + 25}}$; 2.9 beats/min/sec, 0.7 beats/min/sec,

0.2 beats/min/sec, 179 beats/min

79. 160π ft²/sec   81. −27 mph/decade; 19 mph

83.

$$(1.42)\left[\dfrac{(3t^2 + 80t + 550)(14t + 140) - (7t^2 + 140t + 700)(6t + 80)}{(3t^2 + 80t + 550)^2}\right],$$

or $\dfrac{1.42(140t^2 + 3500t + 21{,}000)}{(3t^2 + 80t + 550)^2}$; 31,312 jobs/year/month

85. −400 sports watches/(dollar price increase)

87. True   89. True

Using Technology Exercises 9.6, page 644

1. 0.5774   3. 0.9390   5. −4.9498

7. 5,414,500 people/year; 2,513,600 people/year

Exercises 9.7, page 652

1. $3e^{3x}$ 3. $-e^{-t}$ 5. $e^x + 2x$ 7. $x^2e^x(x + 3)$

9. $\dfrac{e^x(x - 1)}{x^2}$ 11. $3(e^x - e^{-x})$ 13. $-\dfrac{2}{e^w}$

15. $6e^{3x-1}$ 17. $-2xe^{-x^2}$ 19. $-\dfrac{3e^{1/x}}{x^2}$

21. $25e^x(e^x + 1)^{24}$ 23. $\dfrac{e^{\sqrt{x}}}{2\sqrt{x}}$ 25. $e^{3x+2}(3x - 2)$

27. $\dfrac{2e^x}{(e^x + 1)^2}$ 29. $16e^{-4x} + 9e^{3x}$ 31. $6e^{3x}(3x + 2)$

33. $y = 2x - 2$ 35. $\dfrac{5}{x}$ 37. $\dfrac{1}{x + 1}$ 39. $\dfrac{8}{x}$ 41. $\dfrac{1}{2x}$

43. $-\dfrac{2}{x}$ 45. $\dfrac{8x - 5}{4x^2 - 5x + 3}$ 47. $\dfrac{1}{x(x + 1)}$ 49. $x(1 + 2\ln x)$

51. $\dfrac{2(1 - \ln x)}{x^2}$ 53. $\dfrac{3}{u - 2}$ 55. $\dfrac{1}{2x\sqrt{\ln x}}$

57. $\dfrac{2\ln x}{x}$ 59. $\dfrac{3x^2}{x^3 + 1}$ 61. $\dfrac{(x\ln x + 1)e^x}{x}$

63. $-\dfrac{1}{x^2}$ 65. $\dfrac{2(2 - x^2)}{(x^2 + 2)^2}$ 67. $y = x - 1$

69. $-0.1694, -0.1549, -0.1415$; the percentage of the total population relocating was decreasing at the rate of 0.17%/year in 1970, 0.15%/year in 1980, and 0.14%/year in 1990.

71. **a.** 70,000; 353,700   **b.** 37,800/decade; 191,000/decade

73. **a.** 181
   b. 0/decade; −27/decade; −38/decade; −32/decade
   c. 52/decade

75. 12.9%/year; 10.9%/year; 9.3%/year; and 7.9%/year

77. **a.** 70°F   **b.** −14.7°F   **c.** 30°F

79. **a.** −1.64¢/bottle; −1.34¢/bottle
   b. $231.87/bottle; $217.03/bottle

81. 1.8/$1000 output/decade; −0.11/$1000 output/decade;
   −0.23/$1000 output/decade; −0.13/$1000 output/decade

83. 0.0580%/kg; 0.0133%/kg

85. **a.** 14.42%/year; 2.4%/year

87. False

Using Technology Exercises 9.7, page 656

1. 5.4366   3. 12.3929   5. 0.1861

7. **a.**

   b. 4.2720 billion/half century

9. **a.** 153,024; 235,181
   b. 634; 18,401

11. **a.** 69.63%   **b.** 5.09%/decade

Exercises 9.8, page 663

1. **a.** $C(x)$ is always increasing because as the number of units x produced increases, the amount of money that must be spent on production also increases.
   b. 4000

3. **a.** $1.80; $1.60   **b.** $1.80; $1.60

5. **a.** $100 + \dfrac{200{,}000}{x}$ **b.** $-\dfrac{200{,}000}{x^2}$

   c. $\overline{C}(x)$ approaches $100 if the production level is very high.

7. $\dfrac{2000}{x} + 2 - 0.0001x$; $-\dfrac{2000}{x^2} - 0.0001$

9. **a.** $8000 - 200x$ **b.** $200, 0, -200$ **c.** $40

11. **a.** $-0.04x^2 + 600x - 300{,}000$
   b. $-0.08x + 600$ **c.** $200; -40$
   d. The profit increases as production increases, peaking at 7500 units; beyond this level, profit falls.

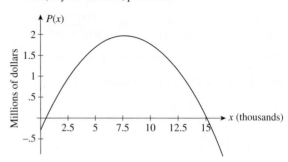

13. **a.** $600x - 0.05x^2$; $-0.000002x^3 - 0.02x^2 + 200x - 80{,}000$
   b. $0.000006x^2 - 0.06x + 400$; $600 - 0.1x$;
      $-0.000006x^2 - 0.04x + 200$
   c. 304; 400; 96

   d.
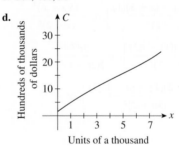

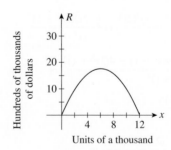

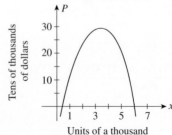

15. a. $0.000002x^2 - 0.03x + 400 + \dfrac{80{,}000}{x}$

b. $0.000004x - 0.03 - \dfrac{80{,}000}{x^2}$

c. -0.0132; 0.0092; the marginal average cost is negative (average cost is decreasing) when 5000 units are produced and positive (average cost is increasing) when 10,000 units are produced.

d.

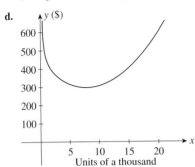

17. a. $\dfrac{50x}{0.01x^2 + 1}$ **b.** $\dfrac{50 - 0.5x^2}{(0.01x^2 + 1)^2}$

c. \$44,380; when the level of production is 2000 units, the revenue increases at the rate of \$44,380 per additional 1000 units produced.

19. 1.21   **21.** 0.288

23. a. $100xe^{-0.0001x}$ **b.** $100(1 - 0.0001)e^{-0.0001x}$ **c.** 0\$/pair

25. False

Chapter 9 Concept Review Questions, page 667

1. $f(x)$; L; a

2. a. L^r **b.** $L \pm M$ **c.** LM **d.** $\dfrac{L}{M}$; $M \neq 0$

3. a. L; x **b.** M; negative; absolute

4. a. right   **b.** left   **c.** L; L

5. a. continuous   **b.** discontinuous   **c.** every

6. a. a; a; $g(a)$ **b.** everywhere   **c.** $Q(x)$

7. a. $[a, b]$; $f(c) = M$ **b.** $f(x) = 0$; (a, b)

8. a. $f'(a)$ **b.** $y = f(a) + m(x - a)$

9. a. $\dfrac{f(a + h) - f(a)}{h}$ **b.** $\displaystyle\lim_{h \to 0} \dfrac{f(a + h) - f(a)}{h}$

10. a. 0   **b.** nx^{n-1} **c.** $cf'(x)$ **d.** $f'(x) \pm g'(x)$

11. a. $f(x)g'(x) + g(x)f'(x)$ **b.** $\dfrac{g(x)f'(x) - f(x)g'(x)}{[g(x)]^2}$

12. a. $g'[f(x)]f'(x)$ **b.** $n[f(x)]^{n-1}f'(x)$

13. marginal cost; marginal revenue; marginal profit; marginal average cost

14. a. $e^{f(x)}f'(x)$ **b.** $\dfrac{f'(x)}{f(x)}$

Chapter 9 Review Exercises, page 668

1. -3 **2.** 2   **3.** -21 **4.** 0   **5.** -1

6. The limit does not exist.   **7.** 7   **8.** $\frac{9}{2}$ **9.** 1   **10.** $\frac{1}{2}$

11. 1   **12.** 1   **13.** $\frac{3}{2}$ **14.** The limit does not exist.

15.

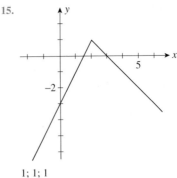

1; 1; 1

16.

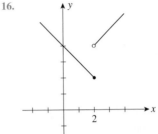

4; 2; the limit does not exist.

17. $x = 2$ **18.** $x = -\frac{1}{2}, 1$ **19.** $x = -1$ **20.** $x = 0$

21. a. 3; 2.5; 2.1   **b.** 2   **22.** 4

23. $\dfrac{1}{x^2}$ **24.** $\frac{3}{2}$; $y = \frac{3}{2}x + 5$

25. -4; $y = -4x + 4$ **26. a.** Yes   **b.** No

27. $15x^4 - 8x^3 + 6x - 2$ **28.** $24x^5 + 8x^3 + 6x$

29. $\dfrac{6}{x^4} - \dfrac{3}{x^2}$ **30.** $4t - 9t^2 + \frac{1}{2}t^{-3/2}$ **31.** $-\dfrac{1}{t^{3/2}} - \dfrac{6}{t^{5/2}}$

32. $2x - \dfrac{2}{x^2}$ **33.** $1 - \dfrac{2}{t^2} - \dfrac{6}{t^3}$ **34.** $4s + \dfrac{4}{s^2} - \dfrac{1}{s^{3/2}}$

35. $2x + \dfrac{3}{x^{5/2}}$ **36.** $\dfrac{(2x - 1)(1) - (x + 1)(2)}{(2x - 1)^2}$, or $-\dfrac{3}{(2x - 1)^2}$

37. $\dfrac{(2t^2 + 1)(2t) - t^2(4t)}{(2t^2 + 1)^2}$, or $\dfrac{2t}{(2t^2 + 1)^2}$

38. $\dfrac{(t^{1/2} + 1)\frac{1}{2}t^{-1/2} - t^{1/2}(\frac{1}{2}t^{-1/2})}{(t^{1/2} + 1)^2}$, or $\dfrac{1}{2\sqrt{t}\,(\sqrt{t} + 1)^2}$

39. $\dfrac{(x^{1/2} + 1)(\frac{1}{2}x^{-1/2}) - (x^{1/2} - 1)(\frac{1}{2}x^{-1/2})}{(x^{1/2} + 1)^2}$, or $\dfrac{1}{\sqrt{x}\,(\sqrt{x} + 1)^2}$

40. $\dfrac{(2t^2 + 1)(1) - t(4t)}{(2t^2 + 1)^2}$, or $\dfrac{1 - 2t^2}{(2t^2 + 1)^2}$

41. $\dfrac{(x^2 - 1)(4x^3 + 2x) - (x^4 + x^2)(2x)}{(x^2 - 1)^2}$, or $\dfrac{2x(x^4 - 2x^2 - 1)}{(x^2 - 1)^2}$

42. $3(4x + 1)(2x^2 + x)^2$

43. $8(3x^3 - 2)^7(9x^2)$, or $72x^2(3x^3 - 2)^7$

44. $5(x^{1/2} + 2)^4 \cdot \dfrac{1}{2}x^{-1/2}$, or $\dfrac{5(\sqrt{x} + 2)^4}{2\sqrt{x}}$

45. $\dfrac{1}{2}(2t^2 + 1)^{-1/2}(4t)$, or $\dfrac{2t}{\sqrt{2t^2 + 1}}$

46. $-2t^2(1 - 2t^3)^{-2/3}$, or $-\dfrac{2t^2}{(1 - 2t^3)^{2/3}}$

47. $-4(3t^2 - 2t + 5)^{-3}(3t - 1)$, or $-\dfrac{4(3t - 1)}{(3t^2 - 2t + 5)^3}$

48. $-\dfrac{3}{2}(2x^3 - 3x^2 + 1)^{-5/2}(6x^2 - 6x)$, or
$-9x(x - 1)(2x^3 - 3x^2 + 1)^{-5/2}$

49. $(2x + 1)e^{2x}$ **50.** $\dfrac{e^t}{2\sqrt{t}} + \sqrt{t}e^t + 1$ **51.** $\dfrac{1 - 4t}{2\sqrt{te^{2t}}}$

52. $\dfrac{e^x(x^2 + x + 1)}{\sqrt{1 + x^2}}$ **53.** $\dfrac{2(e^{2x} + 2)}{(1 + e^{-2x})^2}$ **54.** $4xe^{2x^2 - 1}$

55. $(1 - 2x^2)e^{-x^2}$ **56.** $3e^{2x}(1 + e^{2x})^{1/2}$ **57.** $(x + 1)^2e^x$

58. $\ln t + 1$ **59.** $\dfrac{2xe^{x^2}}{e^{x^2} + 1}$ **60.** $\dfrac{\ln x - 1}{(\ln x)^2}$

61. $\dfrac{x - x \ln x + 1}{x(x + 1)^2}$ **62.** $(x + 2)e^x$ **63.** $\dfrac{4e^{4x}}{e^{4x} + 3}$

64. $\dfrac{(r^3 - r^2 + r + 1)e^r}{(1 + r^2)^2}$ **65.** $\dfrac{1 + e^x(1 - x \ln x)}{x(1 + e^x)^2}$

66. $\dfrac{(2x^2 + 2x^2 \cdot \ln x - 1)e^{x^2}}{x(1 + \ln x)^2}$

67. $2\left(x + \dfrac{1}{x}\right)\left(1 - \dfrac{1}{x^2}\right)$, or $\dfrac{2(x^2 + 1)(x^2 - 1)}{x^3}$

68. $\dfrac{(2x^2 + 1)^2(1) - (1 + x)2(2x^2 + 1)(4x)}{(2x^2 + 1)^4}$, or $-\dfrac{6x^2 + 8x - 1}{(2x^2 + 1)^3}$

69. $(t^2 + t)^4(4t) + 2t^2 \cdot 4(t^2 + t)^3(2t + 1)$, or $4t^2(5t + 3)(t^2 + t)^3$

70. $(2x + 1)^3 \cdot 2(x^2 + x)(2x + 1) + (x^2 + x)^2 3(2x + 1)^2(2)$, or
$2(2x + 1)^2(x^2 + x)(7x^2 + 7x + 1)$

71. $x^{1/2} \cdot 3(x^2 - 1)^2(2x) + (x^2 - 1)^3 \cdot \dfrac{1}{2}x^{-1/2}$, or
$\dfrac{(13x^2 - 1)(x^2 - 1)^2}{2\sqrt{x}}$

72. $\dfrac{(x^3 + 2)^{1/2}(1) - x \cdot \frac{1}{2}(x^3 + 2)^{-1/2} \cdot 3x^2}{x^3 + 2}$, or $\dfrac{4 - x^3}{2(x^3 + 2)^{3/2}}$

73. $\dfrac{(4x - 3)\frac{1}{2}(3x + 2)^{-1/2}(3) - (3x + 2)^{1/2}(4)}{(4x - 3)^2}$, or
$-\dfrac{12x + 25}{2\sqrt{3x + 2}(4x - 3)^2}$

74. $\dfrac{(t + 1)^3\frac{1}{2}(2t + 1)^{-1/2}(2) - (2t + 1)^{1/2} \cdot 3(t + 1)^2(1)}{(t + 1)^6}$, or
$-\dfrac{5t + 2}{\sqrt{2t + 1}(t + 1)^4}$

75. $2(12x^2 - 9x + 2)$ **76.** $-\dfrac{1}{4x^{3/2}} + \dfrac{3}{4x^{5/2}}$

77. $\dfrac{(t^2 + 4)^2(-2t) - (4 - t^2)(2)(t^2 + 4)(2t)}{(t^2 + 4)^4}$, or $\dfrac{2t(t^2 - 12)}{(t^2 + 4)^3}$

78. $4e^{-2x}(x - 1)$ **79.** $\dfrac{e^x(1 - e^x)}{(1 + e^x)^3}$ **80.** $\dfrac{1}{x}$ **81.** $-\dfrac{9}{(3x + 1)^2}$

82. $2(15x^4 + 12x^2 + 6x + 1)$

83. $2(2x^2 + 1)^{-1/2} + 2x\left(-\dfrac{1}{2}\right)(2x^2 + 1)^{-3/2}(4x)$, or $\dfrac{2}{(2x^2 + 1)^{3/2}}$

84. $(t^2 + 1)^2(14t) + (7t^2 + 1)(2)(t^2 + 1)(2t)$, or $6t(t^2 + 1)(7t^2 + 3)$

85. 0   **86.** -2

87. a. $(2, -25)$ and $(-1, 14)$
    b. $y = -4x - 17$; $y = -4x + 10$

88. a. $\left(-2, \frac{25}{3}\right)$ and $\left(1, -\frac{13}{6}\right)$
    b. $y = -2x + \frac{13}{3}$; $y = -2x - \frac{1}{6}$

89. $y = -\dfrac{\sqrt{3}}{3}x + \dfrac{4}{3}\sqrt{3}$ **90.** $y = 112x - 80$

91. $y = -\dfrac{1}{e^2}(2x - 3)$ **92.** $y = \dfrac{1}{e}$

93. $-\dfrac{48}{(2x - 1)^4}$; $\left(-\infty, \dfrac{1}{2}\right) \cup \left(\dfrac{1}{2}, \infty\right)$ **94.** 20

95. a. $C'(x)$ gives the instantaneous rate of change of the total manufacturing cost C in dollars with respect to the quantity produced when x units of the product are produced.
    b. Positive
    c. $20

96. a. 20,430   **b.** 225 cameras/year

97. a. 15%; 31.99%   **b.** 0.51%/year; 1.04%/year

98. 200 subscribers/week

99. 75 years; 0.07 year/year

100. a. $-0.02x^2 + 600x$ **b.** $-0.04x + 600$
    c. 200; the sale of the 10,001st phone will bring a revenue of $200.

101. a. $2.20; $2.20   **b.** $\dfrac{2500}{x} + 2.2$; $-\dfrac{2500}{x^2}$

102. a. $2000x - 0.04x^2$; $-0.000002x^3 - 0.02x^2 + 1000x - 120,000$;
    $0.000002x^2 - 0.02x + 1000 + \dfrac{120,000}{x}$
    b. $0.000006x^2 - 0.04x + 1000$; $2000 - 0.08x$; $-0.000006x^2 - 0.04x + 1000$; $0.000004x - 0.02 - \dfrac{120,000}{x^2}$
    c. 934; 1760; 826
    d. -0.0048; 0.010125; at a production level of 5000, the average cost is decreasing by 0.48¢/unit; at a production level of 8000, the average cost is increasing by 1.0125¢/unit.

103. a. 1175, 2540, 3289   **b.** 4000

104. a. $9/unit   **b.** $8/unit/week   **c.** $18/unit

Chapter 9 Before Moving On, page 671

1. 2   2. **a.** 0   **b.** 1; no

3. $-1; y = -x$ 4. $6x^2 - \dfrac{1}{x^{2/3}} - \dfrac{10}{3x^{5/3}}$ 5. $\dfrac{4x^2 - 1}{\sqrt{2x^2 - 1}}$

6. $-\dfrac{2x^2 + 2x - 1}{(x^2 + x + 1)^2}$ 7. $-\dfrac{1}{2(x + 1)^{3/2}}; \dfrac{3}{4(x + 1)^{5/2}}; -\dfrac{15}{8(x + 1)^{7/2}}$

8. $\dfrac{e^{\sqrt{x}}}{2\sqrt{x}}$ 9. $1 + \ln 2$

CHAPTER 10

Exercises 10.1, page 686

1. Decreasing on $(-\infty, 0)$ and increasing on $(0, \infty)$

3. Increasing on $(-\infty, -1)$ and $(1, \infty)$ and decreasing on $(-1, 1)$

5. Decreasing on $(-\infty, 0)$ and $(2, \infty)$ and increasing on $(0, 2)$

7. Decreasing on $(-\infty, -1)$ and $(1, \infty)$ and increasing on $(-1, 1)$

9. Increasing on $(20.2, 20.6)$ and $(21.7, 21.8)$, constant on $(19.6, 20.2)$ and $(20.6, 21.1)$, and decreasing on $(21.1, 21.7)$ and $(21.8, 22.7)$

11. **a.** Positive   **b.** Positive   **c.** Zero   **d.** Zero
    e. Negative   **f.** Negative   **g.** Positive

13. Increasing on $(-\infty, \infty)$

15. Decreasing on $(-\infty, \frac{3}{2})$ and increasing on $(\frac{3}{2}, \infty)$

17. Decreasing on $\left(-\infty, -\dfrac{\sqrt{3}}{3}\right)$ and $\left(\dfrac{\sqrt{3}}{3}, \infty\right)$ and increasing on $\left(-\dfrac{\sqrt{3}}{3}, \dfrac{\sqrt{3}}{3}\right)$

19. Increasing on $(-\infty, -2)$ and $(0, \infty)$ and decreasing on $(-2, 0)$

21. Increasing on $(-\infty, \infty)$

23. Decreasing on $(-\infty, 3)$ and increasing on $(3, \infty)$

25. Decreasing on $(-\infty, 2)$ and $(2, \infty)$

27. Decreasing on $(-\infty, 1)$ and $(1, \infty)$

29. Decreasing on $(-\infty, \infty)$ 31. Increasing on $(-1, \infty)$

33. Increasing on $(-4, 0)$; decreasing on $(0, 4)$

35. Increasing on $(0, 2)$; decreasing on $(-\infty, 0)$ and $(2, \infty)$

37. Increasing on $(0, e)$; decreasing on (e, ∞)

39. Decreasing on $(-\infty, 0)$ and $(0, \infty)$

41. Relative maximum: $f(0) = 1$; relative minima: $f(-1) = 0$ and $f(1) = 0$

43. Relative maximum: $f(-1) = 2$; relative minimum: $f(1) = -2$

45. Relative maximum: $f(1) = 3$; relative minimum: $f(2) = 2$

47. Relative minimum: $f(0) = 2$ 49. (a)   51. (d)

53. Relative minimum: $f(2) = -4$

55. Relative maximum: $h(3) = 15$ 57. None

59. Relative maximum: $g(0) = 5$; relative minimum: $g(2) = 1$

61. Relative maximum: $f(0) = 0$; relative minima: $f(-1) = -\frac{1}{2}$ and $f(1) = -\frac{1}{2}$

63. Relative minimum: $F(3) = -5$; relative maximum: $F(-1) = \frac{17}{3}$

65. Relative minimum: $g(3) = -7$ 67. None

69. Relative maximum: $f(-3) = -4$; relative minimum: $f(3) = 8$

71. Relative maximum: $f(1) = \frac{1}{2}$; relative minimum: $f(-1) = -\frac{1}{2}$

73. Relative maximum: $f(1) = 2e^{-1}$

75. Relative minimum: $f(1) = 1$

77.

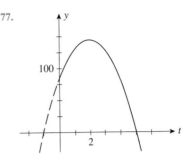

Rising in the time interval $(0, 2)$; falling in the time interval $(2, 5)$; when $t = 5$ sec

79. The number of subscribers is increasing.

81. Rising on $(0, 33)$ and descending on $(33, T)$ for some positive number T.

83. f is decreasing on $(0, 1)$ and increasing on $(1, 4)$. The average speed decreases from 6 A.M. to 7 A.M. and then picks up from 7 A.M. to 10 A.M.

85. **a.** Increasing on $(0, 10)$ **b.** Sales will be increasing.

87. Increasing on $(0, 4)$; decreasing on $(4, 5)$. The cash in the fund will be increasing from 2005 to 2045 and decreasing from 2045 to 2055.

89. **a.** 30%; 41.4%

91. Increasing on $(0, 4.5)$ and decreasing on $(4.5, 11)$; the pollution is increasing from 7 A.M. to 11:30 A.M. and decreasing from 11:30 A.M. to 6 P.M.

93. **b.** 4505/year/year; 272 cases/year/year

95. **a.** $0.0021t^2 - 0.0061t + 0.1$
    b. Decreasing on $(0, 1.5)$ and increasing on $(1.5, 15)$. The gap (shortage of nurses) was decreasing from 2000 to mid-2001 and is expected to be increasing from mid-2001 to 2015.
    c. $(1.5, 0.096)$. The gap was smallest ($\approx 96{,}000$) in mid-2001.

97. True   99. True   101. False   105. $a = -4; b = 24$

107. **a.** $-2x$ if $x \neq 0$ **b.** No

Using Technology Exercises 10.1, page 693

1. **a.** f is decreasing on $(-\infty, -0.2934)$ and increasing on $(-0.2934, \infty)$.
   b. Relative minimum: $f(-0.2934) = -2.5435$

3. **a.** f is increasing on $(-\infty, -1.6144)$ and $(0.2390, \infty)$ and decreasing on $(-1.6144, 0.2390)$.
   b. Relative maximum: $f(-1.6144) = 26.7991$; relative minimum: $f(0.2390) = 1.6733$

5. **a.** f is decreasing on $(-\infty, -1)$ and $(0.33, \infty)$ and increasing on $(-1, 0.33)$.
   b. Relative maximum: $f(0.33) = 1.11$; relative minimum: $f(-1) = -0.63$

7. **a.** f is decreasing on $(-\infty, 0.40)$ and increasing on $(0.40, \infty)$.
   b. Relative minimum: $f(0.40) = 0.79$

9. **a.**

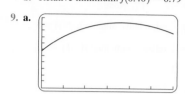

   b. Increasing on $(0, 3.6676)$ and decreasing on $(3.6676, 6)$

11. Increasing on $(0, 4.5)$ and decreasing on $(4.5, 11)$; 11:30 A.M.; 164 PSI

Exercises 10.2, page 704

1. Concave downward on $(-\infty, 0)$ and concave upward on $(0, \infty)$; inflection point: $(0, 0)$

3. Concave downward on $(-\infty, 0)$ and $(0, \infty)$

5. Concave upward on $(-\infty, 0)$ and $(1, \infty)$ and concave downward on $(0, 1)$; inflection points: $(0, 0)$ and $(1, -1)$

7. Concave downward on $(-\infty, -2), (-2, 2)$, and $(2, \infty)$

9. **a.** Concave upward on $(0, 2), (4, 6), (7, 9)$, and $(9, 12)$ and concave downward on $(2, 4)$ and $(6, 7)$
   b. $(2, \frac{5}{2}), (4, 2), (6, 2)$, and $(7, 3)$

11. (a)   13. (b)

15. **a.** $D_1'(t) > 0, D_2'(t) > 0, D_1''(t) > 0$, and $D_2''(t) < 0$ on $(0, 12)$
    b. With or without the proposed promotional campaign, the deposits will increase; with the promotion, the deposits will increase at an increasing rate; without the promotion, the deposits will increase at a decreasing rate.

17. (c)   19. (d)

21. **a.** Between 8 A.M. and 10 A.M., the rate of change of the rate of change of the number of smartphones assembled is increasing; between 10 A.M. and 12 noon, it is decreasing.
    b. At 10 A.M.

23. At the time t_0, corresponding to its t-coordinate, the restoration process is working at its peak.

29. Concave upward on $(-\infty, \infty)$

31. Concave downward on $(0, \infty)$; concave upward on $(-\infty, 0)$

33. Concave upward on $(-\infty, 0)$ and $(3, \infty)$; concave downward on $(0, 3)$

35. Concave downward on $(-\infty, 0)$ and $(0, \infty)$

37. Concave downward on $(-\infty, 4)$

39. Concave downward on $(-\infty, 3)$; concave upward on $(3, \infty)$

41. Concave upward on $\left(-\infty, -\frac{\sqrt{6}}{3}\right)$ and $\left(\frac{\sqrt{6}}{3}, \infty\right)$; concave downward on $\left(-\frac{\sqrt{6}}{3}, \frac{\sqrt{6}}{3}\right)$

43. Concave downward on $(-\infty, 1)$; concave upward on $(1, \infty)$

45. Concave upward on $(-\infty, 0)$ and $(0, \infty)$

47. Concave upward on $\left(-\infty, \frac{5}{2}\right)$; concave downward on $\left(\frac{5}{2}, \infty\right)$

49. Concave upward on $(0, \infty)$; concave downward on $(-\infty, 0)$

51. Concave upward on $(-\infty, -1)$ and $(1, \infty)$; concave downward on $(-1, 0)$ and $(0, 1)$

53. $(0, -2)$ 55. $(1, -20)$ 57. $(0, 1)$ and $\left(\frac{2}{3}, \frac{11}{27}\right)$

59. $(0, 0)$ 61. $(1, 2)$ 63. $\left(-\frac{\sqrt{2}}{2}, 2e^{-1/2}\right); \left(\frac{\sqrt{2}}{2}, 2e^{-1/2}\right)$

65. $\left(e^{-3/2}, -\frac{3}{2}e^{-3}\right)$ 67. Relative maximum: $f(1) = 5$ 69. None

71. Relative maximum: $f(-1) = -\frac{7}{3}$, relative minimum: $f(5) = -\frac{115}{3}$

73. Relative maximum: $g(-3) = -6$; relative minimum: $g(3) = 6$

75. None   77. Relative minimum: $f(-2) = 12$

79. Relative maximum: $g(1) = \frac{1}{2}$; relative minimum: $g(-1) = -\frac{1}{2}$

81. Relative minimum: $f(1) = \frac{1}{e}$

83. Relative minimum: $f(0) = 0$

85.

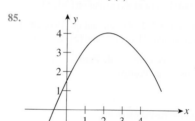

87.

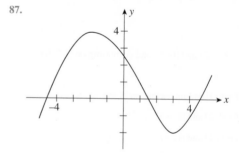

89.

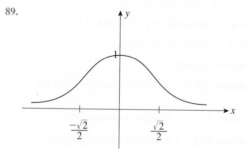

91. **a.** N is increasing on $(0, 12)$.
    b. $N''(t) < 0$ on $(0, 6)$ and $N''(t) > 0$ on $(6, 12)$
    c. The rate of growth of the number of help-wanted advertisements was decreasing over the first 6 months of the year and increasing over the last 6 months.

93. $f(t)$ increases at an increasing rate until the water level reaches the middle of the vase at which time (corresponding to the inflection point) $f(t)$ is increasing at the fastest rate. The water rises faster when the vase is narrower. After that, $f(t)$ increases at a decreasing rate until the vase is filled.

95. **b.** The rate of increase of the average state cigarette tax was decreasing from 2001 to 2007.

97. **b.** The rate was increasing.

99. **a.** 1.43%; 5.80%

103. **a.** $-0.6t^2 + 3.28t + 1.31$; $-1.2t + 3.28$
    c. (2.73, 14.93); the rate was increasing least rapidly in late September of 2006.

105. **a.** 506,000; 125,480
    b. The number of measles deaths was dropping from 1999 through 2005.
    c. April 2002; approximately −41 deaths/year/year

107. **b.** 44   109. False   111. True

Exercises 10.3, page 719

1. Horizontal asymptote: $y = 0$

3. Horizontal asymptote: $y = 0$; vertical asymptote: $x = 0$

5. Horizontal asymptote: $y = 0$; vertical asymptotes: $x = -1$ and $x = 1$

7. Horizontal asymptote: $y = 3$; vertical asymptote: $x = 0$

9. Horizontal asymptote: $y = 0$

11. Horizontal asymptote: $y = 0$; vertical asymptote: $x = 0$

13. Horizontal asymptote: $y = 0$; vertical asymptote: $x = 0$

15. Horizontal asymptote: $y = 1$; vertical asymptote: $x = -2$

17. None

19. Horizontal asymptote: $y = 1$; vertical asymptotes: $t = -4$ and $t = 4$

21. Horizontal asymptote: $y = 0$; vertical asymptotes: $x = -2$ and $x = 3$

23. Horizontal asymptote: $y = 2$; vertical asymptote: $t = 2$

25. Horizontal asymptote: $y = 1$; vertical asymptotes: $x = -2$ and $x = 2$

27. None   29. f is the derivative function of the function g.

31.
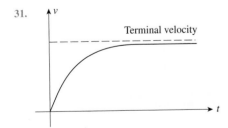

33.
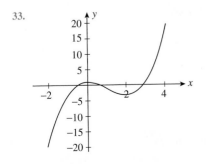

35.
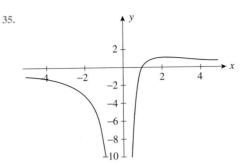

37.
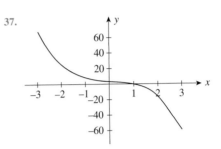

39.
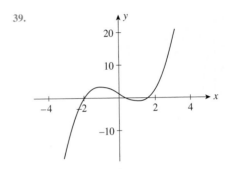

41.
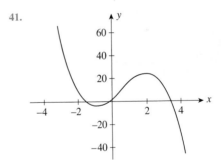

43.
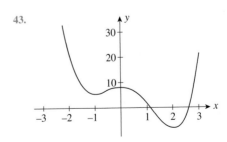

45.
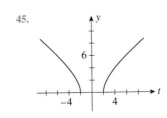

47.

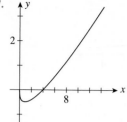

49.

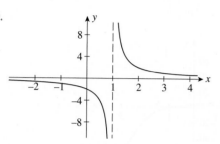

51.

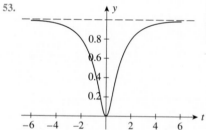

53.

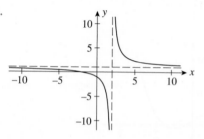

55.

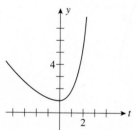

57.

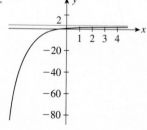

59.

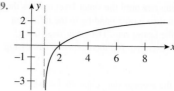

61. a. $x = 100$ **b.** No

63. a. $y = 0$
   b. As time passes, the concentration of the drug decreases and approaches zero.

65.

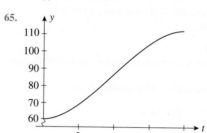

67.

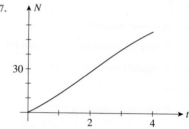

69.

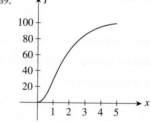

71.

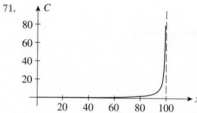

73. a. 30   **b.** $N'(x) = \dfrac{297{,}000e^{-x}}{(1 + 99e^{-x})^2}$

   c.

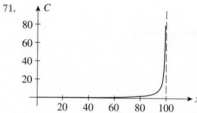

3000

Using Technology Exercises 10.3, page 725

1.

3.
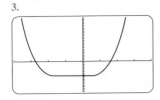

5. −0.9733; 2.3165, 4.6569   7. 1.5142   9. −0.7680; 1.6873

11.
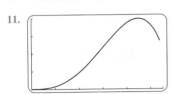

Exercises 10.4, page 733

1. None   3. Absolute minimum value: 0

5. Absolute maximum value: 3; absolute minimum value: −2

7. Absolute maximum value: 3; absolute minimum value: $-\frac{27}{16}$

9. Absolute minimum value: $-\frac{41}{8}$

11. No absolute extrema   13. Absolute maximum value: 1

15. Absolute maximum value: 5; absolute minimum value: −4

17. Absolute maximum value: 14; absolute minimum value: 5

19. Absolute maximum value: 19; absolute minimum value: −1

21. Absolute maximum value: 16; absolute minimum value: −1

23. Absolute maximum value: 3; absolute minimum value: $\frac{5}{3}$

25. Absolute maximum value: $\frac{17}{2}$; absolute minimum value: 5

27. Absolute maximum value ≈ 1.04; absolute minimum value: −1.5

29. No absolute extrema

31. Absolute maximum value: 1; absolute minimum value: 0

33. Absolute maximum value: 0; absolute minimum value: −3

35. Absolute maximum value: 2; absolute minimum value: $\frac{2}{e}$

37. Absolute maximum value: $2e^{-3/2}$; absolute minimum value: −1

39. Absolute maximum value: 3 − ln 3; absolute minimum value: 1

41. 144 ft   43. 17.72%

45. $f(6) = 3.60, f(0.5) = 1.13$; the number of nonfarm, full-time, self-employed women over the time interval from 1963 to 1993 reached its highest level, 3.6 million, in 1993.

47. $3600   49. 6000   51. 3333

53. **a.** $0.0025x + 80 + \dfrac{10{,}000}{x}$ **b.** 2000   **c.** 2000   **d.** Same

57. 533   59. In 7.72 years; $160,208

61. **a.** 2 days after the organic waste was dumped into the pond
    b. 3.5 days after the organic waste was dumped into the pond

71. **a.** 2000; $105.8 billion
    b. 1995; $7.6 billion

75. $R = r; \dfrac{E^2}{4r}$ watts

79. **a.** −70 mg/day; −43 mg/day
    b. At $t = 1$
    c. 125 mg

81. False   83. False

87. **c.**
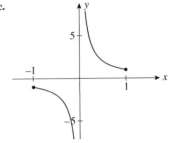

Using Technology Exercises 10.4, page 739

1. Absolute maximum value: 145.9; absolute minimum value: −4.3834

3. Absolute maximum value: 16; absolute minimum value: −0.1257

5. Absolute maximum value: 11.8922; absolute minimum value: 0

7. Absolute maximum value: 2.8889; absolute minimum value: 0

9. **a.**

11. **b.** approximately 1145

Exercises 10.5, page 746

1. 25 ft × 25 ft

3. 750 yd × 1500 yd; 1,125,000 yd²

5. $10\sqrt{2}$ ft × $40\sqrt{2}$ ft

7. $\frac{16}{3}$ in. × $\frac{16}{3}$ in. × $\frac{4}{3}$ in.

9. 5.04 in. × 5.04 in. × 5.04 in.

11. 18 in. × 18 in. × 36 in.; 11,664 in.³

13. $r = \dfrac{36}{\pi}$ in.; $l = 36$ in.; $\dfrac{46{,}656}{\pi}$ in.³

15. $\frac{2}{3}\sqrt[3]{9}$ ft × $\sqrt[3]{9}$ ft × $\frac{2}{5}\sqrt[3]{9}$ ft

17. 250; $62,500; $250

19. 85; $28,900; $340   21. 60 miles/hr

23. $w ≈ 13.86$ in.; $h ≈ 19.60$ in.

25. $x = 2250$ ft   27. 2.68

29. 440 ft; 140 ft; 184,874 sq ft   31. 45; 44,445

Chapter 10 Concept Review, page 751

1. **a.** $f(x_1) < f(x_2)$ **b.** $f(x_1) > f(x_2)$

2. **a.** increasing   **b.** $f'(x) < 0$ **c.** constant

3. **a.** $f(x) \leq f(c)$ **b.** $f(x) \geq f(c)$

4. **a.** domain; $= 0$; exist   **b.** critical number
   c. relative extremum

5. **a.** $f'(x)$ **b.** > 0 **c.** concavity
   d. relative maximum; relative extremum

6. $\pm\infty; \pm\infty$ 7. $0; 0$ 8. $b; b$

9. **a.** $f(x) \leq f(c)$; absolute maximum value
   b. $f(x) \geq f(c)$; open interval

10. continuous; absolute; absolute

Chapter 10 Review Exercises, page 752

1. **a.** f is increasing on $(-\infty, \infty)$ **b.** No relative extrema
   c. Concave down on $(-\infty, 1)$; concave up on $(1, \infty)$
   d. $\left(1, -\frac{17}{3}\right)$

2. **a.** f is increasing on $(-\infty, \infty)$ **b.** No relative extrema
   c. Concave down on $(-\infty, 2)$; concave up on $(2, \infty)$ **d.** $(2, 0)$

3. **a.** f is increasing on $(-1, 0)$ and $(1, \infty)$ and decreasing on
   $(-\infty, -1)$ and $(0, 1)$
   b. Relative maximum value: 0; relative minimum value: -1
   c. Concave up on $\left(-\infty, -\frac{\sqrt{3}}{3}\right)$ and $\left(\frac{\sqrt{3}}{3}, \infty\right)$; concave down
   on $\left(-\frac{\sqrt{3}}{3}, \frac{\sqrt{3}}{3}\right)$
   d. $\left(-\frac{\sqrt{3}}{3}, -\frac{5}{9}\right); \left(\frac{\sqrt{3}}{3}, -\frac{5}{9}\right)$

4. **a.** f is increasing on $(-\infty, -2)$ and $(2, \infty)$ and decreasing on
   $(-2, 0)$ and $(0, 2)$
   b. Relative maximum value: -4; relative minimum value: 4
   c. Concave down on $(-\infty, 0)$; concave up on $(0, \infty)$ **d.** None

5. **a.** f is increasing on $(-\infty, 0)$ and $(2, \infty)$; decreasing on $(0, 1)$
   and $(1, 2)$
   b. Relative maximum value: 0; relative minimum value: 4
   c. Concave up on $(1, \infty)$; concave down on $(-\infty, 1)$ **d.** None

6. **a.** f is increasing on $(1, \infty)$ **b.** No relative extrema
   c. Concave down on $(1, \infty)$ **d.** None

7. **a.** f is decreasing on $(-\infty, \infty)$ **b.** No relative extrema
   c. Concave down on $(-\infty, 1)$; concave up on $(1, \infty)$ **d.** $(1, 0)$

8. **a.** f is increasing on $(1, \infty)$ **b.** No relative extrema
   c. Concave down on $(1, \frac{4}{3})$; concave up on $(\frac{4}{3}, \infty)$
   d. $\left(\frac{4}{3}, \frac{4\sqrt{3}}{9}\right)$

9. **a.** f is increasing on $(-\infty, -1)$ and $(-1, \infty)$
   b. No relative extrema
   c. Concave down on $(-1, \infty)$; concave up on $(-\infty, -1)$
   d. None

10. **a.** f is decreasing on $(-\infty, 0)$ and increasing on $(0, \infty)$
    b. Relative minimum value: -1
    c. Concave down on $\left(-\infty, -\frac{\sqrt{3}}{3}\right)$ and $\left(\frac{\sqrt{3}}{3}, \infty\right)$; concave up
    on $\left(-\frac{\sqrt{3}}{3}, \frac{\sqrt{3}}{3}\right)$
    d. $\left(-\frac{\sqrt{3}}{3}, -\frac{3}{4}\right); \left(\frac{\sqrt{3}}{3}, -\frac{3}{4}\right)$

11. **a.** f is increasing on $(-\infty, 3)$ and decreasing on $(3, \infty)$
    b. Relative maximum value: e^3
    c. Concave up on $(-\infty, 2)$; concave down on $(2, \infty)$
    d. $(2, 2e^2)$

12. **a.** f is decreasing on $(0, e^{-1/2})$ and increasing on $(e^{-1/2}, \infty)$
    b. Relative minimum value: $-\frac{1}{2}e^{-1}$
    c. Concave down on $(0, e^{-3/2})$; concave up on $(e^{-3/2}, \infty)$
    d. $(e^{-3/2}, -\frac{3}{2}e^{-3})$

13.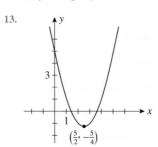

14.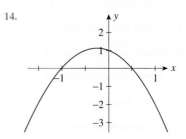

15.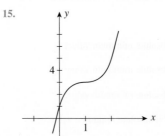

16.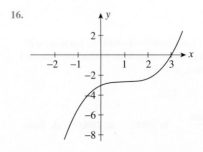

17.

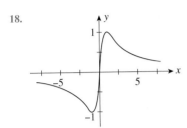

18.

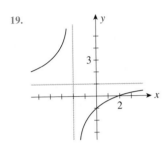

19.

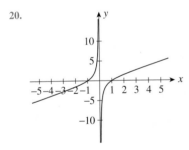

20.

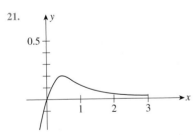

21.

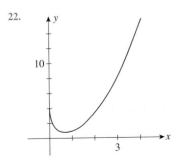

22.

23. Vertical asymptote: $x = -\frac{3}{2}$; horizontal asymptote: $y = 0$

24. Vertical asymptote: $x = -1$; horizontal asymptote: $y = 2$

25. Vertical asymptotes: $x = -2$, $x = 4$; horizontal asymptote: $y = 0$

26. Vertical asymptote: $x = 1$; horizontal asymptote: $y = 1$

27. Absolute minimum value: $-\frac{25}{8}$

28. Absolute minimum value: 0

29. Absolute maximum value: 5; absolute minimum value: 0

30. Absolute maximum value: $\frac{5}{3}$; absolute minimum value: 1

31. Absolute maximum value: -16; absolute minimum value: -32

32. Absolute maximum value: $\frac{1}{2}$; absolute minimum value: 0

33. Absolute maximum value: $\frac{8}{3}$; absolute minimum value: 0

34. Absolute maximum value: $\frac{215}{9}$; absolute minimum value: 7

35. Absolute maximum value: $\frac{3}{e}$; absolute minimum value: $-6e^2$

36. Absolute maximum value: ln 2; absolute minimum value: 0

37. Absolute maximum value: $\frac{1}{2}$; absolute minimum value: $-\frac{1}{2}$

38. No absolute extrema

39. a The sign of $R_1'(t)$ is negative; the sign of $R_2'(t)$ is positive. The sign of $R_1''(t)$ is negative and the sign of $R_2''(t)$ is positive.
  b. The revenue of the neighborhood bookstore is decreasing at an increasing rate while the revenue of the new branch of the national bookstore is increasing at an increasing rate.

40. The rumor spread initially with increasing speed. The rate at which the rumor is spread reaches a maximum at the time corresponding to the t-coordinate of the point P on the curve. Thereafter, the speed at which the rumor is spread decreases.

41. $4000

42. a. $16.25t + 24.625$; sales were increasing.
  b. 16.25; the rate of sales was increasing from 2002 to 2005.

43. (100, 4600); sales increase at an increasing rate until $100,000 is spent on advertising; after that, any additional expenditure results in increased sales but at a slower rate of increase.

44. a. Decreasing on (0, 21.4); increasing on (21.4, 30)
  b. The percentage of men 65 years and older in the workforce was decreasing from 1970 until mid-1991 and increasing from mid-1991 through 2000.

45. (267, 11,874); the rate of increase is lowest when 267 calculators are produced.

47. a. 13.0%, 22.2%

48. a. $I'(t) = -\dfrac{200t}{(t^2 + 10)^2}$

  b. $I''(t) = \dfrac{-200(10 - 3t^2)}{(t^2 + 10)^3}$; concave up on $\left(\sqrt{\dfrac{10}{3}}, \infty\right)$;

  concave down on $\left(0, \sqrt{\dfrac{10}{3}}\right)$

c.

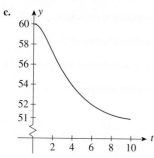

d. The rate of decline in the environmental quality of the wildlife was increasing the first 1.8 years. After that time, the rate of decline decreased.

49. 168   **50.** 3000

51. a. $0.001x + 100 + \dfrac{4000}{x}$ **b.** 2000

52. 10 A.M.

53. a. Decreasing on $(0, 12.7)$; increasing on $(12.7, 30)$
  b. 7.9
  c. The percentage of women 65 years and older in the workforce was decreasing from 1970 to Sept. 1982 and increasing from Sept. 1982 to 2000. It reached a minimum value of 7.9% in Sept. 1982.

55. 74.07 in.³   **56.** Radius: 2 ft; height: 8 ft

57. 1 ft × 2 ft × 2 ft   **58.** 20,000 cases

59. $a = -4$; $b = 11$ **60.** $c > \frac{3}{2}$

62. a. $f'(x) = 3x^2$ if $x \neq 0$ **b.** No

Chapter 10 Before Moving On, page 755

1. Decreasing on $(-\infty, 0)$ and $(2, \infty)$; increasing on $(0, 1)$ and $(1, 2)$

2. Rel. max: $(1, 4e^{-1})$; inflection point: $(2, 8e^{-2})$

3. Concave downward on $\left(-\infty, \frac{1}{4}\right)$; concave upward on $\left(\frac{1}{4}, \infty\right)$; $\left(\frac{1}{4}, \frac{83}{96}\right)$

4.

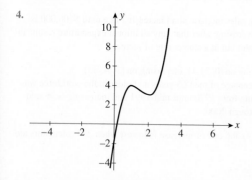

5. Abs. min. value: -5; abs. max. value: 80

6. $r = h = \dfrac{1}{\sqrt[3]{\pi}}$ (ft)

CHAPTER 11

Exercises 11.1, page 766

5. b. $y = 2x + C$
  c.

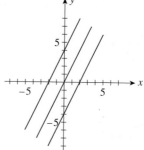

7. b. $y = \frac{1}{3}x^3 + C$
  c.

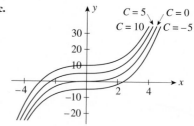

9. $6x + C$ **11.** $\frac{1}{4}x^4 + C$

13. $-\dfrac{1}{3x^3} + C$ **15.** $\frac{3}{5}x^{5/3} + C$

17. $-\dfrac{4}{x^{1/4}} + C$ **19.** $-\dfrac{1}{x^2} + C$

21. $\frac{2}{3}\pi t^{3/2} + C$ **23.** $3x - 2x^2 + C$

25. $\dfrac{1}{3}x^3 + \dfrac{1}{2}x^2 - \dfrac{1}{2x^2} + C$ **27.** $5e^x + C$

29. $x + \frac{1}{2}x^2 + e^x + C$ **31.** $x^4 + \dfrac{2}{x} - x + C$

33. $\frac{2}{7}x^{7/2} + \frac{4}{5}x^{5/2} - \frac{1}{2}x^2 + C$ **35.** $\frac{2}{3}x^{3/2} + 4\sqrt{x} + C$

37. $\frac{1}{9}u^3 + \frac{1}{3}u^2 - \frac{1}{3}u + C$ **39.** $\frac{2}{3}t^3 - \frac{3}{2}t^2 - 2t + C$

41. $\dfrac{1}{3}x^3 - 2x - \dfrac{1}{x} + C$ **43.** $\frac{1}{3}s^3 + s^2 + s + C$

45. $e^t + \dfrac{t^{e+1}}{e + 1} + C$ **47.** $\dfrac{1}{2}x^2 + x - \ln|x| - \dfrac{1}{x} + C$

49. $\ln|x| + \dfrac{4}{\sqrt{x}} - \dfrac{1}{x} + C$ **51.** $\frac{3}{2}x^2 + x + \frac{1}{2}$

53. $x^3 + 2x^2 - x - 5$ **55.** $x - \dfrac{1}{x} + 3$ **57.** $x + \ln|x|$

59. $\sqrt{x}$ **61.** $e^x + \frac{1}{2}x^2 + 2$ **63.** Branch A

65. $s(t) = \frac{4}{3}t^{3/2}$ **67.** 3370 **69.** 5000 units; $34,000

71. a. $0.0029t^2 + 0.159t + 1.6$ **b.** $4.16 trillion

73. a. $-0.125t^3 + 1.05t^2 + 2.45t + 1.5$ **b.** 24.375 million

75. a. $-1.493t^3 + 34.9t^2 + 279.5t + 2917$ **b.** $9168

77. a. $3.133t^3 - 6.7t^2 + 14.07t + 36.7$ **b.** 103,201

79. a. $4.096t^3 - 75.2797t^2 + 695.23t + 3142$ **b.** $3766.05

81. 21,960   **83.** $-t^3 + 96t^2$; 59,400 ft

85. a. $0.75t^4 - 5.9815t^3 + 14.3611t^2 + 26.632t + 108$
b. $321.25 million

87. a. $-0.0000124t^3 + 0.00186t^2 - 0.186t + 9.3$
b. 7030   **c.** 6610

89. $\frac{1}{2}k(R^2 - r^2)$ **91.** $9\frac{7}{9}$ ft/sec²; 396 ft   **93.** 0.924 ft/sec²

95. a. $\dfrac{16\sqrt{2}}{3} t^{3/2} - 8t^4$ **b.** 2.2 in.

97. True   **99.** True

Exercises 11.2, page 778

1. $\frac{1}{5}(4x + 3)^5 + C$ **3.** $\frac{1}{3}(x^3 - 2x)^3 + C$

5. $-\dfrac{1}{2(2x^2 + 3)^2} + C$ **7.** $\frac{2}{3}(t^3 + 2)^{3/2} + C$

9. $\frac{1}{10}(x^2 - 1)^{10} + C$ **11.** $-\frac{1}{5}\ln|1 - x^5| + C$

13. $\ln(x - 2)^2 + C$ **15.** $\frac{1}{2}\ln(0.3x^2 - 0.4x + 2) + C$

17. $\frac{1}{3}\ln|3x^2 - 1| + C$ **19.** $-\frac{1}{2}e^{-2x} + C$

21. $-e^{2-x} + C$ **23.** $-\frac{1}{2}e^{-x^2} + C$

25. $e^x + e^{-x} + C$ **27.** $2\ln(1 + e^x) + C$

29. $2e^{\sqrt{x}} + C$ **31.** $-\dfrac{1}{6(e^{3x} + x^3)^2} + C$

33. $\frac{1}{8}(e^{2x} + 1)^4 + C$ **35.** $\frac{1}{2}(\ln 5x)^2 + C$

37. $2\ln|\ln x| + C$ **39.** $\frac{2}{3}(\ln x)^{3/2} + C$

41. $\frac{1}{2}e^x - \frac{1}{2}\ln(x^2 + 2) + C$

43. $\frac{2}{3}(\sqrt{x} - 1)^3 + 3(\sqrt{x} - 1)^2 + 8(\sqrt{x} - 1) + 4\ln|\sqrt{x} - 1| + C$

45. $\dfrac{(6x + 1)(x - 1)^6}{42} + C$

47. $4\sqrt{x} - x - 4\ln(1 + \sqrt{x}) + C$

49. $-\frac{1}{252}(1 - v)^7(28v^2 + 7v + 1) + C$

51. $\frac{1}{2}[(2x - 1)^5 + 5]$ **53.** $e^{-x^2 + 1} - 1$ **55.** 17,341,000

57. $21,000 - \dfrac{20,000}{\sqrt{1 + 0.2t}}$; 6858   **59.** $\dfrac{250}{\sqrt{16 + x^2}}$

61. $30(\sqrt{2t + 4} - 2)$; $14,400\pi$ ft²

63. $\dfrac{65.8793}{1 + 2.449e^{-0.3277t}} + 0.3$; 56.22 in.   **65.** $\dfrac{r}{a}(1 - e^{-at})$

67. True

Exercises 11.3, page 788

1. 4.27

3. a. 6

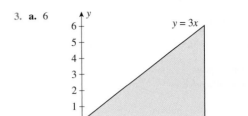

b. 4.5   **c.** 5.25   **d.** Yes

5. a. 4

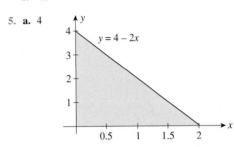

b. 4.8   **c.** 4.4   **d.** Yes

7. a. 18.5   **b.** 18.64   **c.** 18.66   **d.** $\approx 18\frac{2}{3}$

9. a. 25   **b.** 21.12   **c.** 19.88   **d.** ≈ 19.9

11. a. 0.0625   **b.** 0.16   **c.** 0.2025   **d.** ≈ 0.2

13. 4.64   **15.** 0.95   **17.** 9400 sq ft

Exercises 11.4, page 799

1. 6   **3.** 8   **5.** 12   **7.** 9   **9.** $\ln 2$ **11.** $17\frac{1}{3}$ **13.** $18\frac{1}{4}$

15. $e^2 - 1$ **17.** 6   **19.** 6   **21.** $\frac{56}{3}$ **23.** $\frac{4}{3}$ **25.** $\frac{45}{2}$

27. $\frac{7}{12}$ **29.** $\ln 4$ **31.** 56   **33.** $\frac{256}{15}$ **35.** $\frac{2}{3}$ **37.** $\frac{8}{3}$

39. $\frac{39}{2}$ **41. a.** $4100   **b.** $900   **43. a.** $2800   **b.** $219.20

45. a. $0.86t^{0.96} + 0.04$ **b.** $4.84 billion   **47.** $10,133\frac{1}{3}$ ft

49. a. $0.2833t^3 - 1.936t^2 + 5t + 5.6$ **b.** 12.8%   **c.** 5.2%

51. 15,477   **53.** 149.14 million   **55.** $\frac{23}{15}$

57. False   **59.** False

Using Technology Exercises 11.4, page 802

1. 6.1787   **3.** 0.7873   **5.** -0.5888 **7.** 2.7044

9. 3.9973   **11.** 46%; 24%   **13.** 333,209   **15.** 3,761,490

Exercises 11.5, page 809

1. 10   **3.** $\frac{19}{15}$ **5.** $\frac{484}{15}$ **7.** $\sqrt{3} - 1$ **9.** $\frac{1031}{5}$

11. $\frac{32}{15}$ **13.** $\frac{272}{15}$ **15.** $e^4 - 1$ **17.** $\frac{1}{2}e^2 + \frac{5}{6}$

19. 0   **21.** $\ln 4$ **23.** $\frac{1}{3}(\ln 19 - \ln 3)$

25. $2e^4 - 2e^2 - \ln 2$

27. $\frac{1}{2}(e^{-4} - e^{-8} - 1)$ 29. 6   31. $\frac{1}{2}$ 33. $2(\sqrt{e} - \frac{1}{e})$

35. 5   37. $\frac{17}{3}$ 39. -1 41. $\frac{13}{6}$ 43. $\frac{1}{4}(e^4 - 1)$

45. 120.3 billion metric tons

47. \$2.24 million   49. \$40,339   51. \$3.24 billion/year

53. **a.** 160.7 billion gal/year
    b. 150.1 billion gal/year/year

55. \$297.9 million   57. \$6$\frac{1}{3}$ million

59. 0.071 mg/cm³   61. \$14.78

63. 80.7%   71. Property 5

73. 0   75. **a.** -1 **b.** 5   **c.** -13

77. True   79. False   81. True

Using Technology Exercises 11.5, page 813

1. 7.71667   3. 17.56487   5. 10,140   7. 60.45 mg/day

Exercises 11.6, page 820

1. 108   3. $\frac{2}{3}$ 5. $\frac{8}{3}$ 7. $\frac{3}{2}$ 9. 3   11. $\frac{10}{3}$

13. 27   15. $2(e^2 - e^{-1})$ 17. $\frac{38}{3}$ 19. $\frac{10}{3}$ 21. $\frac{19}{4}$

23. $12 - \ln 4$ 25. $e^2 - e - \ln 2$ 27. $\frac{5}{2}$ 29. $\frac{22}{3}$ 31. $\frac{3}{2}$

33. $e^3 - 4 + \frac{1}{e}$ 35. $\frac{125}{6}$ 37. $\frac{1}{12}$ 39. $\frac{71}{6}$ 41. 18

43. S is the additional revenue that Odyssey Travel could realize by switching to the new agency; $S = \int_0^b [g(x) - f(x)]\, dx$

45. Shortfall $= \int_{2010}^{2050} [f(t) - g(t)]\, dt$

47. **a.** $A_2 - A_1$
    b. The distance car 2 is ahead of car 1 after T sec

49. $840 - \int_0^{12} f(t)\, dt$ 51. 42.8 billion metric tons

53. 57,179   55. True   57. False

Using Technology Exercises 11.6, page 825

1. **a.**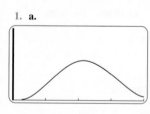

3. **a.**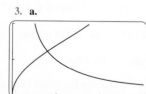

b. 1074.2857   **b.** 0.9961

5. **a.**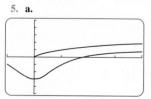

7. **a.**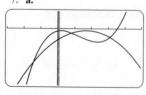

b. 5.4603   **b.** 25.8549

9. **a.**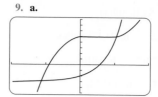

11. **a.**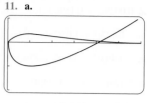

b. 10.5144   **b.** 3.5799

13. 207.43

Exercises 11.7, page 835

1. \$11,667   3. \$6667   5. \$11,667

7. **a.** 1257/month   **b.** \$48,583   9. \$199,548

11. Consumers' surplus: \$13,333; producers' surplus: \$11,667

13. \$824,200   15. \$148,239   17. \$43,788

19. \$47,916   21. \$142,423   23. \$24,780

25. **a.** 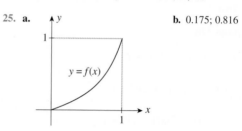   **b.** 0.175; 0.816

27. **a.** 0.31; 0.49   **b.** college teachers

Using Technology Exercises 11.7, page 838

1. Consumers' surplus: \$18,000,000; producers' surplus: \$11,700,000

3. Consumers' surplus: \$33,120; producers' surplus: \$2880

5. Investment A

Chapter 11 Concept Review, page 840

1. **a.** $F'(x) = f(x)$ **b.** $F(x) + C$

2. **a.** $c \int f(x)\, dx$ **b.** $\int f(x)\, dx \pm \int g(x)\, dx$

3. **a.** unknown   **b.** function   4. $g'(x)\, dx$; $\int f(u)\, du$

5. **a.** $\int_a^b f(x)\, dx$ **b.** minus

6. **a.** $F(b) - F(a)$; antiderivative   **b.** $\int_a^b f'(x)\, dx$

7. **a.** $\dfrac{1}{b-a}\displaystyle\int_a^b f(x)\, dx$ **b.** area; area   8. $\int_a^b [f(x) - g(x)]\, dx$

9. **a.** $\int_0^{\bar{x}} D(x)\, dx - \bar{p}\,\bar{x}$ **b.** $\bar{p}\,\bar{x} - \int_0^{\bar{x}} S(x)\, dx$

10. **a.** $e^{rT}\int_0^T R(t)e^{-rt}\, dt$ **b.** $\int_0^T R(t)e^{-rt}\, dt$

11. $\dfrac{mP}{r}(e^{rT} - 1)$ 12. $2\int_0^1 [x - f(x)]\, dx$

Chapter 11 Review Exercises, page 841

1. $\frac{1}{4}x^4 + \frac{2}{3}x^3 - \frac{1}{2}x^2 + C$ 2. $\frac{1}{12}x^4 - \frac{2}{3}x^3 + 8x + C$

3. $\dfrac{1}{5}x^5 - \dfrac{1}{2}x^4 - \dfrac{1}{x} + C$ 4. $\frac{3}{4}x^{4/3} - \frac{2}{3}x^{3/2} + 4x + C$

5. $\frac{1}{2}x^4 + \frac{2}{5}x^{5/2} + C$ 6. $\frac{2}{7}x^{7/2} - \frac{1}{3}x^3 + \frac{2}{3}x^{3/2} - x + C$

7. $\frac{1}{3}x^3 - \frac{1}{2}x^2 + 2\ln|x| + 5x + C$ 8. $\frac{1}{3}(2x+1)^{3/2} + C$

9. $\frac{3}{8}(3x^2 - 2x + 1)^{4/3} + C$ 10. $\frac{(x^3+2)^{11}}{33} + C$

11. $\frac{1}{2}\ln(x^2 - 2x + 5) + C$ 12. $-e^{-2x} + C$

13. $\frac{1}{2}e^{x^2+x+1} + C$ 14. $\frac{1}{e^{-x}+x} + C$ 15. $\frac{1}{6}(\ln x)^6 + C$

16. $(\ln x)^2 + C$ 17. $\frac{(11x^2-1)(x^2+1)^{11}}{264} + C$

18. $\frac{2}{15}(3x-2)(x+1)^{3/2} + C$ 19. $\frac{2}{3}(x+4)\sqrt{x-2} + C$

20. $2(x-2)\sqrt{x+1} + C$ 21. $\frac{1}{2}$ 22. -6 23. $\frac{17}{3}$

24. 242 25. -80 26. $\frac{132}{5}$ 27. $\frac{1}{2}\ln 5$ 28. $\frac{1}{15}$

29. 4 30. $1 - \frac{1}{e^2}$ 31. $\frac{e-1}{2(1+e)}$ 32. $\frac{1}{2}$

33. $f(x) = x^3 - 2x^2 + x + 1$ 34. $f(x) = \sqrt{x^2 + 1}$

35. $f(x) = x + e^{-x} + 1$ 36. $f(x) = \frac{1}{2}(\ln x)^2 - 2$

37. **a.** It gives the distance Car A is ahead of Car B.
    b. $t = 10$, $\int_0^{10}[f(t) - g(t)]\,dt$

38. **a.** It gives the amount by which the revenue of Branch A exceeds that of Branch B.
    b. $t = 10$, $\int_0^{10}[f(t) - g(t)]\,dt$

39. -4.28 40. $\$6740$

41. **a.** $-0.015x^2 + 60x$ **b.** $-0.015x + 60$

42. $V(t) = 1900(t-10)^2 + 10,000$; $\$40,400$

43. **a.** $0.05t^3 - 1.8t^2 + 14.4t + 24$ **b.** $56°F$

44. **a.** $-0.01t^3 + 0.109t^2 - 0.032t + 0.1$ **b.** 1.076 billion

45. 3.375 ppm   46. $3000t - 50,000(1 - e^{-0.04t})$; $16,939$

47. $15,000\sqrt{1 + 0.4t} + 85,000$; $112,659$

48. $26,027$ 49. $\frac{240}{5-x} - 30$

50. $\$3100$ 51. 37.7 million

52. **a.** $205.89 - 89.89e^{-0.176t}$ **b.** $\$161.43$ billion

53. 15 54. $\frac{1}{2}(e^4 - 1)$

55. $\frac{2}{3}$ 56. $\frac{9}{2}$

57. $e^2 - 3$ 58. $\frac{3}{10}$ 59. $\frac{1}{2}$

60. $234,500$ barrels   61. $\frac{1}{3}$ 62. $26°F$

63. 49.7 ft/sec   64. $67,600$/year   65. $\$270,000$

66. Consumers' surplus: $\$2083$; producers' surplus: $\$3333$

67. $\$197,652$ 68. $\$174,420$ 69. $\$505,696$

70. **a.**

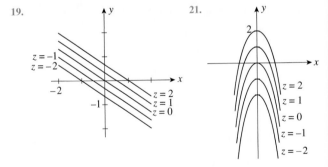

b. 0.1017; 0.3733 **c.** 0.315

71. $90,888$

Chapter 11 Before Moving On, page 844

1. $\frac{1}{2}x^4 + \frac{2}{3}x^{3/2} + 2\ln|x| - 4\sqrt{x} + C$ 2. $e^x + \frac{1}{2}x^2 + 1$

3. $\sqrt{x^2 + 1} + C$ 4. $\frac{1}{3}(2\sqrt{2} - 1)$ 5. $\frac{9}{2}$

CHAPTER 12

Exercises 12.1, page 853

1. $f(0, 0) = -4$; $f(1, 0) = -2$; $f(0, 1) = -1$; $f(1, 2) = 4$; $f(2, -1) = -3$

3. $f(1, 2) = 7$; $f(2, 1) = 9$; $f(-1, 2) = 1$; $f(2, -1) = 1$

5. $g(1, 4) = 12$; $g(4, 1) = 16$; $g(0, 4) = 2$; $g(4, 9) = 56$

7. $h(1, e) = 1$; $h(e, 1) = -1$; $h(e, e) = 0$

9. $g(1, 1, 1) = e$; $g(1, 0, 1) = 1$; $g(-1, -1, -1) = -e$

11. All real values of x and y

13. All real values of u and v except those satisfying the equation $u = -v$

15. All real values of r and s satisfying $rs \geq 0$

17. All real values of x and y satisfying $x + y > -5$

19.   21.

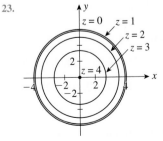

23.

25. $\sqrt{x^2 + y^2} = 5$ 27. b   29. No   31. 9π ft^3

33. **a.** 24.69 **b.** 81 kg

35. **a.** $-\frac{1}{5}x^2 - \frac{1}{4}y^2 - \frac{1}{5}xy + 200x + 160y$
   b. The set of all points (x, y) satisfying $200 - \frac{1}{5}x - \frac{1}{10}y \geq 0$,
   $160 - \frac{1}{10}x - \frac{1}{4}y \geq 0, x \geq 0, y \geq 0$

37. **a.** $-0.005x^2 - 0.003y^2 - 0.002xy + 20x + 15y$
   b. The set of all ordered pairs (x, y) for which
   $20 - 0.005x - 0.001y \geq 0$
   $15 - 0.001x - 0.003y \geq 0, x \geq 0, y \geq 0$

39. **a.** The set of all ordered pairs (P, T), where P and T are positive
   numbers
   b. 11.10 L

41. $7200 billion   43. 103

45. **a.** $1798.65; $2201.29   **b.** $2509.32

47. 40.28 times gravity

49. The level curves of V have equation $\dfrac{kT}{P} = C$ (C, a positive constant).

   The level curves are a family of straight lines $T = \left(\dfrac{C}{k}\right)P$ lying in

   the first quadrant, since k, T, and P are positive. Every point on the
   level curve $V = C$ gives the same volume C.

51. False   53. False   55. False

Exercises 12.2, page 866

1. **a.** 4; 4
   b. $f_x(2, 1) = 4$ says that the slope of the tangent line to the curve of
   intersection of the surface $z = x^2 + 2y^2$ and the plane $y = 1$ at
   the point $(2, 1, 6)$ is 4. $f_y(2, 1) = 4$ says that the slope of the tangent line to the curve of intersection of the surface $z = x^2 + 2y^2$
   and the plane $x = 2$ at the point $(2, 1, 6)$ is 4.
   c. $f_x(2, 1) = 4$ says that the rate of change of $f(x, y)$ with respect to x
   with y held fixed with a value of 1 is 4 units/unit change in x.
   $f_y(2, 1) = 4$ says that the rate of change of $f(x, y)$ with respect to y
   with x held fixed with a value of 2 is 4 units/unit change in y.

3. $f_x = 2; f_y = 3$ 5. $g_x = 6x; g_y = 2$ 7. $f_x = -\dfrac{4y}{x^3}; f_y = \dfrac{2}{x^2}$

9. $g_u = \dfrac{2v}{(u + v)^2}; g_v = -\dfrac{2u}{(u + v)^2}$

11. $f_s = 3(2s - t)(s^2 - st + t^2)^2; f_t = 3(2t - s)(s^2 - st + t^2)^2$

13. $f_x = \dfrac{8x}{3(2x^2 + y^2)^{1/3}}; f_y = \dfrac{4y}{3(2x^2 + y^2)^{1/3}}$

15. $f_x = ye^{xy+1}; f_y = xe^{xy+1}$

17. $f_x = \ln y + \dfrac{y}{x}; f_y = \dfrac{x}{y} + \ln x$ 19. $g_u = e^u \ln v; g_v = \dfrac{e^u}{v}$

21. $f_x = yz + y^2 + 2xz; f_y = xz + 2xy + z^2; f_z = xy + 2yz + x^2$

23. $h_r = ste^{rst}; h_s = rte^{rst}; h_t = rse^{rst}$ 25. $f_x(2, 1) = 5; f_y(2, 1) = 8$

27. $f_x(2, 1) = 1; f_y(2, 1) = 3$ 29. $f_x(2, 1) = 1; f_y(2, 1) = -2$

31. $f_x(1, 1) = e; f_y(1, 1) = e$

33. $f_x(1, 0, 2) = 0; f_y(1, 0, 2) = 8; f_z(1, 0, 2) = 0$

35. $f_{xx} = 2y; f_{xy} = 2x + 3y^2 = f_{yx}; f_{yy} = 6xy$

37. $f_{xx} = 2; f_{xy} = f_{yx} = -2; f_{yy} = 4$

39. $f_{xx} = \dfrac{y^2}{(x^2 + y^2)^{3/2}}; f_{xy} = f_{yx} = -\dfrac{xy}{(x^2 + y^2)^{3/2}};$
   $f_{yy} = \dfrac{x^2}{(x^2 + y^2)^{3/2}}$

41. $f_{xx} = \dfrac{1}{y^2}e^{-x/y}; f_{xy} = \dfrac{y - x}{y^3}e^{-x/y} = f_{yx};$
   $f_{yy} = \dfrac{x}{y^3}\left(\dfrac{x}{y} - 2\right)e^{-x/y}$

43. **a.** 7.5; 40   **b.** Yes

45. $p_x = 10$—at $(0, 1)$, the price of land is changing at the rate of
   $10/ft^2$/mile change to the right; $p_y = 0$—at $(0, 1)$, the price of land is
   constant/mile change upward.

47. Complementary commodities

49. $30/unit change in finished desks; $-$25/unit change in unfinished
   desks. The weekly revenue increases by $30/unit for one additional
   finished desk produced (beyond 300) when the level of production of
   unfinished desks remains fixed at 250; the revenue decreases by
   $25/unit when one additional unfinished desk (beyond 250) is pro-
   duced and the level of production of finished desks remains fixed at
   300.

51. **a.** 20°F   **b.** -0.3°F

53. $\dfrac{\partial N}{\partial x} \approx 1.06; \dfrac{\partial N}{\partial y} \approx -2.85$

55. 0.039 L/kelvin; -0.014 L/mm of mercury. The volume increases by
   0.039 L when the temperature increases by 1 kelvin (beyond 300 K)
   and the pressure is fixed at 800 mm of mercury. The volume decreases by 0.014 L when the pressure increases by 1 mm of mercury
   (beyond 800 mm) and the temperature is fixed at 300 K.

61. True   63. False

Using Technology Exercises 12.2, page 870

1. 1.3124; 0.4038   3. -1.8889; 0.7778   5. -0.3863; -0.8497

Exercises 12.3, page 877

1. $(0, 0)$; relative maximum value: $f(0, 0) = 1$

3. $(1, 2)$; saddle point: $(1, 2, 4)$

5. $(8, -6)$; relative minimum value: $f(8, -6) = -41$

7. $(1, 2)$ and $(2, 2)$; saddle point: $(1, 2, -1)$; relative minimum value:
   $f(2, 2) = -2$

9. $\left(-\frac{1}{3}, \frac{11}{3}\right)$ and $(1, 5)$; saddle point: $\left(-\frac{1}{3}, \frac{11}{3}, -\frac{319}{27}\right)$; relative minimum
   value: $f(1, 5) = -13$

11. $(0, 0)$ and $(1, 1)$; saddle point: $(0, 0, -2)$; relative minimum value:
   $f(1, 1) = -3$

13. $(2, 1)$; relative minimum value: $f(2, 1) = 6$

15. $(0, 0)$; saddle point: $(0, 0, -1)$

17. $(0, 0)$; relative minimum value: $f(0, 0) = 1$

19. $(0, 0)$; relative minimum value: $f(0, 0) = 0$

21. 200 finished units and 100 unfinished units; $10,500

23. Price of land ($200/ft²) is highest at $(\frac{1}{2}, 1)$

25. (0, 1) gives desired location.

27. $r = \dfrac{130}{3\pi}$ in., $l = \dfrac{130}{3}$ in.; $\dfrac{2{,}197{,}000}{27\pi}$ in.³

29. $10'' \times 10'' \times 5''$; 500 in.³

31. $30'' \times 40'' \times 10''$; $7200   33. False

35. False   37. True

Chapter 12 Concept Review, page 880

1. xy; ordered pair; real number; $f(x, y)$

2. Independent; dependent; value

3. $z = f(x, y)$; f; surface

4. $f(x, y) = k$; level curve; level curves; k

5. Constant; x 6. Slope; $(a, b, f(a, b))$; x; b

7. $\leq$; (a, b); $\leq$; domain

8. Domain; $f_x(a, b) = 0$ and $f_y(a, b) = 0$; exist; candidate

Chapter 12 Review Exercises, page 881

1. $0, 0, \frac{1}{2}$; no   2. $e, \dfrac{e^2}{1 + \ln 2}, \dfrac{2e}{1 + \ln 2}$; no

3. $2, -(e + 1), -(e + 1)$

4. The set of all ordered pairs (u, v) such that $u \neq v$ and $u \geq 0$

5. The set of all ordered pairs (x, y) such that $y \neq -x$

6. The set of all ordered pairs (x, y) such that $x \leq 1$ and $y \geq 0$

7. The set of all triplets (x, y, z) such that $z \geq 0$ and $x \neq 1$, $y \neq 1$, and $z \neq 1$

8. $z = x^2 + 2y$

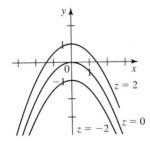

9. $z = y - x^2$

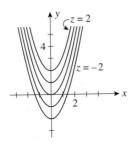

10. $z = \sqrt{x^2 + y^2}$

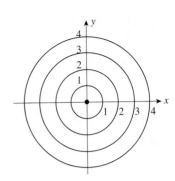

11. $z = e^{xy}$

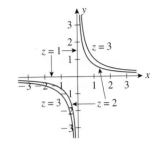

12. $f_x = 2xy^3 + 3y^2 + \dfrac{1}{y}$; $f_y = 3x^2y^2 + 6xy - \dfrac{x}{y^2}$

13. $f_x = \sqrt{y} + \dfrac{y}{2\sqrt{x}}$; $f_y = \dfrac{x}{2\sqrt{y}} + \sqrt{x}$

14. $f_u = \dfrac{v^2 - 2}{2\sqrt{uv^2 - 2u}}$; $f_v = \dfrac{uv}{\sqrt{uv^2 - 2u}}$

15. $f_x = \dfrac{3y}{(y + 2x)^2}$; $f_y = -\dfrac{3x}{(y + 2x)^2}$

16. $g_x = \dfrac{y(y^2 - x^2)}{(x^2 + y^2)^2}$; $g_y = \dfrac{x(x^2 - y^2)}{(x^2 + y^2)^2}$

17. $h_x = 10y(2xy + 3y^2)^4$; $h_y = 10(x + 3y)(2xy + 3y^2)^4$

18. $f_x = \dfrac{e^y}{2(xe^y + 1)^{1/2}}$; $f_y = \dfrac{xe^y}{2(xe^y + 1)^{1/2}}$

19. $f_x = 2x(1 + x^2 + y^2)e^{x^2+y^2}$; $f_y = 2y(1 + x^2 + y^2)e^{x^2+y^2}$

20. $f_x = \dfrac{4x}{1 + 2x^2 + 4y^4}$; $f_y = \dfrac{16y^3}{1 + 2x^2 + 4y^4}$

21. $f_x = \dfrac{2x}{x^2 + y^2}$; $f_y = -\dfrac{2x^2}{y(x^2 + y^2)}$

22. $f_{xx} = 6x - 4y$; $f_{xy} = -4x = f_{yx}$; $f_{yy} = 2$

23. $f_{xx} = 12x^2 + 4y^2$; $f_{xy} = 8xy = f_{yx}$; $f_{yy} = 4x^2 - 12y^2$

24. $f_{xx} = 12(2x^2 + 3y^2)(10x^2 + 3y^2)$;
$f_{xy} = 144xy(2x^2 + 3y^2) = f_{yx}$;
$f_{yy} = 18(2x^2 + 3y^2)(2x^2 + 15y^2)$

25. $g_{xx} = \dfrac{-2y^2}{(x + y^2)^3}$; $g_{xy} = \dfrac{2y(x - y^2)}{(x + y^2)^3} = g_{yx}$;
$g_{yy} = \dfrac{2x(3y^2 - x)}{(x + y^2)^3}$

26. $g_{xx} = 2(1 + 2x^2)e^{x^2+y^2}$; $g_{xy} = 4xye^{x^2+y^2} = g_{yx}$; $g_{yy} = 2(1 + 2y^2)e^{x^2+y^2}$

27. $h_{ss} = -\dfrac{1}{s^2}$; $h_{st} = h_{ts} = 0$; $h_{tt} = \dfrac{1}{t^2}$

28. $3; 3; -2$

29. $(2, 3)$; relative minimum value: $f(2, 3) = -13$

30. $(8, -2)$; saddle point: $(8, -2, -8)$

31. $(0, 0)$ and $(\frac{3}{2}, \frac{9}{4})$; saddle point: $(0, 0, 0)$; relative minimum value: $f(\frac{3}{2}, \frac{9}{4}) = -\frac{27}{16}$

32. $(-\frac{1}{3}, \frac{13}{3})$, $(3, 11)$; saddle point: $(-\frac{1}{3}, \frac{13}{3}, -\frac{445}{27})$; relative minimum value: $f(3, 11) = -35$

33. $(0, 0)$; relative minimum value: $f(0, 0) = 1$

34. $(1, 1)$; relative minimum value: $f(1, 1) = \ln 2$

35. $k = \dfrac{100\,m}{c}$

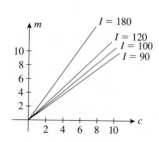

36. **a.** $R(x, y) = -0.02x^2 - 0.2xy - 0.05y^2 + 80x + 60y$
    b. The set of all points satisfying $0.02x + 0.1y \le 80$, $0.1x + 0.05y \le 60$, $x \ge 0$, $y \ge 0$
    c. 15,300; the revenue realized from the sale of 100 16-speed and 300 10-speed electric blenders is $15,300.

37. Complementary

38. The company should spend $11,000 on advertising and employ 14 agents to maximize its revenue.

39. 337.5 yd $\times$ 900 yd

Chapter 12 Before Moving On, page 882

1. All real values of x and y satisfying $x \ge 0$, $x \ne 1$, $y \ge 0$, $y \ne 2$

2.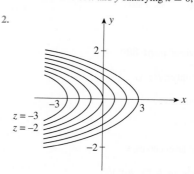

3. 14; 29; at the point $(1, 2)$, $f(x, y)$ increases at the rate of 14 units for each unit increase in x with y held constant at a value of 2; $f(x, y)$ increases at the rate of 29 units per unit increase in y with x held fixed at 2.

4. $f_x = 2xy + ye^{xy}$; $f_{xx} = 2y + y^2e^{xy}$; $f_{xy} = 2x + (xy + 1)e^{xy} = f_{yx}$; $f_y = x^2 + xe^{xy}$; $f_{yy} = x^2e^{xy}$

5. Rel. min. value: $f(1, 1) = -7$

INDEX

How–To Technology Index

Basic Rules of Differentiation

1. $\dfrac{d}{dx}(c) = 0 \qquad (c, \text{ a constant})$

2. $\dfrac{d}{dx}(u^n) = nu^{n-1}\dfrac{du}{dx}$

3. $\dfrac{d}{dx}(u \pm v) = \dfrac{du}{dx} \pm \dfrac{dv}{dx}$

4. $\dfrac{d}{dx}(cu) = c\dfrac{du}{dx} \qquad (c, \text{ a constant})$

5. $\dfrac{d}{dx}(uv) = u\dfrac{dv}{dx} + v\dfrac{du}{dx}$

6. $\dfrac{d}{dx}\left(\dfrac{u}{v}\right) = \dfrac{v\dfrac{du}{dx} - u\dfrac{dv}{dx}}{v^2}$

7. $\dfrac{d}{dx}(e^u) = e^u\dfrac{du}{dx}$

8. $\dfrac{d}{dx}(\ln u) = \dfrac{1}{u} \cdot \dfrac{du}{dx}$

Basic Rules of Integration

1. $\displaystyle\int du = u + C$

2. $\displaystyle\int kf(u)\, du = k\int f(u)\, du \qquad (k, \text{ a constant})$

3. $\displaystyle\int [f(u) \pm g(u)]\, du = \int f(u)\, du \pm \int g(u)\, du$

4. $\displaystyle\int u^n\, du = \dfrac{u^{n+1}}{n+1} + C \qquad (n \neq -1)$

5. $\displaystyle\int e^u\, du = e^u + C$

6. $\displaystyle\int \dfrac{du}{u} = \ln|u| + C$

Formulas

Equation of a Straight Line

a. point-slope form: $y - y_1 = m(x - x_1)$
b. slope-intercept form: $y = mx + b$
c. general form: $Ax + By + C = 0$

Compound Interest

$$A = P(1 + i)^n \qquad (i = r/m, \ n = mt)$$

where A is the accumulated amount at the end of n conversion periods, P is the principal, r is the interest rate per year, m is the number of conversion periods per year, and t is the number of years.

Effective Rate of Interest

$$r_{\text{eff}} = \left(1 + \frac{r}{m}\right)^m - 1$$

where r_{eff} is the effective rate of interest, r is the nominal interest rate per year, and m is the number of conversion periods per year.

Future Value of an Annuity

$$S = R\left[\frac{(1 + i)^n - 1}{i}\right]$$

Present Value of an Annuity

$$P = R\left[\frac{1 - (1 + i)^{-n}}{i}\right]$$

Amortization Formula

$$R = \frac{Pi}{1 - (1 + i)^{-n}}$$

Sinking Fund Payment

$$R = \frac{iS}{(1 + i)^n - 1}$$

The Number of Permutations of n Distinct Objects Taken r at a Time

$$P(n, r) = \frac{n!}{(n - r)!}$$

The Number of Permutations of n Objects, Not All Distinct

$$\frac{n!}{n_1! \, n_2! \cdots n_m!}, \quad \text{where } n_1 + n_2 + \cdots + n_m = n$$

The Number of Combinations of n Distinct Objects Taken r at a Time

$$C(n, r) = \frac{n!}{r!(n - r)!}$$

The Product Rule for Probability

$$P(A \cap B) = P(A) \cdot P(B|A)$$

Bayes' Formula

$$P(A_i \,|\, E) = \frac{P(A_i) \cdot P(E \,|\, A_i)}{P(A_1) \cdot P(E \,|\, A_1) + P(A_2) \cdot P(E \,|\, A_2) + \cdots + P(A_n) \cdot P(E \,|\, A_n)}$$

Expected Value of a Random Variable

$$E(X) = x_1 p_1 + x_2 p_2 + \cdots + x_n p_n$$

List of Applications

BUSINESS AND ECONOMICS

(continued)

List of Applications (*continued*)

SOCIAL SCIENCES